SINTERING OF CERAMICS – NEW EMERGING TECHNIQUES

Edited by **Arunachalam Lakshmanan**

Sintering of Ceramics – New Emerging Techniques
Edited by Arunachalam Lakshmanan

Published by InTech

Janeza Trdine 9, 51000 Rijeka, Croatia

Publishing Process Manager Sasa Leporic

Technical Editor Teodora Smiljanic

Cover Designer InTech Design Team

Additional hard copies can be obtained from orders@intechopen.com

Sintering of Ceramics – New Emerging Techniques, Edited by Arunachalam Lakshmanan
p. cm.
ISBN 978-953-51-0017-1

For EU product safety concerns:
IN TECH d.o.o., Prolaz Marije Krucifikse Kozulić 3, 51000 Rijeka, Croatia,
info@intechopen.com or visit our website at intechopen.com.

Meet the editor

Dr Arunachalam Lakshmanan obtained his Bachelor and Master Degree in Physics with Distinction at the Annamalai University. He joined Bhabha Atomic Research Centre in 1971 and served as a scientific officer untill 1990. He obtained his PhD degree on Thermoluminescence at the Bombay University in 1981. In 1982 and 1983 he worked as a Scientific Associate at CERN, Geneva, Switzerland. For a period of 2 years, Dr Lakshmanan worked as a AvH fellow at the University of Wuppertal, Germany. He was a visiting Professor in several universities in Japan, France and South Korea. He was an Editorial Board Member in Radiation Protection Dosimetry Journal, UK. From 1990 to 2007, Dr. Lakshmanan was head of the Radiation Dosimetry Section at the Indira Gandhi Centre for Atomic Research. In 2008, Dr. Lakshmanan published a book titled "Luminescence of Display Phosphors- Phenomena and Applications". Presently he is working as a Physics Professor at the Saveetha Engineering College, Chennai, India. His research areas include Synthesis of Luminescence Phosphors and Exploiting Zero Point Energy for power production.

Contents

Preface

Pottery and ceramics have been an important part of human culture for thousands of years. From prehistoric storage jars to tiles on the space shuttles, pottery and ceramics have played a key role in innumerable human endeavors. The art of ceramics is one of the oldest known, dating to prehistoric times.

Once heated (fired) to between 350^0 and 800^0C, the clay is converted to ceramic and will never dissolve again. Ceramic products are used as structural including bricks, pipes, floor and roof tiles, Refractories, such as kiln linings, gas fire radiants, steel and glass making crucibles, Whitewares, including tableware, wall tiles, pottery products and sanitary ware or technical, such items include tiles used in the Space Shuttle program, gas burner nozzles, biomedical implants etc. Frequently, the raw materials do not include clay. *Calcination* refers to thermal treatment which removes volatile fraction. Calcination treatment prior to sintering improves the densification of ceramics. For instance, in the case of hydroxyapatite, a bioceramic, calcining at 900 °C prior to sintering at 1250 °C results in a higher bending strength (about 55MPa) with finer grain size.

The chapters covered by this book on Sintering of Ceramics include emerging new techniques on sintering which include Spark plasma sintering, Magnetic Pulsed compaction Low Temperature Co-fired Ceramic technology for the preparation of 3-dimesinal circuits, Microwave sintering of thermistor ceramics, Synthesis of Bio-compatible ceramics, Dielectrics and opto-electronic materials, Sintering of Rare Earth Doped Bismuth Titanate Ceramics prepared by Soft Combustion, nanostructured ceramics, alternative solid-state reaction routes yielding densified bulk ceramics and nanopowders, Sintering of intermetallic superconductors such as MgB_2, impurity doping in luminescence phosphors using soft techniques etc. Other advanced sintering techniques such as radiation thermal sintering for the manufacture of thin film solid oxide fuel cells, are also covered.

Spark plasma sintering (SPS) is a form of sintering where both external pressure and an electric field are applied simultaneously to enhance the densification of the amorphous or metallic/ceramic powder compacts. This form of densification uses lower temperatures and shorter amount of time than typical sintering. The theory behind it is that there is a high-temperature or high-energy plasma that is generated between the gaps of the powder materials; materials can be metals, inter-metallic, ceramics, composites and polymers. It is a novel technique used to sinter ceramics, metals and composites within a few minutes and to obtain densities greater than 99.9%.

Selective laser sintering (SLS) is an additive manufacturing technique that uses a high power laser (for example, a carbon dioxide laser) to fuse small particles of plastic, metal (direct metal laser sintering), ceramic, or glass powders into a mass that has a desired 3-dimensional shape.

A *microwave* furnace can sinter dental ceramics within 1/12th the time of a conventional oven. Microwave has the added advantage of producing better mechanical strength and resistance to low temperature degradation. In order to be effective, the ceramic powders to be sintered are to be surrounded by microwave susceptors such as ferric oxide. Microwave is the best choice for sintering dental ceramics such as Zirconia. Dental laboratories, overloaded with requests, usually take seven or more days to process the permanent dental restoration with a conventional oven.

In magnetic pulsed compaction (MPC), it is possible to apply a high pressure (up to 5 Gpa) to a ceramic powder for a short period of time (~ 500 μs) which results in improved density.

Vacuum sintering. ZrO_2 based ceramic materials have unique properties (high abrasion resistance, strength and breakdown viscosity), and are being used in different technologies. A basic characteristic of the fabrication of such materials is that during traditional sintering at 1600^{0}C in air to obtain high densities causes a Tetragonal → Monoclinic transformation on cooling, which destroys the sample, due to grain enlargement. However sintering at vary low partial oxygen pressure (10^{-6} Pa) stabilizes the high temperature tetragonal phases of ZrO_2 similar to oxide additives (CaO, MgO, Y_2O_3, etc).

Recycling green building materials using *cold-bonding technique* incorporates the principles of cement chemistry which results in the reduction of energy consumption and CO_2 emission.

Arunachalam Lakshmanan
Saveetha Engineering College, Thandalam, Chennai, India

Part 1

New Sintering Techniques

1

Microwave Fast Sintering of Ceramic Materials

Romualdo R. Menezes[1], Pollyane M. Souto[2]
and Ruth H.G.A. Kiminami[2]
[1]Department of Materials Engineering, Federal University of Paraiba
[2]Department of Materials Engineering, Federal University of São Carlos
Brazil

1. Introduction

Microwaves are electromagnetic waves with wavelengths ranging from 1m to 1 mm, which correspond to frequencies between 0.3 and 300 GHz. This frequency range lies just above radio waves and just below visible light on the electromagnetic spectrum (Katz, 1992). The possibility of processing ceramics by microwave heating was discussed over 50 years ago by Von Hippel (1954a), and experimental studies on microwave processing of ceramics were started in the mid 1960s by Tinga and Voss (Tinga & Voss, 1968). Since then, the results of many investigations into microwave sintering and joining of ceramics have been reported (Bykov et al., 2001). Activity in this field began to accelerate in the mid-1970s because of a shortage of natural gas, prompting the investigation of microwave heating and sintering of several ceramic materials in the late 1970s and 1980s.

While most of today's industrial microwave applications involve the relatively low-temperature processing of food, wood, rubber, polymers, etc., interest in high-temperature microwave processing of materials has been growing. In recent years, microwave heating has been widely employed in the sintering and joining of ceramics (Bykov et al., 2001; Huang et al., 2009).

Most of the reposts published in the literature assert that microwave-driven processes are faster than conventional heating processes. This faster speed is manifested as a reduction in the densification time of ceramic powder compacts, often allied to lower sintering temperatures (Bykov et al., 2001). In general, the kinetics of synthesis and sintering reactions are reportedly augmented by two or three orders of magnitude or even more when conventional heating is substituted for microwave radiation (Oghbaei & Mirzaee, 2010).

Microwave heating is a process whereby microwaves couple to materials, which absorb the electromagnetic energy volumetrically and transform it into heat. This differs from conventional methods in which heat is transferred between objects through the mechanisms of conduction, radiation and convection. Because the material itself generates the heat, heating is more volumetric and can be very rapid and selective (Sutton, 1989). Thus, microwave sintering techniques allows for the application of high heating rates, markedly shortening the processing time.

Microwave processing eliminates the need for spending energy to heat the walls of furnace or reactors, their massive components and heat carriers. Hence, the use of microwave processing methods significantly reduces energy consumption, particularly in high-temperature processes, since heat losses escalate considerably as processing temperatures increase. However, the advantages of using microwave energy in high-temperature processes are by no means limited to energy savings. In many cases, microwave processing can improve the product quality (Bykov et al., 2001).

High heating rates not only shorten processing time and reduce energy consumption. Many high-temperature processes involve a sequence of steps that follow each other as the temperature rises. These sequences occur in multistage thermally activated processes, in which separate stages are characterized by different values of activation energy. Some of these stages may have a negative effect on the properties of the final product. In such cases, rapid heating may be vital in reducing the effects of undesired intermediate stages of the process. An example of such a multistage process is the sintering of ceramics. Various diffusion processes, such as surface, grain boundary and bulk diffusion, determine mass transport in different sintering stages.

Harmer and Brook (1981) postulated that rapid sintering produces a finer grained microstructure. This theory applies to all rapid sintering techniques, including microwave sintering. The theory of rapid sintering is based on the assumption that densification and grain growth are thermally activated processes and that the activation energy for grain growth is lower than for densification.

In fast firing, the objective is to enhance the densification rate in detriment to the coarsening rate by a rapidly approaching to the sintering temperature. Because coarsening mechanisms (e.g., surface diffusion and vapor transport) usually prevail over densification mechanisms (e.g., lattice and grain-boundary diffusion) at lower temperatures, it has been suggested that rapid heating to higher temperatures can be beneficial to achieve high density allied to fine grain size. In this case, the shorter time spent at lower temperatures serves to reduce the extent of coarsening, while the driving force for densification is not decreased significantly (Menezes & Kiminami, 2010), resulting in high densification and fine microstructures, which is a factor of paramount importance in the sintering of nanostructured ceramic and composite materials.

However, various fundamental problems are usually encountered when sintering materials by direct microwave heating. Most of the research on material processing by microwaves is based on conventional low-frequency (2.45 GHz) microwave applicators. However, such applicators do not couple microwave power efficiently to many ceramics at room temperature, and poor microwave absorption characteristics make initial heating difficult. Thermal instabilities may occur, which can lead to the phenomenon of thermal runaway; i.e., the specimen overheats catastrophically. The temperature gradients inherent in volumetric heating can lead to severe temperature non-uniformities, which, at high heating rates, may cause non-uniform properties and cracking.

These problems have led researchers to develop hybrid heating techniques that combine direct microwave heating with infrared heat sources. Increasing the temperature (with radiant heat) is a common method used by many researchers to couple microwaves with poorly absorbing (low-loss) materials. Once a material is heated to its critical temperature, microwave absorption becomes sufficient to cause self-heating. This hybrid method can result in more uniform temperature gradients because the microwaves heat volumetrically, and the external heat source minimizes surface heat losses. Hybrid heating can be achieved by using either an independent heat source such, as a gas or electric furnace in combination with microwaves, or an external susceptor that couples with the microwaves. In the latter, the material is exposed simultaneously to radiant heat produced by the susceptor and to microwaves (Clark & Sutton, 1996).

There is growing evidence to support the use of microwave hybrid heating in ceramic sintering and to justify continued research and development for its use in many ceramics manufacturing processes. In this context, this chapter discusses microwave interactions with ceramic materials, dielectric properties of ceramics, and theoretical aspects of microwave sintering, as well as results that highlight the successful application of microwaves to the rapid sintering of ceramic materials.

2. Fundamentals of microwave heating

Microwaves are electromagnetic waves in the frequency band of 300 MHz (3×10^8 cycles/second) to 300 GHz (3×10^{11} cycles/second), which correspond to a wavelength range of 1 m to 1 mm. Within this portion of the electromagnetic spectrum there are frequencies that are used for cellular phones, radar, and television satellite communications (Thostenson & Chou, 1999). Typical frequencies for materials processing are 915 MHz, 2.45 GHz, 5.8 GHz, 22.00 GHz, 24.12 GHz, 28GHz and 60GHz, but only 915 MHz and 2.45 GHz are widely applied ((Katz, 1992; Committee on Microwave Processing of Materials: An Emerging Industrial Technology et al., 1994). The advantages of higher frequencies are more uniform electric field distribution in the cavity and higher power dissipated in dielectric materials. However, microwave apparatus is available on a very limited basis and at extremely high cost.

The interaction of an electric field with a material may elicit several responses, and microwaves can be reflected, absorbed and/or transmitted by the material. In a conductor, electrons move freely in the material in response to the electric field, resulting in electric current. Unless the material is a superconductor, the flow of electrons will heat the material through resistive heating. However, microwaves will be largely reflected from metallic conductors, and therefore such conductors are not effectively heated by microwaves. In insulators, electrons do not flow freely, but electronic reorientation or distortions of induced or permanent dipoles can give rise to heating (Committee on Microwave Processing of Materials: An Emerging Industrial Technology et al., 1994).

Because microwaves generate rapidly changing electric fields, these dipoles change their orientations rapidly in response to the changing fields. If the electric field change takes place close to the natural frequency at which reorientation can occur, the maximum amount of energy is consumed, resulting in optimum heating. In microwave processing terminology, this event is described by the term "well-coupled" material.

It is known that in dielectric materials the external electric field causes a redistribution of internal bound charges, which results in the polarization of the material. A measure of the response of a material to an external electric field is its dielectric permittivity, ε. If the external electric field is alternating (ac), the dielectric response of the material follows it, generally with some lag behind the field changes. To describe this phenomenon quantitatively, a complex dielectric permittivity that depends on the field frequency, ω (ω is equal to $2\pi f$, where f is the frequency in cycles per second), is formally introduced:

$$\varepsilon(\omega)^* = \varepsilon(\omega)' - i\varepsilon(\omega)'' = \varepsilon_o\,(\varepsilon_r(\omega)' - i\varepsilon_r(\omega)'') \quad (1)$$

where $i = (-1)^{1/2}$, ε' is the permittivity (or dielectric constant), ε'' is the dielectric loss factor, ε_o is the permittivity of a vacuum, and ε_r is the relative permittivity (or relative dielectric constant) (ε_r' equals $\varepsilon'/\varepsilon_o$ and ε_r'' equals $\varepsilon''/\varepsilon_o$).

The real component of complex permittivity, ε', is commonly referred to as the dielectric constant. However, because ε' is not constant but may vary significantly with frequency and temperature, it is generally referred to simply as permittivity.

However, microwave propagation in air or in materials depends on the dielectric and magnetic properties of the medium. Thus, the electromagnetic properties of a medium are characterized by complex permittivity and also complex permeability, μ^* (Equation 2). In magnetic materials, the dipoles may be able to couple with the magnetic component of the electromagnetic field and provide an additional heating mechanism. Similarly to the dielectric properties of materials, their magnetic permeability, μ', and magnetic loss, μ'', must be considered in microwave processing,

$$\mu(\omega)^* = \mu\,(\omega)' - i\,\mu\,(\omega)'' \quad (2)$$

where μ' and μ'' are the permeability and magnetic loss factor, respectively.

In an ideal dielectric, the electric charge adjusts instantaneously to any change in voltage or field. In practice, however, there is an inertia-to-charge movement that shows up as a relaxation time for charge transport. Relaxation time, τ, can be defined as the time interval in which a disturbed system is restored to its equilibrium configuration after the application of a microwave field. The behavior of τ determines the frequency and temperature dependencies of ε' and ε'' and varies widely in liquids and solids (Committee on Microwave Processing of Materials: An Emerging Industrial Technology et al., 1994). Thus, when an alternate electric field interacts with a dielectric, the time required for polarization appears as a phase retardation of the charging current. Instead of advancing by 90°, it advances by some angle, δ, other than 90°. This phase shift, which corresponds to a time lag between an applied voltage and induced current, causes loss of current and dissipation of energy in the material (Kingery et al, 1976). The imaginary term in equation (1), ε'', which quantifies the power dissipated in the material, is known as the dielectric loss factor.

Another term sometimes used to quantify the "lossiness" of a material is the loss tangent (Equation 3). The inverse of the loss tangent is the quality factor, Q, which is widely used in waveguide technology (Von Hippel, 1954a).

$$\tan\delta = \frac{\varepsilon''\mu' - \varepsilon'\mu''}{\varepsilon'\mu' + \varepsilon''\mu''} \quad (3)$$

For negligible magnetic loss ($\mu'' = 0$) the loss tangent is expressed as:

$$\tan\delta = \frac{\varepsilon''}{\varepsilon'} = \frac{\varepsilon_r''}{\varepsilon_r'} \tag{4}$$

The charging current for a simple dielectric plate with a sinusoidal applied voltage is given by $I_c = i\omega\varepsilon' E$ and the loss current by $I_l = \omega\varepsilon'' E = I_l = \sigma E$, where σ is the dielectric conductivity. Engineers concerned with dielectric heating will generally use the dielectric constant, ε′, and the dielectric conductivity, σ, to characterize the processed material and express its response at a given frequency.

The power absorbed per unit volume, considering that the electric field is uniform throughout the volume, can be expressed by:

$$P = \frac{1}{2}\left(\omega\varepsilon'' \mathrm{E}^2 + \omega\mu'' \mathrm{H}^2\right) = \omega\left(\varepsilon'' \mathrm{E}^2_{rms} + \mu'' \mathrm{H}^2_{rms}\right) \tag{5}$$

where E_{rms} and H_{rms} are the root mean squares of the internal electric and magnetic fields.

Equation (5) shows that the power dissipated in a ceramic body during microwave processing is directly related to the applied frequency ($\omega = 2\pi f$) and the material's conductivity, σ ($\sigma = \omega\varepsilon''$). For instance, 25 times more power can be dissipated in alumina at 28 GHz, than at 2.45 GHz.

Depending on their electrical and magnetic properties, materials can be divided into three categories according to their microwave absorption properties. Materials with a very low dielectric loss factor (or conductivity) allow microwaves to pass through with very little absorption and are said to be transparent to microwaves. Materials with an extremely high dielectric loss factor, i.e. metals, reflect microwaves and are said to be opaque. Materials with intermediate loss tangents will absorb microwaves (Katz, 1992).

Ceramics with loss factors between the limits of $10^{-2} < \varepsilon'' < 5$ are good candidates for microwave heating. Ceramics with $\varepsilon'' < 10^{-2}$ would be difficult to heat, while those with $\varepsilon'' > 5$ would absorb most of the heating on the surface and not in the bulk.

Absorption energy is quantified by the absorbed power per unit volume and by the attenuation distance, D (or depth of penetration), which expresses the magnitude of the energy attenuation inside the material.

A plane wave varying periodically in time with the frequency $f = \omega/2\pi$ and advancing in the + x direction through space presents a complex propagation factor, γ (Metaxas & Binner, 1990):

$$\gamma = j\omega\sqrt{\varepsilon^* \mu^*} = \alpha + j\beta \tag{6}$$

where α is the attenuation factor and β is the phase factor of the wave.

Equating the real and imaginary parts of Equation (6), the attenuation and phase factors can be expressed in terms of permittivity and permeability (Von Hippel, 1954a):

$$\alpha = \frac{\lambda\omega^2}{4\pi}(\varepsilon'\mu'' + \varepsilon''\mu') \tag{7}$$

$$\beta = \omega\sqrt{\frac{\varepsilon'\mu' - \varepsilon''\mu''}{2}\left(1 + \sqrt{1 + \left(\frac{(\varepsilon'\mu'' + \varepsilon''\mu')}{(\varepsilon'\mu' + \varepsilon''\mu'')}\right)^2}\right)} \tag{8}$$

where λ is the wavelength of the plane wave inside the material.

The attenuation produced by a dielectric is frequently expressed as the attenuation distance (or penetration depth or skin depth), *D*, defined as 1/α, through which the field strength decays to 1/e (0.368) of its original value. For materials with negligible magnetic loss (μ″ = 0), and considering the permeability of the vacuum ($\mu' = \mu_o$):

$$D = \frac{1}{\alpha} = \frac{c}{\omega}\sqrt{\frac{2}{\varepsilon_r'\left(\sqrt{1 + \tan^2\delta} - 1\right)}} \tag{9}$$

where *c* is the speed of light.

Another parameter that can be defined is power penetration depth, D_P, which is the distance at which the power drops to 1/e (0.368) from its value at the surface (Metaxas & Binner, 1990):

$$D_P = \frac{1}{2\alpha} = \frac{D}{2} \tag{10}$$

It is important to differentiate between power penetration depth and electric field attenuation distance or skin depth (*D*), because at frequencies allocated for industrial use in the microwave regime, the power penetration depths could be very small indeed and the size of the ceramic to be treated, particularly when it is fairly lossy, could be many times larger than *Dp*, resulting in unacceptable temperature non-uniformities (Metaxas & Binner, 1990).

The index of absorption, *k*, can be expressed in terms of the attenuation, α, and the phase factor β, according to Equation (11). The reflection coefficient, *R* (the fraction of reflected power), for a normal incidence of the electromagnetic wave on a plane boundary between material and vacuum can also be expressed in terms of loss tangent (Equation 12) (Von Hippel, 1954b; Bykov et al., 2001):

$$k = \frac{\alpha}{\beta} = \sqrt{\frac{\sqrt{1 + \tan 2\delta} - 1}{\sqrt{1 + \tan 2\delta} + 1}} \tag{11}$$

$$R = \frac{1 - \sqrt{2\varepsilon_r'[1 + \sqrt{1 + tg^2\delta}]} + \varepsilon_r'\sqrt{1 + tg^2\delta}}{1 + \sqrt{2\varepsilon_r'[1 + \sqrt{1 + tg^2\delta}]} + \varepsilon_r'\sqrt{1 + tg^2\delta}} \tag{12}$$

The absorption properties of non-metallic materials vary greatly in the microwave frequency range. The loss tangent, tan δ, varies at room temperature from 10^{-4}–10^{-3} (for example in pure alumina and silicon nitride) up to 1 and higher (in carbides, borides, some oxides, and intermetallic compounds). Correspondingly, the penetration depth varies from tens of meters to fractions of a millimeter. For example, at 2.45 GHz and room temperature, graphite and alumina have skin depths of 38 μm and 25 m (according to data in Von Hippel, 1954b; Metaxas & Binner, 1990), respectively.

In general, at room temperature, many ceramics do not absorb microwaves appreciably at 2.45 GHz, and thus present high attenuation distances. However, their absorption can be increased by raising the temperature, adding absorbents (e.g., SiC, carbon, binders), altering their microstructure and defect structure (Agrawal, 1998) by changing their form (e.g., bulk vs. powder), or by changing the frequency of incident radiation (Clark & Sutton, 1996).

The dielectric properties of ceramic materials are highly dependent on temperature, and their loss factor generally increases with temperature, improving the ability of "transparent" materials to couple with microwaves. The rate of temperature increase depends upon the type of ceramic and the operating frequency, but a sharp increase in microwave losses begins at temperatures of about 0.4–0.5 T_m (where T_m is the melting temperature of the material). In this temperature range, the bonds between ions in ionic crystals begin to break, and the electrons in covalent materials begin to populate the conduction bands. Due to this sharp increase in dielectric losses, "transparent" materials couple efficiently with microwaves. This is characteristic of solids with both ionic (Al_2O_3, ZrO_2) and covalent (Si_3N_4, AlN) bonds. For instance, the attenuation distances of alumina and zirconia (Y-stabilized) change from meters at 20°C to 23 and 1 cm at 1200°C, respectively.

When the skin depth is greater than the dimension of the sample, this effect may be neglected. Conversely, when dimensions of samples are greater than its penetration depth penetration of microwave energy will be limited, making uniform heating impossible

Historically, there is a paucity of data on the dielectric properties of most materials as a function of temperature in the microwave range. The data reported in the literature indicate that there is a complex interdependence of dielectric properties on temperature and frequency (Clark & Sutton, 1996). Figures 1 and 2 illustrate the behavior of loss tangent, attenuation distance and reflection coefficient according to complex permittivity data published in the literature (Andrade et al., 1992; Arai et al., 1993; Batt et al., 1995).

Von Hippel (1954b) compiled the first tabulation of measurements for numerous materials over a wide range of frequencies (10 Hz - 10 GHz). The materials he characterized were both organic and inorganic in nature and both liquid and solid. Subsequently, dielectric data on high temperature solid materials, including oxides, nitrides, silicates, rocks and minerals were obtained by Inglesias and Westphal (Inglesias & Westphal, 1967, as cited in Metaxas & Binner, 1990). Since then, four reports have been produced at the same MIT Laboratory for Insulation Research where Von Hippel obtained his results as described by Metaxas & Binner (Metaxas & Binner, 1990). These reports contain much valuable information on the frequency and temperature dependence of the dielectric properties of a large number of ceramic materials (both oxide and some non-oxide), as well as various minerals, many organic compounds and some foodstuffs. The frequency and temperature varied up to 25 GHz and 1000°C, respectively (Metaxas & Binner, 1990).

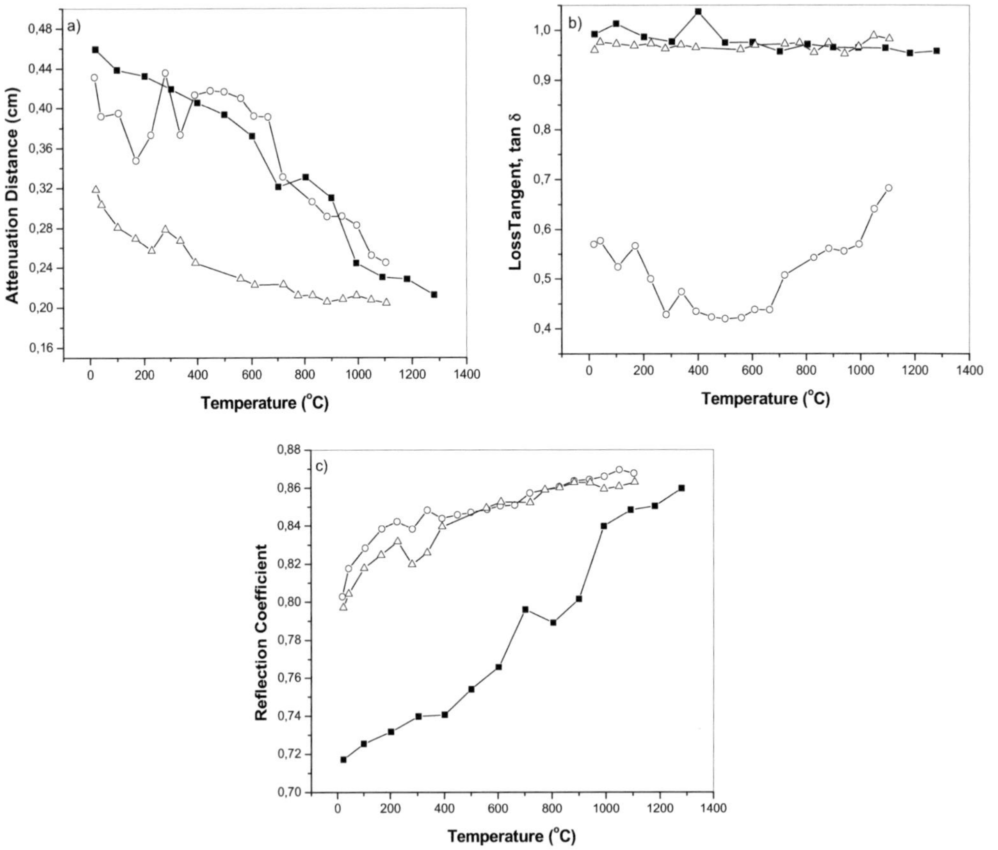

Fig. 1. a) Attenuation distance b), loss tangent, and c) reflection coefficient of SiC.

Studies (Sutton, 1989; Meredith, 1998) showed that when low-loss dielectric materials exhibit an abrupt increase in ε'' with increasing temperature, the potential exists for very rapid changes in temperature. The temperature at which this abrupt change in ε'' occurs is known as critical temperature, *Tc*. Above this temperature, the dielectric loss factor begins to increase, and the material begins to couple with microwaves. Figure 2 illustrates this behavior is when the loss tangent of zirconia increases significantly increases after 600°C.

However, many researchers have shown that thermal runaway in microwave-heated materials often occurs when critical temperature is reached. The sharp increase in microwave absorption with temperature may cause thermal instability, which is commonly known as temperature runaway.

The rapid rise in the dielectric loss factor with temperature is the main issue in thermal runaway (unstable accelerated heating). An increase in local temperature is accompanied by an enhanced microwave energy absorption, which results in local acceleration of heating, a further rise in temperature and the development of thermal runaway. This instability develops when microwave power exceeds some threshold value, which corresponds to an S-type dynamic curve of temperature versus power (Bikov et al., 2001). Below a "critical power level", the material will heat in a stable manner to a steady state value on the lower branch of the response curve. If the power is increased to exceed the upper critical power, the temperature will jump to the upper branch of the temperature-power curve, giving rise to thermal runaway.

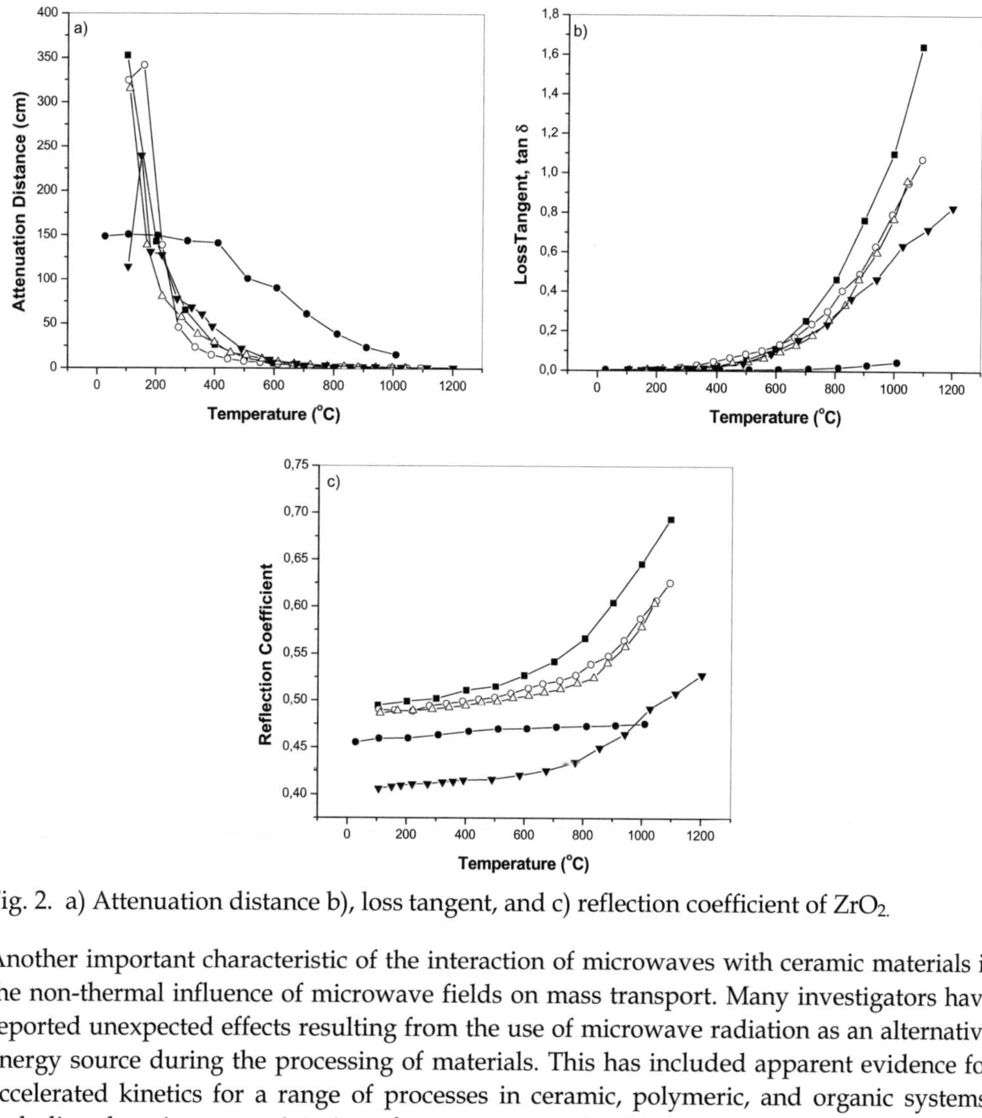

Fig. 2. a) Attenuation distance b), loss tangent, and c) reflection coefficient of ZrO_2.

Another important characteristic of the interaction of microwaves with ceramic materials is the non-thermal influence of microwave fields on mass transport. Many investigators have reported unexpected effects resulting from the use of microwave radiation as an alternative energy source during the processing of materials. This has included apparent evidence for accelerated kinetics for a range of processes in ceramic, polymeric, and organic systems, including the microwave sintering of ceramic materials. This unexpected effect was named "microwave effect". As a general rule, the kinetics of synthesis and sintering reactions of ceramic materials is reportedly augmented by two or three orders of magnitude or even more when conventional heating is substituted for microwave radiation.

In principle, the term "microwave effect", or, more strictly, "microwave field non-thermal effect", should be reserved for the deviations of microwave processes from conventional processes that occur given the identical temperature dynamics of these processes. Over the last decade, considerable effort has been focused on the development of theoretical models for microwave non-thermal enhancement of transport phenomena in solids and experimental justification of these models (Bikov et al., 2001). Studies (Rybakov & Semenov, 1994a, 1994b; Freeman et al., 1994; Booske et al., 1997) observed that, through an amplification and rectification process, the electric fields inside a microwave oven are sufficient to significantly affect the sintering behavior of ceramics, inducing an additional driving force for diffusion in solid materials (Clark & Sutton, 1996). They demonstrated that, under certain conditions, microwave-induced oscillating vacancy fluxes are rectified, leading to directional macroscopic mass transport; and that when the electromagnetic field interacts with the free surfaces of interfaces, an additional driving force for diffusion is created. This driving force has been named "ponderomotive" driving force. This force can compete with different thermochemical driving forces and significantly affect mass transport in solids during sintering.

3. Microwave fast sintering of ceramics

The initial surge in the development of microwave technology was driven by the military needs of World War II. The tremendous effort that went into the development of radar during World War II generated am enormous body of knowledge on the properties of microwaves and related technologies (Committee on Microwave Processing of Materials: An Emerging Industrial Technology et al., 1994). In the post-war years, further development resulted in microwaves used for heating, especially for domestic purposes (Meredith, 1998). The first commercial microwave oven was a large floor-standing model was developed by the Raytheon Company of North America in 1951. Microwave ovens for domestic use were launched on the market in the early 1960s, initiating the mass market for this product. Soon thereafter, industrial microwave applications began to emerge, beginning with rubber extrusion, plastic manufacturing, and the treatment of foundry core ceramics.

Microwave processing of materials is a technology that has proven useful in a number of applications, presenting some important advantages over conventional heating methods. The main benefits of exploiting microwave energy in thermally activated processes stem from the specificity of microwave energy absorption. In contrast to all the other methods commonly used, microwaves allow for volumetric heating of materials. Microwave energy transforms into heat inside the material, which generally result in significant energy savings and shorter processing times. This factor plays a decisive role in most applications that have gained industrial acceptance to date.

Studies on microwave processing of ceramics started in the mid 1960s by Tinga (Tinga & Voss, 1968; Tinga & Edwards, 1968) and, since then, interest in the use of microwaves for heating and sintering ceramics has grown steadily. However, the particular requirements for ceramic powder sintering make this process one of the most challenging applications for microwave processing. These requirements often include high temperatures, high heating rates, uniform temperature distribution, etc. (Menezes et al., 2007). Moreover, during microwave heating, the absorption of electromagnetic energy raises the temperature of the entire sample, but heat loss from the surface causes the near-surface temperatures to become lower than the internal temperatures. Thus, in ceramics with poor thermal conductivity, large temperature gradients can develop in the body. The shape of the ceramic body and the microwave frequency strongly influence the temperature gradients, and achieving sufficiently uniform heating of the body can be difficult. High heating rates can lead to localized heating, which can cause non-uniform properties and even cracking of the sample; as well thermal runaway when the temperature of the sample increases rapidly (Rahaman, 1995).

In microwave sintering, heat is generated internally through the interaction of the microwaves with the atoms, ions, and molecules of the material, which produces an inverse heating profile. Figure 3 presents a schematic diagram of heating by conventional and microwave processes. The microwave heating process is more volumetric, thus allowing higher heating rates. However, most research to date is based on conventional low-frequency (2.45 GHz) microwave applicators, and at this frequency, many ceramics present poor microwave absorption characteristics. Thus, materials, that are transparent to microwaves, such as SiO_2 and Al_2O_3 (ceramics with low-loss, as pure oxides and nitrides), are difficult to heat at room temperature.

Many ceramic materials that are difficult to heat at room temperature possess electrical conductivity or dielectric loss factors that rapidly increase in magnitude as the temperature rises. Thus, these materials will absorb microwave energy if they can be preheated to a suitable temperature using another heat source. This has led to the development of hybrid heating methods (Krage, 1981; Kimrey et al., 1991). Hybrid heating can be achieved by using either an independent heat source, such as a gas or electric furnace in combination with microwaves, or an external susceptor material (high-loss material) that couples with the microwaves. In the latter case, the material is exposed simultaneously to radiant heat produced by the susceptor material and to microwaves (usually, the susceptor material is simply named as "susceptor").

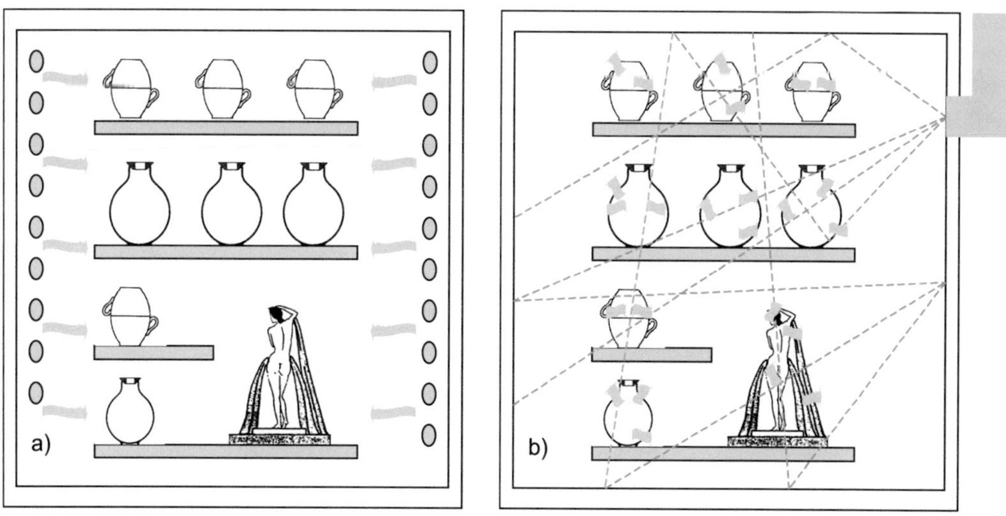

Fig. 3. Heating patterns in: a) conventional, and b) microwave furnaces.

Increasing the temperature (with radiant heat) is a common method many researchers employ to couple microwaves with poorly absorbing (low-loss) materials. Once a material is heated to its critical temperature, *Tc*, microwave absorption becomes sufficient to cause self-heating. This hybrid method can result in more uniform temperature gradients because the microwaves heat volumetrically and the external heat source minimizes surface heat losses. Therefore, one of the most important characteristics associated with the use of microwave hybrid heating (MHH) is the potential to achieve uniform heating throughout the cross-section of a material, as illustrated in Figure 4. In the early 1990s, Dé (Dé et al., 1991a, 1991b) showed that microwave hybrid heating (MHH) could result in samples with no significant density gradient throughout the cross-section and that this phenomenon was enhanced with increased sample size. The consequences of more uniform temperature gradients are homogenous microstructures and improved properties of the final body.

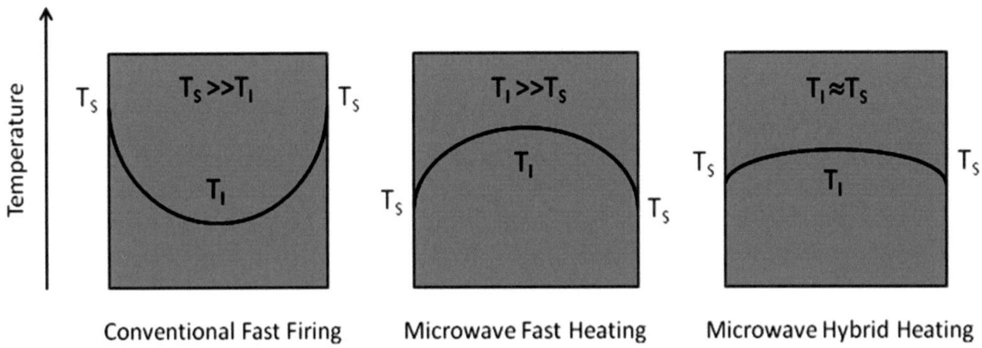

Fig. 4. Temperature distribution in ceramic material heated by conventional fast firing, microwave fast firing and microwave hybrid heating (T_S - Surface temperature, T_I - Internal temperature).

Figure 5 displays homogeneous microstructures of microwave hybrid fast sintered (MHFS) alumina samples (low-loss material). The samples sintered for 20min reached 97% densification and presented internal and surface microstructures with a similar grain size distribution, which reinforces the efficiency of hybrid fast firing in the densification and uniform heating of low-loss ceramic materials. Using hybrid heating, our research group has also microwave fast sintered other ceramic materials, such as mullite, zirconia, ZnO, porcelains, ferrites, and zirconia (Menezes et al., 2005, 2007, 2008, 2010; Souto et al., 2007; 2011), which has yielded similar homogenous microstructures in fast sintering cycles.

On the other hand, our researchers have also observed that, in two conditions, microwave hybrid fast heating can produce non-uniform microstructures. This was verified when susceptor material responded very rapidly to microwaves and the sample presented a very low loss, leading to a heating behavior similar to that of the conventional fast firing. The other condition was detected when processing high-loss materials. These materials coupled very efficiently with microwaves and their temperature rose very rapidly without similar heating of the susceptor. However, both conditions can be controlled by using different susceptors or by optimizing the sintering cycle. Figure 6 depicts the homogeneous microstructure of microwave hybrid fast sintered Ni-Zn ferrite (high-loss material). Using heating rates of up to 100oC/min up to 1000oC allied with lower heating rates at higher temperatures it was possible to reach high densification, 99% and uniform microstructures. Grain size distribution in the region near the surface and in the core of the samples was practically equal, highlighting the uniform temperature distribution inside the material, notwithstanding their high loss and rapid sintering.

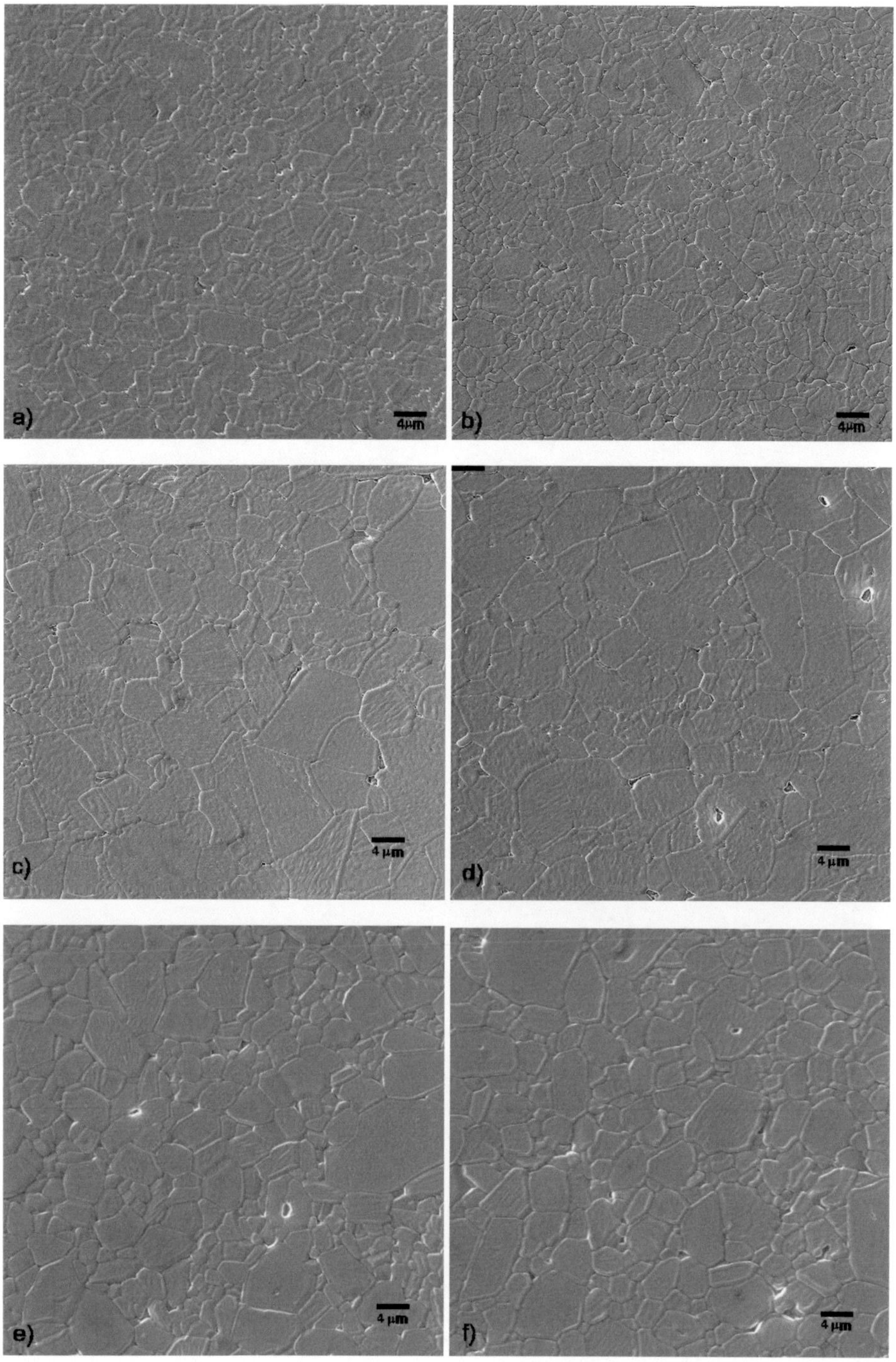

Fig. 5. SEM micrographs of MHFS alumina (Alcoa - A1000SG): at 1550°C (100°C/min up to 1100°C and 50°C/min up to 1550°C, soak time of 30 min – density of 95.3%) a) interior, b) surface; using 1.5kW for 25min (density of 96.6%), c) interior, d) surface; using 1.8kW for 20 min (density of 97.1%), e) interior, and f) surface.

An outstanding example of the use of microwave hybrid heating was the sintering of, ZrO_2/8% Y_2O_3 and zirconia/12% CeO_2 at 2.45 GHz by Janney (Janney et al., 1992). Although these materials could be sintered readily at 28 GHz, attempts to use 2.45 GHz, which involves lower equipment costs, were unsuccessful until SiC rods were inserted into the insulation surrounding the specimens, in what is referred to as the "picket fence" arrangement (Figure 7). The microwave energy initially heated the SiC rods, which resulted in uniform heating of the zirconia, and hence, in a homogenous coupling with microwaves after the critical temperature was reached.

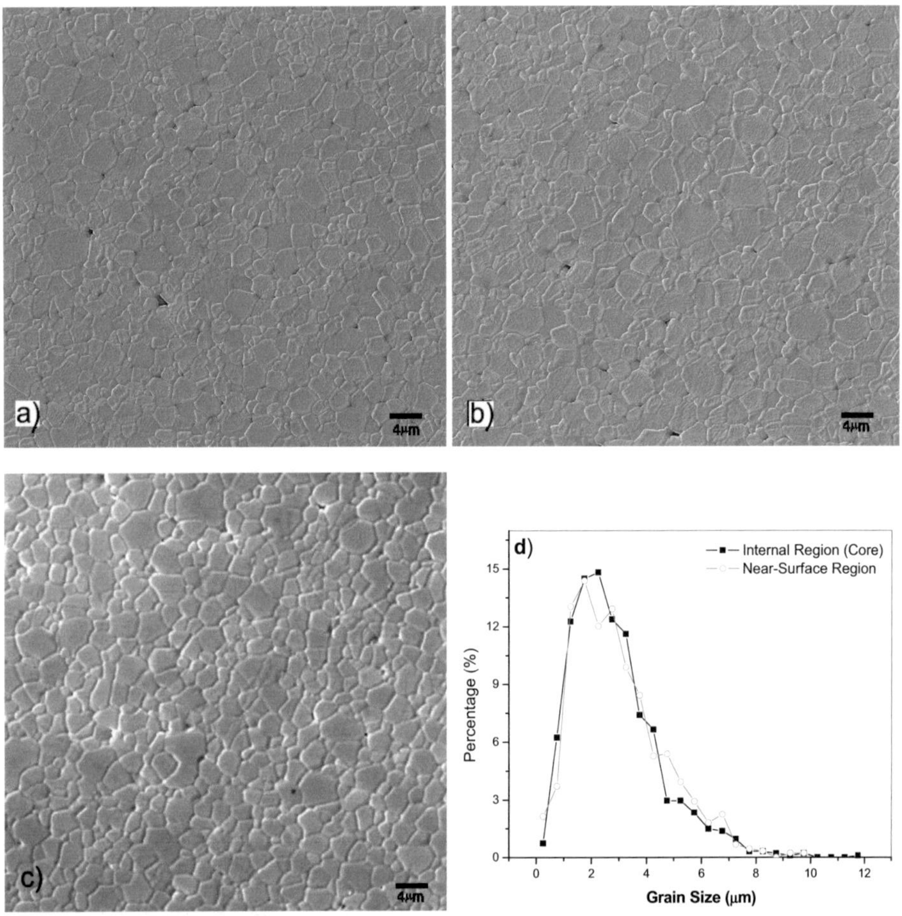

Fig. 6. SEM micrographs of homogeneous microstructures of Ni-Zn ferrite ($Ni_{0,5}Zn_{0,5}Fe_2O_4$): microwave hybrid fast sintered at 1220°C (100°C/min up to 1000°C and 50°C/min up to 1200°C, soak time of 20 min - density of 99%): a) sample interior (core), b) sample surface, c) sintered conventionally at 1200°C (5°C/min, soak time of 120min – density of 95.4%), and d) internal and near-surface grain size distribution of microwave-sintered sample.

Inside the insulation system, the thermal uniformity provided by susceptors plays an important role in reducing thermal runaways during fast sintering. Catastrophic overheating (thermal runaway) of processed material was observed in a large variety of ceramics (such as Al_2O_3, SiO_2, Fe_2O_3, ZrO_2, etc.), and was a serious problem in the processing of materials whose dielectric properties are abruptly altered with rising temperature. Studies (Spotz et al., 1995) demonstrated that the insulation normally used to contain heat inside the cavity can exacerbate the problem of thermal runaway. However, the data also suggested that thermal runaway may be minimized or controlled by hybrid heating, thus allowing low-loss ceramic materials to be processed using rapid firing.

However, care must be taken in designing the shape and size of the susceptor arrangement, because some susceptor materials, e.g., SiC, have high loss factors and therefore low penetration depths. Raising temperature leads to higher dielectric losses and significantly decreased penetration depths (see Figure 1). In this condition, it may not be possible to determine what the fraction of the sintering resulted from the interaction with microwaves and what fraction was sintered conventionally (by radiation of the susceptor material), as illustrated in Figure 8.

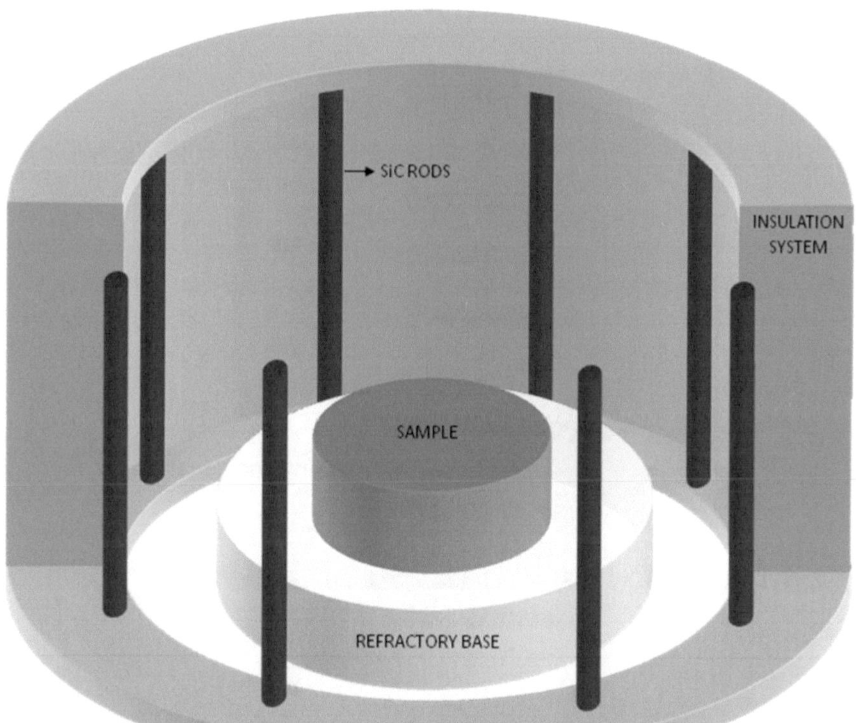

Fig. 7. Illustration of the "picket fence" arrangement.

Nanostructured materials have received much attention in recent years, due to their potential unusual physical and mechanical properties, such as superplasticity in ceramics at elevated temperatures, high hardness and mechanical strength, transparency for usually opaque materials, etc. However, one of the most challenging goals of modern ceramic technologies is still the manufacture of dense ceramic parts with submicrometric or nanostructured grains. Thus, controlled grain growth during the final steps of densification is an extremely important processing concern that requires the development of processing routes to effectively suppress grain growth during densification.

When ceramic materials are processed by traditional sintering techniques, the high temperatures required to fully densify ceramic powders result in large grain sizes due to Ostwald ripening. This makes it extremely difficult to obtain dense materials with nanometric and submicrometric grain sizes. Thus, although suitably uniform nanophase powder materials are becoming increasingly available, the fabrication of fully dense nanostructured products still offers challenges. In this context, the microwave fast sintering appears as an alternative for the sintering and densification of ceramics with suppressed grain growth.

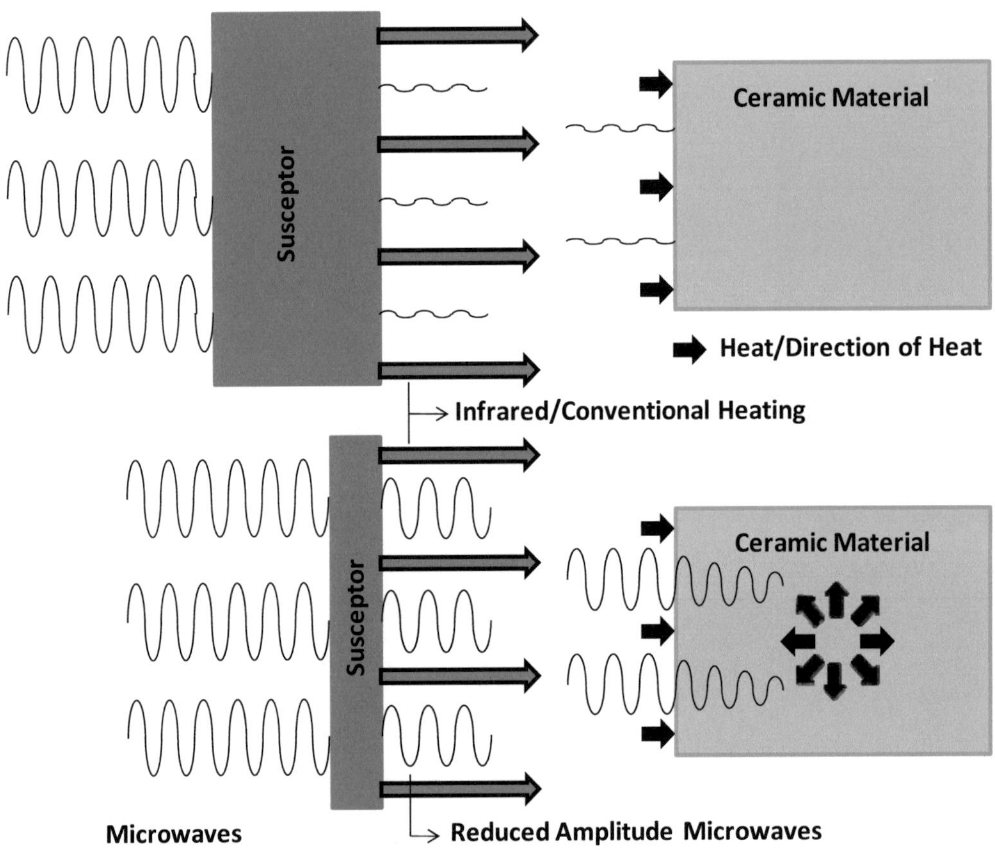

Fig. 8. Microwave hybrid heating arrangement/effect of susceptor material volume on the final heating pattern.

As mentioned early, the objective of fast firing is to enhance the densification rate in detriment to the coarsening rate by a rapidly approaching to the sintering temperature. Because coarsening mechanisms (e.g., surface diffusion and vapor transport) commonly prevail over densification mechanisms (e.g., lattice and grain-boundary diffusion) at lower temperatures, it has been suggested that rapid heating to higher temperatures can be beneficial to achieve high density allied to fine grain size. In this case, the shorter time spent at lower temperatures serves to reduce the extent of coarsening while the driving force for densification is not decreased significantly.

The authors' research group employed microwave hybrid fast heating to sinter submicrometric alumina (Sumitomo, AKP 53 - average particle size of 0.2 μm and surface area of 10.4m²/g). Rapid densification, 98.9%, suppression of grain growth, and submicrometric grain size microstructures were achieved (SEM micrographs in Figure 9)

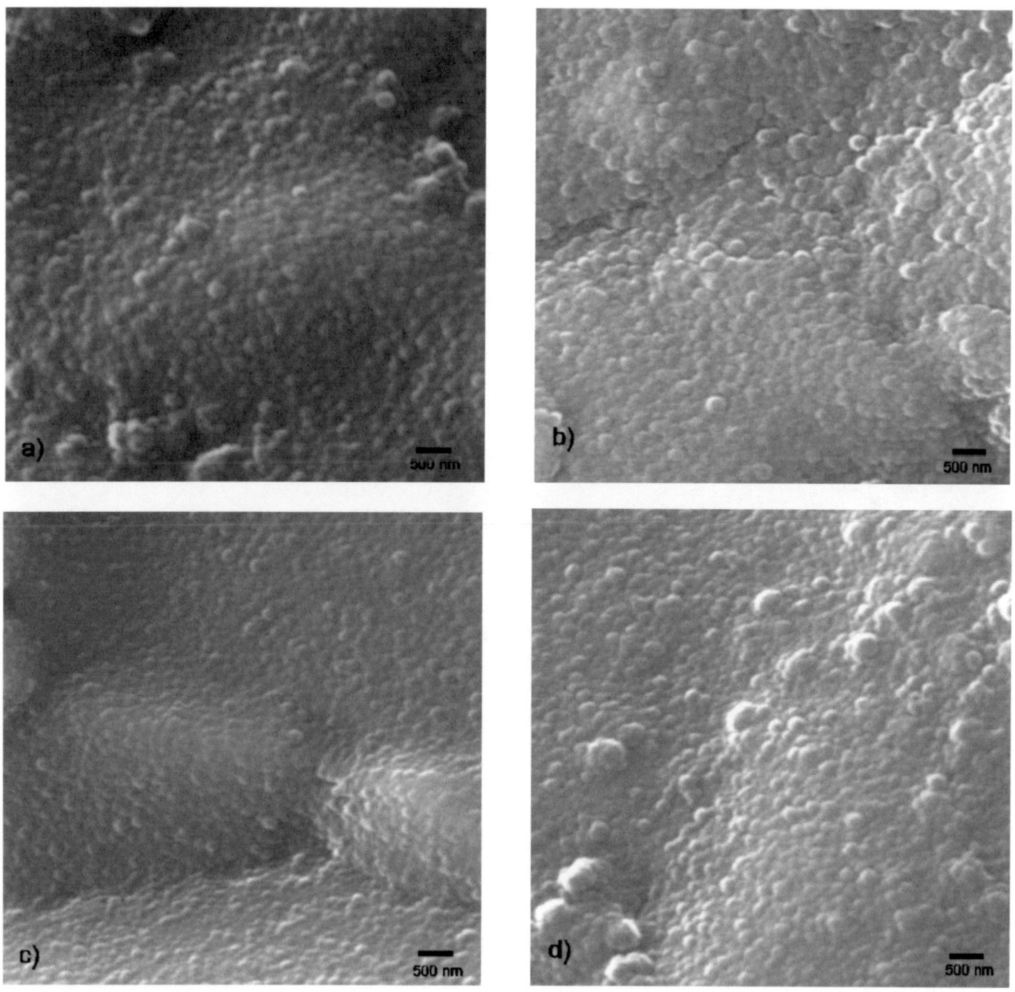

Fig. 9. SEM micrographs of the fracture surface of MHFS alumina (Sumitomo AKP - 53) processed at 2.45GHz with power level of 1.8kW for 26min (98.9% density): a) and c) grains in a internal region, b) and d) grains in a near-surface region.

Figure 10 shows samples of these alumina bodies sintered in a multimode furnace operating at 2.45 GHz. Microwave fast sintered alumina bodies were devoid of cracks, which is a reliable indicator of homogeneous temperature distribution in the bodies. In contrast, conventionally fast fired samples (50°C/min) presented cracks and non-uniform microstructures.

A alumina-zirconia nanocomposites were also rapid-sintered using the heating system developed for fast heating of alumina samples. Alumina (Sumitomo AKP 53)/ zirconia (yttria-stabilized zirconia powder, 3YSZ, Nanostructured & Amorphous Materials Inc., 58–76nm particle size) nanocomposites containing up to 5 vol.% zirconia, reached densities of approximately 99% in 35 min. Uniform microstructures and suppression of grain growth were obtained using rapid hybrid sintering. Figure 11 presents SEM micrographs of sintered nanocomposite containing 5%vol.% zirconia.

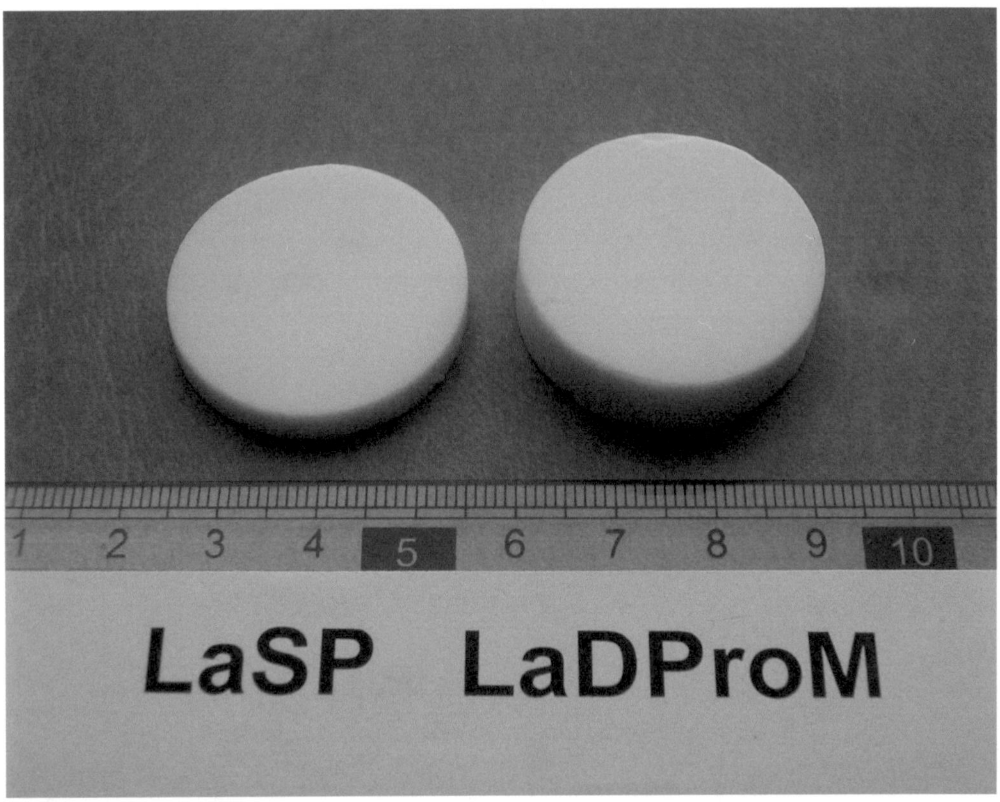

Fig. 10. Alumina samples processed by microwave hybrid fast sintering.

Grain growth suppression was also achieved in the processing of submicrometric zirconia (Zirconia Tosoh TZ-3Y, Tosoh Corporation, $D_{50} \approx 0{,}4\mu m$), which yielded a high densification of 99% a fine-grained microstructure with an average grain size of 0.6μm. The micrographs (Figure 12) of the internal and near-surface regions are practically identical, showing intergranular fracture, uniform grain size distribution, and no visible cracks, pores or abnormal grain growth.

Fig. 11. SEM micrographs of the fracture surface of MHFS nanocomposite alumina (Sumitomo AKP - 53)/zirconia (Nanostructured & Amorphous Materials, 3YSZ) (3 vol.% zirconia) processed at 2.45GHz with power level of 1.8kW for 35min (99% density).

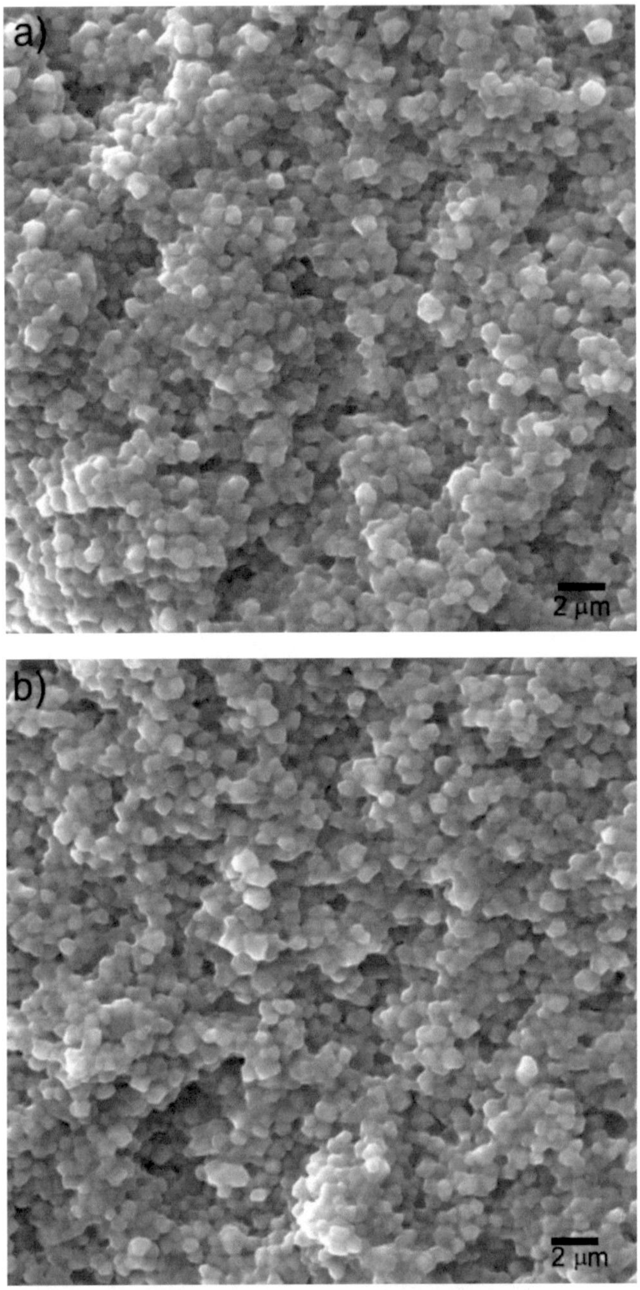

Fig. 12. SEM micrographs of the fracture surface of MHFS zirconia (Tosoh TZ-3Y) processed at 2.45GHz with power level of 1.8kW for 20min (99% density), a) internal region, b) near-surface region.

It is difficult to process zirconia in a 2.45GHz microwave furnace because the dielectric properties of this material change rapidly with temperature (see Figure 2) and its thermal conductivity is very low. Therefore, the use of hybrid heating is very important to produce uniform heating, thus avoiding thermal runaway and cracks. The rapid sintering of zirconia specimens without cracked, allied to the high densities achieved (> 97%), indicate a highly uniform distribution of temperature in the processed bodies. The non-uniform distribution of temperature in zirconia hinders its densification and may lead to thermal runaway may.

The results obtained by our research group indicate the successful use of microwave hybrid fast sintering in the densification of ceramic bodies and its effectiveness in suppressing grain growth in the final stages of sintering. The rapid densification achieved, accompanied by only minor grain growth, seems to be an indication of the microwave effect. However, regardless of the presence or absence of the microwave effect, the results of our research group indicate that microwave hybrid heating is a great potential method for rapid sintering of low and high-loss ceramic materials, yielding uniform microstructures and allowing for suppressed grain growth (when necessary).

4. Conclusions

Studies by our research group confirm that microwave hybrid fast sintering can be employed successfully to sinter low and high dielectric loss ceramics, producing highly dense ceramics and suppressing grain growth in the final stages of sintering. Controlled heating schedules and the use of suitably designed susceptor materials yield uniform microstructures without stress-related cracking or the development of incipient fusion, hot spots or thermal runaway. Thus, successful microwave fast sintering depends on the strict control of sintering conditions, striking an adequate balance between power and time or temperature and time to obtain highly dense and homogeneous microstructures.

5. Acknowledgements

The authors gratefully acknowledge the Brazilian agencies FAPESP (proc. 07/59564-0) and CNPq (Proc. 577201/2008-5/ and 303388/2009-9) for their financial aid.

6. References

Agrawal, D. K. (1998). Microwave processing of ceramics. *Current Opinion in Solid State & Materials Science*, Vol. 3, pp. (480-485), 1359-0286.

Andrade, O. M., Iskander, M. F. & Bringhurst, S. (1992). High temperature broadband dielectric properties measurement techniques. In: *Microwave Processing of Materials III, (Materials Research Society Symposium Proceedings)* Vol. 269, Beatty, R. L., Sutton, W. H. & Iskander, M. F., pp. (527-539), Materials Research Society, 1-55899-164-6, Pittisburg.

Arai, M., Binner J. G. P, Carr, G. E. & Cross, T. E. (1993). High temperature dielectric property measurements of engineering ceramics. In: *Microwaves: Theory and Application In Materials Processing II, Ceramics Transactions*, Vol. 36, Clark, D. E., Tinga, W. R. & Laia Jr, J. R., pp (483-492), The American Ceramic Society, 0-944904-66-1, Westerville.

Batt, J. Binner, J. G. P., Cross, T. E., Greenacre, N. R., Hamlyn, M. G., Hutcheon, R. M., Sutton, W. H. & Weil, C. M. (1995). A parallel measurement programme in high temperature dielectric property measurements: An update. In: *Microwaves: Theory and Application in Materials Processing III, Ceramics Transactions*, Vol. 59, Clark, D. E., Folz, D. C., Oda, S. J. & Silberglitt, R. pp. (243-258), The American Ceramic Society, 1-57498002-5, Westerville.

Booske, J, Cooper, R. F., Freeman, S. A., Meng, B, Ribakov, K. I. & Semenov, V. E. (1997). Thermal and nonothermal interactions between microwave fields in ceramics. In: *Microwaves: Theory and Application in Materials Processing IV, Ceramics Transactions*, Vol, 80, Clark, D. E., Sutton, W. H. & Lewis, D. A., pp. 143-151), The American Ceramic Society, 1-57498025-4, Westerville.

Bykov, Y. V., Rybakov K. I. & Semenov, V. E. (2001). High-temperature microwave processing of materials. *Journal of Physics D: Applied Physics*, Vol. 34, pp. (R55–R75), 0022-3727.

Clark, D. E. & Sutton, W. H. (1996). Microwave Processing of Materials, *Annual Review of Materials Science*, Vol. 26, pp. (299-331), 0084-6600.

Committee on Microwave Processing of Materials: An Emerging Industrial Technology, National Materials Advisory Board, Commission on Engineering and Technical Systems, National Research Council, (1994). *Microwave Processing of Materials*, National Academy Press, 0-309-05027-8, Washington.

Dé, A., Ahmad, I., Whitney, D. & Clark, D. E. (1991b). Microwave (hybrid) heating of alumina at 2.45 GHz: I. Microstructural uniformity and homogeneity. In: *Microwaves: Theory and Application In Materials Processing II, Ceramics Transactions*, Vol. 21, Clark, D. E., Gac. F. D. & Sutton, W. H., pp (319-328), The American Ceramic Society, 0-944904-43-2, Westerville.

Dé, A., I. Ahmad, E. D. Whitney, &. Clark. D. E. (1991a). Effect of Green Microstructure and Processing Variables on the Microwave Sintering of Alumina. In: *Microwave Processing of Materials II, (Materials Research Society Symposium Proceedings)*, Vol. 189, B. Snyder Jr., W. B., Sutton, W. H., Islanker, M. F. & Johnson, D. L., pp. (283-288), Materials Research Society, 1-55899-078-X, Pittsburgh.

Freeman, S. A., Booske, J. H., Cooper, R. F. & Meng, B. (1994). Microwave Radiation Effects on Ionic Current in Ionic Crystalline Solids. In: *Microwave Processing of Materials IV, (Materials Research Society Symposium Proceedings)*, Vol. 347, Iskander, M. F., Lauf, R. J. & Sutton, W. H., pp. (347-385), Materials Research Society, 1558992472, Pittsburgh.

Harmer, M. P. & Brook, R. J. (1981). Fast firing – Microstructural benefits. *Transaction of the British Ceramic Society*, Vol. 80, pp. (147-148), 0307-7357.

Huang, Z., Gotoh, M. & Hirose, Y. (2009). Improving sinterability of ceramics using hybrid microwave heating. *Journal of Materials Processing Technology*, Vol. 209, pp. (2446–2452), 0924-0136.

Janney, M. A., Calhoun, C. D. & Kimrey. H. D. (1992). Microwave Sintering of Solid Oxide Fuel Cell Materials, I: Zirconia-8 mol% Yttria. *Journal of the American Ceramic Society*, Vol. 75, pp. (341—346), 1551-2916.

Katz, J. D. (1992) Microwave Sintering of Ceramics. *Annual Review of Materials Science*, Vol 22, pp. (153-170), 0084-6600.

Kimrey, H. D., Kiggans, J. O., Janney, M. A. & Beaty, R. L. (1991). Microwave sintering of zirconi-toughened alumina composites. In: *Microwave Processing of Materials II,*

(Materials Research Society Symposium Proceedings), Vol. 189, B. Snyder Jr., W. B., Sutton, W. H., Islanker, M. F. & Johnson, D. L., pp. (243-256), Materials Research Society, 1-55899-078-X, Pittsburgh.

Kingery, W. D. , Bowen, H. K. & Uhlmann, D. R. (1976). *Introduction to Ceramics*, 2nd ed., John Wiley & Sons, 0-471-47860-1, New York.

Krage, M. (1981). Microwave sintering of ferrites. *American Ceramic Society Bulletin*, Vol. 60, pp. (1232-1234), 0002-7812.

Menezes, R. R. & Kiminami, R.H.G.A. (2008). Microwave sintering of alumina–zirconia Nanocomposites. *Journal of Materials Processing Technology*, Vol. 203, pp. (513–517), 0924-0136

Menezes, R. R., Souto, P. M., Fagury-Neto, E., Kiminami, R. (2005). Microwave sintering of ceramics, *Proceedings of the fourth world congress on microwave and radio frequency application*, 9780978622206, Austin, Novemeber, 2004.

Menezes, R. R., Souto, P. M., Kiminami, R. H. G. A. (2007). Microwave hybrid fast sintering of porcelain bodies. *Journal of Materials Processing Technology*, Vol. 190, pp. (223–229), 0924-0136.

Menezes, R. R., Souto, P. S. & Kiminami, R. H. G. A. (2010) Microwave Fast Sintering of Submicrometer Alumina. *Materials Research* Vol. 13, pp. (345-350, 1516-1439.

Meredith, R. (1998). *Engineers' Handbook of Industrial Microwave Heating*, The Institution of Electrical Engineers, 0-85296-916-3, London.

Metaxas, A. C., Binner, J. G. P. (1990). Microwave Processing of Ceramics, In: *Advanced Ceramic Processing and Technology, Volume 1*, Binner, J. G. P, pp. (285-362), Noyes Publications, 0-8155-1256-2, Park Ridge.

Oghbaei, M. & Mirzaee, O. (2010). Microwave versus conventional sintering: A review of fundamentals, advantages and applications. *Journal of Alloys and Compounds*, Vol. 494 pp. (175–189), 0925-8388.

Rahaman, M. N. (1995). *Ceramic Processing and Sintering*, Marcel Dekker Inc., 0-8247-9573-3, New York.

Rybakov, K. I. & Semenov, V. E. (1994a) A non-thermal vacancy-drift mechanism of plastic deformation of grains in ceramics during microwave sintering. In: *Microwave Processing of Materials IV, (Materials Research Society Symposium Proceedings)*, Vol. 347, Iskander, M. F., Lauf, R. J. & Sutton, W. H., pp. (661-666), Materials Research Society, 1558992472, Pittsburgh.

Rybakov, K. I. & Semenov, V. E. (1994b). Possibility of Plastic Deformation of an Ionic Crystal Due to the Nonthermal Influence of a High-Frequency Electric Field. *Physical Review B*, Vol. 49, pp. (64–68), 1098-0121.

Souto, P. M., Menezes, R. R. & Kiminami, R. H. G. A. (2007). Microwave hybrid sintering of mullite powders. *American Ceramic Society Bulletin*, Vol. 86, pp. (9201–9206), 1945-2705.

Souto, P. M., Menezes, R. R., & Kiminami, R.H.G.A. (2011). Effect of Y_2O_3 additive on conventional and microwave sintering of mullite, *Ceramics International*, Vol. 37, pp.(241–248), 0272-8842.

Spotz, M. S., Skamser, D. J. & Johnson, D. L. (1995). Thermal Stability of Ceramic Materials in Microwave Heating. *Journal of the American Ceramic Society*, Vol. 78, pp. (1041-1048), 1551-2916.

Sutton, W. H. (1989). Microwave processing of ceramic materials. *American Ceramic Society Bulletin*, Vol. 68, pp. (376-386), 0002-7812.

Thostenson, E. T. & Chou, T. W. (1999). Microwave processing: fundamentals and applications. *Composites: Part A*, Vol. 30, pp. (1055–1071), 1359-835X.

Tinga, W. R. & Voss, A. G. (1968). *Microwave Power Engineering*, Academic, New York.

Von Hippel, A. R. (1954a) *Dielectric and Waves*, John Wiley & Sons, New York.

Von Hipple, A. R. (1954b). *Dielectric Materials and Applications*. MIT Press, Cambridge.

2

Synthesis of Diamond Using Spark Plasma Sintering

Faming Zhang and Eberhard Burkel
Physics of New Materials,
University of Rostock, Rostock
Germany

1. Introduction

Carbon is a complex system rich in polymorphs due to its ability to form *sp*^{1}-, *sp*^{2}-, and *sp*^{3}-hybridized C-C bonds. One of the most outstanding achievements in carbon science was the synthesis of diamond from graphite under high-pressure and high-temperature (HPHT) conditions (Giardini et al., 1960). Since diamond particles and films have now been obtained by many other methods including detonation (Vereschagin et al., 1994), combustion flames (Hirose et al., 1990) and chemical vapour deposition (CVD) with RF plasma (Watanabe et al., 1992) or microwave plasma (Kobashi et al., 1988) etc., the HPHT method is still the most popular commercial method for the diamond synthesis. Nowadays, man-made diamond plays an indispensable role in modern industry for abrasives, tool coatings, microelectronics, optics and other applications.

Spark plasma sintering (SPS), commonly also defined as field assisted sintering (FAST) or pulsed electric current sintering (PECS) is a novel pressure assisted pulsed electric current sintering process utilizing ON-OFF DC pulse energizing. Due to the repeated application of an ON-OFF DC pulse voltage and current in powder materials, the spark discharge point and the Joule heating point (local high temperature-state) are transferred and dispersed to the overall specimen (Zhang & Burkel, 2010). The SPS process is based on the electrical spark discharge phenomenon: a high energetic, low voltage spark pulse current momentarily generates high localized temperatures, from several to ten thousand degrees between the particles resulting in high thermal and electrolytic diffusion. During SPS treatment, powders contained in a die can be processed for diverse novel bulk material applications, for example nanostructured materials, functional gradated materials, hard alloys, biomaterials, porous ceramics and alloys etc (Zhang & Burkel, 2010). The SPS has been used to prepare diverse advanced materials; nevertheless, it is still a new technique for the diamond synthesis.

In the year 2004, during the study of the thermal stability of multi-walled carbon nanotubes (MWCNTs) under various SPS conditions, Zhang et al first found that under SPS conditions of 1500 ^{o}C at very low pressure (80 MPa) carbon nanotubes were unstable and transformed into diamonds without any catalysts being involved (Zhang, Shen et al., 2005). The transformation mechanism involves the breakage of some C–C bonds, the formation of carbon nano-onions (multilayer fullerenes), and the nucleation and growth of the diamond phase within the onion cores (Shen, Zhang et al., 2006). The effect of the high temperatures and sparking plasmas is akin to the electron or ion beam irradiations or high pressures. As a result, diamond forms from the intermediate nano-onions when localized high energy conditions are satisfied (Shen, Zhang et al., 2006). Additionally, the low purity MWCNTs (60% purity) with a very cheap price also can be used as starting materials for the direct synthesis of diamond by the SPS technique (Zhang, Shen et al., 2006). It is postulated that the spark plasmas play a key role to provide most of the energy required in these diamond transitions. It is seen as an important evidence for the presence of spark plasmas during the SPS process. These studies indicate that the SPS has a potential to be used as an alternative method for diamond generation. But it needs further investigation to promote the SPS method to be used as a large-scale synthetic diamond production technique instead of the present hydrostatic HPHT method.

This chapter will focus on the synthesis of diamond using the SPS technique. We aim to promote the SPS method to be used as a large-scale synthetic diamond production technique as an alternative to the present HPHT method. In the first part of this chapter, the thermal stability of carbon nanotubes, C60 and graphite under the pulsed DC field of SPS and AC field of conventional sintering will be studied. In the second part of this chapter, the application of catalysts in the diamond synthesis by the SPS will be investigated. In the last part of this chapter, the factors that influence the diamond growth in the SPS, including carbon modifications and atmospheres (Vacuum, Ar) will be studied in order to increase the diamond sizes and transition rates.

2. Major raw materials and methods

- The MWCNTs with purity above 95.0% were obtained from Shenzhen Nanotech Port, Ltd., China. The C60 powders with purity of 99.5% were obtained from SES Research, Huston, USA. The graphite powders with purity of 99.0% were purchased from Alfa Aesar, Germany.
- The catalyst powders of Fe-35Ni, Ni etc. were purchased from Alfa Aesar, Germany. They were prepared by gas atomization method with 99.0% purity.
- The SPS experiments were conducted using a Model HP D-125/5 FCT spark plasma sintering system (FCT systeme GmbH, Rauenstein, Germany) installed in our Tycho Sinter Lab at the University of Rostock, Germany. The powders were pressed into a ф20 mm graphite die for the SPS treatment to form disk-shaped samples with thickness of 5-8 mm.
- The sintered samples were etched in a boiling solution of concentrated H_2SO_4 (90 vol.%) and HNO_3 (10 vol.%) for 2 h. The etched samples were washed using deionized water repeatedly, and dried in a vacuum oven.
- The phase identification of the etched carbon samples was performed using high-energy X-ray diffraction with energies of 80-100 keV at beamline BW5 (DESY/HASYLAB Hamburger Synchrotron Laboratory). The carbon samples were also analyzed by a Renishaw-2000 Laser Raman spectroscopy system with a He–Ne laser excited at 512 nm to identify the diamond phase.
- Scanning electron microscope (SEM, Zeiss Supra 25, Germany) and transmission electron microscope (TEM, Zeiss-Libra120, Germany) operating at 120 keV, were employed to characterize the products following the SPS treatment.

- The comparison investigation on the stability of the carbon materials was conducted by the in-situ high temperature X-ray diffraction at the MAX80/F2.1 high-pressure beamline of Helmholtz Centre Potsdam at HASYLAB/DESY Hamburg.

3. Stability of carbon nanotubes, C60 and graphite under various fields

The transformation of carbon nanotubes to diamond at very low pressure under SPS has been observed for the first time by Zhang, Shen et al. (2004). Recently, Inam et al reported that multiwall carbon nanotubes were not preserved for ceramic matrices that require high sintering temperatures (>1600°C) and longer processing times (>13 min) in the SPS (Inam et al., 2010). Zhang et al proposed that the spark plasmas may play a key role to provide most of the energy required in this diamond transition, and it provided an indirect way to validate the existence of the plasmas during the SPS. Due to the still on-going arguments about whether the spark plasmas actually occur during the SPS process (Anselmi-Tamburini et al., 2005, Hulbert et al., 2008), it needs further investigations on this point. In this part, we used such an indirect way to prove the presence of plasmas during the SPS. The thermal stability and phase transitional behaviour of carbon nanotubes, C60 and graphite were investigated under the SPS (pulsed DC field). For a comparison study, these carbon materials were also studied using the in-situ high temperature (AC field) synchrotron radiation X-ray diffraction.

3.1 Stability and phase transformation of carbon materials under pulsed DC field

The pure MWCNTs were SPSed at 1500 °C under pressure of 80 MPa for a holding time of 20 min. Figure 1(a) shows the synchrotron radiation-high energy X-ray diffraction patterns of the raw MWCNTs and the spark plasma sintered (SPSed) MWCNTs. The raw MWNCTs show a main diffraction peak at 3.43 Å corresponding to the CNTs (002) plane spacing, and weak peaks at 2.10 and 1.70 Å corresponding to the CNTs (100) and (004) plane spacing, respectively. After SPS processing, the MWCNTs diffraction peaks are still present in the sintered MWCNTs compacts, but the peaks of the CNTs (002), (100) and (004) are stronger than those in the raw MWCNTs. It indicates that the SPS process improved the crystallinity of the MWCNTs (Zhang, Mihoc et al., 2011). Additionally, new peaks were detected in the sample centered at 2.05, 1.23, 1.06 and 1.76 Å corresponding to the cubic diamond (ICDD No. 65-537) (111), (220), (311) and n-diamond (ICDD No. 43-1104) (200) plane spacing, respectively. Figure 1(b) shows the Raman spectra of the raw MWCNTs and the SPSed MWCNTs. The result of the raw MWCNTs show that their D band appeared at 1344 cm^{-1} and G band appeared at 1569 cm^{-1}. After SPS processing, the D peak shifted to 1333 cm^{-1} corresponding to the cubic diamond but there was still a weak peak at 1344 cm^{-1} belonging to the un-reacted MWCNTs, the G band shifted to 1566 cm^{-1} relating to the sp^2 bonded carbon vibrations. The results of the X-ray diffraction and Raman spectroscopy confirmed the diamond formation in the MWCNTs sample after SPS at 1500 °C under 80 MPa for 20 min.

Figure 2 shows the SEM micrographs of the raw MWCNTs and the spark plasma sintered MWCNTs at 1500 °C under 80 MPa for 20 min. The fibrous structures of the raw MWCNTs can be observed in the Figure 2(a). But these structures have disappeared, and some diamond crystals are found in the samples after SPS. Figure 2 (b) shows one diamond crystal with particle size of 35 μm around. In the background of this diamond crystal, no MWCNTs are found.

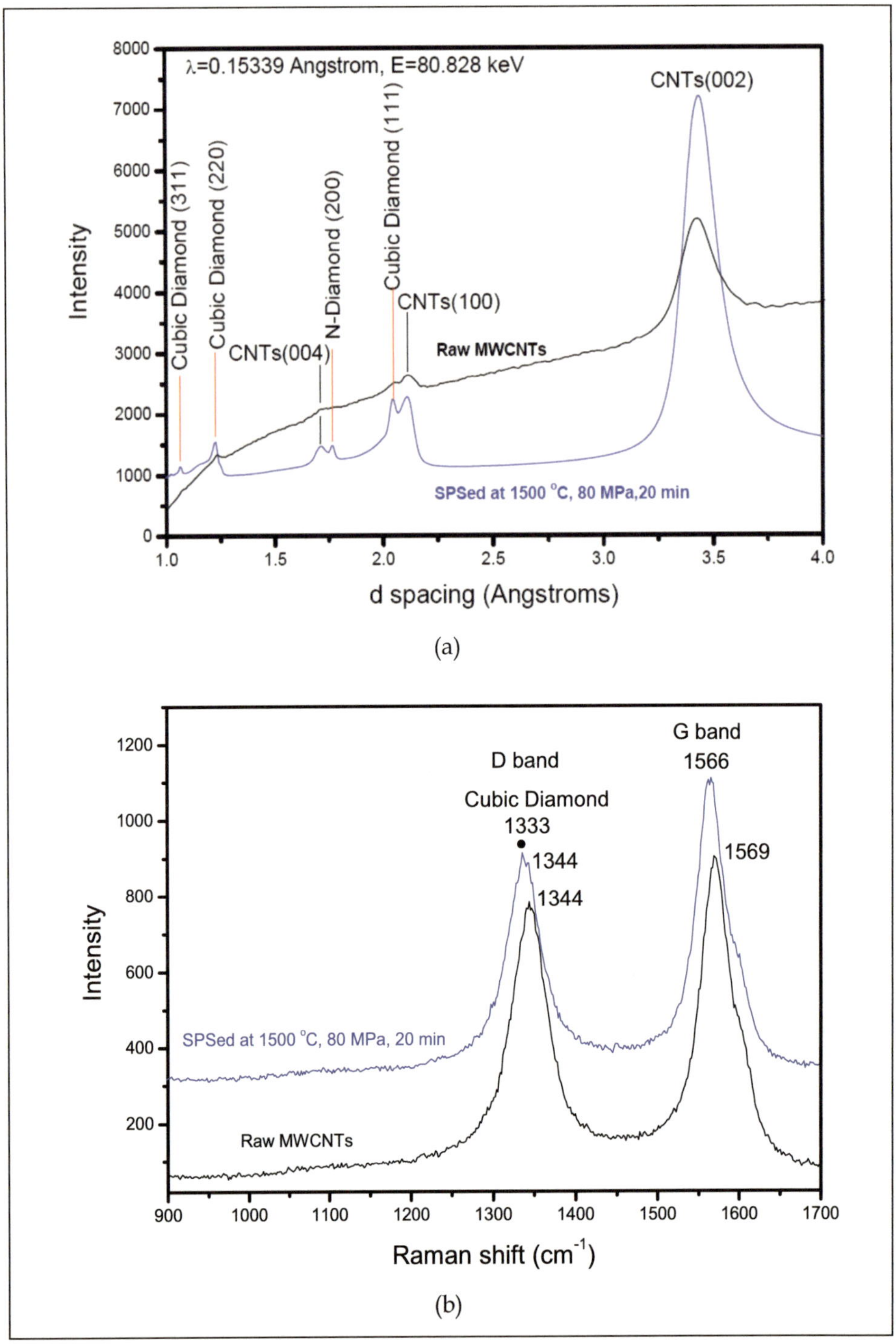

Fig. 1. Synchrotron radiation-high energy X-ray diffraction patterns (a) and Raman spectra (b) of the raw MWCNTs and the spark plasma sintered MWCNTs at 1500 °C, 80 MPa for 20 min.

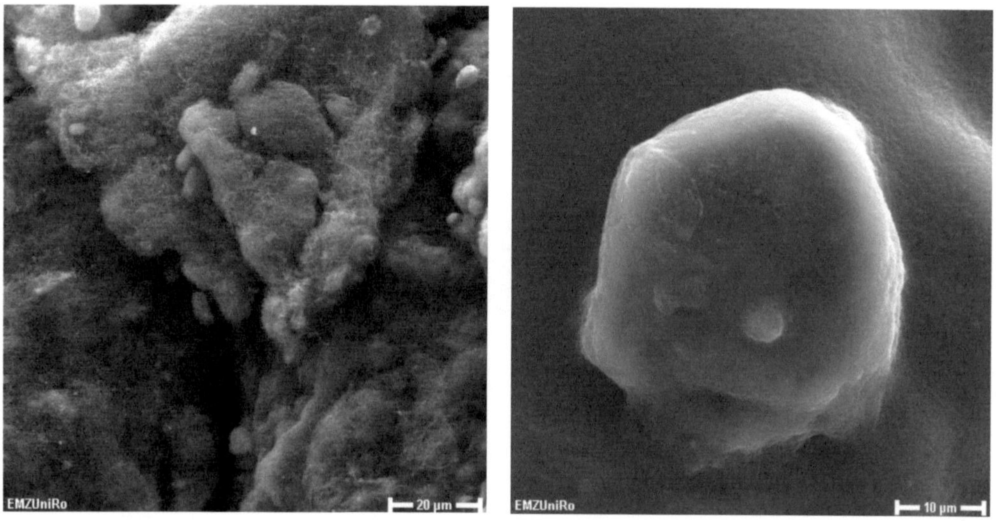

(a) (b)

Fig. 2. SEM micrographs of the raw MWCNTs (a) and the spark plasma sintered MWCNTs at 1500 °C, 80 MPa for 20 min showing one diamond crystal (b).

Figure 3(a) shows the synchrotron radiation-high energy X-ray diffraction patterns of raw C60 and the SPSed C60 at 1500 °C, 80 MPa for 20 min. The raw C60 exhibits diffraction peaks at d spacing of 5.01, 4.28, 4.11, 3.18, 2.9, 2.74 Å belonging to C60 (110), (112), (004), (114), (300), (006) planes (ICDD No. 47-0787), respectively. The C60 after SPS shows the cubic diamond diffraction peaks at d spacing of 2.06 and 1.23 Å and a broad graphite peak. The C60 diffraction peaks disappeared indicating the C60 has completely transformed into diamond and graphite phases after the SPS processing. Figure 3(b) shows the Raman spectra of the raw C60 and the SPSed C60. The raw C60 shows a sharp peak appearing at 1460 cm^{-1}, and two weak, broad peaks centered at 1568 and 1515 cm^{-1}. After SPS processing, it shows the cubic diamond peak at 1333 cm^{-1} and graphite peak at 1558 cm^{-1}, but the C60 peak at 1460 cm^{-1} disappeared. It is consistent with X-ray diffraction results that the C60 has completely transformed into diamond and graphite phases after SPS at 1500 °C under 80 MPa for 20 min.

Figure 4 shows the SEM micrographs of the raw C60 and the spark plasma sintered C60 at 1500 °C under 80 MPa for 20 min. The raw C60 powders are nano-particle agglomerates and show bundles of C60 in the Figure 4(a). Some diamond crystals with sizes from 2 to 8 µm can be observed in the Figure 4(b). The structures of C60 are not noticeable in the background of the diamond crystals.

Figure 5 (a) shows the synchrotron radiation diffraction patterns of the raw graphite and the SPSed graphite at 1500 °C under 80 MPa for 20 min. The raw graphite sample presents Graphite-3R and Graphite-2H diffraction peaks those are centered at 3.348 Å [G-3R(003)], 1.674 Å [G-3R(006)], 1.228 Å [G-3R(110)] (ICDD No. 26-1079), and 2.138 Å [G-2H(100)] , 2.039 Å [G-2H(101)], 1.16 Å [G-2H(112)] (ICDD No. 41-1487). However, the diamond phase is not found in the graphite samples after the SPS processing. Only, an increased intensity in the

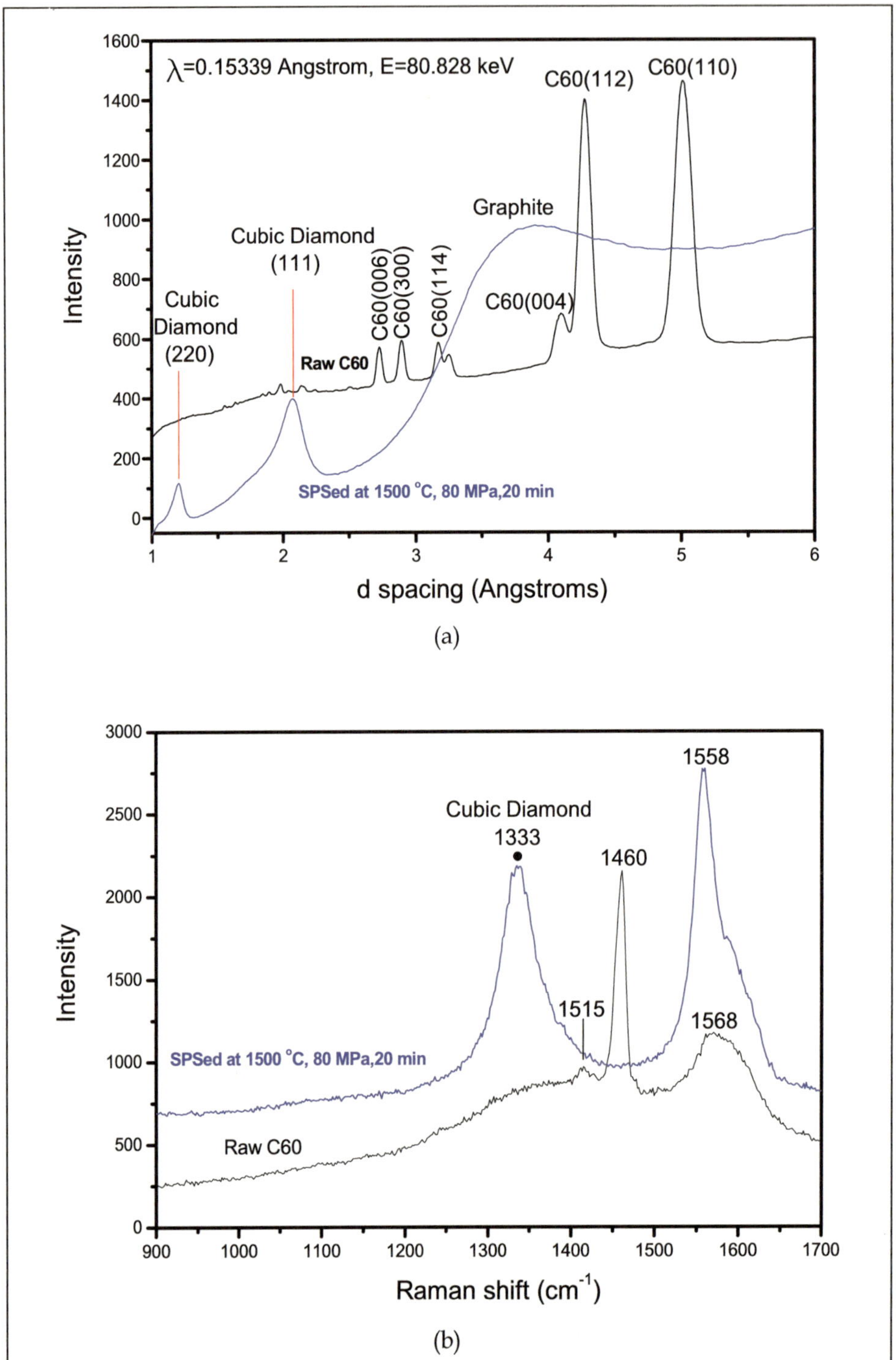

Fig. 3. Synchrotron radiation-high energy X-ray diffraction patterns (a) and Raman spectra (b) of the raw C60 and the spark plasma sintered C60 at 1500 °C, 80 MPa for 20 min.

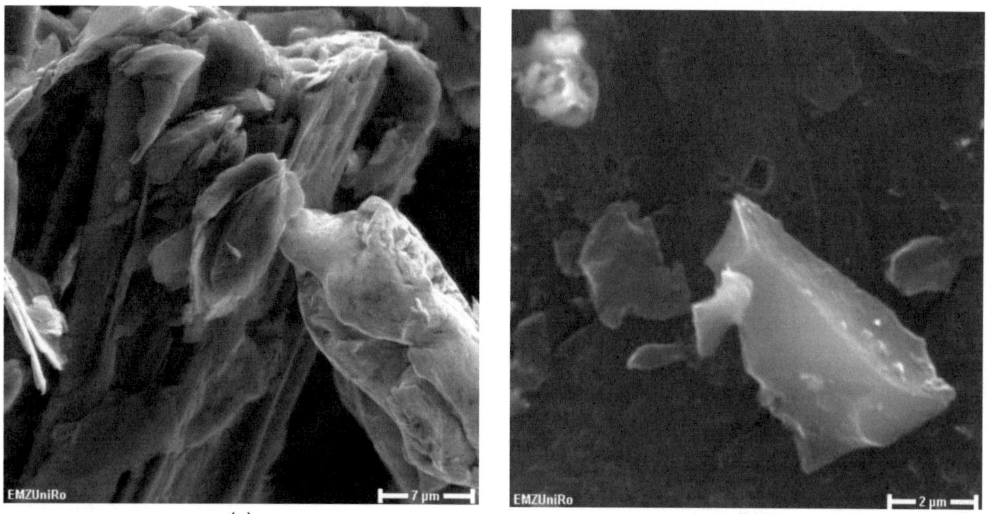

(a) (b)

Fig. 4. SEM micrographs of the raw C60 (a) and the spark plasma sintered C60 at 1500 °C, 80 MPa for 20 min (b).

graphite peaks indicating the improved crystallinity is visible. Figure 5(b) shows the Raman spectra of the raw graphite and the SPSed graphite. The raw graphite shows a sharp peak at 1579 cm^{-1}, and a weak peak at 1350 cm^{-1}. After SPS processing, the intensity of the peak at 1350 cm^{-1} has improved, but there is no diamond peak in the Raman spectra. The X-ray diffraction and Raman spectroscopy results confirmed that there is no diamond conversion from pure graphite after SPS at 1500 °C under 80 MPa for 20 min.

Fig. 6 shows the SEM micrographs of the raw graphite and the spark plasma sintered graphite at 1500 °C under 80 MPa for 20 min. The raw graphite shows the typical layered structure as shown in Figure 6(a). After the SPS of MWCNTs and C60 at the identical condition, there is no presence of diamond in the sample (Figure 6b). The sample shows the similar structure as the raw graphite. The SEM results agree well with the XRD and Raman results and confirmed that there is no diamond conversation from pure graphite after SPS at 1500 °C under 80 MPa for 20 min.

3.2 Stability and phase transformation of carbon materials under AC field

Figure 7 (a) shows the synchrotron radiation-in situ X-ray diffraction patterns of the pure MWCNTs at 80 MPa under different temperatures. In the in-situ sintering furnace (AC filed) of the MAX80/F2.1 high-pressure beamline. The combining peak of MWCNT and graphite has shifted to lower energy values. It indicates the thermal expansion of the nanotubes and graphite planes with the increase of temperature. The boron nitride (BN) peaks are from the container of the powder sample during the in-situ high temperature X-ray experiments. However, there is no diamond formation at or below the temperature of 1500 °C under 80 MPa. This means that the MWCNTs are dynamically stable at this temperature 1500 °C

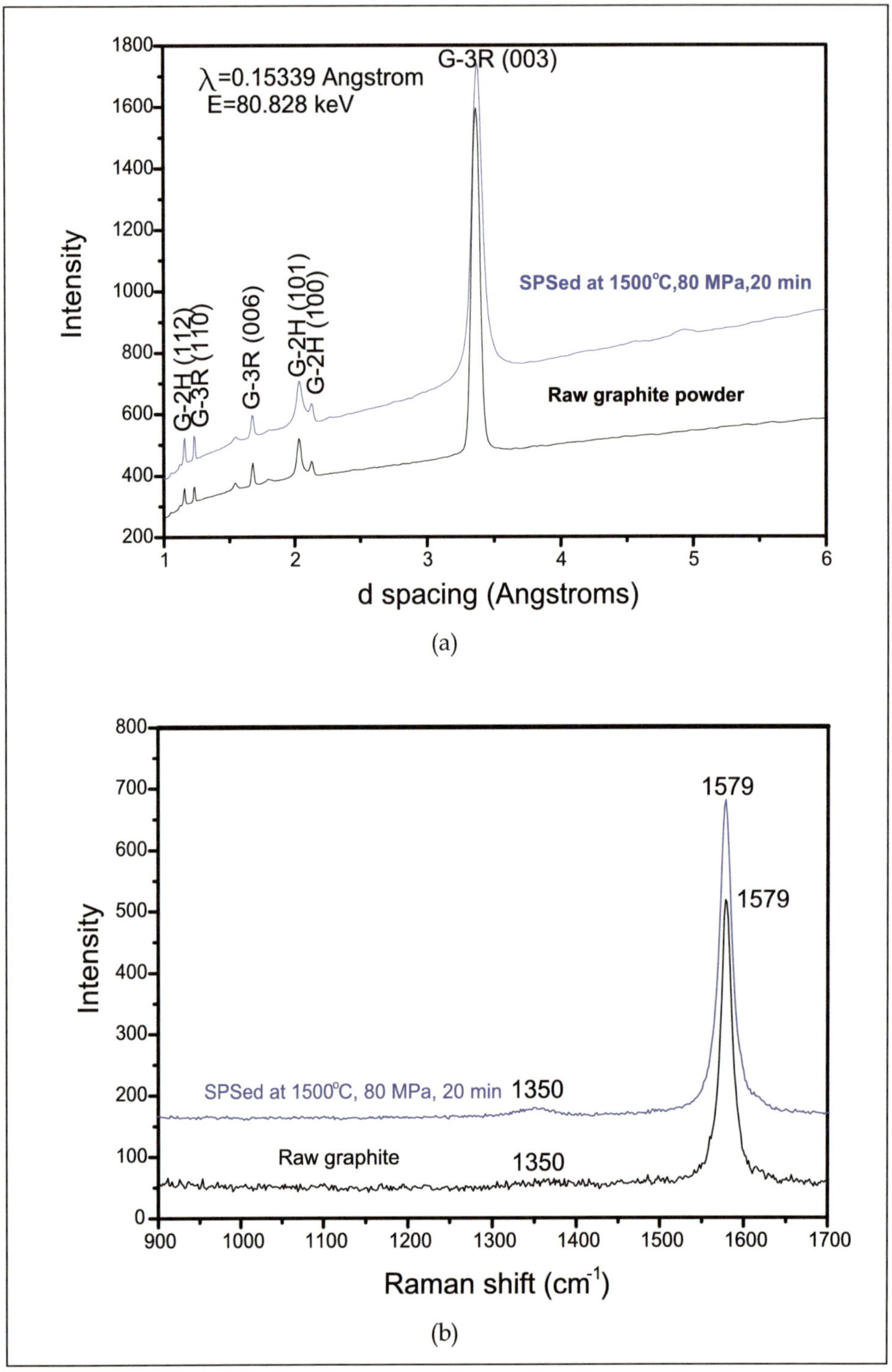

Fig. 5. Synchrotron radiation-high energy X-ray diffraction patterns (a) and Raman spectra (b) of the raw graphite and the spark plasma sintered graphite at 1500 °C, 80 MPa for 20 min.

(a) (b)

Fig. 6. SEM micrographs of the raw graphite (a) and the spark plasma sintered graphite at 1500 °C, 80 MPa for 20 min (b).

under 80 MPa in a non-oxygen atmosphere during the AC sintering. Figure 7 (b) shows the synchrotron radiation-in situ X-ray diffraction patterns of the pure C60 at 80 MPa under different temperatures. There is no diamond formation in the C60 sample. It shows that the C60 is stable below the temperature of 900 °C. However, the C60 is unstable above that temperature point. The C60 (110) peak disappeared above temperature of 900 °C and C60 (112) peak disappeared above temperature of 990 °C. It is found that the graphite is very stable in the in-situ high temperature X-ray experiments at or below 1500 °C under 80 MPa.

3.3 Mechanisms

Synchrotron radiation-high energy X-ray diffraction was used to identify the diamond phase in the carbon samples after SPS. In order to confirm the diamond formation, Raman spectroscopy was also used to identify the formation of sp^3 bonded diamonds. By using the high energy X-ray diffraction and Raman spectroscopy, the cubic diamond phases were identified and confirmed in the SPSed MWCNTs and C60 samples. The n-diamond was also found in the SPSed MWCNTs sample. The n-diamond is a new kind of carbon allotrope, which is a metallic form of carbon with face-centred cubic structure. It is a metastable and intermediate phase, can decompose slowly at room temperature, and has been synthesized accidentally by various processes (Zhang, Mihoc et al., 2011). It is noted that the n-diamond can also be synthesized by the SPS process. The standard d spacings of the cubic diamond (111), (220) and (311) planes are centered at 2.059 Å, 1.261 Å and 1.075 Å (ICDD No. 65-537). The cubic diamond in the SPSed MWCNTs centered at 2.05, 1.23 and 1.06 Å, and in the SPSed C60 appeared at 2.06 and 1.23 Å spacing. The diffraction peaks of the synthesized diamond from MWCNTs and C60 are very close to the standard diamond diffraction data, but there is a little shift. The diamond peak shifts are due to the existence of residual stress

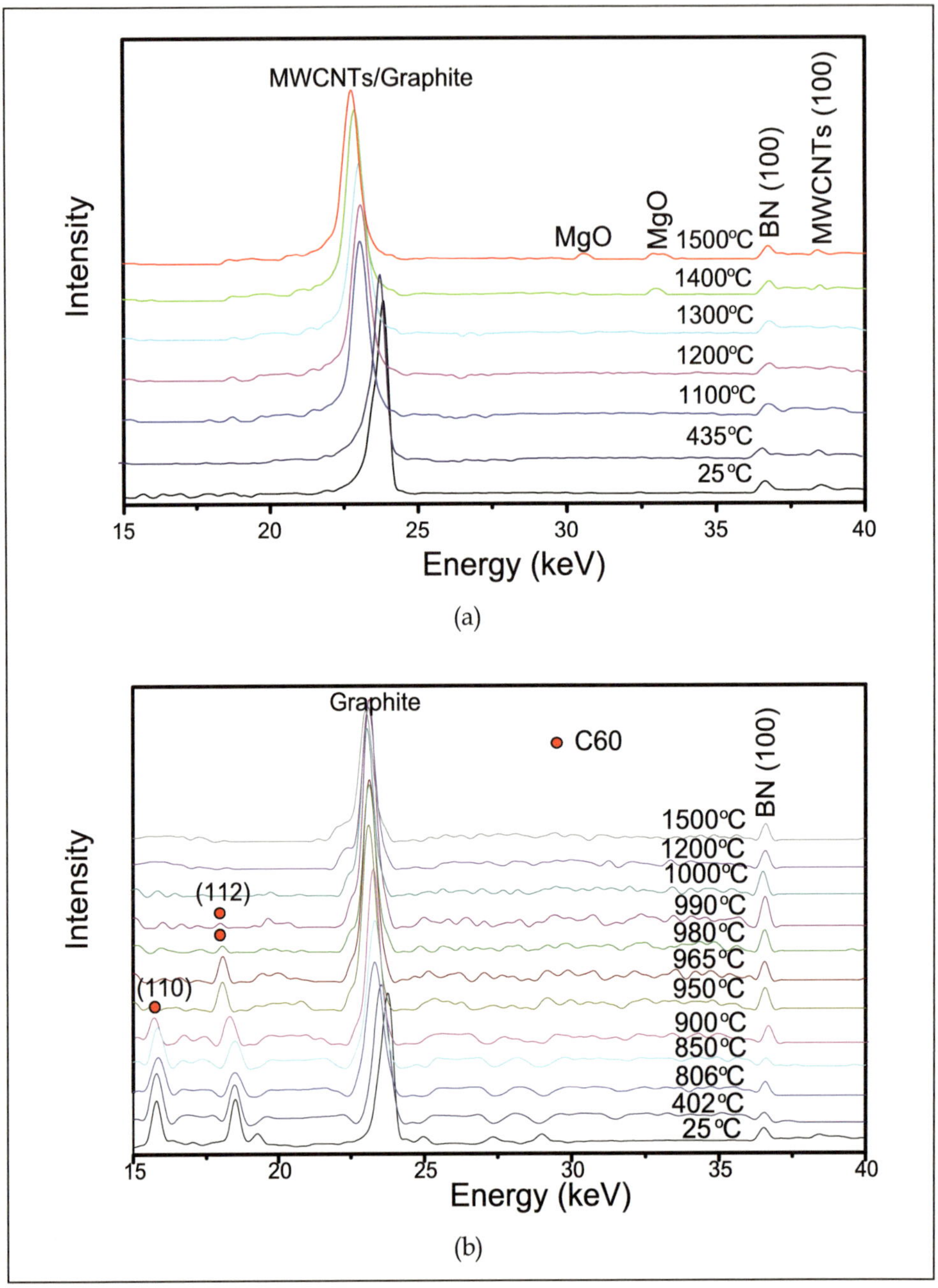

Fig. 7. Synchrotron radiation-in situ X-ray diffraction patterns of the pure MWCNTs (a) and C60 (b) at 80 MPa under different temperatures.

in the synthesized diamonds from MWCNTs and C60 by using the SPS. The residual stress of the diamond is because of the stress that remains after the original cause of the stresses (uniaxial forces, heat gradient) has been removed after the SPS processing. In this study, an uniaxial force of 80 MPa was applied and there generally existed some temperature gradients during the operation of the SPS. Therefore, the diamond peaks in the SPSed MWCNTs and C60 have shifted a little. Combining the results of the Raman spectroscopy, the formation of diamond phases in these MWCNTs and C60 samples is confirmed. It is found that there are no C60 peaks in the X-ray diffraction and Raman results of the SPSed C60 sample, but there are strong unreacted MWCNTs peaks in the SPSed MWCNTs sample, and there are no diamond phases in the SPSed graphite sample. There exists a high activation barrier from the graphite, MWCNTs and C60 to diamond, the exact height of which is unknown. The results this study indicate that the activation barrier between the C60 and diamond is lower than that of the MWCNTs with diamond, and the barrier between MWCNTs and diamond is lower than that of the graphite with diamond. The graphite is the most stable crystalline modification of carbon among the MWCNTs, C60 and graphite allotropes under the SPS processing.

The SPS is a remarkable technique to synthesize and consolidate a large variety of materials. The process typically uses moderate uniaxial pressures usually below 100 MPa in combination with a pulsing on-off DC current during its operation. There are many mechanisms proposed to account for the enhanced sintering abilities of the SPS process; for example, field assisted diffusion, spark impact pressure, plasma cleaning of particle surfaces, Joule's heating, local melting and evaporation especially in metallic systems, surface activation on particles and electron wind force (Zhang, Mihoc et al., 2011). The one that draws the most controversy of these mechanisms involves the presence of momentary plasma. In this study, the diamond converted from the MWCNTs and C60 without any catalysts being involved in the SPS. However, the parallel investigations by using the synchrotron radiation in-situ high temperature X-ray diffraction show that there is no diamond formation in the MWCNTs and C60 samples in the AC sintering at the same pressure (80 MPa) and temperature (1500 ^{o}C). What is the phase transition mechanism from MWCNTs and C60 to diamond in the SPS? Such a clear, significant difference in the products is due to the special sintering principle of SPS. It is a field activated sintering technique based on an DC electric spark discharge phenomenon, i.e. a high energy and low voltage spark pulse DC momentarily generates sparking plasma between particles, which causes localized high temperatures. It is an electric AC heating in the in-situ high temperature experiments. Without plasma effect, it would need 8000-10000 ^{o}C at pressure of 80 MPa to get diamond from the MWCNTs and C60, as we calculated. Therefore, super-high pressures (5-10 GPa) are required for the diamond formation in the hydrostatic HPHT technique. Since the SPS only needs MPa level pressure, it is believed that the plasma plays the key role for the diamond transformation from the MWCNTs and C60. The high current, low voltage, momentary pulsed plasma discharge have generated highly localized Joule's heating up to a few thousand degrees Celsius between particles in few minutes. The current density in the SPS is typically on the order of 10^2 A/cm^2 and is highly concentrated at the inter-granular contact or interface (Yang et al., 2010). The momentary pulsed plasma provided energy equivalent to thousand degrees to help the nano-carbon across their activation barriers to the diamond phase. It leads to the transformation of mainly sp^2 bonded MWCNTs and C60 to sp^3 bonded diamonds. Despite the on-going argument about whether the spark plasmas actually occur during the SPS process, our present study, regarding generating diamond under such a low pressure, suggests that such spark plasmas indeed take place during SPS of these nano-carbon materials with excellent electrical conductivities and high surface areas. The plasmas generated very high localized temperatures up to about 8000-10000 ^{o}C and dramatically reduced the pressures required for diamond formation from the GPa to the MPa level. Eventually, this research provided some new indirect evidences for the presence of plasmas during the SPS operation. Therefore, we take plasma into consideration in the thermodynamic analysis (Zhang, 2005; Zhang, Mihoc et al., 2010). The total energy for the diamond formation:

$$Q = \Delta H_T + Q_P + \Delta H_M \,,$$

where Q is Total Energy, ΔH_T is the Energy due to temperature difference, Q_p is the Energy due to pressure difference, ΔH_M is the Energy due to plasma effect. The enthalpy of plasma:

$$H = H_E + H_K + H_D + H_I \,,$$

where H is the plasma contribution, H_E is the kinetic contribution, H_K is the excitation contribution, H_D is the dissolution contribution, H_I is the electrolytic contribution. Then,

$$dS = \frac{\delta Q}{T} \,,$$

$$\Delta Q(T) - \Delta Q(T_0) = \Delta S(T_0 - T) \,,$$

Where T is the temperature, T_0 is the starting temperature, ΔQ is the difference of mol free energy, ΔS *is* the difference of mol entropy. Only when $\Delta Q(T)<0$, MWCNTs and C60 can be transformed into diamond. So, we can get an equation:

$$T > T_0 + \frac{\Delta Q(T_0)}{\Delta S}$$

The effect of the plasmas in the SPS has increased the entropy ΔS of the whole SPS system resulting in a lower sintering temperature T for the diamond formation. Diamond were converted from MWCNTs and C60 at 1500 °C under very low pressure of 80 MPa. The SPS is a marvelous process to prepare a wide range of advanced materials. The technique significantly uses uniaxial pressures normally 30-100 MPa to integrate a on-off DC current while its running. In this study, the diamond conversion in the SPSed MWCNTs and C60 samples without any catalysts being involved has validated the high localized temperatures between particles. This is due to the presence of momentary plasmas during SPS of these electrically conductive and high surface area nano-carbon materials. The plasmas have increased the entropy of the whole SPS system resulting in milder conditions for the diamond formation.

In summary, the thermal stability of MWCNTs, C60 and graphite has been investigated under the pulsed DC field in a SPS furnace. Cubic diamond and n-diamond have been converted from pure MWCNTs; cubic diamond has been converted from pure C60, both without catalysts being involved by the SPS at conditions of 1500 °C, 80 MPa for 20 min. There was no notice of diamond formation in the case of pure graphite sample processed by SPS at this condition. The graphite is the most stable crystalline modification of carbon among the MWCNTs, C60 and graphite allotropes under the SPS. The parallel investigations by using the synchrotron radiation in-situ high temperature (AC field) X-ray diffraction show that there is no diamond formation in the MWCNTs and C60 samples at the same pressure (80 MPa) and temperature (1500 °C). Their phase transitional mechanism from MWCNTs and C60 to diamond indicated the high localized temperatures between particles due to the presence of momentary plasmas during the SPS process. The plasmas have increased the entropy of the SPS system resulted in milder conditions for the diamond formation.

4. Diamond synthesis with catalysts by the SPS

In the HPHT method, the involved solvent catalysts could decrease the energy barrier and affect the rate of a kinetics reaction for diamond nucleation and contribute to the formation of diamond from graphite. Besides being able to reduce the transforming temperature and pressure from graphite to diamond, they can also affect the quality and crystal form of the diamond (Zhang, Adam et al., 2011). It is indicated that the solvent catalysts may have the same effects to promote diamond growth from MWCNTs and graphite in the SPS method In this part, the catalysts were involved in the SPS diamond synthesis with carbon modifications of MWCNTs and graphite and the effects of catalysts were investigated.

4.1 Carbon nanotubes with FeNi catalyst

Currently preferred metal catalyst materials are Fe-Ni alloys, such as Fe-35Ni, Fe-31Ni-5Co, Fe-30Ni, and other INVAR alloys, where Fe-35Ni being the most preferred and more readily available (Zhang, Adam et al., 2011). In order to increase the transitional rate of diamond, the Fe35Ni alloy powders were chosen as catalysts for diamond synthesis from MWCNTs by the SPS method here. The starting MWCNTs materials have an external and internal diameter of ca. 40 nm and 20 nm, respectively. The MWCNTs/FeNi samples were sintered in the SPS furnace at various temperatures under 70 MPa for 20 min.

Figure 8 (a) shows the Raman spectra of the starting MWCNTs and the spark plasma sintered MWCNTs/Fe35Ni samples at 1100-1500 ^{o}C before etching. The diamond band (D band) of the starting MWCNTs appeared at 1345 cm^{-1}. After spark plasma sintering at 1100 ^{o}C, the D band has shifted to 1342 cm^{-1}. It was found that the characteristic Raman shift of the cubic diamond phase appeared at 1333 cm^{-1} in the 1200-1500 ^{o}C sintered samples. The D band shifted from the starting 1345 cm^{-1} to 1333 cm^{-1} indicating the diamond formation above temperatures of 1200 ^{o}C. The broad peak of Raman spectra at 1333 cm^{-1} in the 1200-1500 ^{o}C sintered samples is due to the existence of un-reacted MWCNTs. The G band is due to the E_{2g} mode of graphite band (G band), relating to the sp^2 bonded carbon vibrations in a 2-dimensional graphitic hexagonal lattice. The G bands appeared at 1570 cm^{-1} in the starting MWCNTs and 1100 ^{o}C sintered sample, has shifted to 1574 cm^{-1} (1200 ^{o}C), 1572 cm^{-1} (1300 ^{o}C), 1576 cm^{-1} (1400 ^{o}C), 1561 cm^{-1} (1500 ^{o}C), which implied the vibrations of sp^2 bonded carbon during the SPS. The Raman results indicated that diamonds are converted from the MWCNTs/Fe35Ni at temperatures of 1200-1500 ^{o}C.

Additionally, all the samples were etched in boiling acid to remove the FeNi catalysts and the un-reacted MWCNTs. Figure 8(b) shows the X-ray diffraction patterns of the starting MWCNTs, Fe35Ni catalyst, and the spark plasma sintered MWCNTs/Fe35Ni samples at 1100-1500 ^{o}C after etching with a CuKα radiation Bruker-XRD. The starting MWCNTs show the (002) plane at 2θ of 25.86 degree without Ni and La catalysts peaks. The Fe35Ni catalysts show diffraction peaks at 2θ of 43.60 and 50.79 degree. After etching of the obtained MWCNTs/Fe35Ni samples, no obvious Fe35Ni diffraction peaks were detected in the 1100-1500 ^{o}C sintered samples. It indicated that the FeNi catalysts have been completely removed from the carbon samples by the boiling acid treatment. The peak at 2θ of 42.90 degree in the raw CNTs has shifted to 43.37 degree in the 1100 ^{o}C sintered sample.

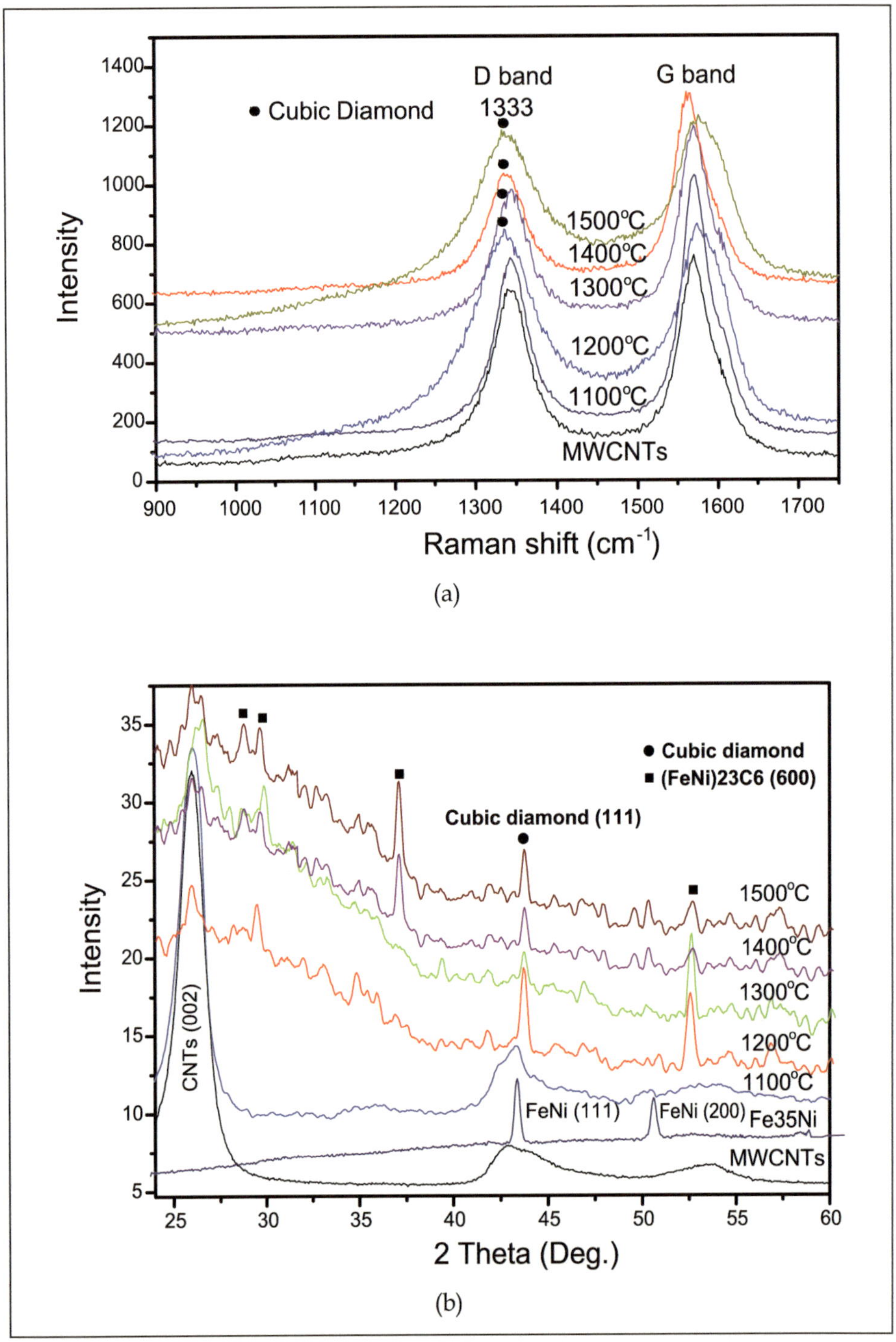

Fig. 8. Raman spectra of the starting MWCNTs and the spark plasma sintered MWCNTs/Fe35Ni samples at 1100-1500 °C before etching (a), and X-ray diffraction patterns of the starting MWCNTs, Fe35Ni catalyst, and the sintered samples after etching.

It is noted that this sample is in the transitional stage from CNTs to diamond. It is identified that the cubic diamond peak at 2θ of 43.95 with d spacing of 0.26 nm in the samples of 1200-1500 °C. There is still a broad CNTs (002) peak indicating there are some un-reacted and un-removed CNTs in the samples. Additionally, haxonite $(Fe,Ni)_{23}C_6$ peaks are found in the XRD results of 1200-1500 °C. With temperature increase from 1200 to 1500 °C, the haxonite peaks get stronger and stronger. This is due to the reaction between the FeNi catalysts and the MWCNTs at high temperatures. The Raman and XRD results confirmed the diamond formation in the 1200-1500 °C sintered samples.

The SEM micrographs of the spark plasma sintered MWCNTs/FeNi samples at 1100 °C and 1200 °C after etching are shown in Figure 9. Compared with the starting CNTs, the MWCNTs in the 1100 °C sample were almost melted and adhered together, but the tubular structure of the CNTs is noticeable in the growing diamond (Figure 9a). After sintered at 1200 °C, diamond crystals with sizes of 10-40 μm are observed in the samples (Figure 9b). These diamond crystals are in shape of hexahedron. Some flake-like carbons are noticed in the sample, as indicated by circles. The higher magnification SEM micrograph shows that the diamond crystals without residual CNTs left on their surface (Figure 9c). The high magnification micrograph in Figure 9 (d) indicated the layer-by-layer texture on the diamond crystals. The SEM micrographs of the spark plasma sintered MWCNTs/FeNi samples from 1300 °C to 1500 °C after etching were also studied (Zhang, Adam et al., 2011).. The particle size of the diamond crystals did not increase with the increase of sintering temperatures. A 1500 °C synthesized diamond with crystal size about 20 μm also showed clear layer-by-layer textures. On the matrix of this diamond crystal, there are many flake-like carbons. Such flake-like carbons were found in all these samples from 1300 to 1500 °C. The carbon flakes in these higher temperature sintered samples are similar to those in the 1200 °C sintered one. These flakes also showed layer-by-layer microstructures. The SEM observations agree well with the Raman and XRD results.

Figure 10 shows the TEM micrographs and selected area diffraction patterns of the spark plasma sintered MWCNTs/Fe35Ni sample at 1200 °C after etching. There are some mono-crystal and poly-crystal diamonds in the samples (Figure 10a-d). The monocrystalline diamond also shows the layer-by-layer structures (Figure 10a), which is consistent with the results of the SEM. The selected area diffraction pattern of the crystal in the bottom of the Figure 10 (a) confirmed the diamond is mono-crystal along [110] direction (Figure 10b). The poly-crystal diamond is in size of tens of micrometers (Figure 10c). The selected area diffraction pattern with diffraction rings were calculated and confirmed the diamonds are cubic poly-crystals (Figure 10d). The Raman, XRD, SEM, and TEM identification results have confirmed the diamond formation in MWCNTs with Fe35Ni as catalysts at a lower temperature of 1200 °C.

The Raman, XRD, SEM and TEM results confirmed that monocrystalline and polycrystalline diamonds were synthesized in the MWCNTs/Fe35Ni sample after SPS at temperatures above 1200 °C. The SEM results showed better diamond crystal shapes in the 1200 °C sintered samples. The TEM with selected area diffraction patterns showed the existence of diamond mono-crystals and poly-crystals in the 1200 °C sintered samples. Higher temperatures (1300-1500 °C) did not lead to larger diamond crystals as seen in the SEM images. The temperature 1200 °C is the optimal SPS temperature for the MWCNTs/Fe35Ni samples. This temperature 1200 °C for the diamond synthesis is much lower than that of the MWCNTs without catalyst (1500 °C) in our previous research (Zhang, Shen et al., 2005).

Fig. 9. SEM micrographs of the spark plasma sintered MWCNTs/Fe35Ni samples at 1100 °C (a), 1200 °C (b-d) after etching, exhibiting the growing process of diamond.

It indicates that the FeNi catalysts are effective to enhance the diamond conversion from MWCNTs in the SPS technique. The melting point of Fe35Ni alloy is about 1460 °C as measured in its phase diagram. During the SPS processing, the melting point of this Fe35Ni powder has been decreased due to the pulsed current induced powder activation, and the applied pressures. There is usually some temperature difference between the mold surface and the actual temperature in the SPS sample. The temperature measurement design in the FCT spark plasma sintering system allowed a very accurate temperature control since the temperature difference between the centre of the sample and the controlling pyrometer was always below 5 °C (Vanmeensel et al., 2005). The catalysts of Fe35Ni alloy powders were melted at SPS temperature of 1200 °C, which was noticed during the SPS of the MWCNTs/Fe35Ni sample ; so that, it reduced the SPS temperature to 1200 °C and the pressure to 70 MPa for the diamond synthesis as well as increased the diamond transition rate using the MWCNTs as carbon sources. In general, milder conditions were realized for the diamond synthesis by using the Fe35Ni catalysts in this study. In the HPHT method, the carbon-carbon diagram for the diamond synthesis is crucial (Novikov, 1999). In this SPS method, there will be a new carbon-carbon diagram for the diamond synthesis, which can predict the optimal temperature, pressure regions for the diamond synthesis in this SPS method.

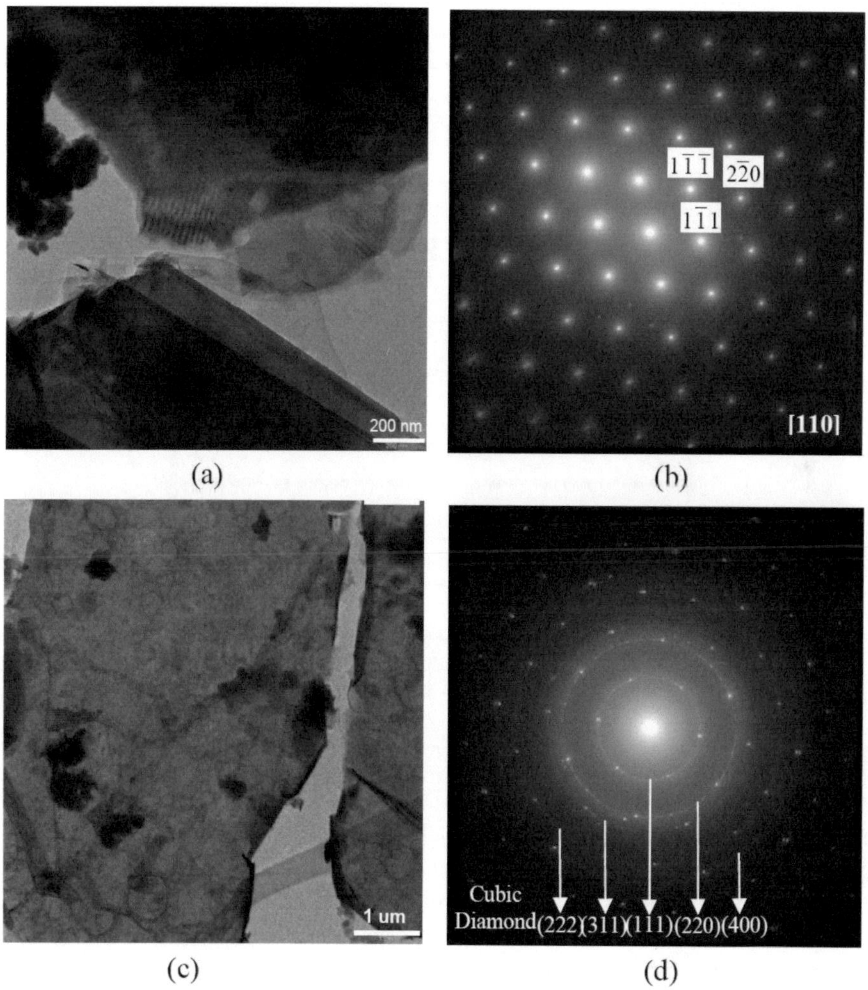

Fig. 10. TEM micrographs with selected area diffraction patterns (SADP) of the 1200 °C spark plasma sintered MWCNTs/Fe35Ni sample after etching, showing the monocrystalline diamond (a, b) and polycrystalline diamond (c, d).

A layer-by-layer structure of diamond crystals was found in the SEM and TEM micrographs. Our previous research revealed the initial diamond growth mechanism from MWCNTs to diamond in SPS without catalysts, that is from CNTs to intermediate phase carbon onions and finally to diamond. The diamond crystals in the samples without FeNi catalysts also show the similar layer-by-layer structures. Many flake-carbons with layers structures were found in the samples of 1200-1500 °C. These indicate that the diamonds were grown up from these carbon flakes. Based on the above analysis, a model for the growth of diamond crystals during the SPS is proposed in Figure 11. The direction of pressure during the SPS is in two axial directions, but not in six directions as the HPHT six-anvil press. Therefore, the diamonds were easier to growth in the direction without pressure. As a result, the MWCNTs were grown to layered diamond flakes vertically to the direction of pressure. Finally, several diamond flakes reacted together and formed a three-dimensional diamond crystal. The growth mechanism of diamond from MWCNTs is a layer-by-layer growth model in the SPS method. This model is available for the MWCNTs to diamond with and without catalysts. This mechanism will be constructive and helpful for the large diamond crystals synthesis by using the SPS technique.

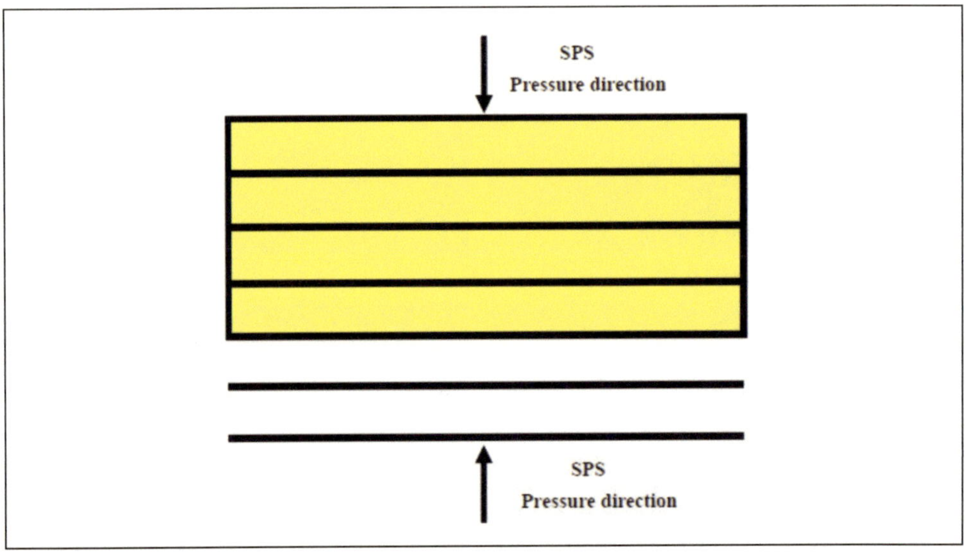

Fig. 11. Schematic illustration of the growth model of diamond crystals from MWCNTs in SPS.

In summary, a Fe35Ni solvent catalyst has been involved to synthesize diamond from MWCNTs by using the SPS technique. Cubic diamond crystals were synthesized from the MWCNTs/Fe35Ni mixtures at lower SPS temperature of 1200 °C under pressure of 70 MPa. In the sample, well-crystallized diamond mono-crystals and poly-crystals consisted particle sizes ranged 10-40 μm. The Fe35Ni catalysts achieved an effective reduction of the SPS temperature to 1200 °C and the SPS pressure to 70 MPa for the diamond synthesis, as well as an increment of diamond transition rate from MWCNTs in the SPS. A model was also proposed to describe the diamond growth and revealed as a layer-by-layer growth mechanism.

4.2 Graphite with various catalysts

Graphite has been used as the main carbon source for the diamond synthesis in the HPHT technique. In this study, the Ni, MnNi, MnNiFe and AlCuFe quasicrystal powder were tested as the catalysts for the diamond synthesis from the graphite by the SPS technique. Each catalyst was weighted and mixed with the graphite powder in the mass ratio of 4:6. The mixture powders were mixed homogeneously by high energy ball milling for 5 hours.

Fig. 12. SEM micrographs of the spark plasma sintered graphite samples at 1300 °C under 50 MPa with Ni (a), AlCuFe quasicrystal (b), MnNi (c), and MnNiFe powder catalysts (d) exhibiting the different diamonds.

Then, the mixtures were subjected to the SPS machine. The sintered samples were tested at SPS temperatures of 1200-1500 °C and under pressures of 10-80 MPa. The results show that diamond crystals can be converted from graphite at the SPS temperature of 1300 °C for holding time of 20 min under the pressure of 50 MPa.

Figure 12 shows the SEM micrographs of the spark plasma sintered graphite samples at 1300 °C for 20 min under 50 MPa with Ni, AlCuFe quasicrystal, MnNi, and MnNiFe powder catalysts after etching. It is interesting that we got diamond nano- and micro-rods with the Ni catalysts from the graphite (Figure 12a) by the SPS. The diameter of the diamond rods are from 80 nm to 2 μm. It is noted that the SPS also can be used as a new method to synthesize diamond nano- and micro-rods. The AlCuFe quasicrystal was introduced as catalyst for the graphite to diamond in the SPS. The diamond crystals with good diamond shapes from 1 to 3 μm were converted from the graphite with the AlCuFe catalyst. It indicates that the AlCuFe quasicrystal powder also can be used as catalyst for the conversion from graphite to diamond in the SPS. With MnNi catalysts, some short rod like diamond crystals are found in the sample (Figure 12c). However, in the sample of graphite/MnNiFe, there are no good diamond crystals visible (Figure 12d).

Additionally, the diamond phase was identified by the X-ray diffraction. Figure 13 shows the synchrotron radiation-high energy X-ray diffraction patterns of the spark plasma sintered graphite samples at 1300 °C under 50 MPa with Ni, AlCuFe quasicrystal , MnNi, and MnNiFe powder catalysts after etching. The Graphite/Ni sample shows strong cubic diamond peaks at d-spacing of 2.06, 1.26 and 1.07 Å. The Graphite/AlCuFe samples show

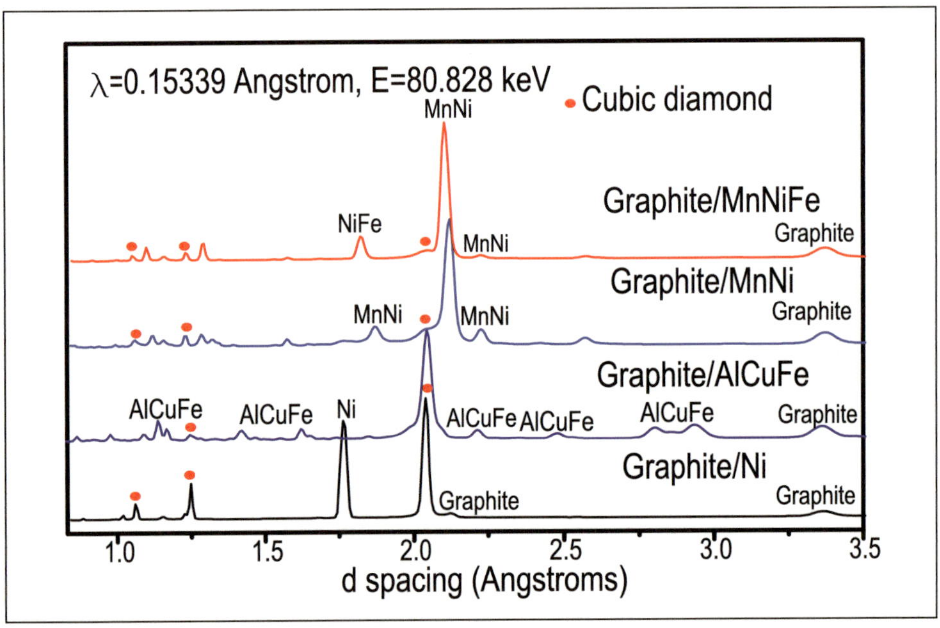

Fig. 13. Synchrotron radiation-high energy X-ray diffraction patterns of the spark plasma sintered graphite samples at 1300 °C under 50 MPa with Ni (a), AlCuFe quasicrystal (b), MnNi (c), and MnNiFe powder catalysts (d).

the strong diamond peaks at d-spacing of 2.06 and 1.26 Å. The Graphite/MnNi and Graphite/MnNiFe samples showed weak diamond peaks at d-spacing of 2.06, 1.26 and 1.07 Å. The XRD results agree well with the SEM results that good crystalline diamonds have been successfully synthesized from the graphite with catalysts of Ni and AlCuFe.

5. Factors affecting the diamond growth in the SPS

In this part, the factors that could affect the diamond growth will be studied in order to increase the diamond crystal size and transition rate. The factors including carbon modifications and atmospheres (Vacuum, Ar) will be studied and discussed.

5.1 Large diamond crystals from the C60 by the SPS

For the carbon modification selection, we will use the C60 as the carbon source for the diamond generation in the SPS. Our previous work has shown that C60 can be converted into diamond under the same SPS conditions as carbon nanotubes are converted to diamond (1500 oC, 80 MPa). Since the C60 has a higher sp^3 hybridization fraction than carbon nanotubes, it makes the transformation of C60 into diamond easier. Therefore, the C60 may be able to increase the diamond size in the SPS process. In this study, the diamond synthesis from the C60 was studied in the SPS (Zhang, Ahmed et al., 2011).

The C60 powders were spark plasma sintered (SPSed) at different temperatures under a pressure of 50 MPa. Figure 14 (a) shows the Raman spectra of the raw C60 and the SPSed C60 samples after etching. The raw C60 shows a sharp peak appeared at 1460 cm^{-1} and two weak broad peaks centered at 1568 and 1413 cm^{-1}. The cubic diamond peaks can also be detected at 1333 cm^{-1} in the Raman spectra taken for the samples processed in the temperature range from 1150 ^{o}C to 1500 ^{o}C under 50 MPa. However, the diamond band of the sample sintered at 1150 ^{o}C is very broad having the lowest height. Its graphite band at 1568 cm^{-1} is at the same value as that of the raw C60. It indicates that there is only a small fraction of diamond in this 1150 ^{o}C SPSed sample. With an increase in temperature to 1200 ^{o}C, 1300 ^{o}C and 1500 ^{o}C, the diamond band at 1333 cm^{-1} gets sharper and sharper, as well as the graphite band is shifted to a higher value of 1576 cm^{-1}. The result of the 1300 ^{o}C SPSed C60 shows the Raman spectra similar to the 1500 ^{o}C SPSed sample. Figure 14 (b) shows the XRD results of the raw C60 and the SPSed C60 samples after etching. In the 1150 ^{o}C sintered C60 sample, we found very weak diamond peaks. The C60 after SPS at temperatures above 1200 ^{o}C show the cubic diamond diffraction peaks at d spacing of 2.06 and 1.26 Å and a broad graphite peak. The C60 diffraction peaks disappeared indicate that the C60 has completely transformed into diamond and graphite phases after the SPS processing at temperatures from 1200-1500 ^{o}C.

The SEM micrographs of the C60 samples SPSed from 1150 to 1500 ^{o}C after etching are shown in Figure 15. There are few fine diamond crystals in the 1150 ^{o}C SPSed sample (Figure 15a). Some diamond crystals with hexagonal, tetragonal or triangular shapes are found in the micrograph of the 1200 ^{o}C SPSed sample (Figure 15b). The particle sizes of the diamond crystals are from tens of micrometers up to 200 μm. The diamond crystals with perfect hexahedron shapes are clearly observed in the 1300 ^{o}C sintered sample (Figure 15c). The diamond sizes range from 100 to 250 μm, they are larger than those of the sample sintered at 1200 ^{o}C. The SEM micrographs of the 1500 ^{o}C sintered C60 samples show that the big diamond crystals are almost melted. There are many small diamond crystals below 4 μm on big crystals.

It is obvious that the diamond crystal sizes do not increase with the increase in temperature; however, well-defined diamond crystals are created. A processing temperature of 1300 °C is the best for the phase transformation of C60 directly to diamond, according to this study.

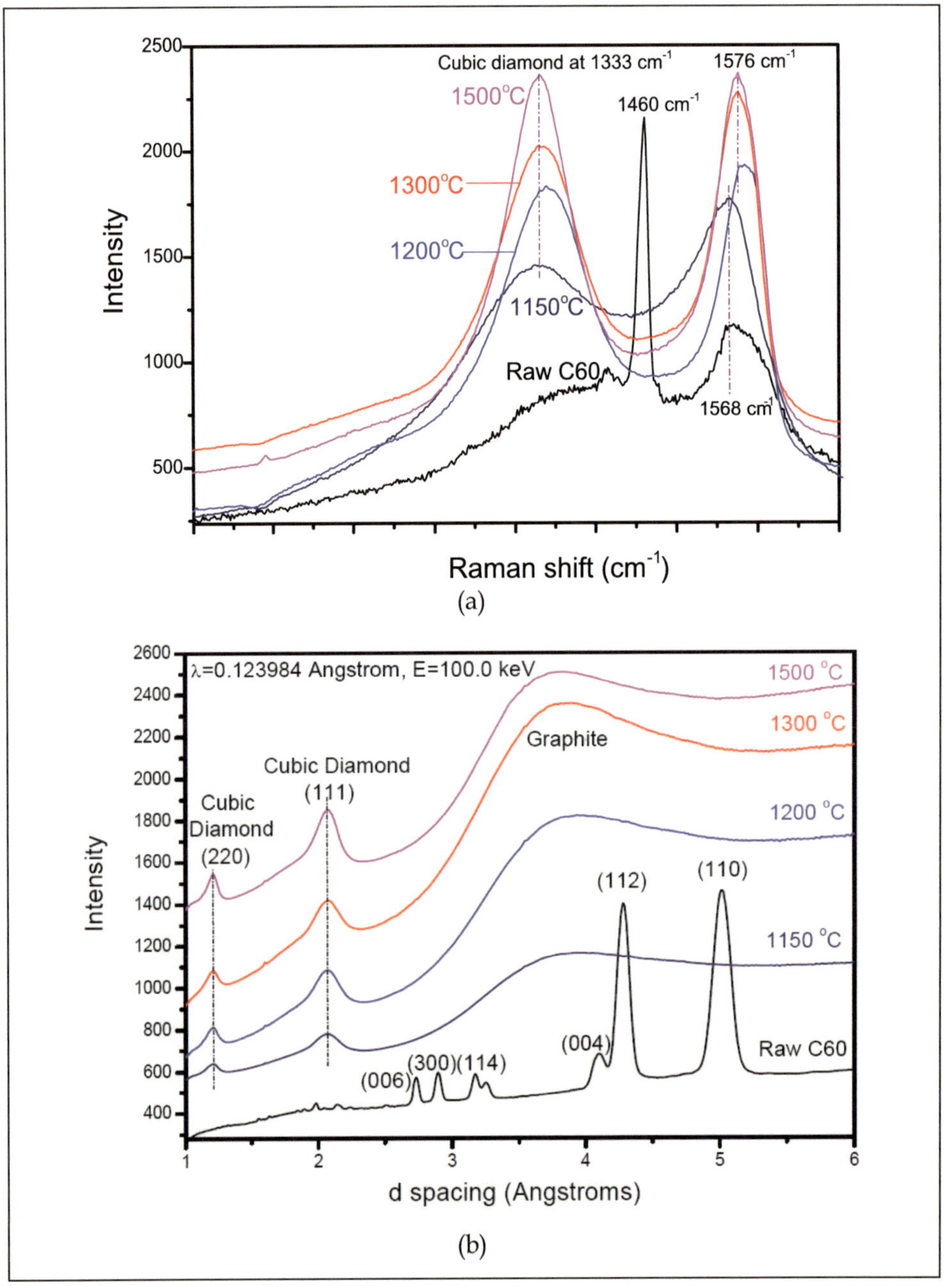

Fig. 14. Raman spectra (a) and Synchrotron radiation-high energy X-ray diffraction patterns (b) of the raw C60 and the spark plasma sintered C60 at different temperatures under 50 MPa pressure.

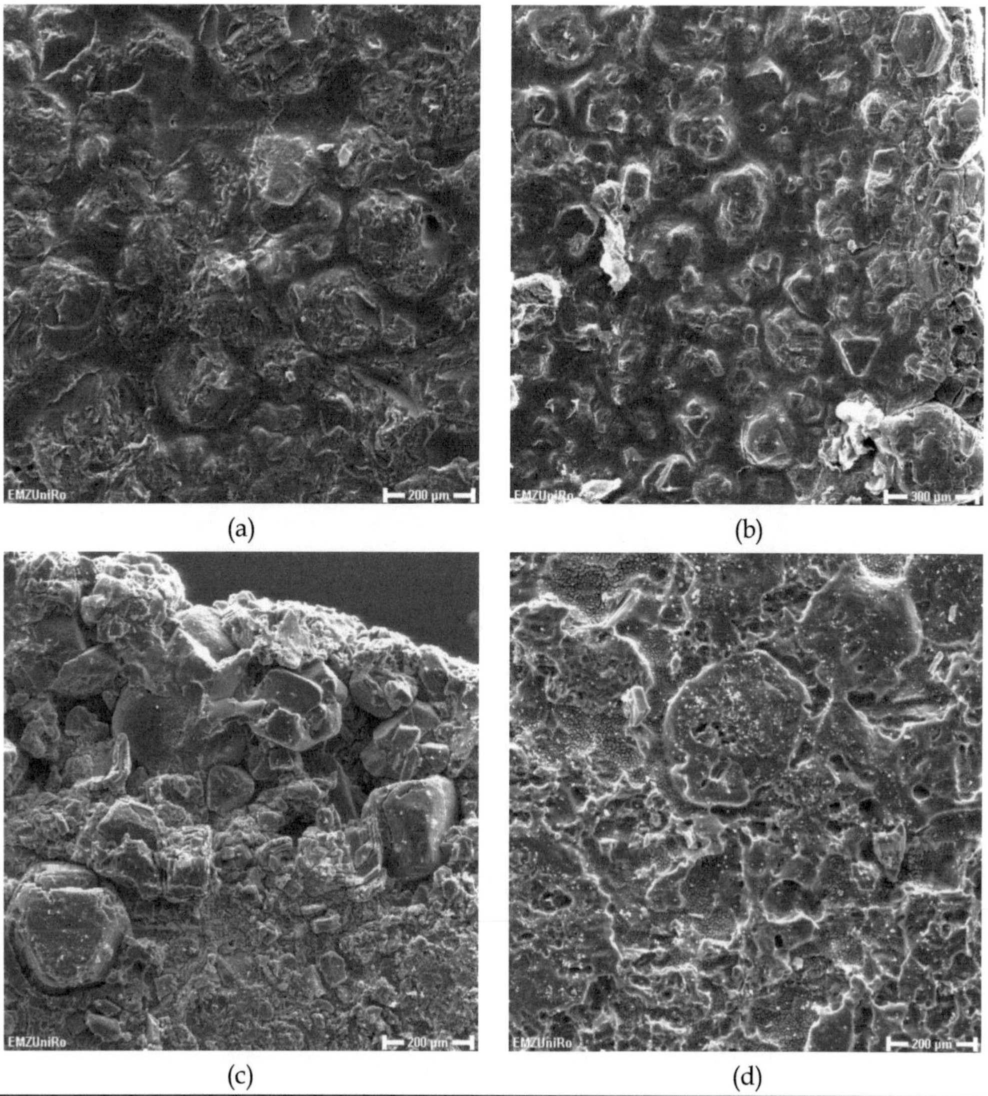

Fig. 15. SEM micrographs of spark plasma sintered C60 samples at 1150 °C(a), 1200 °C (b), 1300 °C (c), and 1500 °C (d) after etching, showing the diamond formation.

The particle size of the diamond crystals made from C60 is up to 250 µm. It is a very large size for such conversion without any catalyst being involved in the process. The carbon atoms in C60 are sp^2 hybridized with a high fraction of sp^3 hybridized structure due to angular strain. It is difficult to transform the planar sp^2 structure to the diamond sp^3 network. The C60 can be considered as a folded graphite sheet with the predominant hybridization sp^3 in the pentagons. This makes the transformation of C60 into diamond easier. A dense assembly of C60 spheroids, where 48 out of 60 carbon atoms have quasi-tetrahedral coordination, is sterically fairly close to that of the diamond (Regueiro et al., 1992). It implies that a small rearrangement of the atoms of C60 can result in the change of its structure. However, it still needs solely superhigh pressure or high pressure and high temperatures for the phase transition from C60 to diamond (Zhang, Ahmed et al., 2011). We got diamond from C60 by the SPS at only 50 MPa and temperatures above 1150 °C. Well-crystallized diamonds with particle sizes up to 250 μm are obtained at 1300 °C and no further growth in particle size is seen beyond this temperature.

5.2 Effects of the CNTs diameters on the diamond sizes by the SPS

In the previous research, it was proposed that the CNTs transformed to carbon nano-onions, and the nucleation and growth of the diamond phase within the onion cores (Shen, Zhang et al., 2006). A model for the diamond nucleation at the internal surface of carbon onion was established, as shown in Fig.16. Based on this model, the energy need for the nucleation of diamond at the internal surface of carbon onion can be formulized (Zhang, 2005):

$$\Delta G = \Delta G_V + \Delta G_S + \Delta G_E$$

Where ΔG — Difference of Free energy for the nucleation of diamond ;

ΔG_V— Difference of Volume Free energy for the nucleation of diamond ;
ΔG_S— Difference of Surface Free energy for the nucleation of diamond ;
ΔG_E— Difference of Elastic strain energy for the nucleation of diamond.

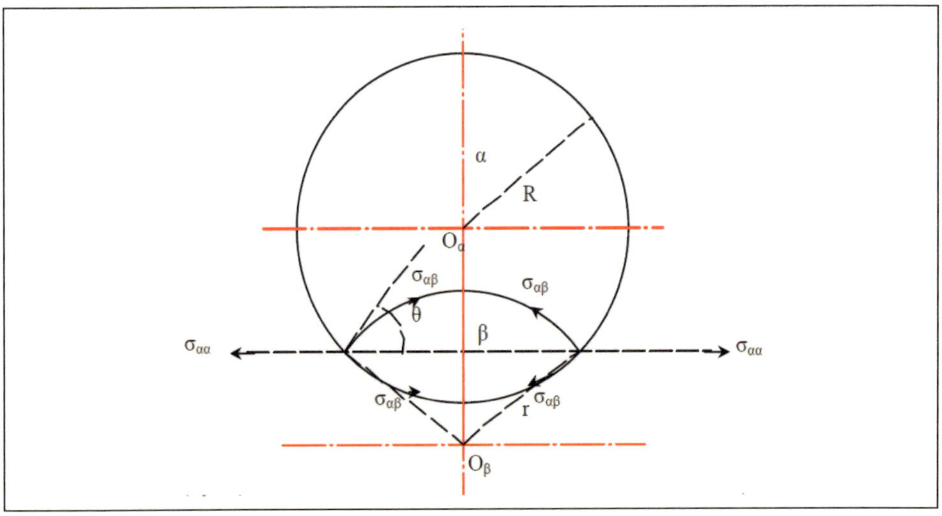

Fig. 16. Model for diamond (*β*) nucleation at the internal surface of carbon onion (*α*).

$$\Delta G_V = \frac{\pi \Delta g_V}{3}\left\{r^3\left(2-3\cos\theta+\cos^3\theta\right)+R^3\left(2-3\sin\theta+\sin^3\theta\right)\right\}$$

$$\Delta G_S = 2\pi\sigma_{\alpha\beta}\left\{r^2(1-\cos\theta)+R^2(1-\sin\theta)\right\}$$

$$\Delta G_E = V\Delta g_E = \Delta g_E\left\{\frac{\pi r^3}{3}\left(2-3\cos\theta+\cos^3\theta\right)+\frac{\pi R^3}{3}\left(2-3\sin\theta+\sin^3\theta\right)\right\}$$

Nuclear powder of diamond from carbon onion ΔG,

$$\Delta G = (\frac{\pi \Delta g_V}{3} + \Delta g_E)\begin{Bmatrix} r^3\left(2 - 3\cos\theta + \cos^3\theta\right) + \\ R^3\left(2 - 3\sin\theta + \sin^3\theta\right) \end{Bmatrix}$$
$$+2\pi\sigma_{\alpha\beta}\left\{r^2(1-\cos\theta) + R^2(1-\sin\theta)\right\}$$

The above equation shows the that decrease of R leads to the decrease of ΔG. The value of R is the inner radius of CNTs. It means the decrease of the diameter of CNTs could decrease the energy for the diamond nucleation ΔG. The lower ΔG leads to higher transition rates and larger particle sizes of diamond. This was validated by the following experiments.

The MWCNTs with diameters from 10-100nm were mixed with Fe35Ni catalyst at a weight ratio of 1:1 by ball milling, and pressed into a graphite die for SPS treatment (1200 °C, 50 MPa, 20 min) to form disk-shaped samples. The raw materials used in this experiment was shown in Table 1 with internal diameters of 10-20 nm, 20-40 nm, 40-60 nm and 60-100 nm.

Main range of diameter of MWCNTs	10-20 nm	20-40 nm	40-60 nm	60-100 nm
Length	5-15 μm	5-15 μm	5-15 μm	5-15 μm
Purity	>95%	>95%	>95%	>95%
Ash	<0.2 wt%	<0.2 wt%	<0.2 wt%	<0.2 wt%
Specific surface area	40-300 m^2/g	40-300 m^2/g	40-300 m^2/g	40-300 m^2/g
Amorphous carbon	<3%	<3%	<3%	<3%

Table 1. The raw materials those used in this experiments.

The SPSed samples after etching were subjected to the SEM analysis. Figure 17 shows the SEM micrographs of the diamond crystals obtained from different MWCNTs. It really shows various diamond sizes. The 10-20 nm MWCNTs generated 15-30 μm diamond crystals. The 20-40 nm MWCNTs generated 10-20 μm diamond. The 40-60 nm MWCNTs produced 7-10 μm diamonds, and the 60-100 nm MWCNTs formed 4-10 μm diamonds. The results are summarized in Table 2. It indicates that smaller diameters MWCNTs could produce larger diamond particles.

Main range of diameter of MWCNTs	10-20 nm	20-40 nm	40-60 nm	60-100 nm
Diamond particle size	15-30 μm	10-20 μm	7-15 μm	4-10 μm

Table 2. The diamond particle size from various MWCNTs.

The Raman results also confirmed the higher transition rate in the 10-20 nm MWCNTs. Therefore, the experiments validated the theoretical assumption. Consequently, the MWCNTs have been transformed into diamond under SPS conditions in presence of a FeNi catalyst at pressures of 50 MPa and temperatures of 1200 °C. The diamond particle size depends on the diameter of the MWCNTs. The MWCNTs with diameters from 60 to 100nm produced diamond particle from 4 to 10 μm, while in the sample of MWCNTs with diameters from 10 to 20 nm generated diamond particle sizes from 15 to 30 μm.

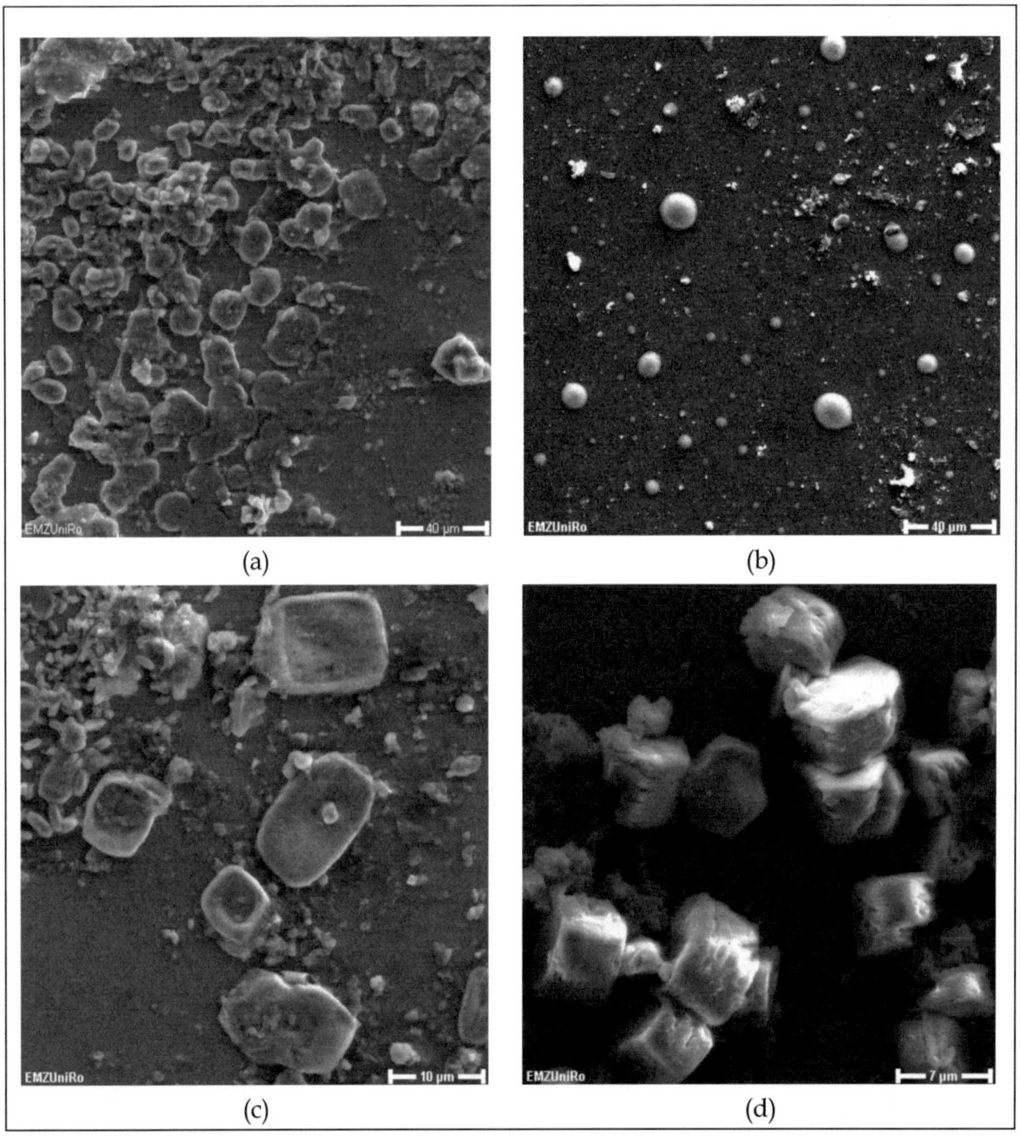

Fig. 17. SEM micrographs of the SPSed MWCNT/FeNi samples with various internal nanotube diameters of 10-20 nm (a), 20-40 nm (b), 40-60 nm (c) and 60-100 nm (d) showing the different diamond sizes.

5.3 Effect of atmospheres on the diamond growth in the SPS

In the SPS process, we usually use the vacuum atmosphere. We proposed that the plasma plays the key role for the diamond transition from various carbon modifications. Based on the theory of plasma physics, the gases like H_2, Ar, O_2, and so on can be used as plasma generating gases which can enhance the plasmas generation (Zheng et al., 2009). Therefore, in this study, the effect of the atmosphere on the diamond growth in the SPS was studied. The MWCNTs/FeNi mixture powders were spark plasma sintered in vacuum and Ar gas atmospheres at 1200 °C under 10 MPa which is the minimum pressure of the Model HP D-5 FCT SPS system. The sintered samples were etched and examined by the SEM and Raman spectroscopy.

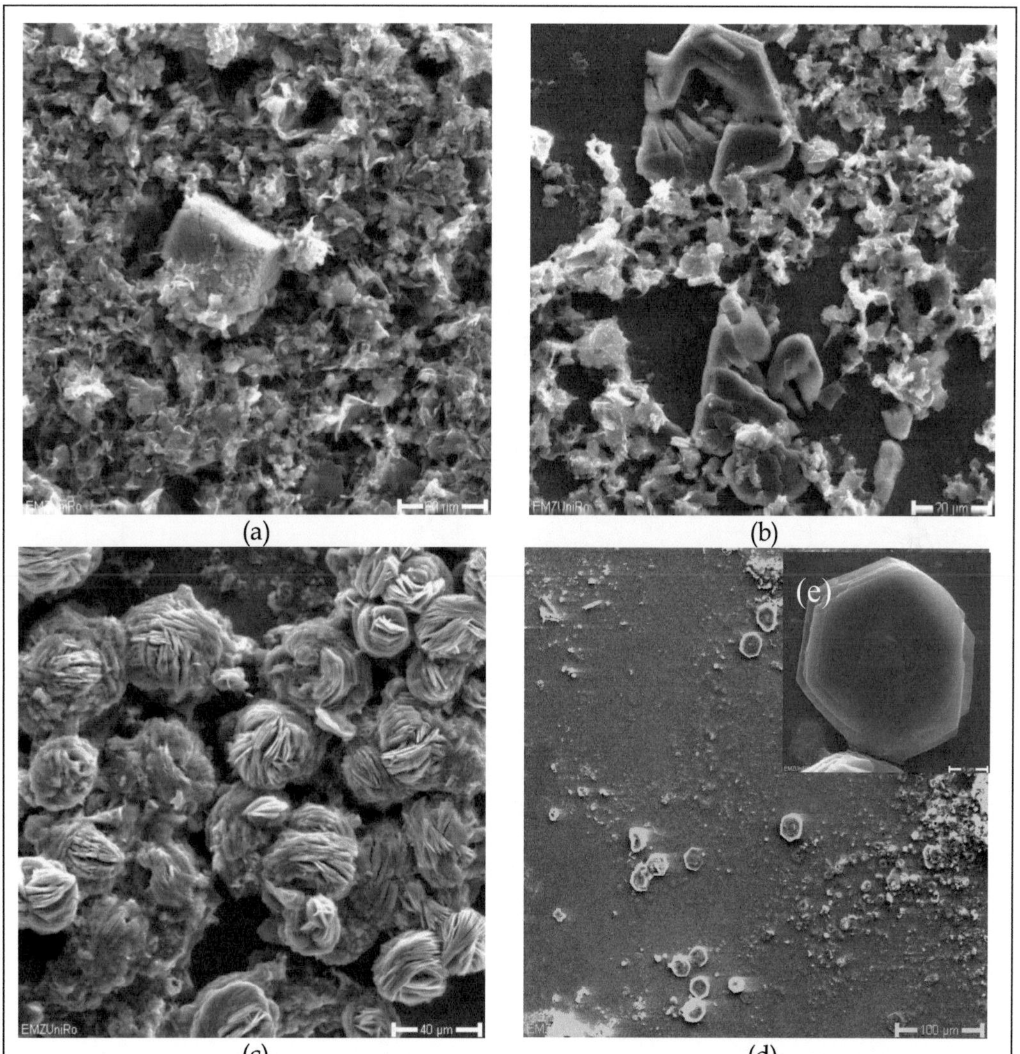

Fig. 18. SEM micrographs of the MWCNTs/FeNi samples SPSed at 1200 °C under 10 MPa in vacuum (a, b) and Ar gas (c-e).

Figure 18 shows the SEM micrographs of the MWCNTs/FeNi samples SPSed at 1200 °C under 10 MPa in vacuum and Ar gas. There are only a few poor quality diamond crystals created in the vacuum atmosphere of the SPS (Figure 18 a, b). In the case of the SPS treatment in Ar gas atmosphere, some flower-like structured carbon are observable (Figure 18c). This indicates that sparking plasmas may have happened and generated such flower-like structured carbon. It is exciting that some high quality diamond crystals with hexahedron structures are found in the sample (Figure 18 d). A high magnification SEM micrograph of one perfect diamond crystal is shown in Figure 18 (e).

Figure 19 shows the Raman spectra of the MWCNTs/FeNi samples SPSed at 1200 °C under 10 MPa in vacuum and Ar gas. Both of them show the cubic diamond peak at 1332 cm^{-1} Raman shift. The vacuum atmosphere SPSed sample shows a very weak diamond peak; however, the Ar gas atmosphere SPSed sample exhibits a strong diamond peak. Both of the G bands centered at the same Raman shift. But it is much sharper in the vacuum atmosphere SPSed sample than that in the Ar gas. The Raman results are accordingly to the SEM results that the Ar gas atmosphere in the SPS promotes the diamond formation. The SPS pressure that we used is very small (10 MPa), so that the pressure effect is negligible for the diamond conversion. The diamonds have been generated by the SPS from such low pressure in vacuum and Ar gas atmospheres. It means that the plasma played the key role for the diamond formation. It provided another evidence for the existence of plasma during the SPS that such a low pressure diamond formation. The Ar gas atmosphere enhanced plasma generation and promoted the diamond transition.

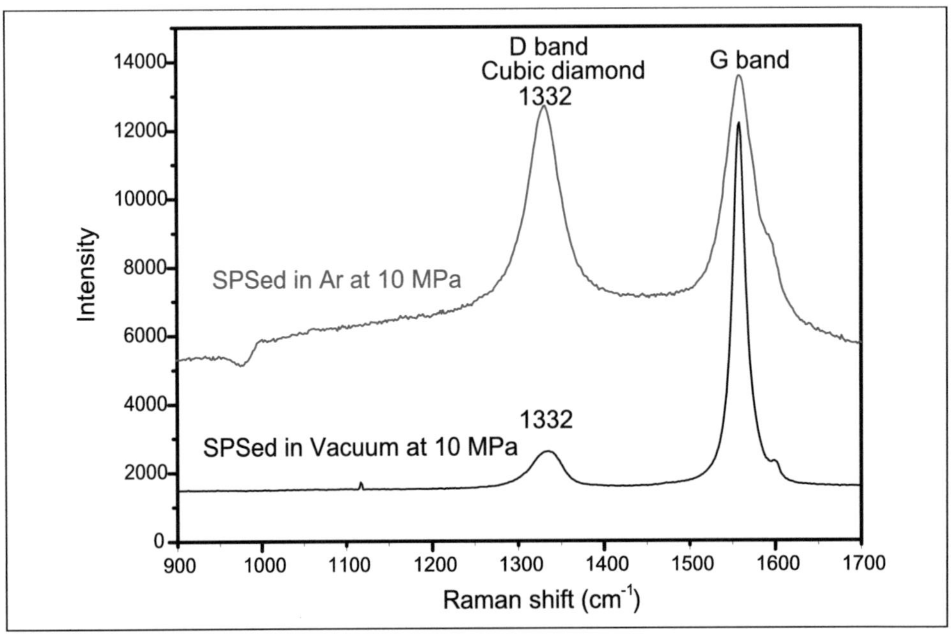

Fig. 19. Raman spectra of the MWCNTs/NiFe samples SPSed at 1200 °C in vacuum and Ar gas.

In this part, we investigated the factors affecting the diamond growth in the SPS. It is found that the C60 is able to increase the diamond size in the SPS process. Due to the high price of the C60, it is suggested that the C60 can be used as a doping catalyst material to promote the diamond transition. The diamond particle size depends on the internal diameter of the MWCNTs. Smaller diameter MWCNTs generated larger diamond particles. It is suggested to choose the smaller diameter MWCNTs when the MWCNTs are used as carbon source for the diamond synthesis in SPS. The Ar gas atmosphere enhanced plasma generation and promoted the diamond transition. It indicates that we select the Ar gas atmosphere to enhance the plasma effect in the SPS.

In this chapter, the diamond synthesis using the SPS was investigated and discussed. The diamond synthesis from pure carbon nanotubes has been covered by China Patent ZL 200410044157.0. (Shen, Zhang, 2004). The diamond synthesis from all the carbon modifications with catalysts as well as pure C60, graphene to diamond has been protected by the Deutsches Patent P162-11 (Zhang, Burkel et al. 2011). The SPS is a new technique for the diamond synthesis. It still needs further investigations to promote the SPS method to be used as a large-scale synthetic diamond production technique.

6. Conclusions and outlook

The thermal stability of MWCNTs, C60 and graphite has been investigated under the pulsed DC field in the SPS furnace. Cubic diamond and n-diamond have been converted from pure MWCNTs; cubic diamond has been converted from pure C60 without catalysts being involved by the SPS at conditions of 1500 °C and 80 MPa for 20 min. There was no notice of diamond formation in the case of pure graphite sample processed by SPS at this condition. The graphite is the most stable crystalline modification of carbon among the MWCNTs, C60 and graphite allotropes under the SPS. The parallel investigations by using the synchrotron radiation in-situ high temperature X-ray diffraction show that there is no diamond formation in the MWCNTs and C60 samples at the same pressure (80 MPa) and temperature (1500 °C). Their phase transition mechanisms from MWCNTs and C60 to diamond indicated the high localized temperatures between particles due to the presence of momentary plasmas during the SPS process. The thermal dynamic analysis reveals that the plasmas have increased the entropy of the SPS system resulting in milder conditions for the diamond formation.

Catalysts were involved in the SPS diamond synthesis with carbon modifications of carbon nanotubes and graphite. A Fe35Ni solvent catalyst has been incorporated to synthesize diamond from MWCNTs by using the SPS. Cubic diamond crystals were synthesized from the MWCNTs/Fe35Ni mixtures at lower SPS temperature of 1200 °C under pressure of 70 MPa. Well-crystallized diamond mono-crystals and poly-crystals with particle sizes ranging around 10-40 μm are synthesized. The Fe35Ni catalysts achieved an effective reduction of the SPS temperature to 1200 °C and the SPS pressure to 70 MPa for the diamond synthesis, as well as an increment of the diamond transition rate from MWCNTs in the SPS. A model was also proposed to describe the diamond growth and revealed as a layer-by-layer growth mechanism. The Ni, MnNi, MnNiFe and AlCuFe quasicrystal powder were used as the catalysts for the diamond synthesis from graphite by the SPS. Diamond crystals have been converted from the graphite at the SPS condition of 1300 °C and 50 MPa for 20 min. Diamond nano- and micro-rods (80 nm-2 μm) have been obtained with the Ni catalysts from the graphite by the SPS. Diamond crystals with good diamond shapes from 1 to 3 μm have been converted from the graphite with the AlCuFe catalyst.

The factors affecting the diamond growth including carbon modifications and atmospheres have been studied in order to increase the diamond crystal size and transition rate. Well-crystallized diamonds with particle sizes up to 250 μm are obtained at 1300 °C by using the fullerene C60 as the carbon source. The mechanism analysis indicates that the high sp^3 hybrid fraction in the C60 leads to its transformation to diamond at lower pressures and temperature during the SPS application. It is suggested that the C60 can be used as doping catalyst materials to promote the diamond transition. A model for the diamond nucleation at the internal surface of carbon onion has been established. Based on the model, the energy need for the nucleation of diamond at the internal surface of carbon onion has been formulized. It is postulated that the decrease of diameter of CNTs could decrease the energy for the diamond nucleation. The results show that the MWCNTs with diameters from 60 to 100nm produced diamond particle from 4 to 10 μm, while in the sample of MWCNTs with diameters from 10 to 20 nm generated diamond particle sizes from 15 to 30 μm. The transition rate has been increased in the 10-20 nm MWCNTs. The experiments validated the theoretical assumption. The effect of the atmosphere on the diamond growth in the SPS was studied. The MWCNTs/FeNi mixture powders were spark plasma sintered in vacuum and Ar gas atmospheres at 1200 °C under the minimum pressure of 10 MPa. The diamonds have been generated by the SPS from such a low pressure in vacuum and Ar gas atmospheres. High quality diamond crystals with hexahedron structures are created in the Ar gas atmosphere of the SPS. It provided another evidence for the existence of plasma during the SPS because of such a low pressure diamond formation. The Ar gas atmosphere enhanced the plasma generation and promoted the diamond transition.

What makes the SPS so interesting for the diamond synthesis is the fact that only relatively low pressures of 10-80 MPa are needed. The SPS possesses a wide range of sintering temperatures from a few hundreds up to 2000 °C, controllable heating rates which can be set to several hundred degrees per minute for extremely rapid processing, as well as the capacity to process large samples up to 100 mm in diameter and 20 mm thick. As we know, the HPHT technique only can prepare very small samples in order to achieve the several GPa level high pressures. Therefore, the SPS is a highly efficient and energy saving technique for diamond synthesis. The investigation in this chapter indicate that the SPS, a pulsed electric field processing technique, has great potential to be used as an alternative and novel method for high-efficiency diamond generation. The diamond generation at such low pressures also provided some indirect evidences for the presence of plasmas during the SPS operation. However, it still needs further investigations to promote the SPS method to be used as a large-scale synthetic diamond production technique. The future highlights will be the development of diamond purification methods to get high purity diamond crystals from the SPSed carbon compacts and the synthesis of millimetre sized diamond crystals and achieving higher diamond transitional rate by using the SPS technique. The mechanical properties (hardness, Elastic modulus) and functional properties including electrical, thermal, optical, magnetic properties etc. of the SPSed diamond particles and the SPSed carbon samples with in-situ formed diamonds need to be investigated. The in-situ synthesis of diamond/ceramics or diamond/metals composites from CNTs/ceramics or CNTs/metals by the SPS is also a very interesting subject for the future researches.

7. Acknowledgements

The authors acknowledge the financial support from the DFG-Deutschen Forschungsgemeinschaft (German Research Foundation) with grant No. BU 547/10-1 and the HASYLAB/DESY project with grant No. II-20090264. We also thank Dr. Christian Lathe, Dr. Martin von Zimmermann, Dr. Jozef Bednarcik at HASYLAB/DESY, Hamburg for their supports in the F2.1 and BW5 experiments, and Dr. Furqan Ahmed at the University of Erlangen-Nürnberg for his support in the Raman spectroscopy experiments.

8. References

Anselmi-Tamburini, U.; Gennari, S.; Garay, J. E.; Munir, Z. A. (2005). Fundamental investigations on the spark plasma sintering/synthesis process: II. Modeling of current and temperature distributions. *Mater. Sci. Eng. A.* 394: 139-148

Giardini, A. A.; Tydings, J. E.; Levin, S. B. (1960). A very high pressure-high temperature research apparatus and the synthesis of diamond. *American Mineralogist.* 45(1-2): 217-221

Hirose, Y.; Amanuma, S.; Komaki, K. (1990). The Synthesis of High-Quality Diamond in Combustion Flames. *Journal of Applied Physics.* 68(12):6401-6415

Hulbert, D. M.; Anders, A.; Dudina, D. V.; Andersson, J.; Jiang, D.; Unuvar, C.; Anselmi-Tamburini, U.; Lavernia, E. J.; Mukherjee, A. K. (2008). The absence of plasma in `spark plasma sintering'. *J. Appl. Phys.* 104(3): 033305-7

Kobashi, K.; Nishimura, K.; Kawate, Y.; Horiuchi T. (1988). Synthesis of diamonds by use of microwave plasma chemical-vapor deposition-morphology and growth of diamond Films. *Physical Review B.* 38(6):4067-4084

Novikov, N.V. (1999). New trends in high-pressure synthesis of diamond. *Diamond and Related Materials.* 8(8-9):1427-1432

Regueiro, M. N.; Monceau, P.; Hodeau, J.L. (1992). Crushing C60 to diamond at room temperature. *Nature.* 355: 237-239

Shen, J.; Zhang, F.; Sun J. F.; Wang, G. (2004). Low pressure synthesis of diamonds from carbon nanotubes by a new process. *China Patent.* ZL 200410044157.0

Shen, J.; Zhang, F.; Sun, J. F.; Zhu, Y. Q.; McCartney, D. G. (2006). Spark plasma sintering assisted diamond formation from carbon nanotubes at very low pressure. *Nanotechnology.* 17: 2187-2191

Vanmeensel, K.; Laptev, A.; Hennicke, J.; Vleugels, J.; Van der Biest, O. (2005). Modelling of the temperature distribution during field assisted sintering. *Acta Materialia.* 53: 4379-4388

Vereschagin, A. L.; Sakovich, G. V.; Komarov, V. F.; Petrov, E. A. (1994). Properties of ultrafine diamond clusters from detonation synthesis. *Diamond and Related Materials.* 3(1-2):160-162

Watanabe, I.; Matsushita, T.; Sasahara, K. (1992). Low-temperature synthesis of diamond films in thermoassisted Rf plasma chemical vapor-deposition. *Japanese Journal of Applied Physics Part 1.* 31(5A):1428-31

Yang, K.; He, J.; Su, Z.; Reppert, J. B.; Skove, M. J.; Tritt, T. M.; Rao, A. M. (2010). Inter-tube bonding, graphene formation and anisotropic transport properties in spark plasma sintered multi-wall carbon nanotube arrays. *Carbon*. 48:756-762

Zhang, F. (2005). Strengthening and toughening mechanisms of carbon nanotubes reinforced WC nanocomposites synthesized by spark plasma sintering. *Ph.D thesis*. Harbin Institute of Technology

Zhang, F., Shen, J.; Sun, J.; Zhu, Y. Q.; Wang, G.; McCartney, G. (2005). Conversion of carbon nanotubes to diamond by a spark plasma sintering. *Carbon*. 43:1254-1258

Zhang, F.; Shen, J.; Sun, J. F.; McCartney, D. G. (2006). Direct synthesis of diamond from low purity carbon nanotubes. *Carbon*. 44 : 3136-3138

Zhang, F.; Burkel. E. (2010). Novel titanium manganese alloys and their macroporous foams for biomedical applications prepared by field assisted sintering. In: *Biomedical Engineering, Trends, Researches and Technologies*, Anthony N. Laskovski, pp. 203-224. InTech, ISBN 978-953-7619, Rijeka, Croatia

Zhang, F.; Adam, M.; Otterstein, E.; Burkel E. (2011). Pulsed electric field induced diamond synthesis from carbon nanotubes with solvent catalysts. *Diamond and related Materials*. 20: 853–858

Zhang, F.; Mihoc, C.; Burke E. (2010). Spark plasma sintering assisted carbon conversion from various modifications to diamond. *Conference Proceedings of Materials Science and Technology (MS&T) 10*, pp. 2312-2317, ISBN 978-0-87339-756-8, Houston, TX, USA, October 17-21, 2010

Zhang, F.; Mihoc, C.; Ahmed, F.; Latte, C.; Burkel, E. (2011). Thermal stability of carbon nanotubes, fullerene and graphite under spark plasma sintering. *Chemical Physics Letters*. 510:109-114

Zhang, F.; Ahmed, F.; Holzhüter, G.; Burkel, E. (2011). Growth of diamond from fullerene C60 by spark plasma sintering. *Journal of Crystal Growth*. doi:10.1016/j.jcrysgro.2011.11.01494

Zhang, F.; Burkel, E.; Rott, G. (2011). Verfahren zur synthese von diamanten. *Deutsches Patent*. P162-11

Zheng, Z.; Liao, L.; Yan, B.; Zhang, J. X.; Gong, H.; Shen, Z. X.; Yu, T. (2009). Enhanced field emission from argon plasma-treated ultra-sharp *a*-Fe_2O_3 nanoflakes. *Nanoscale Res Lett*. 4:1115–1119

3

The Effects of Sintering Temperature Variations on Microstructure Changes of LTCC Substrate

Rosidah Alias
TM Research & Development Sdn. Bhd.
Malaysia

1. Introduction

The successful development and commercialization of high performance ceramic materials has attracted much attention especially for multilayer substrates using the Low Temperature Co-fired Ceramic (LTCC) technology. This technology has become a popular technology for automobiles and wireless communications due to the advantages of the excellent combination of electrical, thermal, mechanical and chemical stability for a wide range of applications, thus allowing preparation of 3-dimensional circuits incorporating passive components within a multilayer construction (Matters-Kammerer et al., 2006; Zhou et al., 2008). This approach also allows the presence a number of interfaces and thus reduction of the overall substrate size and cost can be realized (Lo and Duh, 2002; Chen et al., 2004 and Zhu et al., 2007). The circuits are capable of withstanding sintering during processing temperatures up to 1000 °C. For telecommunication applications the usage of ceramic is implemented in Telecom control station and power supply circuits for the capability to dissipate excess heat and maintain dimensional control stability of the ceramic package. This is important where back-up power is required to maintain operation during primary power outages when cooling is restricted (Barlow and Elshabini, 2007). Another important parameter for wireless communication devices is the requirement to have low dielectric loss ($\tan \delta \sim 10^{-3}$ or less) for higher processing speed, higher dielectric constant ($\varepsilon' > 10$) for miniaturization of the devices and higher integration density (≥ 3 g/cm^3) (Kume et al., 2007; Long et al., 2009). For this reason, it is important to prepare high quality LTCC substrate/package whose properties are strongly dependent on microstructure, phase purity and sintering temperature (Xiang et al., 2002). Therefore the microstructure must be carefully controlled to get dense and fine grained ceramics in order to improve their properties and reliability in many applications (Hsu et al., 2003).

The starting point of the LTCC technology is the development of LTCC tape materials containing a glass-ceramic system that usually needs to show good compatibility with the paste system which acts as a conductive track for RF signal transportation from one location on the circuit to another. It should also have low energy loss in microwave applications to make sure high circuit performance can be achieved (Wang et al., 2009). One of the most important processes in LTCC manufacturing of multilayer LTCC substrates involves co-firing of metal and glass-ceramic. The main problem in the development of materials for LTCC integrated modules arises from the chemical incompatibility of the different materials within the module during the sintering process (Valant and Suvorov, 2000). So, mismatch in sintering kinetics, sintering stress, density variations between tape layers and non-uniform shrinkage of the individual tapes should be taken into consideration to avoid some defects such as delamination, blister, and camber form during a multilayer LTCC process. Therefore, the processes that result in the above mentioned defects must be completely understood before the processing parameters and material compositions are optimized.

Thick-film technology is a method whereby conductive, resistive and dielectric pastes are applied to a ceramic substrate. There are various available ways of producing thick-films such as dip coating, spin coating, screen printing and chemical vapor deposition technique down to 25 μm thickness. However the screen printing method is simpler, more convenient and the most cost effective method to transfer the desired thick-film pattern onto the substrate in order to realize interconnect films with a thickness ranging from 3-15 μm which depends upon the screen printing parameter. Highly conductive metals such as copper, silver and gold are typical electrode materials in LTCC components and modules. A silver (Ag) conductor is very attractive because the price of silver is much cheaper than that of others and can be fired in air. It also has some advantages such superior electrical conductivity and good thermodynamic stability below 200 °C. Above 200 °C it deoxidize to metallic silver (Imanaka, 2005; Bangali et al., 2008; Jean et al., 2004). For the LTCC process the dielectric ceramics and silver should be cofired simultaneously so it is important to carried out the sintering process below the melting point of the metal electrode such as Ag (961 °C) to reduce sintering mismatch between each other which form some defects for the whole substrate (Feteira and Sinclair, 2008; Chang and Jean, 1998; Long et al., 2009). With regard to the co-firing process technology and to get detailed understanding of LTCC process behavior, the mechanism of low temperature sintering was studied by Valant et al., in 2006 using $BaTiO_3$ whose original sintering temperature was about 1250 °C -1300 °C. They used a small amount of Li_2O as a sintering aid and found that small amount of sintering aid about 0.3 wt% was able to reduce the sintering temperature to 820 °C and ceramics with more than 95% of relative density could be produced.

The study of transition metal elements especially controlling the properties of silver is of great importance from the view point of industrial applications. It is not a simple task and demands thorough understanding of structure ⇔ property ⇔ processing relationship (Despande et al., 2005). The overall behavior of silver metal is influenced by the nature of atoms, the size of the grain and the nature of grain boundaries in addition to controlled chemical composition and preparation conditions. So the material specifications/characteristics of the thick-film can fulfill conductor paste requirements in a wide range of applications. Typically, a thick-film conductor paste consists of metal powder as a major component and glass and/or oxide (minor) dispersed in an organic medium (Sergent and Harper, 1995). The metal powder forms a continuous film on the substrate upon firing. The role of the glass and /or oxide in the conductor paste is to help the metal film adhere to the substrate. The investigation of the preparation of LTCC packages using glass + ceramic and glass ceramic system with high electrical conductivity such as gold and silver has been conducted by several researchers such as Shimada et al., 1983, Tummala, 1991, Imanaka and Kamehara, 1992, Jean et al., 2001 and Bangali et al.,

2008. Most of their studies have been performed on multiphase material systems and focused crystallization behavior of the systems. Wang and Zhou (2003) have studied the relation of densification and dielectric properties of glass ceramics with different compositions selected from the $CaO-B_2O_3-SiO_2$ system. They have found that too high sintering temperature will ruin the properties and sintering with a heating speed of 10 °Cmin^{-1} at 850 °C for 60 min is an optimal technology for CBS system. They also notice that too much B_2O_3 will damage the dielectric properties of the glass ceramics (Wang and Zhou, 2003). More recently, Hsi and co-workers have further studied the relation of silver and LTCC. They demonstrate the easy method to decrease effectively the diffusivity of silver ions in LTCC dielectrics by adding 5 wt % of SiO_2 (Hsi et al., 2011). However, not much literature has been found to investigate the microstructure changes of silver conductor and glass-ceramic substrate by changing the sintering temperature and this situation may not be very helpful in understanding the surface properties of the printed conductor and the microstructure of glass-ceramic system in this project. Nevertheless, the results obtained in this work can provide useful findings for researchers and industries.

There are a few factors that determine the microstructure for both printed conductor and LTCC substrate such as the quality of the raw materials and the sintering temperature. The parameters of microstructure are the grain size, grain size distribution, pore size and pore size distribution and the density. The size, shape and also the distribution of grain size and pore size of the conductor and glass-ceramics will vary with preparation conditions and techniques. The objective of this work is to investigate the effect of sintering temperature variations on the microstructure of both metal and ceramic materials of the LTCC substrate for microwave applications.

In this chapter, the work is constructed as follows; firstly, a simple transmission line pattern on the eight layers substrate is fabricated using a standard LTCC technology. Printed patterns were dried in a box oven for 10 minutes at about 70 °C to remove the organic solvent slowly before cofiring process in order to avoid blister, void formation, delamination, cracks and camber effect. These undesirable defects are usually formed due to the different shrinkage rates of LTCC tape and thick-film paste during firing (Hsueh and Evans, 1985; Hrovat et al., 2009). The film was then fired at various sintering temperatures of 800 °C to 900 °C at 25 °C increments. Generally at peak firing temperature the film achieves desired electrical properties and the ceramic composite also has undergone changes in physical properties. Further, sample characterizations were done such as crystal structure determination, linear shrinkage, density measurement and microstructure observation for both metal and LTCC substrate. After that sample analysis and detail discussion is carried out to explain the research findings.

2. Experimental procedure

In the following sections, the various process stages will be described. The effect of process variables on the physical properties and microstructure of the finished product has always been a subject of great importance. The factors that have the most influence on these properties are the purity of the constituent oxides, their proportions and homogeneity in the powder mix and the control of temperature and atmosphere during sintering.

The key stages in the fabrication of glass-ceramics are sintering process of various components together. During these processes the constituent atoms redistribute themselves in such a way as to minimize the free energy of the system. This involves a considerable movement of ions, their inter-diffusion to form a new phase, the minimization of the internal surface area and increase in grain size.

2.1 LTCC multilayer substrate preparation

In the present work, the commercial Ferro A6S was chosen and cut into 8 pieces for the required dimension (21 mm x 21 mm) using a die cutting machine (ATOM SE 20C). The multilayer stack was prepared by standard LTCC process flow as shown in Fig. 1. The thickness of the green tape is about 100 μm for each layer; the green tape was punched and screen printed with CN 33-391 for the surface and inner conductor and CN 33-407 for via fill. The printed pattern was then stacked using a manual stacker plate. The stacked substrate was then laminated by using an isostatic laminator system under pressure and temperature of 3000 psi (21 MPa) and 70 °C respectively to ensure that the layers of the stack are well adhered to each other and to form a compact multilayer substrate. The laminated substrate was placed in a tube furnace and fired using a sintering profile as suggested by the tape manufacturer. The substrates were sintered in the temperature range of 800 °C to 900 °C at 25 °C increments. This sintering is the most critical process step because during the firing step the material becomes a compact ceramic LTCC substrate and its properties are determined. Through sintering, a transformation from the original porous compact to a dense ceramic takes place (Kingery, 1976). At a high temperature, the particles are in an increased contact and the particles grow together to form crystallite grains.

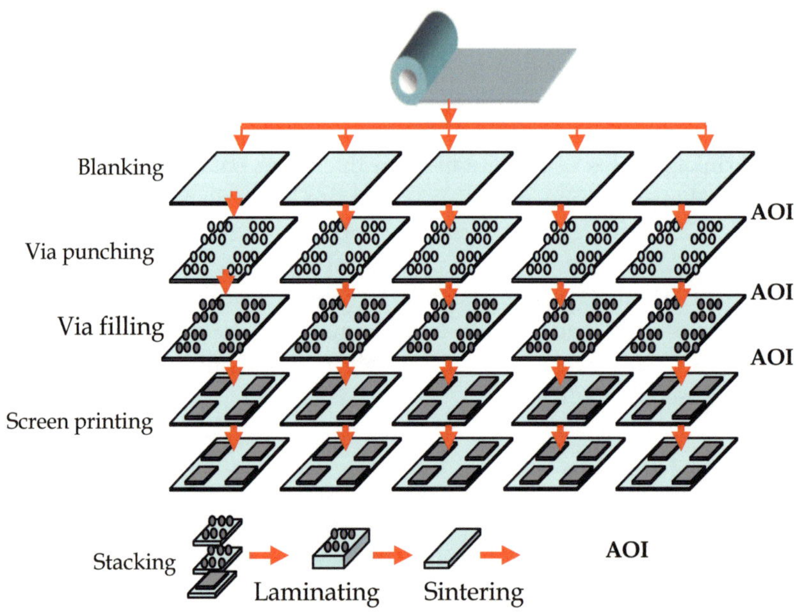

Fig. 1. LTCC Multilayer Fabrication Process Flow.

A typical practice is to heat up the furnace to 450 °C with a gradient of about 2-5 °C/min for 1 hour to completely remove the organic solvent. In the next 2 hours the temperature is raised to about 850 °C at which the sintering process of the composite material is started. The temperature remains constant for 10 minutes to finish the sintering. The last stage is the cooling period which takes about 3 hours depending on the size of the product. This firing condition is selected because of the practical experience on thick-film circuit and suggested by the tape manufacturer.

2.2 Substrate characterizations

2.2.1 Crystal structure determination

The structural characteristics of the LTCC material and metal surface were measured using a Pan Analytical Diffractometer system operating at 45 kV and 40 mA and employing CuK_α radiation with λ = 1.54060 Å. The scanning measurements were 2θ in a range from 2° to 80°, in steps of 0.05° of 2θ and a counting time of 25 s per step. The average crystallite size of all the samples is determined from the full width at half maximum (FWHM) of the (1 1 1) reflection peak in the XRD patterns by using the Debye Scherrer formula shown in equation (1), (Klug & Alexander, 1974).

$$D = \frac{0.94\lambda}{\beta\cos\theta} \tag{1}$$

where λ is the wavelength of the incident x-ray; β is the full width at half maxima; θ is the Bragg's diffraction angle.

2.2.2 Shrinkage

The most important point to bear in mind in the sintering process is controlling the variation in dimensional changes. One of the methods to control the dimensional changes in terms of shrinkage is through the control part as shown in Fig. 2. The steps are as follows:

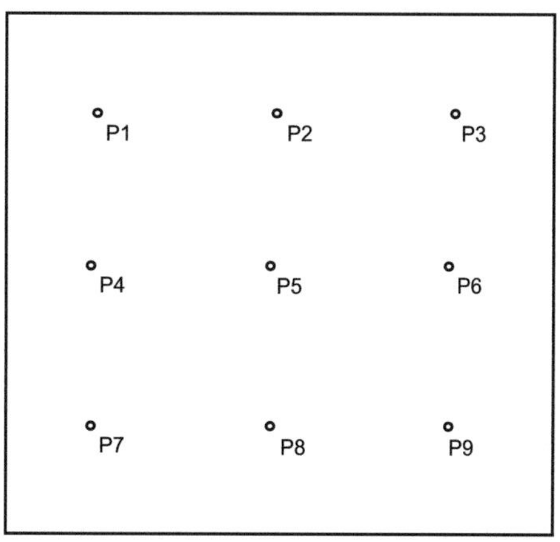

Fig. 2. Control part (Ferro).

1. The control part study main point is to study shrinkage variation. Measurements are taken before and after firing. Measure thickness (Z) of each laminate midway between each via and record data. There will be eight (8) measurements total. Next, measure the average width and length of each laminate by measuring from corner to adjacent corner and record data.
2. Next, measure the distance between each via. Record data, there are 12 measurements needed. Via holes 1,2 - 2,3 - 4,5 - 5,6 - 7,8 - 8,9 are for the X direction. Via holes 1,4 - 2,5 - 3,6 - 4,7 - 5,8 - 6,9 are for the Y direction.
3. The shrinkage and density of sample test pattern are calculated using equation (2).

$$\text{Shrinkage} = \left(\frac{\text{Length}_{\text{before fired}} - \text{Length}_{\text{after fired}}}{\text{Length}_{\text{before fired}}} \right) \text{x}100\% \qquad (2)$$

For the substrate, linear shrinkage was measured along the compaction direction and the diametrical shrinkage from the geometry of the substrate. Repeatability and consistency of the shrinkage percentage must be the top criteria when designing the LTCC product because the shrinkage of LTCC substrate depends on the reactivity of the co-fired material containing ceramic oxide, glass, metal, organic solvent and also the firing conditions such as temperature, time and ambient air. Better reproducibility increases the uniformity of finished product characteristics and therefore increases the process yield. It is not an easy task because all process parameters (lamination, binder burnout, sintering) and material properties (high temperature reactivity, thermal expansion, etc.) must be matched (Rabe et al., 2005). Furthermore, the material quality of the finished product and process conditions also must be optimized in micro and macro structures in order to make sure work in progress are homogeneous at each process step (Imanaka, 2005). In order to achieve desired shrinkage data, the process engineer must establish control of the critical process variable. In commercial production, the designed shrinkage is generally between 12-16 % for XY direction and 20-25 % for Z direction.

2.2.3 Microstructure observation

The surface morphology and composition of the LTCC material and the printed material were observed by using a FEI NOVA Nano SEM 230 machine equipped with an Energy Dispersive X-ray (EDX) detector to reveal the microstructure of the resulting product. Most of the samples were imaged several times, with at least three pictures in each case from different areas of the sample. The thickness and width of conductive traces were measured by an optical microscope and SEM through the cross section view respectively. The average grain size was calculated using the line intercept method. The EDX spectrum is used to identify elements within a sample.

2.2.4 Density measurement

The properties of the final ceramic composite materials depend on the sintered density of the whole substrate. A stacked and laminated LTCC substrate before firing consists of a relatively porous compact of oxides in combination with a polymer solvent. During sintering the organic solvent evaporates and the oxides react to form crystallites, or grains of the required composition, the grains nucleating at discrete centers and growing outwards until the boundaries meet those of the neighboring crystallites. During this process, the density of the material rises; if this process were to yield perfect crystals meeting at perfect boundaries the density would rise to the theoretical maximum, i.e. the x-ray density, which is the material mass in a perfect unit crystal cell divided by the cell volume. In practice imperfections occur and the sintered mass has microscopic voids both within the grains and at the grain boundaries. The resulting density is referred to as the sintered density. The density of the sample was measured using the Archimedes principle shown in equation (3);

$$\rho = \left(\frac{W_a}{W_w} \right) \rho *_w \tag{3}$$

where W_a = weight of sample in air
W_w = weight of sample in water
$\rho*_w$ = density of water = 1 gcm^{-3}

The percentage theoretical density ($\%D_{th}$) was calculated using this formula:

$$\%D_{th} = (\text{measure density / theoretical density}) \times 100\% \tag{4}$$

The theoretical density of Calcium Boron silicate ($CaB_2Si_2O_8$) was calculated using equation below:

$$\rho = \frac{N_C A}{V_C N_A} \tag{5}$$

Where N_C = No. of molecules per unit cell, A = molecular weight, V_C = volume of one molecules and N_A = Avogadro number. So the theoretical density of $CaB_2Si_2O_8$ was calculated by taking the molecular weight of $CaB_2Si_2O_8$ to be 245.87g. Since orthorhombic structure of $CaB_2Si_2O_8$ have four formula units per unit cell, the molecular weight of one cell is (4) x (245.87) = 983.48. The volume of the orthorhombic structure unit cell is a x b x c. The volume of a mole of the material is therefore N_A x a x b x c, where N_A is Avogadro's number. The unit cell edge *a, b and c* (Å) of $CaB_2Si_2O_8$ = 8.7500, 8.0100 and 7.7200 therefore a x b x c = 541.08 Å^3. As 1Å^3 = 10^{-24}cm^3, A x a x b x c is therefore:

$$[6.023 \times 10^{23}][541.08 \times 10^{-24}] = 325.92 \text{ cm}^3$$

so, theoretical density is mass/volume equals to 983.48 g /325.89 = 3.018 g/cm^3.

The theoretical density ρ_x was calculated as above. The percentage of porosity of the sample was calculated using the relation as below:

$$P = 1 - \frac{\rho}{\rho_x} \tag{6}$$

where ρ is the measured density of the sample.

3. Results and discussion

3.1 Phase analysis

Results of the crystallization behaviour of glass-ceramic Ferro A6S substrate sintered at various sintering temperatures are presented in Fig. 3. It can be seen that the small intensity peaks which appeared for the 800 °C sintering belongs to crystalline phases of Calcium Silicates with various stoichiometric ratios Ca_2SiO_4, and Ca_3SiO_5. Increasing the temperature to 825 °C and 850 °C, the two phases of Calcium silicate phase are still maintained but accompanied by a new phase of Calcium Borate (CaB_2O_4). As the sintering temperature increases the formation of Danburite ($CaB_2Si_2O_8$) start to appear in small intensities while crystalline phases of Ca_2SiO_4, Ca_3SiO_5 including $CaSiO_3$ and CaB_2O_4 phase are maintain in this composite system. They are the stable phases when crystallization is complete. Due to the high peak intensity of $CaSiO_3$ at $2\theta \approx 29.9572$, it can be noted that the main phase in this system is Calcium Silicate $CaSiO_3$. The above results were consistent with the work done by Chang and Jean (1999). However, from their studies, they concluded that no crystalline phase was detected at temperatures below 850 °C. The crystallite size as a function of sintering temperature was calculated from the most intense peak (3 1 1) by using the Scherrer equation (equation 1) and tabulated in Table 1.

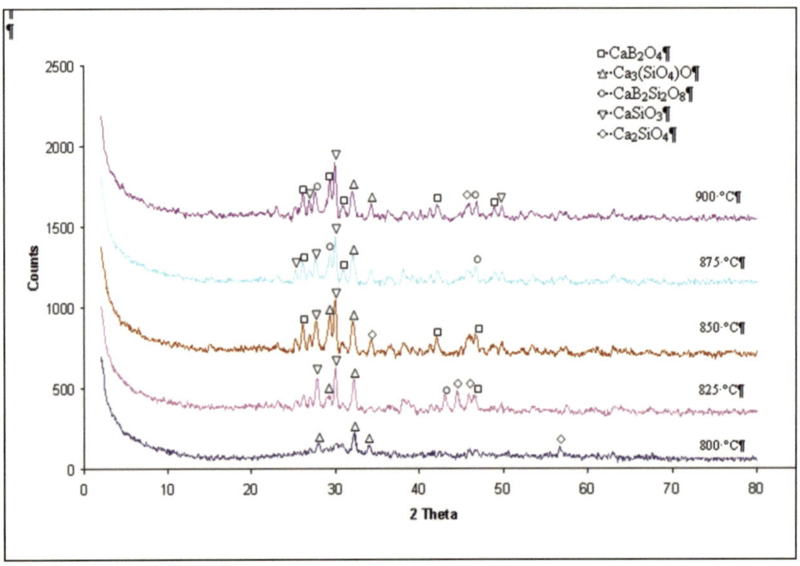

Fig. 3. XRD pattern of the LTCC substrate sintered at various sintering temperature.

Generally sintering temperature plays an important role in determining the crystallite size of the silver paste. However the range of sintering temperature 800 °C and 825 °C in this study did not play an important role in the variation of the crystallite size evolution due to the stable growth of the grains. However, increasing sintering temperature slightly variation of crystallite size was obtained. Thus a further investigation of sintering temperature whether increase or decrease is necessary in order to see the significance different of the crystallite size.

Sintering Temperature (°C)	2 Theta	$D_{spacing}$	Crystallite size (D_p)
800	32.2207	2.77597	29.2565
825	32.1571	2.78132	29.2518
850	32.0970	2.78638	19.4982
875	32.0726	2.78845	25.0677
900	31.9778	2.7965	17.5432

Table 1. Parameter of XRD.

3.2 Physical measurements

3.2.1 Density

The physical data of the LTCC substrate fired at various temperatures are tabulated in Table 2. The relationship of density with sintering temperatures is shown by Fig. 4. The results reveal decreasing trend of the density with sintering temperature with the highest density being 2.992 g/cm^3 at 800 °C and the lowest density is 2.806 g/cm^3 at 900 °C. The densities of all the samples were between 92 to 99 % of theoretical density (3.018 g/cm^3). This trend seems to contradict the normal expected phenomenon that increasing sintering temperature should increase the density. As is well known the sintering process of a ceramic based material is the sintering of the powder compact into the final material. During this step, the porosity decreases and the microstructure of the material develop; this determines its final performance. During the sintering of a homogeneous material the porosity induced during the preparation of the green compact gradually decreases, depending on the powder morphology, agglomeration, the presence of liquid phase sintering and the sintering condition itself. However, the sintering of heterogeneous materials in the LTCC substrate, reactive sintering occurs in the concurrent process of reaction and densification during sintering. A variety of reactions are possible: oxidation-reduction, phase transition or solid solution formation. In this way reaction caused by impurities, additives or other product formed during heating which are often included in the normal sintering process may imply some sort of reactive sintering which usually generates additional porosity. This sintering process is complicated because the phase changes are involved. Thus the understanding of material behaviors such as binder burnout, densification mechanisms of LTCC, pore evolution and deformation of suspended LTCC is important in optimizing the fabrication process for multilayer LTCC substrates as well as for tailoring new LTCC systems (Kemethmuller et al., 2007).

Sample	Density (g/cm^3)	Porosity (%)	Shrinkage (X)/%	Shrinkage (Y)/%	Shrinkage (Z)/%
800	2.992	0.92	14.13	13.28	23.13
825	2.873	4.97	14.35	13.67	22.74
850	2.910	3.64	14.41	14.00	21.56
875	2.860	5.29	15.06	15.15	21.16
900	2.806	7.08	14.53	13.99	21.85

Table 2. Physical properties of LTCC tape samples fired at various sintering temperature.

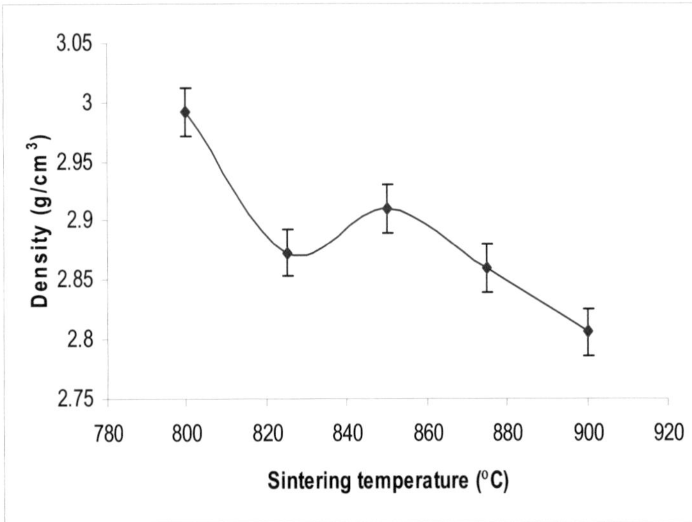

Fig. 4. The plot of the density with the variation of sintering temperature for substrate with metal conductor.

According to the Ferro guideline the tape A6S used in this work is a crystallizing Ca-B-Si-O system and CN33-391 used as a thick-film material (Ferro; Muralidhar et al., 1992). Both of them typically consist of glass and/or oxide dispersed in organic medium. Owing to the coexistence of ceramic filler and glass in LTCCs, liquid phase sintering can be considered; due to the large amount of glass, viscous sintering can be assumed, so the liquefaction of glass has a dominant role. The presence of these mechanisms increased the volume and size of pores inside the material. The density drop due to porosity is based on the same weight for all the samples i.e.; for $\rho = \frac{M}{V}$; M does not change but V increases to accommodate higher porosity (more pores) within the samples. However in the present case, the porosity present inside the substrate cannot support the decreasing trend of density with increasing sintering temperature. So, it is believed that the main reason of the decreasing trend of the density with increasing sintering temperature could be due to the chemical reaction e.g. between some components such as silver in the system with the oxygen. The chemical reaction can result in an expanded volume for the system. Referring back to the XRD analysis in Fig. 3, increasing sintering temperature the phases of Calcium Silicate ($CaSiO_3$) become more stable with the highest intensity for each samples. This phase has the lowest theoretical density (2.92 g/cm^3) among other phases. This finding was consistent with the results found by Erol and his co-workers in 2009 from their studies on the influence of the binder on the properties of sintered glass-ceramics produced from industrial wastes (Erol et al., 2009). They concluded that the densities of the sintered glass-ceramic samples changed depending on the amount of crystalline phase. The higher the content of crystalline phases with high density, the higher the measured density; the higher the amount of crystalline phase with the lower density, the lower the measured density will be obtained. So this might be a reason why the decreasing density was obtained in this present work.

3.2.2 Linear shrinkage

In general, dimensional control is a fundamental problem in ceramic processing. In this co-fired package the dimensional control, reproducibility and consistency of the shrinkage value to exact tolerance are strongly required and critical. Better reproducibility increases the uniformity of finished product characteristics and thereby increases the process yield. Each process step must complement the preceding and subsequent steps to achieve reproducibility. The main factor governing the reproducibility is the control of shrinkage. This rigid dimensional control in terms of shrinkage is the key to manufacturing high quality, high yield LTCC devices especially when the dimensions of lines and vias continuously decrease. Furthermore, repeatable shrinkage results are also needed for accurate positioning of circuit features such as vias, landing pads, cavities, surface mount component placement and post-fired printing and testing alignment (Sawhill, 1988). Dimensional uncertainties can result in disregistries of the laminated substrate due to the difficult control of the shrinkage amount (Raj and Cannon, 1999). So, to achieve desired properties and desired dimension, the process engineer must establish control of critical process variables.

It is clear that the linear shrinkage as tabulated in Table 2 increases up to 875 °C and drops slightly at 900 °C. It was due to the reactivity of the cofired material components such as ceramic oxide, glass, metal and organic solvent. This reaction also depends on the firing condition such as temperature, time and ambient atmosphere (Rabe et al., 2005). Typically, the shrinkage of the LTCC substrate across its width (XY) will be nearly identical while for the Z direction it has big variations. In the commercial practice, the standard values for XY and Z shrinkage are about 14-17 % and 20-25 % respectively. In this work the shrinkage was about 13-15 % for XY and 21-24 % for Z direction (Table 2). Chiang et al. (2011) has noted that in his work the shrinkage in the thickness direction is always larger than that in-plan shrinkage due to more organic material accumulated on the top surface of the green tape caused by unidirectional drying. However for the XY direction the better particle packing is believed to reduce shrinkage during sintering.

3.3 Microstructure

Microstructural characterization of ceramic packaging materials plays an important role in the understanding and improvement of most material properties such as thermal, electrical and mechanical properties (Pinckney and Beall, 2008). The properties of LTCC material are determined not only by the chemical composition and crystal structure but are also governed by the microstructural features such as density, grain size, porosity, intra- and inter-granular pore distribution, phases, crystalline morphology, crystallography and the chemistry of the interfaces. Such a microstructural arrangement can produce property inhomogeneities. The inhomogeneities usually occur during sample preparation. Brook (1988) has start earlier investigation on the role of inhomogeneities in the sintering process of alumina. Since early work on the subject, abnormal grain growth in alumina has often been attributed to inhomogeneities. He also noted that there are two principal kinds of inhomogeneities that may exist in a green or sintered body; extrinsic and intrinsic. Extrinsic inhomogeneities are associated with imperfect processing such as chemical segregation during drying, agglomeration or defect in powder consolidation. Intrinsic inhomegeneities are associated with anisotropy in the material. In a later investigation by Cho et al., (2000) and Sone et al., (2001) has shown that the impurity inhomogeneities in alumina would also produce abnormal grain growth. So, the desirable properties at particular frequencies can be improved by carefully controlling of the microstructure where the change of microstructure is the main issue in engineering materials.

3.3.1 Grain growth

One of the microstructural aspects normally studied is the grain growth. To date there exists a fairly detailed understanding of how grains of a sintered material can grow. According to Kingery (1976) the normal grain growth is the process by which the average grain size of the material increases continuously without a change the graphical shape of the grain size distribution during the heat treatment. Grain growth in polycrystalline materials is conventionally considered to be the results of the migration of grain boundary in response to the driving force provided by the system energy associated with the reduction in the total interface area by interface migration (Dillon and Rohrer, 2009). Since mass is conserved, some grains get smaller and disappear while others grow larger. In grain growth, the fundamental process is transferring an atom across the boundary from one grain to another as shown in Fig. 5. In this respect, it is well known that sintering is an interfaced related process in terms of driving force (Jo et al., 2006). This curvature-drive motion of grain boundary can lead to a normal behavior.

The occurrence of grain growth involves the elimination of free surface, i.e. the elimination of pores during sintering and is one of the ways in which there occurs a spontaneous decrease in the free energy during heating of a system with large interfacial areas (Pampuch, 1976). Interface energy is associated with the boundary between individual grains. A decrease in the free energy of such a system is also possible when the area of interfaces between the grains (grain boundary) decreases because of the excess energy possessed by these boundaries. The decrease in the area of the grain boundaries means an increase in the size of the grains. This is possible when some grains grow at the expense of others as the consequence of the movement of the grain boundaries toward their centers of their curvature. The boundary velocity has been observed to be linearly dependent on the curvature. This phenomenon can be explained in terms of the capillary force. The driving force for the grain growth process is the difference between the fine grain material and the larger grain size product resulting from the decrease in grain boundary area and the total boundary energy (Kingery, 1976). Grain growth must be controlled by even significant defects of the texture of the poly-crystal. It may be stopped by the presence of foreign inclusions (Shackelford, 1992). Besides the solid phase inclusions, the most important influence is exerted by pores (inclusion of gaseous phase). This explains the fact that, in some cases, grain growth does not take place at all, or wherever it is observed, it takes place only after the elimination of most of the pores, or after a considerable decrease in the volume fraction of pores on sintering.

Atoms can move across the grain boundary as well as within crystals when heated at high temperature. Equal numbers of atoms cross a plane boundary in opposite directions. However, atoms on a concave-surfaced boundary are likely to have more neighbors, and hence less energy than atoms on a convex-surfaced boundary. This brings about the chemical potential of the atoms. This causes a movement of the boundary towards the center of its curvature (See Fig. 5). Owing to this movement the grain growth takes place i.e. an increase in the size of the grains with many neighbors at the expense of grains with few neighbors.

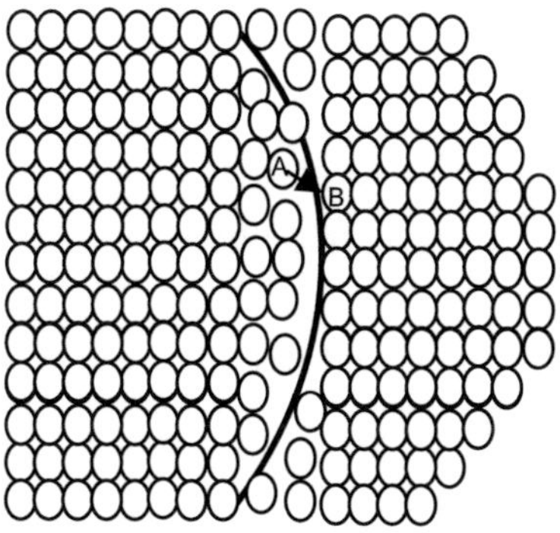

Fig. 5. Classical picture of a grain boundary.

Across the curved grain boundaries, there is the difference in the chemical potential resulting from the stresses which should be directed from the concave side of the curved interface to its convex side. Thus atoms on the convex side are in a state of compression and those on the concave side are in tension. This brings about a change in the chemical potential of the atoms (Rahaman, 1995). The gradient of the chemical potential constitute the driving force for the process of transferring atoms from the convex to the concave side across the curved grain boundary; this causes the movement of the boundary towards the centre of its curvature i.e. in a manner increasing the radius of the curvature. Owing to this movement, grain growth takes place, i.e. an increase in the size of the grains with many neighbors at the expense of grains with few neighbors. The rate of growth is proportional to the driving force and the driving force is proportional to the total amount of grain boundary energy. The driving force for each crystal to grow or shrink is given by the free energy difference between the atoms on the convex side and its difference across the interface amounts which given by the equation 7,

$$\Delta G = \gamma \overline{V}\left(\frac{1}{r_1} + \frac{1}{r_2}\right) \tag{7}$$

Where ΔG is the change in the free energy on going across the curved interface, γ is the boundary energy $\overline{V}$ is the molar volume and r_1 and r_2 are the principal radii of the curvature.

3.3.2 Glass-ceramic tape microstructure

Glass-ceramics formed by the controlled nucleation and crystallization of a glass precursor. As mention in the previous section, the Ferro A6S tape system consists of ceramic filler in glass matrix. The formation of a liquid phase is helpful to improve the development of microstructure behavior of the substrate. For glass-ceramics based on internal nucleation and grain growth, a general evolutionary pattern is observed in the crystallization cycle: amorphous phase separation and/or precipitation of primary crystalline nuclei, nucleation and growth of metastable crystalline phases and approach to a stable crystalline structure (Pinckney and Beall, 2008). Amorphous phase separation is generally the first stage in glass-ceramic formation. This phase is highly unstable as a glass and will precipitate primary crystalline nuclei on heating at temperature near the annealing point of the host glass. The next nucleation stage usually involves the heterogeneous nucleation and growth of metastable crystalline phases on the primary crystalline nuclei, resulting in a fine-grained metastable solid solution assemblage. Finally with increasing temperature, this metastable phase assemblage breaks down into stable crystalline phases by means of crystal phase transformation, reaction between metastable phases or a combination of several of these mechanisms.

The effects of sintering temperature on microstructure of glass-ceramic tape are shown in Fig. 6. It is clearly seen that there is inhomogeneous microstructure for the entire sample with some sample showing a big size pore meaning that probably only a few crystallite sizes present in the bulk samples. This feature of the microstructure could be due to the agglomeration within the sample. In the LTCC process, some particles are densely packed and some particles are loosely packed due to the agglomeration of particle at some places. The presence of agglomeration is a common problem in ceramics processing and influenced the microstructure behavior of the whole substrate. As mentioned by Lange (1984) and Hirata et al., (2009), when a laminated substrate or powder compact is heated, the inhomegeneity of the packing provides the different densification rate and the produce a microstructure which is usually not uniform when agglomeration is severe. An inhomogeneous distribution of particles leads to an inhomogeneous liquid distribution such that there is no driving force for redistribution of the liquid, so the densification rate is not homogeneous and the microstructure development also becomes inhomogeneous.

Based on their research, Deng et al., (2007) studied the microstructure of porous ZrO_2 ceramics and found that the agglomeration resulted in localized non-uniform shrinkage and the large pores are believed to originate from the large interagglomerated particles and greater microstructure non-uniformity. The agglomerated tape however leads to lower densities with large shrinkage deviations in particular direction giving a poor quality (Raj and Cannon, 1999). Agglomeration promotes uneven sintering which sometimes results in a mechanically weak and porous product. Thus, to achieve a high density material and good microstructure development, the agglomeration needs to be controlled (Forrester et al., 2008).

The densification process of the glass-ceramic composite can be described by the conventional three-stage liquid phase sintering; particle rearrangement, dissolution and precipitation and solid state sintering. The presence of glass in this composite system, acts as a sintering aid which produces liquid phase formation at a temperature lower than the sintering temperature and may considerably increase the rate of sintering. The viscous liquid which occurs during the sintering process may promote an additional diffusion mechanism of dissolution/precipitation, particle rearrangement and capillary forces and finally affect the densification at high temperature. Compared to the solid state sintering, liquid phase sintering enhances densification by two mechanisms; the densification is caused by mass transport via lattice diffusion from grain boundaries and grain boundary diffusion that should be encouraged to obtain high density when the sintering temperature is increased.

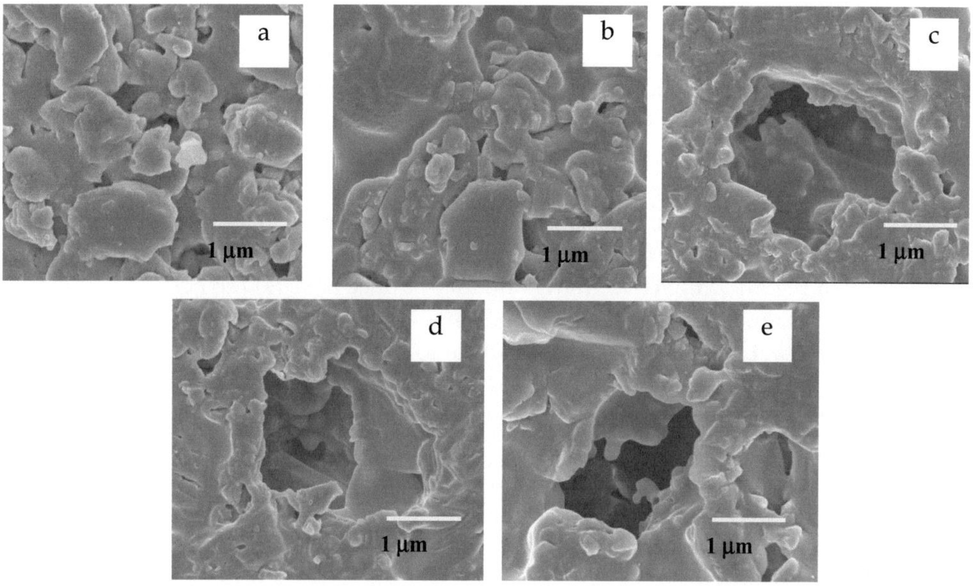

Fig. 6. SEM micrograph of laminated sample sintered at a) 800 °C, b) 825 °C, c) 850 °C, d) 875 °C and e) 900 °C.

The whole densification rate at the final stage, is a summation of the developed local internal defect such as the presence of pores that are no longer large enough to prevent grain growth and this is the major process going on in addition to the final densification. With the increasing grain size, the densification rate decreases as the distance of the defects to the grain boundaries increases. Grain growth also gives pore coalescence where smaller pores are merged together into larger ones: this also reduces the densification rate and explains the density results obtained. In the case of the sintering process of glass ceramic material, if crystallization occurs before densification, the viscosity of samples will be increased. It is due to the contribution of glass composition into a crystalline phase structure, resulting in the reduction of viscous flow of the system. As a result, densification through viscous flow sintering will not occur properly and a porous body will be formed (Banijamali et al., 2009).

3.3.3 Silver grain microstructure

In the LTCC technology, the microstructure changes involves the combination of a substrate and conductor material where LTCC tape materials contain a glass binder and organic solvent for tape casting purpose, thick-film conductor also contain glass frit as adhesion element with the substrate via formation of a vitreous bond (Kuromitsu et al., 1994). However, knowledge about the interaction of ceramic filler and amorphous matrix is very limited. Moreover the understandings of how an amorphous matrix influence the crystal grain growth in the metal film is not yet clear (Liu and Shen, 2004). According to Yajima and Yamaguchi (1984) the densification process of the substrate and printed film are depends upon sintering where the conductor and the substrate consisting of $CaO\text{-}B_2O_3\text{-}SiO_2$ sinter simultaneously. So the liquid phase formed is believed to help densification of metal powder to a denser packing of grains (see Fig. 7).

The SEM micrograph observation of microstructural changes of silver with sintering temperature is shown in Fig. 7. As observed here, the distribution of silver grain size is not homogenous; some grains have grown pretty well and some grains have not. There are relatively small grains about 2-4 μm in size coexisting with many large grains about 8-12 μm. It was also noted that the grain morphology are non-uniform/irregular shape. There are many curved boundaries and there is little obvious faceting or shape anisotropy in certain areas. Verhoeven (1975), has stated that generally two factors influence the shape of grains; the requirement to fulfill space and to be in a state of metastable equilibrium particularly with regard to the total grain boundary surface area. So the net result is that the three dimensional grains are irregular polyhedron with curved faces. Kang et al., (1985) has revealed that the grain shape changed mainly due to the anisotropic grain growth of grains rather than by contact flattening as suggested by Kingery's model. The distribution of the glass-frit diffusion is an inhomogeneous manner; the distribution of the metal grain will be varied. The quality of silver thick-film is dependent on the properties of the bottom substrate. Rane et al., (2003) reported that the microstructure changes in thick film that occurs during firing depends primarily on the sintering of silver powders as well as on the glass bonded interaction presuming that the substrate material has a slight influence on the sintering and microstructure. In LTCC materials, the LTCC substrate also plays an important role in the development of the microstructure during sintering since it is a co-firing process where the conductor and the substrate were fired together. Moreover the microstructural changes in the conductor film during the co-firing process also depend strongly on the sintering of metal powders, binder material interaction and the decomposition of the binder present in both paste and LTCC tape materials which lead the formation of cavities / voids on the surface of the fired film (Bangali et al., 2008).

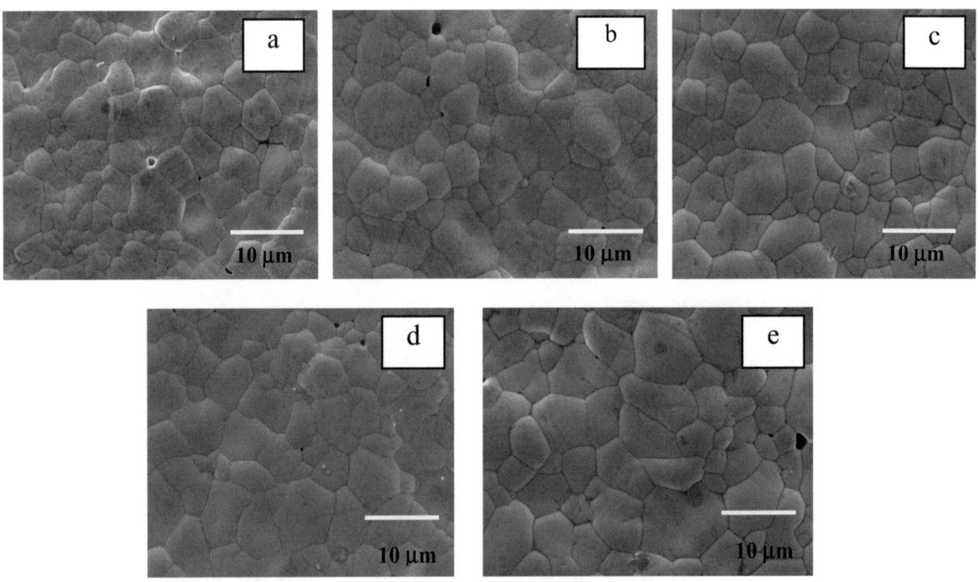

Fig. 7. SEM micrograph of silver surface sintered at a) 800 °C, b) 825 °C, c) 850 °C, d) 875 °C and e) 900 °C.

The tape system as a support substrate for printed metal contain SiO_2 and Ca or a combination of $CaSiO_3$ as observed in XRD peak in the samples would cause some defects in microstructure due to the formation of liquid phase. As reported by Louet et al. (2005), the SiO_2 content can increase the sintering rate due to the formation of an intragranular vitreous phase while the combination of CaO and SiO_2 is responsible for abnormal grain growth. As the contents of SiO_2 and Ca or a combination of $CaSiO_3$ increase at some places, they should increase the grain growth kinetics due to the formation of liquid phase which preferentially fill the spaces between the solid particles. It is indicated that the liquid phase has excellent wettability with $CaSiO_3$. Therefore the silver powder was sintered with liquid glass phase. When the $CaSiO_3$ becomes liquid, chemical reactions such as the migration of silver into the liquid frit would occur on the surface of silver. It also believes that the silver conductive thick-film contains about 90 wt% silver and 10wt % glass frit. So, the presence of the liquid phase gives rise to changes of its atomic mobility and is known to be the cause of abnormal grain growth. During sintering process, the glass phase may decompose into liquid phase providing the capillary force for neighboring particles to come closer and react. If the glass phase is not uniformly distributed, the grain growth is not homogeneous.

The average grain size of sintered body was measured over 200 grains by the linear intercept method. Fig. 8 was plotted with varied sintering temperature for the average grain size and the density of the substrate. As we can see with increase the firing temperature, the silver grains grow in much good shape up to 900 °C and the average grain size increased from ~4.39μm to ~5.60 μm. The work is consistent with the findings by Huang and Tsai (2001), which noted that the grain size increases with increasing sintering temperature.

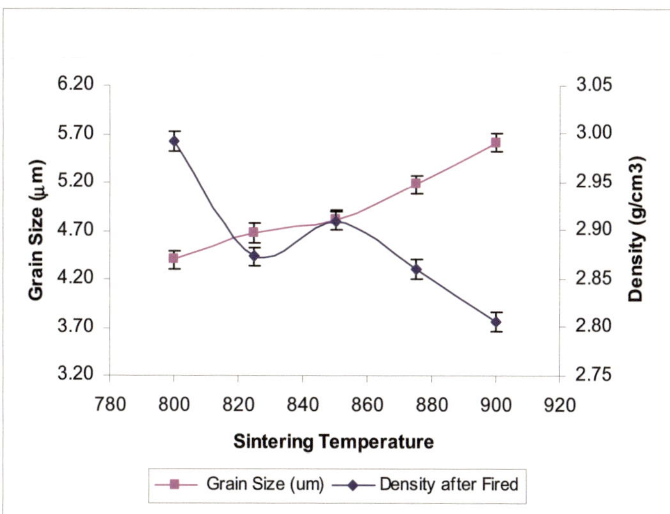

Fig. 8. Average grain size of printed conductor fired at various sintering temperature.

However, the trend of the average grain size (metal surface) and the density (whole substrate) are somewhat opposing trend with increasing sintering temperature. The relation of microstructural changes with decreasing density is not significant in this work because the density is highly extremely extrinsic and macroscopic property while the average grain size is less extrinsic and microscopic average of crystallite size in polycrystalline material.

The average grain size can be increased when increasing the sintering temperature however the density does not always show a similar trend of the average grain size with increase the temperature due to the presence of pores and other defects or stresses within the sample (Alias et al., 2009). The average grain size of a sintered product is influenced by the total shrinkage characteristic during sintering. The glass contained in metal thick-film formed the liquid phase during sintering resulting in faster grain growth speed of the conductor because of its excellent low temperature shrinkage characteristics. The increase of average grain size shows the microstructural evolution of the samples. These can be either be intrinsic effects related to the crystal structure, defect, surface or energy anisotropy of the material or be extrinsic such as pore size, density heterogeneity, partial wetting by the liquid phase and preferential impurity (Feteira and Sinclair, 2008) . The microstructure of printed silver metal can be designed to attain the features such as fine grain size, low porosity and negligible second phase that can improve operation at the particular frequency. The control of microstructure of metal films is very important for a good device performance.

The microstructural evolution of the sintered samples can be described by adapting the sintering mechanism explained by Sameshima et al. (2000). There are 5 stages involved in the process:

1. First initial stage where no grain growth and mass transport of the sintered samples occur by grain boundary diffusion.
2. Second initial stage occurs afterwards where no grain growth occurs and the mass transport of sintered body is by lattice diffusion.
3. First intermediate and final stage later takes place where lattice diffusion occurs without any grain growth of the samples.
4. Second intermediate and final stage where there is grain growth happening and lattice diffusion still playing its role.
5. Final intermediate stage where grain boundary diffusion occur with no grain growth.

Microstructure evolution has a large influence on the material properties generally and can be characterized by abnormal grain growth behavior where a few selected grains grow by replacing the other grains (Ubach et al., 2004). The main cause of the abnormal growth is due to the presence of anisotropy of either the surface energy or the strain density (Paik et al., 2003). Anisotropy of the grain boundary energy has been considered to be origin of the abnormal grain growth observed in bulk materials. It is possible that this may apply to the silver film. The silver grain growth mechanism could be evaluated according to its activation energy (Q). Coble's theory in Coble (1961) mentioned that from the behavior of particle growth, the activation energy of metal grain growth can be predicted using Arrhenius equation as below:

$$d\ln k/dT = Q/RT^2 \tag{8}$$

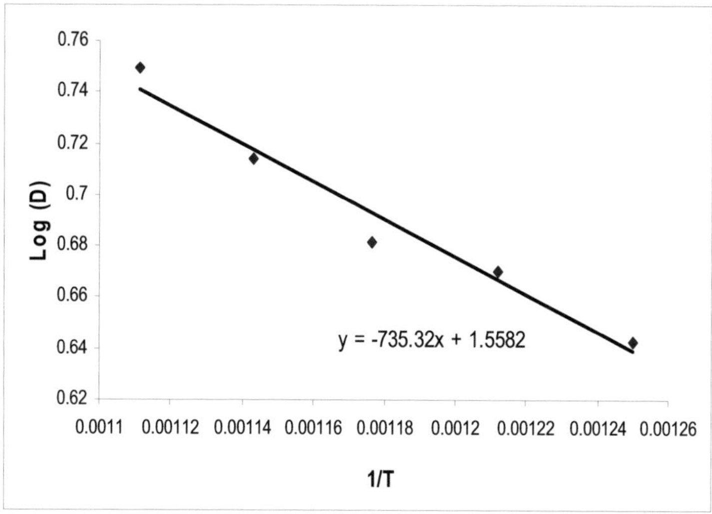

Fig. 9. Plots of log D versus the reciprocal of absolute temperature (1/T).

where k is the specific reaction rate constant, Q is the activation energy, T is the absolute temperature, and R is the ideal gas constant. The value of k however can directly be related to grain size according to Jarcho et al. (1976), which resulted in the equation below:

$$LogD = (-Q / 2.303R)1/T + A \tag{9}$$

Where T is the absolute temperature, A is the intercept and D is the grain size. By using equation 9, one can obtain a best fitted straight-line plot of grain size where a plot of log D versus the $1/T$ as shown in Fig. 9. From the equation (9), we obtained the slope of the line which is $-Q/2.303R$ and the value of the activation energy of metal grain growth (Q) can be calculated from the Arrhenius plot which is 14.079kJ/mol which might be ascribed to grain boundary grooving, precipitation, impurity drag or interfacial reaction between the films and the substrate. The obtained values is considered too low compared to the activation energy calculated by Wu et al. (2011) from their studies on the behavior of ZnO-doped silver thick-film and the silver grain growth mechanism.

3.3.4 Anisotropic grain growth

A closer look for sample sintered at 800 °C and 825 °C as shown in Fig. 10, we clearly see pronounced layered-structure grain growth/faceted grain microstructure. The large relatively faceted grains have differed orientations between each other. For the oriented grain growth the preferred direction is the same for all directions. So in this case, the anisotropic structure occurs for a restricted number of grains and can be due to the crystallographic effects. For all cases of the direction growth of anisotropic grain is random. Growth anisotropy is produced naturally during the process of grain growth (Uematsu et al., 1997).

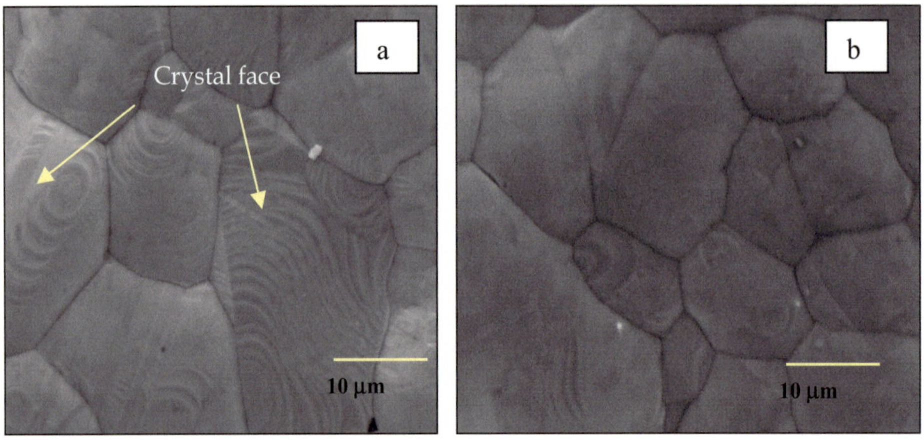

Fig. 10. Anisotropic grain microstructure for sample sintered at a) 800 °C and b) 825 °C.

There are some investigations on the anisotropic grain growth as well as the presence of faceting grain of alumina interface in the presence of glass (Simpson and Carter, 1990). The faceting of alumina interface in the presence of a glass is a factor affecting the grain boundary movement during sintering. The geometry and movements of facet during liquid phase sintering process are expected to influence directly the final microstructure of the grain boundaries that ultimately control the properties of the alumina compact.

For a faceted grain in a liquid matrix, the dihedral angle is not uniquely defined because of the torque on a facet, contrary to the case of a rounded grain. When two faceted grains are in contact with a certain angle of crystallographic orientation in a liquid matrix, the shape of the contact boundary in equilibrium must be that of the minimum interfacial energy. These abnormal grains originate from heterogeneities in the chemical composition, particle size and packing density. In the LTCC process, the substrate also plays an important role in the development of microstructure during sintering since it is a co-firing process where the conductor and the substrate are sintered simultaneously. So, the nuclei of abnormal grains may be formed during different stages of sintering. From XRD peak observation for samples sintered at 800 °C and 825 °C, a few phases including $CaSiO_3$ and $Ca_3(SiO_4)O$ were present suggesting the nucleation rate is not zero. The results found was consistent with the work of Rohrer (2001) which noted that at lower temperature, the anisotropy grain of the surface energy is expected to be higher. The wollastonite phase ($CaSiO_3$) becomes a liquid phase before it decomposes. The liquid phase exists preferentially, filling spaces in small pores and particle contact. The capillary force that is generated by the liquid phase can drag the nearby particle close together. The distribution of the liquid phase may not be uniform because the amount of liquid is small. According to Kang et al, (1991), liquid phase, even in a small amount forms due to the presence of impurities, grains are faceted and abnormal grain growth occurs. Such heterogeneities may trigger the formation of nuclei of abnormal grain. Because the nuclei are small, they have a greater chance to grow at their own pace until they encounter other abnormal grains (Chen and Tuan, 2000). However, when a liquid phase forms due to a high concentration of impurities, the grain shape is faceted and abnormal grain growth occurs at low annealing temperatures. The edges of faceted grains become rounded with an increase in annealing temperature and normal grain growth occurs because of an increased contribution of entropy (a reduction of self free energy). Critical amounts of CaO and SiO_2 to induce and to suppress abnormal grain growth were also measured. According to the results, when the amounts of CaO and SiO_2 are within their solubility limits, normal grain growth occurs. On the other hand, abnormal grain growth occurs if the amount is above the solubility limit.

With increasing temperature, the faceted grain microstructure becomes less prominent. It might be due to the presence of boundary defect. As studied by Maclaren et al. (2003), all faceted microstructure being limited by a paucity of steps, ledges or boundary dislocation to serve as site for atom attachment or detachment. As the sintering temperature increases the faceted or anisotropic microstructure are limited. Although the faceted grain microstructure observed is not large, the effects on surface morphology are significant. Furthermore it should be pointed out that there is geometry limits to observable anisotropies for perpendicular surface in any equilibrium experiment.

4. Conclusion

As a conclusion, the performance of the laminated sample strongly depends on the starting raw materials and processing method; good technical skills and handling procedures are required to minimize the defect/issue occurring in the LTCC multilayer process. As with other advanced electronic ceramics, the quality of raw materials will have a marked effect on the final properties of a ceramic product. The raw material selection includes characterization of purity, particle size distribution, surface area, consistency and cost. The processing of a ceramic material must be optimized with respect to microstructural characteristics, because only by this way we can produce ceramics with properties that approach the level of intrinsic dielectric properties. The chosen raw material is the first key to controlling the microstructure and making a consistent product.

The present study is able to explain the effect of the different sintering temperatures on the whole substrate properties and printed thick-film. The microstructural evolution at various sintering temperatures was carefully tracked. The sintering temperature therefore plays a vital role in forming microstructures, porosity and shrinkages of the as-prepared materials. This is more complex than the simple sintering of metal grains. In order to understanding better the microstructure evolution, the study of samples fired should be carried out at the different stages of the sintering process. This work will be very helpful to predict how the microstructural factors and material properties of the LTCC system would affect the microwave-frequency performance of relevant devices. It should be mentioned that the results obtained from this work is not always be consistent with observations from other works.

5. Acknowledgement

The authors wish to thank Telekom Malaysia for their funding support under project IMPACT (RDTC/100745) and Assoc. Prof. Dr. Mansor Hashim, Mr. Ismayadi Ismail and Mrs. Sabrina Mohd. Shapee for the guidance and support for this research work.

6. References

Alias, R., Ibrahim, A., Shapee, S.M., Ambak, Z., Yusoff, Z.M. and Saad, M.R. (2009). Microstructure and Thermal Diffusivity of Al_2O_3-SiO_2-Based Substrate with Varied Sintering Temperatures. *Journal of the Australian Ceramic Society,* Vol. 45, No. 2, pp. 80-84.

Bangali, J., Rane, S., Phatak, G. & Gangal, S. (2008). Silver Thick-Film Paste for Low Temperature Co-fired Ceramic: Impact of Glass Frit Variation. *Soldering and Surface Mount Technology,* Vol. 20, No. 3, pp. 41-46.

Banijamali, S., Eftekkhari Yekta, B., Rezaie, H.R. & Marghussian, V.K. (2009). Crystallization and Sintering Characteristics of CaO-Al_2O_3-SiO_2 Glasses in the Presence of TiO_2, CaF_2 and ZrO_2. *Thermochimica Acta,* Vol. 488, No. 1-2, pp. 60-65.

Barlow, F.D. & Elshabini, A. (2007). *Ceramic Interconnect Technology Handbook.* CRC Press, New York.

Brook., R. J., Tuan, W.H and Xue, L. A (1988) Critical Issues and future directions in sintering science Ceramic Transaction 1, pp; 811-823.

Chang, C.-R. & Jean, J.-H. (1998). Effect of Silver Paste Formulation in Camber Development during the Co-Firing of Silver Based LTCC Package. *J. Am. Ceram. Soc.,* Vol. 81, No. 11, pp. 2805-2814.

Chang, C.-R. & Jean, J.-H. (1999). Crystallization Kinetics and Mechanism of Low-Dielectric, Low Temperature, Cofirable CaO-B_2O_3-SiO_2 Glass Ceramics. *J. Am. Ceram. Soc.,* Vol. 82, No. 7, pp. 1725-1732.

Chen, C-.S., Chiou, C.-C., Chen, C.-S. & Lin, I.-N. (2004). Microwave Dielectric Properties of Glass-MCT Low Temperature Cofireable Ceramics. *J. European Ceramic,* Vol. 24, pp. 1795-1798.

Chen, C.-Y. & Tuan, W.-H. (2000). Effect of Silver on the Sintering and Grain Growth Behavior of Barium Titanate. *J. Am. Ceram. Soc.,* Vol. 83, No. 12, pp. 2998-2992.

Chiang, M.-J., Jean, J.-H. & Lin, S.-C. (2011). The Effect of Anisotropy Shrinkage in Tape-Cast Low Temperature Co-Fired Ceramics on Camber Development of Bilayers Laminates. *J. Am. Ceram. Soc.,* Vol. 94, No. 3, pp. 683-686.

Cho, S.-J., Kim, K.H., Kim, D.J and Yoon, K.J. (2000). Abnormal Grain Growth at the Interface Centrifugally Cast Alumina Bilayer during Sintering, *J. Am. Ceram. Soc.,* No. 83, pp. 1773-1776.

Coble, R.L. (1961). Sintering Crystalline Solids. I. Intermediate and Final State Diffusion. *J. Appl. Phys.,* Vol. 32, No. 5, pp. 787-792.

Deng, Z.-Y., Fereira, J.M., Tanaka, Y. & Isoda, Y. (2007). Microstructure and Thermal Conductivity of Porous ZrO_2 Ceramics. *Acta Materialia,* Vol. 55, pp. 3663-3669.

Despande, V., Kshirsagar, A., Rane, S., Seth, T., Phatak, G.J., Mulik, U.P. and Amalnerkar, D.P. (2005). Properties of Lead-Free Conductor Thick-Films of a Co-Precipitated Silver Palladium Powders. *Materials Chemistry and Physics,* No. 93, pp. 320-324.

Dillon, S.J. & Rohrer, G.S. (2009). Mechanism for the Development of Anisotropy Grain Boundary Character Distribution during Normal Grain Growth. *Acta Materialia,* Vol. 57, pp. 1-7.

Erol, M., Kucukbayrak, S and Ersoy-Mericboyu, A. (2009). The Influence of the Binder on the Properties of Sintered Glass-Ceramics Produced from Industrial Wastes. *Ceramics International,* No. 35, pp. 2609-2617.

Ferro Corporation; Ferro Design Guideline in www.ferro.com.

Feteira, A. & Sinclair D.C. (2008). Microwave Dielectric Properties of Low Firing Temperature $Bi_2W_2O_9$. *J. American Ceram. Soc.*, Vol. 91, No. 4, pp. 1338-1341.

Forrester, J.S., Goodshaw, H.J., Kisi, E.H., Suaning, G.J. & Zobec, J.S. (2008). Effect of Milling on the Sintering Behavior of Alumina. *J. Am. Ceram. Soc.*, Vol. 44, No. 1, pp. 47-52.

Hirata, Y., Hara, A. & Aksay, A. (2009). Thermodynamics of Densification of Powder Compact. *Ceramics International*, Vol. 35, pp. 2667-2674.

Hrovat, M., Balavic, D, Kita, J., Holc, J, Cilensek, J. and Drnovsek, S. (2009). Thick-Film Thermistors and LTCC Materials; The Dependence of the Electrical and Microstructural Characteristics on the Firing Temperature. *J. of Europ. Ceram. Soc.*, 29, pp. 3265-3271.

Hsi, C.-S, Chen, Y.-R and Hsiang, H.-I (2011), Diffusivity of Silver Ions in the Low Temperature Co-fired Ceramic (LTCC) Substrate. *J. Mater. Sci*, Vol. 46, Issue 13, pp. 4695-4700.

Hsu, Y.-F, Wang, S.-F & Cheng, T.-W. (2003). Effects of Additives on the Densification and Microstructural Evolution of Fine θ-Al_2O_3 Powder. *Mat. Sc. & Eng. A*, Vol. 362, Issue 1-2, pp. 300-308.

Hsueh, C. H. and A. G. Evans. (1985). Residual Stress and Cracking in Metal/Ceramic Systems for Microelectronics Packaging, *J. Am. Ceram. Soc.*, No. 68, Vol. 3, pp. 120–127

Huang, C.-L. & Tsai, J.-T. (2001). Effects of Sintering Temperature on CaO-Li_2O-Sm_2O_3-TiO_2 Microwave Dielectric Ceramic. *Proc. Natl. Sci. Counc. ROC (A)*, Vol. 25, No. 5, pp. 317-321.

Imanaka, Y. & Kamehara, N. (1992). Influence of Shrinkage Mismatch between Copper and Ceramics on Dimensional Control of the Multilayer Ceramic Circuit Board. *J. Ceram. Soc. Jpn. Int. Ed.*, Vol. 100, pp. 558-561.

Imanaka, Y. (2005). *Multilayered Low Temperature Co-fired Ceramic (LTCC) Technology*. Springer, New York.

Jarcho M., Bolen C.H., Thomas M.B., Bobick J., Kay J.F., Doremus R.H. (1976) Hydroxylapatite Synthesis and Characterization in Dense Polycrystalline Form. *Journal of Materials Science*. Vol.11, No. 11, pp. 2027-2035.

Jean, J.-H and Fang, Y-C., Dai S. X. and. Wilcox, D. L. Sr. (2001) Devitrification Kinetics and Mechanism of K_2O–CaO–SrO–BaO–B_2O_3–SiO_2 Glass-Ceramic. *J. Am. Ceram. Soc.*, Vol. 84, No. 6, pp. 1354–1360.

Jean, J.-H., Chang, C.-R. & Lei, C.-D. (2004). Sintering of Crystallizable CaO-B_2O_3-SiO_2 Glass with Silver. *J. American Ceram. Soc.*, Vol. 87, No. 7, pp. 2630-2633.

Jo, W., Kim, D.-Y. & Hwang, N.-M. (2006). Effect of Interface Structure on the Microstructural Evolution of Ceramics. *J. Am. Ceram. Soc.*, Vol. 89, No. 8, pp. 2369-2380.

Kang, S.-J.L., Kaysser, W.A., Petzow, G. & Yoon, D.N. (1985). Growth of MO Grains around Al_2O_3 Particles during Liquid Phase Sintering. *Acta Metall.*, Vol. 33, pp. 1919-1926.

Kang, S.-J.L., Kim, K.-H. & Yoon, D.N. (1991). Densification and Shrinkage during Liquid Phase Sintering. *J. Am. Ceram. Soc.*, Vol. 74, No. 2, pp. 425-427.

Kemethmuller, S., Hagymasi, M., Stiegelschmitt, A. & Roosen, A. (2007). Viscous Flow as the Driving Force for the Densification of Low Temperature Co-Fired Ceramic. *J. Am. Ceram. Soc.*, Vol. 90, No. 1, pp. 64-70.

Kingery, W.D., Bowen, H. K., & Uhlman, D.R. (1976). *Introduction to Ceramics*, (2nd Ed.), John Wiley and Sons, New York.

Klug, H.P. & Alexander, L.E. (1974). *X-Ray Diffraction Procedures for Polycrystalline and Amorphous Materials*. John Wiley and Sons, New York.

Kume, S., Yasuoka, M., Lee, S.-K., Kan, A., Ogawa, H. And Watari, K. (2007). Dielectric and Thermal Properties of AIN Ceramics. *J. Europ. Ceram. Soc.*, Vol. 27, pp. 2967-2971.

Kuromitsu, Y., Wang, S.F., Yoshikawa, S. & Newnham, R. E. (1994). Evolution of Interfacial Microstructure between Barium Titanate and Binary Glass. *J. American Ceram. Soc*, Vol. 77, No. 3, pp. 852-856.

Lange, F.F. (1984). Sinterability of Agglomerated Powders. *J. Am. Ceram. Soc.*, Vol. 67, No. 2, pp. 83-89.

Liu, Z.-J. & Shen, Y.G. (2004). Effect of Amorphous Matrix on the Grain Growth Kinetics in Two-Phase Nanostructured Film: A Monte Carlo Study. *Acta Materiala*, Vol. 52, pp. 729-736.

Lo, C.-L. & Duh, J.-G. (2002). Low Temperature Sintering and Microwave Dielectric Properties of Anorthite-Based Glass-Ceramic. *J. Am. Ceram. Soc.*, Vol. 85, No. 9, pp. 2230-2235.

Long, Y., Wang, Y., Wu, W., Wang, D. & Li, Y. (2009). Sintering and Microwave Dielectric Properties of $LiNB_{0.63}Ti_{0.4625}O_3$ Ceramics with the B_2O_3-SiO_2 Liquid Phase Additives. *J. American Ceram. Soc.*, Vol. 92, No. 11, pp. 852-856.

Louet, N., Gono, M and Fantozzi, G. (2005). Influence of the Amount of Na_2O and SiO_2 on the Sintering Behavior and the Microstructural Evolution of a Bayer Alumina Powder, *Ceramics International*, Vol. 31, pp. 981-987.

MacLaren, I., Cannon R. M., Gulgun, M. A., Voytovych, R., Nicoletta P.-P.,¶ Scheu, C. Taffner, U. and Ruhle M. (2003). Abnormal Grain Growth in Alumina: Synergistic Effects of Yttria and Silica, *J. Am. Ceram. Soc.*, Vol. 86, No. 4, pp. 650–659.

Matters-Kammerer, M., Mackens, U., Reimann, K., Pietig, R., Hennings, D., Schreinemacher, B., Mauczok, R. , Gruhlke, S. & Martiny, C. (2006). Material Properties and RF Applications of High *k* and Ferrite LTCC Ceramics. *Microelectronics Reliability*, Vol. 46, pp. 134–143.

Muralidhar, S.K., Shaikh, A.S., Robert, G.J., Hankey, D.L., Leandri, D.J. & Vlach, T.J. (1992). *Low Dielectric, Low Temperature Fired Glass-Ceramic.* U.S Patent 5,164,342.

Paik, P.-M., Park, Y.-J, Yoon, M.-S., Lee, J.-H. & Joo, Y.-C. (2003). Anisotropy Grain Boundary Energies as Cause of Abnormal Grain Growth in Electroplated Copper Film. *Scripta Materiala*, Vol. 48, pp. 683-688.

Pampuch, R. (1976). *Ceramic Materials: An Introduction to their Properties.* Elsevier, Scientific Publishing Company, NY.

Pinckney, L. R. and Beall, G.H. (2008). Microstructural Evolution in Some Silicate Glass-Ceramics: A Review. *J. Am. Ceram. Soc.*, Vol. 91, No. 3, pp. 773-779.

Rabe, T., Schiller, W.A., Hochheimer, T., Modes, C. & Kipka, A. (2005). Zero Shrinkage of LTCC Self-Constrained Sintering. *Int. J. Appl. Ceram. Tech.*, Vol. 2, No. 5, pp. 374-382.

Rahaman, M.N. (1995). *Ceramic Processing and Sintering.* Marcel Dekker, Inc. NY.

Raj, P.M. & Cannon, R. (1999). Anisotropy Shrinkage in Tape-Cast Alumina: Role of Processing Parameter and Particle Shape. *J. Am. Ceram. Soc.*, Vol. 82, No. 10, pp. 2619-2625.

Rane, S. B., Khanna,.P.K., Seth, T., Phatak, G.J., Amalnerkar, D.P and Das, B.K. (2003). Firing and Processing Effects on Microstructure of Fritted Silver Thick-film Electrode Materials for Solar Cells. *Materials Chemistry and Physics,* Vol. 82, pp. 237-245.

Rohrer, G. S. (2001), The Anisotropic of Metal Oxide Surface Properties; In *Oxide Surfaces,* D.P. Woodruff, pp. 485-513, Elsevier Science, UK.

Sameshima, S., Higashi, K. & Hirata, Y. (2000). Sintering and Grain Growth of Rare-Earth-Doped Ceria Particles. *J. Ceram. Processing,* Vol. 1, No. 1, pp. 27-33.

Sawhill, H.T. (1988). Materials Compatibility and Co-firing Aspect in Low Temperature Co-fired Ceramic Packages. *Ceram. Eng. Sc. Proc.*, Vol. 9, No. 11-12, pp. 1603-1617.

Sergent, J.E. & Harper, C.A. (1995). *Hybrid Microelectronics Handbook,* (2nd Ed.) McGraw-Hills Inc., New York.

Shackelford, J.F. (1992). *Introduction of Materials Science for Engineers.* Macmillan Publishing Company, USA.

Shimada, Y., Utsumi, K., Suzuki, M., Takamizowa, H., Nitta, M. and Watari, T. (1983). Low Firing Temperature Multilayer Glass Ceramic Substrate. *IEEE Trans. Compon. Hybrids Manuf. Tech.*, Vol. 6, No. 4, pp. 382-388.

Simpson, Y.K and Carter, B.C. (1990). Faceting Behaviour of Alumina in the Presence of a Glass. *J. Am. Ceram. Soc.,* Vol. 73, No. 8, pp. 2391-2398.

Sone, T.-W,Han, J.-H., Hong, S.-H and Kim, D.-Y .(2001). Effect of Surface Impurities on the Microstructure Development during Sintering of Alumina. *J. Am. Ceram. Soc.*, Vol. 84, pp. 1386-1388.

Tummala, R.R. (1991). Glass-Ceramic Packaging in the 1990's. *J. Am. Ceram. Soc.*, Vol. 74, pp. 895-908.

Ubach, R.L.J.M., Schreurs, P.J.G. & Geers, M.G.D. (2004). Microstructure Evolution of Tin-Lead Solder. *IEEE Trans. on Components and Packaging Technologies,* Vol. 27, No. 4, pp. 635-641.

Uematsu, K., Ishaka, S., Shinohara, N. & Okumiya, M. (1997). Grain Oriented Microstructure of Alumina Ceramics Made Through the Injection Moulding Process. *J. Am. Ceram. Soc.*, Vol. 80, No. 5, pp. 1313-1315.

Valant, M. & Suvorov, D. (2000). Microstructure vs. Dielectric Property Correlation in the Stoichiometric Sillinites. *Korean J. Crystallography,* Vol. 11, No. 4, pp. 191-194.

Valant, M., Suvorov, D., Pullar, R.C., Sarma, K. & Alford, N. M. (2006). A Mechanism for Low Temperature Sintering. *J. Europ. Ceram. Soc.*, Vol. 26, Issue 13, pp. 2777-2783.

Verhoeven, J.D. (1975). *Fundamental of Physical Metallurgy.* John Wiley, New York.

Wang, S.-F., Wang, S.J., Wang, Y.-R., Hsu, Y.-F., Chen, L.-Y. & Tsai, J.-S. (2009). Effect of SiO_2 Addition on the Microstructure and Microwave Dielectric Properties of Ultra Low Fire $Tite_3O_8$ Ceramics. *Ceramics International,* Vol. 35, pp. 1813-1817.

Wang, S.-H and Zhou, H.-P .(2003). Densification and Dielectric Properties of CaO-B_2O_3-SiO_2 System Glass-Ceramics. *Material Science and Engineering B,* Vol. 99, pp. 597-600.

Wang, X. & Atkinson, A. (2011). Microstructure Evolution in Thin Zirconia Films: Experimental Observation and Modeling. *Acta Materialia,* Vol. 59, pp. 2514–2525.

Wu, S.P., Zhao, Q.Y., Zheng, L.Q. and Ding, X. H. (2011). Behaviors of ZnO-doped Silver Thick-Film and Silver Grain Growth Mechanism. *Solid State Science,* Vol. 13, pp. 548-552.

Xiang J.-H., Yong, H. & Xie, Z.-P. (2002). Study of Gel-Tape-Casting Process of Ceramic Materials. *Mat. Sc. & Eng. A,* Vol. 323, Issue 1-2, pp. 336-341.

Yajima, K. & Yamaguchi, T. (1984). Sintering and Microstructure Development of Glass Bonded Silver Thick-Film. *Journals of Material Science,* Vol. 19, pp. 77-84.

Zhou, D., Wang, H., Yao, X. & Phang, L.-X. (2008). Dielectric Behavior and Co-Firing with Silver Monoclinic $BiSbO_4$ Ceramic. *J. Am. Ceram. Soc.,* Vol. 91, No. 4, pp. 1380-1383.

Zhu, H., Liu, M., Zhou, H., Li, L. & Lv, A. (2007). Study on Properties of CaO-SiO_2-B_2O_3 System Glass-Ceramic. *Materials Research Bulletin,* Vol. 42, Issue 6, pp. 1137-1144.

4

Microwave Sintering of Thermistor Ceramics

Aimin Chang[1], Huimin Zhang[1], Qing Zhao[1] and Bo Zhang[1,2]
[1]*Xinjiang Key Laboratory of Electronic Information Materials and Devices, Xinjiang Technical Institute of Physics and Chemistry. CAS, Urumqi*
[2]*Graduate School of the Chinese Academy of Sciences, Beijing*
PR China

1. Introduction

Microwave sintering is a new sintering technology developed in the middle to late period of the 1980's, which is characterized by fast densification for ceramic materials[1]. In recent years, microwave heating has been well employed in the field of sintering and joining of ceramics as a result of its advantages against conventional methods. These ceramic materials include oxides, mixed oxides, non-oxides, composite ceramics, etc[2]. In addition, because ceramics have low thermal conductivities and are processed at high temperatures, many researchers have attempted to take advantage of volumetric heating for sintering, chemical vapor infiltration (CVI), and pyrolysis of polymeric precursors[3]. Now it has been found that the microwave sintering can also be applied as efficiently and effectively to thermistor ceramics as well as many other ceramics. This chapter compares advantages of microwave sintering with conventional sintering and presents some applications in thermistor ceramics.

2. Theoretical aspects of microwave sintering

Microwave is the name given to electromagnetic radiation 1 m to 1 mm in wavelength that corresponds to a frequency of about 300MHz to 300 GHz. Microwave heating is a process in which the materials couple with microwaves, absorb the electromagnetic energy volumetrically, and transform into heat. This is different from conventional methods where heat is transferred between objects by the mechanisms of conduction, radiation and convection[4]. In conventional heating, the material's surface is first heated followed by the heat moving inward. This means that there is a temperature gradient from the surface to the inside. However, microwave heating generates heat within the material first and then heats the entire volume[5].

2.1 The principle of microwave sintering

Microwave sintering makes use of dielectric loss of materials in microwave electromagnetic fields to heat ceramic matrix to the sintering temperature fast. The microwave sintering effect mainly comes from the continuous changes of material polarization in alternating electromagnetic field. The continuous changes of dipoles in materials can produce strong vibration and friction, therefore, generating heat to achieve sintering[1].

2.2 Advantages of microwave sintering comparison to conventional methods

Microwave sintering has achieved worldwide acceptance due to its significant advantages against conventional sintering methods, especially in ceramic materials. A summary of these are[2, 4]: (1) The observation of more rapid reaction and/or sintering during microwave processing of ceramics had led to speculation that microwave processing results in enhanced diffusion; (2) Since microwaves penetrate the dielectric with absorption and heat generation throughout the work piece, microwave sintering is much more efficient and lower processing costs;(3)Generally higher density and better grain distribution can be achieved through microwave sintering;(4) Microwave sintering can significantly shorten the sintering time leading to consume much lower energy than conventional sintering.

3. Application of microwave sintering in thermistor ceramics

Microwave sintering is a new sintering technology which can densify ceramic materials at a very rapid rate and at a substantially lower temperature than the conventional sintering process[6]. The following demonstrates microwave sintering advantages against conventional sintering for some oxide thermistor ceramics, including NTC thermistor ceramics and PTC thermistor ceramics.

3.1 Microwave sintering of NTC thermistor ceramics

Because of integrity and uniformity of microwave heating, the NTC thermistors of depending on uniformity and densification may show uniform microstructure during microwave sintering. A dense microstructure is necessary to obtain a good reproducibility of the electrical characteristics of the ceramic[7]. During microwave sintering, in addition to heating, the rise in temperature is accompanied by an increase in grain diameter and a consequent decrease in the porosity during densification[8].Consequently, microwave sintering can enhance the densification and improve uniformity of NTC thermistors.

3.1.1 Brief introduction of NTC thermistors

Negative temperature coefficient (NTC) thermistors are thermally sensitive resistors whose resistance decreases with increasing temperature. Their resistivity can be expressed by the following Arrhenius equation[9]: $\rho=\rho_0\exp(B/T)$, where ρ_0 is the resistivity of the material at infinite temperature, T is the absolute temperature and B is the B constant, which is sometimes called the coefficient of temperature sensitivity. In fact, B constant indicates a sensitivity to temperature excursions and is given by $B=q/k$, where q is the activation energy for electrical conduction, and k is the Boltzmann constant[10]. They are widely used in various industrial and domestic applications, e.g., elements for the suppression of in-rush current, for temperature measurements and controls, and for compensation for other circuit elements[11].

3.1.2 Microwave sintering of NTC thermistor ceramics

Jin et al.[12, 13] successfully calcined and sintered $Mn_{0.43}Ni_{0.9}CuFe_{0.67}O_4$ samples by microwaves. Fig.1 compares the XRD patterns of powders calcined by microwave treatment and conventional heating, respectively. As can be seen, phase analysis of conventionally calcined $Mn_{0.43}Ni_{0.9}CuFe_{0.67}O_4$ powders shows metallic oxide as the main crystalline phase and a very minor amount of spinel phase. These results indicate that oxides were still not fully reacted. However, for the microwave heating, the complete spinel phase has obtained when powders were calcined at 650°C. Consequently, the Microwave sintering leads to a lower calcination temperature than conventional methods. In the conventional sintering processes, extremely high sintering temperatures and long holding times (several hours) under air atmosphere are applied in the fabrication of thermistor products to achieve the highest density and minimum porosity. Fig.2 shows the SEM micrographs from the fracture surface of sintered samples. Although the same sintering temperature and holding time (1000°C, 40 min) were applied, the microstructures of conventional and the microwave sintered samples were different. It can be seen from the Fig.2(a) and 2(b) that the microwave-sintered samples for both nearsurface and interior regions are practically identical, showing intergranular fracture and uniform grain size. Additionally, most pores are located at the isolated grain boundaries. In contrast, as shown in Fig.2(c) and 2(d), the conventionally sintered samples show exaggerated grain growth and many isolated and closed pores. The exaggerated grain growth may indicate a very high mobility of grain boundaries[14]. The variation in densification parameter of the sintered samples as a function of holding time at 1000°C is shown in Fig.3. It can be observed that the green compacts are well densified when sintered at 1000°C for 40 min, showing a densification of above 0.82 and porosity of below 4.5%. Microwave-sintered samples exhibit enhanced densification for the same holding time compared with the conventionally sintered samples. That is because the densification is a thermally activated process and the activation energy for microwave densification is lower than the value for conventional sintering[15, 16]. Fig.4 compares the B constant and resistivity (ρ) of twenty components by microwave and conventional sintering. Microwave sintering can obtain the components with well uniformity of B value and resistivity, of which the B_{avg} is 1930K (deviation of 1.47%) and resistivity ρ_{avg} is 135Ωcm (deviation of 4.45%). However, the B_{avg} is 1720K (deviation of 1.47%) and resistivity ρ_{avg} is 78Ωcm (deviation of 25.34%) for the conventionally sintered components. Therefore, microwave sintering is more effective in promoting uniformity and consistency of components compared to conventional sintering.

There is an increasing need for thermistors with high-precision and exchange-type in various industrial and domestic applications, such as temperature measurements and controls. In order to obtain NTC thermistor with low B value (2100 K), high-precision, exchange-type and fine homogeneity, $Mn_{0.43}Ni_{0.90}CuFe_{0.67}O_4$ NTC thermistor material precursors were prepared by the polymerized complex method using microwave sintering and conventional sintering[17]. The obtained results indicated that the particles microwave calcined at 650°C were small and uniform. The ceramics mainly consisted of rich-Cu phase and poor-Cu phase. The homogeneity of components is significantly increased by microwave sintering. Moreover, the rate of finished products is increased to 85% from 30% compared with the one of conventional sintering. The sintering activity and electrical properties of CoMnNiO based ceramic were investigated by Zhang et al.[18]. CoMnNiO NTC thermistor materials prepared by coprecipitation were sintered by microwave and conventional sintering. Compared with the conventional sintering, the samples prepared by microwave sintering have homogeneous density microstructure and fine grain size. The resistance deviation of components is 5% and 3% for conventional and microwave sintering, respectively. Electrical properties indicate that microwave sintering improves uniformity and consistency of components. In conclusion to this, microwave sintering method has potential superiority on synthesis NTC ceramic materials.

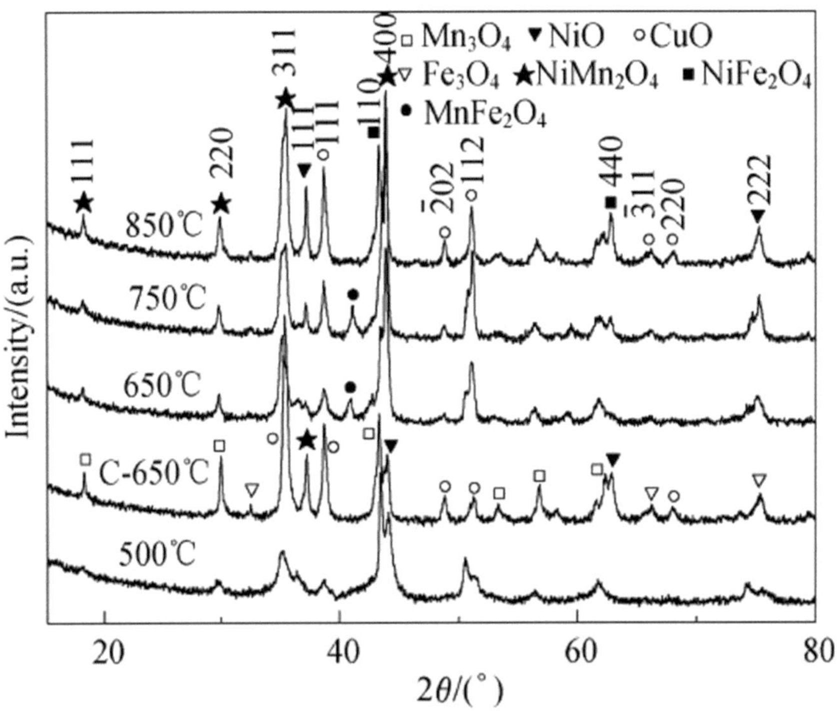

Fig. 1. XRD patterns of $Mn_{0.43}Ni_{0.9}CuFe_{0.67}O_4$ powder calcined at different temperature 500-850°C: microwave sintering with dwelling time of 30min; C-650°C: conventionally calcined at 650°C[12].

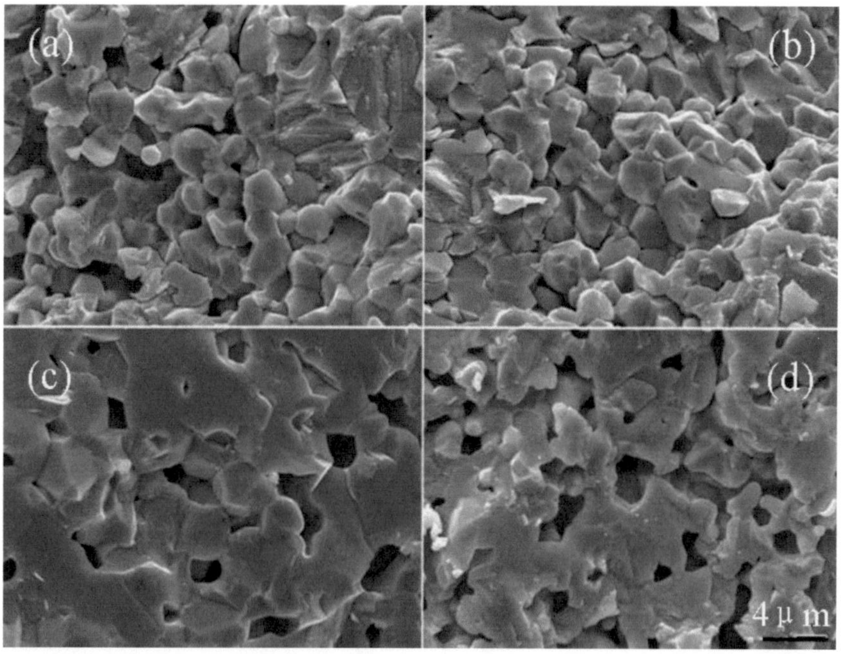

Fig. 2. SEM images of the fractures of samples: (a) near-surface region, microwave; (b) interior region, microwave;(c) near-surface region, conventional; (d) interior region, conventional[13].

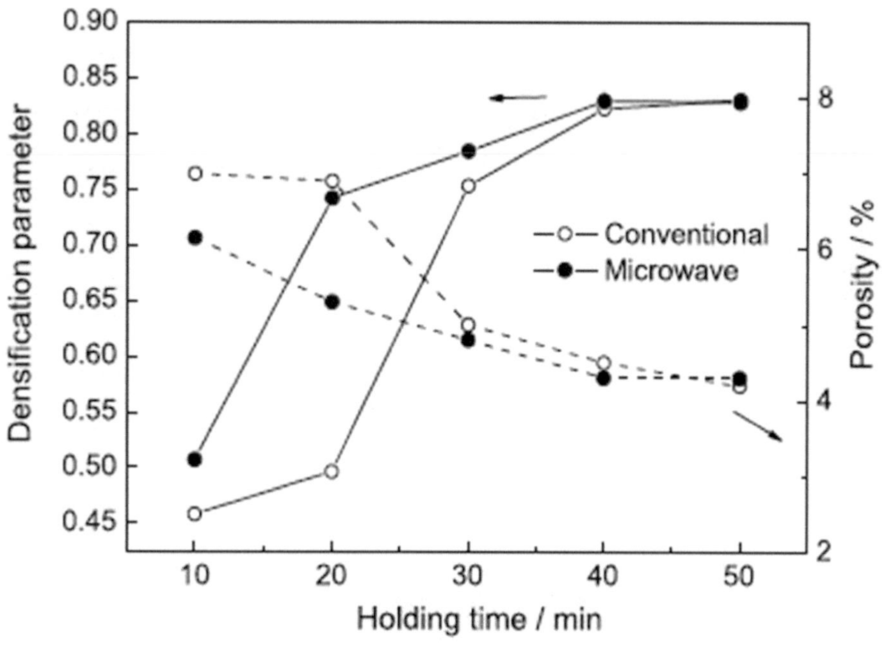

Fig. 3. The densification parameter and porosity of samples sintered for various holding time[13].

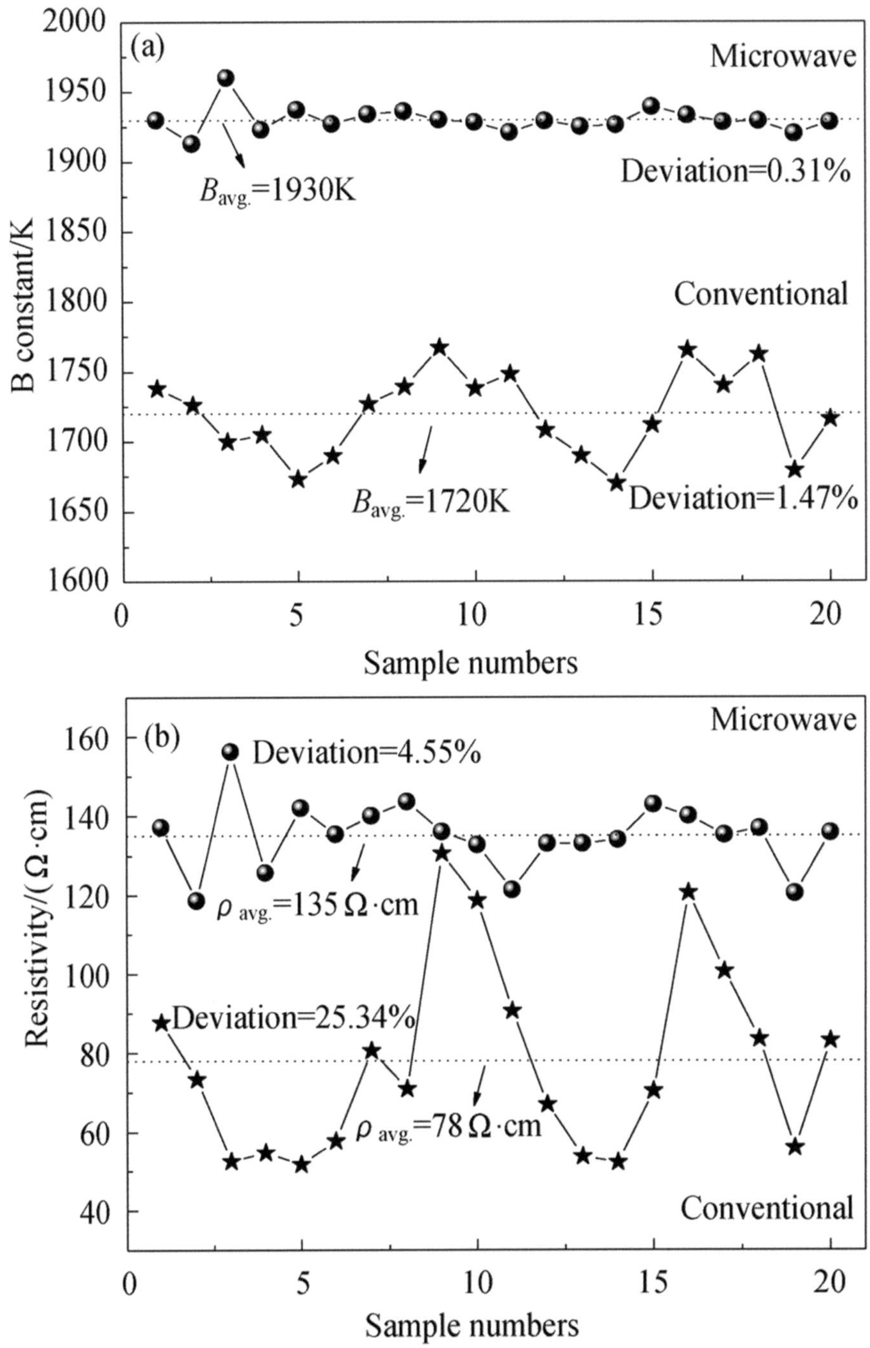

Fig. 4. Distribution of (a) B constant, (b) ρ for twenty components[12].

3.2 Microwave sintering of PTC thermistor ceramics

Positive temperature coefficient (PTC) thermistors are characterised by an increase in the electrical resistance with temperature. They are widely used in various industrial and domestic applications, such as electronic and electrical temperature control.

Chang et al. have studied the orientational growth of grains in doped $BaTiO_3$ PTCR materials by microwave sintering[19]. Fig.5 exhibits typical microstructures of sintered

(a)

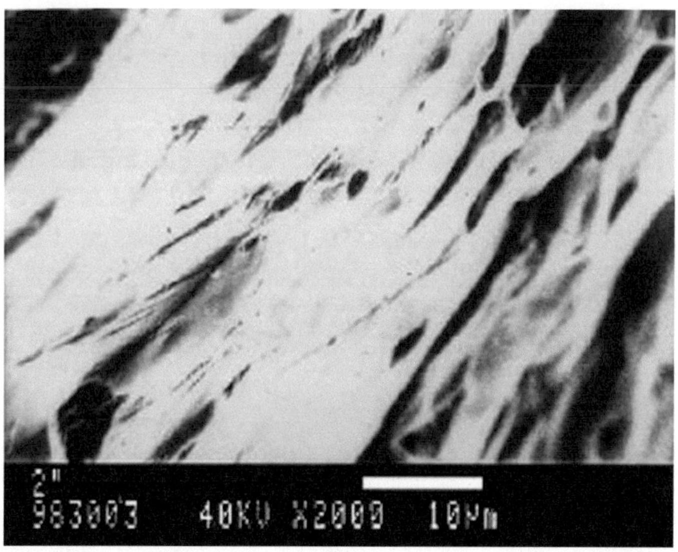

(b)

Fig. 5. The microstructure of the samples: (A) conventional sintering; (B) microwave sintering[19].

sample. The conventional-sintered sample shows ecliptic spherical shape, whereas the microwave-sintered sample shows an orientational strip-like microstructure. The growth of grains in microwave-sintered BTO is obvious orientational, which is very different from the familiar conventional-sintered BTO. The XRD (Fig.6) analysis indicates that the perovskite structure of microwave-sintered BTO is tetragonal, whereas the conventional-sintered sample is cubic. The reason for this abnormal phenomenon can perhaps be attributed to the polarisation of BTO as a typical piezoelectric material in the microwave field, and also to the enhancement of grain diffusion by the microwave-sintered.

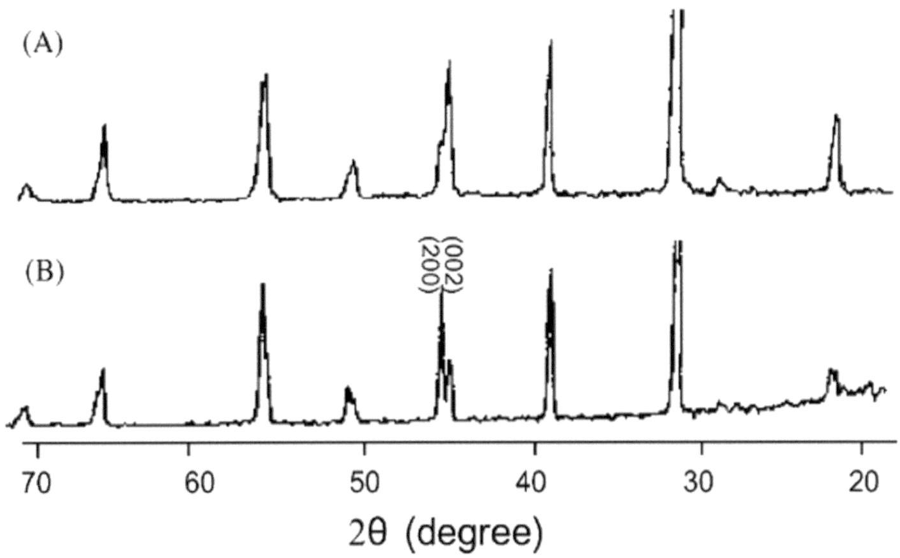

Fig. 6. XRD patterns of doped BTO samples: (A) conventional sintering; (B) microwave sintering[19].

3.3 The study of insulator materials for the microwave sintering of oxide thermistor ceramics

CoMnNiO series NTC and $BaTiO_3$-based PTC ceramics, having good dielectric coupling characteristic, can more easily couple with microwaves, therefore, generating heat to achieve sintering. However, due to the problems of thermal cracks during the microwave sintering and especially in the rapid cooling procedure, the production rate of oxide thermistor ceramics by microwave sintering is very low.

Chang et al. have successfully prepared $MgA1_2O_4$-$LaCrO_3$ insulator materials which mix $MgA1_2O_4$ spinel materials with superior thermal properties and $LaCrO_3$ perovskites materials with high efficiency absorption of microwave[20].The design of this kind insulating system solves the problems of thermal cracks often occur in the microwave sintering of many oxide ceramics, such as CoMnNiO series NTC ceramics and $BaTiO_3$-based PTC ceramics. Additionally, the $MgA1_2O_4$-$LaCrO_3$ insulator materials can produce a homogeneous heating for the samples in the microwave sintering, and grain crack-free and dense ceramic samples. Adopting $MgAl_2O_4$-$LaCrO_3$ susceptor in the temperature range of

1100-1200°C, microwave sintering of large-sized (50rum in diameter and 50mm in thickness) crack-free Co-Mn-Ni spinel oxide NTC ceramics has been achieved successfully[21]. Besides, $BaTiO_3$-based PTC ceramics (30mm in diameter) were microwave sintered at 950°C for 50min by the adoption of $MgAl_2O_4$-$LaCrO_3$ susceptor[20].

4. Summary

The fundamentals, applications of microwave sintering to thermistor ceramics and $MgA1_2O_4$-$LaCrO_3$ insulator materials for the microwave sintering of oxide thermistor ceramics are reviewed in this chapter. The advantages of microwave sintering against conventional sintering methods for thermistor ceramics are summarized as follows:(1) Microwave sintering consumes much lower energy than conventional sintering; (2) Higher heating rates can be attained and thus the sintering time reduces by using microwave sintering;(3) Generally higher density and better grain distribution can be achieved through microwave sintering. What's more, microwave sintering can enhance the densification rate and hence improve the uniformity and consistency of thermistors to a greater extent than the conventional sintering.

In the past year and half, significant developments and advances have taken place in the field of microwave sintering of ceramics. In spite of these successes, microwave processing has yet to see wide spread application in the ceramic industry. Additionally, more research must be conducted to have a full understanding of the process. It can be predicted that there is a great future for microwave sintering for the successful commercialization for specialty ceramics.

5. References

[1] H.S. Hao, L.H. Xu, Y. Huang, X.M. Zhang, Z.P. Xie, Kinetics mechanism of microwave sintering in ceramic materials, Science in China Series E: Technological Sciences, 52 (2009) 2727-2731.

[2] J.D. Katz, Microwave sintering of ceramics, Annual Review of Materials Science, 22 (1992) 153-170.

[3] E.T. Thostenson, T.W. Chou, Microwave processing: fundamentals and applications, Composites Part A: Applied Science and Manufacturing, 30 (1999) 1055-1071.

[4] M. Oghbaei, O. Mirzaee, Microwave versus conventional sintering: A review of fundamentals, advantages and applications, Journal of Alloys and Compounds, 494 (2010) 175-189.

[5] P. Yadoji, R. Peelamedu, D. Agrawal, R. Roy, Microwave sintering of Ni-Zn ferrites: comparison with conventional sintering, Materials Science and Engineering B, 98 (2003) 269-278.

[6] I.N. Lin, W.-C. Lee, K.-S. Liu, H.-F. Cheng, M.-W. Wu, On the microwave sintering technology for improving the properties of semiconducting electronic ceramics, Journal of the European Ceramic Society, 21 (2001) 2085-2088.

[7] K. Park, S. Kim, J.G. Kim, S. Nahm, Structural and electrical properties of MgO-doped $Mn_{1.4}Ni_{1.2}Co_{0.4-x}Mg_xO_4$ ($0 \le x \le 0.25$) NTC thermistors, Journal of the European Ceramic Society, 27 (2007) 2009-2016.

[8] A. Chatterjee, T. Basak, K. Ayappa, Analysis of microwave sintering of ceramics, Aiche Journal, 44 (1998) 2302-2311.

[9] K. Park, J. Lee, The effect of ZnO content and sintering temperature on the electrical properties of Cu-containing $Mn_{1.95-x}Ni_{0.45}Co_{0.15}Cu_{0.45}Zn_xO_4$ ($0 \leq x \leq 0.3$) NTC thermistors, Journal of Alloys and Compounds, 475 (2009) 513-517.

[10] Y. Luo, X. Liu, G. Chen, Effect of Y_2O_3 addition on the electrical properties of $BaTiO_3$-based NTC thermistors, Materials Letters, 60 (2006) 1011-1013.

[11] K. Park, J. Lee, S.J. Kim, W.S. Seo, W.S. Cho, C.W. Lee, S. Nahm, The effect of Zn on the microstructure and electrical properties of $Mn_{1.17-x}Ni_{0.93}Co_{0.9}Zn_xO_4$ ($0 \leq x \leq 0.075$) NTC thermistors, Journal of Alloys and Compounds, 467 (2009) 310-316.

[12] X.J. Jin, A.M. Chang, H.M. Zhang, D.Y. Zhang, Preparation and Microwave Sinterability of $Mn_{0.43}Ni_{0.9}CuFe_{0.67}O_4$ NTC Thermistor Materials by Pechini Method, Journal of Inorganic Materials, 24 (2009) 1013-1018.

[13] X. Jin, A. Chang, H. Zhang, D. Zhang, A Comparison Study of Sinterability and Electrical Properties for Microwave and Conventional Sintered $Mn_{0.43}Ni_{0.9}CuFe_{0.67}O_4$ Ceramics, Journal of Materials Science & Technology, 26 (2010) 344-350.

[14] S.-G. Wang, A.-M. Chang, H.-M. Zhang, Q. Zhao, Preparation and characterization of $Mn_{0.43}Ni_{0.9}CuFe_{0.67}O_4$ by a polymerized complex method, Materials Chemistry and Physics, 110 (2008) 83-88.

[15] J. Wang, J. Binner, B. Vaidhyanathan, N. Joomun, J. Kilner, G. Dimitrakis, T.E. Cross, Evidence for the Microwave Effect During Hybrid Sintering, Journal of the American Ceramic Society, 89 (2006) 1977-1984.

[16] K.H. Brosnan, G.L. Messing, D.K. Agrawal, Microwave Sintering of Alumina at 2.45 GHz, Journal of the American Ceramic Society, 86 (2003) 1307-1312.

[17] X.J. Jin, A.M. Chang, H.M. Zhang, D.Y. Zhang, Microwave sinter of $Mn_{0.43}Ni_{0.90}CuFe_{0.67}O_4$ thermistor materials prepared by polymerized complex method, Electronic Components and Materials,28(2009)30-34.

[18] ZHANG Hui min, CHANG Ai min, ZHAO Qing, Study on the Sintering and Electric Characters of the CoMnNiO-Based NTC Thermistor, Micronanoelectronic Technology,7/8(2007)86-90.

[19] A. Chang, J. Jian, The orientational growth of grains in doped $BaTiO_3$ PTCR materials by microwave sintering, Journal of materials processing technology, 137 (2003) 100-101.

[20] Chang, A. M., W. Yang, et al. $MgAl_2O_4$-$LaCrO_3$ insulator materials for the microwave sintering of electronic oxide ceramics, Journal of Inorganic Materials, 17(2002): 713-718.

[21] Chang, A. M., J. W. Zhuang, et al. Microwave sintering of large-sized NTC thermistor ceramics, Microwave and Radio Frequency Applications, (2003)191-198.

5

Cold-Bonding Technique – A New Approach to Recycle Innocuous Construction Residual Soil, Sludge, and Sediment as Coarse Aggregates

Chih-Ta Tsai
Sustainable Environment Research Center
National Cheng Kung University
Taiwan

1. Introduction

This chapter mainly illustrates that the mechanism and process of cold-bonding technique as well as using of three various innocuous recycling resources, construction residual soil, granite sludge, and lime sludge, to produce recycling coarse aggregates through the cold-bonding technique.

1.1 Output and source of the innocuous recycling resources in Taiwan

The attention of shortage of the primitive aggregate has been received in Taiwan. Although the natural resource in Taiwan is rather lack, the innocuous recycling resources are quite plentiful (e.g., construction residual soil, granite sludge, lime sludge, reservoir sediments, and so on). Using above-mentioned resources to manufacture the recycling green building materials is a good means in light of several aspects, including the reduction of waste, recycling of resource, and low carbon society. The Taiwan government has been devoted to developing the sustainability of society and has promulgated some acts to achieve this goal (Hsieh et al., 2012). Certainly, use of above-mentioned resources is one of the crucial issues because the resources and space for storing waste in Taiwan is so limited. Moreover, the global warming (or so-called climate change) is one of hot issues, in which how to reduce CO_2 has received much attention.

For the promotion and facility of reuse and recycling of construction residual soil, which are generally sorted as 9 categories (i.e. B1 to B7) in Taiwan as shown in Table 1 (Industrial Technology Research Institute [ITRI], 1996). Herein the construction residual soils of B1, B2-1, B2-2, and B5 categories were immediately reused through uncomplicated process (like crushing and sieving, etc.), but the reuse rate of B2-3, B3, B4, B6, and B7 categories is extremely low due to their poor properties. According to the evaluation by the Construction and Planning Agency Ministry of the Interior, R.O.C. the construction residual soil of about 40 millions m^3 is generated every year as shown in Table 2, and the B2-3, B3, B4, B6, and B7 categories with extremely low reuse rate accounted for more than 50 %.

Code	Properties of construction residual soil
B1	Rocks, gravels, crushed rocks, or sand
B2-1	Blended soil, gravels, and sand (soil <30 % by vol.)
B2-2	Blended soil, gravels, and sand (30 %<soil<50 % by vol.)
B2-3	Blended soil, gravels, and sand (soil>50 % by vol.)
B3	Silt
B4	Clay
B5	Brick and concrete blocks
B6	Sediment or soil contains >30 % moisture by wt.
B7	Bentonite from continuous walls construction

Table 1. Categories of construction residual soil in Taiwan.

	B1 (m^3)	B2-1 (m^3)	B2-2 (m^3)	B2-3 (m^3)	B3 (m^3)	B4 (m^3)	B5 (m^3)	B6 (m^3)	B7 (m^3)	Total (m^3)
2002	9,745,339	3,157,653	2,660,776	3,859,634	6,253,732	4,641,081	286,941	175,606	131,461	30,912,223
2003	6,491,354	6,256,473	4,009,632	6,125,236	6,927,483	4,852,074	995,285	803,362	424,588	36,885,487
2004	4,109,131	8,417,495	6,705,485	6,530,775	8,408,148	5,177,942	809,651	612,185	107,955	40,878,767
2005	2,178,436	8,148,354	6,974,924	9,182,229	8,409,695	5,884,677	1,095,015	485,395	78,657	42,437,382
2006	1,749,982	7,975,805	6,272,035	9,352,248	7,340,871	5,571,844	1,370,277	917,939	46,155	40,597,156
2007	3,099,089	6,538,887	4,829,821	9,762,949	6,524,040	4,483,718	1,636,861	1,294,205	31,871	38,201,441
2008	3,079,980	7,420,172	4,499,688	8,924,265	6,560,130	3,821,069	1,895,799	538,005	55,848	36,794,956
2009	2,419,110	5,494,537	3,982,374	8,289,454	3,818,809	1,917,679	1,393,881	529,180	237,392	28,082,416
2010	1,862,649	5,984,624	3,840,541	11,238,137	3,732,365	2,533,991	1,653,261	948,366	680,717	32,474,651
Total	34,735,070	59,394,000	43,775,276	73,264,927	57,975,273	38,884,075	11,136,971	6,304,243	1,794,644	327,264,479
(%)	10.61	18.15	13.38	22.39	17.72	11.88	3.40	1.93	0.55	100.00

Table 2. Output of the construction residual soil every year in Taiwan (ITRI, 1996).

The granite sludge of about 300,000 to 500,000 ton was generated from cutting and grinding granite in Taiwan. The common treatment and final disposal technology of granite sludge were solidification and landfill in the past, therefore its recycling amount is only approximately 50,000 ton every year. The granite sludge should not be classified as an industrial waste, but rather a recycling resource. On the basis of environmental protection aspects and increase economic benefits, the cold-bonding technique was adopted to recycle granite sludge as coarse aggregates.

The China Petrochemical Development Corporation (CPDC) An-Shun site that was a decommissioned chloroalkaline and pentachlorophenol manufacturing plant in Tainan, Taiwan includes 115,000 m^3 of chloroalkaline plant, 40,000 m^3 of pentachlorophenol plant, 47,000 m^3 of vegetation area, 27,000 m^3 of grass area, and 155,000 m^3 of seawater storage pond as shown in Fig. 1 (Chao et al., 2008). The highest mercury concentration was found in the chloroalkaline plant with a level as high as 3,370 mg/kg in soil, way above the soil control standard of 20 mg/kg. The worst dioxin contamination was found in the pentachlorophenol plant with levels as high as 64,100,000 ng-I-TEQ/kg in soil, whereas the soil control standard is 1,000 ng-I-TEQ/kg. But there is a large amount of about 90,000 m^3 of uncontaminated and innocuous lime sludge in vegetation area. The CPDC expected to recycle the above-mentioned uncontaminated and innocuous lime sludge as recycling coarse aggregates through using cold-bonding technique for the remediation and reconstruction of CPDC An-Shun site in the future.

Fig. 1. The surrounding location of CPDC An-Shun site (Chao et al., 2008).

1.2 Mechanism of cold-bonding technique

Based on the purposes of green building materials (i.e. reduction of waste and CO_2 footprint, energy conservation, lightening of material, and so on), it is a critical issue for building and construction department to treat the wastes properly and encourage the recycling of resources. In spite of many investigators (Chen et al., 2010; Hung & Hwang, 2007) indicate that the sintering technique has been successfully applied to recycle above-mentioned resources as lightweight aggregates. But the energy consumption and CO_2 emission of sintering process are too much to be extensively adopted. A new approach, the cold-bonding technique (Cai et al., 2010, 2012 & Tsai et al., 2011, 2012) incorporates the principles of the cement chemistry (Mehta, 1986; Mindess & Young, 1981) and composite material (Gibson, 1994), was developed to recycle these resources as recycling coarse aggregates. Consequently, the main difference between cold-bonding and sintering technique is the reduction of energy consumption and CO_2 emission.

The cold-bonding recycling coaese aggregate was regarded as a fiber reinforced concrete or a cement-based composite that is the original concept for developing cold-bonding recycling coarse aggregates. In which cement, blast-furnace slag (BF slag), and fly ash are regarded as cementitious materials or binders, the construction residual soil, granite or lime sludge is as a filler (i.e. aggregate), and the glass fiber is as a reinforcement. In view of the fundamental principle of concrete materials (Dowling, 1993; Skalny & Mindess,

1989) that is the higher packing density of component materials of concrete, the higher will be the properties of concrete. To ensure characteristics of cold-bonding recycling coarse aggregates are acceptable, the cement-based composites were granulated as the recycling coarse aggregates with a higher stress of greater than 28 MPa after proportioning and mixing. These mixture proportions and conditions of granulation will be illustrated in subsequent sections.

2. Constituent materials

The constituent materials of cold-bonding recycling coarse aggregates mainly include 1) cementitious materials: cement, BF slag, and fly ash, 2) recycling resources: the innocuous construction residual soil, granite sludge, and lime sludge, 3) other materials: the recycling glass fibers and superplasticizer.

2.1 Cementitious materials

The type I Portland cement produced by Universal Cement Corporation, BF slag provided by CHC Resources Corporation, and class F fly ash supplied by Taiwan Power Station are employed to produce the cold-bonding recycling coarse aggregates. These cementitious materials conform to the related American Society for Testing and Material (ASTM) standards and their physical properties as well as chemical compositions are shown in Tables 3.

	Item	Cement	BF slag	Fly ash
Physical properties	Specific gravity	3.15	2.90	2.14
	Surface area (m^2/g)	2,970	4,350	3,110
	Time of initial setting (min)	135	-	-
	Time of final setting (min)	377	-	-
Chemical compositions (%)	SiO_2	22.16	35.56	49.86
	Al_2O_3	5.63	14.34	37.89
	Fe_2O_3	2.17	0.33	3.18
	CaO	67.35	50.23	6.04
	MgO	-	5.66	-
	SO_3	2.08	0.95	0.66
	f-CaO	0.08	-	-
	TiO_2	0.25	0.44	1.20
	Na_2O	0.31	-	-
	K_2O	0.15	0.09	0.44
	Loss on ignition	0.51	0.31	6.08

Table 3. Physical properties and chemical compositions of cementitious materials.

2.2 Recycling resources

There were four various construction residual soils (i.e. B2-3, B3, B4, and B6 categories) employed to make the cold-bonding recycling coaese aggregates as shown in Fig. 2.

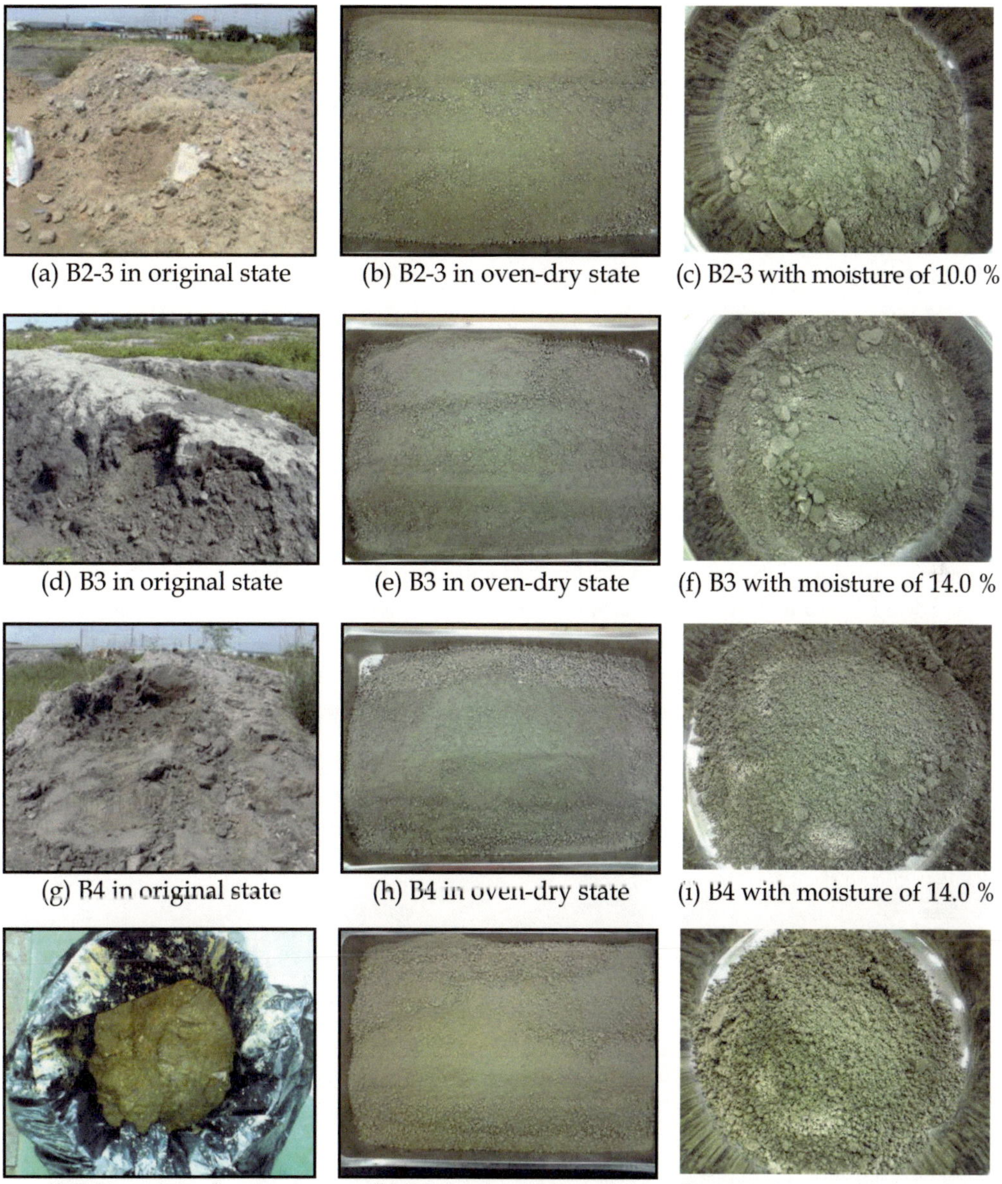

(a) B2-3 in original state (b) B2-3 in oven-dry state (c) B2-3 with moisture of 10.0 %

(d) B3 in original state (e) B3 in oven-dry state (f) B3 with moisture of 14.0 %

(g) B4 in original state (h) B4 in oven-dry state (i) B4 with moisture of 14.0 %

(j) B6 in original state (k) B6 in oven-dry state (l) B6 with moisture of 14.0 %

Fig. 2. Photos of four construction residual soils in various moisture states.

In which B2-3, B3, and B4 categories were provided from a local collecting plant of construction wastes in Tainan as well as the construction residual soil of B6 category was acquired from a construction site near Shida Road in Taipei, Taiwan. For engineering purposes, these construction residual soils of B2-3, B3, B4, and B6 categories were classified as SP (i.e. poorly graded sand), ML (i.e. silt), SM (i.e. silty sand), and CL (i.e. lean clay), respectively according to Unified Soil Classification System (ASTM, 2006). Their specific gravity as well as chemical compositions are shown in Tables 4.

	Construction residual soil				Granite sludge		Lime sludge
	B2-3	B3	B4	B6	A	B	
CaO (%)	1.96	2.29	2.16	0.27	12.08	6.39	55.76
SiO_2 (%)	75.70	73.20	69.46	69.52	55.49	70.63	21.70
Al_2O_3 (%)	14.83	16.77	18.68	21.29	11.04	14.76	8.45
Fe_2O_3 (%)	3.01	3.71	4.17	4.08	10.90	2.72	2.57
MgO (%)	N.D.	N.D.	0.85	0.70	8.08	N.D.	6.07
SO_3 (%)	0.08	0.14	0.23	0.02	0.09	0.07	3.81
K_2O (%)	2.65	2.92	3.43	3.17	1.56	4.69	N.D.
TiO_2 (%)	0.54	0.67	0.72	0.70	0.22	0.29	0.39
V_2O_5 (%)	N.D.	N.D.	N.D.	N.D.	N.D.	N.D.	N.D.
Specific gravity	2.57	2.64	2.54	2.870	2.90	2.70	2.62

Table 4. The specific gravity and chemical compositions of recycling resources.

In order to accelerate the settling of sludge for facilitating the recycling of water, the flocculants were added into the water reclamation pond in masonry plants. Therefore there were two various granite sludges (see Fig. 3), A granite sludge does not contain any flocculants and B granite sludge contains a few flocculants, were provided from the Stone and Resource Industry R&D Center in Hualien, Taiwan and their compositions were shown in Table 4.

(a) A in original state (b) A in oven-dry state (c) A with moisture of 19.5 %

(d) B in original state (e) B in oven-dry state (f) B with moisture of 20.0 %

Fig. 3. Photos of granite sludge in various moisture states.

The lime sludge (see Fig. 4) was provided from the CPDC An-Shun site that was a decommissioned chloroalkaline and pentachlorophenol manufacturing plant in Tainan, Taiwan (Chao et al., 2008). The specific gravity and chemical compositions of lime sludge are shown in Tables 4.

(a) A in original state (b) A in oven-dry state (c) A with moisture of 45.0 %

Fig. 4. Photos of lime sludge in various moisture states.

In Taiwan, the government continuously makes great efforts to promote the development of recycling materials (like green building materials, etc.) produced with the recycling wastes or resources over the past decade. For protecting the safety and health of users and purchasers, the stringent control standards were establish to prevent any noxious components from incorporating into the recycling resources. In other words, any recycling wastes or resources must confirm to the stringent control standards before they were used to manufacture any recycling products. The toxicity characteristic leaching procedure (TCLP) test acts an impotant and decisive role in the above-mentioned control standards. The TCLP results of recycling resources were below the criteria of general enterprise wastes and green building materials (Chiang, 2007) in Taiwan as shown in Table 5.

	Cr (mg/L)	Cd (mg/L)	Pb (mg/L)	Cu (mg/L)	As (mg/L)	Hg (mg/L)
B2-3 construction residual soil	<0.02	<0.02	<0.02	<0.02	0.0018	N.D.
B3 construction residual soil	<0.02	<0.02	<0.02	<0.02	N.D.	N.D.
B4 construction residual soil	<0.02	<0.02	<0.02	<0.02	N.D.	N.D.
B6 construction residual soil	<0.02	<0.02	<0.02	<0.02	0.0015	N.D.
A granite sludge	0.087	0.180	0.049	0.005	0.003	N.D.
B granite sludge	1.270	0.080	0.032	0.006	0.009	N.D.
Lime sludge	0.313	0.001	0.026	0.007	0.079	N.D.
Criteria of general enterprise wastes	5.0	1.0	5.0	15.0	5.0	0.2
Criteria of green building material	1.5	0.3	0.3	0.15	0.3	0.005

Table 5. TCLP results of recycling resources.

2.3 Other materials

Type G superplasticizer, a carboxylate-based, was purchased from local factory in Taiwan and the characteristics of superplasticizer as shown in Table 6. The glass fibers were recycled from printed circuit board (PCB) wastes by Much Fortune Technology Co., Ltd. In fact, the glass fibers also should be regarded as a recycling resource.

Item	
Specific gravity	1.09
Water reduction (%)	30.00
Solid content (%)	25.70
pH value	2.81

Table 6. The characteristics of carboxylate-based superplasticizer.

3. Mixture proportions design

This section principally introduce that a local mixture proportion method in Taiwan, densified mixture design algorithm (DMDA), is employed to design and prepare the cement-based composite for producing the cold-bonding recycling coarse aggregates, the logic and procedure of DMDA, how to design mixture proportions of cold-bonding recycling coarse aggregates by utilization of DMDA, and the mixture proportions of cold-bonding recycling coarse aggregates.

3.1 Design logic of densified mixture design algorithm, DMDA

The densified mixture design algorithm (DMDA) was developed by National Taiwan University of Science and Technology since 1992 is a mixture proportion method for cement-based composite (e.g. mortar, plain concrete, high performance, self-consolidating concrete and so on), in which the paste volume method and the current dry loose density (i.e. unit weight) method are incorporated together for obtaining the optimized cement paste through densely packing aggregates (Chang et al., 2001; Tsai, 2005; Tsai et al., 2006). The most important design logic of DMDA is the achievement of "least void" through the utilization of fly ash (to fill the void between blended aggregates) and the cement paste (to fill the rest of the void). The utilization of fly ash (in addition to the cement paste) to fill the void between blended aggregates will increase the density of cement-based composite (Tsai et al., 2006). And the addition of the super plasticizers (SP) is helpful to solve the potential problem of tangling or balling of fibers (Tsai et al., 2009, 2010). Thus the workability of the cement-based composite with the aid of SP is ensured as a result.

3.2 Design consideration of cold-bonding recycling aggregate

In conventional mixture design, the cement-based composite workability is decided by the water amount and the compressive strength whereas the durability is decided by the water-to-cement ratio (w/c) (American Concrete Institute [ACI], 1991). The workability can be improved by increasing the water amount (Neville, 2000) and the strength can be increased by increasing the cement content. However, too much cement paste will cause chemical shrinkage, and the shrinkage rate or expansion rate is in direct proportion to the water and cement amount due to the hydration of the cement (Hwang & Lu, 2000; Mather, 2000). Besides, ordinary cement-based composite contains water at least 20 % by its volume, and hence drying shrinkage will be unavoidable. So the durability of cement-based composite is destroyed due to disintegration and crack formation. To avoid these problems cement-based composite mixture designed with low water amount and low cement content is proposed.

Durability design should be considered for improving both the fresh and hardened stages of the cement-based composite and should finally extend their service life. First and foremost the cement-based composite mixture design should have a very low water amount (Neville, 2000) so as to minimize the shrinkage rate or the expansion rate (Hwang & Lu, 2000). Then, the cement-based composite must be designed to satisfy the construction needs such as with zero or low slump for cold-bonding recycling aggregate or roller compacted concrete, with high slump for self- consolidating concrete or high performance concrete, type of construction work, and the required final finished result. In plastic stage, the cement-based composite is designed to prevent the occurrence of plastic shrinkage cracks due to excess water evaporation from its surface. A certain amount of fibers should be included in the cement-based composite to absorb energy and in the case of crack formation, to stop their propagating (Rossi et al., 1987). The addition of pozzolanic materials (i.e. BF slag and fly ash) is necessary to help the self-healing of cracks if they are generated (Tsai et al., 2009). A strict standard operation procedure for mixture proportion, material selection, trial batch, quality control and curing are required to lower the possibility of crack formation.

The DMDA was adopted to design the intended cement-based composites for producing the cold-bonding recycling coarse aggregates. In order to minimize the shrinkage rate or the expansion rate and ensure the durability of cold-bonding recycling coarse aggregate, a very low water-to-cementitious ratio of 0.20 was selected to design mixture proportions of cement-based composite. And a total of 39.4 kg glass fibers (the volume=0.02 m) was added to reach the intended design value of 2.0 % by volume of the cement-based composite for preventing such cracks and enhancing the toughness as well as volume stability (Rossi et al., 1987; Tsai et al., 2009, 2010) of cold-bonding recycling coarse aggregate. In view of cement has the most energy consumption and CO_2 emission in all constituent materials of cold-bonding recycling coarse aggregate as well as the above-mentioned disadvantages, the amount of cement is limited to less than 200 kg/m^3. There were three various amount of cement (i.e. 50, 100 and 200 kg/m^3) designed for every recycling resource to magnify the application cold-bonding recycling coarse aggregates in the future.

3.3 Mixture design procedure of cold-bonding recycling aggregate by DMDA

The following steps can be used to provide computational basis for designing the cement-based composite mixture to produce the cold-bonding recycling coarse aggregates employing the DMDA procedure.

(1) Select proper material resource and gather material information.

This is an important step for the mix design of cement-based composite mixture for producing the cold-bonding recycling coarse aggregates. The basic quality information of the ingredients of cement-based composite is necessary for the purpose of quality control.

(2) Obtain the maximum dry loose density (i.e. unit weight) by iteratively packing recycling resources, BF slag, and fly ash in filler system.

(2-a) Fill fly ash with BF slag and then obtain:

$$\alpha = \frac{W_{slag1}'}{W_{slag1}' + W_{flyash}'} \tag{1}$$

where α is the ratio at maximum dry loose density as fly ash is filled with BF slag; W_{slag1}' is the weight of BF slag (kg) in filler system; W_{flyash}' is the weight of fly ash (kg).

(2-b) Fill recycling resource with the blend of fly ash and BF slag under fixed α, and obtain:

$$\beta = \frac{W_{slag1}' + W_{flyash}'}{W_{slag1}' + W_{flyash}' + W_{\mathrm{Re}}'} \tag{2}$$

where β is the ratio at maximum dry loose density as recycling resource were filled with the blend of fly ash and BF slag; W_{Re}' is the weight of recycling resource (kg).

(3) Select the volume of glass fiber (η) added into cement-based composite.

(4) Calculate the least void, V_{υ} :

$$V_{\upsilon} = 1 - \sum \frac{W_i'}{\gamma_i} - \eta \tag{3}$$

where W_i' (kg/m^3) and γ_i (kg/m^3) are the weight and density of i constituent material in filler system, respectively.

(5) Assign a lubricating paste thickness (t) and calculate the volume of cement paste.

$$V_p = nV_{\upsilon} \tag{4}$$

where n is a multiplier for lubricating paste; V_p is the volume of cement paste.

(6) Calculate the factor of volume variation (υ) (Tsai, 2005):

$$\upsilon = \frac{1 - \eta - nV_{\upsilon}}{1 - \eta - V_{\upsilon}} \tag{5}$$

(7) Calculate the weight of recycling resource, BF slag, fly ash, and glass fiber in filler system, respectively:

$$W_{fiber} = \eta \times \gamma_{fiber} \tag{6}$$

$$W_{flyash} = \upsilon \times W_{flyash}' \tag{7}$$

$$W_{slag1} = \upsilon \times W_{slag1}' \tag{8}$$

$$W_{\mathrm{Re}} = \upsilon \times W_{\mathrm{Re}}' \tag{9}$$

where γ_{fiber} is the density of glass fiber (kg/m³); W_{fiber}, W_{flyash}, W_{slag1}, and W_{Re} are weights of glass fiber, fly ash, BF sand, and recycling resource, respectively in the cement-based composite mixture (kg/m³).

(8) Calculate the amount of cement, BF slag and mixing water in paste system:

$$V_P = n \cdot V_V = \frac{W_{water}}{\gamma_{water}} + \frac{W_{cement}}{\gamma_{cement}} + \frac{W_{slag2}}{\gamma_{slag}} \tag{10}$$

If ξ is the ratio of replacing cement with BF slag by weight, then:

$$\xi = \frac{W_{slag2}}{W_{cement} + W_{slag2}} \tag{11}$$

where W_{water}, W_{cement}, and W_{slag2} are weights of water, cement, and BF slag in paste system, respectively (kg/m³); γ_{water}, γ_{cement}, and γ_{slag} are densities of water, cement, and BF slag, respectively (kg/m³).

Substitute Equation 11 into Equation 10 to obtain:

$$V_p = \frac{\left(\frac{W_{water}}{W_{cement}}\right) W_{cement}}{\gamma_{water}} + \frac{W_{cement}}{\gamma_{cement}} + \frac{\left(\frac{\xi}{1-\xi}\right) W_{cement}}{\gamma_{slag}} \tag{12}$$

If the water-to-cementitious material ratio (w/cm) is λ, then:

$$w / cm = \lambda = \frac{W_{water}}{W_{cement} + W_{flyash} + W_{slag1} + W_{slag2}} \tag{13}$$

Using Equations 11 and 13, Equation 12 can be used to solve for W_{cement}:

$$W_{cement} = \frac{V_p - \frac{\lambda}{\gamma_w}(W_{flyash} + W_{slag1})}{[\frac{\lambda}{\gamma_{water}} + \frac{1}{\gamma_{cement}} + \frac{\xi}{1-\xi}(\frac{\lambda}{\gamma_{water}} + \frac{1}{\gamma_{slag}})]} \tag{14}$$

The calculated W_{cement} can be substituted both into Equation 11 and Equation 13 to obtain W_{slag2} and W_{water}, respectively.

(9) Determine the dosage of SP and amount of water

The dosage of SP is determined by its quality and the total water content. Under fixed total amount of water and w/cm ratio, the SP dosage can be estimated according to past experience as shown in Equation 15 and Fig. 5 (Chang et al., 2009).

$$SP\ (\%) = 4.59 \times 10^{-5} \times X^2 - 1.81 \times 10^{-2} \times X + 1.98 \tag{15}$$

where X is total amount of water (kg/m^3).

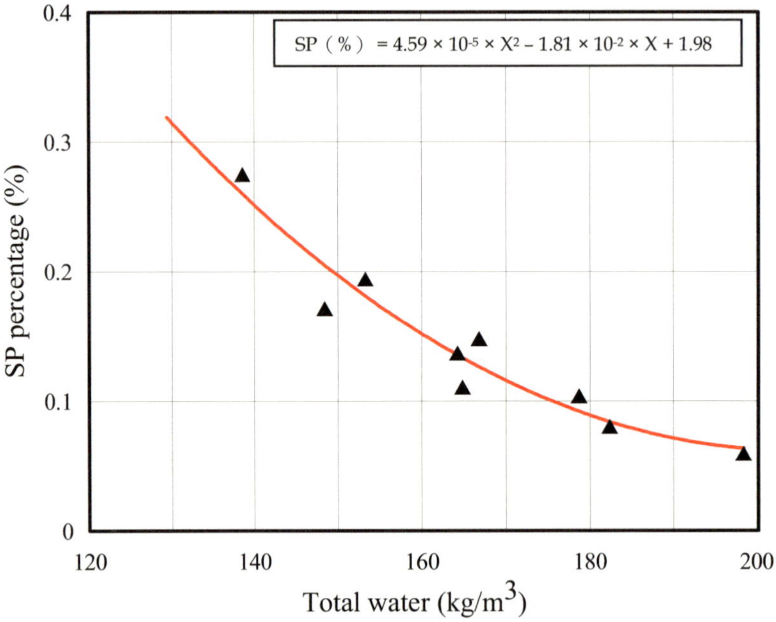

Fig. 5. The SP dosage for cold-bonding recycling coarse aggregates (Chang et al., 2009).

3.4 Mixture of cold-bonding recycling aggregate

According to the procedure of DMDA described in Section 3.3, the mixture proportions of cement-based composite with various recycling resources for producing the cold-bonding recycling coarse aggregates are shown in Table 7. For instance, B2-3-200 represents a cement-based composite contains the construction residual soil of B2-3 category and a cement amount of 200 kg/m^3; B-100 represents a cement-based composite with B granite sludge and a cement amount of 100 kg/m^3; L-50 represents a cement-based composite with lime sludge and a cement amount of 50 kg/m^3.

4. Granulation by cold-bonding technique

Dowling (1993) and Skalny & Mindess (1989) indicated that the higher packing density of component materials of cement-based composite, the higher properties of cement-based composite will be. In order to ensure that characteristics of cold-bonding recycling coarse aggregates are able to satisfy with the related criteria, a higher stress was adopted to the cement-based composites as recycling coarse aggregates. The various methods, molds, and conditions of granulation are discussed in this section.

4.1 Method of granulation

Three various methods were conducted to granulate the cement-based composites as cold-bonding recycling coarse aggregates. They are: 1) spirally push method; 2) immediately squeeze out method; 3) press ingot method. The granulation energy of spirally push method is too small to adequately form the recycling aggregates (see Fig. 6) due to the intended cement-based composite contains lower moisture to minimize the shrinkage rate or the expansion rate and ensure the durability of cold-bonding recycling coarse aggregate. On the contrary, while the cement-based composite with higher moisture could be successfully granulated by using spirally push method, but the produced recycling aggregates will not have nice properties.

Mix No.	Mixture proportions (kg/m^3)					
	Cement	BF slag	Fly ash	Recycling resource	Glass fiber	SP+Water
B2-3-50	50	150	235	1656	39.4	87
B2-3-100	100	110	255	1617	39.4	93
B2-3-200	200	20	280	1573	39.4	100
B3-50	50	150	235	1574	39.4	87
B3-100	100	110	255	1537	39.4	93
B3-200	200	20	280	1496	39.4	100
B4-50	50	150	235	1528	39.4	87
B4-100	100	110	255	1491	39.4	93
B4-200	200	20	280	1450	39.4	100
B6-50	50	150	235	1615	39.4	87
B6-100	100	110	255	1578	39.4	93
B6-200	200	20	280	1535	39.4	100
A-50	50	100	390	1403	39.4	108
A-100	100	70	345	1446	39.4	103
A-200	200	20	252	1522	39.4	95
B-50	50	100	385	1308	39.4	107
B-100	100	70	358	1324	39.4	106
B-200	200	20	252	1412	39.4	95
L-50	50	120	315	976	39.4	97
L-100	100	85	279	999	39.4	93
L-200	200	20	204	1047	39.4	85

Table 7. Mixture proportions of cold-bonding recycling aggregates.

Regardless of the cement-based composite contains how much moisture, the immediately squeeze out method will not successfully achieve the granulation of cold-bonding recycling coarse aggregate. This is due to the water within cement-based composite was drained out as shown in Fig. 7, like consolidation in geotechnical engineering (Holitz & Kovacs, 1981), during the process of granulating recycling coarse aggregates. Therefore the immediately squeeze out method also cannot be employed to form the cold-bonding recycling coarse aggregates.

(a) A commercial spirally-push machine (b) Photo of improper recycling aggregates

Fig. 6. Spirally push method.

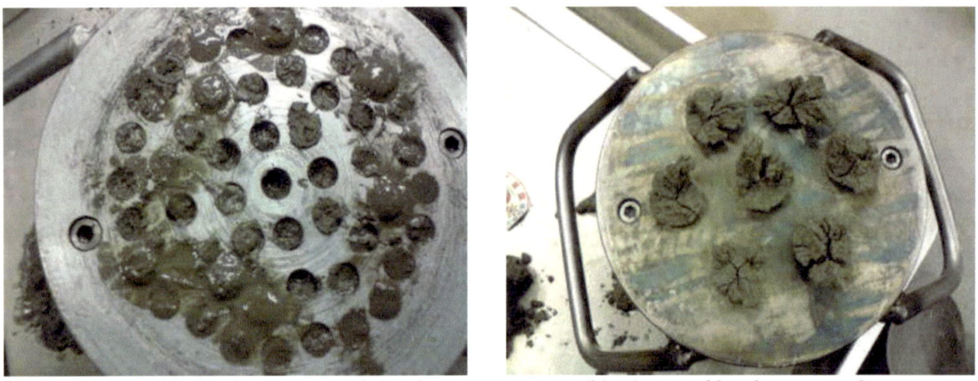

(a) The water within cement-based composite was drained out (b) Photo of broken recycling aggregates

Fig. 7. Immediately squeeze out method.

Finally the press ingot method (see Fig. 8) was developed and successfully granulated the cold-bonding recycling coarse aggregates. Fig. 8 also shows the procedure of press ingot method and the procedure is described as follows:

Step 1: To fill the mixed cement-based composite into the mold.
Step 2: To set up the pestle into the mold.
Step 3: To press and form the recycling aggregates.
Step 4: To take off the recycling aggregates from the mold.

4.2 Optimum moisture of granulation

In the original designing concept of cement-based composites mixture proportions for producing the recycling coarse aggregates, the recycling resources were regarded as fillers (i.e aggregates of concrete). But the natures of recycling resources are quite different from primitive aggregates (e.g. particle shape, gradation, absorption, and so on). The purpose of exploring the optimum moisture of recycling resource for granulating aggregate is to avoid two issues: 1) with lower moisture: the recycling coarse aggregates can not be adequately granulated; 2) with higher moisture: the cold-bonding recycling coarse aggregates don't have only sufficient durability (Mehta, 1986; Mindess & Young, 1981; Neville, 2000) but the redundant water also will be drained out during forming recycling aggregates (like consolidation in geotechnical engineering). The drained water will result in the excessively high water-to-cement ratio (w/c) or water-to-cementitious materials (w/cm) around the surface of the recycling aggregate. And the excessively high w/c or w/cm will immensely affect the strength, hardness, abrasion resistance, soundness, permeability of the cold-bonding recycling coarse aggregate (Mehta, 1986; Mindess & Young, 1981).

(a) Step 1: To fill the mixed cement-based composite into the mold

(b) Step 2: To set up the pestle into the mold

(c) Step 3: To press and form the recycling aggregates

(d) Step 4: To take off the recycling aggregates from the mold

Fig. 8. The procedure of press ingot method.

The higher adopted stress is not always better for granulating cold-bonding recycling coarse aggregates. Because the water has the incompressible nature. While the too high stress is adopted for granulation, the water certainly will be drained out. Then the cold-bonding recycling coarse aggregate will generate cracks due to tensile stress in capillary pores caused by absorption of the drained water during unloading process. Therefore the proposed stress of granulation by using press ingot method is 35.0 to 42.0 MPa. The corresponding optimum moistures of recycling resources are shown in Table 8 and the blended cement-based composites contain the recycling resources with such moisture are shown in Fig. 2, Fig. 3, and Fig. 4. This result also implies that the cold-bonding technique is able to be applied to handle recycling resources with moisture and reduce the energy consumption and CO_2 emission resulted from the oven-dry process. It is worth mentioning that the optimum moisture of blended recycling resource can be estimated by the proportion and optimum moisture of every constituent recycling resource. For example, a blended recycling resource is composed of 40 % B2-3 construction residual soil and 60 % lime sludge, its optimum moisture will be approximate 31 %.

Type of recycling resource	Optimum moisture
B2-3 construction residual soil	10.0 %
B3 construction residual soil	14.0 %
B4 construction residual soil	14.0 %
B6 construction residual soil	14.0 %
A granite sludge	19.5 %
B granite sludge	20.0 %
Lime sludge	45.0 %

Table 8. The optimum moistures of recycling resources.

4.3 Improvement of particle shape

The press ingot method was developed and successfully granulated the cement-based composites as cylindrical cold-bonding recycling coarse aggregates with five various diamemters of 24, 18, 12, 8, and 5 mm as shown in Fig.9. Mindess & Young (1981) and Mehta (1986) indicated that aggregate shap affects the workability of fresh concrete through their influence on cement paste requirments. The ideal aggregate particle is one that is close to spherical in shape. But the spherical shape is unfavorable for mechanical properties of concrete. Whereas both of the above contentions, the cylindrical cold-bonding recycling coarse aggregates were expected to be rounder (i.e. to smooth the four corners of cylinder). Only the molds of press ingot method with three various diamemters of 24, 18, and 12 mm was improved. Because the corner effects of cylindrical cold-bonding recycling coarse aggregate with diamemter of 8 or 5 mm are very limited. After improving particle shape, the molds and recycling coarse aggregates are shown in Fig 10.

(a) Diameter = 24 mm (b) Diameter = 18 mm (c) Diameter = 12 mm

(d) Diameter = 8 mm (e) Diameter = 5 mm

Fig. 9. Photos of cold-bonding recycling coarse aggregates with various diameter.

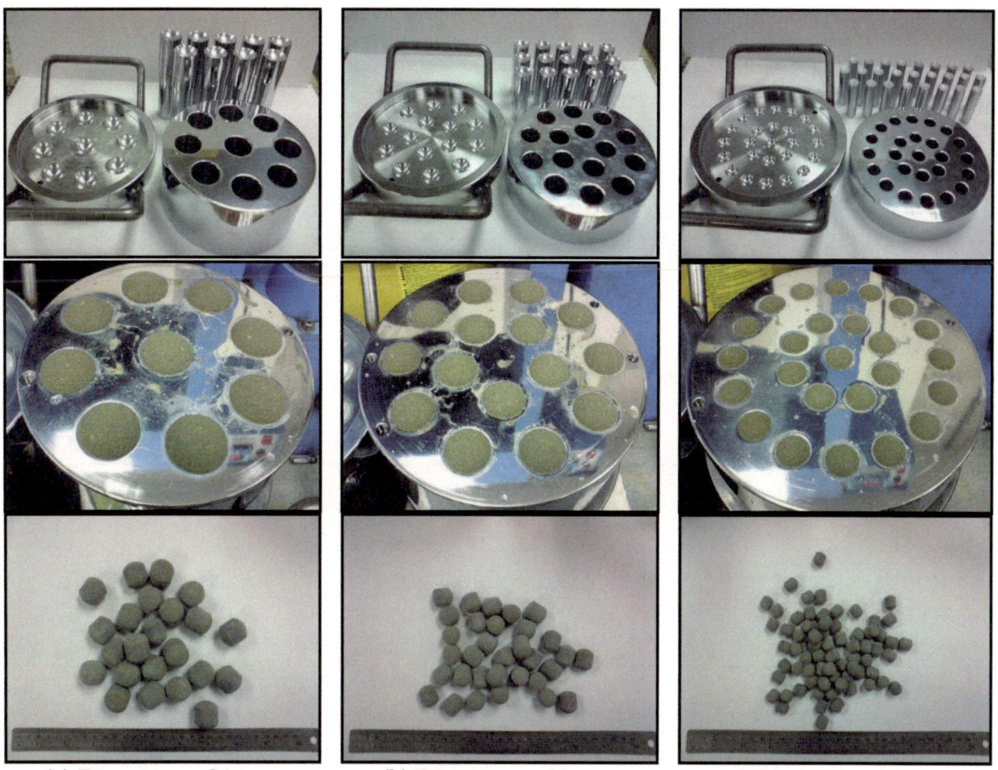

(a) Diameter = 24 mm (b) Diameter = 18 mm (c) Diameter = 12 mm

Fig. 10. Photos of molds and recycling aggregates after improvement of particle shape.

5. Characteristics of cold-bonding recycling coarse aggregates

The promotion of up-to-date green building materials was often impeded by the lack of relevant standards or specifications cited to verify their characteristics (Chang et al., 2009). To avoid such obstruction affecting the application of cold-bonding recycling coarse aggregates in the future, the characteristics of cold-bonding recycling coarse aggregates were certified in accordance with ASTM C33. The aggregate gradation significantly affects on workability, strength, durability, and economy of concrete (Mehta, 1986; Mindess & Young, 1981; Tsai et al., 2006). Therefore the gradation of cold-bonding recycling coarse aggregate was established by the above-mentioned five various diameter particles according to the mean of upper and lower limit from size number 56 and 6 recommended by ASTM C33.

5.1 Basic properties

After granulating cold-bonding recycling coarse aggregates by using the press ingot method, these recycling aggregates were cured in saturated limewater at the temperature of 23± 2.0 °C according to ASTM C192. They were conducted teste of basic properties include specific gravity in oven-dry (OD) and saturated surface dry (SSD) states, absorption, bulk density (i.e. unit weight) and voids according to relevant ASTM standards after the age of 28 days and these results were shown in Table 9. The result indicates that the specific gravity of

Mix No.	Basic properties				
	Specific gravity (OD)	Specific gravity (SSD)	Absorption (%)	Unit weight (kg/m³)	Voids (%)
B2-3-50	1.80	2.07	14.9	1,191	33.9
B2-3-100	1.81	2.07	14.2	1,192	34.2
B2-3-200	1.81	2.08	14.9	1,176	35.0
B3-50	1.73	2.02	17.0	1,144	33.9
B3-100	1.73	2.01	16.4	1,153	33.4
B3-200	1.73	2.02	16.7	1,103	36.3
B4-50	1.77	2.05	15.8	1,165	34.2
B4-100	1.79	2.06	15.3	1,175	34.4
B4-200	1.80	2.08	15.5	1,141	36.6
B6-50	1.84	2.13	15.8	1,207	33.9
B6-100	1.83	2.11	15.2	1,215	33.6
B6-200	1.83	2.11	15.2	1,201	34.5
A-50	1.66	2.01	21.3	1,084	34.7
A-100	1.69	2.03	20.2	1,101	34.9
A-200	1.74	2.08	19.3	1,133	34.9
B-50	1.55	1.89	22.1	1,033	33.3
B-100	1.58	1.91	21.2	1,045	33.9
B-200	1.65	1.99	20.0	1,088	34.1
L-50	1.23	1.72	40.0	837	32.0
L-100	1.21	1.72	41.6	827	31.6
L-200	1.20	1.73	44.6	811	32.4
Standards	ASTM C127	ASTM C127	ASTM C127	ASTM C29	ASTM C29

Table 9. Basic properties of cold-bonding recycling coarse aggregates.

cold-bonding recycling coarse aggregate is lighter than primitive aggregate, and the optimum moisture of recycling resource for granulation directly influences the absorption of recycling aggregate. The more optimum moisture of recycling resource, the higher properties of cement-based composite will be.

Other characteristics of ASTM C33 (e.g. contents of chert, clay lumps and friable particles, materials less than 75 μm, and coal and lignite, abrasion, and soundness, etc.) of recycling coarse aggregate were conducted, too. The test results show that the other characteristics of recycling coarse aggregates satisfy the specification of ASTM C33 except the soundness of L-200 (see Table 10). The reason causing the soundness of L-200 is much higher than the ASTM C33 criterion of 12 % may be attributed to the fact that L-200 contains too calcium components to have adequate sulfate resistance (Mangat & Khatib, 1995). This result also implies that using lime sludge or other recycling resources with abundant calcium components to produce cold-bonding recycling coarse aggregates should choose the mixture proportions with cement amount lower than 200 kg/m^3 to ensure their sulfate resistance. Table 11 shows the comparisons of properties of cold-bonding recycling coarse aggregate between before and after improvement of particle shape. The results indicate that the unit weight, voids, and abrasion of cold-bonding recycling coarse aggregate have the significant advancement.

Mix No.	Other characteristics of ASTM C33 (%)					
	Clay lumps and friable particle	Chert (<2.4 sp gr SSD)	Materials <75μm	Coal and lignite	Abrasion	Soundness
B2-3-50	N.D.	N.D.	0.45	N.D.	45.7	8.68
B2-3-100	N.D.	N.D.	0.56	N.D.	40.7	4.93
B2-3-200	N.D.	N.D.	0.42	N.D.	35.2	1.55
B3-50	N.D.	N.D.	0.53	N.D.	48.9	11.16
B3-100	N.D.	N.D.	0.49	N.D.	44.3	8.12
B3-200	N.D.	N.D.	0.59	N.D.	39.2	4.62
B4-50	N.D.	N.D.	0.33	N.D.	46.7	11.66
B4-100	N.D.	N.D.	0.42	N.D.	42.8	8.47
B4-200	N.D.	N.D.	0.44	N.D.	38.8	4.94
B6-50	N.D.	N.D.	0.41	N.D.	43.8	8.23
B6-100	N.D.	N.D.	0.41	N.D.	40.2	4.68
B6-200	N.D.	N.D.	0.35	N.D.	34.5	1.47
L-50	N.D.	N.D.	0.53	N.D.	38.8	7.65
L-100	N.D.	N.D.	0.50	N.D.	37.4	9.87
L-200	N.D.	N.D.	0.48	N.D.	37.8	81.47
Criteria of ASTM C33	2.0 to 10.0	8.0	1.0	0.5 to 1.0	50.0	12.0

Table 10. ASTM C33 other characteristics of cold-bonding recycling coarse aggregates.

Mix No.	Before improvement of particle shape			After improvement of particle shape		
	Unit weight (kg/m³)	Voids (%)	Abrasion (%)	Unit weight (kg/m³)	Voids (%)	Abrasion (%)
B2-3-50	1,191	33.9	45.7	1,205	32.9	39.7
B2-3-100	1,192	34.2	40.7	1,216	33.3	35.4
B2-3-200	1,176	35.0	35.2	1,219	33.4	28.5
B3-50	1,144	33.9	48.9	1,153	33.3	44.2
B3-100	1,153	33.4	44.3	1,173	33.1	38.7
B3-200	1,103	36.3	39.2	1,182	32.9	31.7
B4-50	1,165	34.2	46.7	1,163	33.3	42.1
B4-100	1,175	34.4	42.8	1,182	32.9	36.6
B4-200	1,141	36.6	38.8	1,205	33.0	31.1
B6-50	1,207	33.9	43.8	1,215	33.4	38.2
B6-100	1,215	33.6	40.2	1,219	33.5	34.7
B6-200	1,201	34.5	34.5	1,230	32.8	27.0

Table 11. ASTM C33 other characteristics of cold-bonding recycling coarse aggregates.

5.2 Mechanical properties

Referring to ASTM C39, the cold-bonding recycling coarse aggregate was conducted single particle compressive strength test at the age of 3, 7, 10, 14, 28, 56, and 91-day, respectively. The result indicates that the higher the cement amount, the higher will be the single particle compressive strength test of cold-bonding recycling coarse aggregate as shown in Fig. 11, Fig. 12, and Fig. 13. And the single particle compressive strength increases with the increase of curing age due to the fact that the contribution of hydration of cement and pozzlanic reaction (Dinajar et al., 2008; Malhotra, 1990).

Fig. 11 shows that the cold-bonding recycling coarse aggregates using of B3 and B4 construction residual soil have worse performances. The reason may be attributed to the fact that the B3 and B4 construction wastes are belong to silt and clay with worse properties, respectively. Especially, the single particle compressive strength of recycling aggregate with B granite sludge is significantly higher than A granite sludge as shown in Fig 12. Due to B granite contains some flocculants which are helpful for cement-based composite (like the mechanism of polymer concrete). The addition of flocculants contributes the densification of capillaries and interface within cement-based composite, which will enhance the bonding strength within the recycling coarse aggregate. The cold-bonding recycling coarse aggregate using of lime sludge has the highest single particle compressive strength. The reason causing the single particle compressive strength of recycling aggregate with lime sludge higher than construction residual soil and granite sludge may be attributed to the fact that lime does not only contain a little hydration but also activate the pozzlanic reaction.

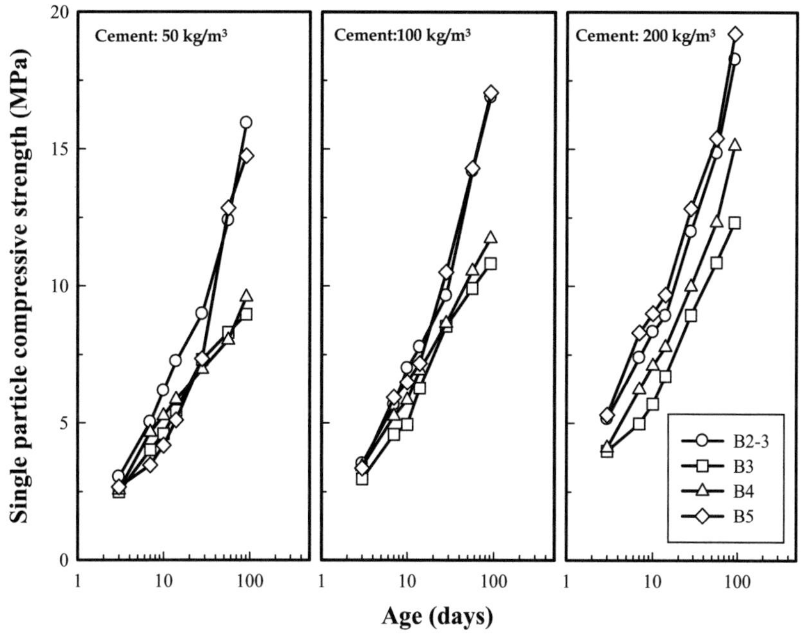

Fig. 11. The single particle compressive strength growth of recycling coarse aggregates with construction residual soil.

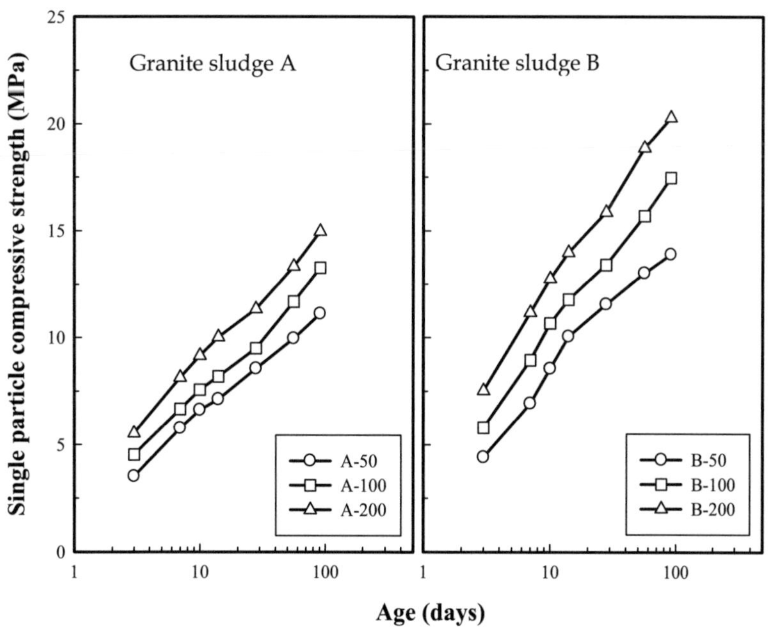

Fig. 12. The single particle compressive strength growth of recycling coarse aggregates with granite sludge.

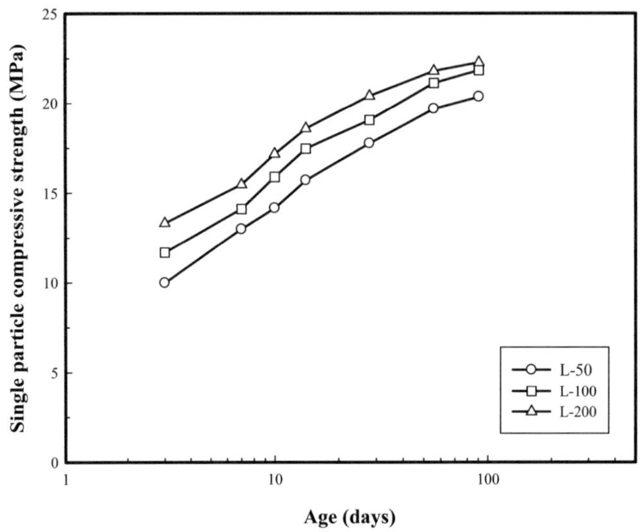

Fig. 13. The single particle compressive strength growth of recycling coarse aggregates with lime sludge.

5.3 Comparisons of various aggregates

According to an approach to estimate energy consumption, CO_2 emission, and prime cost of sintering recycling aggregates (Shiao et al., 2002), the cold-bonding recycling aggregates also was evaluated and compared with primitive and sintering aggregate. Table 6 shows the comparisons of properties, energy consumption, CO_2 emission, and prime cost of three various aggregates (primitive, sintering, and cold-bonding recycling aggregates). The results show that the recycling aggregate produced by using cold-bonding technique can reduce about 65 % CO_2 footprint than using sintering technique (Shiao et al., 2002). The prime cost of sintering recycling aggregate is 5 to 6 times higher than cold-bonding recycling aggregate. Even if the prime cost of cold-bonding recycling aggregate is lower than the primitive aggregate in Taiwan. The gradation of cold-bonding recycling aggregate is controllable and its single particle compressive strength at 91-day is 1.5 to 3 times higher than sintering recycling aggregate.

Item	Aggregate type		
	Primitive	Sintering	Cold-bonding
Gradation	Incontrollable	Single size	Controllable
Single particle compressive strength (MPa)	60.0 to 500.0	<7.0	10 to 22
Energy consumption	Low	High	Low
CO_2 emission (kg/m^3)	-	62.89	20.59
Cost (NTD/ m^3)	850	4,000	650

Table 12. Comparisons of various type aggregates.

6. Conclusion

1. The DMDA is appropriate to design the cement-based composite with recycling resources (i.e. construction residual soil, granite sludge, lime sludge) and glass fibers recycled from printed circuit board (PCB) wastes for producing the cold-bonding recycling coarse aggregates.
2. The press ingot method was developed and successfully granulated the cold-bonding recycling coarse aggregates and the procedure also was established. The proposed stress of granulation by using press ingot method is 35.0 to 42.0 MPa. The cold-bonding technique is able to be applied to handle recycling resources with moisture and reduce the energy consumption and CO_2 emission resulted from the oven-dry process. It is worth mentioning that the optimum moisture of blended recycling resource can be estimated by the proportion and optimum moisture of every constituent recycling resource.
3. The gradation of cold-bonding recycling coarse aggregate was controllable and established by the five various diameter particles according to requirements engineering or relevant standards and specifications. The unit weight, voids, and abrasion of cold-bonding recycling aggregate have the significant advancement after improvement of particle shape.
4. The recycling aggregate produced by using cold-bonding technique can reduce about 65 % CO_2 footprint than using sintering technique and the prime cost of cold-bonding recycling aggregate is 5 to 6 times lower than sintering recycling aggregate. Even if the prime cost of cold-bonding recycling aggregate is lower than the primitive aggregate in Taiwan. The single particle compressive strength at 91-day is 1.5 to 3 times higher than sintering recycling aggregate.
5. The developed cold-bonding recycling coarse aggregate could increase the reuse and recycling of wastes or recycling resources, reduce the energy consumption and CO_2 footprint, and diminish the impact on the environment and future generations. Using of lime sludge and cold-bonding technique to produce recycling coarse aggregates could not only has the above-mentioned benefits, but also these recycling aggregates would be applied to the remediation and reconstruction of CPDC An-Shun site in the future.

7. Acknowledgment

The author greatly appreciates the grant from Architecture and Building Research Institute, Ministry of The Interior, R.O.C. The continuous supports and technical assistances of Distinguished Professor Juu-En Chang, Professor Chao-Lung Hwang, Associate Professor Lung-Sheng Li, Dr. Gordon Tung-Chin Kung, Mr. Chien-Chih Chang and The Performance Experiment Center, Building Research Institute, Ministry of The Interior, R.O.C. are gratefully acknowledged. Also, the author would like to thank the financial support from National Cheng Kung University and the local factories (i.e. UG Enterprise Co., Ltd., and Sun Tech Enterprise Co., Ltd.) in Taiwan.

8. References

ACI Committee 211.1-91. (1991). *Standard Practice for Selecting Proportions for Normal, Heavyweight and Mass concrete*. American Concrete Institute, Farmington Hills, Michigan, USA.

ASTM D2487-06. (2006). *Standard Practice for Classification of Soils for Engineering Purposes (UnifiedSoil Classification System)*, American Society for Testing and Material, Philadelphia, Pennsylvania, USA.

Cai, Z.D.; Zhang, J.Z.; Li, L.S.; Gong, D.Q. & Zhang, Z.E. (2010). Using of Sediment from Foundation Engineering to Produce Recycling Cold-bonding Coarse Aggregates. *Chinese Journal of Geotechnical Engineering*, Vol.32, Supp.2, pp. 619-622, ISSN 1000-4548

Cai, Z.D.; Zhang, J.Z.; Li, L.S. & Zhang, Z.E. (2012) Feasibility of Using Construction Residual Soils to Develop Cold-pressing Recycling Coarse Aggregates. *Journal of Building Materials*, ISSN 1007- 9629 (Accepted for publication)

Chang, P.K.; Peng, Y.N. & Hwang, C.L. (2001). A Design Consideration for Durability of High-performance Concrete. *Cement and Concrete Composites*, Vol.23, No.4-5, pp. 375-380, ISSN 0958-9465

Chang, J.E; Chen, W.C.; Tsai, C.T.; Kung, G.T.C.; Chang, C.C.; Yu. N.; Che, M.D.; Liao C.T.; Wu, C.T. & Liang Y.Y. (2009). *The Report of Technology Development and Promotion of Recycling Green Building Materials Project (II)*. Architecture and Building Research Institute, Ministry of The Interior, R.O.C., Taipei, Taiwan

Chao, C.C.; Lin, K.H.; Lee, W.J.; Chang, J.E. & H.J. Chang. (2008). An Integrated Approach for Site Remediation and Re-Development. *Proceedings of Dioxin 2008*, Birmingham, UK.

Chen, H.J.; Wang, S.Y. & Tang, C.W. (2010). Reuse of Incineration Fly Ashes and Reaction Ashes for Manufacturing Lightweight Aggregate. *Construction and Building Materials*, Vol.24, No.1, pp. 46-55, ISSN 0950-0618

Chiang, C.M. (Ed.). (2007). *Evaluation Manual for Green Building Material 2007 New Edition*, Architecture and Building Research Institute, Ministry of The Interior, R.O.C., ISBN 978-986-008683-6, Taipei, Taiwan

Dinajar, P.; Babu, K.G. & Santhanam, M. (2008). Durability Properties of High Volume Fly Ash Self Compacting Concretes. *Cement Concrete Composites*, Vol.30, No.10, pp. 880-886 ISSN 0958-9465

Dowling, N.E. (1993). *Mechanical Behavior of Materials: Engineering Methods of Deformation, Fracture, and Fatigue*, Prentice-Hall, Inc., ISBN 0-13-579046-8, Englewood Cliffs, New Jersey, USA

Gibson, R.F. (1994). *Principles of Composite Material Mechanics*, McGraw-Hill, Co., ISBN 0-07-113335-6, Singapore

Holitz, R.D. & Kovacs, W.D. (1981). *An Introduction to Geotechnical Engineering*, Prentice-Hall, Inc., ISBN 0-13-484394-0, Englewood Cliffs, New Jersey, USA

Hsieh, T.T.; Chiang, C.M.; Ho, M.C. & Lai, K.P. (2012). The Application of Green Building Materials to Sustainable Building for Environmental Protection in Taiwan. *Advanced Materials Research*, Vols.343-344, pp. 267-272, ISSN: 1022-6680

Hung, M.F. & Hwang, C.L. (2007). Study of Fine Sediments for Making Lightweight Aggregate. *Waste Management and Research*, Vol.25, No.5, pp. 449-456, ISSN 0734-242X

Hwang, C.L. & Lu, H.C. (2000). The Effect of Paste Amount and Water Content on the Volume Stability of Concrete. *Journal of the Chinese Civil and Hydraulic Engineering*, Vol.12, No.3, pp. 621-626, ISSN 1015-5856

Industrial Technology Research Institute. (1996). In: *Taiwan Construction Residual Soil Information Service Center*, Available from http://140.96.175.34

Malhotra, V.M. (1990). Durability of Concrete Incorporating High-volume of Low-calcium (ASTM Class F) Fly Ash. *Cement Concrete Composites*, Vol.12, No.4, pp. 271-277 ISSN 0958-9465

Mangat, P.S.; & Khatib, J.M. (1995). Influence of Fly Ash, Silica Fume, and Slag on Sulfate Resistance of Concrete. ACI Materials Journal, Vol.92, No.5, pp. 542552 ISSN 0889-325X

Mather, B. (2000). Use Less Cement. *Concrete International*, Vol.22, No.11, pp. 55-56 ISSN 0162-4075

Mehta, P.K. (1986). *Concrete: Structure, Properties and Materials*, Prentice-Hall, Inc., ISBN 0-13-167115-4, Englewood Cliffs, New Jersey, USA

Mindess, S. & Young, J.F. (1981). *Concrete*, Prentice-Hall, Inc., ISBN 0-13-167106-5, Englewood Cliffs, New Jersey, USA

Neville, A.M. (2000). Water and Concrete: A Love-Hate Relationship. *Concrete International*, Vol.22, No.12, pp. 34-38 ISSN 0162-4075.

Rossi, P.; Acker, P. & Malier, Y. (1987). Effect of Steel Fibers at Two Stages: the Material and the Structure. *Materials and Structures*, Vol.20, pp. 436-439, ISSN 1359-5997

Shiao, J.P.; Hwang, C. L. & Pan, C.P. (2002). *The Economic Evaluation of the Application of the Lightweight Aggregate Concrete on RC Structure*, Architecture and Building Research Institute, Ministry of The Interior, R.O.C., ISBN 9-57-013048-2, Taipei, Taiwan

Skalny, J. & Mindess, S. (Eds.). (1989). *Materials Science of Concrete II*, The American Ceramic Society, Inc., ISBN 0-94-490437-8, Westerville, Ohio, USA

Tsai, C.T. (2005). The Study on the Re-deduction and Application of Densified Mixture Design Algorithm for Cement-Based Composites. *PhD Dissertation*, Department of Construction Engineering, National Taiwan University of Science and Technology, Taipei, Taiwan.

Tsai, C.T.; Chang, C.C.; Kung, G.T.C. & Chang, J.E. (2011). The Application of Construction Wastes and Cold-bonding Technique in Recycling Coarse Aggregates. *Proceedings of the 7th International Symposium on Cement & Concrete (ISCC2010) and the 11th International Conference on Advance in Concrete Technology and Sustainable Development*, pp. 624-629, ISBN 978-7-119-06783-4, Jinan, Shandong, P.R. China, May 9-12, 2010

Tsai, C.T.; Chang, C.C., Li, L.S. & Kung, T.C. (2012). Recycling Lime Sludge from CPDC An-Shun Site as Coarse Aggregates through Cold-bonding Technique. *Advanced Materials Research*, Vols.343-344, pp. 283-288, ISSN: 1022-6680

Tsai, C.T.; Li, L.S.; Chang, C.C. & Hwang, C.L. (2009). Durability Design Consideration and Application of Steel Fiber Reinforced Concrete in Taiwan. *Arabian Journal for Science and Engineering*, Vol.34, No.1B, pp. 57-79, ISSN 1319- 8025

Tsai, C. T.; Kung, G.T.C. & Hwang, C. L. (2010). Use of High Performance Concrete on Rigid Pavement Construction for Exclusive Bus Lanes. *Construction and Building Materials*, Vol.24, No.5, pp. 732-740, ISSN 0950-0618

Tsai, C.T.; Li, L.S. & Hwang, C.L. (March 2006). The Effect of Aggregate Gradation on Engineering Properties of High Performance Concrete. In: *Journal of ASTM International*, 09.03.2006, Available from http://www.astm.org

6

Advanced Sintering Techniques in Design of Planar IT SOFC and Supported Oxygen Separation Membranes

Vladislav Sadykov et al.*
Boreskov Institute of Catalysis, Novosibirsk State University, Novosibirsk, Russia

1. Introduction

Thin film solid oxide fuel cells (SOFC) operating in the intermediate temperature (IT) range are now considered as promising for distributed, mobile, standby or auxiliary power generation. At present one of the most important scientific aims in design of solid oxide fuel cells is to lower the operating temperatures to 600-800°C. In this temperature range, majority of problems inherent to SOFC operating at high (950-1000°C) are alleviated. Thus, cations interdiffusion and solid state reactions between electrolyte and electrodes are hampered and thermal stresses are decreased which prevent degradation of the functional layers [Yamamoto, 2004]. Hence, design of thin film SOFC requires also elaboration of nano-structured electrodes compatible with electrolytes from chemical and thermophysical points of view and providing a developed three-phase boundary (TPB). In this respect, broad options are provided by design of nanocomposite mixed ionic-electronic conducting (MIEC) functional layers – (Sadykov et al., 2010; Sadykov et al., 2009; Sadykov et al., 2008).

One of the most demanding problem in solid oxide fuel cells design is caused by the necessity of co-sintering of thin layers (electrolyte, functionally graded nanocomposite cathode) to provide required density without degradation of their transport, electrochemical and thermo-mechanical properties. The most developed and cost-effective are methods based upon supporting electrolyte powders with addition of organic binders and dispersants via screen-printing (Souza et al., 1998), tape casting (Kobayashi et al., 2002) or slurry coating technique (Jung et al., 2007). Thus supported «green» layers are sintered at

* Vladimir Usoltsev[1], Yulia Fedorova[1], Natalia Mezentseva[1], Tamara Krieger[1], Nikita Eremeev[1], Marina Arapova[1], Arcady Ishchenko[1], Alexey Salanov[1], Vitaly Pelipenko[1], Vitaly Muzykantov[1], Artem Ulikhin[2], Nikolai Uvarov[2], Oleg Bobrenok[3], Alexander Vlasov[4], Mikhail Korobeynikov[4], Aleksei Bryazgin[4], Andrei Arzhannikov[4], Petr Kalinin[4], Oleg Smorygo[5] and Manfred Thumm[6]
[1]*Boreskov Institute of Catalysis, Novosibirsk State University, Novosibirsk, Russia*
[2]*Institute of Solid State Chemistry and Mechanochemistry, Novosibirsk, Russia*
[3]*Institute of Thermal Physics, Novosibirsk, Russia*
[4]*Budker Institute of Nuclear Physics, Novosibirsk, Russia*
[5]*Institute of Powder Metallurgy, Minsk, Belarus*
[6]*Karlsruhe Inst. Technology, Karlsruhe, Germany*

temperatures ~ 1400°C. As a rule, high sintering temperatures required for electrolyte densification impose restrictions on the nature of electrode materials. Thus, at temperatures of sintering exceeding 1200°C, application of cathode substrates comprised of Sr-substituted perovskites could result in formation of isolating La(Sr)-zirconate layers (Chiba et al., 2008). Similarly, in the case of metal substrates high-temperature sintering is possible only in vacuum or in reducing gas atmospheres (Tucker, 2010). Another approach to supporting thin layers of electrolytes is based upon application of vapor deposition methods operating at substrate temperatures below 1000°C. This prevents formation of undesired phases in the course of synthesis, thus broadening the scope of electrode materials which can be used as electrolyte substrates. Among these methods, the most developed are different versions of Chemical Vapor Deposition (CVD) and Plasma Deposition (either in air or in vacuum) (Yamane & Hirai, 1987; Schiller et al., 2000; Minh, 1993).

Traditional sintering approaches requiring too high temperatures often failed in this respect. In design of supported oxygen-separation membranes, similar problems appear due to necessity of sintering thin dense mixed ionic-electronic conducting nanocomposite layers supported on macroporous ceramic (cermet or metal) substrates. For example to get Gd-doped ceria (GDC) pellet with density 97% from agglomerated commercial powders by simple heating in the furnace, required temperature is 1600-1700°C (Ma et al., 2004). Densification rate of GDC powder was found to increase only after 1100°C being accompanied by grain growth (Zhang & Ma, 2004). By using weakly agglomerated GDC nanopowders (average grain size 10 nm), the density of 94% from the theoretical value was achieved after sintering at 1100°C for several hours, and after 108 hours the relative density increased up to 99%. Increasing sintering temperature to 1200-1300°C allows to obtain 97-99 % density samples after 5 hours dwelling (Ivanov et al., 2007). Sintering of composites of ceria with strontium-doped lanthanum manganites was also studied using conventional techniques. Samples with 95-98% density could be obtained at temperatures 1350-1400°C. Sintering dynamics was found to depend upon samples composition, the best sinterability being observed for samples containing 30-70% ceria (Cutler et al., 2003). Together with application of nanosized powders, a popular approach in conventional sintering aiming at lowering sintering temperature is based upon using special sintering additives. Some of these additives form melts which enhance mass transfer between the particles of materials with a higher melting point. Bismuth oxide melts at 820°C and can substitute rare earth cations in the functional layers improving electrochemical performance. Sintering of GDC was enhanced by adding small amounts of Bi_2O_3: samples of GDC containing 1 wt% bismuth sintered at 1200-1400°C for 2 - 4 h were dense bodies (98-99% of theoretical density) which is achieved at temperatures lower by 250-300°C than required for sintering of undoped GDC (Gil at al., 2007). Moreover, even refractory additives can improve sintering via affecting defect chemistry at interfaces. Thus, the density and grain size of sintered Gd-doped CeO_2 increased with adding up to 2 mol. % Al_2O_3 while decreasing at a higher content of this additive (Lee et al., 2004 ; Liu at al., 2008). Similarly, iron oxide addition via mechanical activation treatment also provides a higher density of GDC (Zhang at al., 2004). The maximum density was obtained by adding 5 mol.% Ga_2O_3 to GDC at sintering temperature 1400°C (Lee et al., 2004). Combination of nanosized freeze-dried powders with addition of 1-3 mol.% cobalt to GDC allowed to get 94% density at 1000° C [Perez-Coll et al., 2003].

For sintering of cathode or membrane perovskite layers, silver or Ag-Pd alloy aids were successfully used (Chen et al, 1992).

Advanced sintering techniques are based upon application of high-energy irradiation that could be focused directly on the sample or even functional layer (as described in selective laser sintering studies (Kumar, 2003). Electron beams as PVD modification were used for deposition of alumina or composite cermets functional layers of carbides/nitrides (Singh & Wolfe, 2005). Only a few works mention e-beam as a promising approach to deposit thin and dense YSZ or GDC electrolyte layer on electrode (mainly NiO/YSZ anode) (Laukaitis et al., 2007; Lemkey et al., 2005). Deposition through the vapor (PVD or CVD) ensures thin uniform layers but, unfortunately, this approach requires high vacuum that increases costs and reduces commercial usage. Nevertheless, e-beam technique was used for direct sintering, including application of high-energy e-beams to produce layered metal composites (Zaeh&Kahnert, 2009), sintering of zirconia and zirconia-corundum composites etc. (Annenkov, 1996). Enhanced sintering of ceramics under e-beams action is explained by dissipation of radiation energy in heterogeneous structures, thermal-diffusion stimulation of mass transfer due to point defect formation and increase of thermal vibrations of the lattice (Annenkov, 1996)

Compared with radiation-induced sintering techniques, microwave heating at frequency 2.45 GHz is more broadly applied for sintering of zirconia-based electroceramics (Alexeff & Meek, 2011; Mazaheri et al., 2008). Zirconia or ceria- based materials have a low microwave absorption at temperatures below 400°C (Jiao, 2011). Jump-like change of their dielectric properties at 400-1000°C requires to use special susceptors with a better absorption for preheating samples and smoothing the heating curve. As a susceptor usually disks of SiC were used to provide so called hybrid microwave sintering. Microwave heating produces samples sintered to density 90-95% of the theoretical value at temperatures lower by 200°C than conventional heating (Mazaheri et al., 2008 ; Charmond et al., 2010). However, the maximum density obtained for doped zirconia samples in these conditions was ca 98%, and to produce fully dense samples the hybrid mode heating was used with dwelling about 1 hour at 1350°C (Charmond et al., 2010). Microwave heating uses also significantly higher heating rates 5-50° (and even 200°)/min as compared with conventional sintering carried out at a heating rate 2-5°/min with dwelling at the maximum temperature for 5-10 hours. This becomes possible due to very uniform heating without space gradients of temperature in the sample. A higher heating rate provides a lower average grain size (2.3 μm at 5°/min vs 0.9 μm at 50°/min) (Mazaheri et al., 2008 ; Charmond et al., 2010). Despite all these apparent advantages, up to date these advanced techniques have not been systematically applied for sintering functional layers comprised of mixed ionic –electronic conductors –perovskites and their nanocomposites with different electrolytes as required in design of thin film IT SOFC and oxygen separation membranes.

This work presents results of research aimed at filling such a gap and providing verification of these advanced sintering techniques (e-beam and microwave heating) for such an application. Functional layers were comprised of oxides with ionic conductivity (doped ceria, doped zirconia and δ-Bi_2O_3, doped La silicate with apatite structure) and mixed ionic-electronic conductivity (perovskites and their nanocomposites with electrolytes). Effect of sintering parameters on the real structure/microstructure and functional properties of these materials was elucidated. Experimental samples of button-size fuel cells and oxygen separation membranes were manufactured using advanced sintering techniques and successfully tested showing promising performance.

2. Experimental

2.1 Powder synthesis

Nanocrystalline complex oxides with ionic conductivity (fluorite-type electrolytes GDC, YSZ, δ-Bi_2O_3 stabilized by Er or Y+Sm) and mixed ionic-electronic conductivity (rare-earth manganites, nickelates, ferrites, cobaltites and their solid solutions including those doped by Sr and/or Bi) were synthesized by modified polymerized polyester citric acid-ethylene glycol precursors (Pechini) and mechanical activation routes. Synthesis procedures and structural features of these complex oxides are described in details elsewhere (Kharlamova et al., 2011; Sadykov et al., 2011). Nanocomposites were prepared by ultrasonic dispersion of the mixture of powders in isopropanol (solvent) using a T25 ULTRA-TURRAX (IKA, Germany) homogenizer with addition of polyvinyl butyral. Detailed compositions of complex oxide and their composites prepared for sintering are summarized in Table 1.

Materials	Abbreviation	Chemical composition	Geometry	Sintering techniques*
Cathode	LFN	$LaFe_{0.7}Ni_{0.3}O_{3-\delta}$	Pellets	CH
	LFC	$LaFe_{0.5}Co_{0.5}O_{3-\delta}$	Pellets	CH
Cathode composite	LFN-GDC	$LaFe_{0.7}Ni_{0.3}O_{3-\delta}$ (50 wt.%)-$Ce_{0.9}Gd_{0.1}O_{2-\delta}$ (50 wt.%)	Pellets	CH, RTS
	LFC-GDC	$LaFe_{0.5}Co_{0.5}O_{3-\delta}$ (50 wt.%)-$Ce_{0.9}Gd_{0.1}O_{2-\delta}$ (50 wt.%)	Pellets	CH, RTS
Electrolyte	GDC	$Ce_{0.9}Gd_{0.1}O_{2-\delta}$	Pellets Layer	CH, RTS, MH
	YSZ	$Y_{0.08}Zr_{0.92}O_{2-\delta}$	Pellets	MH
	BE	$Bi_{0.5}Er_{0.5}O_{2-\delta}$	Pellets	CH
	BYS	$Bi_{1.5}Y_{0.3}Sm_{0.2}O_{2-\delta}$	Pellets	CH, MH
Anode	NiO/YSZ	NiO (60%)-$Y_{0.16}Zr_{0.84}O_{2-\delta}$ (40%)	Pellets, Foam-based	CH, RTS CH

* CH- conventional heating, MH- microwave heating, RTS-radiation-thermal sintering

Table 1. Compositions and samples geometry.

2.2 Samples preparation for sintering

Pellets were uniaxially pressed from powders under 5.5 t/cm^2. As a porous metal substrate for composite anode or cathode, a unique type of macroporous planar Ni-Al foam substrate with a high electric and thermal conductivity, thermal stability and corrosion resistance was used (Smorygo et al., 2008; Sadykov et al. 2010). Along with these metal substrates, traditional NiO/YSZ planar anode substrates covered by thin layers of YSZ electrolyte were used for supporting cathode layers (Sadykov et al, 2009, 2011). Functional layers were supported by slip casting or screen-printing on the support surface.

2.3 Sintering techniques

Functional layers and uniaxially pressed pellets were sintered by different techniques: conventional heating (CH) in the furnace at temperatures up to 1300°C (nanocomposites) or 1400°C (electrolytes) with or without sintering aids (Bi or Ag nitrates, etc) in different gas atmospheres (air, Ar, etc); microwave heating (MH) or radiation-thermal sintering (RTS) under electron beam action.

For conventional heating, a high-temperature oven equipped with a quartz reactor specially designed for heating in the argon flow was used.

MH was carried out using a system based on gyrotron with frequency 24 GHz specially designed for heating of materials. Samples were heated by the focused radiation (power 0.5-1.5 kW) up to 1000-1200°C with heating rate 50°/min followed by dwelling at the final temperature for 30 min and then cooling to room temperature.

RTS was carried out on an ILU-6 accelerator that gives electron pulses with a high (2.4 MeV) energy. Temperature was varied in range of 900-1200 °C (heating rate 30-40 °/min) by changing the frequency in the range of 8-20 Hz. Time of treatment was varied in the range of 10-240 min.

2.4 Characterization of the real structure and functional properties

Genesis of the real structure of sintered functional layers and pellets was studied in details by X-ray powder diffraction, high resolution transmission electron microscopy and scanning electron microscopy with local elemental analysis by EDX. XRD patterns were obtained with an ARLX'TRA diffractometer (Thermo, Switzerland) using CuK_{α} monochromatic radiation (λ=1.5418 Å) in 2θ range 5-90°. Transmission Electron Microscopy (TEM) micrographs were obtained with a JEM-2010 instrument (lattice resolution 1.4 Å, acceleration voltage 200 kV). Analysis of the local elemental composition was carried out by using an energy-dispersive EDX spectrometer equipped with Si(Li) detector (energy resolution 130 eV).

Transport properties were characterized by impedance spectroscopy and oxygen isotope heteroexchange. Conductivity was measured in air with a MO-10 Micro-Ohmmeter at 25 Hz frequency with four Ag paste electrodes placed symmetrically at a perimeter of the pellet. The oxygen mobility was characterized by the temperature -programmed oxygen isotope exchange (a static installation with MS control of the gas phase isotope composition at pO_2 4 Torr). Performance of functional layers sintered by using different techniques was tested using respective devices: button-size fuel cells operating with wet H_2–air feeds and reactors equipped with supported membranes operating in the mode of oxygen separation from air into He stream (Sadykov et al, 2010, 2011).

3. Results and discussion

3.1 Effect of sintering stage on the real structure of SOFC materials

Electrolytes

Fresh (after synthesis) powders of GDC and BE are characterized by cubic fluorite -like structure, with some tetragonal distortions for YSZ. After sintering stage (radiation-thermal treatment, microwave or conventional heating) new phases were not revealed and no significant changes in diffraction pattern could be observed (Fig. 1). For fresh BYS sample, diffraction pattern corresponds to rhombohedral $Bi_{0.775}Sm_{0.225}O_{1.5}$ phase [89-4391, 44-0043] (Fig. 2). After sintering, it transforms into cubic fluorite phase of $Bi_{1.5}Y_{0.5}O_3$ type [33-0223] with lattice parameter a =5.511 Å (CH) - 5.507 Å (RTS). The lattice parameters of GDC and YSZ calculated for fresh or sintered samples vary only slightly (for 1-2%). Diffraction peaks are narrowed after sintering reflecting grain growth. Thus, X-ray particle sizes calculated by using the Scherrer equation change from 10.8 nm for fresh YSZ sample to 59.8 nm after radiation treatment. For BYS, variation is smaller -from 66.0 nm for fresh sample to 80 nm for sample after RTS.

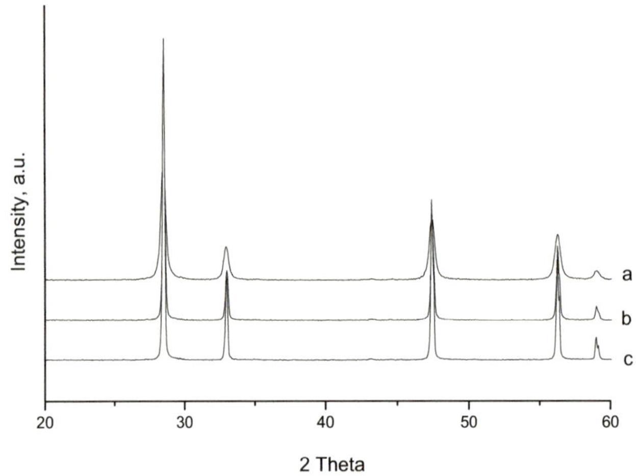

Fig. 1. Effect of sintering on XRD patterns of GDC electrolyte fresh (a), after microwave heating (b) and after radiation-thermal treatment (c).

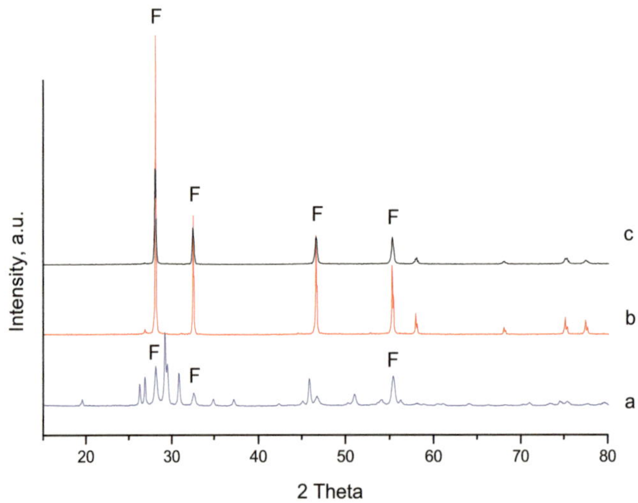

Fig. 2. Effect of sintering on XRD patterns of BYS electrolyte fresh (a), after conventional heating (b) or microwave heating (c).

Fig. 3 presents high resolution TEM images of GDC sample -fresh and after RTS or MH. Fresh sample consists of disorderly stacked nanosized (10-20 nm) domains. Sintering increases domains size up to 50-100 nm and provides their coherent stacking. EDX analysis revealed that the chemical composition of domains in sintered oxide is close to that in fresh sample $Ce_{0.86-0.91}Gd_{0.14-0.09}$. Similar variation of particle size and the mode of stacking after radiation-thermal treatment was observed by TEM for YSZ powder

TEM images for BYS electrolyte are presented in Fig. 4. The 111 lattice spacing observed (3.172 Å) is close to that in $Bi_{1.5}Y_{0.5}O_3$ (PDF#33-223 (Fm3m). The chemical composition of fresh sample at particles edges is characterizes by a higher bismuth content ($Bi_1Y_{0.9}Sm_{0.6}$) as compared with the core ($Bi_{0.4}Y_1Sm_{0.3}$). This segregation disappears after sintering being accompanied by rotational disordering of nanosized domains stacking. Hence, cooperative rhombohedral distortion of fluorite-like BYS structure in fresh sample can be caused by spatial nonuniformity of dopants distribution in particles.

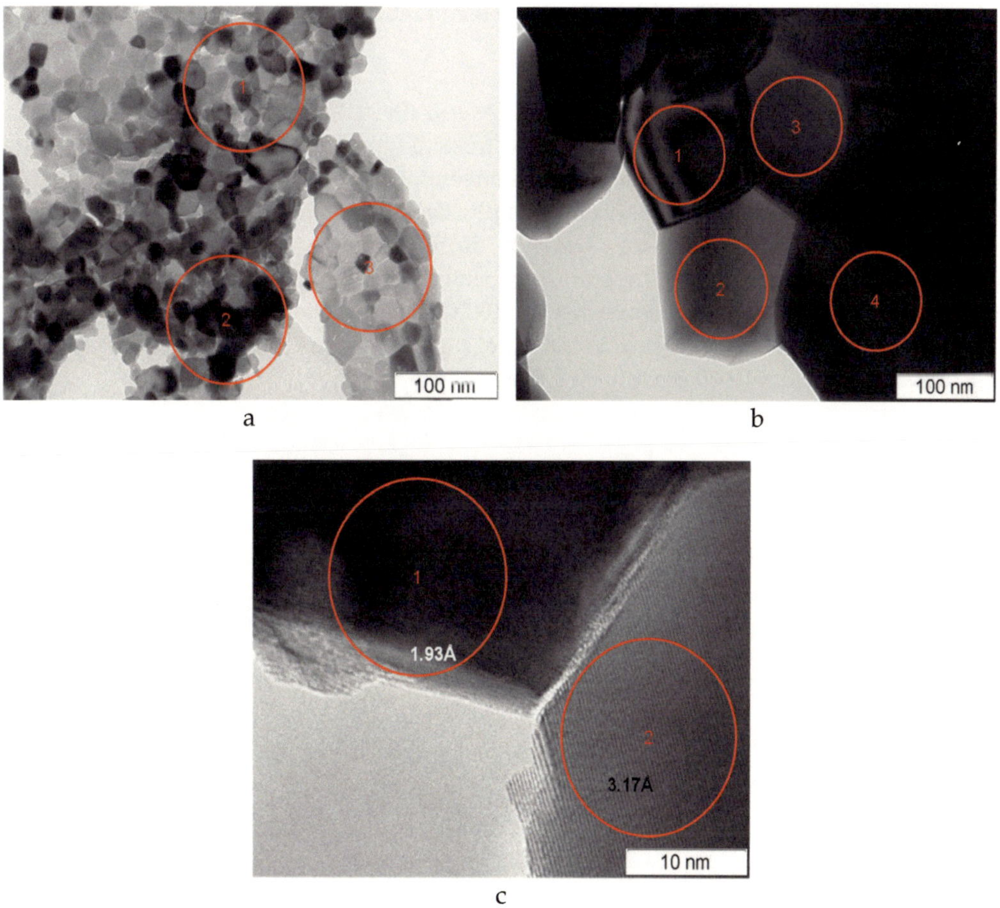

Fig. 3. TEM images of GDC after synthesis (a), radiation-thermal sintering at 1200 ˚C (b) or microwave heating at 1200˚C (c).

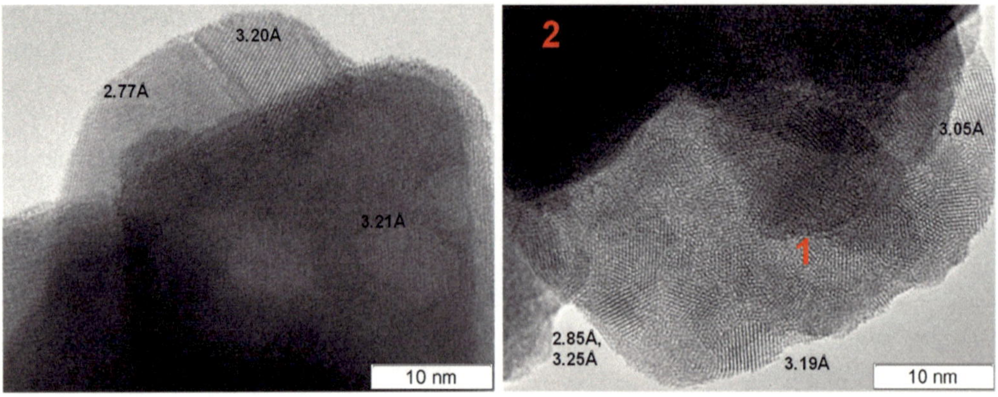

Fig. 4. TEM images of BYS after synthesis (a) and microwave heating at 900°C (b).

Hence, advanced sintering procedures resulted in substantial grain growth even for rather refractory electrolytes (GDC, YSZ) at moderate temperatures.

Cathode nanocomposite materials

After synthesis, complex perovskite oxides LFN and LFC (P) have the rhombohedral -type structure (space group R3-c) , and GDC - fluorite-type (F) cubic structure. The same structures are observed after sintering by conventional heating as well as by radiation treatment (Fig. 5). Some shift of reflections is observed after sintering due to redistribution of cations between P and F phases (Sadykov , 2010). X-ray particle sizes of both phases in composites increase with the temperature of sintering. Thus, for perovskite phases, after radiation-thermal sintering of composites at 1100°C, X-ray particle sizes increase from 20-30 nm for fresh samples to 90-100 nm. Variation of more refractory GDC phase particle size is less pronounced -from 10-20 nm to ~ 40-50 nm.

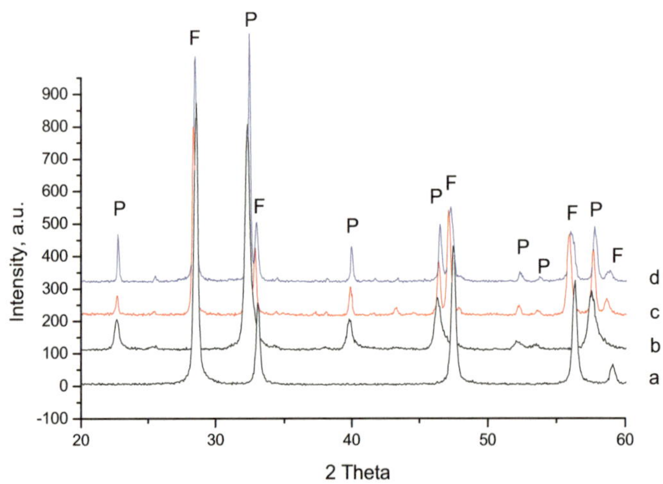

Fig. 5. Effect of sintering on XRD patterns of LFN-GDC composite (a-GDC fresh, b-LFN fresh, c- LFN-GDC after CH 1300°C, d - LFN-GDC, after RTS). P – perovskite-type phase, F – fluorite-type phase.

Fig 6. presents high- resolution TEM images for LFN-GDC cathode composite after radiation-thermal treatment. As judged by TEM data, reasonably uniform spatial distribution of perovskite and fluorite domains is observed. As follows from EDX data, pronounced redistribution of elements between P and F domains takes place. La and Fe from perovskite phase move to fluorite-type oxide while cerium and gadolinium ions migrate to perovskite structure. Nickel has not demonstrated significant mobility. Domain boundaries appear to be rather coherent providing a good epitaxy between P and F phases. Typical domain size for both samples is about 10-50 nm.

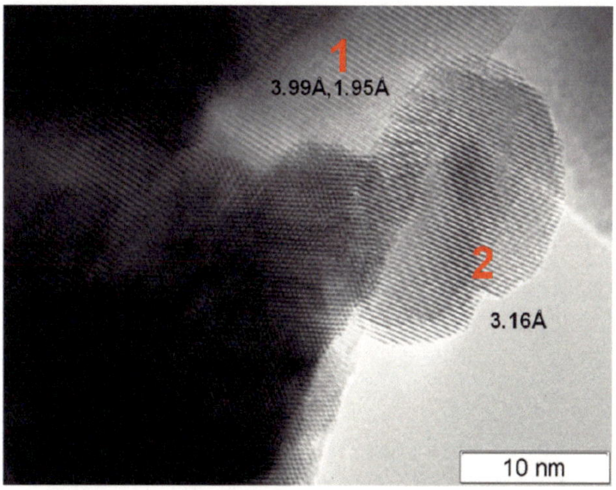

Fig. 6. TEM images of LFN-GDC composite after RTS at 1100° C. 1- LFC domain, 2- GDC domain.

3.2 Microstructure of sintered SOFC materials

Solid electrolytes (GDC, BE)

Microstructure of sintered pellets of solid electrolytes was studied by SEM. Figs. 7. and 8 present typical microstructure of samples after radiation-thermal treatment and microwave heating. Both electrolytes demonstrate a high density without visible cracks. Some closed porosity in observed in BYS (Fig. 8).

The average grain size of GDC oxide after RTS (0.2-0.5 µm) seems bigger than domains of GDC sintered by microwave heating (0.1-0.4 µm). This fact can be explained by very fast diffusion in the solid under e-beam. Thus, microwave heating stimulates local radiation absorption, and only part of this energy dissipates into heat that stimulates grain growth. Moreover, RTS was held for 150 min vs 90 min of MH. For conventional sintering it is a common knowledge that increasing sintering time increases grain size.

Cathode materials (LFN-GDC, LFC-GDC)

Sintering of bulk cathode composites was studied using radiation-thermal sintering and conventional heating techniques. In Figs. 9 and 10 values of density estimated by Archimedes method are presented in dependence on sintering temperature and duration of radiation treatment.

Fig. 7. SEM images demonstrating bulk (cross-section) structure of GDC after radiation-thermal sintering (a) and after microwave heating (b).

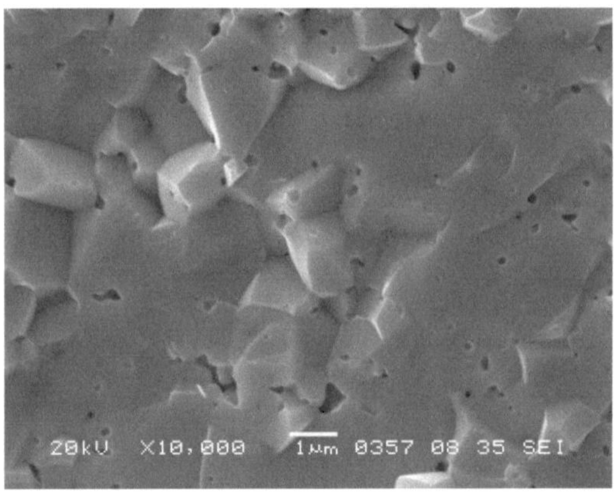

Fig. 8. SEM image demonstrating internal structure of BYS after microwave heating at 900°C.

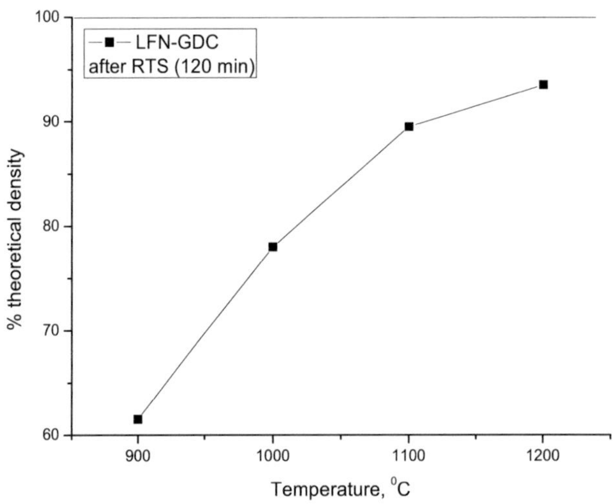

Fig. 9. Influence of temperature on density of LFN-GDC cathode composite.

As follow from Fig.10, duration of radiation-thermal treatment has a small effect on the real density of sample. In fact, after 60 min of RTS no significant increase of density was observed. On the contrary, the temperature strongly affects the density increasing it from 60% at 900° C to 95% at 1200° C.

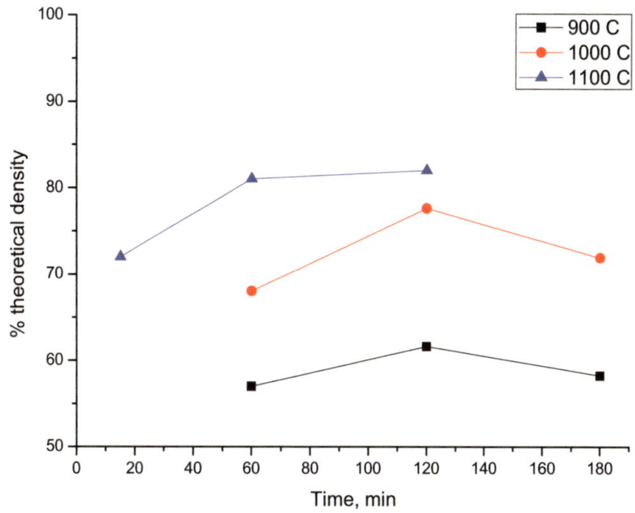

Fig. 10. Effect of sintering time on the density of cathode LFN-GDC composite.

It is worth noting that all samples sintered by RTS have a high mechanical strength even after RTS at 900°C. This fact is obviously explained by formation of crystal-type contacts in the bulk of composite. Nothing of this kind was observed by conventional sintering at the same temperatures and sintering time. Another important fact consists in the level of theoretical density: by using RTS technique we obtained 95 % of theoretical density at 1200° C and 60 min treatment, while by conventional sintering it can be achieved only after keeping at 1300° C for 5 hours.

The internal microstructure of sintered cathodic composite was studied by SEM images of cross-sections in back-scattered electrons (Fig. 11). Dark regions in this image correspond to perovskite-type phase, and gray to GDC. After sintering, interpenetrating structures of perovskite phase (electronic conductor) and fluorite-type GDC (ionic conductor) are observed. LFN-GDC sintered by conventional heating is characterizes by residual porosity (typical size of pores 0.5-1 μm) due to shrinkage of nanosized particles. The same features of microstructure were observed for LFC-GDC composite. Hence, cathodic composites based on lanthanum ferrites/cobaltite with GDC are expected to demonstrate the same features during radiation-thermal sintering: fast shrinkage under e-beam, formation of matrix even at moderate temperatures, which prevents diffusion of La cations into electrolyte typical to high-temperature treatments.

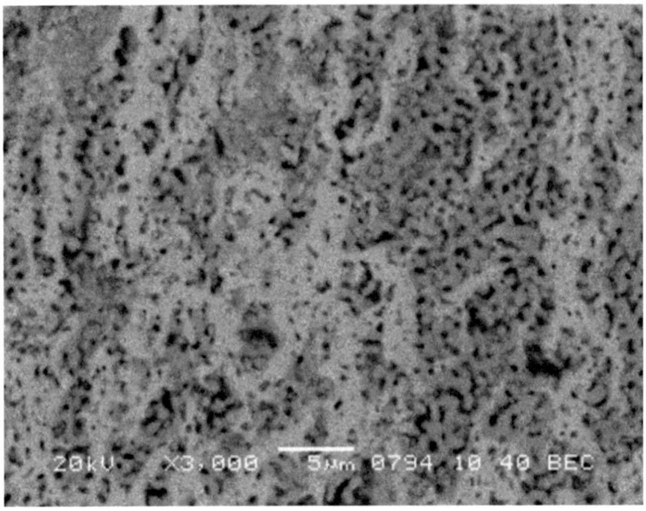

Fig. 11. Internal microstructure by SEM images in back-scattered electrons of LFN-GDC nanocomposite, CH 1300° C.

3.3 Microstructure of functional layers on metallic anodic substrates

Typical SEM images of different electrolyte layers supported on anodic substrate NiO/YSZ/Ni-Al foam are shown in Fig. 12. GDC layer was sintered to complete density by radiation -thermal treatment at 1200°C (Fig. 12a), which is ~ 200°C lower than for conventional sintering. Bi-Er and BYS layer were sintered by this method to a similar density even at much lower (900°C) temperature. Though small (less than 0.5 μm diameter) closed pores were observed, their share is estimated to be below 5%.

For Sr-doped lanthanum silicate electrolyte $La_9SrSi_6O_{26.5}$ known for its poor sinterability (Sadykov et al, 2010), porosity is apparently bigger (Fig. 12d), though the layer is also rather dense. Hence, for the latter type of electrolyte, combination of radiation-thermal treatment with sintering aids could be applied to provide complete layer densification.

Sintering of cathode composites should provide reasonably porous (porosity ~ 30-40%) mechanically strong layers strongly attached to the layer of electrolyte. As follows from Fig. 13a, radiation-thermal treatment (as well as microwave radiation) provides required characteristics combining porosity with developed triple-phase boundary to ensure both developed surface area for oxygen molecules activation and diffusion paths for oxygen ions to enter the electrolyte layer. At the same temperatures cathode layer sintered by conventional heating was only weakly attached to electrolyte.

On the other hand, for nanocomposite mixed ionic-electronic conducting layers supported on the surface of oxygen separation membrane, complete density/absence of porosity is required to provide selective oxygen permeation. As shown in Fig. 13b, Bi-containing composites are sintered to complete density by radiation-thermal treatment at moderate (900°C) temperatures, thus meeting requirement.

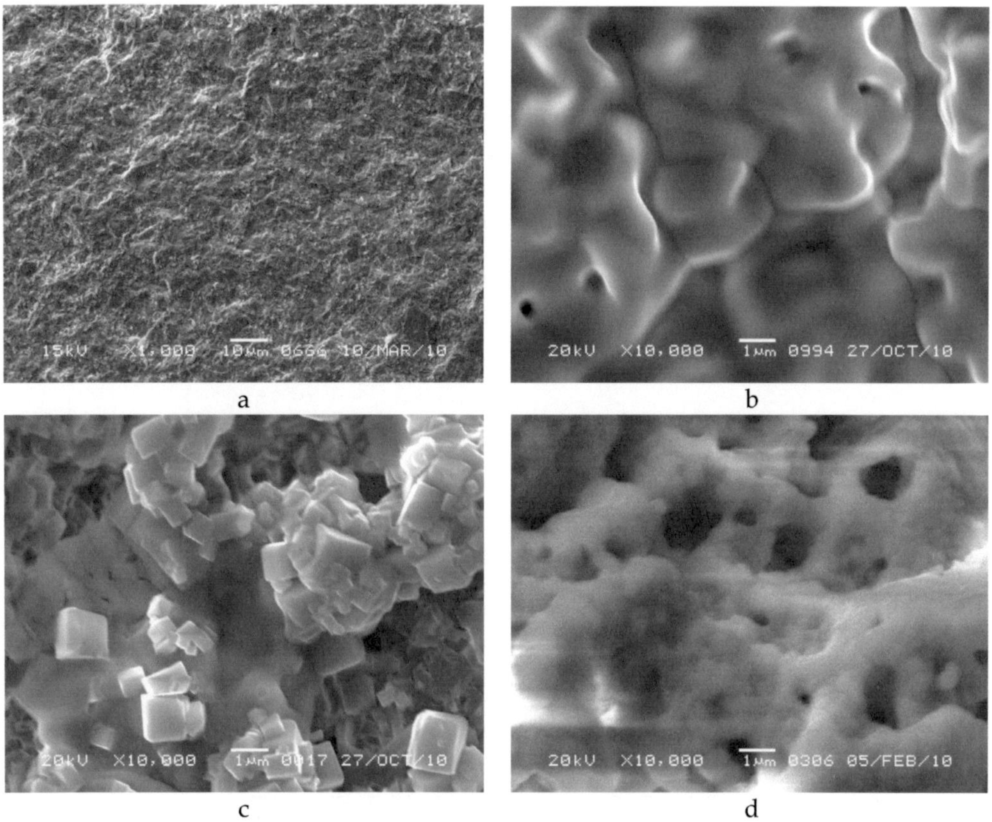

Fig. 12. SEM images of thin (10 microns) layers of electrolytes supported on anodic substrate NiO/YSZ/Ni-Al foam and sintered by radiation-thermal treatment at different temperatures. a- GDC, 1200˚C; b - BE , 900˚C, c-BYS, 900˚C, d- $La_9SrSi_6O_{26.5}$ with apatite structure, 1200˚C.

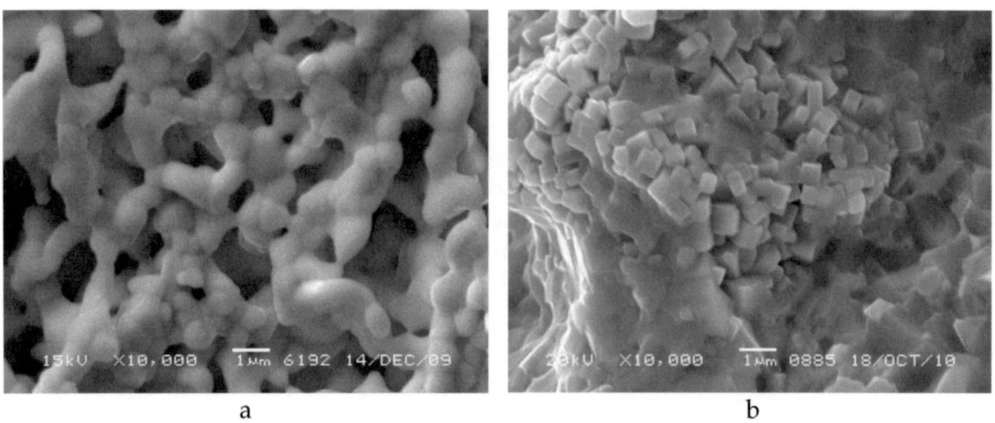

Fig. 13. SEM image demonstrating typical structure of LFN-GDC (a) and LaBiMn -BiEr layers (b) on Ni-Al foam-based substrates after radiation-thermal sintering at 900˚C.

3.4 Electrochemical and transport properties of cathode composite material

Conductivity

Fig. 14 demonstrates that conventional sintering provides a lower specific conductivity even at higher sintering temperature than RTS. While the main factor in controlling conductivity is certainly residual porosity of pellets, note that for the radiation-thermal treatment somewhat higher conductivity was observed after sintering at lower (900°C) temperatures. This can be explained by a loss of oxygen from LFN at higher temperatures leading to conductivity decrease (Sadykov, 2011).

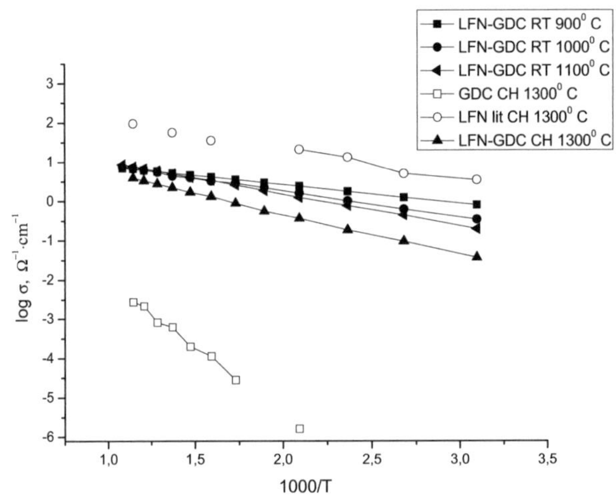

Fig. 14. Effect of sintering mode on specific conductivity of LFN-GDC samples after radiation-thermal sintering and conventional heating compared with literature data (Sadykov, Pavlova et al, 2010).

For LFC-GDC composite (Fig.15) sintering by e-beam at 1100°C for 1 h provides the same level of specific conductivity as prolonged (5 hours) conventional sintering at 1300°C. Surprisingly dwelling under e-beam for 2 hours or more has not improved conductivity. This can be explained by more pronounced interaction between phases in this nanocomposite as compared with LFN-GDC. Indeed, the difference between specific conductivity of LFC and LFC-GDC composite is ~ 3 order of magnitude, which is much bigger than that observed for LFN and LFN-GDC, respectively (Fig. 14).

Oxygen isotope heteroexchange

As a measure of oxygen mobility in perovskite-fluorite nanocomposites, so called dynamic extent of oxygen isotope exchange was shown to be simple and efficient characteristic varying in parallel with the oxygen diffusion coefficient (Sadykov et al, 2009-2010). It is defined as the number of oxygen atoms exchanged up to a given temperature in the temperature -programmed mode of heating the sample with the natural oxygen isotope composition in a static reactor under contact with the gas phase containing initially only $^{18}O_2$. This value can be expressed either in oxygen monolayers (X_S) or as a fraction of all oxygen contained in the bulk of sample (X_V). Fig. 16 presents such dependence for LFN-GDC composites with different ratio of perovskite/fluorite phases. As follows from these results, for composites with a higher content of GDC, already at 700°C more than half of overall amount of bulk oxygen is exchanged, which is clearly possible only in the case of high oxygen mobility in the bulk of composite. The increase of oxygen mobility with the amount of GDC phase, and, hence, the length of perovskite-fluorite interface, agrees with suggestion that such an interface provides a path for fast oxygen diffusion due to its specific structure and composition (Sadykov et al, 2011). The values of X_v for LFN-GDC composites with a high GDC content are close to those for Sr-containing LSFC-GDC composites known for their very high oxygen mobility (Sadykov et al, 2008). Hence, even composites without Sr and Co in complex perovskites could provide a high bulk oxygen mobility combined with a low reactivity with respect to interaction with YSZ if their composition and sintering procedures are properly optimized.

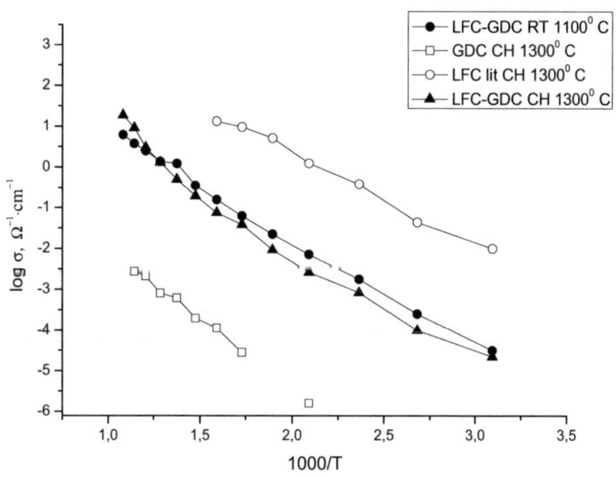

Fig. 15. Effect of sintering mode on specific conductivity of LFC-GD composite after radiation-thermal sintering and conventional heating compared with literature data (Sadykov, Pavlova et al, 2010).

3.5 Some characteristics of SOFC and oxygen separation membranes with co-sintered functional layers

3.5.1 Fuel cells

For assembling cells, a home-made Ni/yttria-doped zirconia (YSZ) anode substrate (the exposed cell surface area 1 cm^2) as well as Ni-Al foam supported NiO/YSZ layers with 10 micron thick 8YSZ layers supported by CVD were used. Cathode slurries made from nano-powders ultrasonically dispersed in isopropanol with addition of butyral resin were deposited on half cells by air spray (perovskite + fluorite 10 microns thick nanocomposite functional layer, such as LSM-ScCeSZ or LSFN-GDC) and by painting (porous thick LSFN cathode layer) followed by drying and sintering at 900-1100°C for 2 h using microwave or e-beam heating (Sadykov et al, 2011). The cell performance was evaluated using air at cathode side and humidified H_2 at anode side with Pt current collectors adding Pt or Ar pastes on the cathode side. For these cells, the typical level of power density at 700°C was in the range of 500 mW/cm^2, which is promising for the practical application. Performance stability was demonstrated for short-term (~ 100 h) testing. No cracking or layers spallation was observed after testing. Decreased sintering temperature allowed to prevent any damages to anodic substrates during cells manufacturing as well as any undesired interaction between YSZ electrolyte and cathode layers leading to formation of isolating pyrochlore layers.

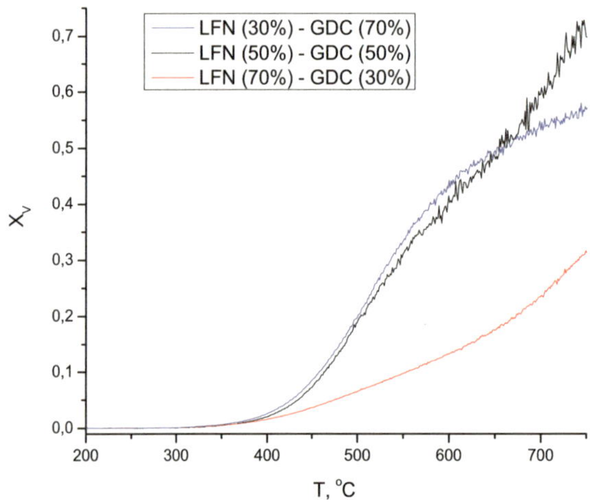

Fig. 16. Temperature dependence of the dynamic extent of exchange X_V for LFN-GDC composites sintered by microwave heating at 1100°C. PO_2 = 4 Torr, heating ramp 5°/min.

3.5.2 Membrane performance

Fig. 17 presents the temperature dependence of oxygen flux through asymmetric supported oxygen separation membranes with Bi-containing perovskite-fluorite functional layers sintered by using microwave radiation (vide supra). As follows from these results, the values of oxygen flux are close to best results obtained with supported membranes in these conditions (Sadykov et al, 2010).

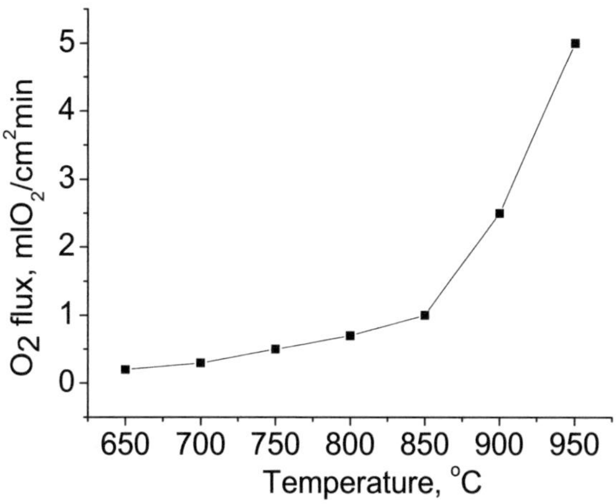

Fig. 17. Temperature dependence of oxygen flux under air/He gradient for supported asymmetric membrane comprised of La-BiMnO-YSmBi layers on binary Ni-Al foam substrate.

4. Conclusions

Advanced sintering techniques based upon radiation-thermal sintering by e-beam action and microwave heating allow to provide required density and consolidation of thin functional layers in design of intermediate temperature solid oxide fuel cells and oxygen separation membranes. Due to decreased temperature and duration of sintering as compared with conventional sintering methods, such negative phenomena as variation of functional layers phase composition, their cracking and damage of metallic substrates were prevented. For oxide mixed ionic-electronic conducting composites advanced sintering provides developed interfaces which act as paths for fast oxygen diffusion required for considered applications. As the results, fuel cells and oxygen separation membranes manufactured using advanced sintering techniques demonstrate performance characteristics promising for the practical applications.

5. Acknowledgements

The authors gratefully acknowledge support from by Integration Project 57 (T09CO-003) SB RAS- NAS of Belarus, Russian Federation Government Grant N 11.G34.31.0033, OCMOL FP7 EC Project, Project 57 of RAS Presidium Program No. 27 and Contract 02.740.11.0852 of the Federal Program "Scientific and Educational Cadres of Russia.

6. References

Annenkov, Yu. (1996). Physical foundations high-temperature electron-beam of ceramics, *Russian Physics Journal*, 39, No. 11, 1146-1159, ISSN: 1573-9228

Alexeff, I. & Meek, T. (2011). The effect of electric field intensity on the microwave sintering of zirconia, *Materials Letter*, 65, 2111-2113, ISSN: 0167-577X

Charmond, S.; Carry, C. & Bouvard, D. (2010). Densification and microstructure evolution of Y-tetragonal zirconia polycrystal powder during direct and hybrid microwave sintering in a single mode cavity, *Journal of European Ceramic Society*, 30, 1211-1221, ISSN: 0955-2219

Chiba, R.; Tabata, Y.; Komatsu. T.; Orui, H.; Nozawa, K.; Arakawa M. & Arai H. (2008). Property change of a $LaNi_{0.6}Fe_{0.4}O_3$ cathode in the initial current loading process and the influence of a ceria interlayer, *Solid State Ionics.*, 178, 1701-1705, ISSN: 0167-2738.

Cutler, R & Meixner D. (2003). Ceria–lanthanum strontium manganite composites for use in oxygen generation systems, *Solid State Ionics,* 159, 9-19, ISSN: 0167-2738.

Gil, V; Tartaj, J; Moure, C. & Duran P. (2007). Rapid densification by using Bi_2O_3 as an aid for sintering of gadolinia-doped ceria ceramics, *Ceramics International,* 33, 471-475, ISSN: 0272-8842.

Ivanov, V.; Khrustov, V; Kotov Yu, Medvedev, A.; Murzakaev, A.; Shkerin, S. & Nikonov, A. (2007). Conductivity and structure features of $Ce_{1-x}Gd_xO_{2-\delta}$ solid electrolytes fabricated by compaction and sintering of weakly agglomerated nanopowders, *Journal of the European Ceramic Society*, 27, 1041-1046, ISSN: 0955-2219.

Yamamoto, O. (2000) Solid oxide fuel cells: fundamental aspects and prospects *Electrochimica Acta*, 45, 2423-35, ISSN 0013-4686

Jiao, Z.; Shikazono, N. & Kasagi, N. (2011). An ultra-fast fabrication technique for anode support solid oxide fuel cells by microwave, *Journal of Power Sources*, 196, 5490-5493, ISSN: 0378-7753.

Jung, H.; Hong, K.-S.; Jung, H.-G.; Kim, H.; Kim, H.-R.; Son, Ji-W.; Kim, J.; Lee, H.-W. & Lee, J.-H. (2007). SOFCs with Sc-Doped Zirconia Electrolyte and Co-Containing Perovskite Cathodes, *Proceedings of the Electrochemical Society.*, 154, 480-485, ISSN: 1945-7111.

Kharlamova, T.; Pavlova, S.; Sadykov V., Bespalko, Y., Krieger, T.; Pelipenko, V.; Belyaev, V.; Muzykantov, V.; Alikina, G.; Okhlupin, Yu.; Uvarov, N. & Smirnova, A. (2011) Nanocomposite Cathode Materials for Intermediate Temperature Solid Oxide Fuel Cells, *ECS Transactions*, 35, 2331-2340, ISSN: 1938-5862

Kobayashi, K.; Takahashi, I.; Shiono, M. & Dokiya M. (2002). Supported $Zr(Sc)O_2$ SOFCs for reduced temperature prepared by electrophoretic deposition, *Solid State Ionics*, 152-153, 591-596, ISSN: 0167-2738.

Kumar,. S. (2003) Selective Laser Sintering:A Qualitative and Objective Approach *Journal of the Minerals,* 55, 43-47, ISSN 1047-4838

Laukaitis, G.; Dudonis, J. & Milčius D. (2007). Formation of Gadolinium Doped Ceria Oxide Thin Films by Electron Beam Deposition, *Materials Science (Medžiagotyra)*, 13, 23-26, ISSN: 1392-1320.

Lee, J.-S.; Choi, K.-H; Ryu, B.-K, Shin, B.-C. & Kim I.-S. (2004). Effects of alumina additions on sintering behavior of gadolinia-doped ceria, *Ceramics International*, 30, 807-812, ISSN: 0272-8842

Lee, J.-S.; Choi, K.-H; Ryu, B.-K, Shin, B.-C. & Kim I.-S. (2004). Effects of gallia additions on sintering behavior of gadolinia-doped ceria, *Materials Research Bulletin*, 39, 2025-2033, ISSN: 0025-5408.

Lemkey, F. & Movchan B. (2005). Synthesis of porous and dense elements of SOFC by electron beam physical vapor deposition (EBPVD), In: *Fuel Cell Technologies: State and Perspectives,* Sammes, N et al. (Ed.), pp. 73-80, Springer, ISBN: 978-1-4020-3496-1, Netherlands.

Liu, Y.; Yin, C.; Wang, L; Li, D; Lian, J.; Hu, J. & Guo, Z. (2008). Properties of a Ceria-Based (C6S2G2) Solid Oxide Electrolyte Sintered with Al_2O_3 Additive, *Science of Sintering,* 40, 13-20, ISSN: 0350-820X

Ma, J.; Zhang, T.; Kong, L; Hing, P. & Chan, S. (2004). $Ce_{0.8}Gd_{0.2}O_{2-\delta}$ ceramics derived from commercial submicron-sized CeO_2 and Gd_2O_3 powders for use as electrolytes in solid oxide fuel cells, *Journal of Power Sources,* 132, 71-76, ISSN: 0378-7753.

Mazaheri, M.; Zahedi, A. & Hejazi, M. (2008). Processing of nanocrystalline 8 mol.% yttria-stabilized zirconia by conventional, microwave-assisted and two-step sintering, *Materials Science and Engineering: A,* 492, 261-267, ISSN: 0921-5093

Minh, N.Q. (1993). Ceramic Fuel Cells, *Ceram. Soc.,* 76, 563-588, ISSN: 0002-7820.

Pérez-Coll, D.; Núñez, P.; Frade, J. & Abrantes C. (2003). Conductivity of CGO and CSO ceramics obtained from freeze-dried precursors, *Electrochimica Acta,* 48, 1551-1557, ISSN: 0013-4686

Sadykov, V.; Kharlamova, T.; Batuev, L.; Mezentseva, N.; Alikina, G.; Muzykantov, V.; Krieger, T.; Pavlova, S.; Zaikovskii, V.; Ishchenko, A.; Zarubina, V.; Rogov, V.; Bobrenok, O.; Uvarov, N.; Kilner, J.; Druce, J. & Smirnova, A. (2008). Design and Characterization of Nanocomposites Based on Complex Perovskites and Doped Ceria as Advanced Materials for Solid Oxide Fuel Cell Cathodes and Membranes, *Mater. Res. Soc. Symp. Proc.,* 1098, 1-6, ISSN: 0272-9172.

Sadykov, V.; Muzykantov, V.; Bobin, A.; Batuev, L.; Alikina, G.; Lukashevich, A.; Boronin, A.; Krieger, T.; Ishchenko, A.; Bobrenok, O.; Uvarov, N.; Smirnova, A. & Vasylyev, O. (2009). Design and characterization of LSM-ScCeSZ nanocomposite as MIEC material for SOFC cathodes and oxygen-separation membranes., *Mater. Res. Soc. Symp. Proc.,* 1126, 1-6, ISSN: 0272-9172.

Sadykov, V.; Pavlova, S.; Zarubina ,V.; Bobin, A.; Alikina, G.; Lukashevich, A.; Muzykantov, V.; Usoltsev, V.; Kharlamova, T.; Boronin, A.; Koscheev, S.; Krieger, T.; Ishchenko, A.; Mezentseva, N.; Salanov, A.; Smirnova, A.; Bobrenok, O. & Uvarov, N. (2009). Design and Characterization of Functionally Graded Cathode Materials for Solid Oxide Fuel Cells, *ECS Trans.,* 25, 2403-2412, ISSN: 1938-6737.

Sadykov, V.; Zarubina, V.; Pavlova, S. et al. (2010). Design of asymmetric multilayer membranes based on mixed ionic–electronic conducting composites supported on Ni–Al foam substrate, *Catalysis Today,* No. 156, (October 2010) pp. 173-180, ISSN: 0920-5861

Sadykov, V.; Pavlova, S.; Kharlamova, T.; Muzykantov, V.; Uvarov, N.; Okhlupin, Yu.; Ishchenko, A.; Bobin, A.; Mezentseva, N.; Alikina G., Lukashevich, A.; Krieger, T.; Larina, T.; Bulgakov, N.; Tapilin, V.; Belyaev, V.; Sadovskaya, E.; Boronin, A.; Sobyanin, V.; Bobrenok, O.; Smirnova, A.; Smorygo O. & Kilner J. (2010) Perovskites and Their Nanocomposites with Fluorite-Like Oxides as Materials for Solid Oxide Fuel Cells Cathodes and Oxygen-Conducting Membranes, In: *Perovskites: Structure, Properties and Uses,* Borowski, M (Ed.), pp. 67-178, Nova Science Publishers, Inc. New York, ISBN: 978-1-61668-525-6, .

Sadykov, V; Mezentseva, N.; Usoltsev, V., Sadovskaya , E., Ishchenko, A., Pavlova, S., Bespalko, Y., Kharlamova, T., Zevak, E., Salanov, A., Krieger, T., Belyaev, V., Bobrenok, O., Uvarov, N., Okhlupin, Yu., Smorygo, O., Smirnova, A.; Singh, P.,

Vlasov, A., Korobeynikov, M., Bryazgin. A., Kalinin, P. & Arzhannikov, A. (2011). Solid oxide fuel cell composite cathodes based on perovskite and fluorite structures, *Journal of Power Sources,* 196, 7104-7109, ISSN: 0378-7753.

Sadykov, V., Usoltsev, V., Fedorova, Yu., Sobyanin, V., Kalinin, P., Arzhannikov, A., Vlasov, A., Korobeynikov, M., Bryazgin; A., Salanov, A., Predtechenskii, M., Bobrenok, O., Ulikhin, A., Uvarov, N., Smorygo, O., Il'yushchenko, A., Ul'yanitskii, V., Zlobin, S. (2011). Design of medium–temperature solid oxide fuel cells on porous supports of deformation strengthened Ni–Al alloy. *Russian Journal of Electrochemistry*, 47, 488–493. ISSN: 1023-1935.

Schiller, G.; Henne, R; Lang, M.; Ruckdaschel, R. & Schaper, S. (2000) Development of vacuum plasma sprayed thin-film SOFC for reduced operating temperature, *Fuel Cells Bulletin,* pp. 7-12, ISSN: 1464-2859.

Singh, J. & Wolfe D. (2005). Nano and macro-structured component fabrication by electron beam-physical vapor deposition (EB-PVD), *Journal Of Materials Science,* 401-26, ISSN: 1573-4803.

Smorygo, O.; Mikutski, V.; Leonov, A.; Marukovich, A. & Vialiuha, Y. (2008) Nickel foams with oxidation-resistant coatings formed by combustion synthesis, *Scripta Materialia,* 58, 910-913, ISSN: 1359- 6454.

de Souza, S.; Visco, S. & de Jonghe, L. (1997). Thin-film solid oxide fuel cell with high performance at low-temperature, *Solid State Ionics.*, 98, 57-60, ISSN: 0167-2738.

Tucker, M. (2010). Progress in metal-supported solid oxide fuel cells: A review, *J. Power Sources,* 195, 4570-4582, ISSN: 0378-7753.

Yamane H. & Hirai T. (1987). Preparation of ZrO_2-film by oxidation of $ZrCl_4$, *Materials Science Letters,* 6, 1229-1235, ISSN: 0261-8028.

Zaeh, M. & Kahnert, M. (2009). The effect of scanning strategies on electron beam sintering, *Prod. Eng. Res. Devel.,* 3, 217-224, ISSN: 0022-1817.

Zhang, T. & Ma, J. (2004). Dense submicron-grained $Ce_{0.8}Gd_{0.2}O_{2-d}$ ceramics for SOFC applications, *Scripta Materialia,* 50, 1127-1130, ISSN: 1359- 6454.

Zhang, T; Ma, J; Kong, L; Chan, S; Hing, P. & Kilner J. (2004). Iron oxide as an effective sintering aid and a grain boundary scavenger for ceria-based electrolytes, *Solid State Ionics,* 167, 203-207, ISSN: 0167-2738

7

Sintering and Properties of Nb_4AlC_3 Ceramic

Chunfeng Hu[1], Qing Huang[1],
Yiwang Bao[2] and Yanchun Zhou[3]
[1]*Ningbo Institute of Material Science and Technology, Chinese Academy of Sciences, Ningbo*
[2]*State Key Laboratory of Green Building Materials, China Building Materials Academy, Beijing*
[3]*Ceramic and Composites, Aerospace Research Institute of Materials and Technology, Beijing, China*

1. Introduction

Layered ternary compounds, $M_{n+1}AX_n$ (where M is an early transition metal, A is an A group element, X is C or N, and n = 1-3), also called the MAX phases, are layered carbides or nitrides crystallizing with hexagonal symmetry structure [1,2]. These ceramics combine the characteristics of metals and ceramics such as high strength and modulus, low density, good electrical and thermal conductivity, easy machinability, damage tolerance, and resistance to thermal shock and high temperature oxidation. To date, more than 50 M_2AX compounds (not list for brevity), five M_3AX_2 compounds (Ti_3SiC_2, Ti_3GeC_2, Ti_3AlC_2, Ta_3AlC_2, and Ti_3SnC_2), and seven M_4AX_3 (Ta_4AlC_3, Ti_4AlN_3, Ti_4SiC_3, Ti_4GeC_3, Nb_4AlC_3, V_4AlC_3, and Ti_4GaC_3) were identified. For 413 phases, it has been determined that there are two kinds of atomic stacking sequences along [0001] direction. Ti_4AlN_3, Ti_4SiC_3, Ti_4GeC_3, α-Ta_4AlC_3, Nb_4AlC_3, and V_4AlC_3 have the same Ti_4AlN_3-type crystal structure with atomic arrangement of AB$\underline{A}$BACB$\underline{C}$BC along [0001] direction. Only β-Ta_4AlC_3 was determined to have the AB$\underline{A}$BABA$\underline{B}$AB atomic arrangement along [0001] direction. In detail, the difference of atomic arrangements between β-Ta_4AlC_3 and α-Ta_4AlC_3 lay in the diversity of atomic positions. The atomic positions of β-Ta_4AlC_3 were described as Ta1 at (4*f*) (1/3, 2/3, 0.05524), Ta2 at (4*f*) (2/3, 1/3, 0.16016), Al at (2*c*) (1/3, 2/3, 1/4), C1 at (2*a*) (0, 0, 0), and C2 at (4*e*) (0, 0, 0.11125). While, the atomic positions of α-Ta_4AlC_3 were determined as Ta1 at (4*f*) (1/3, 2/3, 0.05453), Ta2 at (4*e*) (0, 0, 0.15808), Al at (2*c*) (1/3, 2/3, 1/4), C1 at (2*a*) (0, 0, 0), and C2 at (4*f*) (2/3, 1/3, 0.10324).

Ti_4AlN_3-type Nb_4AlC_3 was firstly discovered by heat treating Nb_2AlC at 1700°C, and the crystal structure was determined using a combined technique of *ab initio* calculation and high resolution scanning transmission electron microscopy. Additionally, the single phase Nb_4AlC_3 could be synthesized by hot pressing and spark plasma sintering. The microstructure, electrical, thermal, and mechanical properties of as-prepared Nb_4AlC_3 were systematically described.

2. Crystal structure

Figure 1 compares the projection of atoms (Figs. 1(a) and (b)) and the Z-contrast image of Nb_4AlC_3 (Fig. 1(c)), wherein the layer stacking sequence of Nb and Al atoms along the [0001] direction was directly shown. In one Nb-C slab, the number of Nb layers was four, i.e., the regular atomic arrangement was four Nb atoms layers per Al atoms layer alternately stacking along the [0001] direction. The Nb layers were separated by close packed Al atom (0001) planes. The Z-contrast image confirmed that the Nb_4AlC_3 crystallized in the Ti_4AlN_3-type crystal structure. It basically consisted of two units: nonstoichiometric $NbC_{0.75}$ slab and Al atomic plane. The atom arrangement of Nb and Al was AB<u>A</u>BACB<u>C</u>BC [3].

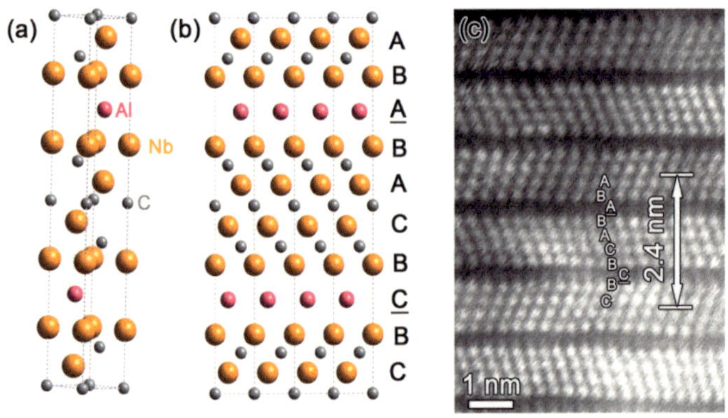

Fig. 1. (a) and (b) Atom arrangement of Nb_4AlC_3. (c) High-resolution Z-contrasting TEM image after FFT filtering of Nb_4AlC_3 [3].

3. Hot pressing

3.1 Synthesis procedure

Commercial powders of niobium (99%, -200 mesh), aluminum (99%, -300 mesh), and graphite (99%, -200 mesh) were used as starting materials. Firstly, the molar ratio of Nb : Al : C = 4 : 1.3 : 2.7 was selected for investigating the reaction path of Nb_4AlC_3. Excess Al and less graphite were used because Al might lose during high temperature processing and C-deficient existed in most of Al-containing MAX phases. The powders were dryly mixed in a resin jar, ball milled for 12 hours, and then sieved. The mixed powders were uniaxially pressed at 5 MPa to form the green compacts in a BN-coated graphite die. Afterwards, the green compacts were heated to 1500, 1550, 1600, 1650, and 1700°C, with a heating rate of 15°C/min in a flowing Ar atmosphere. The samples were held at target temperatures for 60 minutes under a pressure of 5 MPa, and then cooled down to room temperature with the furnace cooling rate. After composition optimization, single-phase dense Nb_4AlC_3 was prepared using the starting materials with the molar ratio of Nb : Al : C = 4 : 1.1 : 2.7. The green compact was held at 1700°C for 60 minutes under a pressure of 30 MPa.

Temperature	Phase compositions
1500°C	NbC, Nb_2AlC, Nb_4AlC_3, C, Nb_2Al, Al_3Nb, Nb_3Al_2C
1550°C	NbC, Nb_2AlC, Nb_4AlC_3, Al_3Nb
1600°C	NbC, Nb_2AlC, Nb_4AlC_3, Al_3Nb
1650°C	Nb_2AlC, Nb_4AlC_3, Al_3Nb
1700°C	Nb_4AlC_3, Al_3Nb

Table 1. Phase compositions of the samples sintered at the temperatures range from 1500 to 1700°C [4].

Figure 2 shows the X-ray diffraction patterns of the samples sintered at 1500-1700°C using initial powders with the molar ratio of Nb : Al : C = 4 : 1.3 : 2.7. The identified phase compositions of the samples were listed in Table 1. At 1500°C, the phases in the sample were NbC, Nb_2AlC, Nb_4AlC_3, C, Nb_2Al, Al_3Nb, and Nb_3Al_2C (Fig. 2(a)). As the temperature was raised to 1550°C, only NbC, Nb_2AlC, Nb_4AlC_3, and Al_3Nb were detected in the sample (Fig. 2(b)). C, Nb_2Al, and Nb_3Al_2C were completely consumed. The formation of Nb_2AlC was probably associated with the reactions in equations (1) and (2):

$$Nb_2Al + C = Nb_2AlC \tag{1}$$

$$Nb_3Al_2C + NbC = 2Nb_2AlC \tag{2}$$

When the temperature increased to 1600°C, the amount of Nb_4AlC_3 increased with the consumption of Nb_2AlC and NbC (Fig. 2(c)). Possibly, the reaction occurred as following:

$$Nb_2AlC + 2NbC = Nb_4AlC_3 \tag{3}$$

When the temperature reached 1650°C, the diffraction peaks of NbC disappeared. The main crystalline phase was Nb_4AlC_3, together with small quantities of Nb_2AlC and Al_3Nb (Fig. 2(d)). When a higher temperature of 1700°C was used, the final sample contained only Nb_4AlC_3 and Al_3Nb (Fig. 2(e)). All diffraction peaks of Nb_2AlC disappeared. The decomposition reaction could be described as:

$$9Nb_2AlC = 3Nb_4AlC_3 + 2Al_3Nb + 4Nb \tag{4}$$

Figure 3 shows the X-ray diffraction pattern of single phase Nb_4AlC_3. All the diffraction peaks corresponded to Nb_4AlC_3. The crystal structure of Nb_4AlC_3 prepared by the present method was Ti_4AlN_3-type. No impurity phases were detected.

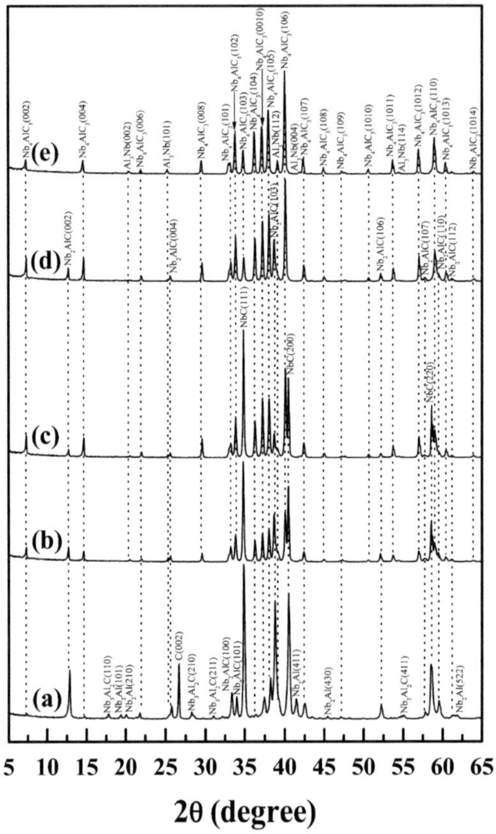

Fig. 2. X-ray diffraction (XRD) patterns of initial powders with the molar ratio of Nb : Al : C = 4 : 1.3 : 2.7 sintered at (a) 1500°C, (b) 1550°C, (c) 1600°C, (d) 1650°C, and (e) 1700°C [4].

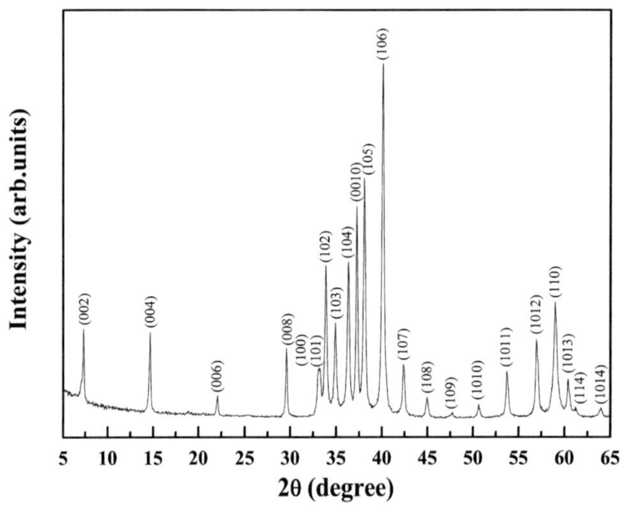

Fig. 3. XRD pattern of Nb_4AlC_3 prepared using initial powders with the molar ratio of Nb : Al : C = 4 : 1.1 : 2.7 [4].

3.2 Microstructure

The etched surface of Nb_4AlC_3 was shown in Fig. 4. Plate-like Nb_4AlC_3 grains distributed irregularly with a few equiaxed grains. The average grain size of Nb_4AlC_3 was 50 μm in length and 17 μm in width.

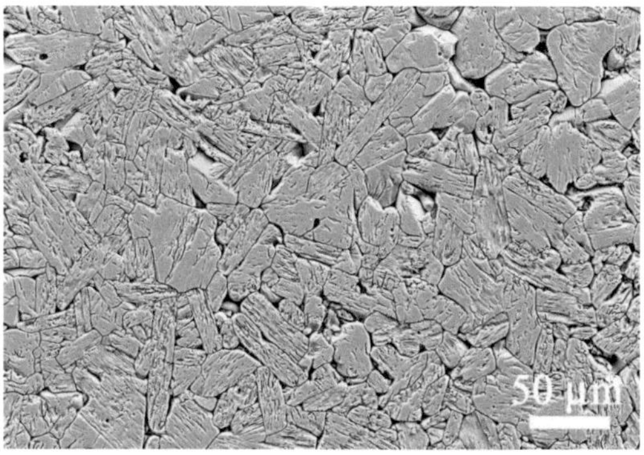

Fig. 4. Scanning electron microscope (SEM) micrograph of etched surface of Nb_4AlC_3 [4].

3.3 Properties evaluation

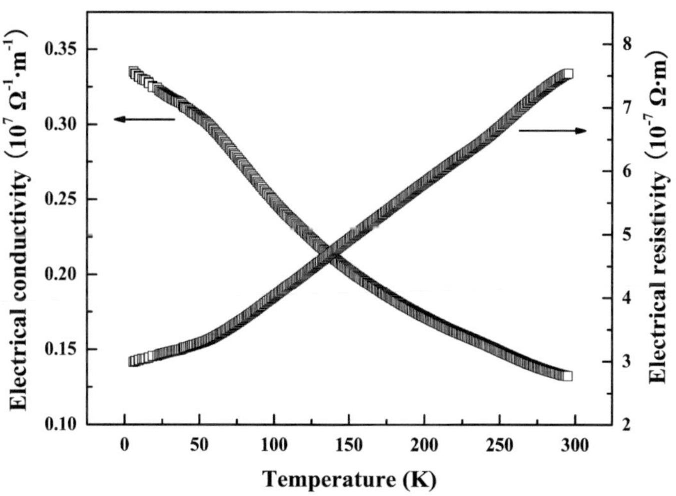

Fig. 5. Temperature dependences of electrical conductivity and electrical resistivity of Nb_4AlC_3 [4].

Figure 5 shows the electrical conductivity and electrical resistivity of Nb_4AlC_3 in the temperature range of 5-300 K. With the increasing temperature, the electrical conductivity of Nb_4AlC_3 decreased from 3.35×10^6 Ω^{-1} m^{-1} to 1.33×10^6 Ω^{-1} m^{-1}. The electrical resistivity of Nb_4AlC_3 increased linearly above 50 K, indicating a metallic characteristic. Fitting the resistivity in the temperature range from 50 to 300 K, the temperature dependence of electrical resistivity was obtained with a coefficient of determination, r^2, of 0.99:

$$\rho(\mu\Omega \cdot m) = \rho_0(1-\beta\Delta T) = 7.133[1-0.0025(273.15-T)] \quad (5)$$

where ρ_0 was the electrical resistivity at 273.15 K (μΩ·m), T the absolute temperature (K), and β the temperature coefficient of resistivity (K^{-1}). The temperature coefficient of resistivity was 0.0025 K^{-1}.

The thermal expansion coefficient of Nb_4AlC_3 was measured as 7.2 × 10^{-6} K^{-1} in the temperature range of 200-1100°C. Figure 6 shows the temperature dependences of molar heat capacity and thermal conductivity of Nb_4AlC_3. The molar heat capacity increased linearly with increment of temperature, which fitted a third-order polynomial. The molar heat capacity of Nb_4AlC_3 approached to a plateau above 1227°C. At room temperature, the molar heat capacity of Nb_4AlC_3 was determined as 158 J·(mol·K)$^{-1}$. A least square fitting the temperature dependence of thermal conductivity for Nb_4AlC_3 was described as:

$$\lambda = 11.6 + 0.0064T \qquad (6)$$

with r^2 of 0.99. At 25°C, the thermal conductivity of Nb_4AlC_3 was deduced to 13.5 W·(m·K)$^{-1}$. Up to 1227°C, the thermal conductivity of Nb_4AlC_3 increased to 21.2 W·(m·K)$^{-1}$. The total thermal conductivity was associated with both the electron and phonon contributions:

$$\lambda_{total} = \lambda_{electron} + \lambda_{phonon} \qquad (7)$$

in which $\lambda_{electron} = L_0\sigma T$ (Wiedmann-Franz Law), where σ was the electrical conductivity at the selected temperature T, and $L_0 = 2.45\times10^{-8} W\cdot\Omega\cdot K^{-2}$. At room temperature, the calculated result was 9.6 W·(m·K)$^{-1}$, about 71% of total thermal conductivity. Therefore, the electrons mainly contributed to the conductivity at 25°C.

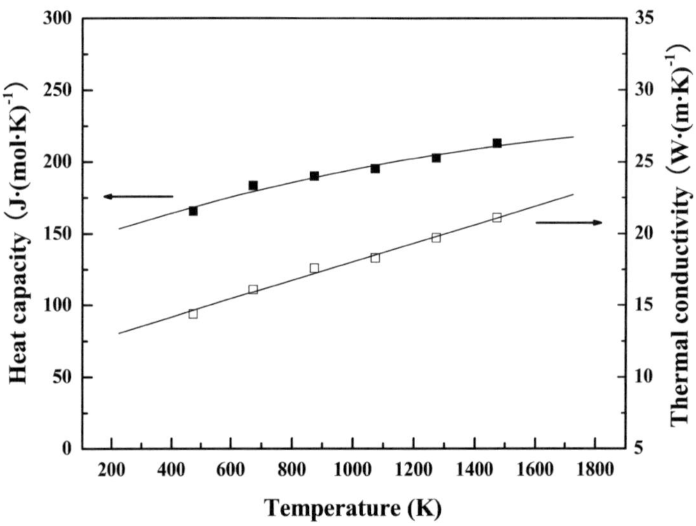

Fig. 6. Temperature dependences of molar heat capacity and thermal conductivity of Nb_4AlC_3 [4].

Figure 7 shows the indentation load dependence of Vickers hardness of Nb_4AlC_3. The insets were the Vickers indentation on the polished surface of Nb_4AlC_3 under a load of 10 N. With increasing load from 3 to 200 N, the hardness gradually decreased from 6.2 to 2.6 GPa. At a lower load, the bigger scatter was seen due to the anisotropic nature of grains. Above 50 N, the hardness value converged to 2.6 GPa. Therefore, the intrinsic hardness of Nb_4AlC_3 was 2.6 GPa. The morphology of the indent produced by a load of 10 N showed that no cracks initiated and propagated from the diagonals, and the material was pushed out around the indent (Fig. 7(a)). The microscale plasticity was associated with infragrain multiple basal-plane slips between microlamellae, intergrain sliding, lamellae or grain push out, and microfailures at the ends of the constrained shear-slips (Fig. 7(b)). In addition, the zigzag crack propagation was observed in an individual Nb_4AlC_3 grain (Fig. 7(c)). Additionally, the measured shear strength of Nb_4AlC_3 was 116 MPa. The low shear strength implied good damage tolerance and easy machinability of Nb_4AlC_3.

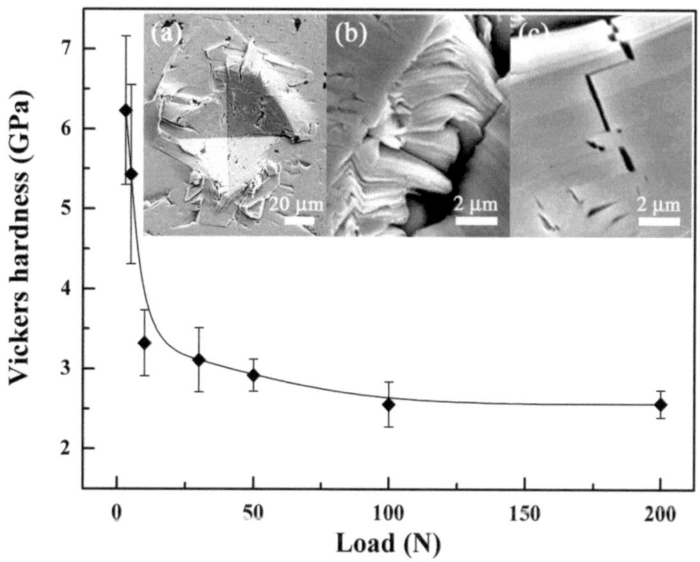

Fig. 7. Vickers hardness of Nb_4AlC_3 as a function of indentation loads. The insets show (a) push-out, (b) delamination, kink, and basal plane slips, and (c) zigzag crack propagation in a grain in an indent under a load of 10 N [4].

Figure 8(a) shows the temperature dependences of normalized Young's moduli of Nb_4AlC_3, Nb_2AlC, β-Ta_4AlC_3, and Ta_2AlC. The temperature dependences of mechanical damping, Q^{-1}, of Nb_4AlC_3, Nb_2AlC, β-Ta_4AlC_3, and Ta_2AlC were shown in Fig. 8(b). Below 1400°C, a slight linear decrease of Young's modulus of Nb_4AlC_3 was observed with increasing temperature. Whereas, a break was seen at a temperature between 1400 and 1500°C. Similar turning points were also observed at 1200-1300°C for Nb_2AlC, 800-900°C for β-Ta_4AlC_3, and 800-900°C for Ta_2AlC (Fig. 8(a)). Corresponding to the accelerated decrease of Young's modulus, the mechanical damping of Nb_4AlC_3 also started to increase at 1400°C (Fig. 8(b)). Above 1500°C, more rapid decrease of Young's modulus of Nb_4AlC_3 was observed with increasing temperature. The rapid decrease of Young's modulus started at 1300°C, 900°C, and 900°C, respectively, for Nb_2AlC, β-Ta_4AlC_3, and Ta_2AlC. The higher critical temperature for the rapid decrease of Young's modulus of Nb_4AlC_3 indicated that this material could be used at much higher temperatures. The Young's modulus of Nb_4AlC_3 could retain up to 1580°C with a loss of 21%. The modulus loss of Nb_4AlC_3 was 16% at 1400°C and 21% at 1580°C (Fig.

8(a)). Therefore, the flexural strength of Nb_4AlC_3 might still possess a high value up to 1580°C. Figure 9 displays the temperature dependence of flexural strength for Nb_4AlC_3. Obviously, the flexural strength of Nb_4AlC_3 could retain up to 1400°C without any degradation.

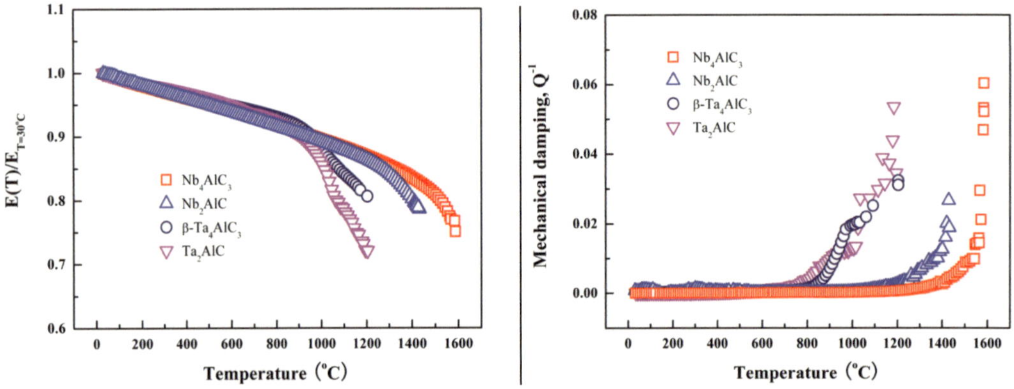

Fig. 8. Temperature dependences of (a) normalized Young's moduli and (b) mechanical damping, Q^{-1}, of Nb_4AlC_3, Nb_2AlC, β-Ta_4AlC_3, and Ta_2AlC [4].

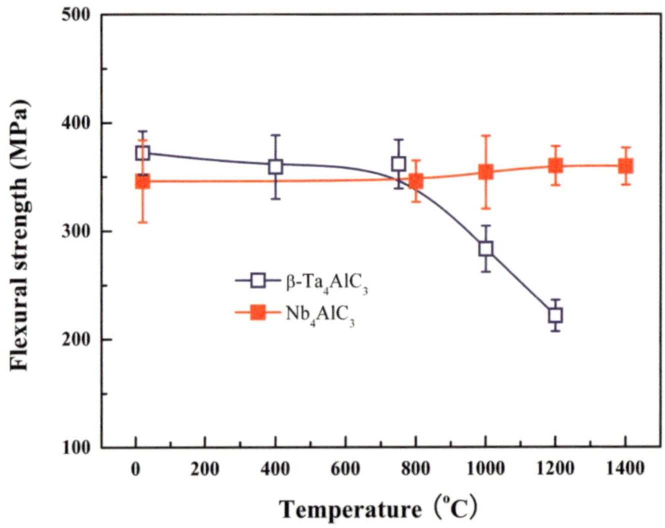

Fig. 9. Temperature dependent flexural strength of Nb_4AlC_3 and β-Ta_4AlC_3 [4].

4. Spark plasma sintering

4.1 Synthesis procedure

Commercial powders of niobium (45 μm, 99.9%), aluminum (30 μm, 99.9%), and carbon black (20 nm, 99%) were used for investigating the synthesis of Nb_4AlC_3 using the SPS technique. For investigating the reaction path, niobium, aluminum, and carbon black powders with a molar ratio of 4 : 1.5 : 2.7 were weighed using an electrical balance with an accuracy of 10^{-2} g. The powders were put into an agate jar and milled for 12 hours using ethanol as the dispersant. After milling, the mixed powders were dried in air and sieved using a 100 mesh sieve. The obtained mixture was put into a graphite die with a diameter of 20 mm. A layer of carbon sheet (~0.2 mm thickness) was put in the inner of die for lubrication. A layer of heat isolation carbon fiber was used to wrap the die for inhibiting the rapid heat diffusion. The mixture was firstly cold pressed as a compact green. Then the green together with the die was heated in a spark plasma sintering facility (100 kN SPS-1050, Syntex Inc., Japan). The sintering temperature was measured by an optical pyrometer focusing on a hole in the wall of die. From ambient temperature to 700°C, it took 5 minutes to heat the sample. Between 700 and 1400°C, a heating speed of 50°C/min was adopted. Above 1400°C, the heating speed was set as 10°C/min. The annealing temperatures were selected as 800, 1000, 1200, 1400, and 1600°C, respectively. The vacuum degree was 7-10 Pa. The holding time was 2 minutes.

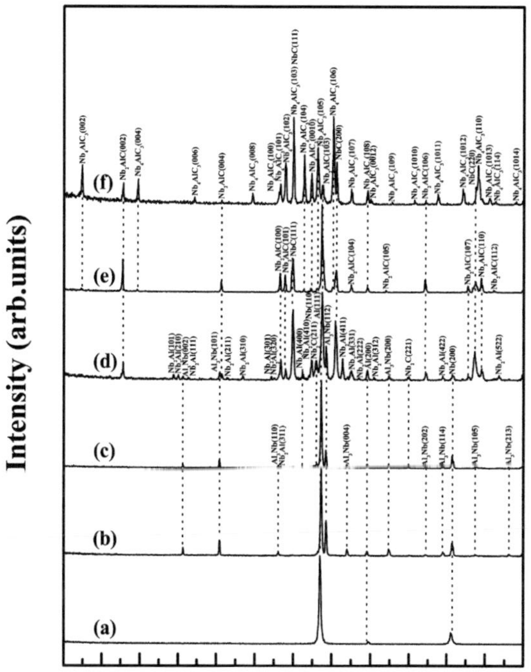

Fig. 10. XRD patterns of samples sintered at different temperatures using Nb, Al, and carbon black mixture powders with the molar ratio of 4 : 1.5 : 2.7: (a) ambient temperature, (b) 800°C, (c) 1000°C, (d) 1200°C, (e) 1400°C, and (f) 1600°C [5].

Figure 10 shows XRD patterns of samples sintered from ambient temperature to 1600°C using Nb, Al, and carbon black mixture powders with the molar ratio of 4 : 1.5 : 2.7. The XRD data of initial mixture powder was shown in Fig. 10(a). Carbon black could not be detected, which might be due to the fine structure. When the temperature was increased to 800°C, Al_3Nb was detected in the sample by XRD (Fig. 10(b)). The melting point of aluminum was 660°C, which meant that the melting aluminum probably combined niobium to form Al_3Nb during the heating process:

$$3Al(l) + Nb(s) \rightarrow Al_3Nb(s) \tag{8}$$

When increasing the temperature up to 1000°C, Nb_2Al and Nb_2C appeared in the sample with the consumption of Al_3Nb (Fig. 10(c)). The reaction could be possibly described as:

$$Al_3Nb(s) + 5Nb(s) \rightarrow 3Nb_2Al(s) \tag{9}$$

Additionally, Nb_2C was formed due to the reaction between Nb and carbon black:

$$2Nb(s) + C(s) \rightarrow Nb_2C(s) \tag{10}$$

As temperature increased to 1200°C, the diffraction analysis showed that new phases of NbC and Nb_2AlC became the main phases in the sintered sample (Fig. 10(d)). The amounts of Nb_2Al and Nb_2C also increased with the consumption of Nb, Al, and Al_3Nb. Probably, the formation of NbC was ascribed to the reaction:

$$Nb(s) + C(s) \rightarrow NbC(s) \tag{11}$$

Nb_2AlC was probably formed due to the reaction between Nb_2Al and carbon black. The reaction equation was described as:

$$Nb_2Al(s) + C(s) \rightarrow Nb_2AlC(s) \tag{12}$$

When the temperature increased up to 1400-1600°C, the existed phases in the samples were only NbC, Nb_2AlC, and Nb_4AlC_3. In hot pressing, it was found that Nb_3Al_2C existed in the sample sintered by hot pressing at 1500°C. Due to the initial composition difference (Nb : Al : C = 4 : 1.3 : 2.7), it might be due to the kinetics of phase formation. At 1400°C, Nb_4AlC_3 was detected in the prepared sample (Fig. 10(e)). Probably, it was ascribed to the following two equations:

$$Nb_2AlC(s) + Nb_2C(s) + C(s) \rightarrow Nb_4AlC_3(s) \tag{13}$$

and

$$Nb_2AlC(s) + 2NbC(s) \rightarrow Nb_4AlC_3(s) \tag{14}$$

It was supposed that carbon black has been completely consumed at this temperature. When the temperature rose to 1600°C, Nb_4AlC_3 became the main phase with the consumption of Nb_2AlC and NbC (Fig. 10(f)).

In order to get single phase Nb_4AlC_3, the sintering temperature was further increased. Figure 11 shows the effect of annealing temperatures on the synthesis of Nb_4AlC_3. When the annealing temperature was 1620°C, only Nb_4AlC_3 and Nb_2AlC could be detected in the sample (Fig. 11(a)). NbC has been completely consumed during the reaction process (Eq. (14)). When increasing the temperature as 1650°C, less Nb_2AlC could be detected in the sample by XRD (Fig. 11(b)). However, Al_3Nb appeared again:

$$9Nb_2AlC(s) \rightarrow 3Nb_4AlC_3(s) + 2Al_3Nb(s) + 4Nb(s) \tag{15}$$

When the annealing temperature was increased up to 1665°C, more Al_3Nb was formed and a small quantity of NbC also appeared in the sample (Fig. 11(c)). NbC was from the decomposition of Nb_4AlC_3 due to the loss of Al. Up to 1680°C, the amount of NbC increased and Al_3Nb disappeared in the sample (Fig. 11(d)). The optimized annealing temperature in present work was 1650°C.

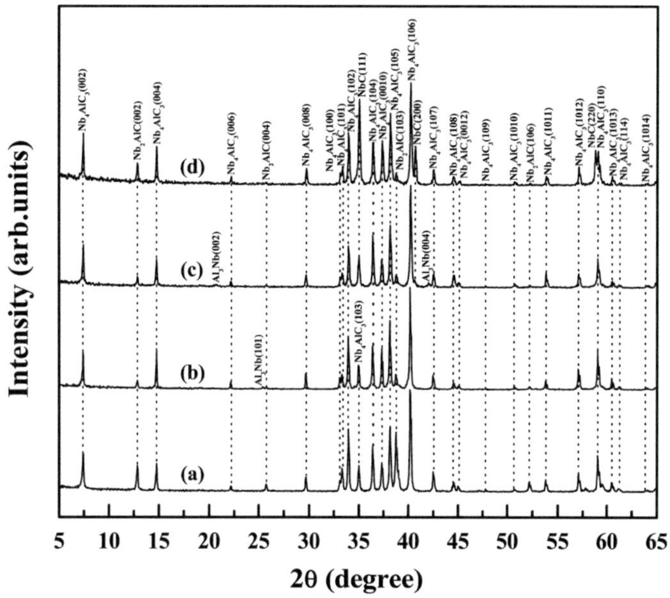

Fig. 11. Effect of annealing temperature on the synthesis of Nb_4AlC_3: (a) 1620°C, (b) 1650°C, (c) 1665°C, and (d) 1680°C [5].

In order to investigate the effect of initial composition on the synthesis of Nb_4AlC_3, the initial mixture powders with different molar ratios of Nb : Al : C = 4 : 1.1 : 2.7, 4 : 1.3 : 2.7, 4 : 1.4 : 2.7, and 4 : 1.5 : 2.7 were selected and sintered at the optimized temperature of 1650°C. Figure 12 shows the XRD patterns of sintered samples. With the increasing Al content in the initial compositions, the amount of NbC in the sample decreased continuously (Figs. 12(a)-(d)). The optimized composition for synthesizing Nb_4AlC_3 by hot pressing in a flowing argon atmosphere was Nb : Al : C = 4 : 1.1 : 2.7. Because of the high vacuum level in SPS furnace (7-10 Pa), Al element was easier to evaporate at high temperature. Therefore, more Al element was added into the compositions. When the initial composition of Nb : Al : C = 4 : 1.5 : 2.7 was used for preparing Nb_4AlC_3, there were only a small quantity of Nb_2AlC and a trace of Al_3Nb existing in the sample. Therefore, the optimized composition was selected as 4 : 1.5 : 2.7. Additionally, it was hoped to eliminate the impurities of Nb_2AlC and Al_3Nb by modifying the holding time. However, when prolonging the holding time from 2 to 4 minutes, though Al_3Nb has disappeared, a plenty of NbC appeared in the sample and Nb_2AlC couldn't be removed, as shown in Fig. 13. Therefore, the optimized holding time in present work was 2 minutes.

Based on the above investigations, the optimized parameters were used to synthesize dense bulk Nb_4AlC_3. Figure 14 shows the X-ray diffraction pattern of as-prepared Nb_4AlC_3. The primary phase was Nb_4AlC_3 and a few amount of Nb_2AlC and Al_3Nb existed in the sample. The impurities of Nb_2AlC and Al_3Nb were less than 6 wt%.

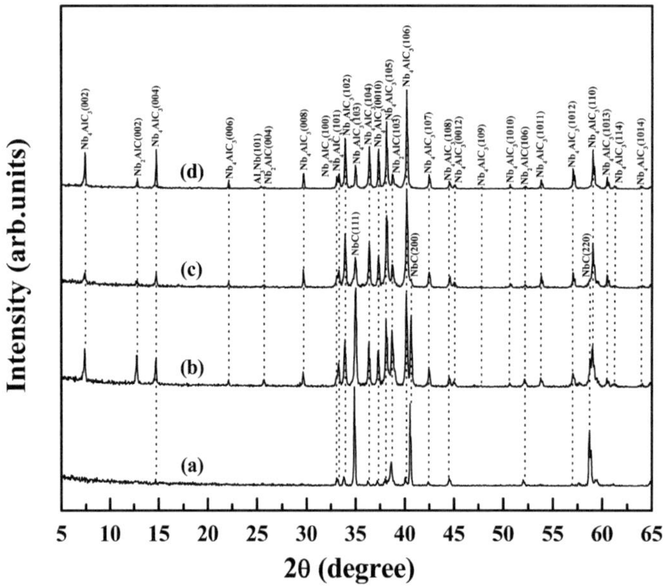

Fig. 12. Effect of initial element compositions (molar ratio of Nb, Al, and C) on the synthesis of Nb_4AlC_3: (a) 4 : 1.1 : 2.7, (b) 4 : 1.3 : 2.7, (c) 4 : 1.4 : 2.7, and (d) 4 : 1.5 : 2.7 [5].

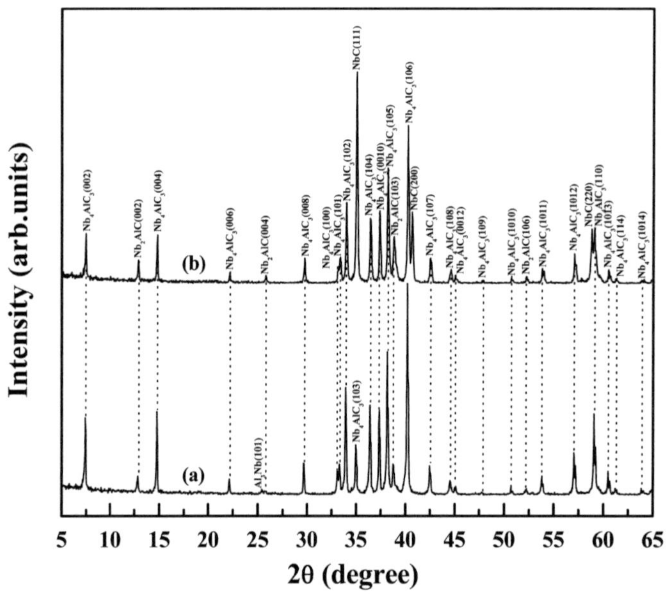

Fig. 13. Effect of holding time on the synthesis of Nb_4AlC_3: (a) 2 minutes and (b) 4 minutes [5].

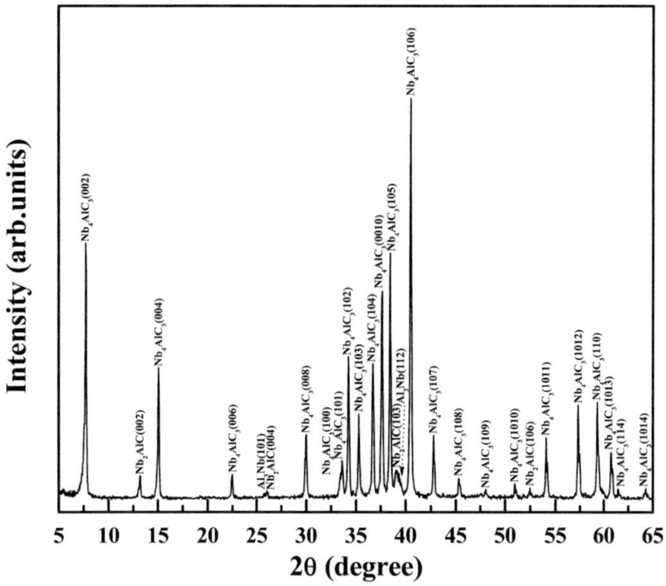

Fig. 14. XRD pattern of dense Nb_4AlC_3 synthesized using the optimized parameters under a pressure of 30 MPa [5].

4.2 Microstructure

Figure 15 shows the etched surface and fracture surface of Nb_4AlC_3. Laminar grains could be clearly observed in the etched surface. The growth of grain did not show the preferable direction, i.e., textured microstructure. The mean grain size was determined as 21 μm in length and 9 μm in width. In the fracture surface, Nb_4AlC_3 grains exhibited the multiplex damage modes, such as transgranular fracture, intergranular fracture, kink bands, and delaminations.

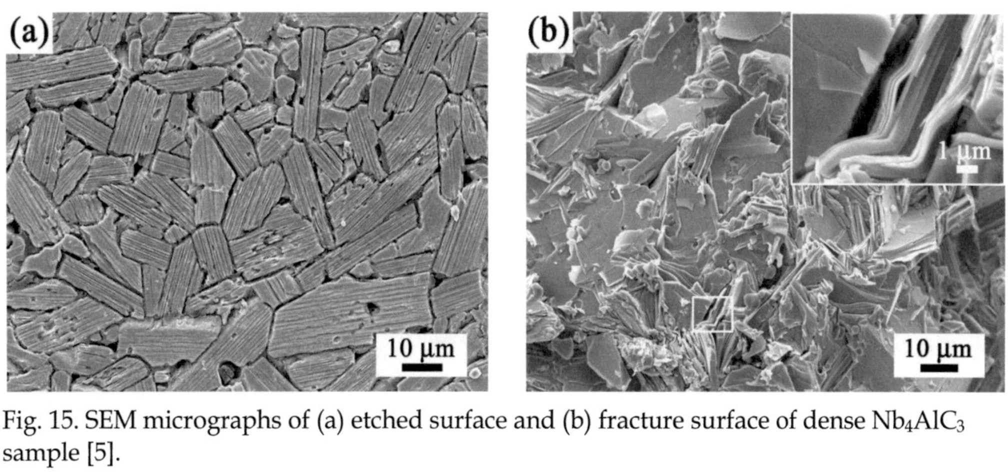

Fig. 15. SEM micrographs of (a) etched surface and (b) fracture surface of dense Nb_4AlC_3 sample [5].

4.3 Properties evaluation

The thermal expansion and technical thermal expansion coefficient of Nb_4AlC_3 sample were shown in Fig. 16. With increasing temperature, the thermal expansion of Nb_4AlC_3 showed the linear change. Fitting the thermal expansion in the temperature range from -128 to 282°C, the temperature dependence of thermal expansion was obtained with a coefficient of determination, r^2, of 0.99:

$$\frac{\Delta L}{L_0} = -0.18204 + 6.5483 \times 10^{-4} T \qquad (16)$$

in which ΔL (m) was the length change at temperature T (K), and L_0 (m) was the length of sample at 5°C (initial room temperature). The technical thermal expansion coefficient, $\alpha_{tech.}$ (K^{-1}), was defined as:

$$\alpha_{tech.} = \frac{1}{L_{145K}} \frac{\Delta L_K - \Delta L_{145K}}{T_K - 145} \qquad (17)$$

in which L_{145K} (m) was the length of sample at -128°C, ΔL_{145K} (m) the length change at -128°C, and ΔL_K (m) the length change at temperature T_K (K). The calculated thermal expansion coefficient at 282°C was 6.7×10^{-6} K^{-1}. The technical thermal expansion coefficient of Nb_4AlC_3 increased from -128 to about -73°C rapidly, which might be attributed to the nonlinear increase of instrument temperature. Above -73°C, the technical thermal expansion coefficient approached a constant.

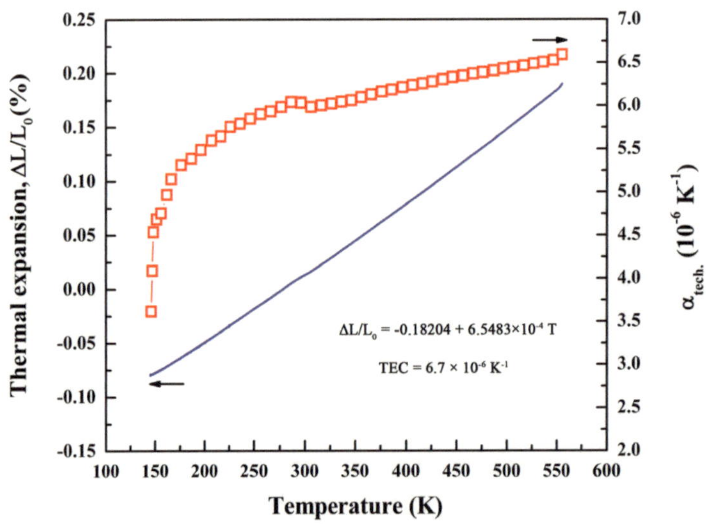

Fig. 16. Temperature dependence of thermal expansion and technical thermal expansion coefficient of Nb_4AlC_3 in a temperature range of -128 and 282°C [5].

Figure 17 shows the temperature dependent electrical conductivity and electrical resistivity of Nb_4AlC_3 in a temperature range of 25-827°C. With increasing temperature, the electrical conductivity decreased corresponding to the increase of electrical resistivity. The measured electrical conductivity of Nb_4AlC_3 at 25°C was $2.25 \times 10^6\ \Omega^{-1}\cdot m^{-1}$, higher than that of hot pressed sample ($1.33 \times 10^6\ \Omega^{-1}\cdot m^{-1}$), which might be due to the existence of Nb_2AlC ($3.45 \times 10^6\ \Omega^{-1}\cdot m^{-1}$) and Al_3Nb. The electrical resistivity of Nb_4AlC_3 increased with a linear rule below 300°C. Fitting the electrical resistivity in the temperature range of 25-300°C, the temperature dependent resistivity could be obtained with a determination coefficient of 0.99:

$$\rho(\mu\Omega \cdot m) = \rho_0(1-\beta\Delta T) = 0.44371[1-0.003048(298-T)] \tag{18}$$

in which ρ_0 was the electrical resistivity at 298 K ($\mu\Omega\cdot m$), T the absolute temperature (K), and β the temperature coefficient of resistivity (K^{-1}). However, the temperature dependent electrical resistivity showed the nonlinear increase above 300°C with the increment of temperature. At 827°C, Nb_4AlC_3 still had a high electrical conductivity of $0.76 \times 10^6\ \Omega^{-1}\cdot m^{-1}$, indicating the excellent high temperature conductive capability.

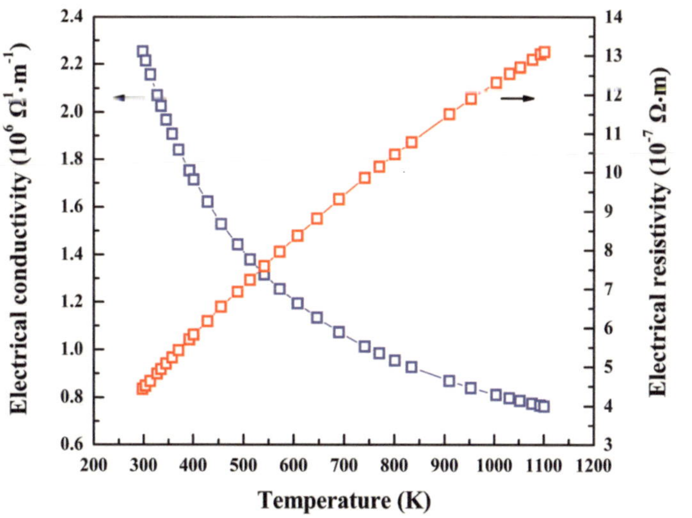

Fig. 17. Temperature dependence of electrical conductivity of Nb_4AlC_3 in a temperature range of ambient temperature and 827°C [5].

The measured Vickers hardness of as-prepared Nb_4AlC_3 was 3.7 GPa, which was close to the value of hot pressed Nb_4AlC_3. Figure 18 displays the three cycles load versus depth-of-microindentation of one Nb_4AlC_3 grain whose basal plane was perpendicular to the surface. Inset was the loops of single cycle indentation on both perpendicular (PE) and parallel (PA) directions, in comparison with that of ZrB_2. The three indentation cycles were all open without reversibility. However, the open scope, i.e., loop area, was decreasing with more cycles, which showed the slight harder behavior. Additionally, the indentation responses were different along PA and PE directions. Obviously, the indentation depth and loop area along PA direction were larger than those along PE direction. It was easier to form kink bands along PA direction because the top surface was unconstrained. Pop-in appeared during the indentation when along PE direction, which was probably due to the delaminations between basal planes. In comparison with the hexagonal ZrB_2, the smaller elastic recovery indicated the more effective energy dispersive capability.

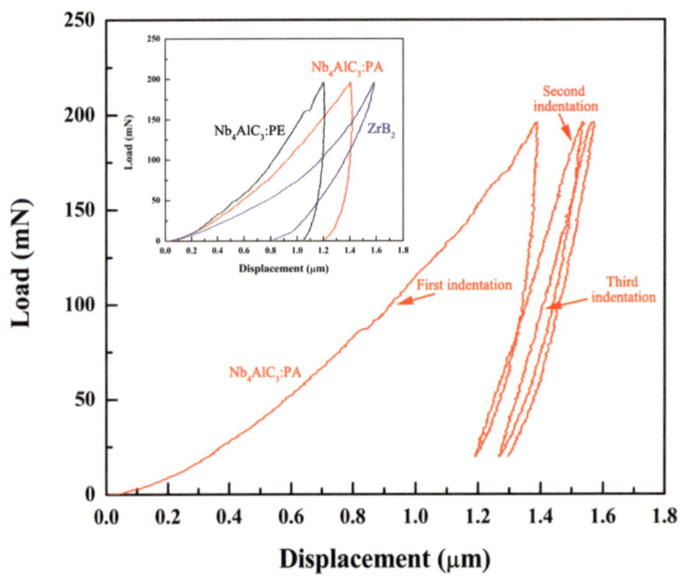

Fig. 18. Typical load vs. depth of indentation response of one Nb_4AlC_3 grain whose basal plane is perpendicular to the surface. Inset is the loops of single indentation on both perpendicular (PE) and parallel (PA) directions, in comparison with that of ZrB_2 (hexagonal structure) [5].

The ambient flexural strength of Nb_4AlC_3 was tested as 455 MPa, higher than that of hot pressed Nb_4AlC_3 (346 MPa), which might be ascribed to the finer grain size. When the samples were tested at 1000 and 1400°C, the flexural strength of Nb_4AlC_3 were 297 and 230 MPa, respectively. The decrease of flexural strength might be ascribed to the existence of Al_3Nb in the samples, which caused the plasticity deformation of bars at high temperatures.

5. Summary

In this chapter, bulk Nb_4AlC_3 ceramic was prepared by an *in situ* reaction/hot pressing method and spark plasma sintering using Nb, Al, and carbon as the starting materials. The reaction path was investigated. Additionally, it was found that when different sintering methods were adopted the final properties of ceramic were different. *Hot pressing*: The thermal expansion coefficient was determined as 7.2×10^{-6} K^{-1} in the temperature range of 200-1100°C. The thermal conductivity of Nb_4AlC_3 increased from 13.5 $W\cdot(m\cdot K)^{-1}$ at room temperature to 21.2 $W\cdot(m\cdot K)^{-1}$ at 1227°C, and the electrical conductivity decreased from 3.35×10^6 to 1.13×10^6 $\Omega^{-1}\cdot m^{-1}$ in a temperature range of 5-300 K. Nb_4AlC_3 possessed a low hardness of 2.6 GPa and high flexural strength of 346 MPa. Most significantly, Nb_4AlC_3 could retain high modulus and strength up to very high temperatures. The Young's modulus at 1580°C was 241 GPa (79% of that at room temperature), and the flexural strength could retain the ambient strength value without any degradation up to the maximum measured temperature of 1400°C. *Spark plasma sintering*: The coefficient of thermal expansion was measured as 6.7×10^{-6} K^{-1} from -128 to 282°C. The electrical conductivity was tested as 0.76×10^6 $\Omega^{-1}\cdot m^{-1}$ at 827°C, showing excellent high temperature conductivity. The Vickers hardness and flexural strength were measured as 3.7 GPa and 455 MPa, respectively. The micro-indentation evaluation indicated the anisotropic response of Nb_4AlC_3 grains, reflecting the anisotropic crystal structure. Additionally, the flexural strength could remain a high value of 230 MPa up to 1400°C.

Acknowledgements: This work is supported by the "Chunlei" program in Ningbo Institute of Material Technology and Engineering in China.

6. References

[1] H. Nowotny, "Struktuchemie Einiger Verbindungen Der Ubergangsmetalle Mit Den Elementen C, Si, Ge, Sn," *Prog. Solid State Chem.*, 2, 27 (1970).

[2] M. W. Barsoum, "The $M_{N+1}AX_N$ Phases: A New Class of Solids; Thermodynamically Stable Nanolaminates," *Prog. Solid State Chem.*, 28, 201-81 (2000).

[3] C. F. Hu, F. Z. Li, J. Zhang, J. M. Wang, J. Y. Wang, and Y. C. Zhou, "Nb_4AlC_3: A New Compound Belonging to the MAX Phases," *Scripta Mater.*, 57, 893-6 (2007).

[4] C. F. Hu, F. Z. Li, L. F. He, M. Y. Liu, J. Zhang, J. M. Wang, Y. W. Bao, J. Y. Wang, and Y. C. Zhou, "*In Situ* Reaction Synthesis, Electrical and Thermal, and Mechanical Properties of Nb_4AlC_3," *J. Am. Ceram. Soc.*, 91, 2258-63 (2008).

[5] C. F. Hu, Y. Sakka, H. Tanaka, T. Nishimura, and S. Grasso, "Low Temperature Thermal Expansion, High Temperature Electrical Conductivity, and Mechanical Properties of Nb_4AlC_3 Ceramic Synthesized by Spark Plasma Sintering," *J. Alloys Compd.*, 487, 675-81 (2009).

8

Effect of Magnetic Pulsed Compaction (MPC) on Sintering Behavior of Materials

Soon-Jik Hong[1], Md. Raihanuzzaman Rumman[1]
and Chang Kyu Rhee[2]
[1]*Kongju National University,*
[2]*Korea Atomic Energy Research Institute*
Republic of Korea

1. Introduction

Sintering is one of the most frequently used processing techniques in the studies of material science that is used to produce density-controlled materials and components from metal or/and ceramic powders by applying thermal energy. Hence, sintering is categorized in the synthesis or processing element among the basic elements of materials science and engineering. As material synthesis and processing have become immensely vital in recent years for materials development, the importance of sintering is increasing as a material processing technology. Sintering is, in fact, one of the oldest technologies, originating in the prehistoric era with the firing of pottery. The production of tools from sponge iron was also made possible by sintering. Nevertheless, it was only after the 1940s that sintering was studied fundamentally and scientifically. Since then, remarkable developments in sintering science have been made. One of the most important and beneficial uses of sintering in the modern era is the fabrication of sintered parts of all kinds, including powder-metallurgical parts and bulk ceramic components (German, 1996).

Sintering mostly aims to produce sintered parts with reproductive and, if possible, designed microstructure through control of sintering variables (Chiang et al., 1996; Green et al., 1989). Microstructural control means the control of grain size, sintered density, and size and distribution of other phases including pores. In most cases, the final goal of microstructural control is to prepare a fully dense body with a fine grain structure.

Now in order to improvise the conditions of sintering, a number of researches have been carried out throughout the world. By sintering conditions, we mean the parameters involved in the process, namely, sintering time, temperature, rate of temperature increase, pressure involved in the process, or even the microstructural behavior in greater sense. This chapter focuses on one of these recently developed techniques called Magnetic Pulse Compaction, which may be used as an additional process on materials, prior to sintering. In special cases, based on commercial requirements, this process can stand alone, without even having the necessity of sintering.

In addition to the detailed description of Magnetic Pulse Compaction, a number of material fields have also been tried to cover where this process can either be used, or has successful contribution compared to other conventional methods. Several experimental results and reports have been discussed, especially on ceramics and hard-to-compress materials. In each case, we have assayed to delineate the differences that this process is introducing, the contributions it has, or the necessity to incorporate this in the chain of processes. Finally, we made some concluding remarks based on real term experimental results about the advantages of this process, which may aid certain streams of material research. Applications of MPC, both as a stand-alone process and prior to sintering have been highlighted throughout the entire chapter.

2. Magnetic pulse compaction

In the commercial sector there is a constant demand for high density, net shape parts that come with an affordable price. At the moment automotive parts, such as powertrain gears for high- performance applications, are typically machined from forged and wrought blanks. Due to high machining costs, these components are much more expensive than conventional press and sintered powder metallurgy (PM) parts. Although PM hot pressing and extrusion techniques can be used effectively to produce bulk materials, these processes are costly due to numerous intermediate steps involved in the process (Kim, 1998; Seaok et al., 2004). Beside, in static compaction methods like cold pressing, hot isostatic pressing etc., the particles often have slow movement with the range of 10^{-3} to 10^{-4} m/s of velocity, resulting in grain growth of nanopowders during the static process (Lee et al., 2004). One of the ways to produce high relative density powder product is magnetic pulse compaction (MPC), which is a kind of dynamic magnetic compaction (DMC) technique. Having superiority to other compaction methods in energy control and forming efficiency, DMC is a kind of high-energy rate method which can be used to produce composite powder, ceramic, metal etc. (Min et al., 2010). While using MPC, it is possible to apply a high pressure (up to 5 GPa) to powders within a short period of time (~500μs). This ultra high pressure in 500μs brings about an enhanced deformation and microstructural rearrangement to these compacted powders, resulting in the improved green density (J.G. Lee et al., 2010).

In the static compaction methods, in general, the particles move slowly with the range of 10^{-3} to 10^{-4} m/s of velocity, whereas in MPC process it is possible to achieve the high speed movement of particle over 10-100m/s owing to very high pressure, more than 1GPa (Lee et al., 2004). During MPC, since effect time of the magnetic force is too short, it can be considered that there is an impulsive force acting in MPC and it is well known that impulsive forces are very high, which are capable of breaking the activation energy of the compacted particle to move them from their places in the bulk.

Magnetic Pulse Compaction (MPC) is a new PM process that can produce components like these at costs comparable to the single-sinter, single- sinter route. In addition, MPC technology is also expected to find use in applications where special shapes, characterized by radially symmetric geometries and/or long length/diameter (L/Ds), are required.

Magnetic forces have been used for more than two decades in high rate metal forming and powder compaction. The same electro-magnetic based pulse forces are used in MPC to realize net shape powder consolidation. The powder-pressing route makes use of transmitted impact energy in a process analogous to driving a nail into a board with a hammer. While almost any material can be compacted to full density using a sufficiently large impact pressure, the important benefits of magnetic powder compaction when compared to die pressing are higher green densities and higher aspect ratios.

The basic working principle of the Magnetic Pulse Compaction process is shown in Figure 1. Powders are filled in a conductive container (armature) placed in the bore of a high field coil. The coil is pulsed with a high current to produce a magnetic field in the bore using the Biot-Savart law, which in turn, induces currents in the armature using Ohm's law.

Biot-Savart Law:

$$\vec{B}(t) = \frac{\mu_0 I(t)}{4\pi} \int \frac{d\vec{l} \times \vec{r}}{r^2} \tag{1}$$

where μ_0 is the permeability of free space, r is the distance from the element to the point *P*, and $\hat{r}$ is a unit vector pointing from dl toward point *P*. We find the total field at *P* by integragrating this expression over the entire current distrubition. Now, Faraday's law of induction states that the emf (ε) induced in a circuit is directly proportional to the time rate of change of magnetic flux through the circuit.

Faraday's Law:

$$\varepsilon = -\iint \frac{d\vec{B}}{dt} dydx \tag{2}$$

Hence, current (I) is induced using the Ohm's Law

Ohm's Law:

$$I(t) = \frac{1}{R}(V - \varepsilon(t)) \tag{3}$$

where *V* is terminal voltage, *R* is resistance of external resistor and ε is the induction magnetic field. The induced currents interact with the applied magnetic field to produce an inwardly acting magnetic force that collapses the tube, thereby compacting the powder, using Lorentz force.

Lorentz Force:

$$\vec{F} = I(t) \int dx \times \vec{B} \tag{4}$$

where I(t) and $\vec{B}$ are induction current and magnetic field, respectively. The armature is launched into the powders with a large kinetic energy within a few microseconds of the compaction cycle. The powders are pressed to full density via the transmitted impact energy, with the entire compaction occurring in less than one millisecond.

The process steps of MPC are similar to conventional PM compaction technique, and include tooling for compaction, powder filling, part extraction and sintering, as well as the optional steps of sizing and finishing. In most commercial applications, the powder filling and compaction are done in air medium at room temperature. The filling can also be carried out in special environments, such as in an inert gas or other cover gases, for special applications. The powders can also be compacted at elevated temperatures with suitable system modifications. The *MAGNEPRESSTM* system used in the DMC process consists of four subsystems - pulsed power system, electromagnetic field coil, material handling capabilities, and programmable logic controller (PLC) systems. The power supply has a modular design and has been designed to accommodate expanding energy needs. High production rates up to ten parts per minute can be achieved with this system. Material handling capabilities, such as powder filling and transportation of filled cassettes, can be easily integrated into the system for a given application.

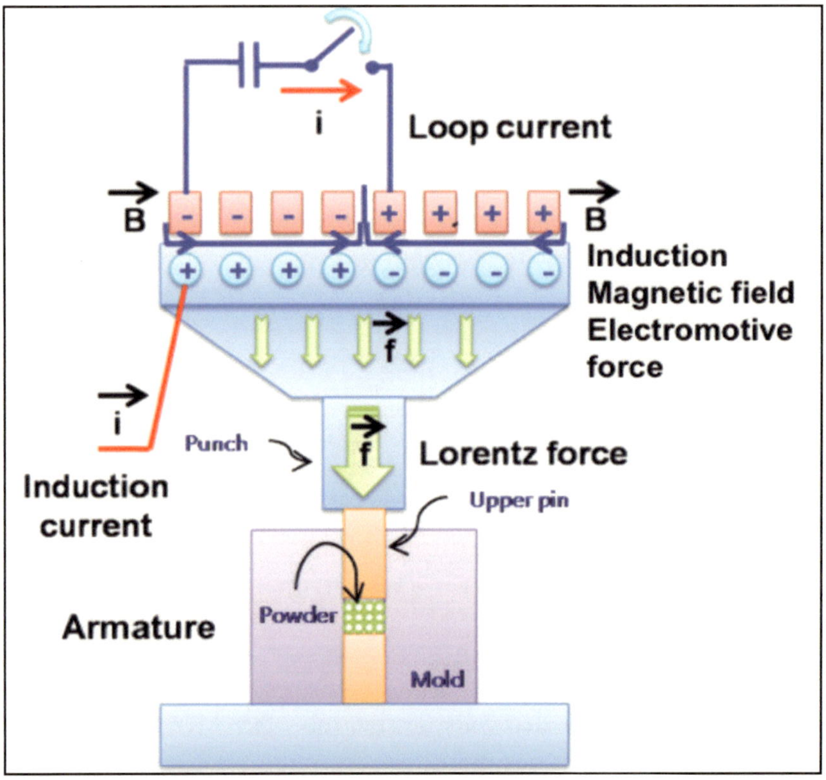

Fig. 1. Working principle of Magnetic Pulse Compaction

3. Effects of MPC on different types of materials

3.1 Ceramics: Effects of MPC on alumina (Al_2O_3)

It is of great importance to know how ceramic nanopowders behave under Magnetic Pulse Compaction, and their consequent effects after sintering. So, a number of experiments were carried out on Al_2O_3 nanopowders, and their microstructural behavior, characteristics and mechanical properties were studied.

In the present investigation, dynamic compaction by magnetic pulsed compaction (MPC) and precompaction were conducted in order to prepare the Al_2O_3 bulks. The starting powder was α-Al_2O_3 (purity of 99.8%) with an average powder particle size of 50~200 nm.

The Al_2O_3 powder was formed into the shape of a disk by magnetic pulsed compaction (MPC). Table 1 demonstrates the compaction arrangements, the apparent features and final density of the bulks fabricated by the combination of precompaction, MPC, and sintering processes. In order to improve the density and properties, the starting powder was precompacted in a die under 110 MPa, 220 MPa, and 330 MPa, and then each precompacted sample without separation from the die was MPC-ed at room temperature. The obtained density of the MPC-ed specimen increased with increasing MPC pressure. The pressure required to consolidate the nanopowders is related to the force required to push the particles together. In order to push two particles together, the applied force must be equal to or greater than the resisting force. The ceramic powders are not expected to plastically deform during compaction like metals. In hard ceramics, a plastic deformation of the particles or the formation of a particle-to-particle contact is so difficult that the stored strain energy by the compaction pressure could not be readily relaxed, which results in the formation of cracks in the materials compacted at high pressure.

Experimental conditions	Density (%)
Uniaxial static compaction (110 MPa) + Sintering (1450 °C for 3 h)	**90.0**
MPC (0.5 GPa) + Sintering (1450 °C for 3 h)	**90.0**
MPC (1.25 GPa) + Sintering (1,450 °C for 3 h)	**92.0**
MPC (1.8 GPa) + Sintering (1,450 °C for 3 h)	**90.0**
Pre-compaction (110 MPa) + MPC (1.25 GPa) + Sintering (1450 °C for 3 h)	**93.0**
Pre-compaction (220 MPa) + MPC (1.25 GPa) + Sintering (1450 °C for 3 h)	**94.5**
Pre-compaction (330 MPa) + MPC (1.25 GPa) + Sintering (1450 °C for 3 h)	**94.5**

Table 1. Consolidation conditions and relative densities of sintered bulks by a combination of MPC and precompaction processing.

In addition, it is clear that density also increases with increasing precompaction pressures from 110 MPa to 220 MPa and then becomes stagnant at 94.5 % for precompaction pressure beyond 330 MPa. The highest density of 94.5 % was achieved in the sample precompacted at 220 MPa, while the density of the sample without precompaction was 92%. The means that precompaction of the powder before MPC improves the final density of the sintered bulk. This may be due to the higher initial packing density introduced by precompaction pressure, as well as the higher MPC pressure. The increased green density may be due to better packing associated with small particles filling the voids between the larger ones. Therefore, the pressure may play a role in the initial stage through particle rearrangement and distribution of agglomerates as well as in the later stages of densification. Additionally, it was reported that a high initial density is not only effective to enhance the subsequent densification during sintering, but also capable of limiting the rapid grain growth (Bernache-Assolant et al., 1993). Therefore, it is necessary to very carefully control the strain energy during the fabrication of ceramic compacts while using high pressures, like in MPC.

Fig. 2 shows the variation in Vickers hardness as a function of MPC pressureand pre-compaction condition during consolidation. With increasing MPC pressure, the hardness of the bulk increases and the value is much higher than that of the uniaxial compaction. In addition, the difference in porosity in terms of size in different regions can be another important factor leading to the variation. The homogeneously distributed ultra-fine grains and higher density of the MPC-ed bulk might be the reasons for the increased hardness. The hardness of the bulk also increases with increasing precompaction pressure. This suggests that particle rearrangement during precompaction occurs at lower pressures.

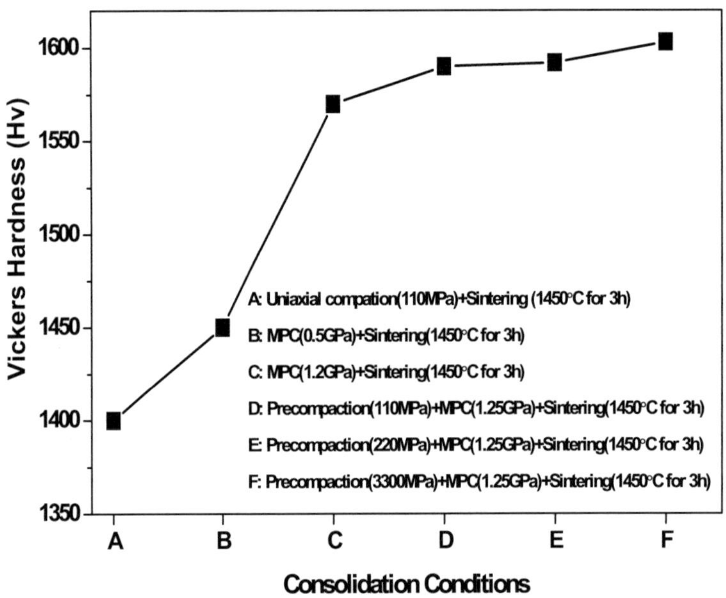

Fig. 2. Changes in Vickers hardness of sintered bulks as a function of different consolidation conditions.

This improved hardness with increasing precompaction pressure can be associated with the enhanced density of the bulk. These results clearly indicate that the precompaction prior to MPC and sintering is an efficient approach to promote higher density and Vickers hardness. Besides, the structural characteristics, which can be controlled by the consolidation and processing, such as porosity, internal stress, etc., may play an important role in determining the properties. However, it is ambiguous to distinguish the contribution of the property enhancement between the grain size effect and other effects such as pores or strains accompanying the consolidated samples.

Fig.3 represents the SEM micrographs (a and b) of the sintered bulks as a function of MPC pressure and precompaction, and TEM micrographs of the uniaxially compacted specimen and MPC-ed specimen (c and d). The microstructure of a commercial Al_2O_3 plate (Fig. 3(a)) shows coarse grains and porous structures along the boundaries. The microstructure of the MPC-ed and sintered bulk exhibits small grains with fine pores. The average grain size measured for 0.5 GPa MPC-ed pressure without precompaction was 0.74μm as shown in Fig. 3(b). On the contrary, the size range of the commercial Al_2O_3 plate was nearly 5.5 μm, which is much higher than the MPC-ed one. Furthermore, the relative density of the sample precompacted at 220 MPa was the highest at 94.5%, and most of the resolved porosity appeared to be isolated at the grain interstices. Moreover, to distinguish the effect of the compaction on the microstructural characteristics between the uniaxial and MPC pressures, specimens were closely examined by TEM.

Fig. 3 (c and d) shows TEM micrographs of the uniaxially compacted specimen and MPC-ed specimen. A larger grain size with an average of 820 nm together with a considerable porous structure within the particles was observed in the case of uniaxial compaction (c) compared to the specimen compacted by MPC. The higher MPC pressure significantly influenced the decreased grain size as evident in figure 3 (d). Besides, the microstructure indicates a homogeneous dispersion of finer grains varying in size from 380 to 840 nm with an average of 600 nm.

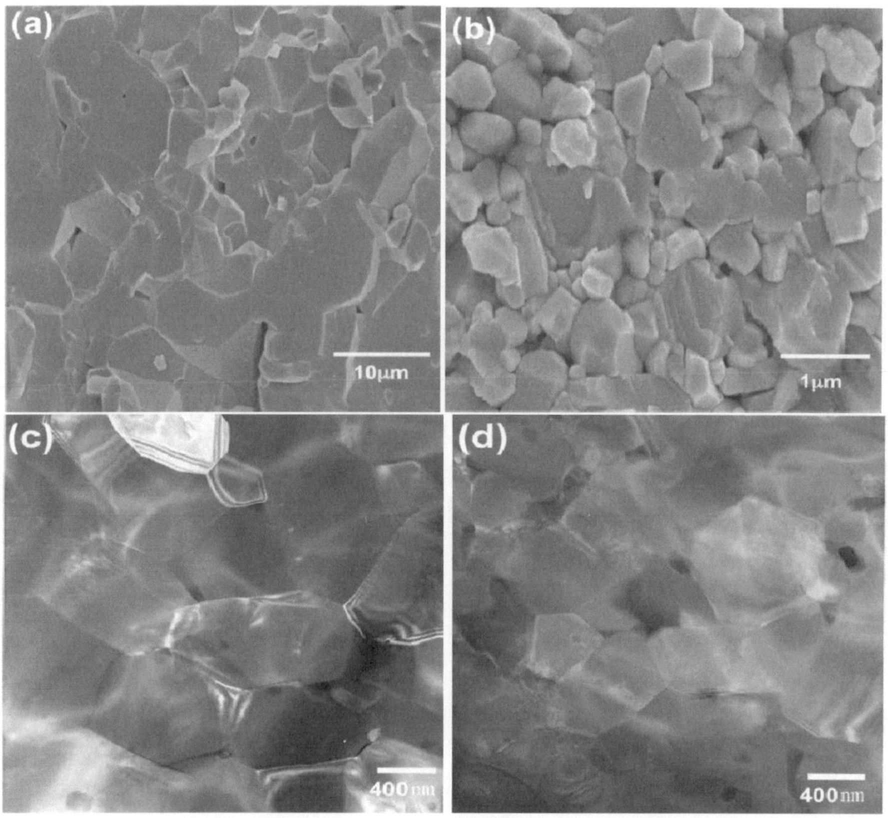

Fig. 3. SEM fracture surface of the sintered bulks as a function of MPC pressure and precompaction process; (a) Commercial Al_2O_3, (b) 0.5 GPa MPC, and bright field transmission electron images taken from (c) Uniaxial compaction (330 MPa) + Sintering (450 °C for 3h) and (d) MPC (1.25 GPa) + Sintering (450 °C for 3h).

The reliability of the consolidated bulk is significant to the generation and propagation behavior of the microcracks, and the indentation test is considered as one of the most effective and convenient methods to determine the fracture of materials. Fig.4 shows representative SEM micrographs of the cracks formed from the corner of the indenter. Despite the secondary cracking, a long crack of 50 μm formed from each corner of the indenter and easily propagated without tortuousness in a spire Al_2O_3 plate as shown in Fig. 4(a). However, the sintered specimens after the MPC show a higher decrease in the crack length (about 40 μm) compared to that of the spire Al_2O_3 plate (an example is shown in figure 4 (b)). In order to identify the fracture mechanism of failure during the breakdown voltage testing, the fracture surface of the bulks was also studied using high magnification SEM micrographs as shown in Fig.4(c and d). Fig. 4(c) displays the fully formed hole and melted grain during testing. However, the observed fracture surface was basically different for the precompacted specimen (Fig.4 (d)). The micrograph shows a shallow hole. The hole formed in the bulk was smaller than that of the bulks without precompaction due to the higher density. The microstructure is coarser as a result of the high temperature during the voltage loading. However, it was difficult to observe any severe cracks, which were easily observed in the samples without precompaction. A fracture analysis demonstrated that bulks with a lower density are more prone to failure than the higher density ones; the lower the density, the greater is the probability to develop cracks and holes in it, and consequently, the lower the load it can withstand before fracture.

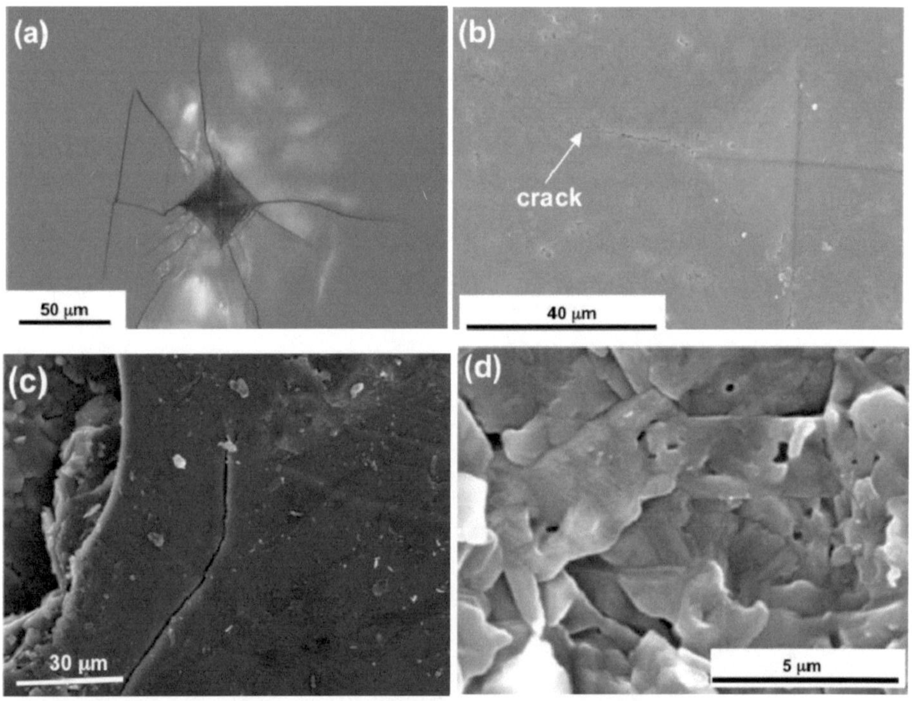

Fig. 4. SEM micrographs of cracks in sintered (a) commercial Sapphire, (b) 0.5 GPa MPC-ed Al_2O_3 bulk, and SEM fracture surface of the sintered bulks after breakdown voltage testing as a function of compaction conditions; (c) commercial, (d) 220 MPa uniaxial precompaction+1.25 GPa.

Another reason is that the increased breakdown voltage in the MPC-ed and precompacted specimens is related to the grain size of sintered bulks. Though the commercial Al_2O_3 showed a high density, a breakdown voltage decreased due to the coarse grain size. The intrinsic defects like the F, F+ center in a single crystalline Al_2O_3 act as heat generation sites during passing the high power wave by a second electron formation. As a result, cracks can be formed at those sites. Alternatively, the grain boundary can absorb the second electron. Therefore, it can prevent the heat generation. In this study, it is expected that the specimen consolidated by the combination of MPC and sintering process showed higher breakdown voltage values than that of the commercial Al_2O_3 due to fine microstructure. Finally, these research results suggest the possible fabrication of the Al_2O_3 bulk showing improved mechanical properties and breakdown voltage due to the ultra-fine microstructure and higher density by the combination of precompaction, MPC, and the sintering process.

It should be noted that the high strength and low dielectric constants of ceramics are attractive for application as a microwave window, but the statistical failure of ceramics severely limits their use. Especially, sapphire has been the best material for a microwave window because of its high strength, absorbed powder and good tolerance to radiation damage. A microwave window includes a transparent disk made of glass or ceramic that lets microwaves through and an attachment device made of metal that is used to attach the transparent disk. Specially processed and mounted sapphire windows have been shown to provide a significant improvement in microwave/radio frequency powder transmission capabilities compared to the current technology.

3.2 Effects of MPC on zirconia (ZrO_2)

Now similar experiments were carried out on ZrO_2. ZrO_2 nanopowders of nanocrystalline polygonal particles with an average size of 50~100 nm were used. In case of specimens where the changes in densification behaviors were studied due to addition of Y_2O_3 as sintering additives, ZrO_2 powders were mixed with Y_2O_3 with a binder (water in this study) to form a solution through mechanical mixing while the rest of the specimens were prepared without Y_2O_3.

Densification experiments were performed on ZrO_2 powders using a die of superalloy 718 with 20 mm internal diameter and 100 mm external diameter, on a punch driven Magnetic Pulsed Compaction machine. Most of the experiments began by cold pressing a powder sample weighing ~10 g by loading up to 0.3~3.2 GPa range and holding the pressure for less than 6 microseconds. The cold compaction pressure was chosen so that a crack-free compact with a known low relative density could be produced. Cold compaction pressures up to 3.2 GPa were used in the set of experiments to study the effect of cold compaction on the hot pressing behavior.

The change in the behavior of the density and shrinkage as a function of varying MPC-ed pressure and 1450°C sintering temperature is delineated in Fig. 5(a & b) for the bulks having standard composition. Fig. 5(a) indicates that both densities of the MPC-ed specimen after first and second sintering increased with increasing MPC pressure from 0.5 GPa to 2.0 GPa, although depending on PVA condition, cracks occurred at times. It was possible to attain a density of 99% on sintered samples after the second sintering. It is due to the fact that high

MPC-ed pressure can break-up the weak agglomerates in the powder during compaction, consecutively making the pores smaller and size distribution narrower (Skandan et al., 1994). The increased green density is due to better packing associated with the smaller particles filling the voids between the bigger ones. The MPC pressure plays a role during the initial stage through particle re-arrangement and distribution of agglomerates as well as during the later stages of densification through plastic or super-plastic deformation. The removal of smaller pores requires much lower sintering temperature and/or shorter sintering period compared to the larger ones. Therefore, a high initial green density is not only effective to enhance subsequent densification during sintering, but also can potentially reduce the duration and sintering temperature significantly (Bruch, 1962; Occhionero & Halloran, 1984). Increasing green density is found to delay the onset of enhanced grain growth during the later stages of sintering (Bruch, 1962).

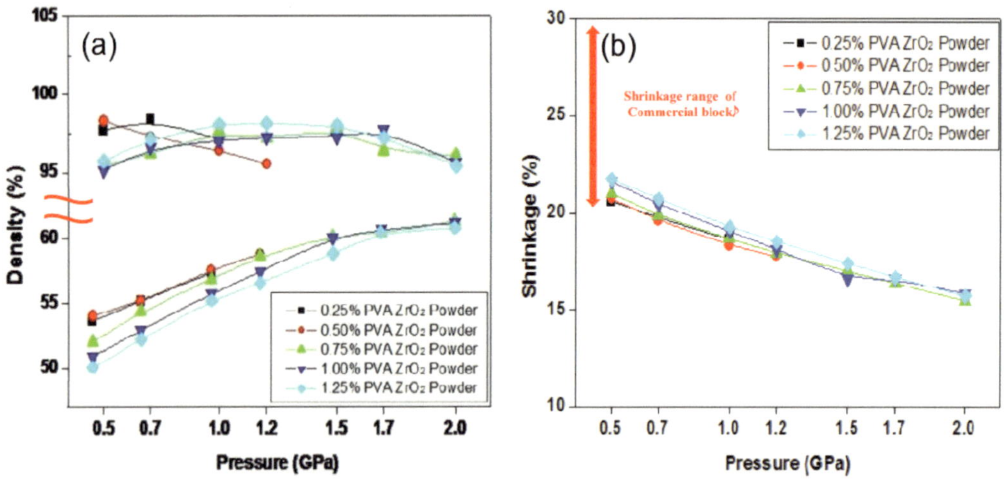

Fig. 5. Variation in (a) relative density and (b) shrinkage, as a function of MPC pressure at 1450° C.

During sintering process, on a microstructural scale, the bonding between particles occurs as cohesive necks grow at the particle contacts. The bonds between contacting particles enlarge and merge as sintering progresses. At each contact, a grain boundary grows to replace the solid-vapor interface, resulting in two particles to coalesce into single particle. Now, while MPC process is being used prior to sintering, particles on a first level, form bonds between contacting particles and initiates particle coalescing to a certain extent, which is then further progressed during sintering, requiring less sintering temperature and time. This interparticle neck growth, with a slight loss of surface area, can occur in powder compacts while improving properties. The densification parameter of the sintering process depends also on the pore elimination process in powder compacts. Now, during this short period of time while MPC process is being executed, pores tend to diminish with increasing pressure until certain limit, which in turn also helps the consequent sintering process to act faster. Apart from the contribution in particle coalescing and pore reduction, the pressure implied during MPC process also helps in shrinking the particles to a much greater extent compared to sintering process alone, given the same time. However, beyond certain extent, this gradually increasing pressure can produce residual stress so high, that it can cause formation of cracks within and along the particles and thus prevents the required neck growth.

Fig. 6 shows an SEM image sequence obtained from consolidated ZrO_2 bulks at a pressure range of 1.7 and 1.9 GPa, and sintered at 1450°C. It can also be delineated that the grain size is decreasing with increasing pressure and has reached a minimum, ranging 2~3 μm at 1.9 GPa pressure, with temperature kept constant at 1450 °C; which is also in good agreement with results reported elsewhere (Li et al., 2006; Hong et al., 2007, Kim et al., 2011). Furthermore, in the sintered bulks, as shown in fig 6(a), pores can be observed although they are not widely dispersed and homogeneously distributed. As pressure tends to increase, pores are filled up and remarkably decreased as shown in fig 6 (c).

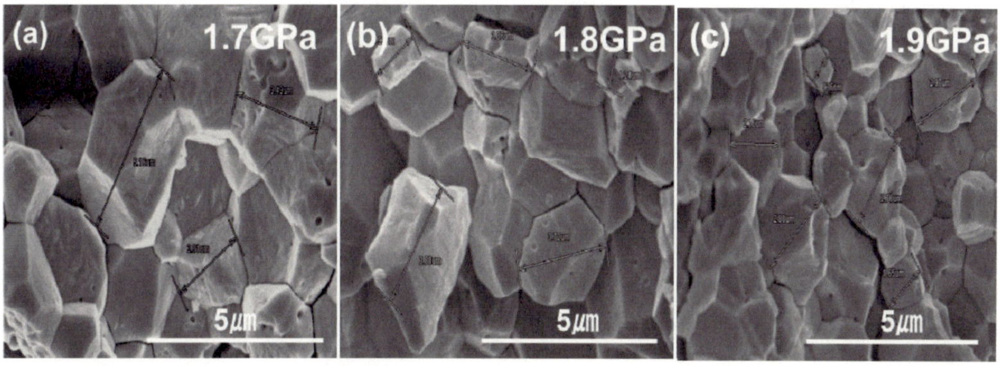

Fig. 6. Microstructures of consolidated bulks with varying pressure a) 1.7 GPa, b) 1.8 GPa, c) 1.9 GPa at 1450 °C.

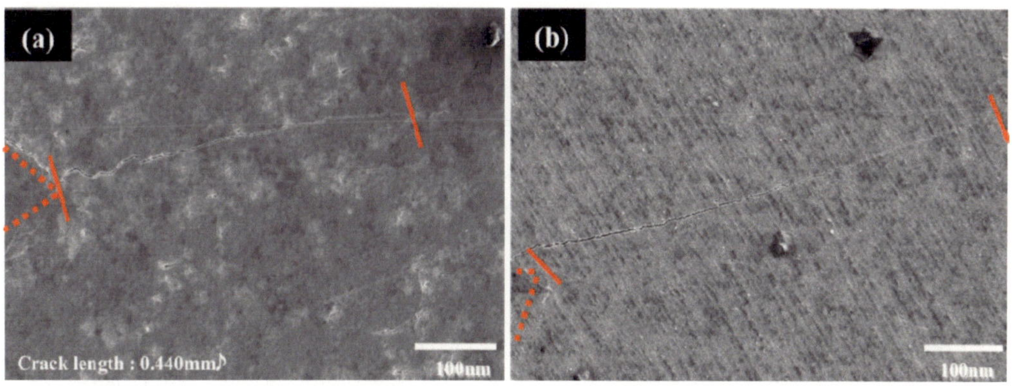

Fig. 7. Comparison of crack length between MPC-ed and commercial bulks (a) MPC-ed block, (b) Commercial block.

The effect of MPC on crack length behavior of sintered ZrO_2 bulk has been compared with commercially available bulks, as shown in Figure 7. It is clearly seen that using MPC on ceramics can result in having smaller crack lengths compared to commercial bulks that haven't gone through this process. This behavior in crack length evidently suggests a resistance against fracture, external force or wear. It also reduces chances of having secondary cracking, or even propagation of crack that may result in failure.

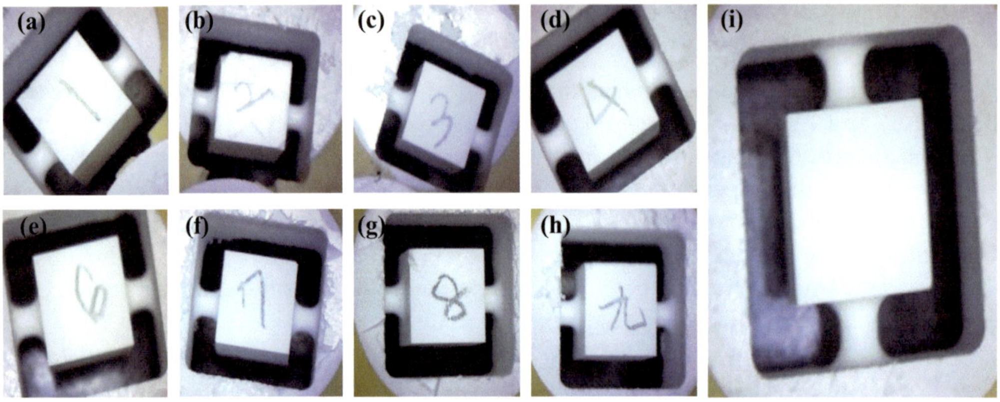

Fig. 8. Formability evaluation of sintered bulks (a) 0.75%PVA, 0.5GPa, (b) 0.75%PVA, 1.2GPa (Edge Crack) (c) 0.75%PVA, 1.5GPa (High Hardness), (d) 1.00%PVA, 0.7GPa (Good Quality) (e) 1.00%PVA, 1.2GPa (Good Quality), (f) 1.25%PVA, 0.7GPa (High Hardness) (g) 1.25%PVA, 1.5GPa (Crack), (h) 1.25%PVA, 1.7GPa (i) 1.00%PVA, 1.0GPa (optimum Quality).

In ceramics, it is also important to know how well or when materials can be machined. The term machinability refers to the ease with which a metal can be machined to an acceptable surface finish. Materials with good machinability require little power to cut, can be cut quickly, easily obtain a good finish, and do not wear the tooling much; such materials are termed as machining free materials. The factors that typically improve a material's performance often degrade its machinability. Therefore, to manufacture components economically, engineers are challenged to find ways to improve machinability without having to compromise with performance.

Machinability can be difficult to predict because machining has so many variables. Two sets of factors are the condition of work materials and the physical properties of work materials. The condition of the work material includes eight factors: microstructure, grain size, heat treatment, chemical composition, fabrication, hardness, yield strength, and tensile strength. Physical properties are those of the individual material groups, such as the modulus of elasticity, thermal conductivity, thermal expansion, and work hardening. Other important factors are operating conditions, cutting tool material and geometry, and the machining process parameters.

Now figure 8 shows the conditions of the sintered bulks, where the machinability as well as the formability of the sintered bulks was evaluated. The formability of most of the bulks was close to perfect. It could also be possible to machine the preforms right after using MPC, which can be highly efficient for industrial purpose.

The MPC process is ideally suited to produce net shape parts with cylindrical symmetry, thin walled tubes, high aspect ratio components and parts with internal features. Shapes and sizes of typical parts produced so far from powders range from 12.7 mm diameter x 76.2 mm long to 127.0 mm diameter x 25.4 mm long. Figure 9 shows some typical MPC-ed ZrO_2 targets (a) and their application as dental blocks (b).

One of the applications being developed is the production of zirconia dental blocks, which are being widely used in dental industries. Special applications of ceramics that go through

Magnetic Pulse Compaction are of great importance, due to the remarkable reduction of shrinkage in sintered bulks. While commercial bulks have considered double step sintering due to machining requirements in ceramics, it was found that simple MPC-ed bulks are capable of going through conventional machining. This could lead to an easier sintering approach altogether, if MPC process is applied in ceramics.

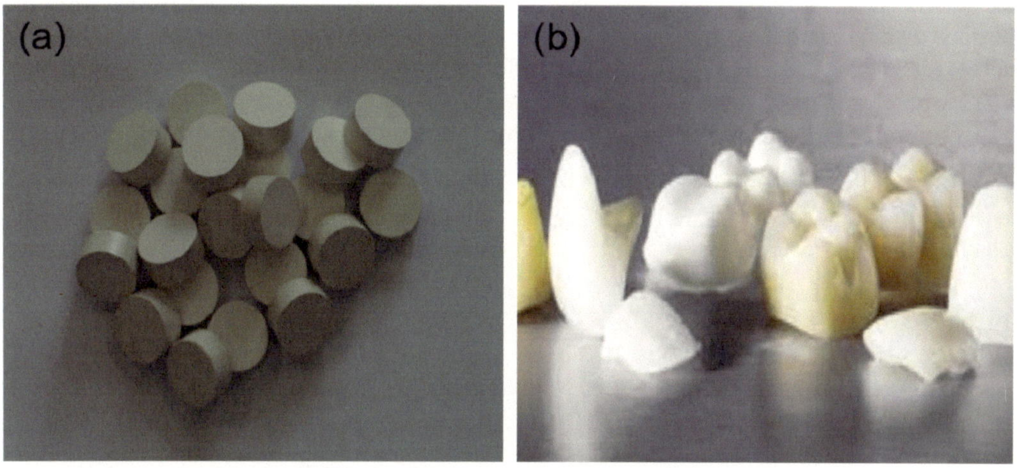

Fig. 9. (a) ZrO_2 metalizing targets of the sintered bulks (1450°C for 2hrs) and
(b) their application as dental blocks compacted by MPC

3.3 Hard materials: Effects of MPC on sintered tungsten carbide-cobalt (WC-Co)

A number of alloy powders, namely ceramics, have gone through the combined MPC and sintering process. At this point, it is necessary to know how hard-to-compress materials react to similar processing conditions. Hard metal is the term used to signify a group of sintered, hard, wear-resisting materials based on the carbides of one or more of the elements; tungsten, tantalum, titanium, molybdenum, niobium and vanadium, bonded with a metal of lower melting point, usually cobalt. Tungsten carbide is however the most widely used one. These materials are commonly referred to as cemented carbides or simply as carbides as, for example, carbide tools. By varying the carbide particle size, the amount of binder metal, and the sintering conditions, the properties such as wear resistance, impact strength, resistance to cratering, and hot hardness may be optimized for a given application.

For example, in the case of a wire drawing die, wear resistance is a major requirement, but for a cutting tool, especially if subject to intermittent loading, high impact strength is required. Figure 10 shows the as-pressed density achieved in WC-Co. In most cases, high initial densities were achieved after MPC, with results getting higher when consecutive pressures were applied. The process parameters, used for some alloys, indicated in Figure 10, are not optimized; and higher densities than shown are possible.

The WC-Co MPC preforms were sintered at conventional PM sintering temperatures and were easily heat treatable. Their densities and hardness after sintering are illustrated in Figure 11 (a) and (b), respectively. It can be seen that hardness values greater than 1450 Hv were achieved in MPC processed WC-Co alloys. Figure 11 (a) illustrates the changes in relative density for both WC-7.5wt%Co and WC-12wt%Co samples after magnetic pulsed compaction and sintering. Before and after sintering, samples of both compositions showed similar pattern with increasing pressure, and successfully reached a maximum of 97% dense state for WC-12wt% sample after sintering, which is higher than most commercial bulks. Here, the conventional explanation of the sintering mechanism of WC-Co comprises of some of the WC dissolving into the cobalt binder phase, migrating and re-precipitating on the surface of the original WC (Schwartzkopf & Kiefer, 1960). The final product consists of a three-dimensional skeleton of WC grains with cobalt as a binder phase matrix (Silva et al., 2001; Cha et al., 2001). In hard alloys such as WC-Co or TiC, even a slight difference in hardness value with respect to their constituents' content, pressure or grain size can mean a change to their mechanical behavior.

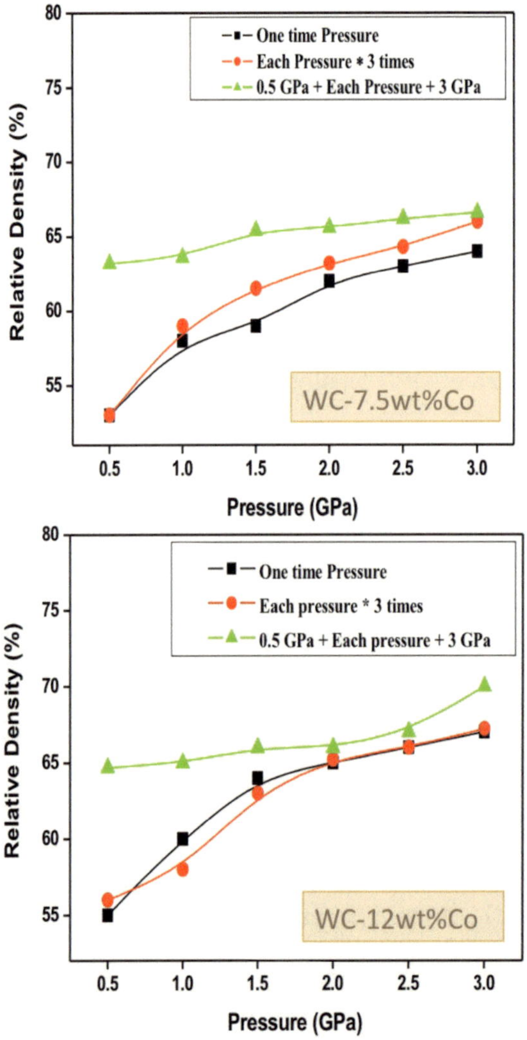

Fig. 10. Changes in initial density with varying MPC-ed pressure on a) WC-7.5wt%Co and b) WC-12wt%Co at room temperature.

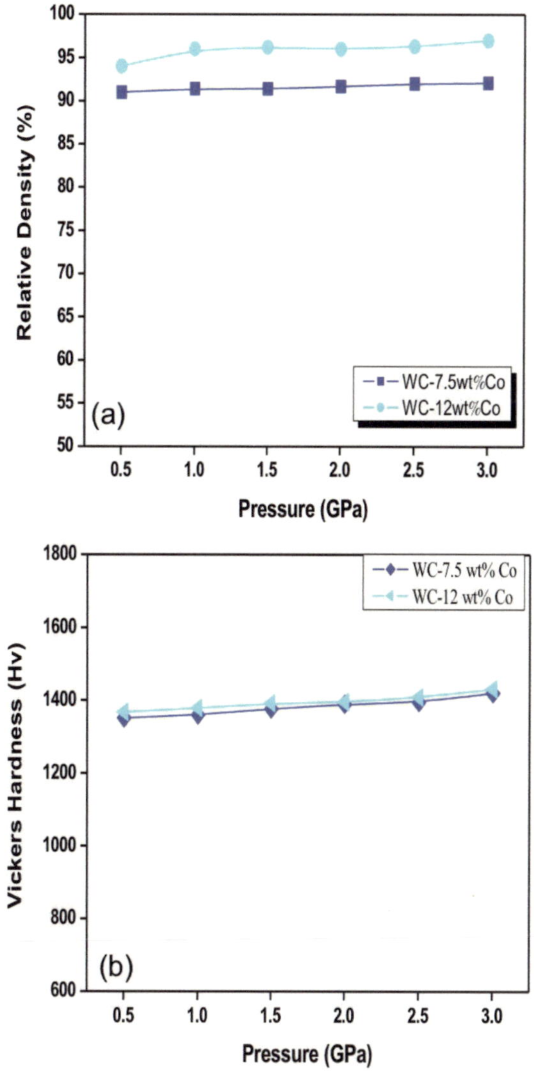

Fig. 11. Changes in (a) relative density, (b) hardness of WC-7.5wt%Co and WC-12wt%Co samples with MPC-ed pressure.

Fig 12 shows the changes in microstructure of sintered samples with varying MPC-ed pressure and Co content. It can easily be comprehended that these fine grained WC got more homogenously distributed in Co binder phase with increased pressure and Co content, at an elevated sintering temperature of 1450ºC. It may have as well resulted in having these samples display a higher relative density and hardness for high magnetic pulsed compaction, especially after sintering, even at a lower operating temperature. Now, it is comparatively easier for finer grained WC material to sinter at lower temperature and become more homogeneous. The first stage in this mechanism is the spreading of binder phase on the WC grains. This happens within minutes after the temperature reaches 1000ºC for fine-grained WC. For coarse-grained WC it happens at ~1100ºC, even though the W-Co-

C system has a ternary eutectic at ~1275°C. The second stage is that Co coated WC particles agglomerate into the Co liquid matrix; this happens faster for the fine-grained WC. The next stage is the forming of a network of agglomerates. The capillary forces close the pores and contract the sample towards an equidimensional shape in competition with the gravitational force, which forces the WC particles to sediment vertically. The last stage is the slight growth of the original WC grains from the tungsten and carbon dissolved in the cobalt.

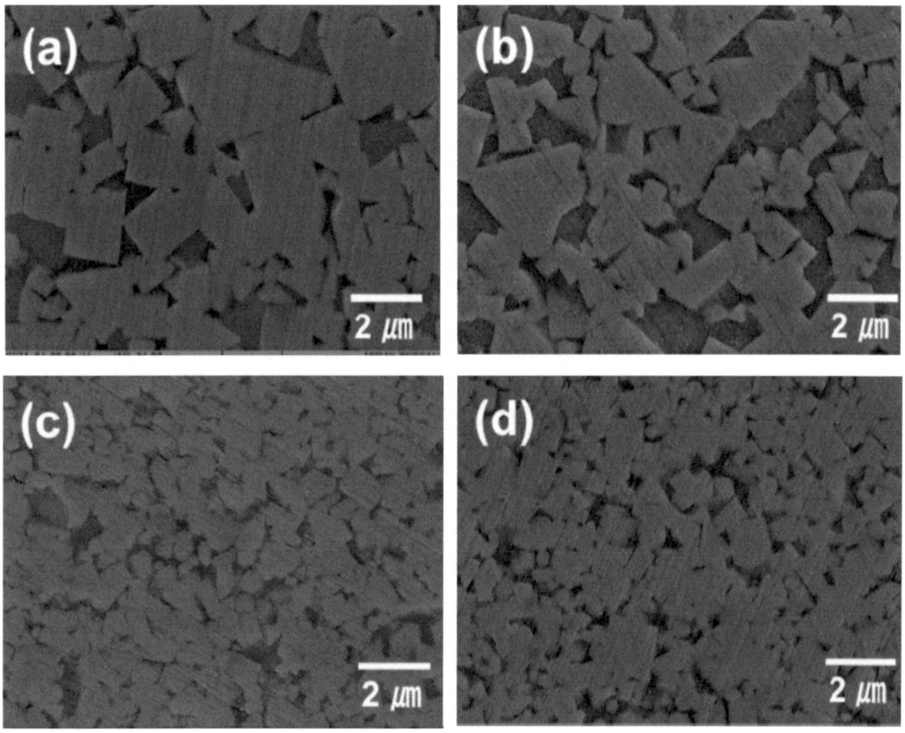

Fig. 12. Changes in microstructure of sintered samples (a) WC-7.5wt%Co at 1 GPa, (b) WC-7.5wt%Co at 3 GPa, (c) WC-12wt%Co at 1 GPa and (d) WC-12wt%Co at 3 GPa.

4. Conclusion

This chapter has mainly focused on the unique advantages of using Magnetic Pulse Compaction as part of commercial material production line. Through most of the observed results, it can be said that this process adds special value to materials and alloys in properties like densification, shrinkage and other microstructural aspects. While MPC is being employed, high initial and final density, and reduced shrinkage have been observed in most of our studies mentioned throughout this chapter. This has resulted in illustrating better microstructural arrangement in many ways. The unique feature of this process also highlights the fact that compaction is performed within fraction of a second with options of varying the pressures. The process can be carried out both at room and elevated temperatures, making it possible to employ this method on a wide range of materials.

A number of applications have already been discussed in this chapter. There are however, more fields of application that MPC can cover. For example, even transmission ring gear of AGMA 9 (American Gear Manufacturers Association) rating for automotive powertrain applications has recently found interest in using MPC. The objectives of these efforts are to replace the cast and machined gear with a lower cost net shape PM gear and to increase the power density through enhanced material properties. In addition, a large number of industrial drilling parts, made of hard materials like WC-Co or WC-Ni-Co, such as drill bits, go through an initial phase of MPC. Initial development efforts have focused on achieving the target density; developing tooling that can survive the dynamic compaction conditions, and producing the desired geometry. With a number of applications now almost ready for commercial implementation, new applications are continuously being sought and developed for the MPC process.

5. References

Green, D.J., Hannink, R.H.J., & Swain, M.V. (1989). *Transformation Toughening of Ceramics.* CRC Press, ISBN 0-8493-6594-5, Boca Raton.

Kim, H.S. (1998). Yield and compaction behavior of rapidly solidified Al–Si alloy powders. *Mater. Sci. Eng. A,* Vol. 251, No. 1-2, (August 1998), pp. (100-105).

Seaok, H.K., Lee, J.C., & Lee, H.I. (2004). Extrusion of spray-formed Al–25Si–X composites and their evaluation. *J. Mater. Proc. Technol.* Vol. 160, (March 2005), pp. (354-360).

Lee, G.H., Rhee C.K., Lee, M.K., Kim, W.W., & Ivanov, V.V. (2004). Nanostructures and mechanical properties of copper compacts prepared by magnetic pulsed compaction method. *Mater. Sci. Eng. A,* Vol. 375-377, (July 2004), pp. (604-608).

Li, M., Yu, H.P., & Li, C.F. (2010). Microstructure and mechanical properties of Ti6Al4V powder compacts prepared by magnetic pulse compaction. *Trans. Nonferrous Met. Soc. China.* Vol. 20, No. 4, (April 2010), pp. (553-558).

Lee, J.G., Lee, M.K., Hong, S.J., Lee, H.W., Pyun, S.P , & Rhee, C.K. (2009). Consolidation of mixed diamond and cobalt granule powders by magnetic pulsed compaction. *Materials Letters.* Vol. 64, No. 1, (January 2010), pp. (35-37).

Bernache-Assolant, D., Soustells, M., Monty, C., Pastor, H., & Chaix J.M. (January 1993). *Chimie Physique du frottage.* Hermes, Paris.

Skandan, G., Hahn, H., Roddy, M., & Cannon, W.R. (1994). Ultrafine-Grained Dense Monoclinic and Tetragonal Zirconia. *J. Am. Ceram. Soc.* Vol. 77, No. 7 (March 1995), pp. (1706-1710).

Bruch, C.A. (1962). Sintering Kinetics for the High Density. Alumina Process. *Am. Ceram. Soc. Bull.* Vol. 41, No. 12, (1962), pp. (799-806).

Occhionero, M.A., & Halloran, J.W. (1984). The influence of green density upon sintering. *Proceedings of the Sixth International Conference on Sintering and Related Phenomena,* ISBN: 0306416662, Notre Dame, 1984.

Li, S., Zheng, W.T., & Jiang, Q. (2006). Size and pressure effects on solid transition temperatures of ZrO_2. *Scripta Materialia.* Vol. 54, No. 12, (June 2006), pp. (2091-2094).

Hong, S.J., Lee, G.H., Rhee, C.K., Kim W.W., & Lee, K.S. (2007). Synthesis and characterization of nanoparticles of ZnO by levitational gas condensation. *J. Mater. Sci. and Eng. A.* Vol. 449-451, (March 2007), pp. (401–406).

Kim J.H., Rumman M.R., Rhee C.K., Lee J.G., Lee M.K. & Hong S.J. (2011). Fabrication and Densification Behavior Analysis of Metalizing Targets Using ZrO2 Nanopowders by Magnetic Pulsed Compaction. *Materials Transactions*, Vol. 52, No. 6 (2011) pp. 1156 to 1162

Schwartzkopf P., & Kiefer, R. (1960). *Cemented Carbides*, Macmillan, New York.

Silva, A.G.P., Schubert, W.D., & Lux, B. The role of the binder phase in the WC-Co sintering. *Mat. Res.* Vol. 4, No. 2, (April 2001), pp. (59- 62), ISSN 1516-1439.

Cha, S.I., Hong, S.H., Ha, G.H., & Kim, B.K. (2001). Microstructure and mechanical properties of nanocrystalline WC-10Co cemented carbides. *Scripta Materialia.* Vol. 44, No. 8-9, (May 2001), pp. (1535–1539).

German, R.M. (January 1996). *Sintering Theory and Practice*, Wiley Interscience, ISBN: 047105786X, New Jersey.

Chiang, Y.M., Birnie, D.P., & Kingery, W.D. (April 1996). *Physical Ceramics: Principles for Ceramic Science and Engineering*. Wiley Interscience, ISBN 0-4715-9873-9.

Part 2

Bio-Ceramics

9

New Challenges in the Sintering of HA/ZrO_2 Composites

Cristina Vasconcelos*
University of Azores/Department of Technological Sciences and Development
Physics Department of FCT-UNL/ Centre of Physics and Technological Research
Portugal

1. Introduction

The search for improved the human living standards and longevity in recent decades has led to a strong development of areas related to life sciences. With increasing life expectancy of the population, the constant appeal for the development of materials to be used clinically in the replacement and regeneration of damaged tissues or organs has also allowed the growth of the multidisciplinary field of biomaterials, which is based on the combination of life sciences with materials science and engineering.

Bioceramics, used initially as alternatives to metals in order to increase the biocompatibility of implants, have become a diverse class of biomaterials, presently including three basic types: bioinert high-strength ceramics, bioactive (or surface reactive) and bioresorbable ones. These are the ceramics, which can be used inside the human body without rejection to replace various diseased or damaged parts of the musculoskeletal system. In the last 50 years, several advances in many specialty bioceramics such as alumina, zirconia (ZrO_2), calcium phosphates and bioactive glasses have made significant contributions to the development of the present health care industry, improving the quality of human life. Recent developments in bioceramics research are, however, focused on bioactive and bioresorbable ceramics, i.e. hydroxyapatite [HA, $Ca_{10}(PO_4)_6(OH)_2$] and calcium phosphates as they exhibit superior biological properties over other materials. However, the great challenge is the reproduction of structures, properties and functionalities of parts of the human skeleton that result from thousands of years of evolution. The major constituent of bone is HA so it attracts major interest for employing in prosthetic applications due to the similarity of its chemical composition and crystallography to those of mineralized bone of human tissues. In addition the formation of chemical bond with the host tissue offers HA a greater advantage in clinical applications over most other bone substitutes. However, in spite of chemical similarities, mechanical performance of synthetic HA is very poor compared to bone. In fact, its poor mechanical strength makes it unsuitable for load-bearing. In most applications of biomedical materials the mechanical properties are especially important, as well as the chemical reactivity of their surfaces. To overcome these limitations and to meet the requirements for the self-load bearing, HA has been incorporated with other compounds such as mullite (Clifford et al., 2001), ZrO_2 (Y.M. Kong et al., 2005; Sung & D.H.

* Corresponding Author

Kim, 2003; W. Li & Gao, 2003), alumina (J. Li et al., 1995), and apatite (Agathopoulos et al., 2003), among others. However, the combination of HA and ZrO_2 have attracted a great attraction for applications to bone tissues exposed under high friction and high impact due to the potential mechanical properties such as high fracture toughness and hardness as well as bioinertness of the ZrO_2 component. A higher strength and fracture toughness can be significantly enhanced either by stress-induced tetragonal (t-ZrO_2) to monoclinic (m-ZrO_2) phase transformation toughening, or by a deflection toughening mechanism. The mechanical behavior is strongly dependent on the structure of the material. Moreover there are basic principles that govern the behavior and properties of a material, which are described as a ratio of it internal structure, processing and their properties. When one aspect of this relationship is changed the others will be affected accordingly.

The processing of materials is another key parameter that will have implications on the final properties of the biomaterial. Many researchers have observed that the mechanical strength and fracture toughness of HA based ceramics can be improved by the use of different sintering techniques which include hot isostatic pressing (HIP), spark plasma sintering (SPS) and microwave sintering. Also a sintering aid based on a low sintering temperature may be helpful if an easy and cost-effective sintering technique has to be used. On the other hand, to minimize the thermal decomposition and improve the densification of the HA/ZrO_2 composites, various atmospheres assisted sintering can be employed. Moreover, the addition of a low melting secondary phase to achieve liquid phase sintering for better densification, incorporation of sintering additives to enhance densification through grain boundary strengthening, and use of nanoscale ceramic powders from sol-gel process for better densification contributed to desirable mechanical and biological properties.

2. Constrained sintering by rigid inclusions: Solid-state sintering of ceramic-composites

In nature the combination of various materials is a constant search for an optimization of properties and functions. By analyzing the different tissues and organs that constitute the human body is possible to verify that they typically result from a combination of different types of biological materials. As an example, natural bone is an ideal nanocomposite, which consists of approximately 30% of matrix material and 70% of nanosized mineral (matrix material- collagen fibers (polymer) plus mineral - HA crystals (ceramic)).

In the development of biomaterials that strategy is also widely used. A variety of materials that result from the combination of two (or more) materials, with composition and structure and different properties that display the higher end of its individual components are composites. One of the most important characteristics of composites is the possibility of modifying the properties with the change of only one of several processing variables such as size, shape, distribution and orientation of the constituents, among others. The principal material of a composite that involves the "reinforcement" is called matrix, with regard to the extent in the body, and the second component (e.g. SiC particles or fibers, ZrO_2 particles, etc.) is usually referred as the filler (or inclusion).

When attempting to join different materials, several factors need to be considered to ensure the integrity of the resulting structure. These include the physical and chemical properties of each component (reactivity, coefficient of thermal expansion, tensile and compression strength, etc.) and the processing conditions (temperature and atmosphere).

One of the most popular paths to the formation of polycrystalline ceramic composites involves powder compaction into a porous body which would receive sintering treating rather than melting or other methods. The reasons for that are two: First, ceramics usually melt at high temperatures, which makes melting difficult and inefficient. Second, ceramics are brittle which is not suitable for processing by thermo-mechanical forming. Sintering commonly refers to the process by which a system of powder particles in contact is consolidated to form dense polycrystalline aggregates through physical and chemical changes when subjected to appropriate temperatures. Almost all aspects of the sintering process have been addressed in many books and publications (Exner, 1980; Exner & Arzt, 1996; German, 1996; Kingery et al., 1976; Kingery & Berg, 1955; Coble & Gupta, 1967).

Sinterization is one of the most important manufacturing and delicate steps that a ceramic body is subject. The most important because it determines, in large part, the final properties of the ceramic, such as microstructure, mechanical properties and final crystalline phases and in the case of bioceramics, their behavior in service. Is one of the more delicate because the achievement of the desired final properties is dependent of the processing parameters such as material properties (e.g. melting temperature, T_m), impurities, temperature and time, initial compact density, initial particle size distribution, and applied pressure. Usually a self-supporting compact is obtained from powders that are compacted uniaxially or isostatically to form a "*green body*", followed by a densification stage involving sintering at appropriate temperatures. So, temperature, time, initial particle size of the material and sintering atmosphere, nature of additives, etc. are the most important factors which can influence the sintering process and the resulting microstructure.

The heat-assisted treatment of sintering usually improves the mechanical strength of the material through the formation of a solid bond between the particles. This process is generally accompanied with densification (i.e. elimination of the pores) and strengthening of the compact, a process driven by interfacial energy (γ). From a macroscopic point of view, during sintering not only dimensional changes of the products occurs, i.e. the compacts usually contract as the material densifies, but also their shape may vary as a consequence of anisotropic sintering contraction, phase transformations, etc. Moreover, anisotropic shrinkage behavior of composite powders can cause formation of coalesced pores and other microstructural changes in sintered compacts. Hence, based on the application and the expected properties of the final product, there are different techniques to control the mechanism and stages of the sintering process.

The driving force to bond individual particles together, during sintering, is the reduction of free energy of surfaces of powder compacts. This can be achieved by the elimination of pores and reduction of grain boundaries via grain growth. Since γ_{sv}(surface free energy of the solid-vapor interface) is normally greater than γ_{ss} (surface free energy of the solid-solid interface), the solid-vapor interfaces tend to be replaced by solid-solid interface when enough energy is provided. Through the driving force, the neck growth between adjacent particles results in a bonded network polycrystalline microstructure with the pores trapped inside, as shown in Fig.1a.

Through further heating of the compact, the residual pores become smaller in size and number, and isolated from each other (Fig.1b). These isolated pores could completely disappear in the final stages of sintering depending on processing parameters.

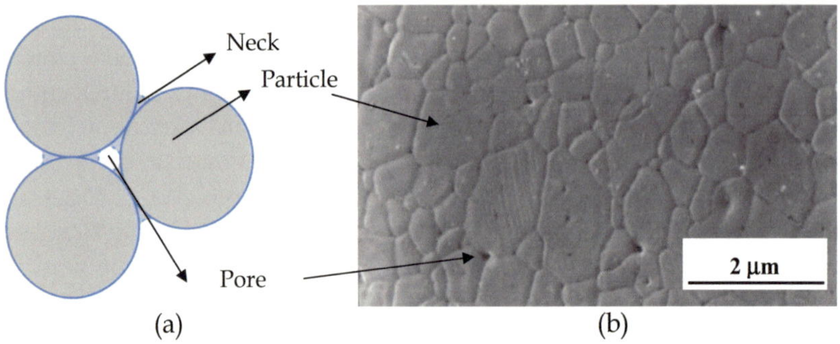

Fig. 1. (a) Sintering model geometry of sphere particles. (b) SEM microstructure of a HA compact, sintered at 1100 °C for 1 h.

During sintering, material moves through mechanisms of viscous flow or diffusion (or both) in order to eliminate porosity and reduce the γ_{sv}. Diffusion is among the most important phenomena observed in sintering of ceramic materials. Diffusive mass transport takes place when there is a gradient in the chemical potential and when the species in question has sufficient mobility. There are a number of competing paths for mass transport during ceramic sintering (Garay, 2010), such as grain boundary diffusion, volume diffusion and surface diffusion. Usually surface diffusion, leads to coarsening, which is the growth of the neck between particles leading to the reduction of the specific surface area. The other transport mechanisms, normally volume diffusion and grain boundary diffusion lead to densification.

Most of ceramic particles, including oxides, need elevated temperatures to be sintered to high levels of densification. Densification of covalent-ionic ceramics without additives is extremely difficult due to the low diffusivity in the solid state following from the nature of bonding in such materials. The sinter temperature is usually at $T > 0.5T_m$ [K]. Hence, the high temperature ensures that the atoms have sufficient mobility to diffuse to the pores and densify ceramic powders. On the other hand, abnormal grains grow due to Ostwald ripening (German, 1996) when traditional sintering techniques are used, can also occur. Temperature, however, is not the only factor that affects the densification process. A high "*green density*", uniform packing, small particle size, spherical particles and narrow size distribution, for narrow-pore size distribution (Yeh & Sacks, 1990) are also important. In the formation of polycrystalline matrix composites, the presence of second-phase inclusions (e.g. particles) leads to a drastic reduction in the matrix densification rate. Thus considerable difficulties are often encountered in the formation of polycrystalline matrix ceramic composites by conventional, pressure less sintering.

Recent developments in bioceramic composites produced a new set of challenges for sintering theory, particularly when differential densification occurs due to constrained sinterization, i.e parts of a structure densify at different rates and temperatures than other parts (Green et al., 2008) because matrix densifies around rigid inclusions. Inclusions and heterogeneities not only reduce the rate of densification, but also cause differential sintering which generates defects that reduce the strength of the final composite due to differential shrinkage. It was shown that the large inclusions can effectively retard the matrix densification and affect the composite microstructure characteristics (Sudre & Lange, 1997).

Also, differences in rates of sinterization between inclusions and matrix may develop stresses which cause sintering defects such as cracks and isolated pores. In fact, the matrix sintering is inhibited when it tries to densify and contract around the inclusions. In this situation, the mean stress will be compressive on the inclusion and tensile on the matrix, thereby opposing densification. The internal stress field will also depend on the shape of the second phase and its volume fraction. It was also reported that the densification rate decreases with increasing volume fraction of inclusions and that densification apparently stops before the theoretical density is reached. It was shown (Brinker & Scherer, 1990) that when inclusions inhibit densification, coarsening mechanisms such as surface diffusion are favored over sintering. Although several improvements in the mechanical properties (e.g. toughness) of ceramic matrices can be achieved by introducing certain reinforcing particles, they tend to inhibit shrinkage so that the full density of the compact may never be achieved, unless a pressurized process, such as hot-pressing or other approaches is applied.

The success in powder consolidation is intimately related to the control of the competition between densification and coarsening. The growth of larger particles, as shown in Fig.2, and shrinkage of small ones (coarsening) is observed often during sintering of crystalline materials. Control of grain growth is thus an essential aspect of controlling the densification-coarsening competition, and producing ceramics with desirable microstructures.

The volume fraction of a particulate second phase present in a composite is an important factor in sintering and resultant shrinkage, and the shape and size of the reinforcing phase may also influence the sintering behavior of the matrix material. The effect of heterogeneities, especially, rigid inclusions, on the matrix densification has been a subject of several studies (Bordia & Scherer, 1988; Scherer, 1987). It was shown that at the first stage of densification, the volume fraction, shape, and distribution of the inclusions affects the morphology of the network formed by the differential densification of the matrix. The matrix properties determine whether, in later stages of densification, the interstitial porous regions can reach full density or not. It was concluded that grain growth in the dense regions of the matrix limits the final densification of the composite.

Many properties of composites are predicted on the basis of the rule of mixtures, which is a method based on an assumption that a composite property is the volume weighed average of the phases (matrix and dispersed phase) properties.

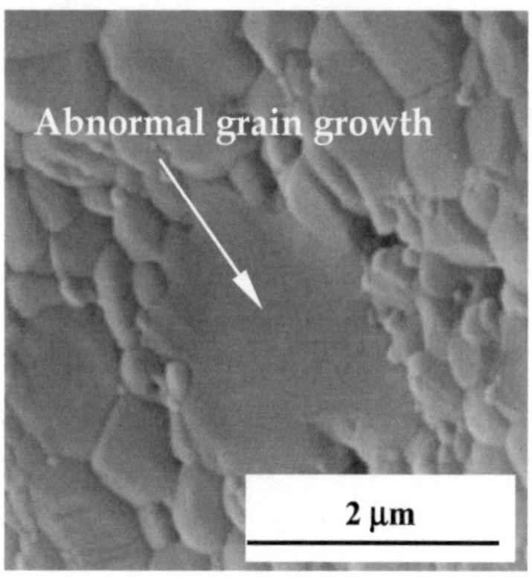

Fig. 2. SEM microstructure of a HA compact after sintering at 1300 °C for 2 h.

Description of sintering of particle-reinforced composites, for a large extent of densification and if no interfacial reaction occurs between the matrix and reinforcement, can be made in terms of a simple rule of mixtures approach (Edrees et al., 1998), as shown by equation (1):

$$\left(\frac{\Delta V}{V_0}\right)_c = \left(\frac{\Delta V}{V_0}\right)_m (1-p) \tag{1}$$

where the subscripts c and m refer to composite and matrix, respectively, and where p is the volume fraction of the filler; therefore, the shrinkage of the composite after sintering is equal to the shrinkage that would be found in the isolated matrix times the matrix volume fraction of the composite.

3. Drawbacks in conventional sinterization of HA/ZrO_2 composites

The inherent low strength and fracture toughness of synthetic HA have restricted its use to high load resistant implants (bones) due to its inferior mechanical properties (especially fracture toughness) compared to cortical bone (LeGeros et al., 1993). For example, its fracture toughness (K_{Ic}) is 1 MPa.m$^{1/2}$, below the minimum of the range of values for human bones: 2–12 MPa.m$^{1/2}$. Moreover, the Weibull modulus (n) is low in wet environments (n = 5–12) (DeWith et al., 1981) which indicates a weak structure of the HA implants. On the other hand, synthetic HA is more isotropic with a larger grain size than the biological HA (Cao & Hench, 1996). Further, bone is a complex composite of an organic (collagen) and an inorganic (biological apatite) component. Therefore, the synthesis of composites made of HA have been put forward, in which the addition of second phase into HA for improving its fracture toughness was widely used and investigated (Kong et al., 1999; H.W. Kim et al., 2003; Shen et al., 2001; Adolfsson et al., 2005; Kumar et al., 2005; Rao et al., 2002; Wu et al., 1998). For example, both the fracture toughness and flexural strength of HA can be improved substantially by the addition of ZrO_2 (Yu et al., 2003). It was widely reported that that HA/ZrO_2 composites show significantly higher mechanical properties, in particular bending strength, micro-hardness and Young's modules have proved to be bigger than those of pure HA (Miao et al., 2004). The incorporation of ZrO_2 into HA enhances its mechanical properties and will not affect its biocompatibility. On the other hand, implants made of the HA/ZrO_2 composites are highly biocompatible with no adverse reactions when used in rabbit mandibles (Vaidhyanathan et al., 1997). Further, the Y_2O_3 addition into ZrO_2 can stabilize the tetragonal phase at room temperature (YSZ), and the tetragonal phase plays a major role in the fracture toughness increase. The toughening mechanisms include transformation and crack deflection toughening. Transformation toughening mechanism is an important mechanism in ceramic composites (Garvie et al., 1975). A great deal of energy is absorbed due to the volume expansion during the transformation of t-ZrO_2 to m-ZrO_2 so that the fracture toughness is enhanced. On the other hand, the ZrO_2 particles dispersed in matrix would disturb the crack tip stress field and make the crack deflect and/or curve, then the driving force of crack propagation decreased and newborn surface area of crack increased, therefore the toughness was enhanced (Faber & Evans, 1983). The change of crack propagation mode, the increase of crack length and the formation of newborn surface area of crack at crack tip will consume energy. The main factor causing crack deflection is the internal stress caused by the interface effect, especially the interaction between internal stress/interface and cracks. The reason for the improvement of fracture toughness by this mechanism is the growth of surface area of crack and the change of the stress field distribution due to the crack propagating by deflection.

In general, the mechanical properties decrease significantly with increasing content of porosity and grain size, while a high crystallinity, a low porosity and small grain size tend to give a higher stiffness, a higher compressive and tensile strength and a greater fracture toughness. Recently it were obtained HA/YSZ composites with improved mechanical properties, flexural strength of ~155 MPa and fracture toughness of ~2.1 MP $m^{1/2}$, due to the contribution of 25 wt% YSZ component (Sung et al., 2007). The increased fracture toughness would result from the stress-induced tetragonal to monoclinic phase transformation in the YSZ component of HA/YSZ composites. In spite of a better mechanical behaviour, HA/YSZ composites must display uniform microstructures with a high degree of dispersion and without decomposition of the HA, during the sintering process. There are, however, some problems concerning the ceramic processing of HA/ZrO_2 composite that should be solved. These problems are related to the influence synthesis and sintering conditions on phase stability of HA and *t*-ZrO_2. The conventional mechanical mixing of HA and YSZ powders has been reported for the preparation of HA/YSZ composites through sintering (H.W. Kim et al., 2005; Kumar et al., 2005; Rapacz-Kmita et al., 2006). However, this mechanical mixing would cause low sintering density and/or non-uniform YSZ phase distribution in the sintered HA matrix due to the large particle size and/or segregation of YSZ particles, as shown in both micrographs of Fig.3.

Non-uniform distribution of the YSZ phase in the HA matrix, as shown in Fig.3, would seriously deteriorate the mechanical properties of HA/YSZ composites. Rather, the addition of ZrO_2 particles reduces the size of HA grains, ZrO_2 can thus act as an effective grain growth inhibitor to the HA grains. However, such composites cannot be sintered at temperatures same as the sintering of pure HA, while the sintering temperature for pure HA does not exceed the 1000 °C, the specimens made of HA–20 vol.% YSZ must be necessarily heated up to 1100 °C to obtain near full dense compacts (Khalil et al., 2007). Furthermore, the higher sintering temperature of HA/ZrO_2 composite, frequently, causes an extreme grain coarsening, also illustrated in Fig.2, and HA phase decomposition, leading to a considerable deterioration of mechanical properties. Also, at high temperature the calcium, which is the main component of the HA, diffuses into YSZ and changes its tetragonal structure into the stable cubic phase, hindering the transformation toughening mechanism (Wu & Yeh, 1988). Furthermore, calcium diffusion also favours the decomposition of HA, leading to the formation of β-tricalcium phosphate (β-TCP, $Ca_3(PO_4)_2$) and calcium zirconate (CZ, $CaZrO_3$). The presence of such phases in the composite drastically reduces it mechanical behaviour (LeGeros, 1993). The formation of other phases among HA also interferes with the sinterability of the composites due to differential shrinkage, leading to overall reduced density as compared to that of pure HA. For instance, the sinterability of a HA–20wt.%ZrO_2 (PSZ; 3 mol% Y_2O_3-doped) composite was affected by the differential shrinkage due to HA thermal decomposition (Wu & Yeh, 1988). Therefore, if there is no reaction between HA and ZrO_2, the strength and toughness of HA are indeed improved significantly. HA and ZrO_2 readily react with each other above 1000 °C (Shen et al., 2001). However, reducing the temperature, the densification is usually difficult unless very special processes, such as HIP, SPS, or other approaches are employed. Hence, another disadvantage of such materials is related to its processing. It is difficult to densify the HA/ZrO_2 by pressureless sintering. However, despite the use of a cold isostatic pressing, a poor densification (apparent porosity ~30%) have been obtained in HA–20-vol%-ZrO_2 composites, after sintering in air at 1350°C for 1 h, (H. W. Kim et al., 2002a) (Fig. 4).

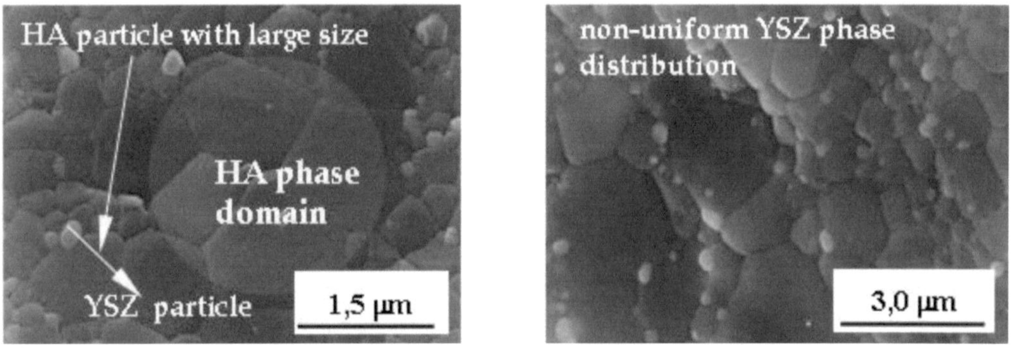

Fig. 3. SEM microstructures of HA–15 vol.% YSZ, sintered at 1300°C, 1h.

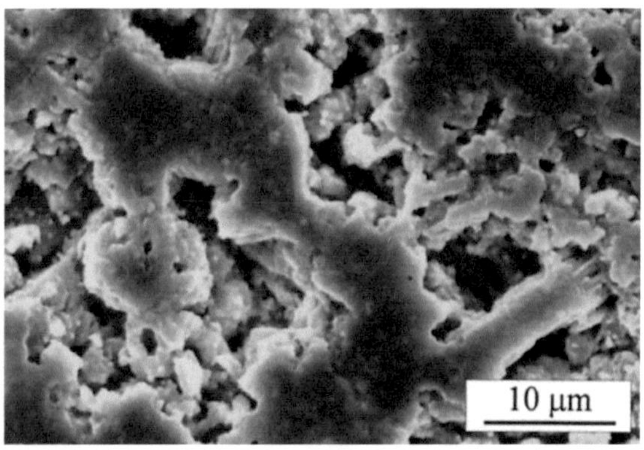

Fig. 4. SEM images showing the microstructure of HA–20- vol%-ZrO_2 composite after sintering at 1350°C for 1 h (source: H. W. Kim et al., 2002a).

A number of approaches have been tried to prepare HA/ZrO_2 composites containing higher fraction of ZrO_2 (20-40 vol%) (Rao & Kannan, 2002; J. Li et al., 1996). However, in these composites, the major problem encountered were the densification of the composites and the decomposition of HA to β-TCP. Actually, the decomposition of HA to β-TCP is accelerated at the higher sintering temperatures (>1300°C). Much effort has been exerted to improve the densification an overcome the reaction between HA and ZrO_2. However, in most cases, hot-pressing (Kong et al., 1999) or HIP (Takaki et al., 1992) processes are still necessary to obtain HA-ZrO_2 composites with high densities. Nevertheless, there is still a desire for a novel technique to produce dense bodies that use lower temperatures in order to facilitate consolidation but that removes any phase decomposition.

4. Enhancements in the sinterability of HA/ZrO_2 ceramic composites

To overcome the problem of grain growth, unconventional sintering and densification techniques have been proposed. The literature shows that dense HA bodies can be produced by a range of thermal treatments in addition to pressureless air sintering such as conventional HIP, microwave and SPS. However, the application of high temperatures during processing may cause decomposition of the principal components.

To minimize these reactions, some efforts have been made toward reducing the sintering temperature and holding time. Simple economical routes to microstructural improvement are therefore worth investigating. One alternative is the use the sol-gel process to synthesize a large variety of compositions carefully doped with additional phases, allowing obtain HA nanosized particles and a greater control of the morphology and microstructure of the derived HA/ZrO_2 compacts.

4.1 Hot Isostatic Pressing (HIP)

Dense ceramic-ceramic composites are usually obtained by pressing and conventional sintering of powders using pressure assisted methods, such as hot pressing, hot isostatic pressing, etc.

The high sintering temperatures and long sintering times required for the consolidation of HAP/ZrO_2 powders often result in excessive grain coarsening and decomposition of the HAP into second phases including phase transformation of *t*-ZrO_2 to cubic phase during sintering in air, which are characteristic for conventional sintering methods and results in the deterioration of the mechanical properties of such ceramics.

The advantages of the hot pressing technique are the enhancement of the densification kinetics and the limiting of grain growth, while the disadvantages are the limited geometry of the end product and the expensive equipment required. Ahn et al., 2005, studied the effect of nano ZrO_2 reinforcement on the strength of hot pressed HA/ZrO_2 composites containing 1.5 to 8 wt% ZrO_2. They observed the highest strength and hardness at 3wt% ZrO_2 which subsequently decreased on higher ZrO_2 loading. They also reported that lower volume fraction of ZrO_2 addition will help to retain both HA and *t*- ZrO_2 phase, which may be due to the combined effect of matrix constraint and uniform dispersion of fine ZrO_2 particles. However, there are still controversial results in the literature about the phase stability of these composites prepared by pressure sintering methods (e.g HIP, hot pressing). Indeed, some reaction has been reported between HA and ZrO_2 by some authors (Veljovic et al., 2007) but not by others (Li et al., 1996). However, there is a general conformity that HA/ZrO_2 composites were considerably stable especially after HIP (Wu & Yen, 1988), even though traces of reaction may be expected due to the increased contact area between HA matrix and dispersed ZrO_2 particles (Adolfsson et al., 2000). On the other hand, a partial reaction between HA and ZrO_2 was observed in the composites prepared by hot pressing but still much less than observed in sintering in air (Evis et al., 2005, 2007). Therefore, the sintering environment and statistical effects on the ZrO_2 distribution within matrix are important parameters that affect the thermal stability of HA and ZrO_2 in the composites. Hot pressing of HA was found to allow the occurrence of densification at temperatures much lower than during conventional sintering (Halouani et al., 1994; Veljovic et al., 2009)

4.2 Microwave sintering developments

It is usually accepted that an important issue in ceramics microstructural development is the interaction between densification and coarsening. To control it, parameters such as temperature, pressure, sintering time, and heating rate must be optimized. Rapid heating (RH) has produced high density compacts when compared with slow heating, for similar densities (D. J. Chen & Mayo, 1996). However, differential sintering that causes differential densification is one of the problems most often encountered in RH.

In this context, microwave sintering has emerged in recent years as an alternative technique to overcome the problems of RH. Because it is a noncontact technique, and the heat is transferred to the body via electromagnetic waves, large amounts of heat can be transferred to material's interior, minimizing the effects of differential sintering. Conventional furnaces heat samples by surface heating transference and, thus depending on the rate of heating, a large thermal gradient from the surface to the centre can be generated within a powder compact. However, volumetric heating via microwave radiation ensures an uniform heating and almost no temperature gradients, which yield higher heating rates and must lower sintering time. Therefore, this technique provides a series of benefits, such as great microstructure control, no limit of the geometry of the bodies, improved mechanical properties and reduced manufacturing costs due to energy savings, lower temperatures of sintering and shorter processing times.

Microwave sintering of ceramic materials with significant time and energy savings has been widely investigated. Microwave sintering of HA was first reported in the 90's. Fang et al., 1995, showed that for transparent HA, the total processing time from start to finish of the sintering process was ~ 20 min for microwave sintering while the same was about 4 h in the case of conventional sintering. Hydroxyapatite ceramics with tailored mechanical properties have also been fabricated by this technique (Rodriguez-Lorenzo et al., 2003). Further it has been shown that HA samples microwave sintered showed better densification, higher density and certainly higher hardness and fracture toughness than samples conventionally sintered at the same temperature (Veljovic et al., 2010). The addition of ZrO_2, however, can reduce the sinterability of the composite and therefore does not reach a high density. At ~20 years ago, only ~50%-78% of relative density was achieved in conventional sintering of HA/ZrO_2 at 1100-1400° C for 3 h (Wu & Yeh, 1988). Then, a few years later, it was reported an achieved relative density of 93% in microwave sintered HA/10%ZrO_2 composites at 1200°C, for only half hour (Fang et al., 1993). Very recently, a compared study between microwave and conventional sintering was performed on HA/ZrO2 composites (Declan et al., 2010). The effect of microwave heating on green bodies has been investigated in order to understand how microwave energy may affect the physical and mechanical properties of the resultant densified composites. The main difference between the two methods is that materials with different microstructures are formed. The higher levels of interconnected open porosity in microwave sinterized HA-ZrO_2 composites are considered to be useful in promoting osteo-integration.

4.3 Spark Plasma Sintering (SPS)

Spark plasma sintering is a newly developed process which makes possible sintering at low temperatures and short periods by charging the voids between powder particles with electrical energy and effectively applying high temperature spark plasma generated briefly. It is regarded as a rapid sintering method similar to microwave sintering. Therefore, it is capable of sintering ceramic powders quickly to a high density at a relatively lower temperature, compared to the conventional sintering method. Due to the applied electric field, the diffusion rate increases and therefore, powder can be densified at much lower temperature with shorter holding time (only few minutes). However, expensive equipment required limits its generalized employ. Nevertheless, Hap ZrO_2 (Kumar et al., 2005) composites have already been prepared by SPS technique and the results showed improved properties compared to their pressureless sintered composites. For example, HA–40 wt.% ZrO_2 composites were sintered at different temperatures by SPS. After sintering at 1200 °C for 5 min, the relative density of the composite increased up to 93% (Miao et al., 2004). The respective microstructure is presented in Fig.5. One can see that the t-ZrO_2 phase (white spots) is uniformly dispersed in the HA matrix, and the composite is also very dense, since large pores are not found.

Fig. 5. SEM micrograph of the HA-YSZ composite prepared by spark plasma sintering at 1200 °C for 5 min (Source: Miao et al., 2004).

4.4 Sol-gel synthesis and powder consolidation on the sinterability of HA/ZrO_2 composites

A currently paradigm of synthesizing and processing of powder compacts emphasizes the tailored assembly of particles, from the atomic scale to the macroscopic scale. One of the trends is prepare finer powder for the ultimate processing and better sintering to achieve dense matrices, with uniform microstructures and high degree of dispersion of the ZrO_2 phase. More is the fineness, more is the surface area, which increases the "reactivity" of powders; The sol-gel process is a very efficient method to synthesize a large variety of compositions carefully doped with additional phases, allowing obtain HA nanosized particles and a greater control of the morphology and microstructure of the following HA/ZrO_2 compacts. Thus, while suitably uniform nanophase powder materials are becoming increasingly available, challenges remain in the fabrication of fully dense nanostructured products. Another important challenge is the production of nanostructures with excellent intergranular distribution of ZrO_2 phase.

Sol-gel chemistry have gained much attention in glass and ceramics fields since the 60ths, as an alternative to the conventional melting techniques, when bulk inorganic gels were formed and converted through low-temperature heat treatments to porous and/or dense glasses or polycrystalline matrixes (Bouquin et al., 1987; Livage, 1998; Brinker & Scherer, 1990). A wide variety of glasses and ceramics with unique properties (materials "*à la carte*") can be prepared through sol-gel processing. The possibility to control a number of parameters of the final product such as homogeneity, (from the chemical composition to desired final characteristics, namely the mechanical properties, purity, microstructure (e.g. porosity, surface area, crystalline phases), and sintering temperature are among the advantages of sol-gel processing methods over the conventional ones. The resulting oxide materials have low melting and sintering temperatures, as opposed to the much higher melting temperatures required if they are produced via melting production routes. In addition, by controlling the rheological properties of the *sol* and *gel*, the final product could be shaped into various forms such as fibers, bulks, thin films coatings and powders. The most considerable constraint in the case of industrial sol-gel processing is the cost of raw materials, due to the high quality of the chemicals as precursors.

The expression sol-gel (from a liquid solution to a solid bulk gel) was applied to a colloidal solution (suspension), which is followed by a gel phase. This definition is now changed to the hydrolysis and polycondensation of a precursor and subsequent formation of a gel (Dimitriev et al., 2008). Upon destabilization of the colloidal suspension (sol), a gel which is a rigid network of the sol aggregated particles, is formed. In other words, gel is a transition state between the solid and liquid. The dispersed particles in the sol are small enough to form a stable suspension because of the Brownian motion. By varying the sol conditions, such as dehydration or pH, the colloidal dispersion (sol) transforms into the gel stage.

Although the concept of sol-gel processing covers a large variety of materials, it is specifically suitable to those obtained from hydrolysable metal alkoxides or salts. Hence, metal alkoxides $M(OR)_n$ are the most common precursors for the sol-gel synthesis of metal oxides. The precipitation of metal oxide particles from solutions was one of the earliest routes to produce ceramics through sol-gel processing.

Considering the numerous applications of HA in biomedical fields, development of various synthesis methods is currently a major issue. A number of synthesis techniques using various sources of Ca and P have been developed which includes wet chemical method (precipitation), hydrothermal synthesis procedure, continuous precipitation, thermal deposition, sol-gel and solid state reaction method (Feng et al., 2005; Thamaraiselvi et al., 2006). However, such techniques can lead to the formation of Ca-deficient apatite structure. Moreover, the presence of β-TCP, $Ca_4P_2O_9$ or $Ca_4(PO_4)_2O$ along with the HA phase in the subsequent sintered material is indicative of a Ca/P ratio lower than the stoichiometric value of pure HA (Ca/P = 1.67). Ca/P ratios higher than 1.67 would indicate the presence of CaO with the HA phase (LeGeros et al., 1993). The final composition of dense HA after sintering depend on the sintering temperature, the experimental conditions and Ca/P molar ratio of the HA preparation prior to sintering (LeGeros et al., 1993). In some cases, a well-crystallized HA phase was only developed while approaching a sintering temperature of 1200 °C.

The sol-gel approach provides significantly easier conditions for the synthesis of HA. A currently paradigm of synthesizing and processing emphasizes the tailored assembly of particles, from the atomic scale to the macroscopic scale. Thus, nanosized particles with high surface area and improved reactivity can be easily obtained by the sol-gel route. Technologically, there are significant benefits from the lower sintering temperatures of HA nanosized particles: possibility of avoiding sintering aids, phase decomposition, deleterious interfacial interactions, and undesirable phase transformations. Compared to the conventional methods, the most attractive features of sol-gel process include *(a)* molecular-level homogeneity can be easily achieved through the mixing of two liquids; *(b)* the homogeneous mixture containing all the components in the correct stoichiometry and ensures a much higher purity; and, *(c)* a lower heat treatment temperature to form polycrystalline ceramics is usually achieved without resorting to a high excessive temperature. As a result, several synthesis routes have been proposed for sol-gel synthesis of HA and HA/YSZ as well as different mixing conditions with various molar ratios among reactants (Han et al., 2004; Bogdanoviciene et al., 2006).

A number of calcium and phosphorous precursor combinations were employed for sol-gel synthesis of HA, namely $Ca(NO_3)_2$, CaO, $Ca(OH)_2$ and Ca alkoxide as a Ca sources, triethyl phosphite [$P(OC_2H_5)_3$] (TEP) and triethyl phosphate [$PO(OC_2H_5)_3$], (P_2O_5) and H_3PO_4 as a P-source. Once more, chemical activity and the temperature required forming the HA structure depends largely on the chemical nature of the precursors. One of most promising combination for the precursors yielding stoicheometric HA with Ca/P ratio of 1.67, is calcium nitrate and TEP using alcohol at the hydrolysis stage (Vasconcelos & Barreto, 2011).

In the development of ceramic sol-gel composites, the different components of the sol may be tailored so that they do not react with each other to form new components. In particular, in HA/ZrO_2 (PSZ; Y_2O_3-doped) composites, ZrO_2 particles can be dispersed into a HA sol before gelation, leading to a composite with good homogeneity and intimate contact between the components.

The gel composite slurry is typically dried at temperatures under 200°C, and sintered at temperatures lower than the corresponding calcined ceramics. Due to process synthesis, ZrO_2 particles served as nucleation sites during HA precipitation, so HA crystals were formed on the surfaces of ZrO_2 particles (Vasconcelos & Barreto, 2011). This phenomenon provided a more intimate mixing in these binary composites. Coarsening (i.e. growth of larger particles and pores, and shrinkage of finer ones), frequently observed during sintering of crystalline materials, results in overall grain growth which is in direct competition with the densification processes. However, in the case of sintering of dried gels derived amorphous powders, these competing processes are absent (there is no grains), therefore all the pores shrink during viscous sintering and the body densifies to its full density.

4.5 Nanostructural design and improvement of the mechanical behavior in HA/ZrO_2 ceramic composites

Fabrication of full dense ceramics composites is not easy. One of the main reasons lies to shrinkage heterogeneities and detrimental grain growth after the density of around 90% of theoretical density, when the continuous network of pores (in second stage of sintering) disintegrates to the closed ones (in final stage of sintering) and leads to an accelerated grain growth in conventionally sintered specimens, in special if the filler is not either homogenously distributed. On the other hand, the retention of t-ZrO_2 in HA derived composites has to be controlled carefully until is used for implant devices. It has been recognized that the level of toughening is complexly dependent on the microstructure of such composites (i.e., volume fraction, size, shape, location and size distribution of ZrO_2 (Heuer et al., 1982). It can be seen (Fig.6) that toughness (K_{Ic}) of HA-based ZrO_2 composites increased with the ZrO_2 content up to 20wt% (Chen et al., 2008). The poor mechanical behavior of the composites with 40 to 60 wt% ZrO_2 is correlated with the presence of high porosities (up to 16%) and phase distribution heterogeneities which tended to retard the densification.

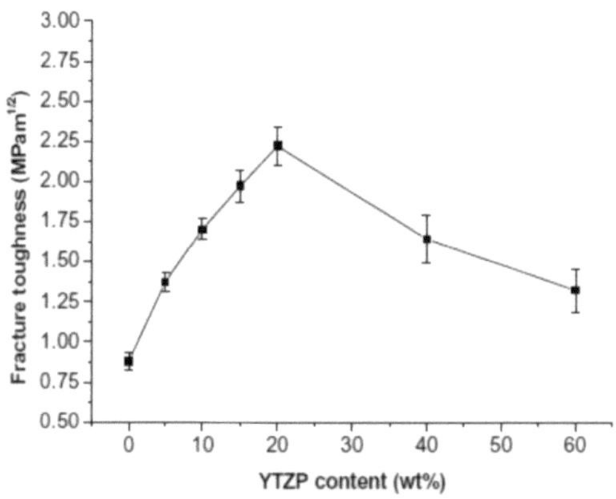

Fig. 6. Fracture toughness vs. ZrO_2 content of HA-based ZrO_2 composites conventionally sintered at 1400°C for 2 hours (Source: Chen et al., 2008).

Homogeneous as well as fine dispersions of ZrO_2 particles in the HA matrix can be obtained by chemical mixing of the constituents in the solution through the sol–gel processing (Vasconcelos & Barreto, 2011). The fine grain size and uniform microstructure of both phases within the composite satisfy the requirements for a toughness improvement; moreover, the toughness can be improved many folds due to the presence of tetragonal phase within the HA matrix, with a potential of 100% of transformation. However, such arguments can changes with the phase composition, crystallinity, crystallite size and pore morphology, specific surface area and subsequent shrinkage.

The sol-gel technology offers processing advantages and gives flexibility in tailoring the composite chemistry to obtain the desired properties. Moreover, the processing conditions, composition, retention of the t-phase of ZrO_2 the calcination temperature and additives (H. W. Kim et al., 2002b) also control the morphology of the powders and their sintering behavior. An important challenge is the production of composite nanostructures with homogeneous microstructural distribution of ZrO_2 phase. Fig.7. illustrates the SEM of a sintered (950 °C for 1 h) HA/10 mol% ZrO_2 composite (Vasconcelos & Barreto, 2011). As can be seen, the image reveals that ZrO_2 particles are present as both intergranular and intragranular in the HA matrix. However, the presence of intragranular particles is much less compared to intergranular particles. The network of intergranular grain is thus the dominant in the toughening mechanism. The average grain size of HA varies from 0.5 mm to 2mm and is quite homogeneous in the entire matrix. ZrO_2 particles are smaller in size (50-

150 nm) and are trapped at grain boundaries between HA grains. This tailored microstructure provide a more intimate mixing in binary composites, yielding a higher dispersion, allowing ZrO_2 particles to be present mostly at grain boundaries, without agglomerates. Thus, the prepared samples were fully densified with very small isolated voids.

The principal merit of the microstructure observed in these particular composites, obtained by the sol-gel route, is the adequate relative grain size ratio and phase distribution between the both phases, allowing ZrO_2 particles to be present mostly at grain boundaries without agglomerates. During the HA grain growth, residual pores were eliminated throughout the sinter of the HA matrix, and thus an equilibrium-point between densification and coarsening were achieved. Simultaneously, ZrO_2 particles, trapped in the grain-boundaries acts as effective abnormal grain growth inhibitor of HA

Other approaches were also used to obtain better mechanical properties. The properties of these materials are determined by their microstructures; therefore, to control their microstructural development and achieve fine microstructures, the sintering parameters must be optimized. For example, partially or fully fluorine-substituted hydroxyapatite (HAF) ceramics (Adolfsson & Hermansson, 2005) have shown a high thermal stability and some level of bioactivity, and they have been used to develop HAF-ZrO_2 composites with aimed improved mechanical behavior (Chen et al, 2008). The HAF-ZrO_2 microstructures exhibited ZrO_2 grains distributed on the grain boundaries and also within the grains of the matrix (Fig.8). Furthermore, with the increase of the ZrO_2 content, more ZrO_2 grains were trapped within the grains.

For mechanical improvement is quite essential that ZrO_2 particles are mainly as tetragonal phase. However, if *m*-ZrO_2 is present on cracked surface, such presence indicates that the transformation toughening is active for toughness enhancement (Vasconcelos & Barreto, 2011). Transformation toughening is induced from the stress-induced transformation experienced by *t*-ZrO_2 particles when interacting with an advancing crack (Fig.9a) (Stevens, 1986).

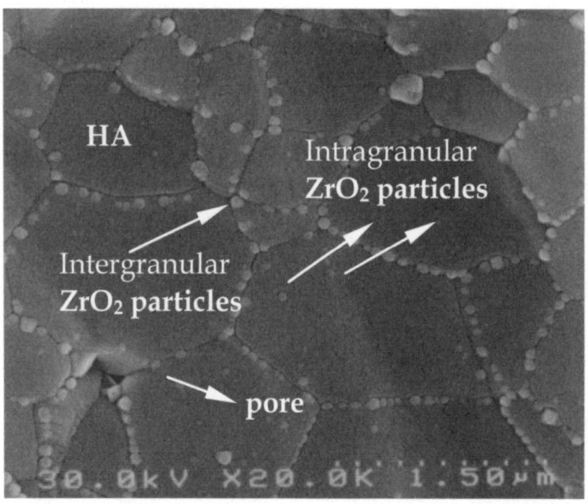

Fig. 7. Tailored microstructure a HA–10 mol%ZrO_2 sol-gel composite (adapted from Vasconcelos & Barreto, 2011).

Moreover, crack deflection toughening by ZrO_2 particles can also contributes to toughening of the composite through crack deflection around the dispersed ZrO_2 particles (Fig.9b) (Vasconcelos & Barreto, 2011). Hence, K_{Ic} changes with the addition of ZrO_2 may be rationalized by the relative predominance of the toughening mechanism, i.e., ZrO_2 phase transformation and/or crack deflection.

Fig. 8. SEM micrographs of the thermally etched surfaces of the HAF-5wt%ZrO_2 composites conventionally sintered at 1400°C (Source: Chen et al, 2008).

In addition, a high degree of densification and homogeneous shrinkage of such composites can be yield if the formation of Ca-deficient HAP phases during sinterization is avoided. Therefore, through sol-gel process the T-onset of decomposition and the densification at low conventional temperatures of ZrO_2-doped HA composites leads to more thermally stable HA-ZrO_2 composite products than those obtained by other (e.g. co-precipitation) routes.

The improvements in sinterability and microstructure had a strong influence on the mechanical properties of the HA-ZrO_2 composites.

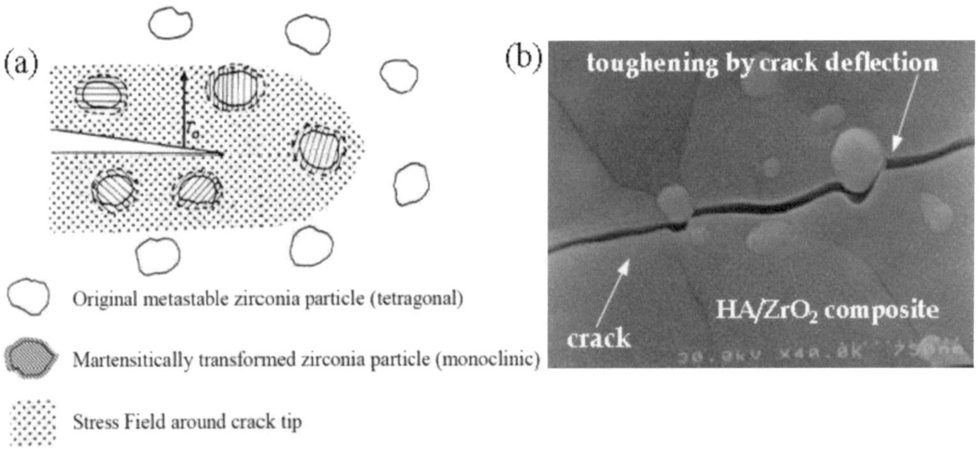

Fig. 9. (a) Stress induced transformation of t-ZrO_2 particles in the elastic stress field of a crack (Stevens, 1986); (b) Indentation crack propagation during indentation fracture of active crack deflection by the ZrO_2 particles (adapted from Vasconcelos & Barreto, 2011).

5. Sintering advances to suppress the decomposition of the HA phase

Apart from the processing techniques to control the external sintering conditions, other approaches are also used to suppress the decomposition of the HA phase. To minimize the thermal decomposition and improve the densification of the composites, various atmospheres assisted sintering have been employed to produce HA/ZrO_2 composites. However, if this chemical interaction exist, unfortunately would affect the phase purity and the microstructure of the HA/ZrO_2 composites, leading to undesirable mechanical and biological properties.

In the spike of that, it was observed that the release of water increases gradually with increasing temperature creating vacancies in HA structure (former OH^- positions is replaced by O^{2-} ions and vacancies) (Adolfsson et al., 1999, 2000). This reaction occurs because HA is a hydrated calcium phosphate material, and begins to dehydroxylate at about 800°C to form oxyhydroxyapatite ($Ca_{10}(PO_4)_6(OH)_{2-2x}O_x$ $_x$, where =vacancy). Further heating, and according to the equilibrium phase diagram of CaO/P_2O_5, HAP will decompose in β-TCP (β-$Ca_3(PO_4)_2$), forming α-TCP when heated to temperatures in excess of 1350°C. It is well established that HA is thermally decomposed into mostly β-TCP, CaO and $H_2O_{(vapour)}$ (Ahn et al., 2001; Shen et al., 2001; Heimann & Vu, 1997), according to the following reactions (2), (3) and (4):

$$Ca_{10}(PO_4)_6(OH)_2 \Leftrightarrow Ca_{10}(PO_4)_6(OH)_{(2-2x)}O_x + xH_2O\uparrow \qquad (2)$$

$$Ca_{10}(PO_4)_6O \Leftrightarrow 2\beta - Ca_3(PO_4)_2 + Ca_4P_2O_9 \qquad (3)$$

$$Ca_{10}(PO_4)_6(OH)_2 \Leftrightarrow 3\beta - Ca_3(PO_4)_2 + CaO + H_2O\uparrow \qquad (4)$$

Another process that removes CaO from HA is dissolution of CaO into *t*-ZrO_2 as seen in reaction (5):

$$CaO + ZrO_{2(tetragonal)} \Leftrightarrow CaO(ZrO_{2(cubic)}) \qquad (5)$$

Some authors (Wu & Yeh, 1998; Rao & Kannan, 2002; Evis et al., 2005;) have reported that the CaO release, due to HA decomposition to β-TCP, enhances the formation of CZ and CaO:ZrO_2 solid solution, according to reaction (6):

$$Ca_{10}(PO_4)_6(OH)_2 + yZrO_{2(tetragonal)} \Leftrightarrow$$

$$3\beta - Ca_3(PO_4)_2 + CaZrO_3 + CaO(ZrO_{2(cubic)})_y + \uparrow(2-2x)H_2O \qquad (6)$$

5.1 Thermal stabilization of HA/ZrO2 composites with water vapor assisted sintering

Since a gaseous species exists on the products side of the decomposition reactions, sintering atmosphere would be expected to influence the decomposition kinetics of HA. Consequently, sintering under vacuum conditions can induce decomposition at lower temperatures, favouring the formation of water vapour. On the other hand sintering in a moist atmosphere can neutralize this effect and avoid or delay the decomposition to some degree.

When a specific fraction of vacancies is created, HA will not be stable and it will decompose. However, if H_2O loss is maintained at a low level, the equilibrium of these reactions is shifted to the left and decomposition into β-TCP, $CaZrO_3$ and $ZrO_{2(cubic)}$ will not occur. In order to prevent water loss, it is recommended the use an atmosphere control for sintering (water vapor, for example), or incorporation of additives (e.g. incorporation of F^- ions) (H. W. Kim et al., 2003) in HA structure. If fraction of vacancies formed is suppressed the temperature stability will be extended.

So far, it has been well recognized that, by pressureless sintering of HA/ZrO_2 composites, it is very difficult to reach full densification in air (Li & Hermansson, 1996; Wu &Yeh, 1988). Therefore, only by introducing pressure water vapor it is possible to densify without any decomposition (Vasconcelos & Barreto, 2011). The obtained HA/ZrO_2 compacts under Water Vapor Assisted Sintering did not contain any phases other than HA and the tetragonal modification of ZrO_2, as revealed by their X-ray powder diffraction patterns (Fig. 10). The presence of an H_2O atmosphere during sintering reaction causes a compensation of the partial vapour atmosphere of water inside the furnace. In this way, vacancies formation in the HA structure through reaction (2) could be effectively avoided by a left shift of the equilibrium of reactions (4) and (6), countering the HA decomposition by means of the Le Châtelier principle.

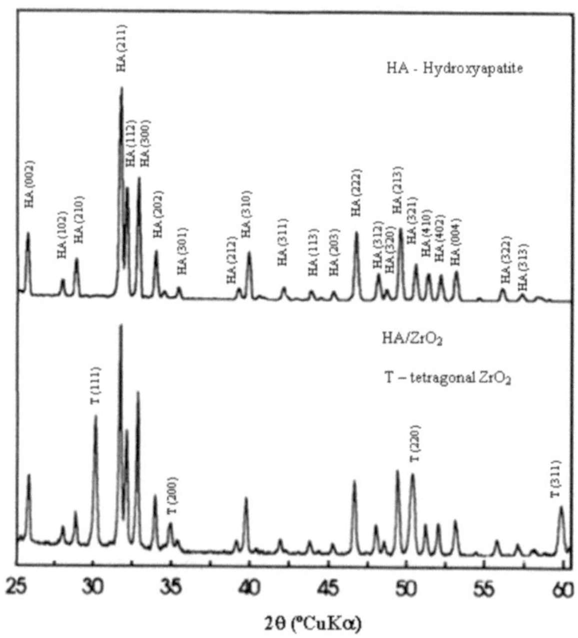

Fig. 10. XRD patterns of HA and HA /10%YSZ powder heated at 950°C, 1 h in H_2O vapour atmosphere (adapted from Vasconcelos & Barreto, 2011).

Similar approaches, like a flowing $H_2O(g)/O_2$ mixtures or hydrothermal sintering conditions have been successful to eliminate or reduce the decomposition of HA or increase the T-onset of decomposition (Ruys et al., 1995). On the other hand, if an intergranular distribution of ZrO_2 particles can be achieved (Fig.4), the decomposition reactions of HA are avoided because diffusion of water from the reaction zone to the surfaces is retarded by the ZrO_2 matrix (intergranular ZrO_2 particles) in boundaries of HA grains, forming a continuous framework (Vasconcelos & Barreto, 2011). Besides, the morphological characteristics of the powders, the good stoichiometry, and their crystallinity are also factors to take into account in sinterability.

5.2 Other approaches

It was reported that one of the main disadvantages of the composite approach applied to HA is related to its processing, leading generally to poor densification. There is an agreement that the addition of ZrO_2 to the HA causes it to decompose at lower temperatures in pressureless sintering, so that several workers have used hot aid pressing to reach higher density. However, it is possible improve densification by applying appropriate sintering additives for HA. The sintering additives should considerably improve composite densification without decomposition of HA. Moreover, the sintering additives could be used as structure stabilizers and/or to control grain size. Thus, several sintering additives (Suchanek et al., 1997) have been used in HA (e.g. sodium, lithium, magnesium, calcium and aluminium fluorides, lithium and sodium phosphates, among others.) Most of them (except NaF and AlF_3) improved densification of HA, not only by influencing the processing conditions, but also by changing chemical properties of the powders (LeGeros, 1991). In a study of HA-ZrO_2 composites sintered without pressure, small additions of CaF_2 (H. W. Kim et al., 2002b) were added as a sintering aid to thermally stabilize the HA-ZrO_2 composites. Thus, due to the substitution of OH- groups by F- ions, Fluorapatite (FHA) was obtained and thereby restrained the decomposition of HA to β-TCP. As a result, dense bodies with high strength and toughness were obtained.

Apart this, other approaches were also used to suppress the decomposition of the HA phase. For instance, (Kong et. al., 1999) added alumina into HA/ZrO_2 composites to reduce the contact areas between the HA phase and the ZrO_2 phase, and hence suppress the interactions between them. However, the introduction of another metal oxide phase might also result in a high level of residual thermal stresses and lead to microcracking of the sintered bodies due to the large difference in the thermal expansion coefficients of the various composite components. In addition, the presence of CaO may cause decohesion of the material due to stresses arriving from formation of $Ca(OH)_2$ and related volume changes (Ababou et al., 1995). However, Heimann & Vu, 1997, have shown that addition of CaO to HA/ZrO_2 composite shifts the chemical equilibrium of the product from β-TCP and tetra calcium phosphate (TTCP) towards HA making it more stable. Additional CaO will be effectively fixed by ZrO_2 acting as a sink for Ca^{2+} ions resulting in the formation of either *t*-ZrO_2 or $CaZrO_3$.

6. Conclusion

Bioceramics, used initially as alternatives to metals in order to increase the biocompatibility of implants, have become a diverse class of biomaterials. HA attracts major interest due to the similarity of its chemical composition and crystallography to mineralized bone of human tissues. However, the mechanical performance of synthetic HA is very poor. In fact, its poor mechanical strength makes it unsuitable for load-bearing. Hence, combination of HA and ZrO_2 have attracted a great attraction since higher strength and fracture toughness can be significantly enhanced either by stress-induced tetragonal (*t*-ZrO_2) to monoclinic (*m*-ZrO_2) phase transformation toughening, or by a deflection toughening mechanism. The high sintering temperatures and long sintering times required for the consolidation of HAP/ZrO_2 powders often result in excessive grain coarsening and decomposition of the HAP into second phases including phase transformation of *t*-ZrO_2 to cubic phase during sintering in air, resulting in the deterioration of their mechanical properties. The sol-gel technology offers processing advantages and gives flexibility in tailoring the composite chemistry to obtain the desired properties yielding a particular composite microstructure with homogeneous intergranular distribution of ZrO_2 phase. The presence of H_2O in the sintering atmosphere prevents the HA decomposition by means of the Le Châtelier principle. However, the main disadvantages of the composite approach applied to HA still is related to its processing, leading generally to poor densification. So, the use of several hot aid pressing techniques to reach higher density is still a possibility. However, it is also possible improve densification and avoid thermal decomposition, by applying appropriate sintering additives for HA.

7. References

Ababou A. & D. Bernache-Assollant, in Proceedings of Fourth EuroCeramics, 1995, edited by A. Ravaglioli (1995), Vol. 8 (Biomaterials), p. 185.

Adolfsson, E.; Nygren, M. & Hermansson, L. (1999). Decomposition mechanisms in aluminum oxide-apatite systems. *J. Am. Ceram. Soc.* Vol. 82, pp. 2909-2912

Adolfsson E.; Alberius-Henning, P. & Hermansson, L. (2000). Phase analysis and thermal stability of hot isostatically pressed zirconia–hydroxyapatite composites. *J. Am. Ceram. Soc.* Vol. 83, pp. 2798-2802

Adolfsson, E. & Hermansson L. (2005). Zirconia-fluorapatite materials produced by HIP, *Biomaterials*, Vol.20, pp. 1263–1267.

Agathopoulos, S.; Tulyaganov, D. U.; Marques, P. A. A. P.; Ferro, M. C.; Fernandes, M. H. V. & Correia, R. N. (2003). The Fluorapatite-Anorthite System in Biomedicine *Biomaterials, Vol.* 24, No.7, pp. 1317-1331

Ahn, E. S.; N. J. Gleason &J. Ying, Y. (2005). The effect of zirconia reinforcing agents on the microstructure and mechanical properties of hydroxyapatite-based composites. *J. Am. Ceram. Soc.*, Vol.88, No.12, pp. 3374-3379

Ahn E. S.; Gleason N. J.; Nakahira, A.; & Ying, J. Y. (2001). Nanostructure Processing of Hydroxyapatite-based Bioceramics. *Nano Lett.*, Vol.1, No.3, pp. 149-153.

Bogdanoviciene, I.; Beganskiene A.; Tonsuaadu K.; Glaser J.; Meyer, H.J. & Kareiva, A. (2006). Calcium hydroxyapatite, $Ca_{10}(PO_4)_6(OH)_2$ ceramics prepared by aqueous sol-gel processing. *Mater. Res. Bull.* Vol.41, pp. 1754-1762.

Bordia, R. K. & Scherer, G. W. (1988). On the Constrained Sintering-III. Rigid inclusions, *Acta Metall*, Vol. 36, No. 9, pp. 2411-2416

Brinker, C. J. & Scherer, G. W. (1990) *Sol-Gel Science - The Physics and Chemistry of Sol-Gel Processing,* Academic Press Inc.

Cao, W. & Hench, L.L. (1996). Bioactive materials, *Ceram. Int.*, Vol.22, pp. 493–507

Chen, D.J. & Mayo, M.J. (1996). Rapid rate sintering of nanocrystallineZrO_2-3mol% Y_2O_3. *J. Am. Ceram. Soc.* Vol.79, pp. 906–912.

Chen, Y.; Dong, Z. & Miao, X. (2008). The Effect of Yttria-Stabilized Zirconia on the Properties of the Fluorine-Substituted Hydroxyapatite Ceramics Prepared by Pressureless Sintering, *Journal of Biomimetics, Biomaterials and Tissue Engineering,* Vol.1 pp. 57-68.

Clifford, A.; Hill, R. G.; Towler, M. R. & Wood D. J. (2001). The Crystallisation of Glasses from the. Ternary CaF_2-$CaAl_2Si_2O_8$-P_2O_5 System, *J. Mater. Sci.,* Vol.36,No.16, pp. 3955-3961

Coble, R.L. & Gupta T.K. In *Sintering and Related Phenomena,* Kuczynski GC, Hooton, Gibbon CF (eds). Gordon and Breach: New York, 1967; 445–441.

Curran D. J.; Fleming T. J.; M. R. Towler & Hampshire, S. (2010). Mechanical properties of hydroxyapatite–zirconia compacts sintered by two different sintering methods. *J Mater S*ci: Mater Med., Vol.21, pp.1109–1120

DeWith, G.; Van Dijk, H. J. A.; Hattu N. & Prijs K. (1981). Preparation, microstructure and mechanical properties of dense polycrystalline hydroxyapatite, *J. Mater. Sci.,* Vol.**16**, pp. 1592-1598

Dimitriev Y.; Ivanova, Y., & Iordanova, R. (2008). History of sol-gel science and technology, *Journal of the University of Chemical Technology and Metallurgy,* Vol.43, No.2, pp. 181-192

Edrees, H. J.; Smitht, A. C. & Hendry, A. (1998). A Rule-of-mixtures Model for Sintering of Particle-reinforced Ceramic-matrix Composites, *Journal of the European Ceramic Society,* Vol.18, pp. 275-278

Evis, Z. & Doremus, R.H. (2007) Hot-pressed hydroxylapatite/monoclinic zirconia composites with improved mechanical properties, *J. Mater. Sci.,* Vol.42, pp. 2426–2431.

Evis, Z.; Ergun, C. & Doremus, R.H. (2005). Hydroxylapatite–zirconia composites: thermal stability of phases and sinterability as related to the CaO–ZrO_2 phase diagram, *J. Mater. Sci.,* Vol.40, pp. 1127-1134

Exner, H. E., Solid-state sintering: critical assessment of theoretical concepts and experimental methods. *Powder Metall. Int.,* 1980, **4**, 203–209.

Exner, H. E. and Arzt, E., Sintering processes. In *Physical Metallurgy, vol.3,* ed. R. W. Cahn and P. Haasen. 4th ed. Elsevier Science, Amsterdam, 1996, pp. 2628–2662.

Faber, K.T. & Evans, A.G. (1983). Crack deflection process—II: experiment, *Acta. Metall.* Vol.31, pp. 577–584.

Fang Y.; Roy, D. M.; Cheng, J.; Roy, R. & Agrawal, D. K. (1993) Ceramic transactions (eds) D. E. Clark et al (Westerville, Ohio: The Am. Ceram. Soc., Inc.) 36 p. 397

Fang Y., Agrawal, D. K.; Roy, D. M. & Roy, R. (1995). Fabrication of Transparent Hydroxyapatite Ceramics by Microwave Processing, *Mater. Lett. Vol.*23, pp. 147-151

Feng, W. Mu-sen. L.; Yu-peng, L. & Yong-xin, Q. (2005). A simple sol-gel technique for preparing hydroxyapatite nanopowders, *Mater. Lett., Vol.* **59,** pp. 916-919

Garay, J.E. (2010). Current-Activated, Pressure-Assisted Densification of Materials, *Annu. Rev. Mater. Res.* Vol.40, pp. 445-468.

Garvie, R.C.; Hughan, P.R. & Pascoe, R.T. (1975). Ceramic steel?, *Nature,* Vol.258, pp. 703–704.

German, R. M., *Sintering Theory and Practice.* John Wiley & Sons, New York, 1996.

Green, D. J.; Guillon, O. & Rodel, J. (2008). Constrained sintering: A delicate balance of scales, *Journal of the European Ceramic Society*, Vol.28, pp. 1451–1466

Halouani R.; Bernache-Assollant, D.; Champion, E. & Ababou, A. (1994). Microstructure and related mechanical properties of hot pressed hydroxyapatite ceramics, *J. Mater. Sci: Mater. Med.*, Vol.5, pp. 563–568.

Han, Y. C.; Li, S.P; Wang, X.Y. & Chen, X. M. (2004). Synthesis and sintering of nanocrystalline hydroxyapatite powders by citric-acid sol-gel combustion method, *Mater. Res. Bull.* Vol. 39, pp. 25-32.

Heimann R. B. & Vu, T. A. (1997). Effect of CaO on thermal decomposition during sintering of composite hydroxyapatite-zirconia mixtures for monolithic bioceramic implants *J Mater. Sci. Lett.*, Vol.16, pp. 437–439

Heuer, A.H.; Claussen, N.; Kriven, W.M. & Ruhle, M. (1982). Stability of tetragonal ZrO_2 particles in ceramic matrices, *J Am Ceram Soc.*, Vol.65, pp. 642–8

Khalil, K.A.; WonKim, S. & Kim, H.Y. (2007). Consolidation and mechanical properties of nanostructured hydroxyapatite-(ZrO_2 + 3 mol% Y_2O_3) bioceramics by high-frequency induction heat sintering, *Material Science and Engineering*, Vol.A 456, pp. 368–372

Kim, H. W.; Koh, Y. H.; Yoon, B. H. & Kim, H. Ee. (2002a). Reaction Sintering and Mechanical Properties of Hydroxyapatite–Zirconia Composites with Calcium Fluoride Additions, *J. Am. Ceram. Soc.*,Vol. 85, No.6, pp. 1634–36

Kim, H. W.; Noh, Y. J.; Koh, Y. H.; Kim, H. Ee. & Kim, H. M. (2002b). Effect of CaF2 on densification and properties of hydroxyapatite– zirconia composites for biomedical applications *Biomaterials*, Vol.23, pp. 4113–4121

Kim, H. W.; Kong, Y. M.; Koh, Y.H.; Kim, H.E.; Kim H. M. & Ko, J. S. (2003). Pressureless sintering and mechanical and biological properties of fluor-hydroxyapatite composites with zirconia, *J. Am. Ceram. Soc.*, Vol.86, pp.2019–2026

Kim, H, W.; Knowles, J. C.; Li, L. H. & Kim, H. E. (2005). Mechanical performance and osteoblast-like cell responses of fluorine-substituted hydroxyapatite and zirconia dense composite. *Journal of Biomedical Materials Research Part A*, Vol.72, No.3, 258 - 268

Kingery, W.D. & Berg M. (1955). Study of the initial stages of sintering solids by viscous flow, evaporation -condensation, and self-diffusion *J. Appl. Phys., Vol.* 26, pp. 1205-1212

Kingery,W. D.; Bowen, H. K. & Uhlmann, D. R. (1976) *Introduction to Ceramics* (*2nd ed.*). John Wiley & Sons, New York

Kong, Y. M.; Kim, S.; Kim, H. E. & Lee, I. S. (1999). Reinforcement of hydroxyapatite bioceramic by addition of ZrO_2 coated with Al_2O_3, *J. Am. Ceram. Soc.*, Vol.82, No.11, pp. 2963–2968

Kong, Y. M.; Bae, C. J.; Lee, S. H.; Kim, H. W. & Kim, H. E. (2005). Improvement in biocompatibility of ZrO_2-Al_2O_3 nano-composite by addition of HA. Biomaterials, *Biomaterials Vol.*26, pp. 509-517

Kumar, R.; Prakash, K.H.; Cheang, P. & Khor, K.A. (2005). Microstructure and mechanical properties of spark plasma sintered zirconia-hydroxyapatite nano-composite powders, *Acta Mater.*, Vol.53, pp. 2327–2335.

LeGeros R. Z., 1991. *Calcium Phosphates in Oral Biology and Medicine, in: Monographs in Oral Science (M. H. Myers, ed.),*Karger Publ. A. G., Zurich.

LeGeros R.Z., J.P. LeGeros, in: L.L. Hench, J. Wilson (Eds.), An Introduction to Bioceramics, Advanced Series in Ceramics, vol. 1, World Scientific, Singapore, 1993, pp. 139–180.

Li, J.; Fartash, B. & Hermansson, L. (1995). Hydroxyapatite-alumina composites and bone-bonding. Biomaterials, *Biomaterials, Vol.*16, pp. 417-422

Li J.; Liao, H. & Hermansson, L. (1996). Sintering of partially stabilized zirconia and partially stabilized zirconia-hydroxyapatite composites by hot isostatic pressing and pressureless sintering, *Biomaterials*, Vol.17 No.18, pp.1787-1790

Li, W. & Gao, L. (2003). Fabrication of HAp-ZrO_2 (3Y) nano-composite by SPS, *Biomaterials Vol.*24, pp. 937-940

Livage, J.; Beteille, F.; Roux, C.; Chatry, M.; & Davidson, P. (1998). Sol-Gel Synthesis of Oxide Materials, *Acta mater.*, Vol. 46, No.3 pp. 743-750

Miao, X.; Chen, Y.; Guo, H. & Khor, K.A. (2004). Spark plasma sintered hydroxyapatite-yttria stabilized zirconia composites, *Ceramics International*, Vol.30, pp. 1793–1796

Nagarajan V. S. & Rao, K. J. (1993). Structural, mechanical and biocompatibility studies of hydroxyapatite-derived composites toughened by zirconia addition, *J. Mater. Chem.* Vol.3, pp. 43–51.

Rao, R. R. & Kannan, T. S. (2002). Synthesis and sintering of hydroxyapatite-zirconia composites, *Mater. Sci. Eng. C*, Vol.20, No. 1-2, pp. 187–193

Rapacz-Kmita, A.; Slosaczyk, A. & Paszkiewicz, Z. (2006). Mechanical Properties of HAP-ZrO_2 Composites, *J. Eur. Ceram. Soc., Vol.*26, pp. 1481-1488

Rodríguez-Lorenzo, L. M. ; Vallet-Regí, M. ; Ferreira, J. M. ; Ginebra, Aparicio, M. P. & Planell, C. (2002). Hydroxyapatite ceramic bodies with tailored mechanical properties for different applications, *J. Biomed Mater Res.* Vol.60, No.1, pp. 159-166

Ruys A. J.; Brandwood, A.; Milthorpe, B. K.; Dickson, M. R.; Zeigler, K. A. & Sorrell, C. C. (1995). The effects of sintering atmosphere on the chemical compatibility of hydroxyapatite and particulate additives at 1200°C, *J. Mater. Sci.: Mater. Med.* Vol.6, pp. 297– 301.

Scherer, G. W. (1987). Sintering with Rigid Inclusions, *J. Am. Ceram. Soc,* Vol.70, No.10, pp. 719-25.

Shen, Z.; Adolfsson, E.; Nygren, M.; Gao, L.; Kawaoka H.; Niihara, K. (2001). Dense hydroxyapatite-zirconia ceramic composites with high strength for biological applications, *Adv. Mater.* Vol.13, pp. 214–216.

Stevens, R. (1986). Zirconia and Zirconia Ceramics, Magnesium Elektron Ltd, p18-19.

Sudre, O. & Lange, F.F. (1997). Effect of Matrix Grain Growth Kinetics on Composite Densification. *J. Am. Ceram. Soc,* Vol.80, No.3, pp.800-802

Sung, Y. M. & Kim, D. H. (2003). Crystallization characteristics of yttria-stabilized zirconia/hydroxyapatite composite nanopowder, *J. Cryst. Growth,Vol.* 254, pp. 411-417

Sung, Y. M.; Shin, Y. K. & Ryu J. J. (2007). Preparation of hydroxyapatite/zirconia bioceramic nanocomposites for orthopaedic and dental prosthesis applications. *Nanotechnology,* Vol.18 No.6, pp. 065602–065607.

Takaki, M.; Mochida, M.; Uchida, N.; Saito, K. & Uematsu, L. (1992). Filter Cake Forming and Hot Isostatic Pressing for TZP-Dispersed Hydroxyapatite Composite, *J. Mater. Sci.: Mater. Med.*, Vol.**3**, pp. 199–203

Thamaraiselvi, T.V.; Prabakaran, K. & Rajeswari, S. (2006) Synthesis of Hydroxyapatite that Mimic Bone Minerology, *Trends Biomater. Artif. Organs, Vol.*19, No.2, pp. 81-83

Vaidhyanathan, B.; Arun Kumar, K.; Shetty, J. N. & Rao, K. J. (1997). Biological evaluation of a zirconia toughened apatitic composite implant, *Bull. Mater. Sci.*, Vol.20, pp. 135–140

Vasconcelos, H.C. & Barreto, M.C. (2011) Tailoring the Microstructure of Sol–Gel Derived Hydroxyapatite/Zirconia Nanocrystalline Composites, *Nanoscale Res Lett* 2011, Vol.**6**, doi:10.1007/s11671-010-9766-z

Veljovic, Dj.; Jokic, B.; Jankovic-Castvan, I.; Smiciklas, I.; Petrovic, R. & Dj. Janackovic, (2007). Sintering behaviour of nanosized HAP powder, *Key Eng. Mater. Vol.*330–332 pp. 259–262.

Veljovic Dj.; Jokic, B.; Petrovic, R.; Palcevskis, E.; Dindune, A.; Mihailescu, I. N. & Janackovic, Dj. (2009). Processing of dense nanostructured HAP ceramics by sintering and hot pressing, *Ceram. Int.* Vol.35, pp. 1407–1413.

Veljovic, Dj.; Zalite, I.; Palcevskis, E.; Smiciklas, I.; Petrovic, R. & anackovic, Dj. (2010). Microwave sintering of fine grained HAP and HAP/TCP bioceramics, *J Ceramics International,* Vol.36, pp.595–603

Wu, J. M. & Yen, T. S. (1988). Sintering of hydroxyapatite-zirconia composite materials, *J. Mater. Sci.*, Vol.23, pp. 3771–3777.

Yeh, T.S. &, Sacks, M.D. (1990). In *Sintering of Advanced Ceramics,* Handwerker CA, Blendell JE, Kaysser WA (ed). pp 309–331.American Ceramic Society: Westerville, OH

Yu, M. Q.; Fan, S. G.; Shen, Q. & Zhang, L. M. (2003), Toughening and Strengthening Mechanism of Zirconia-Alumina Multiphase Ceramics, *Key Engineering Materials,* Vol.249, pp.167-170

10

Ceramics in Dentistry

R. Narasimha Raghavan
Dental Surgeon, Chennai
India

1. Introduction

It is quite usual in dentistry to adopt a material from engineers and adapt it to clinical conditions. A good example of such an instance is dental ceramics. In Dental science, ceramics are referred to as nonmetallic, inorganic structures primarily containing compounds of oxygen with one or more metallic or semi-metallic elements. They are usually sodium, potassium, calcium, magnesium, aluminum, silicon, phosphorus, zirconium & titanium.

As we peep into the dental history, a French dentist De Chemant patented the first porcelain tooth material in 1789. In 1808 Fonzi, an Italian dentist invented a "terrometallic" porcelain tooth that was held in place by a platinum pin or frame. Ash developed an improved version of the platinum tooth in 1837. Dr. Charles Land patented the first Ceramic crowns in 1903.Vita Zahnfabrik introduced the first commercial porcelain in 1963.

Structurally, dental ceramics contain a crystal phase and a glass phase based on the silica structure, characterized by a silica tetrahedra, containing central Si^{4+} ion with four O^{-} ions. It is not closely packed, having both covalent and ionic characteristics. The usual dental ceramic, is glassy in nature, with short range crystallinity. The only true crystalline ceramic used at present in restorative dentistry is Alumina (Al_2O_3), which is one of the hardest and strongest oxides known. Ceramics composed of single element are rare. Diamond is a major ceramic of this type, hardest natural material used to cut tooth enamel. Ceramics are widely used in dentistry due to its dual role - strength and esthetics.

Basically the inorganic composition of teeth and bones are ceramics - Hydroxyapatite. Hence ceramics like hydroxyapatite, wollastonite etc are used as bone graft materials. They have an entire plethora of synthetic techniques like wet chemical, sol-gel, hydrothermal methods etc. Also they are added as bioactive filler particles to other inert materials like polymers or coated over metallic implants. These ceramics are collectively called as bioceramics. There are basically two kinds of bioceramics-inert (e.g. Alumina) and bioactive (hydroxyapatite). They can be resorbable (Tricalciumphosphate) or non-resorbable (Zirconia).

Dental cements are basically glasses. Initially, silicate cements were introduced. They constitute the first dental cement to use glass as its component. The cement powder contains a glass of silica, alumina and fluorides. The liquid, is an aqueous solution of phosphoric acid with buffer salts. Fluoride ions leached out from the set cements are responsible for the anti-cariogenic property. But silicates are discontinued due to low pH during setting reaction that affects the dental pulp.

In cements called glass ionomers, the glass forms the filler and acidic polymers form the matrix. The reaction that proceeds is usually an acid base reaction. Usually they are dispensed as powder containing glass powder, and liquid containing an acid, say, polyacrylic acid. When they are mixed, the acid in liquid etches the glass and reacts with calcium and other ions forming salts and the cement sets to a hard mass. Cements are direct restorative materials - i.e. manipulated and placed onto teeth directly, unlike other restorations that are made outside and fixed to teeth, called indirect restorations. They are highly advantageous as they are quick to set, release fluoride leading to anti-cariogenic action, esthetic and chemically bond to tooth material.

The use of ceramics are encouraged by their biocompatibility, aesthetics, durability and easier customization. The specialty of ceramic teeth is the ability to mimic the natural tooth in colour and translucency along with strength. Ceramics have excellent intraoral stability and wear resistance adding to their durability.

Dental ceramics, since introduction have undergone numerous modifications in terms of chemistry. Ceramics have been able to give heed to the ever changing needs in dentistry. To delve deep into the relevance of ceramic in dentistry, one should understand the physics of forces acting in the oral cavity. The masticatory (chewing) force is the strongest force present here. Other minor forces include that of tongue and periodontal ligament, which do not relate to the use of ceramics in dentistry.

The masticatory force is generated outside oral cavity by basically strong muscles, that move the jaw, open it or close it. Closure of jaw produces two kinds of forces. It is predominantly compressive in nature. Frequently impact kind of force is also experienced. Hence a ceramic has to undergo cycles of these forces indefinitely, without fracture, to result in a successful restoration of lost teeth structures.

In order to have a complete idea of what ceramic means to dentistry, we need to look at the complete range of ceramics used in this discipline.

1.1 Classification of dental ceramics

Classification of ceramics in dentistry is apparently an impossible task due to vast improvements made in the compositions. Nevertheless, the table provided here gives a general idea, say a bird's eye view of ceramics in dentistry.

Microstructural Classification	Category 1: Glass-based systems (mainly silica) Category 2: Glass-based systems (mainly silica) with fillers usually crystalline (typically leucite or a different high-fusing glass) a) Low-to-moderate leucite-containing feldspathic glass b) High-leucite (approx. 50%)-containing glass, glass-ceramics (Eg: IPS Empress) c) Lithium disilicate glass-ceramics (IPS e.max® pressable and machinable ceramics) Category 3: Crystalline-based systems with glass fillers (mainly alumina) Category 4: Polycrystalline solids (alumina and zirconia)

Based on processing technique	1) Powder/liquid glass-based systems 2) Pressable blocks of glass-based systems 3) CAD/CAM systems
Based on Composition	Silicates: These are characterized by amorphous glass phase, containing predominantly silica. Oxide ceramics: It is notable that only oxide ceramics are used in dentistry, since nonoxide ceramics are difficult to process. Oxide ceramics contain a principal crystalline phase like Alumina. Zirconia has very high fracture toughness. Glass ceramics: These are type of ceramics that contains a glass matrix phase & at least one crystal phase. Although classification of dental ceramics based on composition is not much of importance today, due to advances made, it is included for historic importance.
Based on Type	Feldspathic porcelain. Leucite – reinforced porcelain, Aluminous porcelain. Glass infiltrated alumina, Glass infiltrated zirconia. Glass ceramics.
Based on firing temperature	Ultra-low fusing < 850°C Low fusing 850°C - 1100°C Medium Fusing 1101°C - 1300°C High fusing >1300°C
Based on sub-structure metal	Cast Metal, Swaged metal, Glass ceramics. Sintered core ceramics and CAD-CAM porcelain. The various types of metals in metal ceramics include noble alloys like gold alloys, base metals like iron, indium & tin. Pure metals like commercially pure titanium, platinum, gold and palladium alloys and Base metal alloys (nickel, chromium).
Based on reinforcing method	Reinforced ceramic core systems Resin-bonded ceramics Metal–ceramics

Table 1. Classification of dental ceramics.

Material	**wt %**
Silica	~62
Alumina	~18
Boric oxide	~7
Potash (K_2O)	~7
Soda (Na_2O)	~4
Other oxides	~2

Table 2. Typical oxide composition of dental porcelain.

The classification based on microstructure will be dealt with in detail and an idea of classification by processing technique is added in this review. Other classifications are of academic interest.

Basically ceramics are used as indirect restorative materials such as crowns and bridges, Inlays/Onlays and dental implants. Recently ceramic braces are used in orthodontics.

1.2 Crowns and bridges

Crown is technically a "Cap" placed on a tooth to protect it from fracture or sensitivity. On the other hand a bridge is a fixed replacement of missing teeth, with support from adjacent teeth. Both of these function similarly on biological and biomechanical terms, hence discussed together. These are of either *Porcelain fused to metal* (PFMs) or *full ceramics*. In case of PFMs, a metal core is placed in the tooth surface and ceramic is built on it. This is done by initially preparing the metallic portion by conventional casting techniques. Then the ceramic powder is incrementally painted on it and sintered. This is followed by glazing. In case of full ceramics, the wax pattern is prepared for the crown, it is invested and mould space prepared by lost wax technique. The ceramic is fused and typically pressed into the mould space.

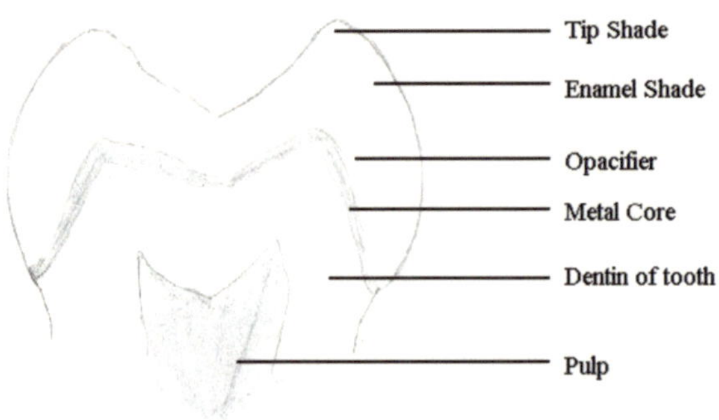

Fig. 1. Ceramic Crowns and their anatomy.

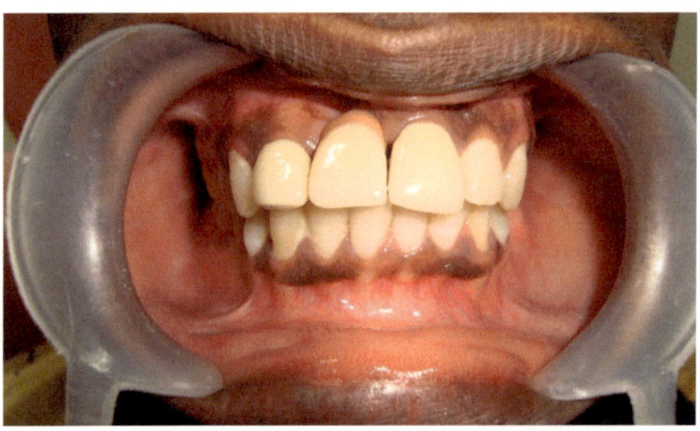

Fig. 2. Clinical picture of crown. Note pink ceramic used of gum recession.

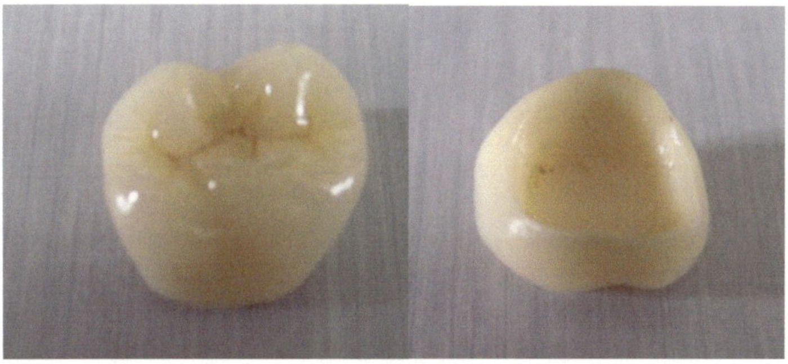

Fig. 3. All Ceramic Crown.

Similarly, ceramic teeth are manufactured in various shades, shapes and sizes to be used in complete dentures. Also, in case of gum recession, in fixed prosthesis, pink coloured ceramics are placed in the lost gum region to make it look natural.

1.2.1 Category 1: Glass-based systems (mainly silica)

Chemical composition of these ceramics is based on silica network and potash feldspar ($K_2O.Al_2O_3.6SiO_2$) or soda feldspar ($Na_2O.Al_2O_3.6SiO_2$) or both. Potassium and sodium feldspar are naturally occurring minerals composed primarily of Potash and soda. The most important property of feldspar is its tendency to form crystalline mineral leucite when melted.

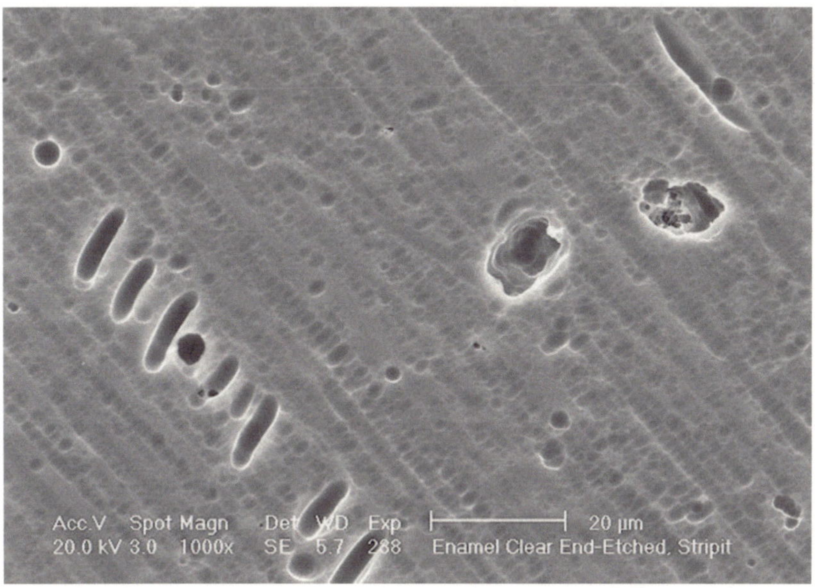

Fig. 4. Scanning electron micrograph of amorphous glass (Russel Giordano et al, 2010).

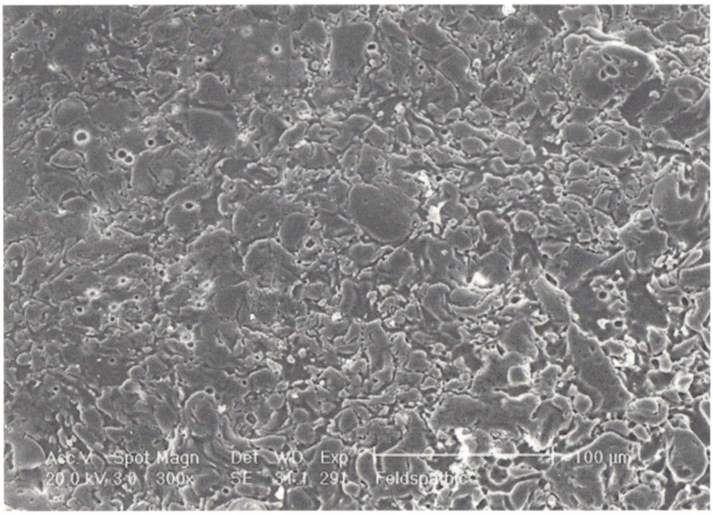

Fig. 5. SEM image of Feldspathic porcelain (Russel Giordano et al, 2010).

Mechanical properties say, flexural strength usually ranges from 60 MPa to 70 MPa. Hence, they can be used as veneer materials for metal or ceramic substructures, as well as for veneers, using either a refractory die technique or platinum foil.

Then, few other components like pigments, opacifiers and glasses are added to control fusion and sintering temperature, thermal properties and solubility. Glass modifiers like boric oxide can be added to reduce viscosity and softening temperature. Pigments are basically metallic oxides - say nickel oxide, manganese oxide, etc. Tin oxide is used for opaquing, iron oxide used for brown, copper oxide for green, titanium oxide for yellow, manganese oxide for purple, cobalt oxide for blue, nickel oxide for brown and rare earth oxides for simulating ultraviolet reflectance of natural teeth in ceramic.

The pigment oxides are also called as colour frits. These are added in appropriate proportions, dictated by intensity of colour required. Then the material is fired and fused to form glass, that is powdered again.

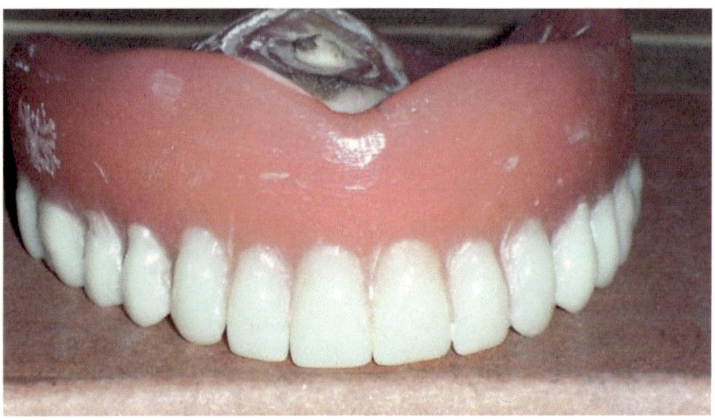

Fig. 6. Acrylic Denture with Ceramic teeth in lab processing stage.

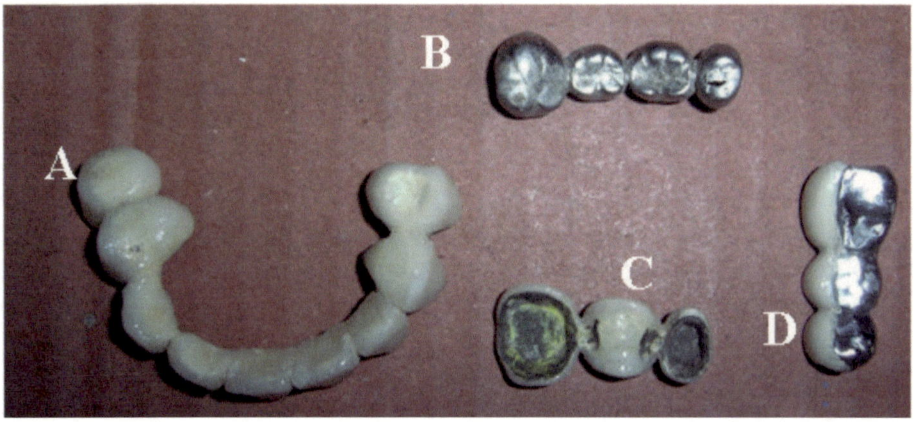

Fig. 7. Different kinds of fixed Dentures A)PFM Bridge, B) Full metal bridge, C)Inner side of PFM bridge, D) Metal bridge with ceramic facing.

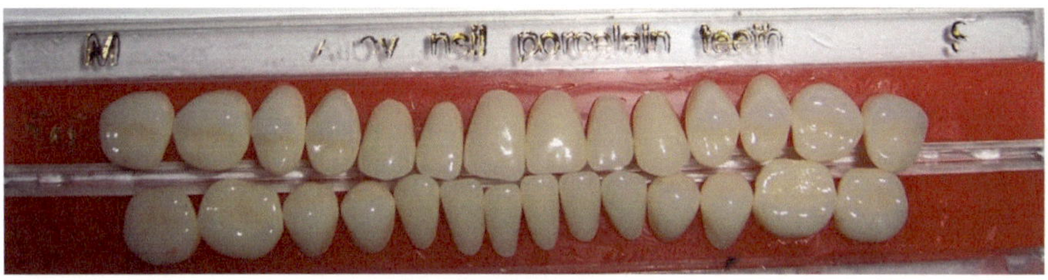

Fig. 8. Porcelain teeth set for removable prosthesis.

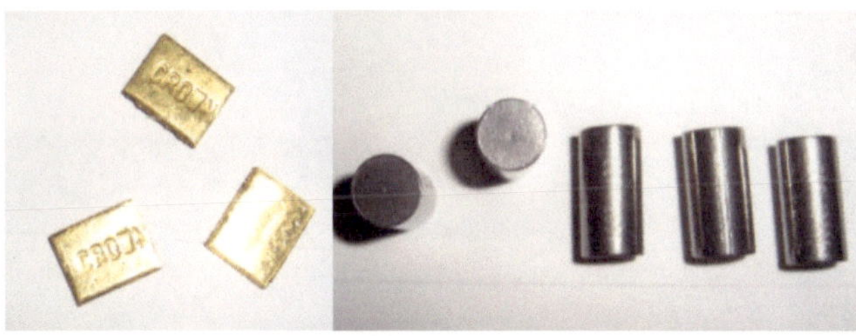

Fig. 9. Representative alloys for PFM cores.

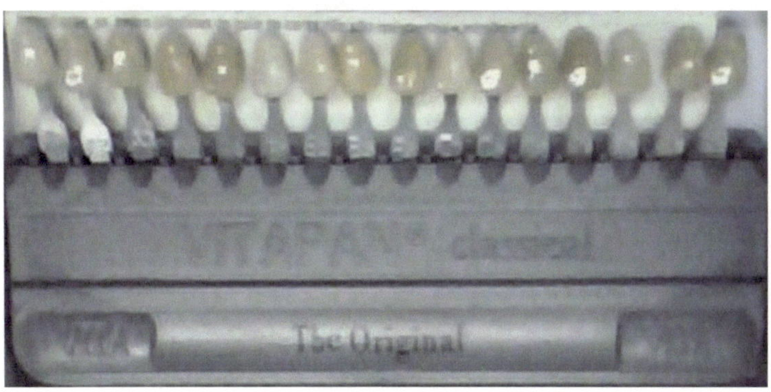

Fig. 10. Commonly used dental shade guide showing corresponding numbers for shades.

These ceramics are strengthened by either development of residual compressive stresses within the surface or by interruption of crack propagation through the material. Residual compressive stresses are introduced by ion exchange and thermal tempering. Interruption of crack propagation is in turn achieved by dispersion of crystalline phases like partially stabilized Zirconia.

Glass ceramics made of a glass matrix phase and at least one crystal phase is produced by controlled crystallization of glass. It is available as castable machinable, pressable and infiltrated forms used in all-ceramic restorations. The first commercially available castable glass ceramic was developed by the Corning Glass Works (Dicor®) in 1950s. It paved way to dental ceramic system relying upon strengthening of glass with various forms of mica. ($SiO_2.K_2O.MgO.Al_2O_3.ZrO_2$, with the addition of some fluorides). Fluorides present in these ceramics are responsible for their nature-like fluorescence. Mould space is created by lost wax method and desired shape is formed. This process followed by coating with veneering porcelain. The noteworthy aspect of this ceramic is the *Chameleon effect* in which some part of color is picked up from adjacent teeth. Here, the ceraming process results in the nucleation and the growth of tetrasilicate mica crystals within the glass. The crystals are needle-like in shape and arrest the propagation of cracks. Mechanical property measurements suggest the flexural strength is in the range of 120–150MPa, may just be adequate for posterior crowns but is not sufficient for the construction of all-ceramic bridges. The passage of light through the material is affected by the crystal size and the difference in the refractive indices of the glass phase and the crystalline phase. If the crystals are smaller than the wavelength of visible light (0.4–0.7mm) the glass will appear transparent.

The refractive index of the small mica crystals is closely matched to that of the surrounding glass phase, such that the tendency for light to scatter is lower than the aluminous porcelains.

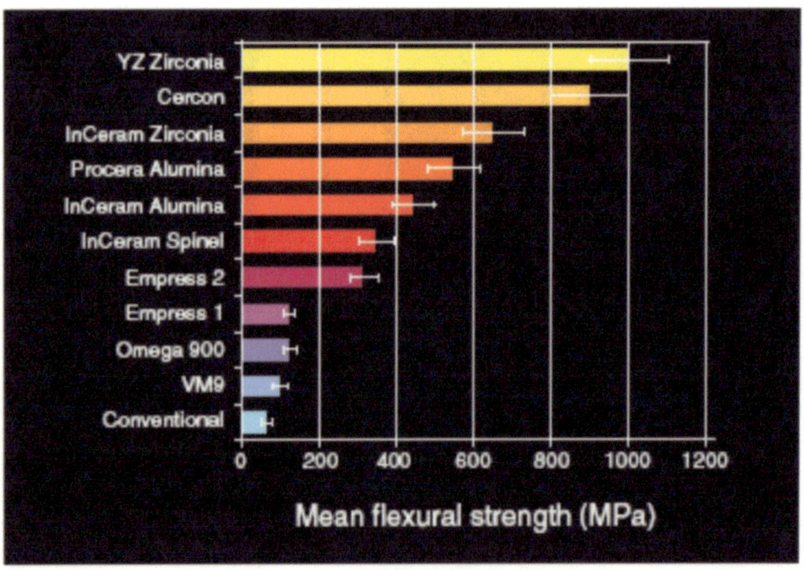

Fig. 11. Flexural strengths of various ceramics (McLaren EA et al, 2005).

The Machinable glass ceramic is another high quality product, which is crystallized during the manufacture and provided as CAD/CAM blanks or ingots. They provide better precision than castable glass ceramic, due to reduction of casting errors. They exhibit similar mechanical properties, to castable forms but are less translucent.

1.2.2 Category 2: Glass-based systems (mainly silica) with fillers usually crystalline (typically leucite or a different high-fusing glass)

This is a modification to category 1, in that varying amounts of other crystals are added or grown. The primary crystal types are either leucite, lithium disilicate, or fluorapatite.

Leucite has been widely used as a constituent of dental ceramics to modify the coefficient of thermal expansion. This is most important where the ceramic is to be fused or baked onto metal (Optec HSP). But in leucite reinforced ceramic system, IPS Empress, leucite has a different role. This material relies on an increased volume of fine leucite particles to increase flexural strength. Leucite is nothing but potassium aluminum silicate mineral with large co-efficient of thermal expansion compared with glasses. The property of Feldspar to form Leucite is exploited in the manufacture of porcelains for metal bonding. Newer generations of materials have much finer leucite crystals (10 μm to 20 μm) and even particle distribution throughout the glass. These materials are less abrasive and have much higher flexural strengths. In Figure 5, a scanning electron micrograph (SEM) of a typical feldspathic porcelain reveals a glass matrix surrounding leucite crystals. These materials are most commonly used as veneer porcelains for metal-ceramic restorations.

Fig. 12. SEM Image of IPS-Empress I - Etched (Russel Giordano et al, 2010).

In the sintering process (Fortress and Optec-HSP), slurry of the ceramic powder is applied to a refractory die (unlike Platinum foil coated die in the procedure of the porcelain jacket crown), dried and subsequently fired in a porcelain furnace. Multiple layers can be built up to develop characterisation. Great skill is required by the dental laboratory technician to get the best aesthetics and appropriate contour. However, the strength of Leucite ceramics are insufficient for all-ceramic restorations.

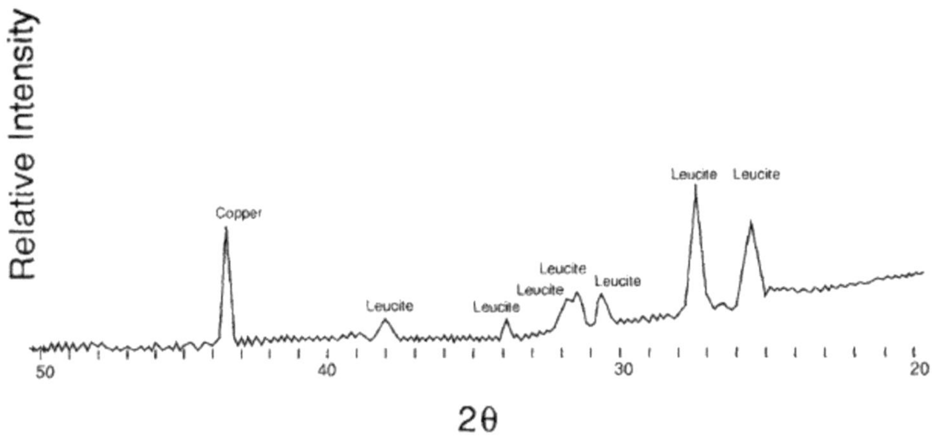

Fig. 13. XRD of Optec HSP (Peter W. PichC et al, 1994).

Lithium Disilicate and Apatite Glass Ceramics are based on SiO_2–Li_2O. (Empress II, Ivoclar-Vivadent). The crystalline phase, lithium disilicate ($Li_2Si_2O_5$) makes up about 70% of the volume of the glass ceramic. Lithium disilicate has microstructure consisting of numerous small plate-like crystals that are interlocking and randomly oriented. This has a reinforcing effect on strength because, the needle-like crystals deflect cracks and arrest the propagation of cracks. A second crystalline phase, consisting of a lithium orthophosphate (Li_3PO_4) of a much lower volume, is present. The mechanical properties of this glass ceramic are far superior to that of the leucite glass ceramic, with a flexural strength in the region of 350–450MPa and a fracture toughness approximately three times that of the leucite glass ceramic. There is a possibility for its use in all-ceramic systems. Also, processing is done by hot pressing technique.

For the alumina-based core systems feldspathic glasses can be used to provide the aesthetic surface layer, as their coefficients of expansion are closely matched. For the leucite glass ceramics the layering ceramic is identical to the core ceramic and so a mismatch in coefficient of expansion does not arise. However, this is not possible for lithium disilicate glass ceramic due to its higher coefficient of expansion. Here emerges a need for new compatible layering ceramic. This new layering ceramic is an apatite glass ceramic. The crystalline phase formed on ceraming is hydroxyapatite ($Ca_{10}(PO_4)_6OH_2$), which is the basic constituent of enamel. Thus, it represents a material that, at least in composition, is the closest match to enamel.

A veneer porcelain made of fluorapatite crystals in an aluminosilicate glass may be layered on the core to create the final morphology and shade of the restoration. The shape and volume of crystals increase the flexural strength to approximately 360 MPa, or about three times that of Empress. This material can be translucent even with the high crystalline content; this is due to the relatively low refractive index of the lithium-disilicate crystals. The material is translucent enough that it can be used for full-contour restorations or for the highest esthetics and can be veneered with special porcelain. Veneer porcelain consisting of fluorapitite crystals in an aluminosilicate glass may be layered on the core to create the final morphology and shade of the restoration. The fluorapatite (fluoride-containing calcium phosphate, Ca5(PO4)3F) crystals contribute to the veneering porcelain's optical properties and CTE, so it matches the lithium-disilicate pressable or machinable material. Both the veneering and lithium-disilicate materials are etchable due to the glassy phase. Initial clinical data for single restorations are excellent with this material, especially if it is bonded. A material with similar properties and structure called 3G OPC is available as a pressable glass-ceramic from Pentron.

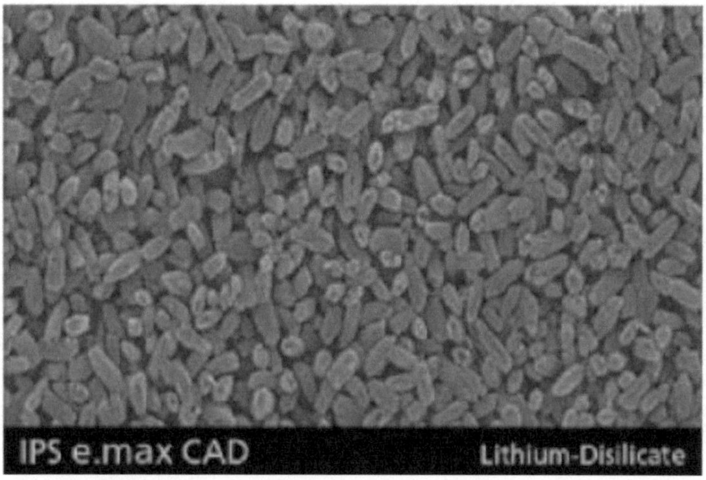

Fig. 14. SEM image of IPS e.max. CAD (McLaren Et al, 2009).

Advantage of PFM is its high strength in clinical service. It uses the sub-structure metal to withstand stresses. Thermal compatibility is good. Lower crack propagation leading to better fracture resistance. But, on the other side of the coin, there is inadequate structure for ceramics due to thickness of metal. This results in reduction of more tooth structure while preparing it for the restoration. Metallic hue can be visible in anterior teeth. Metal will be exposed in case of gingival recession. Patients with allergy to metals can react adversely. As we add more layers between materials, more number of fractures can occur. Bonding failures at porcelain-metal interface can occur due to oxidation of interface metallic surface.

It is good to have a glance into preparation of tooth structure for a full crown, to understand the pros and cons of ceramics. This procedure involves reduction of size of the tooth to accommodate the crown. Briefly, the aim of the procedure is to achieve good retention of restoration on crown, resistance of tooth to fracture, facilitate good chewing, protection of dental pulp while maintaining conservation of tooth structure. Unlike metals that are strong in thin sections, ceramics are strong only in bulk. Hence, when a particular region of crown is involved in mastication, more of tooth material is reduced at that particular region, called functional cusp bevel, to provide bulk to the ceramic. In such cases, amount of tooth reduction will be the sum of clearance for metal and ceramic in PFM and only for ceramic in all ceramic crowns. This definitely projects the tooth conservation in all-ceramic crowns.

In order to overcome the disadvantages of PFMs, and to achieve closer tooth colour match, All-Ceramics were developed. Natural teeth always permit diffuse transmission and regular transmission. Prosthetic teeth must also possess a similar depth of translucency, which is realized in all-Ceramic restorations. Aluminous porcelain, Glass Ceramics, Castable, Machinable and Pressable Glass infiltrated, CAD-CAM and Cercon Zirconia system are few examples of all-ceramic systems.

1.2.3 Category 3: Crystalline-based systems with glass fillers

Aluminous porcelain contains a glass matrix phase and at least 35 vol% of Alumina. It is a commonly used core ceramic and has a thin platinum foil when employed with all ceramic restorations. Aluminous core is stronger than feldspathic porcelain.

Construction of the Procera AllCeram core involves die production from the impression, digitising the geometry of coping using computer software, and transferring this information to a laboratory in Stockholm. The coping is produced by a process, which involves sintering 99.5% pure alumina at 1600–1700°C such that it is completely densified. The coping is then returned to the dental laboratory for building in the crown's aesthetics using compatible feldspathic glasses. The flexural strength of the Al_2O_3 core materials is in the region of 700MPa, and thus similar to that achieved with the In-Ceram Zirconia.

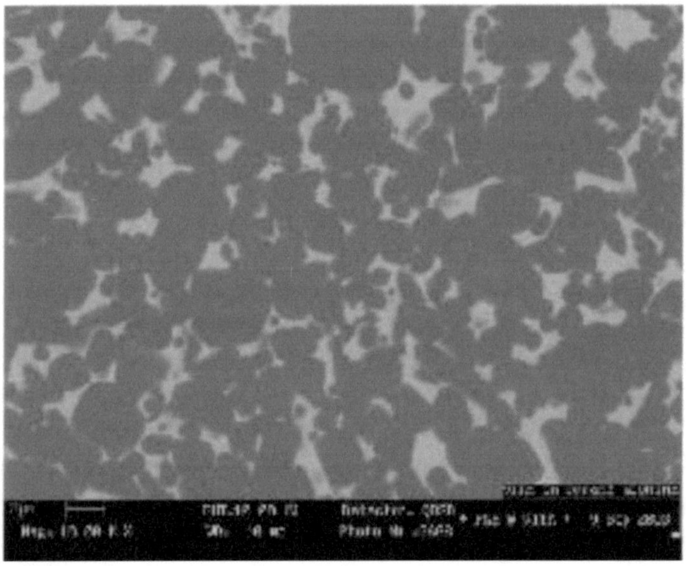

Fig. 15. High density of In-Ceram powder. (McLaren Et al 2009).

Pure alumina cores are produced by Techceram Ltd. In this system the impression can be sent to Techceram Ltd, who will produce a special die onto which the alumina core is deposited using a thermal gun-spray technique. This process produces an alumina core with a density of 80–90%, which is subsequently sintered at 1170°C to achieve optimum strength and translucency. The alumina coping is then returned to the dental laboratory, where the ceramist will develop the final contour and aesthetics using conventional feldspathic glasses. The clinical significance is the better translucency than glass alumina ceramics.

Fig. 16. SEM image of In-Ceram surface (Russel Giordano et al, 2010).

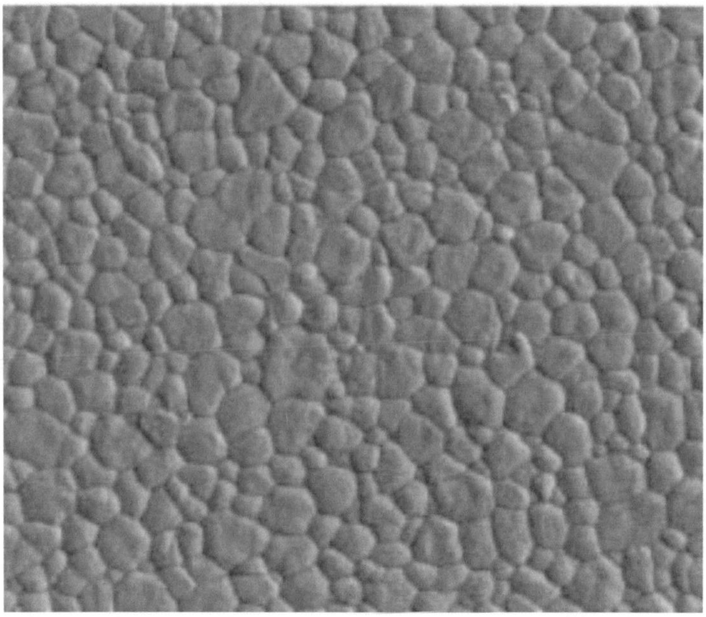

Fig. 17. SEM image of Sintered Zirconia (Lava) (McLaren Et al, 2009).

The use of modern Aluminous Crowns rose in the mid 1960's by McLean. The Nobel Biocare company (Sweden) introduced systems of pressing alumina onto a metal die, removing the pressed shape from the die and then sintering it. They are used to make cores to build up ceramic superstructures for dental implants (CeraOne®), and the second for conventional crowns (Procera®). Here, there is no glassy phase present between the particles. Feldspathic veneering porcelains are then fired onto this core to provide the necessary colour and form for the final restoration.

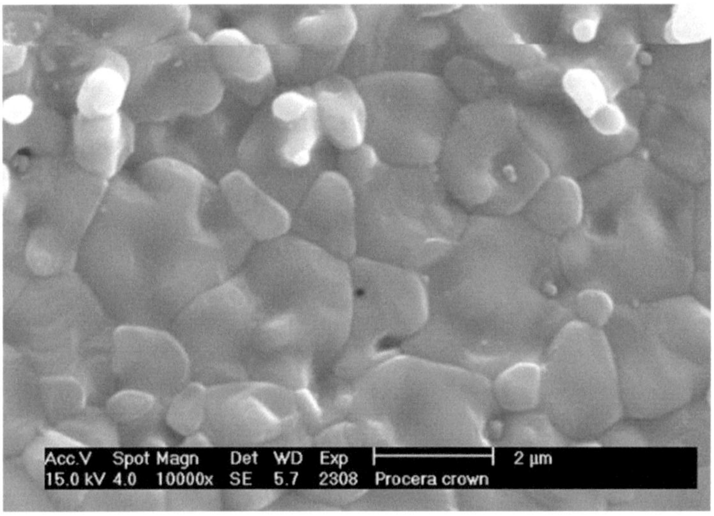

Fig. 18. SEM Image of Procera Crown (Russel Giordano et al, 2010).

The Pressable glass ceramics (IPS Empress 1 & 2) involve pressure molding in the manufacture. Heated ceramic ingot is pressed through a heated tube into a mold, where the ceramic form cools and hardens to the shape of the mold, which is later recovered after cooling. Hot pressing usually occurs over a 45mins at a high temperature to produce the ceramic sub-structure. Then it is stained, glazed or coated by veneering porcelain, according to esthetic needs, which results in translucent ceramic core, moderately high flexural strength, excellent fit & excellent esthetics.

The 1980s witnessed the development of slip casting system using fine grained alumina. The cast alumina was initially sintered and then infiltrated with a Lanthana based glass. Onto this alumina core, a feldspathic ceramic could be baked to provide form, function and esthetics to the crown. Glass infiltrated ceramic is used as one of the 3 core ceramics namely, In-Ceram Spinell, In-Ceram Alumina and In-Ceram Zirconia. They use the technique of slip-cast on a porous refractory die and heated in a furnace to produce a partially sintered coping or framework which is infiltrated with glass at 1100^{0}C for about 4 hrs to strengthen the slip-cast core. They possess high flexural strength and can be successfully cemented using any cement. CAD/CAM involves a technique wherein the internal surface is ground with diamond discs to the dimensions obtained from a scanned image of the preparation.

In-Ceram® (Vident, www.vident.com) consists of a family of all-ceramic restorative materials. The family encompasses a range of strengths, translucencies, and fabrication methodology designed to cover the wide scope of all-ceramic restorations, including veneers, inlays, onlays, and anterior/posterior crowns and bridges. In-Ceram Spinell (alumina and magnesia matrix) is the most translucent with moderately high strength and used for anterior crowns. In-Ceram Alumina (alumina matrix) has high strength and moderate translucency and is used for anterior and posterior crowns. In-Ceram Zirconia (alumina and zirconia matrix) has very high strength and lower translucency and is used primarily for three-unit posterior bridges. In addition, these materials are supplied in a block form for producing milled restorations using a variety of machining systems.

In-Ceram is in a class called interpenetrating phase composites. They consist of at least two phases, which are intertwined and extend continuously from the internal to external surfaces. This class has better mechanical and physical properties relative to the individual components; a tortuous route through alternating layers of both components is required in order for these materials to break.

Interpenetrating phase materials are generally fabricated by first creating a porous matrix; in the case of In-Ceram, it would be a ceramic "sponge." The pores are then filled by a second-phase material, lanthanum-aluminosilicate glass, using capillary action to draw a liquid or molten glass into all the pores to produce the dense interpenetrating material.

The system met with great success as an alternative to conventional metal-ceramics. It uses a sintered crystalline matrix of a high-modulus material (85% of the volume) in which there is a junction of the particles in the crystalline phase. This is different from glasses or glass-ceramic materials in that these ceramics consist of a glass matrix with or without a crystalline filler in which there is no junction of particles (crystals). Slip casting may be used to fabricate the ceramic matrix, or it can be milled from a presintered block. Flexural strengths range from 350 MPa for spinell, 450 MPa for alumina, and up to 650 MPa for zirconia. Clinical studies support In-Ceram Alumina as an almost all purpose ceramic. In-Ceram Alumina had the same survival rates as porcelain-fused-to-metal restorations up to the first molar, with a slightly higher failure rate for the second molar. In-Ceram Zirconia should be used on molars only due to its very high opacity, which is not suitable for anterior esthetics. For anterior teeth, the alumina magnesia version of In-Ceram (called Spinell) is ideal due to its higher translucency.

1.2.4 Polycrystalline solids

Solid-sintered monophase ceramics are formed by directly sintering crystals together without any intervening matrix to form a dense, air-free, glass-free, polycrystalline structure. Several processing techniques allow the fabrication of either solid-sintered aluminous oxide (alumina, Al_2O_3) or zirconium oxide (ZrO_2) framework. The first fully dense polycrystalline material for dental applications was Procera® AllCeram alumina (Nobel Biocare, www.nobelbiocare.com), with a strength of approximately 600 MPa. The alumina powder is pressed and milled on a die and sintered at about 1600°C, leading to a dense coping but with approximately 20% shrinkage.

The use of zirconia has increased rapidly in the past few years. This is Zirconia, partially stabilized with small amounts of other metal oxides. Partially stabilized zirconia allows production of reliable multiple-unit all-ceramic restorations for posterior teeth, since they produce high stress. Zirconia may exist in several crystal types (phases) depending on the addition of minor components. Typically for dental applications, about 3 wt% of yttria is added to pure zirconia.

Zirconia has unique physical characteristics that make it twice as strong and tough as alumina-based ceramics. Values for flexural strength range from approximately 900 MPa to 1100 MPa. There is no direct correlation between flexural strength (modulus of rupture) and clinical performance. Another important physical property is fracture toughness, which has been reported between 8 MPa and 10 MPa for zirconia. This is significantly higher than other dental ceramics. Fracture toughness is a measure of a material's ability to resist crack growth. Zirconia has the apparent physical properties to be used for multiple-unit anterior and posterior FPDs. Clinical reports on zirconia have not shown any problem with the framework, but have shown the chipping and cracking of porcelain. Using a slow-cooling protocol at the glaze bake to equalize the heat dissipation from zirconia and porcelain increased the fracture resistance of the porcelain by 20%. Zirconia may be in the form of blocks that are milled to create the frameworks (CAD/CAM). Mostly, they are fabricated from a porous block, milled oversized by about 25%, and sintered to full density in a 4 - 6 hours cycle. Alternatively, fully dense blocks are milled. However, this approach requires approximately 2 hours of milling time per unit whereas milling of the porous block necessitates only 30 to 45 minutes for a three-unit bridge.

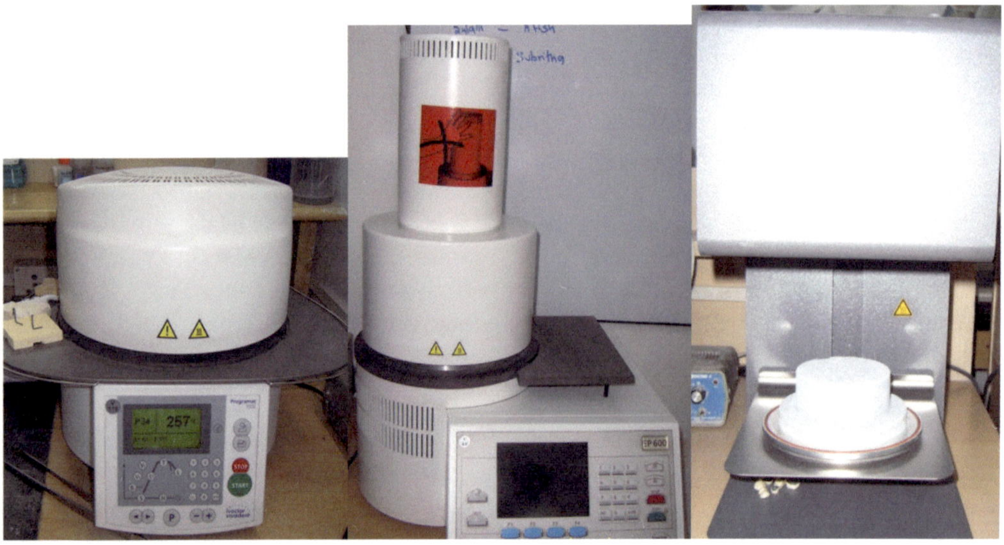

Fig. 19. Dental Ceramic Furnaces.

1.3 Inlays and onlays

Basically, Inlay is an indirect filling placed on teeth. Onlay is a more extensive filling (Involving more that one cusp in case of molars). They work in similar fashion. These are made of many kinds of previously mentioned ceramics. They are fabricated by CAD/CAM technique.

1.4 Implants

Implant is basically a pillar/post placed into bone to act as a replacement to root of a tooth, on which a denture - fixed/removable is placed. Usually these are made from titanium and its alloys. Basic criteria for a material to be used as an implant is strength and bone biocompatibility. Recently Zirconia implants are introduced in to the market. With an increasing consciousness and fear for metal allergy, Zirconia implants are gaining momentum. Their advantage is their non allergic nature. In clinical studies, no difference in performance is reported. Their basic disadvantage over titanium is their brittleness. Also

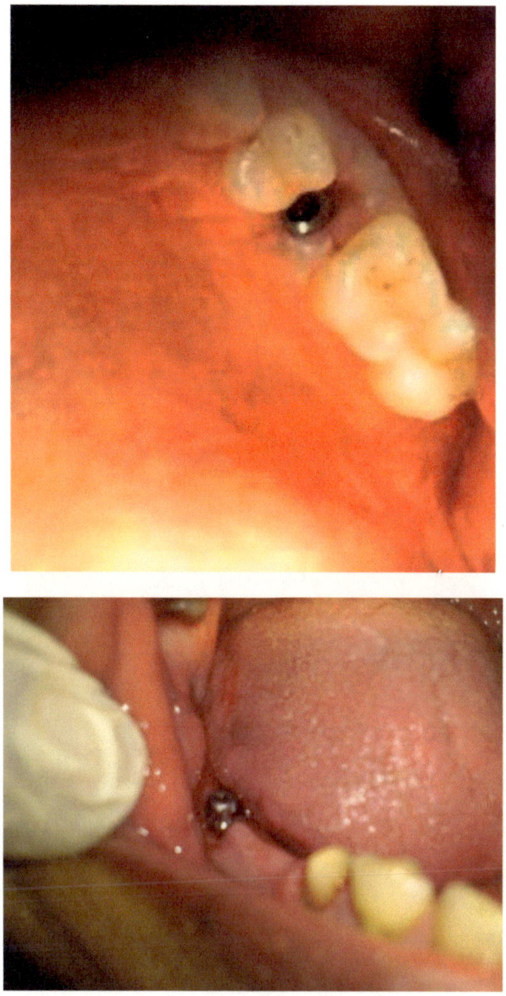

Fig. 20. Picture of titanium implant in place - clinical picture.

hydroxyapatite coated titanium implants are already into the market. These aim at better tissue response and osseointegration. Forces experienced by the implants are basically compressive and bending. Even though alumina has good mechanical properties, it is inert towards bone, hence not used in implant dentistry.

Zirconia implants (Ceraroot® etc) are compressed, sintered and milled to produce the necessary shape and surface texture. In case of dental implants, it is a proven fact that shape and surface texture influence the osseointegration. Right from introduction of implants in dentistry, lots of shapes have been tried. Lots of patented designs are found regarding these designs and surfaces. The researches are focused on chemistry, morphology and surface characteristics for betterment of implant success.

1.5 Braces

The science of braces - called Orthodontics, describes the corrective procedure as unaesthetic metamorphosis to an esthetic result. The aim of ceramic braces is to reverse the above description. The primary use of these brackets is esthetics. Ceramic braces as such do not experience masticatory forces, But are subjected to the orthodontic forces like sliding over wires, torquing etc. Due to its high wear coefficient, when wrongly placed can cause attrition of opposing teeth. They are basically polycrystalline alumina or Zirconia.

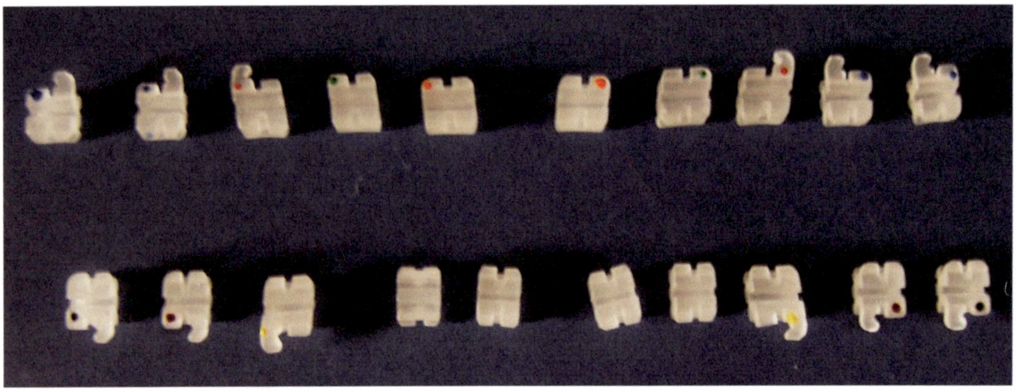

Fig. 21. Ceramic braces, colour codes showing the correct tooth to be cemented to.

2. Classification based on processing techniques

For ease of understanding, classification according to processing techniques is advocated. Generally, there are

1. Powder/liquid glass-based systems
2. Pressable blocks of glass-based systems
3. CAD/CAM systems

This system is more relevant clinically. In spite of same chemistry and microstructure, the processing history determines the properties relevant to clinical performance. Specifically, machined blocks of materials have performed better than powder/liquid versions of the same material.

1. Powder/liquid systems

These can be conventional or slip casted. The conventional construction of a porcelain crown involves compaction, firing and glazing. Briefly compaction involves mixing of powder with water and binder to form a paste by spatulation, brush application, whipping or vibrating, that is aimed at compaction, which is painted over the die that is previously coated with platinum foil. This paste is made from different porcelain powders to mimic the esthetics of natural teeth. Usually, an opaque shade (to mask metal core), a dentin shade and then enamel shade is used. The enamel shade is selected from shade guide matched to patients actual tooth shade.

The objective of condensation techniques is to remove water, resulting in a more compact arrangement with high density of particles that reduces the firing shrinkage. The particle shape and size affect the handling characteristics of the powder and have influence on firing shrinkage. The binder helps to hold the fragile particles together in this so-called green state.

Firing initially involves slow heating of crown in the open entrance to the furnace, to drive off excess water before it forms steam that cracks the ceramic. Dried compact is placed in the furnace and the binders are burnt out. Some contraction occurs in this stage. When the porcelain begins to fuse, continuity is achieved at contact points between the powder particles. The material is still porous, and is usually referred to as being at the low bisque stage. As the higher temperature prevails for longer time, more fusion takes place as the molten glass flows between the particles, resulting in more compaction and filling the voids. A large contraction takes place during this phase (~20%), and the resultant material is apparently non-porous. The high shrinkage is caused by fusion of the particles during sintering, and resultant close contact between particles. Longer sintering will lead to pyroplastic flow and loss of form and will become highly glazed. A very slow cooling rate is employed to avoid cracking or crazing.

The furnaces can be programmed to automate these procedures. Vacuum-firing produces a denser porcelain than air firing, as air is withdrawn during the firing process, resulting in fewer voids and a stronger crown and more predictable shade. Areas of porosity in air fired porcelain alter the translucency of the crown, as they cause light to scatter. Also, air voids become exposed on grinding of the superficial layer, compromising esthetics by giving a rough surface finish.

Glazing is done to eliminate residual surface porosity that might encourage bacterial colonization and its sequel. Glazing results in surface that is smooth, shiny and impervious. To accomplish this, either low fusing glasses are applied to crown after construction and fused, or final firing is done under controlled condition that fuses the superficial layer to make it impervious.

With regard to slip casting, the "slip" is a homogenous dispersion of ceramic powder in water. The water pH adjustment creates a charge on the ceramic particles, which are coated with a polymer to cause the fine suspension in water. In the case of In-Ceram, the slip is applied on a gypsum die to form the underlying core for the ceramic tooth. The water is absorbed by porous gypsum, leading to packing of particles into a rigid network. The alumina core is then slightly sintered in a furnace to create an interconnected porous network. The lanthanum glass powder is placed on the core and glass becomes molten and flows into the pores by capillary action to produce the interpenetrating network. The last step in the fabrication involves application of aluminous porcelain on the core to produce the final form of the restoration. Other powder dispersions, such as those created with zirconia, may be poured into a gypsum mold that removes the water and leads to formation of homogeneous block of zirconia.

2. Pressable ceramics

Pressed ceramic restorations are fabricated using a method described previously, similar to injection molding. Empress restorations and other materials with a similar leucite/glass structure are fabricated in this manner. Pressables may be used for inlays, onlays, veneers, and single-unit crowns.

3. CAD/CAM

Numerous ceramics have found their way into this system, due to its short processing time. Some of them are described here. **Glass/Crystal ceramics** are made from fine-grain powders, producing pore-free ceramics. This was the first material specifically produced for the CEREC system. It has an excellent history of clinical success for inlays, onlays, and anterior and posterior crowns. These blocks are available as monochromatic, polychromatic with stacked shades as in a layered cake, and in a form replicating the hand-fabricated crowns whereas an enamel porcelain is layered on top of dentin porcelain. **Glass/Leucite** is a feldspathic glass with approximately 45% leucite crystal component. **Lithium disilicate** is not initially fully crystallized, which improves milling time and decreases chipping risk from milling. The milled restoration is then heat-treated for 20 - 30 mins to crystallize the glass and produce the final shade and mechanical properties of the restoration. This crystallization changes the restoration from blue to a tooth shade. The microstructural and chemical composition is essentially the same. **Framework** Alumina are fabricated by pressing the alumina-based powder into a block shape. These blocks are only fired to about 75% density. After milling, these blocks are then infused with a glass in different shades to produce a 100% dense material, which is then veneered with porcelain. **Porous Alumina** frameworks may be fabricated from porous blocks of material. Pressing the alumina powder with a binder into molds produces the blocks. The blocks may be partially sintered to improve resistance to machining damage or used as pressed in a fully "green" state (unfired, with binder). The frameworks are milled from the blocks and then sintered to full density at approximately 1500°C for 4 to 6 hours. The alumina has a fine particle size of about 1μm and strength of approximately 600 MPa and is designed for anterior and posterior single units, as well as anterior three-unit bridges. **Porous Zirconia** frameworks milled from porous blocks are fabricated similarly to alumina blocks. As is the case with the alumina block, the milled zirconia framework shrinks about 25% after a 4 - 6 hours cycle at approximately 1300°C to 1500°C. The particle size is about 0.1 μm to 0.5 μm.

3. Conclusions

Any reader of this review, who is well versed in materials science, will be tempted to enter dentistry to introduce newer and better ceramics to dentistry. As one can comprehend, no material is ideal in all aspects, to natural structure of teeth. However, mechanical and biological properties of ceramics have exhibited a significant improvement in the past few decades. Recent surges in all-ceramic crowns for esthetics and durability is a highlight of such improvements. Also, with dentistry, frequently gaining new faces are catered by these ceramic materials. Further scope of research in dental ceramics can be directed to stronger and osseointegrative ceramic implants, more esthetic and strong crowns and veneers, to mention a few. Definitely, future is bound to witness a bigger revolution in field of dental ceramics, with introduction of newer ceramics and nanotechnology for the betterment of dental restorations on the lines of form, function and esthetics, along with improved biocompatibility.

4. Acknowledgements

- Dr. Jacob Mathew Phillip, Senior Lecturer, Tagore Dental College, Chennai.
- Mr. Neamathulla, Dentaffairs Co., Chennai
- Dr. T. A. Srinivasan, Dental surgeon, Chennai.
- Dr. Faizur Rahman, Ankhor Dental Labs, Chennai

5. References

Berg NG, Derand T. A 5-year evaluation of ceramic inlays (CEREC). Swed Dent J. 1997;21(4):121-127.

Brochu JF, El-Mowafy O. Longevity and clinical performance of IPS-Empress ceramic restorations–a literature review. J Can Dent Assoc. 2002;68(4):233-237.

Clarke D. Interpenetrating phase composites. J Am Ceram Soc. 1992;75:739-759.

Della Bona A, Mecholsky JJ Jr, Anusavice KJ. Fracture behavior of Lithia disilicate and leucite based ceramics. Dent Mater. 2004;20(10):956-962.

Dental Ceramics. Chap 26 in Phillip's Science of dental Materials. 10/ed. Edited by Anusavice, 1998. WB Saunders Press.

Giordano R, Pelletier L, Campbell S, et al. Flexural strength of alumina and glass components of In-Ceram. J Dent Res. 1992;71:253.

Guazzato M, Albakry M, Ringer SP, et al. Strength, fracture toughness and microstructure of a selection of all-ceramic materials. Part I. Pressable and alumina glass-infiltrated ceramics. Dent Mater. 2004;20(5):441-448.

Hegenbarth EA. Procera aluminum oxide ceramics: a new way to achieve stability, precision, and esthetics in all-ceramic restorations. Quintessence Dent Technol. 1996;20:21-34.

Hoeland W, Schweiger M, Frank M, et al. A comparison of the microstructure and properties of the IPS Empress 2 and the IPS Empress glass ceramics. J Biomed Mater Res. 2000;53(4):297-303.

Ironside JG and Swain MV. A review and critical issues of dental ceramics. Journal of the Australasian Ceramic Society, 1998; 34(2): 78-91.

Kingery WD, Bowen HK, Uhlmann DR. Introduction to Ceramics. 2nd ed. New York, NY: John Wiley and Sons;1976:1-19.

Kraemer N, Frankenberger R. Clinical performance of bonded leucite-reinforced glass ceramic inlays and onlays after 8 years. Dent Mater. 2005;21(3):262-271.

Marc A Rosenblum & Allan Schulman. A review of All-Ceramic restorations; JADA - 1997 ; 128 ; 297 - 305.

McLaren EA, Giordano RA, Pober R, et al. Material testing and layering techniques of a new two phase all glass veneering porcelain for bonded porcelain and high alumina frameworks. Quintessence Dent Technol. 2003;26:69-81.

McLaren EA, Giordano RA. Zirconia-based ceramics: material properties, esthetics, and layering techniques of a new veneering porcelain, VM9. Quintessence Dent Technol. 2005;28:99-111.

McLaren EA, White SN. Survival of In-Ceram crowns in a private practice: a prospective clinical trial. J Pros Dent. 2000;83(2):216-222.

McLaren EA and Cao PT. Ceramics in Dentistry—Part I: Classes of Materials. Inside Dentistry; Oct 2009: 94-104

Papanagiotou HP, Morgano SM, Giordano RA, et al. In vitro evaluation of low-temperature aging effects and finishing procedures on the flexural strength and structural stability of Y-TZP dental ceramics. J Prosthet Dent. 2006;96(3):154-164.

Peter W. PichC, William J, O Brien,, Carole L. Groh, and Kenneth M. Boenke. Leucite content of selected dental porcelains. Journal of Biomedical Materials Research, 1994; 28: 603-609.

Piwowarczyk A, Ottl P, Lauer HC, et al. A clinical report and overview of scientific studies and clinical procedures conducted on the 3M ESPE Lava All-Ceramic System. J Prosthodont. 2005;14(1):39-45.

Proebster L. Survival rate of In-Ceram restorations. Int J Prosthodont. 1993;6(3):259-263.

Raigrodski AJ, Chiche GJ, Potiket N, et al. The efficacy of posterior three-unit zirconium-oxide-based ceramic fixed partial dental prostheses: A prospective clinical pilot study. J Prosthet Dent. 2006;96(4):237-244.

Russell Giordano and Edward A. McLaren. Ceramics Overview: Classification by Microstructure and Processing Methods. Compendium of Dental Education; Nov-Dec 2010, 31(9): 682-697.

Sailer I, Feher A, Filser F, et al. Five year clinical results of zirconia frameworks for posterior fixed partial dentures. Int J Prosthodont. 2007;20(4):383-388.

Scotti R, Catapano S, D'Elia A. A clinical evaluation of In-Ceram crowns. Int J Prosthodont. 1995;8(4):320-323.

Seghi RR, Daher T, Caputo A. Relative flexural strength of dental restorative ceramics. Dent Mater. 1990;6(3):181-184.

Stappert CF, Guess PC, Chitmongkolsuk S, et al. All-ceramic partial coverage restorations on natural molars. Masticatory fatigue loading and fracture resistance. Am J Dent. 2007;20(1):21-26.

Sukumaran VG and Narasimha Bharadwaj. Ceramics in Dental Applications. Trends Biomater. Artif. Organs, 2006; 20(1):7-11.

Wagner J, Hiller KA, Schmalz G. Long-term clinical performance and longevity of gold alloy vs ceramic partial crowns. Clin Oral Investig. 2003;7(2):80-85.

Part 3

Glass Ceramics / Composites

11

Biocompatible Ceramic – Glass Composite – Manufacturing and Selected Physical – Mechanical Properties

Barbara Staniewicz-Brudnik[1] and Małgorzata Lekka[2]
[1]*The Institute of Advanced Manufacturing Technology, Krakow,*
[2]*The Institute of Nuclear Physics PAS, Krakow,*
Poland

1. Introduction

Biomaterials present specific group of materials, about different composition, structure and properties, which, are accepted by human organism, but some of them (like hydroxyapatite ceramics, bioglass, bioglass - ceramics, modified carbon materials) make connections with alive tissue or take part in its regeneration (Hench L.L. 1998, Krajewski A., Ravaglioli A. 2002).

From historical point of view application of synthetic material for repair of human body are dated from thousand years. It has been discovered, that some Egyptian mummies had dentist denture executed from gold. The first reports about application of ivory as an implementation materials was found in a Greek mythology (Błażewicz S., Stoch L. 2003).

It needs to be mentioned that clinical attempts of refilling of wastes of tissues and bones using a different type of material was initiated before centuries. However, real medical treatment with implementation product manufactured in commercial scale was started in the twentieth century.

In 1902 gold capsules were used for manufactured prostheses of head of femur. From this moment systematical researches on the insertion materials were done. For this group we can classify cobalt - chromium - nickel alloys (Vitallium) that are used in the orthopedics till now (as plates, nails, screws for special applications, dentist implants). Production of polymer materials opened new possibilities.

From 1930 polymetacrylan methylu (PMMA) has been used in the dentistry as a cement fulfills and in the jointing process of metallic bones prostheses, particularly in the case of hip and knee joints.

Application of corundum material by Boutin in the 1972, for manufacturing the elements of joint prostheses was the crucial moment. The following properties have revolutionized the quality of the prostheses: high strength, low friction coefficient, low degree of wear and good biocompatibility (Jaegermann Z., Ślósarczyk A. 2007)

The corundum bioceramics manufactured in different forms (dense, macro and micro porosity) makes many functions during the repair intervention. However, use of many of these materials has experimental character and they are on the clinical stage, there is also another big, commercial group of products that are applied in the clinical practice (Biolox, Synatite, Endobone, Neobone, Cerabone) (Jaegermann Z., Ślósarczyk A. 2007, Jaegermann Z., et al ,2006).

The newest trend in the regenerative medicine area is tissue engineering, which aimes to obtain medicines and decrease number of complications.

Biocompatible alumina composites are a new generation of ceramic glass materials used in tissue engineering (Jaegermann Z. 2005, Szarska S., et al, 2008 Staniewicz - Brudnik B. et al, 2010). Biomaterials substrates (inorganic, polymeric, hybrid) are two - or three dimensional scaffold, which inhabits the cells (eg fibroblasts) by growing them in vitro, and then the resulting product material and cell is implanted in place of the defect (Chen Q. Z., et al 2008, Brovarone C., Verne E. 2006). The main task of such a scaffold is the physical support for cells and the control of their proliferation, differentiation and morphogenesis (Sachlos E., Czernuszka J. T. 2003).

The basic criteria that should have the substrates (Teramoto H., et al, 2005, Staniewicz - Brudnik B., Lekka M., Bączek E. 2010, Czechowska J., Ślósarczyk A. 2011) are formulated as follows:

- the substrate should contain open pores that connect to each other about the right size to inclusion of cells, then tissues and their vascularity;
- they should have appropriate chemical properties (bioactivity, non - toxic) to promote attachment of cells to substrates, their differentiation and multiplication, mechanical properties (tensile strength, torsion, hardness, Young's modulus) close to the natural materials;
- they should be made of materials with controlled biodegradability (biosorption), so that tissue could be replaced after a specified time basis ;
- do not cause adverse reactions (including allergic)
- they should be easily manufactured in various shapes and sizes.

Taking into considerations all these above requirements the substrates were synthesized, which are biocompatible corundum glass system composites. Biocomposites by combining the characteristics of these materials (Al_20_3, glass) allow to achieve unique properties such as high mechanical strength, crack resistance, high biocompatibility and bioactivity (Hee - Gon B., et al, 2008, Abo - Mosallam H.A., et al, 2009).

The aim of our work is to obtain corundum- glass biocomposites - meeting the above criteria, obtained in a simple, inexpensive and energy efficient way.

The usefulness verification of new substrates was based on short -term cultures of fibroblast human skin of CCL 110 line from Prochem company and mouse preosteoblasts MC3T3 - E1 Subclone14 from the same company.

2. The effect of mechanochemical treatment of submicrocrystalline sintered corundum grains on their certain physical – Mechanical properties

2.1 Submicrocrystalline sintered corundum grains

Submicrocrystalline sintered corundum grains is a new generation of alfa type aluminum oxide with ultradispersive structure, which was obtained by transformation of sol - gel process of aluminum oxide by using Mg0, L_2O_3, Nd_2O_3, Y_2O_3 as modificators. Abrasive grains consist of Al_2O_3 plates about 0.5 - 1 μm joint by needle bridge of $MgLaAl_{11}O_{19}$ spinel type. Submicrocrystalline sintered corundum grains have in comparison to conventional corundum materials higher strength (90 MPa, conventional 85 MPa), hardness (20 - 22,5 GPa, conventional 18,5 - 21 GPa) with simultenous increase of fracture toughness. Commercial names of these materials are cubitron, Seeded Gel and Blue Sapphire (Niżankowski Cz. 2002, Markul J. 2008).

The submicrocrystalline sintered corundum grains about 150 granulation (125-150 μm) have microhardness of 21.5 Gpa. They were milled for 10, 15, 20, 25 and 30 hours in the planetary mill Pulverisette 6 type produced by Fritsch Company, in the agate chamber with agate balls in the ethanol addition as a slide agent. The grains samples were removed from the chamber after determined times (10, 15, 20, 25, 30 hours) and taken for the further research procedure. The X - ray research works were conducted on the PW 1710 X - ray diffractometer with cobalt lamp at the range of 2Θ angle from 20° - 90°. The phase identification and calculation of percentage contents and structure parameters were done using EVA program by Brucker Company.

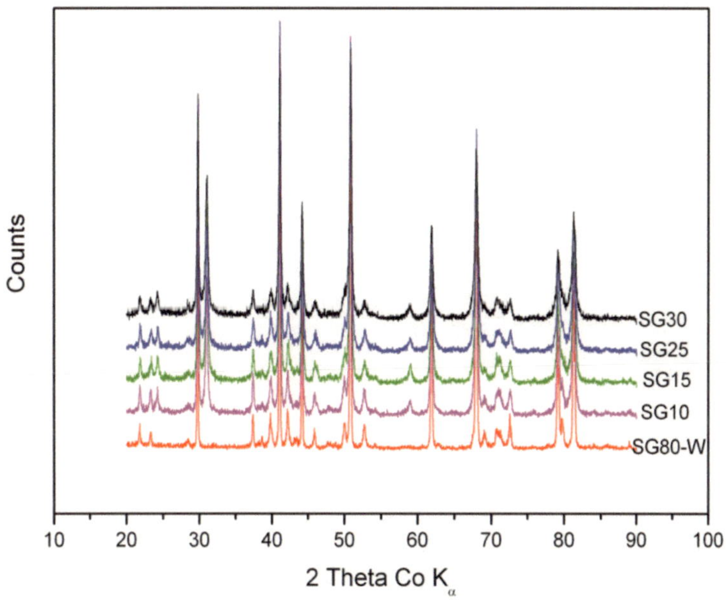

Fig. 1. The X - ray diffraction images of the samples before and after milling for 10, 15, 25, 30 hours.

Basing on this this research in the initial samples of submicrocrystalline sintered corundum the following phases were identified:

α - Al_2O_3 with romboedric structure
kappa - Al_2O_3 with orthoromboedrical structure
δ - Al_2O_3 with orthorombic structure
non - stoichiometric (spinel) compound - $(Mg_{0,03}\ Al_{0,35})\ (Al_{1,68}\ Mg_{0,30})\ O_4$ with cubic structure.

In the milled samples low - temperature quartz (SiO_2) appeared, coming from the used agate balls during the milling process.

The contents of α - Al_2O_3 phase decreased in a benefit of kappa - Al_2O_3, phase, contents of δ - Al_2O_3 was on the same level at different parameters of crystallographic lattice. Lattice parameters of spinel decreased.

Compounds	Time of milling / Structure	Laticce parameters [A°]				
		0	10	15	25	30
Corundum α-Al_2O_3	rombo.	a=4.75970 α=90° b=4.75970 β=90° c=12.9950 γ=120°	a=4.75970 α=90° b=4.75970 β=90° c=12.9950 γ=120°	a=4.75970 α=90° b=4.75970 β=90° c=12.9950 γ=120°	a=4.75970 α=90° b=4.75970 β=90° c=12.9950 γ=120°	a=4.75970 α=90° b=4.75970 β=90° c=12.9950 γ=120°
κ-Al_2O_3	ortorombic	a=4.83400 α=90° b=8.30960 β=90° c=8.93530 γ=90°	a=4.83400 α=90° b=8.30960 β=90° c=8.93530 γ=90°	a=4.83400 α=90° b=8.30960 β=90° c=8.93530 γ=90°	a=4.83400 α=90° b=8.30960 β=90° c=8.93530 γ=90°	a=4.83400 α=90° b=8.30960 β=90° c=8.93530 γ=90°
Magnesium aluminium oxide $(Mg_{0.63}Al_{0.35})(Al_{1.68}Mg_{0.34})O_4$	cubic	a=8.25958	a=8.24575	a=8.17760	a=8.17760	a=8.17760
Aluminium oxide δ-$Aλ_2O_3$	ortorombic	a=8.02684 α=90° b=7.98915 β=90° c=11.63293 γ=90°	a=7.90755 α=90° b=7.95600 β=90° c=11.67196 γ=90°	a=7.99681 α=90° b=7.95600 β=90° c=11.779313 γ=90°	a=7.97367 α=90° b=7.98252 β=90° c=11.82810 γ=90°	a=7.99350 α=90° b=7.95600 β=90° c=11.71100 γ=90°
Low temperature quartz - SiO_2	hexagonal	-	a=4.92000 b=4.92000 c=5.42000	a=4.92000 b=4.92000 c=5.42000	a=4.92000 b=4.92000 c=5.42000	a=4.92000 b=4.92000 c=5.42000

Table 1. Lattice parameters of individual phases of submicrocrystalline sintered corundum determined from X - ray measurement.

The grains composition measurement were done using Sedigraph 5100 X - ray analyzer (Micrometrics Company). Suspensions from research samples were prepared using solution of distillated water and 0,5% sodium pyrophosphate as a sediment liquid. Grain composition was determined in the range of 100 to 0,2 μm. The measurement results are shown as the cumulative and populative curves in Fig. 2 and 3 respectively.

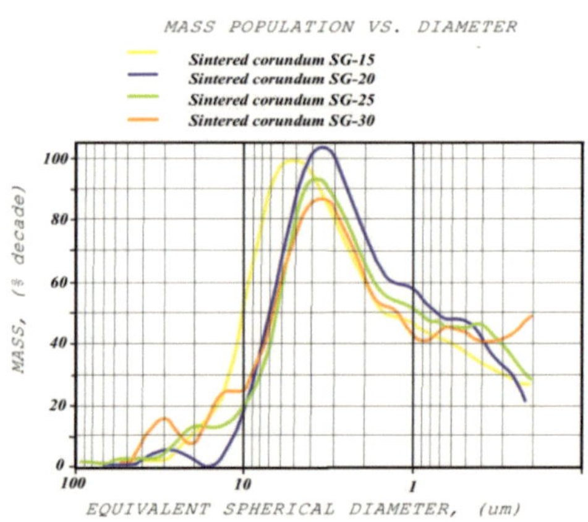

Fig. 2. Populative curves of submicrocrystalline sintered corundum grains after milling for 15,20,25,30 hours.

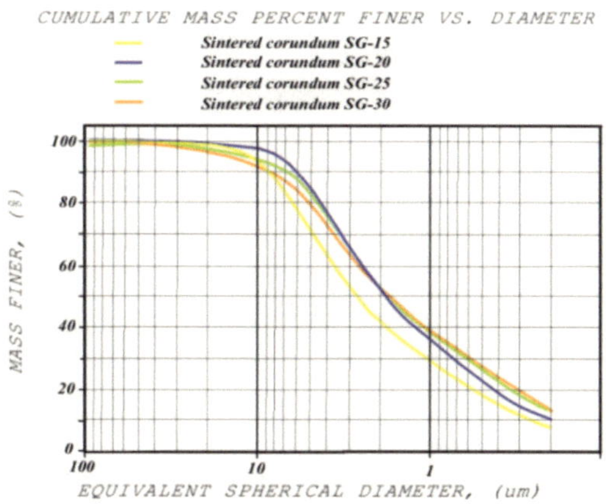

Fig. 3. Cumulative curves of submicrocrystalline sintered corundum grains after milling for 15, 20, 25, 30 hours.

SG - 10 hours - 20 - 1 μm
SG - 15 hours - 10 - 1 μm
SG - 20 hours - 8 - 1 μm
SG - 25 hours - 6 - 1 μm
SG - 30 hours - 6 - 1 μm.

The second population of grains was distinguished on the diagrams that had about 1 μm size with following percentage contents:

SG - 10 hours - 23,6 %
SG - 15 hours - 29,0 %
SG - 20 hours - 35,7 %
SG - 25 hours - 38,9 %
SG - 30 hours - 38,9 %.

The above data indicate that the thickest grains are complementary milled and consequently the contents of ultrafine particles increases. For samples milled for 25 and 30 hours no significant differences in the grains compositions were observed in the entire measuring range. This can be explained by the formation of grains agglomerates, difficult to break even using ultrasonic method.

The specific surface area of samples was determined by S_{BET} physical adsorption of nitrogen at temperatures of liquid nitrogen from the equation Braunaura - Emmett - Teller. For calculations based on the data from the adsorption isotherms the relative pressure at the range p/p_0 0.06 - 0.10 were used. Measurements of specific surface area S_{BET} expressed in m^2/g showed the prolonged milling of samples of submicrocrystalline sintered corundum caused a systematic increase in the specific surface area (from 0.1 m^2/g for the initial sample to 16.4 m^2/g for the sample milled for 30 hours). The greatest effect was achieved after 10 hours milling in relation to the initial sample and after 30 hours of milling in relation to the sample milled for 25 hours and values were 6.4 m^2/g and 3.4 m^2/g respectively.

Sample/fraction below	SG-10	SG-15	SG-20	SG-25	SG-30
40 μm	99.4	100.0	100.0	100.0	100.0
20 μm	95.8	99.3	99.0	97.6	97.2
10 μm	81.7	93.0	97.9	94.0	91.6
5 μm	59.6	71.2	85.0	83.0	79.5
2 μm	34.7	42.3	52.3	53.7	52.1
1 μm	23.6	29.0	35.7	38.9	38.6
0.5 μm	14.8	18.3	22.6	26.9	27.7
0.2 μm	6.1	7.6	10.3	13.1	13.5
Mediane, μm	**3.70**	**2.68**	**1.85**	**1.72**	**1.72**

Table 2. Granulometric analysis of grains (percentage of the fraction below).

Microscopic observations were performed on scanning electron microscope JSM 6460 LV JEOL Company in the range of low and high vacuum, at 20 kV accelerating voltage. SEM and COMPO images were observed at 100x, 400x, 1000x and 2500x magnifications. In these images elongated, irregular grains with sharp edges were present. In the samples milled 10, 15, 20 hours a few large grains and the increased amount of fine grains was observed. In the sample milled for 30 hours cluster of agglomerates appeared.

Fig. 4. SEM image of submicrocrystalline sintered corundum grains:

a. before milling (100x magn.)
b. after 15 hours of milling (2500x magn.)
c. after 20 hours of milling (2500x magn.)
d. after 30 hours of milling (2500x magn.)

2.2 Glass of $Ca0 - Si0_2 - P_20_5 - Na_20$ system

Glass calcium - silicate - phosphate from the system $Ca0 - Si0_2 - P_20_5 - Na_20$ was obtained at 1350^0C by fritting method. Initial materials were:

Calcium carbonate - analytically pure
Sodium phosphate - analytically pure
Silicon dioxide - pure
sodium carbonate - analytically pure.

Three frits were obtained from completely transparent with bluish aquamarine to milky opaque amber colour. Frits were milled in the planetary mill Pulverisette 6 type (Fritsch Company), in the agate chamber with agate balls by 5, 10, 15 and 20 hours with the addition of distilled water. Samples were removed after a certain time subjecting them to further test procedure. A similar procedure was used as in the case of submicrocrystalline sintered corundum grains.

The resulting glass was subjected to chemical analysis on the spectrometer ARL Advant'XP by X - ray fluorescence spectral method. The results of oxide glass compositions were very closed to the calculated theoretically composition, taking into account 10% of the volatility of phosphorus.

Observations of the glassy frits using the scanning electron microscope revealed the presence of elongated grains, irregularly shaped with sharply outlined edges. After prolonged milling a few large grains and clusters of small grains were visible.

Fig. 5. Glass FB3 system after fritting process.

Fig. 6. SEM images of glass FB3 system milled for 5 and 20 hours.

Prolonged milling did not significantly influence the growth of specific surface area of glass system. Only after 20 hours of milling glass system specific surface area was doubled. The reasons for this phenomena might be explained by the acting of large cohesive forces of glass particles.

The densities of glasses were determined by helium method using the pycnometer AccuPyc 330. Analysis of results showed that these glasses belong to a light group of glass (density of 2.65 - 2.70 g/cm^3). Knowing the density and specific surface area and using the formula:

$$2r = \frac{6}{S_{BET} \cdot d} \qquad (1)$$

diameters of the grains were calculated. Only after 20 hours of milling the grain sizes were nearly two times decreased from 2.47 µm to 1.17 µm diameter.

Time milling	Value of specific surface area m^2/g	Average size of grains µm
5 hours	0,9159	2,47
10 hours	0,9553	2,36
15 hours	1,1964	1,89
20 hours	1,9241	1,17

Table 3. Average grains sizes of glass FB3 system after milling by 5, 10, 15, 20 hours and related them specific surface areas.

DTA and DTG studies allowed us to determine the glass temperature of vitrification (FB1 - 538 ⁰C, FB2 - 535 ⁰C, FB3 - 525 ⁰C) and temperature of dilatometric softening point (FB1 697 ᵒC, FB2 700,9 ᵒC, FB3 711,9 ⁰C). DTG curve for the glass FB3 system indicated a normal weight loss about 1 wt.% for temperature of 600 ⁰C. In case of glass FB2 system the mass loss, about 5wt% was far too large. This suggested the presence of internal defects, consisting in incorporating the occluded water and CO_2 particle in the glass structure. This was confirmed by the investigation using the DTA apparatus containing the gas analyzer, indicating volatilization of particles at a temperature of 150 °C, for H_20 and at a temperature of 400 ᵒC for CO_2 particles.

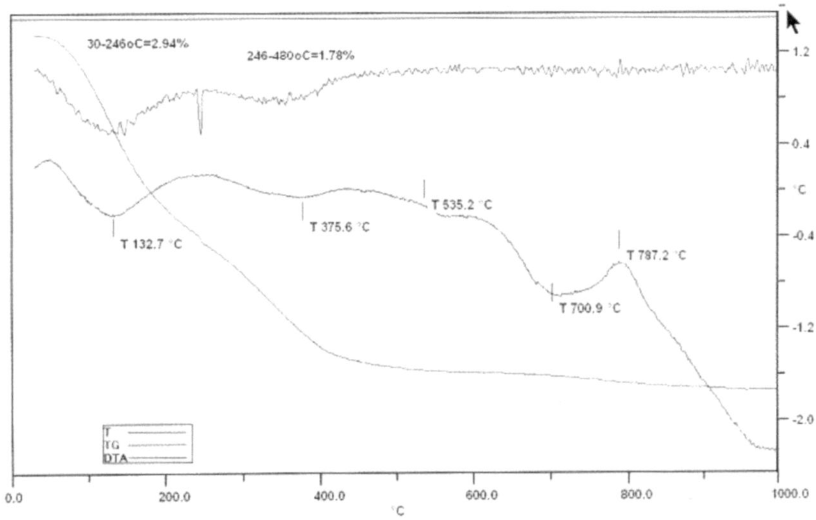

Fig. 7. DTA and DTG of glass FB2 system.

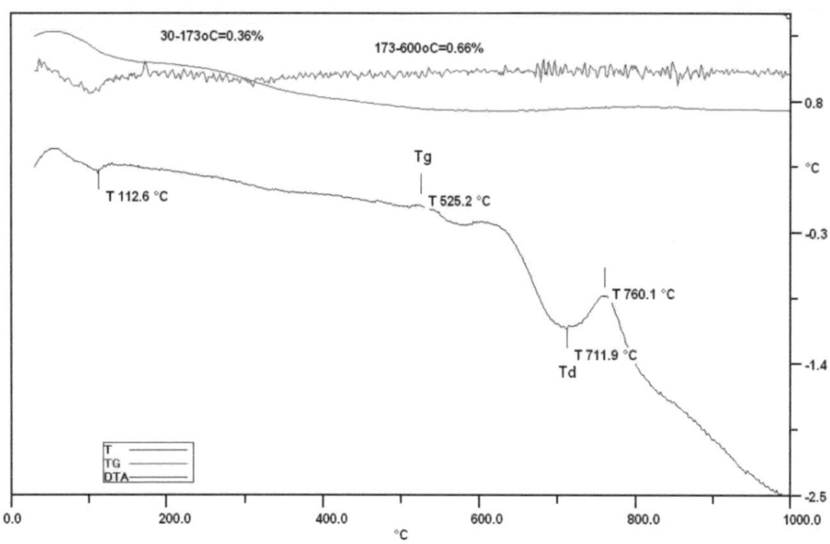

Fig. 8. DTA and DTG of glass FB3 system.

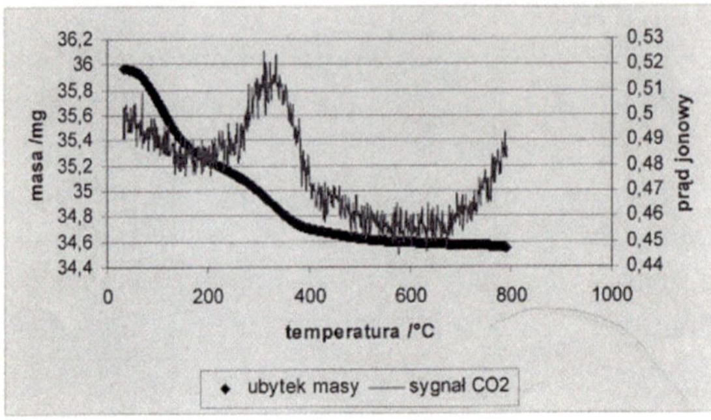

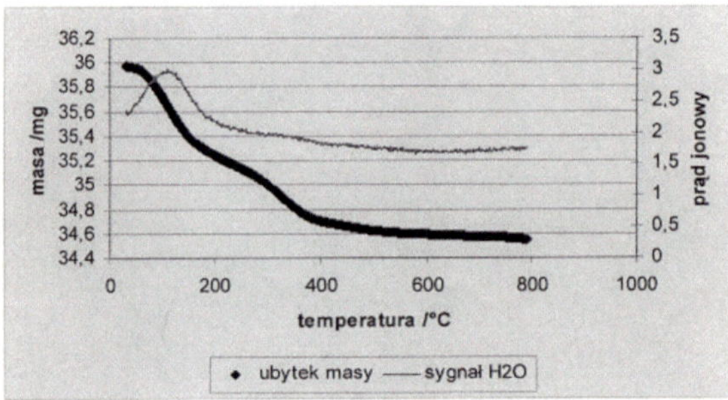

Fig. 9. DTG of glass FB2 system with gas analysis.

Microhardness measurements were carried out using the micro tester FM7 with 100g load. The results show that these are hard glass systems within 5.5 - 6.0 GPa hardness.

Wettability research test of glass system to plate substrates from submicrocrystalline sintered corundum carried out at high temperature microscope MH02 by sessile - drop method showed that the glass FB3 was not gasified, there was no significant change in the dimensions of the swelling and contact angle theta was about 45^0.

a) 110°C a) 904°C a) 1010°C

Fig. 10. Wettability of glass FB3 system on the substrate from submicrocrystalline sintered corundum, a) 110°C, b) 904°C, c) 1010°C.

For further research the submicrocrystalline sintered corundum milled for 20 hours and the glass FB3 system were chosen because of the obtained value of specific surface area of corundum and the properties stability of glass.

3. Biocomposites

The composites containing a matrix from submicrocrystalline sintered corundum and bioglass of CaO - SiO_2 - P_2O_5 - Na_2O system (FB3) in 10, 20 and 30 wt.%, obtained by powder metallurgy technique, in the process of free sintering in air atmosphere, in the electric furnace. Bioglass of the CaO - SiO_2 - P_2O_5 - Na_2O system was obtained by fritting process at 1350 ^{o}C using an electric furnace.

The composites were obtained with two techniques:

- cold pressing and sintering
- cold pressing, isostatic densification and sintering.

Samples of small (Ø10x2) and large (Ø16x5) size were pressed on the screw press, and after they were isostatically densified or not. The heat treatment was performed without the mould in an electric furnace in air, according to the established characteristics of isothermal soak, at maximum temperature at the time of 2 hours.

Phase composition of biocomposite samples was identified using X - ray diffractometer August Siemens Type D 500 Cristal Reflex with copper lamp with monochromatic radiation.

The phases composition were determined for the samples doped with 10 wt.% bioglass admixture isostatic densification or without densification (w1ssc), the identity of the spectra was found in both cases, suggesting the identity of the phase composition of the samples, regardless of the method of obtaining. In samples following phases were identified : alpha and kappa - Al_2O_3 (kappa in the trace amounts), anortit - aluminum calcium silicate - $CaAl_2(SiO_4)_2$, sodium aluminum silicate $NaAlSiO_4$ and quartz SiO_2.

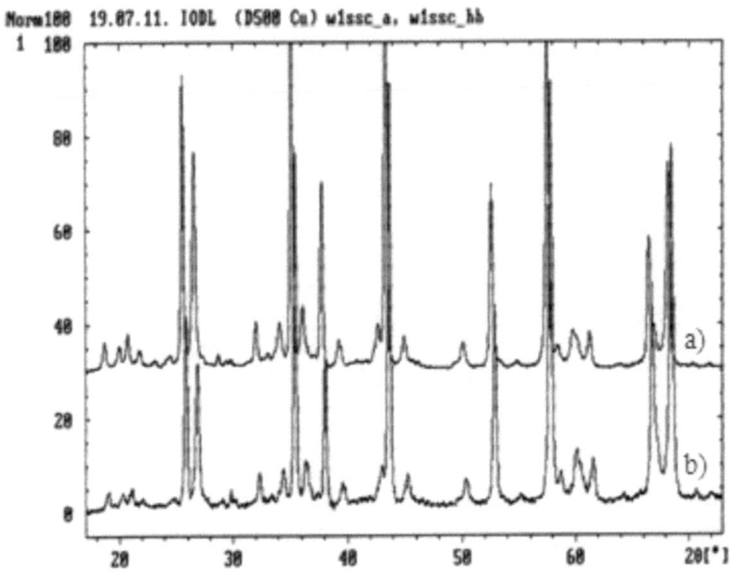

Fig. 11. The XRD spectra for w1ssc samples a) with isostatic densification, b) without densification.

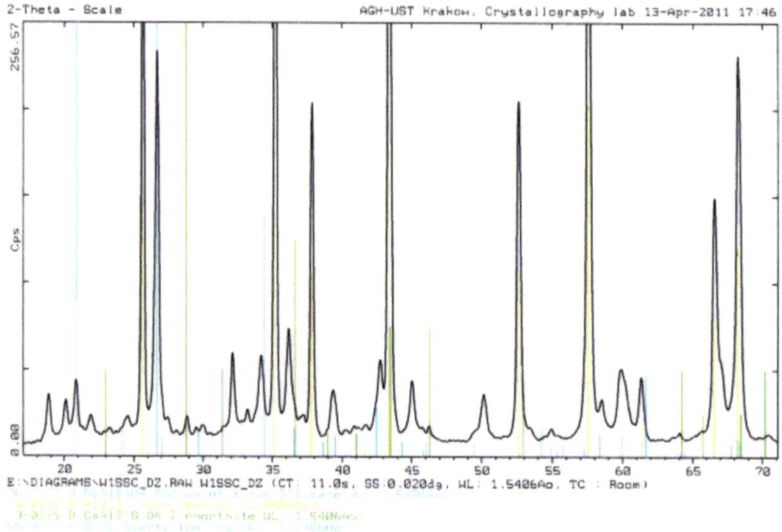

Fig. 12. Identification of XRD spectra for w1ssc isostatically densified sample.

Additionally were made the samples with 20, 30 wt.% glass admixture without isostatic densification. It was found, that the phase composition did not change very much.

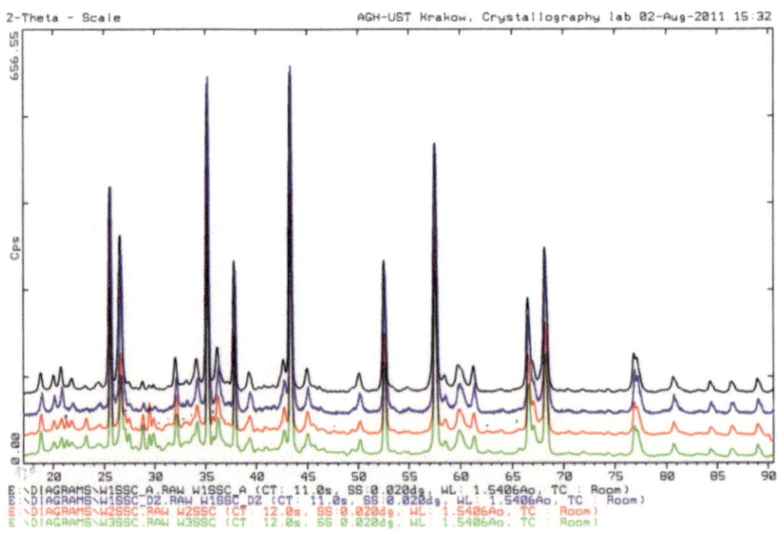

Fig. 13. The XRD spectra for w1ssc, w2ssc, w3ssc samples without densification.

Phase	Strukture	Crystalographic parameters
SiO_2	hexagonal	a = 4,913 c = 5,405
$MgAl_{11}LaO_{19}$	hexagonal	a = 5,582 c = 21,942
CaO	Face - centered cubic	a = 4,797
$NaAlSiO_4$	cubic	a = 7,37
P_2O_5	Face centered orthorhombic	a = 16,3 b = 8,14 c = 5,26 a/b = 2,00246 c/b = 0,64619
Al_2O_3	rhombohedral	a = 4,758 c = 12,991
$CaAl_2(SiO_4)_2$	triclinic	a = 8,21 b = 12,95 c = 14,16 a/b = 0,63398 c/b = 1,09344

Table 4. Crystalographic identification of phase composition.

Microscopic observations with the electron microscopes Jeol JSM 6460 LV type were performed in high vacuum (~ 1,3 x10^{-3}Pa) at 20 kV accelerating voltage, magnification 20x, 100x and 1000x using of BEC image. Microscopic observations carried out with the scanning electron microscope revealed the differences in the microstructure of samples after isostatic densification or without densification. Microstructures of composites contained grains of irregular shape and varying dimensions and pores. The dimensions of grains were in the range below 1 µm to several µm. Leeks (pores) had a varied shape and dimensions of the order of several micrometers. Samples obtained without densification were more porous; but having more smaller pores (mesopores). Isostatically densified samples had dense microstructure with a small amount of larger pores. Photos of samples formed by various techniques are presented in Fig. 14, 15 under magnification 1000x.

a) b)

Fig. 14. The SEM observations of the sample W1ssc, magn. 1000x; a) with isostatic densification, b) without densification.

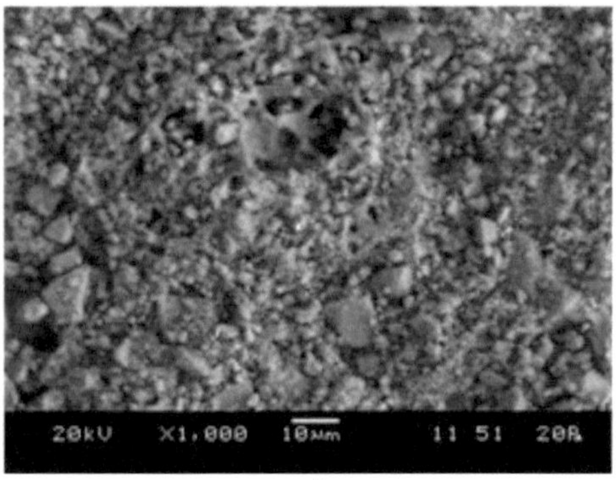

a)

b)

Fig. 15. The SEM observations of the sample W2ssc, magn. 1000x; a) with isostatic densification, b) without densification.

Measurements of the real density (d_{real}) – of powder samples and helium density (on tablets Ø 16x5mm) were performed using helium pycnometer AccuPyc1330 Micrometrics company. Before the relevant measurements of the samples were initially desorbed by a 10 – fold pure helium flushing. Five parallel measurements were done for each sample. The results were used to calculate the closed porosity. Measurements of apparent density (d_{ap}) and total porosity (Pc) were carried out using GeoPyc density analyzer, model 1360 manufactured by Micrometrics. Ten simultaneous measurements were made for each sample. Apparent density (g/cm^3), the volume of pores in the material Vc (cm^3/g) and total porosity (%) were determined.

Sample/ parameter	W1 without dens.	W1 isostat. dens.	W2 without dens.	W2 isostat. dens	W3 without dens.	W3 isostat. dens
d_{real}, g/cm^3	3,6150 ±0,0050	3,6150 ±0,0050	3,5296 ±0,0193	3,5296 ±0,0193	3,4743 ±0,0252	3,4743 ±0,0252
d_{hel}, g/cm^3	3,6684 ±0,0089	3,5457 ±0,0109	3,5010 ±0,0167	3,5237 ±0,0030	3,3942 ±0,0249	3,4319 ±0,0402
$d_{apperend}$, g/cm^3	2,2821 ±0,0054	2,3179 ±0,0052	2,1564 ±0,0070	2,2516 ±0,0097	1,9637 ±0,0044	2,0442 ±0,0052
V_{pores}, cm^3/g	0,166	0,149	0,178	0,160	0,215	0,198
V_{macro}, cm^3/g	0,161	0,146	0,176	0,159	0,214	0,197
V_{mezo},cm^3/g	0,005	0,003	0,002	0,001	0,001	0,001
P, %	36,9	35,9	38,9	36,2	43,5	41,2
S_{BET}, m^2/g	1,99	1,45	1,02	0,88	0,89	0,50

Table 5. Density and porosity of biocomposite samples.

Studies of real density of the sample d_{real} (Table 5) showed that, regardless of the method of preparation, with increasing glass content in the composite the real density decreases (from 3,615 to 3,474 g/cm^3). It resulted from the different density of submicrocrystalline sintered corundum (3,887g/cm^3) and glass FB3 system (2,564 g/cm^3).

The results of determining the apparent density and the skeleton of both open and closed pores became evident that the apparent density decreases with increasing glass content, in both cases of isostatic densification or without densification, but the value of the apparent density of isostatically densified samples was higher than of samples without densification, what resulted from the process of obtaining . Based on density measurements of helium and the actual porosity of the closed set Pz results showed that in the test samples there were present opened pores and closed pores were in negligible volume. Specific surface samples with increasing S_{BET} bioglass content decreased in both cases (samples isostatically densified or without densification). The values of SBET in the 0,5 to approximately 2,0 m^2/g were testified by low content of mesopores which volume ranged from 0,001 to 0,005 cm^3/g.

Visible is that the total porosity increased with increasing bioglass content for both isostatically densified samples (from 0,146 to 0,197cm^3/g) and without densification (from 0,161 to 0,214 cm^3/g) containing mainly macropores with dimensions greater than 0.1 microns, which resulted from an increased quantities of bioglass mainly having opened pores.

Measurements of surface area (S_{BET}) were performed using the multifunctional apparatus (ASAP 2010, Micromeritics American company) for measuring surface area and porosity. The specific surface area was determined by physical S_{BET} nitrogen adsorption at liquid nitrogen temperature (77K) from the equation Brunauer - Emmet - Teller (the theory of multilayer adsorption). Before the measurement surfaces of the test samples were subjected to desorption at temperature 1050°C, in vacuum and flushing with pure helium. Sample degassing time was about 8 hours. Surface degassing state was controlled in automatic mode.

The specific surface area calculations based on data from the adsorption isotherms of the relative pressure range p/p_0 from about 0,06 to about 0,20% and the volume and dimensions of the mesopores were calculated using p/p_0 of 0,97%.

Geometric structures of the biocomposites samples surface were determined using the TOPO 01vP profilometer, by measuring the surface topography parameters (Ra, Rz, Rt), the image of 2D and 3D ,profile material rate and amplitude distributions of the ordinates.

The geometric structure of the surface isostatically densified or without densification samples (w1ssc) have been evaluated by measuring surface topography performed using TOPO 01vP profilometer developed and produced in IZTW. There were defined the basic parameters of roughness (Ra, Rz, Rt). In the isostatically densified sample porosity effect on profile was visible (few large cavities- inequality) .This was confirmed by image analysis of 2D and 3D. Participation of the linear bearing and amplitude distribution of the ordinate was shifted toward negative values.

The profile roughness of sample without densification has more blurred shape, with more small pits and a few large inequalities compared with the profile of the isostatic densification sample. It was testified by the image analysis of 2D and 3D. The amplitude distribution of the ordinate showed the presence of inequality - type cavities.

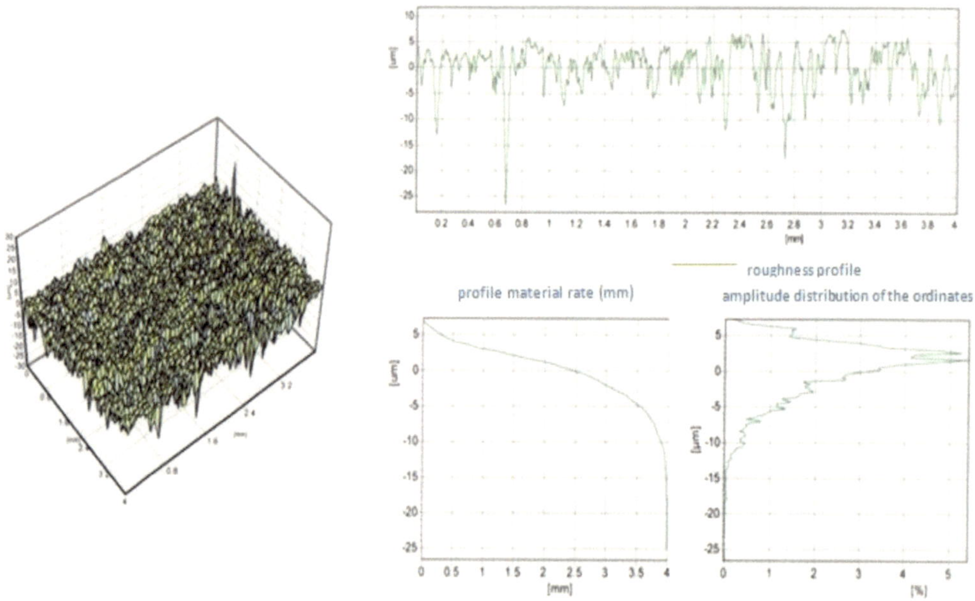

Fig. 16. The image of 2D and 3D, along with a profile material rate and amplitude distribution of the ordinates for isostic densification w1ssc sample.

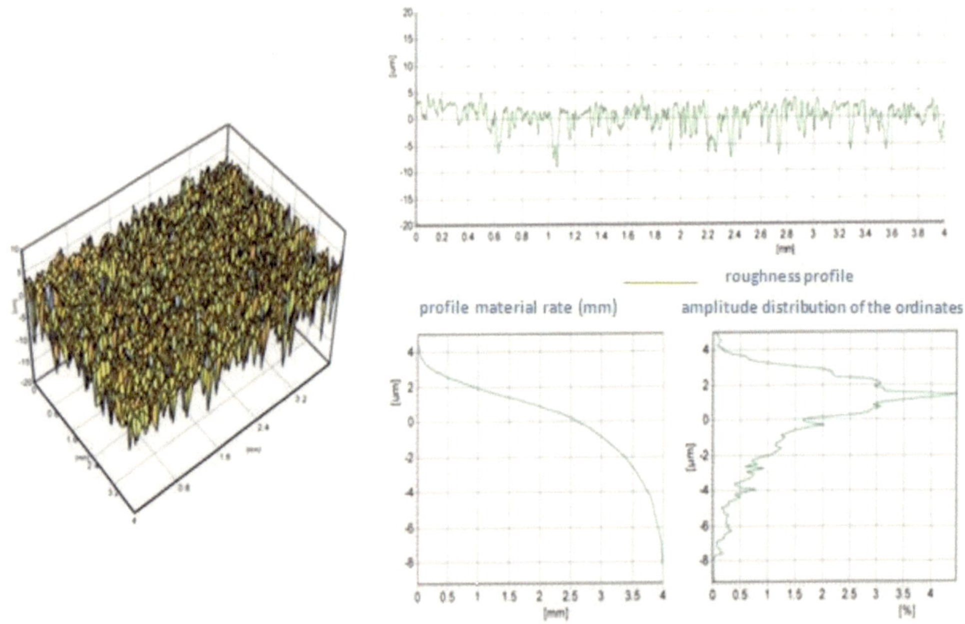

Fig. 17. The image of 2D and 3D, along with a profile material rate and amplitude distribution of the ordinates without densification w1ssc sample.

4. Short term culture of the fibroblast human skin CCL 110 line and mouse preosteoblasts MC3T3 – E1 Subclone 14 on these substrates

The usefulness verification of these research substrates (w1ssc, w2ssc, w3ssc) for cell culture were checked for short - term culture of human skin CCL line and mouse preosteoblasts MC3T3 - E1 Subclone 14 CRL 2594 line.

4.1 Materials and methodology

Bioglass sterilization. The surfaces used for cell growing were w1ssc, w2ssc, and w3ssc. Before seeding the cells, the following protocol for sterilization has been applied. The samples of the bioglass composite (10 mm in diameter and 2 mm in thickness) were immersed in the 70 % alcohol solution for 12 hours. Afterwards, each side of the sample was exposed for 2 hours to UVC light (wavelength of 245 nm), which was provided by a germicidal lamp from a laminar flow chamber (Nuaire Nu 425), at average intensity of 0.1 mW/cm^2 at the working plane. Such sterilized bioglass samples were used immediately for cell growth.

Cell lines. Two types of cell lines were studied:

- human skin fibroblasts (CCL-110, LG Promochem) were cultured in DMEM (Dulbecco's Modified Eagle Medium, Sigma) containing 5 % of fetal bovine serum and 1 % mixture solution of antibiotics (streptomycin, neomycin and penicillin). They were grown at 37°C in an incubator (NUAIRE, USA) providing 95% air / 5% CO_2 atmosphere. Initially, cells were grown in a culture flask (Saarstedt) and when they formed semi-confluent monolayer, they were trypsinized using 0.25% trypsin/EDTA solution (Sigma) and placed into bioglass surfaces for 96 h and 360 h.
- mouse preosteoblasts MC3T3 - E1 Subclone 14 (CRL-2594, LG Promochem). Cells were cultured in Alpha Minimum Essential Medium with ribonucleisides, supplemented with 10% fetal bovine serum (LG Promochem). They were grown at 37°C in an incubator (NUAIRE, USA) providing 95% air/ 5% CO_2 atmosphere. Analogously to fibroblasts, cells were grown in a culture flask (Saarstedt) and when they formed semi-confluent monolayer, they were trypsinized using 0.25% trypsin/EDTA solution (Sigma) and placed into bioglass surfaces for 96 h and 360 h.

Phalloidin staining. To visualize the organization of actin filaments, cells were stained and imaged using fluorescent microscope (the same protocol was applied independently of the cell type). First, cells grown on bioglass surfaces, were fixed with 3.7% paraformaldehyde dissolved in the PBS buffer for 10 minutes, followed by rinsing them twice in the PBS buffer (Phosphate Bufered Saline, Sigma), permeabilization with 0.1% Triton X-100 solution in the PBS buffer. Then, they were again rinsed in the PBS buffer (3x3 minutes). The actin filaments were stained using the solution containing phalloidin labeled with Alexa Fluor 488 (1:200, in PBS buffer, Molecular Probes) for 30 minutes incubation at room temperature. Next the excess of dye was removed by rinsing bioglass surfaces in the PBS buffer. The wet surfaces with stained cells were placed on the microscope slide and immediately imaged.

Fluorescence microscopy. Fluorescence is the ability of organic or inorganic specimen to absorb and subsequent emit of light. The basic function of a fluorescence microscopy is to irradiate the sample with a desired and specific band of wavelengths, and then to separate the much weaker emitted fluorescence from the excitation light. In a properly configured microscope, only the emission light should reach the eye or detector so that the resulting fluorescent structures are superimposed with high contrast against a very dark (or black) background. The limits of detection are generally governed by the darkness of the background, and the excitation light is typically several hundred thousand to a million times brighter than the emitted fluorescence. In the presented studies, the fluorescence microscope was used to visualize the actin filaments, which were stained with phalloidin labeled with Alexa Fluor 488 that absorb blue light ($\lambda = 495$ nm) and emits green fluorescence ($\lambda = 518$ nm). The images were recorded using Olympus 71X microscope equipped with the 100 W mercury lamp, the MWIG2 filter, and digital camera XC 10 (working under Cell^R program).

4.2 Results and discussion

The visualization of the cell shape has been perform through actin filament staining using phalloidin coupled with Alexa - Fluor 488. Although actin filaments are dispersed within entire cell, they are concentrating mainly in cortex layer beneath the plasma membrane, and therefore they can be used for cell-shape visualization (fig. 18). The cell cultures were performed for single cells grown for 96 hours. The images observed for fibroblasts were better due to the higher amount of actin filaments inside as compared to osteoblats. However, independenlty of the cell type, their shape varied from rounded to the well extended (spindle - like) ones.

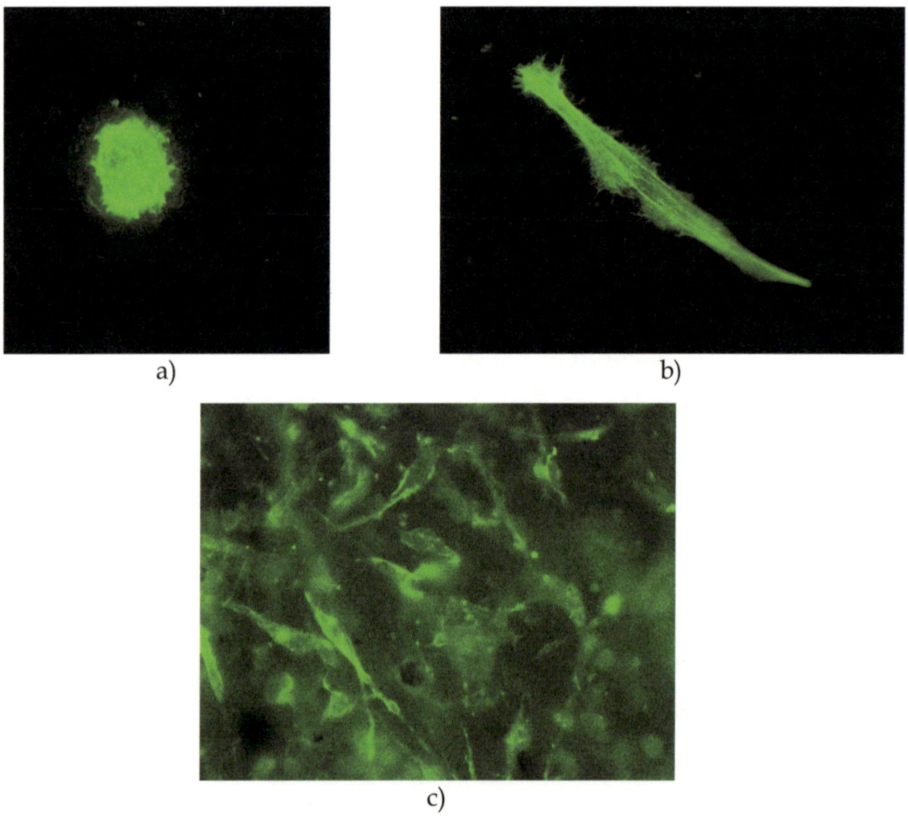

Fig. 18. Fluorescent images recorded after 96 hours of growth on composites (magnitude 600x), a, b – fibroblasts, c – osteoblasts.

The round cells denotes the cells that starts to detach from the investigated surface. In general, cells spread and grow on rigid substrates if the provided conditions enable them to preserve normal functioning (in such conditions fibroblasts turn into spindle-shaped cells) but they retract, and become round on non-biocompatible ones. The table 6 presents the summary of shapes observed for both fibroblasts and osteoblasts cultured on the studed surfaces W1ssc, W2ssc, and W3ssc.

Composite's surface	Isostatic densification	Cell type	Cell shape
W1ssc	Yes	fibroblasts osteoblasts	spindle-like and round spindle-like
W1ssc	No	fibroblasts osteoblasts	round spindle-like
W2ssc	Yes	fibroblasts osteoblasts	spindle-like and round spindle-like
W2ssc	No	fibroblasts osteoblasts	round round
W3ssc	Yes	fibroblasts osteoblasts	round spindle-like
W3ssc	No	fibroblasts osteoblasts	spindle-like spindle-like and round

Table 6. Summary of cellular shapes cultured on the different composite's surfaces.

To estimate the number of cells present on composite surface, the percentage of cells present on a given sample type was calculated as a ratio of cells present on the studied surface divided by the number of cells cultured on a glass coveslip (2500 and 4800 cells on a glass rounded coveslip with diameter of 10 mm for human skin fibroblats and mouse osteoblasts, respectively).

Figure 19 a and b shows the comparison between the percentage of cells attached to composite surfaces (W1ssc, W2ssc, W3ssc) after 96-hour of growth in culture conditions. In case of fibroblasts, the composite surfaces without densifications manifest larger survival level as compared to those proceed after densification. The isostatic densification results in lower biocompability of the surfaces. The observed drop was from 23% to 11%, 7% to 4%, and 15% to 8% for W1ssc, W2ssc and W3ssc composites, respectively. The worst biocompatible properties were observed for W2ssc sample. The ability of cells to grow on W1ssc and W3ssc was comparable.

Mouse osteoblasts behave similarly. The observed increase was from 18% to 24%, 7% to 11%, and 13% to 19% for W1ssc, W2ssc and W3ssc composites, respectively. Analogously as for fibroblasts, the quality of growing conditions for osteoblasts can be order as follows: the worst biocompatible properties were observed for W2ssc composite, then W3ssc, and W1ssc surfaces are placed.

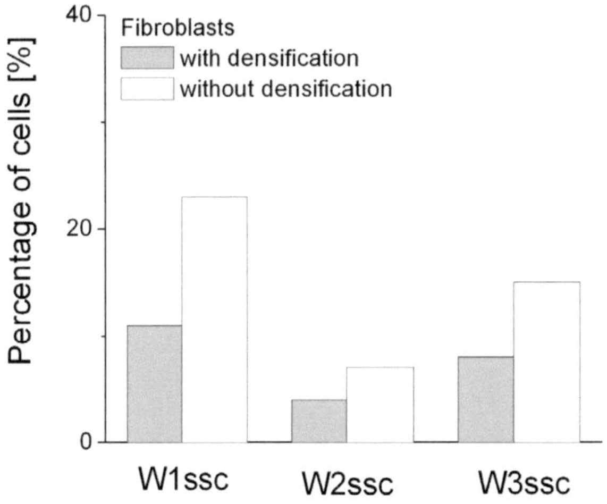

a)

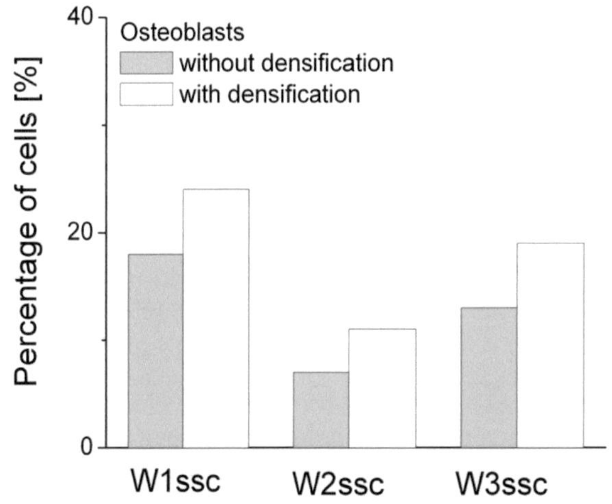

b)

Fig. 19. The percentage of cells (fibroblasts and osteoblasts) present on a surface of composite surfaces.

One of the important conditions that should be provided by biocompatible materials is providing the optimal conditions for cell growth, which can be characterized by the doubling time (Maret D., et. al, 2010, Puttini S., et al, 2009). The chosen cells doubled their amount after 130 and 38 hours for fibroblasts and osteoblasts, respectively, when cultured on Petri dish in optimal conditions. Therefore, this parameter was determined to monitor the changes in their growth. The results are presented in fig. 20 a and b.

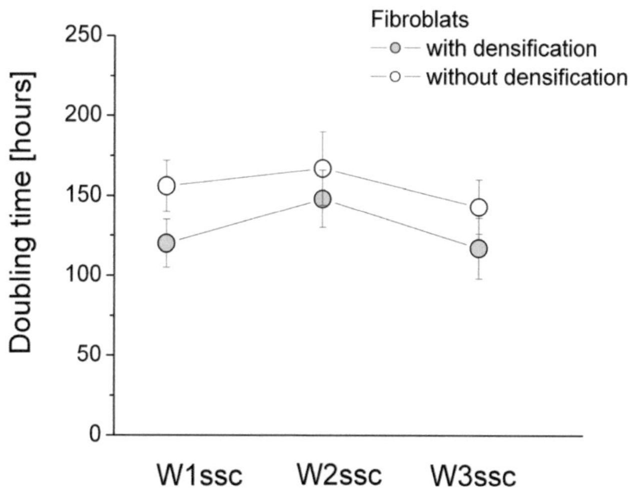

a)

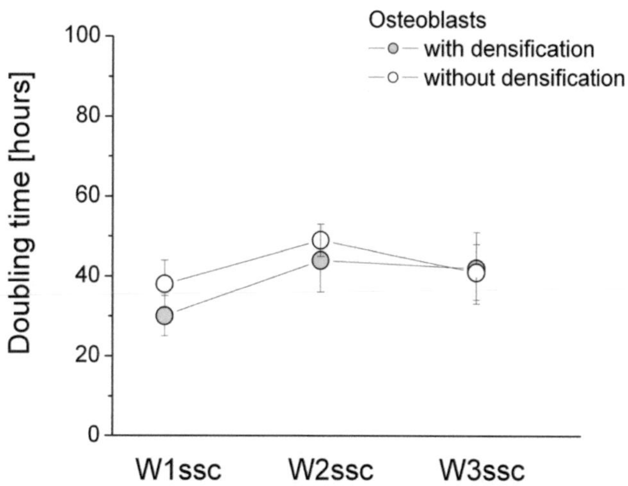

b)

Fig. 20. The percentage of cells (fibroblasts and osteoblasts) present on a surface of composite surfaces.

The doubling time was dependent on the substrate properties. For fibroblasts, the obtained values were 120 ± 15 h, 148 ± 18 h, 117 ± 19 h for composite's surfaces prepared without densification (fig. 20 a). The application of densification step introduces the increase of doubling time to 156 ± 16 h, 167 ± 23 h, and 143 ± 17 h, for W1ssc, W2ssc, and W3ssc, respectively. Longer doubling times correlated well with the lower number of cells present on the corresponding composite's surface. Thus, indicating worse growing conditions for fibroblasts. Analogously, as for fibroblasts, the doubling time was determined for osteoblasts (fig. 20 b). Also in that case, its value showed a relationship with the number of cells on a composite's surfaces, pointing out the best growing conditions in case of W1ssc surfaces.

5. Conclusions

Based on studies conducted so far can be stated that:

- prolonged milling of submicrocrystalline sintered corundum effectively increased the specific surface area of grains from 0.1 m^2/g for the sample which was not milled to the value 16.4 m^2/g for the sample milled for 30 hours. Particle size distribution after milling for 10, 15, 20, 25 and 30 hours of populative and cumulative curves indicated the multimodal shape with two distinctive ranges of particle size depending on the time of milling. Agglomerated particles were observed after 25 and 30 hours of milling.
- X – ray studies revealed the presence of phases α, kappa, δ and Al_2O_3 and non – stoichiometric composition of magnesium aluminum oxide $(Mg_{0.63}Al_{0.35})(Al_{1.68}Mg_{0.30})O_4$ in initial sample. Additionally, in milling samples the silicon dioxide (quartz) coming from the used agate balls from the planetary mill Pulverisette 6 type appeared.
- in the case of glass of calcium – silicon phosphate system prolonged mechanochemical treatment neither had a significant effect on the increase specific surface area (it was only twice after 20 hours milling from 0.9159 m^2/g in the starting glass system to 1.9241 m^2/g after 20 hours milling), nor on the change of size of glass system grains (from 2.47 μm after 5 hours to 1.17 μm after 20 hours).:
- it was found that the phase composition of w1ssc samples (with or without isostatic densification) was identical regardless of the method of manufacturing. It have been identified for w1, w2, w3ssc samples alfa and kappa aluminum oxide and at w1ssc anorthit $CaAl_2(Si0_4)_2$, at w2ssc, w3ssc $NaAlSi_20_6$ additionally;
- observations of biocomposites substrates for scanning electron microscope showed increased porosity in the samples without densification with large number of small pores. Isostatically densified samples had more compact structure, with a small amount of larger pores;
- the difference between total porosity in the samples after isostatic densification and without densification was evident. The samples without densification have a higher porosity with increasing bioglass system A similar trend is maintained in case of isostatically densified composites; with increasing glass content increases the total porosity, apparent and real density decreases. It may be found that the most common pores in all the substrates is far greater than 0,1 micron;
- the analysis of parameters, diagrams and distributions showed significant asymmetry of the distribution in the negative value direction of unevenness for the both cases. It was bigger for not densified sample.
- the studied composite's surfaces (W1ssc, W2ssc, W3ssc) were characterized by distinct biocompatible properties. The growth of two cell types i.e. fibroblasts and osteoblasts revealed cell – type dependent behaviour.. All substrates without densification seemed to provide better growing conditions for fibroblasts and preosteoblasts while isostatic densification induced worser surface properties for cell growth

6. Acknowledgments

This work was partially supported by the project SMW (Single Molecule Workstation), grant agreement number 213717 (NMP4 - SE - 2008 - 213717).

7. References

Abo - Mosallam H.A.; Salama S.N.; Salama S.M. (2009). Formulation and characterization of glass - ceramics based on $Na_2Ca_2Si_3O_9$ - $Ca_5(PO_4)_3F$ - Mg_2SiO_4 - system in relation to their biological activity, *Journal Materials Science: Materials in Medicine*, No. 20, pp. 2385 - 2394

Błażewicz S.; Stoch L. (2003). *Biomateriały t4, in: Biocybernetyka i Inżynieria Biomedyczna*, Akademicka Oficyna Wydawnicza, Exit, ISBN 83 - 876474 - 58 - 3, Warszawa, Poland (in Polish)

Brovarone C. ; Verne E. (2006). Macroporous bioactive glasse-ceramic scaffolds for tissue engineering. *Journal of Materials Science: Materials in Medicine*, Vol.17, No.11, pp. 1069-1078

Chen Q.Z.; Efthymiou A. ; SalihV. ; Boccacini A.R. (2008). Bioglass-derived glass-ceramic scaffolds: study of cell proliferation and scaffold degradation in vitro. *Journal of Biomedical Materials Research*, pp. 1049-1060

Czechowska J.; Ślósarczyk A. (2011). Bioaktywne materiały ceramiczne dla inżynierii tkankowej i medycyny regeneracyjnej. *Szkło i Ceramika*, No. 2, pp. 17 - 22 (in polish)

Hee - Gon B.; Sung - Jin K.; Sang - Yeup Park. (2008). Biocompatibility and the physical properties of bio - glass ceramics in the Na2O - CaO - SiO_2 - P_2O_5 system with CaF_2 and MgF_2 additives, *Journal of Ceramic Processing Research*, Vol. 9, No. 6, pp. 588 - 590

Hench L.L. (1998). Biomaterials: a forecast for the future. *Biomaterials*, No.19, pp. 1419 - 1423

Jaegermann Z. (2005). Porowata bioceramika korundowa, praca doktorska, AGH, Kraków, Poland (in polish)

Jaegermann Z. ; Michałowski S. ; Karaś J. ; Chrościcka A. ; Lewandowska - Szumieł M. (2006). Porowate nośniki korundowe do zastosowania w inżynierii tkankowej. *Szkło i Ceramika*, No.57, pp.16 - 20 (in Polish)

Jaegermann Z. ; Ślósarczyk A. (2007). *Gęsta i porowata bioceramika korundowa w zastosowaniach medycznych*, Uczelniane Wydawnictwo Naukowo - Dydaktyczne AGH, Kraków, Poland (in polish)

Krajewski A. ; Ravaglioli A. (2002). Bioceramics and biological glasses. In: *Integrated Biomaterials Science*, Springer US,

Maret D. (2010) et al. Surface expression of precursor N-cadherin promotes tumor cell invasion. *Neoplasia (New York, NY)* 12, pp. 1066-1080

Markul J. (2008). Synergiczne konstrukcje ściernic ze spoiwem ceramicznym, Wydawnictwo Uczelniane Politechniki Koszalinskiej, Monografia, Nr. 156, ISSN 0239 - 7129, Koszalin, Poland (in Polish)

Niżankowski Cz. (2002). Manufacturing sintered corundum abradants. *Archives of Civil and Mechanical Engineering*, No.2, pp. 53 - 64

Puttini S.; Lekka M.; Saugy D.; Dorchies O.; Incitti T.; Bozzoni I.; Kulik A.J.; Mermod N. (2009). Atomic force microscopy assay of the elasticity of utrophin-expressing muscles. *Molecular Therapy*, Molecular Therapy Editorial Office, Seattle

Sachlos E.; Czernuszka J.T. (2003). Making tissue engineering scaffolds work. Review on the application of solid freeform fabrication technology to the production of tissue engineering scaffolds. *European Cells and Materials*, No.5, pp. 29 - 40

Ślósarczyk A. ; Rapacz - Kmita A.(2004). Bioaktywne ceramiczne materiały kompozytowe. *Ceramika Materiały Ogniotrwałe*, Vol.56, No.4, pp. 144 - 149 (in polish)

Staniewicz - Brudnik B.; Lekka M.; Bączek E.; Wodnicka K.; Wilk W. (2010). The effect of the method of obtaining the submicrocrystalline sintered corundum - bioglass composite matrix system on its microstructure and biocompatibility. *Monograph The Faculty of Mechanical Engineering Lodz University of Technology*, pp. 71 - 80, Lodz, Poland

Staniewicz - Brudnik B.; Lekka M.; Jaworska L.; Wilk W. (2010). Biocompatible glass composite system - some physical-mechanical properties of the glass composite matrix system. *Optica Applicata*, Institute of Physics, Wrocław University of Technology, Wrocław, Poland

Szarska S. ; Staniewicz-Brudnik B.; Lekka M. (2008). The effect of the size of the substrate grain made of submicrocrystalline sintered corundum on the bioglass composite structure and certain physico - mechanical properties of the bioglass. *Optica Applicata XXXVIII*, No.1, pp. 251-258

Teramoto H.; et al. (2005). Resorption of apatite - wollastonite containing glass - ceramic and beta - tricalcium phophate in vivo, *Acta Medica Okayama*, Vol. 59, Issue 5

12

Synthesis and Sintering Studies of Magnesium Aluminum Silicate Glass Ceramic

Shahid-Khan Durrani*, Muhammad-Ashraf Hussain, Khalid Saeed, Syed-Zahid Hussain, Muhammad Arif and Ather Saeed
Materials Division, Directorate of Technology, PINSTECH, P. O. Nilore, Islamabad, Pakistan

1. Introduction

1.1 Ceramic materials

Ceramic materials are complex compounds and containing both metallic and non-metallic elements. Typically, ceramics are very hard, brittle, high melting point materials with low electrical and thermal conductivity, good chemical and thermal stability, and high compressive strengths (Barsoum, 1997; Minh et al., 1995). Ceramics are of tremendous interest primarily because of their wide range of applications in high temperature environments; they are also extensively used in fuel technology (Koshiro et al, 1995), oxygen sensor (Ciacchi, et al. 1994), magnets ceramics (Valenzuela,2005) , all electronic equipments including integrated-chips, capacitors and digital alarms (Miller,& M. R. Miller (2002), telecommunication (Bhargava, A.K. 2005), abrasives (Callister, 2007) , ceramic crystal-glass (Carter, & Norton. (2007), ceramic insulators are widely used in the electrical power transmission system (Chowdhury, 2010), ceramic superconductors (David E. C.&, Brece K.Zoitos, 1992) and other pharmaceuticals industries (Rice, & R. W. Rice, (2002) etc.

Ceramic materials can be classified into four main groups (Rajendran, V. 2004):

i. The amorphous ceramics, which are generally referred to as "glasses".
ii. Crystalline ceramics, which are single phase materials like alumina, or mixtures of such materials.
iii. Bonded ceramics, where individual crystals are bonded together by a glassy matrix, such as clay products.
iv. Cements, these are crystalline, and also amorphous materials.

The structures of ceramics fall into two main groups:

- Simple crystal structures: containing ionic or covalent bonds, or a mixture of the two. Examples are magnesium oxide, which is an ionic compound with cubic structure, and silicon carbide, with covalent bonds and a tetrahedral structure like diamond. Alumina has a close packed hexagonal structure, with a mixture of covalent and ionic bonds,

---

* Corresponding Author

with one-third of the potential aluminum sites vacant in order to suit the valency requirements of the two elements ((Mostaghaci, H. 1996).

- Complex silicate structures**:** The majority of ceramic materials, i.e., derived from clay, sand or cement, contain the silicon in the form of silicates. The arrangements are involving both chains of silicate ions $(SiO_4)^{2-}$, double chains and links in sheet form.
- In advance ceramic, extremely pure microscopic raw materials are used, with strict control of compositions and manufacturing conditions, in order to achieve a more uniform microstructure. Interest in modern ceramic continues to grow because these materials have a unique set of properties which no other family of materials can match. These include high hardness, heat resistance, ability to withstand corrosive atmosphere, resistance to abrasion, ability to sustain compressive loads at elevated temperatures, and an inexpensive and abundant availability of raw materials. Except for their brittleness, the ceramics stand out well against metals and plastics and can be applied as thermal barrier coatings to protect metal structures, wearing surfaces, or as integral components by themselves. These new ceramics are called either "new ceramics or fine ceramics" or frequently called "high-performance ceramics" as described by (Shinroku, 1986). Advance ceramic materials are being used in many new applications and have been proposed for numerous components which may result in significant growth of current and new business (Doremus, 1994; Lewis, 1989; Michael, B. 2007).

In advance ceramic, extremely pure microscopic raw materials are used, with strict control of compositions and manufacturing conditions, in order to achieve a more uniform microstructure. Interest in modern ceramic continues to grow because these materials have a unique set of properties which no other family of materials can match. These include high hardness, heat resistance, ability to withstand corrosive atmosphere, resistance to abrasion, ability to sustain compressive loads at elevated temperatures, and an inexpensive and abundant availability of raw materials. Except for their brittleness, the ceramics stand out well against metals and plastics and can be applied as thermal barrier coatings to protect metal structures, wearing surfaces, or as integral components by themselves. These new ceramics are called either "new ceramics or fine ceramics" or frequently called "high-performance ceramics" as described by (Shinroku, 1986). Advance ceramic materials are being used in many new applications and have been proposed for numerous components which may result in significant growth of current and new business (Doremus, 1994; Lewis, 1989; Michael, B. 2007).

1.1.1 Glass ceramics

Glass ceramic materials are fine grained polycrystalline solid containing residual glass phase formed by melting of glass and forming into products by the way of controlled crystallization of a specially formulated glass (Shackelford, & Doremus, 2008). These are primarily silicates containing oxides such as Al_2O_3, TiO_2, LiO_2, and others. In amorphous form, the glasses are transparent. Glasses can be made to transform into a polycrystalline state by a suitable heat-treatment process, called devitrification. An initiator, such as TiO_2, is added to begin the nucleation of ceramic crystals. The product is called glass ceramic. Glass ceramics are members of ceramic family and an important electroceramic type, were first investigated in the 1940s by Stookey at Corning Glass, (Pannhorst, 1995**)** are extensively exploited as electrical insulators in electronic industries and have a wide variety of technological applications such as microelectronic substrates and packaging, optically transparent components, biomedical implants, catalytic supports, membranes and sensors, as well as for the matrix of composite materials. (Partridge, 1994; Manfredini, et al., 1997). In addition, they have been used in protection layers, sewing-thread in industry, ceramic tiles and also to develop into a very promising substrate material of computer hard disk (Novaes et al., 1994). The universal glass ceramics are Lithia-alumina-silica (Li_2O-ZrO_2-SiO_2-Al_2O_3, LZSA) systems (Montedo et al., 2009) that have zero thermal expansion co-efficient, and hence no thermal shock problem.

The majority commercial glass-ceramic products are formed by highly automated glass-forming processes and converted to a crystalline product by the proper heat treatment. Glass-ceramics can be prepared through a powder processing methods, in which glass frits are sintered and crystallized. The range of potential glass-ceramic compositions is therefore particularly wide, requiring only the ability to form a glass and control its crystallization. Glass-ceramics can provide advantages over conventional glass or ceramic materials, by combining the simplicity and flexibility of forming and inspection of glass with improved and often unique physical properties in glass ceramic. They possess highly uniform microstructures, with crystal sizes (<10mm), this homogeneity indicates that their physical properties are highly reproducible. The glass-ceramics are fully densified with zero porosity. They generally are at least 50% crystalline by volume and often are >90% crystalline (Strand, 1986). Glass-ceramics are also called as pyrocerams, vitrocerams, devitrocerams, sitalls, slag-ceramics, melt-formed ceramics, and devitrifying frits.

1.1.2 Ceramic processing and crystal growth

Glass-ceramic materials are fabricated (green body) by means of either bulk or powder processing methods. In these methods, glass-ceramic products are melted and fabricated to shape in the glass state. Most forming methods are rolling, pressing, spinning, casting, and blowing. The product is then crystallized using a heat treatment designed for that material. This process, known as ceramic processing, usually comprise of a low temperature hold to

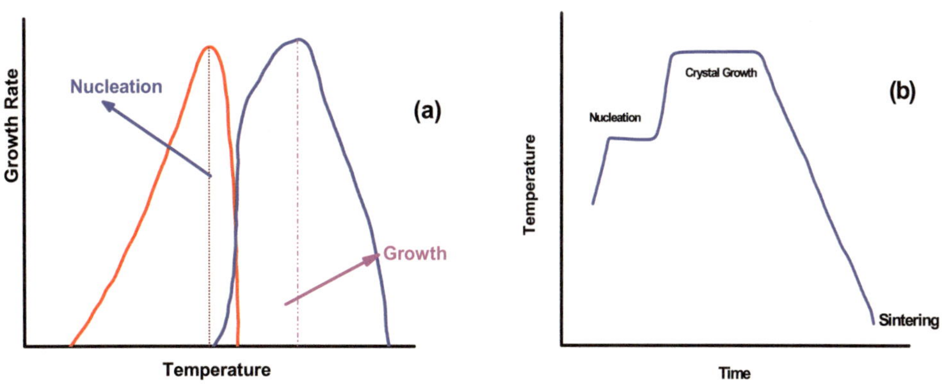

Fig. 1. (a-b): Crystal growing process (a) Nucleation and crystal growth as a function of temperature (b) heat treatment cycle showing crystal growth during sintering as a function of time.

induce nucleation, followed by one or higher temperature holds to promote crystallization and growth of the primary phase. Because crystallization occurs at high viscosity, product shapes are typically preserved with little or no shrinkage (1-3%) or deformation during the ceramming. Nucleation generally begins with phase separation, whereby an amorphous, homogeneous glass unadulterated into two immiscible phases of different compositions. Controlled devitrification is only possible for certain glass compositions and generally takes place in two stages: formation of submicroscopic nuclei, and their growth into macroscopic crystals. These two stages are called nucleation and crystal growth as shown in Figure 1(a-b) (Hölland,et al., 2002; Rincón, 1992).

Nucleation occurs when a small piece of soil forms from the liquid. The solid must achieve a certain minimum critical size before it is stable. Growth of crystal occurs as atoms from the liquid are attached to the tiny solid until no liquid remains. Devitrification means the formation of crystals on or inside of amorphous glass, typically due to a delayed cooling cycle. The presence of local crystalline inclusions supports the glass and makes it more flexible, reducing the presence and severity of micro-cracks and acting as crack stoppers. The heat treatment that supports the growth of these native crystals during the glass formation is called ceramming and it is a two step process. Ceramming is a controlled crystallization of the glass that results in the formation of tiny crystals that are uniformly distributed throughout the body of the glass structure. The size of the crystals, as well as the number and rate of growth is measured by the time and temperature of the ceramming heat treatment. There are two branch of ceramming process; crystal nucleation and crystal growth as shown in Fig. 2(a-c) (Steffestun, & Frischat, G.H. 1993; Durrani et al., 2005; Cattell, M.J. 2006). Each phase take place because the glass body is detained at exact temperature for a definite length of time. Crystals have a propensity to develop in a mixture of glass when it is held at a specific temperature, called the crystal nucleation temperature. This means that when apprehended at the crystal nucleation temperature, multiple seed crystals begin to grow throughout the glass body. The longer the glass is kept at this temperature, the more seed crystals will form. Ideally, a glass ceramic will be strongest when there is a very large number of a small crystal distributed uniformly all over its mass. Once a seed crystal forms, it will also begin to grow larger at this temperature, but somewhat slowly. If the temperature of the glass body is held at the crystal nucleation temperature for a very long time, a very large number of crystals of broadly varying size will form. The initially the seed will be largest, while the crystals that have lately begun to grow, will be the smallest. In order to better control the art of the finished product, the ideal glass ceramic will have crystals of a small and uniform size. Any form of devitrification in a glass structure will produce one degree of opacity. Large crystals are flat to make the glass opaque, while small crystals uniformly scattered throughout the structure have less of an impact on the optical qualities of the finished product. Thus it is of advantage to keep the temperature at the point of maximum seeding for a finite length of time in order to allow several tiny seed crystals to nucleate, and then to stop the nucleation process and support the ones that have formed to grow to appropriate size.

The glass ceramic materials with tailor made properties can be fabricated by controlling nucleation and crystal growth process. Although some glass compositions are self-nucleating, generally some nucleating agents are added to the batch to promote phase separation and internal nucleation. These melt homogeneously into the glass, but promote very fine phase separation on reheating. The isolated phase, which can be a metal, titanate, zirconate, phosphate, sulfide, or halide, is structurally incompatible with the host glass and is usually highly unstable as a glass. Nucleation is followed by one or higher temperature treatments to promote crystallization and development of the desired microstructure. Most commercial glass-ceramic products are formed by highly mechanized glass-forming processes and converted to a crystalline product by proper heat treatment. Glass-ceramic materials have been prepared through powder processing in which glass frits are sintered and then crystallized (Grossman, 1974) the range of potential glass-ceramic compositions is therefore extremely broad, requiring only the ability to form a glass and control its crystallization. The production of glass-ceramics from powdered, using conventional ceramic processes such as spraying, slip-casting, or extrusion, extends the range of possible glass-ceramic compositions by taking advantage of surface crystallization. In these materials, the surfaces of the glass grains serve as the nucleating sites for the crystal phases. The glass composition and processing conditions are selected such that the glass softens proceeding to crystallization and undergoes viscous sintering to full density just before the crystallization process is completed. Given these conditions, the final crystalline microstructure is basically the same as that produced from the bulk process.

Fig. 2. (a-c): Crystal growing in aluminosilicate ceramic materials (a) Aluminum deficient silicalite-I (b) High silica aluminosilicate (c) sodium aluminosilicate.

The precursor glass powders may be produced by various methods, the simplest being the milling of quenched glass to an average particle size of 3-15mm. Sol-gel processes (Yuan, et al., 2010; Chen, et al., 2010; and Lei, B. 2010) in which highly uniform, ultrafine amorphous particles are grown in a chemical solution, may be preferable for certain applications. Such devitrifying frits are employed significantly as sealing frits for bonding glasses, ceramics, and metals. Other applications include co-fired multilayer substrates for electronic packaging, matrices for fiber-reinforced composite materials, refractory cements and corrosion-resistant coatings.

1.1.3 Machinable glass ceramic

Machinable glass ceramics based on magnesium aluminum silicate (MgO-Al_2O_3-SiO_2, MAS) glass ceramic system have technological importance due to their applications for high voltage and in ultra high vacuum (Hattori, T. et al., 1982, Emad M El-Meliegy, 2004). An important group of these materials are mica-containing glass-ceramics. The crystal phase generated in the mica containing glass ceramics is called fluorophlogopite ($KMg_3AlSi_3O_{10}F_2$) receive wide application due to their high machinability, which results in an increased versatility of the products and numerous possibilities for industrial application.These materials not only have strange feature of machinability but also have potential use till date are in electronic and semiconductor industry (precision coil formers & high voltage insulators), laser industry (spacers, cavities and reflectors in laser assemblies), high vacuum industry (thermal breaks in high temperature processing equipment, coil supports), aerospace and space Industry (retaining rings on hinges, windows and doors of NASA's space shuttle, supports and components in several satellite borne systems) and also in nuclear industry (fixtures and reference blocks in power generation units). It is noted (Hench et al., 1993) that glass ceramics can also be used for biomedical purpose including dental materials field. The MAS glass ceramic can be machined into complicated shapes and precision parts with ordinary metal working tools, quickly and inexpensively, and it requires no post firing after machining, no frustrating delays, no expensive hardware, no post fabrication shrinkage, and also no costly diamond tools to meet the required specifications. MAS ceramic material exhibits non-wetting, zero porosity, withstands high temperatures up to 1000°C and has high dielectric strength. MAS materials are being used in nuclear technology (Bozadzhiev L.S.; 2011) in the production of proto type components, used in medicines for the axles of mechanisms providing energy for implanted cardio-stimulators and also used in the production of welding jets or as holders for welded components. MAS have potential application in spacers, headers and windows for microwave tube devices, sample holders for microscopes and aerospace components (Goswami et al., 2002; Baik, et al. 1995; Boccaccini, 1997). Properties of MAS glass ceramics such as hardness, machinability, conductivity depend upon the composition and microstructure. Machining of these materials can be carried out to precise tolerances and surface finish with conventional tools. Factual reason of good machinability character of MAS lies in unique microstructure of inter-locking array of plate like mica crystals, dispersed uniformly throughout glass matrix.

MAS glass ceramic materials have been prepared by controlled crystallization in which a large number of tiny crystals rather than few bigger single crystals have been grown (Margha, et al., 2009). Controlled crystallization or heat treatments generally consist of a two-stage heat treatment, namely a nucleation stage and crystal growth stage. In the nucleation stage, small nuclei are formed within the parent glass. After the formation of stable nuclei, crystallization taking place by growth of a new crystalline phase. The nucleation and crystallization parameters of glasses are very significant in the preparation of glass-ceramics with desired microstructures and properties (Abo-Mosallam, 2009).

Simmons, et al, 1982; studied the effect of fluorine content and its source on the crystallization of MAS materials. Identified predominant crystalline phases in their study were fluorite, norbergite or fluorophlogopite depending on heat treatment, fluorine concentration etc. Common preparation methods have been discussed in literature (Hattori, T. et al. 1982) but due to special category and technological importance of the material, the crucial process has either been vague or missing. Recently, magnesium aluminum silicate (MAS) glass ceramic systems have been synthesized by sintering route (Durrani, et al., 2010 & Hussain, S.Z et al., 2010). Therefore, the preparation of MAS materials acquires special importance for meeting indigenous requirement. In the light of this fact, preparation of machinable MAS glass ceramic was undertaken using sintering route.

1.2 Sintering

Generally, when ceramic powders are formed and then heated (green compact) part, there is a certain temperature below melting point at which they begin to burn, and in most cases there is shrinkage or expansion resulting in densification, phenomenon is called sintering. (Moulson et al. 1992; Rahaman, 2005). The goal of the sintering process is to convert highly porous compacted powder into high strength bodies. Sintering may be considered the process by which an assembly of particles, compacted under pressure or simply confined in a container, chemically bond themselves into a coherent body under the influence of an elevated temperature. The temperature is usually below the melting point of the major constituent. Much of the difficulty in defining and analyzing sintering is based on the many changes within the material that may take place simultaneously or consecutively. In the sintering process the temperature of the granulated sinter compound is raised to temperatures between 1000°C and 1450°C to achieve partial fusion. During the heating and cooling cycle different species react with each other to produce certain phases. Molten material is produced which crystallizes or solidifies into various phases that bond the microstructure together. Therefore, sinter consists of an assembly of various phases of varying chemical composition and morphology. Each of different phases has a unique influence on the sinter quality.

Sintering is a complex process and for any given metal and set of sintering conditions there are likely to be different stages, driving forces and material transport mechanisms associated with the process. Sintering or firing of pure oxide ceramic requires relatively long time and high temperature because the diffusion proceeds in solid state. The complete sintering process is generally considered to occur in three stages: (i) initial stage, (ii) intermediate stage, (iii) and final stage. There is no clear-cut distinction between the stages since the processes that are associated with each stage tend to overlap each other.

The reduction in surface energy can be used to explain the three main stages of solid-state sintering, (Wang, Y., 2008) is shown in Fig. 3(a-c). In the first stage, atoms migrate towards the points of contact between particles to form necks as this filling process reduces the surface area and the surface energy. In the second stage, the grain boundaries grow because, as atoms are removed from the grain boundary and diffuses towards the neck, this causes the centers of particles to mutually converge. In the final stage, the grain is slowly eliminated as grain boundaries merge.

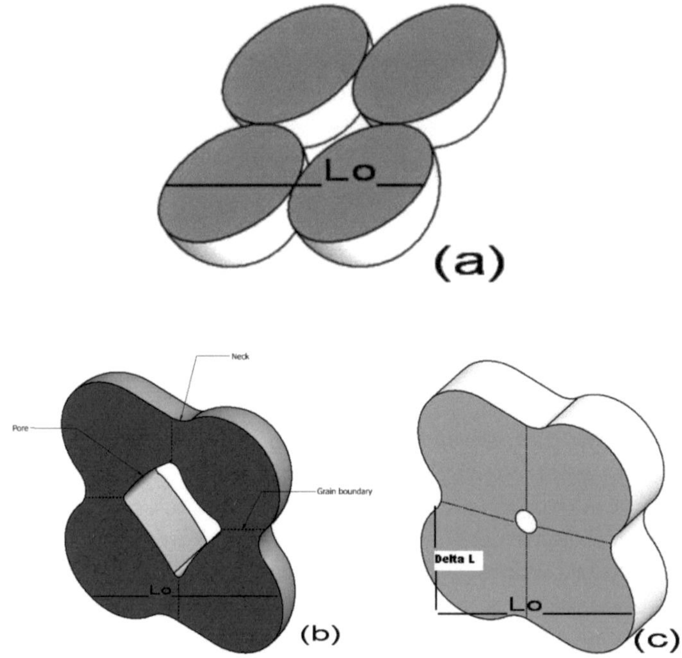

Fig. 3. (a-c): Schematic representation of the neck formation during solid-sintering due to (a) powder compaction, (b) neck growth, and (c) neck growth developed by densification.

The description of such topic is beyond the scope of this chapter, however, a detailed account of these stages has been discussed somewhere else (Richerson, 1992; & German, 1996). The sintering process is an important tool in fabrication of various materials into useful products. Today, it is used to manufacture a wide range of products for consumers, electronics, transportation and biomedical systems, e.g., rocket nozzles, ultrasonic transducers, automobile engines, semiconductor packaging substrate, and dental implants. Sintering is not only used for high temperature materials but also for other materials that can be densified below 1000°C. For example, firing of glass-based substrate and of-screen-printed metallic inks or paste for microelectronic applications. Densification or shrinkage of the sintered part is very often associated with all types of sintering. However, sintering can take place without any shrinkage; expansion or no net dimensional change is quite possible. From the tooling point of view it is preferred to avoid very large amount of dimensional changes. The driving force for solid state sintering is the excess surface free energy. Sintering converts a compacted powder into a denser structure of crystallites joined to one another by grain boundaries. Grain boundaries vary in thickness from 100pm to over 1μm.They may consist of crystalline or vitreous second phases, or may be simply a disordered form of the major phase because of differing lattice orientation in the neighboring grains. Grain boundary is the border between two grains, or crystallites, in a polycrystalline material. Grain boundaries are generally not as dense as the crystals and, in the early stages of sintering at least, allow free diffusion of gas to and from the outside atmosphere. Typically, polycrystalline ceramic's microstructure is developed by solid state sintering as shown in Fig. 4(a-c).

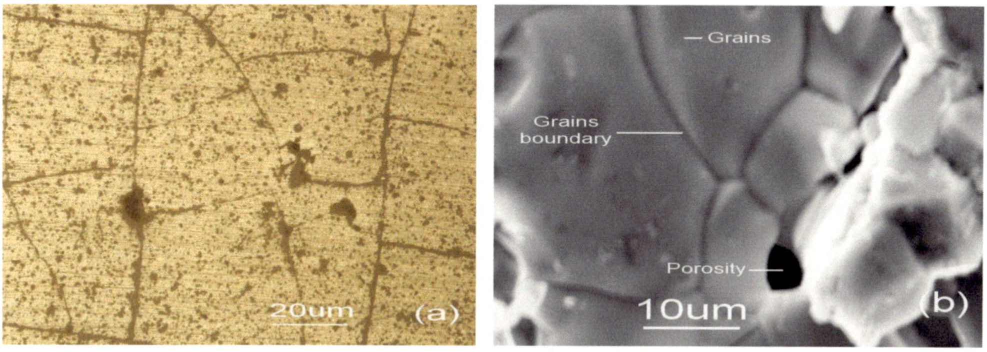

Fig. 4. (a-b): Microstructure of sintered pellet of (a) Metallographic image of magnesium aluminum silicate glass ceramic (MAS-G4) (b) SEM image showing combinations of grains, grain boundaries and porosity.

The polycrystalline ceramic's microstructure consists of numerous grains or crystallites that are defined by the long-range order of the crystal. These grains are separated by a grain boundary that may include single or multiple phases of sintered aid material. A grain boundary is the surface that separates the individual grains and is a narrow zone in which the atoms are not properly spaced. In addition, microstructure is likely to contain a certain amount of porosity left behind during manufacture. Sometime separate phases of material exist within the microstructure as particles as is commonly found in reaction-bonded ceramics (Swab, 2009). During the sintering process, an agent (sintering aid) is added to help in the bonding process and reduced the temperature that is required during the sintering step. These sintering agents lead to a reduction in the mechanical properties of ceramic because they form relatively soft grain boundaries with low melting temperatures. Smaller particles sinter much faster than coarse particles because the surface area is larger and the diffusion distances are smaller. Rate of sintering varies with temperature. Sintering processes can be divided into three large categories (Yu, et al., 2001; Moghadam, et al.1982):

i. Solid Phase Sintering
ii. Liquid Phase Sintering
iii. Gas Phase Sintering

1.2.1 Factors effect the sintering

The most important factors involved during sintering process are "process variables" (temperature, time and furnace atmosphere), "materials variables" (particle size, shape and structure), green density and dimensional changes. Variety of sintering methods are available for sintering the ceramic compacts i.e., standard pressure sintering, reaction-sintering, hot pressing, post-reaction sintering, recrystallization sintering, atmospheric pressure sintering, ultra-high-pressure sintering (Ring, T.A. 1996), chemical vapor deposition (CVD) (Mitchell, (2004) and isostatic hot pressing (HIP) (Olevsky, et al. 2009, Smothers, 2009).

1.3 Aim of work

The present work concerns only about solid phase sintering process. It is very difficult to sinter fully dense state of machinable magnesium aluminum silicate (MAS) glass ceramic materials with fluorophlogopite as the main crystalline phase. The primary objective of the present research work is to provide a simple sintering method for preparation of crystalline magnesium aluminum silicate glass-ceramic body with predominant fluorophlogopite crystal phase, which can be utilized as candidate material for machinable tools acquiring good resistance to attack by acids and alkalies. The variation in sintered densities, mechanism of phase transformation, microstructure changes and thermal expansion co-efficient of MAS glass ceramic was also ascertained. It constitutes a part of our ongoing studies on MAS glass ceramic material in detail and all the results are based on techniques previously applied (Durrani et al., 2010; Hussain et al. 2010).

2. Preparation of MAS glass ceramic by sintering route

Magnesium aluminum silicate glass ceramic specimens (designated as MAS-G1-MAS-G10) were prepared through sintering route using three stage schedules i.e. calcination, nucleation and crystallization process using stiochiometeric amount of inorganic metal salts, such as oxides, hydroxide, fluorides and carbonates in weight percent as shown in **Table 1**. The starting raw materials used in the present study were Silica (SiO_2, Fluka, 99.9%), aluminum oxide (Al_2O_3, BDH, 99.9%), magnesium oxide (MgO, Merck, 99.9%), potassium carbonate (K_2CO_3, Riedel-de Haen 99.9%), boron oxide (B_2O_3, BDH, 99.9%), magnesium fluoride (MgF_2 BDH99.9%), ortho-phosphoric acid (H_3PO_4.BDH, 99.9%), acetone (CH_3COCH_3 BDH, 99.9%), hydrochloric acid (HCl, Marck), sodium carbonate (Na_2CO_3, BDH, 99.9%) and sodium hydroxide pellets (NaOH, Merck). All chemicals were of analytical reagent grade and used without further purification.

In a typical preparation, the initial charge was mixed thoroughly and calcined at 950°C for 24 h with a pre-determined heating schedule. Approximately 4-7wt% MgF_2 was added to calcined charge and milled in a planetary ball mill for 40 h. The median particle size of the ball-milled sample was around 5.3μm. The fine ball milled charge was seasoned in 5% H_3PO_4 acid solution in acetone medium for 72h. Compacts of MAS (61×16×5 mm) were made using hydraulic press of load capacity ~ 25 tons/in^2 (1 ton/in^2 = 15.44 MPa). The compacts were sintered using two-step heating program. In first step the compact was heated up to 600-630°C for 2-4 h to ensure good nucleation and to initiate crystal growth. In second step, heating was carried out with different heating rates in the range of 15-60°C/h up to sintering temperatures of 950-1080°C. The sample was kept at the sintering temperature for a sufficiently long time to achieve the desired crystal growth. In order to measure the phase purity, crystallinity and morphology of synthesized products, the MAS materials were characterized using different analytical techniques i.e., TG/DTA, XRD/XRF and SEM. Phase purity and crystallinity of MAS specimens was determined by X-ray diffraction on a Rigaku Geiger flux instrument using Cu Kα radiation. For measurement of weight loss the combined TG-DTA thermal analysis was performed using Netzsch STA-409 thermoanalyzer. Coefficient of thermal expansion was observed on horizontal high temperature computer-controlled differential dilatometer (Netzsch, DIL, 409) with heating rate of 10°C/min up to 1100°C in static air. The particle size distribution of powder specimens was measured by laser particle size analyzer (SK-Laser Micron PRO-7000S). Density of MAS specimens was measured using Ultra Pycnometer 1000 (Quantachrome). Porosity was measured by radiographic technique using Real-time radiography Instrument (Model Isvolt HS Panta Sei Fert). Micro-structural features, elemental contents and porosity of sintered specimen MAS -G8 were observed using scanning electron microscope (SEM, LEO 4401). The specimen was fully polished, put onto aluminum stud, dried in air and then the specimen was coated with thin gold film for the SEM observation. For the measurement of micro-hardness of sintered specimens, indentation technique using Vickers diamond pyramid indentor on the micro hardness tester was used. Before measurements, the sample surface was polished with 3μ alumina powder to get good reflective surface. The measurement was done on the polished surface by applying 300g load for 15 sec. The effect of K_2CO_3 concentration on sintered density and mechinability of MAS specimens (MAS-K1-MAS-K10) at fixed amount of MgF_2 (4-11%) and nucleation temperature (630°C) was also studied. The sintered specimens were treated with 5% hydrofluoric and hydrochloric acids for 24h at 95°C to observed the effect of these acids. Effect of 5% sodium hydroxide and sodium carbonate was also studied for 6 h duration at 95°C. The samples were weighed for any loss in weight after washing off acids and bases.

Impedance spectroscopy on pellets of MAS glass ceramic, which were sintered at 1040 and 1050°C temperatures was performed in the frequency range of $1 \leq$ frequency $\leq 10^7$ Hz at room temperature, using an alpha-N Analyzer (Novocontrol Germany). The surfaces of both sides of the pellets were cleaned properly and contacts were made by silver paint on opposite sides of the pellet, which were cured at 150°C (423K) for 3 h. Before the impedance experiments, the dispersive behavior of the leads were carefully checked to exclude any extraneous inductive and capacitive coupling in the experimental frequency range. The ac signal amplitude used for all these studies was 0.2 V and WINDETA software was used for data acquisition.

Specimen #	Chemical Composition (wt %)						Calcination		Ball Milling	Soaking
	SiO_2	Al_2O_3	MgO	K_2CO_3	B_2O_3	MgF_2	Temperature (°C)	Time (h)	Time (h)	
MAS -G1	36.62	12.73	16.87	13.93	1.91	17.91	950	24	40	72
MAS -G2	37.57	13.19	16.49	14.26	1.97	16.49	950	24	40	72
MAS -G3	37.72	13.23	16.05	13.98	2.48	16.54	950	24	40	72
MAS -G4	38.51	13.51	16.38	14.27	2.53	14.78	950	24	40	72
MAS -G5	39.28	15.71	15.52	13.59	8.25	7.65	950	24	40	72
MAS -G7	40.67	16.43	14.12	13.43	8.62	6.73	950	24	40	72
MAS -G8	36.34	24.82	13.23	11.32	3.53	10.76	950	24	40	72
MAS -G9	44.32	12.91	15.45	13.62	4.12	9.58	950	24	40	72
MAS -G10	41.75	14.15	16.92	13.54	4.46	9.18	950	24	40	72

MAS-G = Magnesium aluminum silicate glass

Table 1. Chemical composition and reaction conditions for preparation of MAS glass ceramic material by sintering route.

3. Results and discussion

3.1 XRD phase analysis

Experimental results showed the phase stability, thermal stability, compressibility, and sinterability of MAS glass ceramic materials. The crystallinity of MAS glass was studied by

XRD and XRF. The XRD patterns of specimens MAS-G3, MAS-G4 and MAS-G8 sintered at 1040°C for 3h along with standard specimens are presented in Fig. 5. The XRD patterns were indexed and compared with Joint Committee of Powder Diffraction Standard (JCPDS) data cards (McClune, W. F.1989). The predominant phases in the prepared MAS specimens were identified as fluorophlogopite (potassium magnesium aluminum silicate, $KMg_3AlSi_3O_{10}F_2$, JPCPD # 71-1542), sillimanite (aluminum silicon oxide, Al_2SiO_5, JPCPD #.88-0893) and leucite (potassium aluminum silicate $KAlSi_2O_6$ JPCPD # 15-0047). The percentage of these phases in each specimen is given in **Table 2,** whereas crystallographic XRD analysis data are presented in **Table 3.** The well formed diffraction patterns and "d" values confirmed that the predominant crystal phase in MAS-G8 is 57% fluorophlogopite and this fluorophlogopite crystal phase was found near to standard values i.e., 60% fluorophlogopite. It was also observed that specimen MAS-G8 consists of 57%fluorophlogopite crystal phase due to which it possess very good mechinability, but very difficult to be sintered to fully dense state. The leucite (potassium-aluminum-silicate, $KAlSi_2O_6$) crystal phase with large coefficient of thermal expansion is formed when glass ceramic is heated and held at temperatures between 1025°C and 1530°C.

Specimen #	Sintering Conditions			Density (g/cm³)		Phase (%) by XRD	
	Nucleation (°C)	Temperature (°C)	Time (h)	Green	Sintered		
MAS-G1	630	1040	3	1.79	2.18	F= 28	L= 71
MAS-G2	630	1040	3	1.74	2.12	F= 30	L = 69
MAS-G3	630	1040	3	1.91	2.29	F=51	S= 49
MAS-G4	630	1050	3	1.68	2.15	F = 56	S = 43
MAS-G5	630	1040	3	1.65	2.12	F= 28	L= 71
MAS-G6	630	1050	3	1.72	2.14	F=49	S= 42
MAS-G7	630	1060	4	1.64	2.14	F = 50	S = 44
MAS-G8	630	1040	3	1.81	2.38	F= 57	S = 43
MAS-G9	630	1070	2	1.86	2.39	F= 54	S = 42
MAS-G10	630	1080	2	1.74	2.35	F= 51	S = 39
Standard					2.48	F= 60	S = 40

F= Fluorophlogopite L= Leucite and S= Sillimanite

Table 2. Fluorophlogopite phase of magnesium aluminum silicate glass ceramics with different composition.

Crystallographic XRD analysis	Fluorophlogopite	Silliminite
Symmetry	Monoclinic	Orthorhombic
Unit cell volume(Å³)	481.77	331.23
Crystallite size (Å)	184	184
Lattice parameters		
a (Å)	5.308	7.49
b (Å)	9.183	7.67
c (Å)	10.00	5.77

Table 3. XRD phase analysis of MAS-G8 specimen.

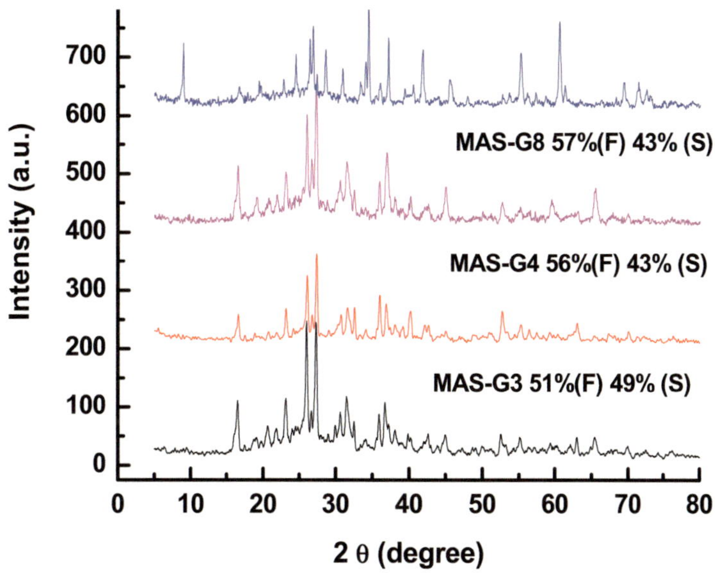

Fig. 5. XRD patterns of magnesium aluminum silicate glass ceramic, specimen MAS-G3, MAS-G4 and MAS-G8 along with standard.

3.2 Thermal analysis

The TG/DTA analysis is proper method to determine the optimal stabilization temperature (Perdomol, et al., 1998; & Izquierdo-Barba, et al., 1999). Thermal analysis TG, DTA of selected specimen the MAS-G8 was carried in the temperature range of 40-1000°C in order to find the crystallization temperature of different crystalline phases as shown in Fig. 6. The total weight loss 9.14% was observed after heating up to 900°C. Since the specimen was not fully dried, some physically bounded water was present at surface of crystals and in micropores, which caused a subsequent loss of mass at the lowest temperature range, i.e., 50-250°C. This process is endothermic as confirmed by DTA. Two peaks in the DTA data were observed. The first peak is endothermic, which is connected with decomposition of water vapors and glass transformation. It can be seen that crystal phase (nucleation) occurs at nearly 732°C, which is evident from the large exothermic peak. The exothermal maxima correspond to the separation of the crystalline phase. The position of the minimum on DTA curve is basically determined by the chemical composition of the separated phase or by its transformation temperature. The exothermal crystallization peak, its position and shape have been characterized by the crystallization process (Bapna & Mueller, H. 1996). In crystallization process [K_2O-Al_2O_3-SiO_2-MgO-B_2O_3-F], first crystal phase to appear is the aluminium borate-mullite solid solution crystal which on increasing temperature, transforms to fluorophlogopite on reaction with the matrix phase (Roy, S. et al. (2011). Heating rate during DTA is around 10°C/min. All the volatiles were completely removed at 900°C. The nucleation temperature of this crystal is a strong function of heating rate as found by (Bapna & Mueller, H. 1996) and observed that with increasing heating rate shifts the nucleation temperature to a higher value. The percent weight loss of other specimens MAS-G2 to MAS-10 is shown in Table 4.

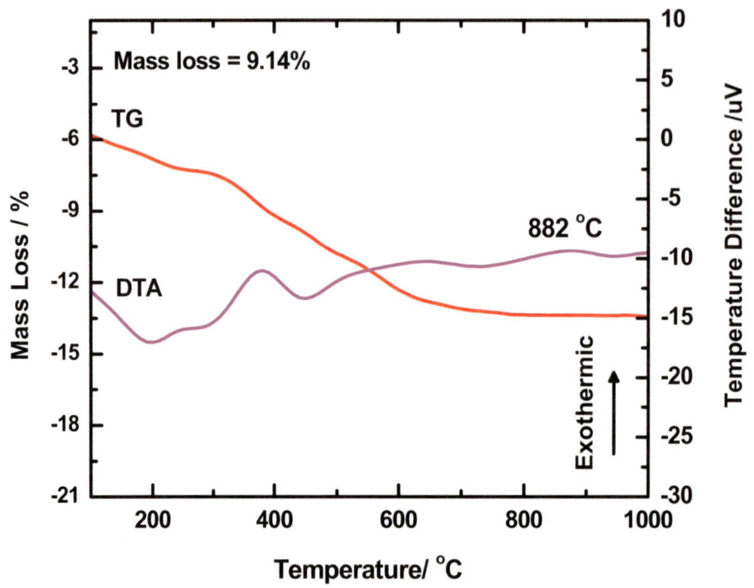

Fig. 6. TG/DTA curves for magnesium aluminum silicate glass ceramic powder, specimen MAS-G8.

3.3 Green compact and thermo dilatometric study

The behavior of densification was measured by dilatometry in order to understand what is occurring during sintering. Dilatometry is a well-known method of studying densification kinetics during the sintering process of ceramic bodies (Holkova, *et al.*, 2003). In dilatometry, the compacted ceramic body undergoes heat treatment in the dilatometer to initiate sintering. Simultaneously, the length of the compacted body is measured as a function of time at a given temperature. The densification can further be assessed by measurement of post-sintering density, whereby as the compacted body shrinks, its density will increase. Isothermal sintering experiments were carried out by heating samples rapidly after binder burnout to the sintering temperature and the data collected at intervals that are set in the control program. An entire densification profile can be obtained from a single sample. The characteristics of pressed powder of MAS-G4 and MAS-G8 during sintering were examined by means of dilatometry. Fig. 7 and Fig.8 display the thermo-dilatometric curves of MAS-G4 and MAS-G8 powders which were calcined at 650°C and then pressed into green pellet respectively. The curve shows a simple thermal shirking or expansion i.e., the change in length during sintering of pellet. There was no change in length on heating up to temperature of 208°C but started at about 557°C. However, a large shrinkage occurred as a result of evolution of decomposed species between 300 and 400°C. The temperature between 400 and 700°C shows the expansion in the material. This expansion is due to nucleation process i.e. decomposition of meat-stable phases of magnesium and aluminum silicates and carbonates and fluorides also the expulsion of organic binder (PVA). Above 850°C, there is gradual increase in dimensional changes (lattice parameters) which are due to crystal growth (development of MAS-G8 crystalline phase). On the basis of dilatometry results, sintering was conducted between 985-1040°C in order to obtain the crystalline MAS-G8 fluorophlogopite phase. Thermal expansion co-efficient (alpha) of MAS-G8 ceramic was measured at different temperature ranges by dilatometric technique. The coefficient of thermal expansion (α = x10^{-5} $^oC^{-1}$) was measured using a dilatometer Fig. 7(a-b) and Fig. 8(a-b) in air at 1000°C for specimens MAS-4 and MAS-G8; the values are -1.563 and -1.093 x10^{-5} $^oC^{-1}$ respectively. The result revealed that α decrease with increase of chemical composition. The detailed coefficient of thermal expansion (α = x10^{-5} $^oC^{-1}$) values are presented in **Table 4.**

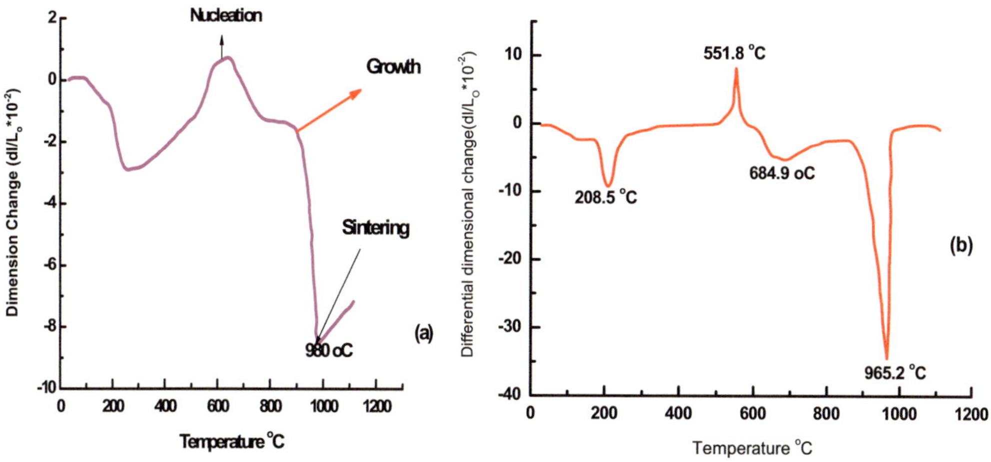

Fig. 7. (a-b): Thermodilatometric (a) and differential dilatometric (b) curves of green compact of magnesium aluminum silicate glass ceramic specimen MAS-G4.

Physico-chemical Properties						
Property	Units	MAS-G2	MAS-G4	MAS-G6	MAS-G8	MAS-G10
Weight loss (DG/DTA)	%	10.15	10.42	9.05	9.14	9.32
Thermal co-efficient (200-1100 ^{o}C)	$^oC^{-1}$	-1.563x10^{-5}	-1.563x10^{-5}	-1.563x10^{-5}	-1.093x10^{-5}	-1.063x10^{-5}
Median diameter of particle	µm	4.4	5.4	5.6	5.3	4.9
Micro hardness	Vicker	449	484	504	527	505
Resistivity	Ω	1.07 x 10^9	1.27 x 10^9	1.35 x 10^9	2.72 x 10^9	2.05 x 10^9
Porosity	%	7-8	7	5	2-3	2-3
Theoretical density	%	79	87	81	93	91

Table 4. Physical properties of MAS glass ceramics.

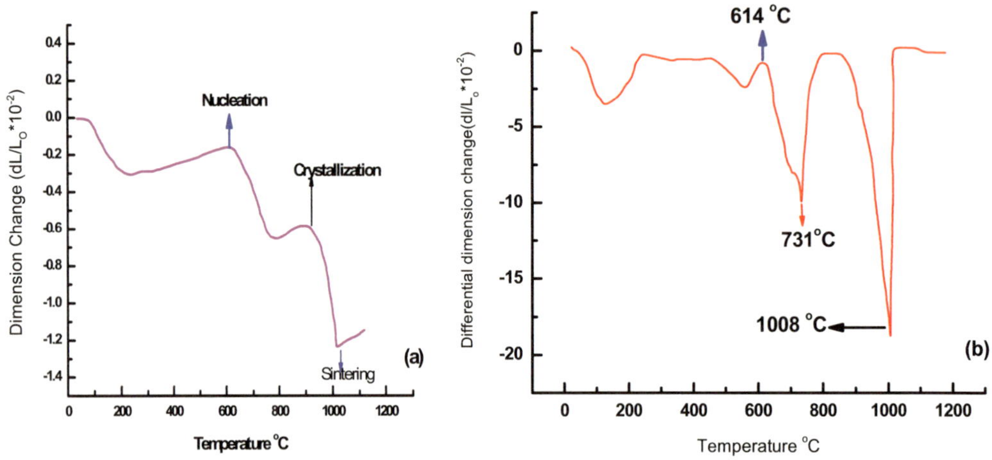

Fig. 8. (a-b): Thermodilatometric (a) and differential dilatometric (b) curves of green compact of magnesium aluminum silicate glass ceramic specimen MAS-G8.

3.4 Sintering MAS glass ceramic

Green solid briquette of dimensions (61x16x5mm) was obtained using hydraulic press. The green body exhibits significant changes of its properties resulting from dehydration at low temperatures, phase changes during dehydroxlation, high-temperature reactions, and densification during sintering. All these changes significantly influence phyisco-mechanical properties of the fired body.It was observed that if the initial charge was not pulverized for sufficiently long time after calcination, the sintered briquette had a lot of open pores on the surface and MAS glass ceramic material looks swelled and deformed as shown in Fig. 9(a-c). The percent relative densities were measured in order to evaluate the material performance. The effect of heating temperature and time duration on density of sintered specimens MAS-G6 to MAS-G10 as a function of sintering temperature is shown in **Table 2**. Insignificant increase in density was observed in sintering temperature range 1040°C to 1060°C with time duration range 2-4h. It was observed that as the sintering temperature was increased beyond 1060°C, excessive thermal energy leads to some rearrangement among the grains as well as some an isotropic crystal growth. In addition, decomposition of fluorophlogopite phase, formed at lower temperature, takes place leading to the development of internal line cracks/voids (Radojic, & Nikolic, 1991) in the material. As a result of this, the material looks swelled, broken, deformed and its density decreases at higher sintering temperatures, Fig. 9(a-c). The important sintering parameters, green and sintered densities of samples MAS-G1 to MAS-G10 are given in **Table 2.** Relative density (93%) was obtained, when the MAS-8 glass ceramic specimen was sintered at 1040°C for 4h. The high sintered density can be attributed due to small degree of agglomeration and fine nature of powder. The results revealed that sintered MAS-G8 glass ceramic is composed of uniformly sized grains and is >93% of theoretical density. Formation of fluorophlogopite phase, good relative density and micro porosity are mainly responsible for producing machinability properties. Fig.10 (a-b) shows photographic, microscopic and scanning electron image of green and sintered MAS-G4 glass specimens whereas XRD pattern of such specimen is shown in Fig. 5.

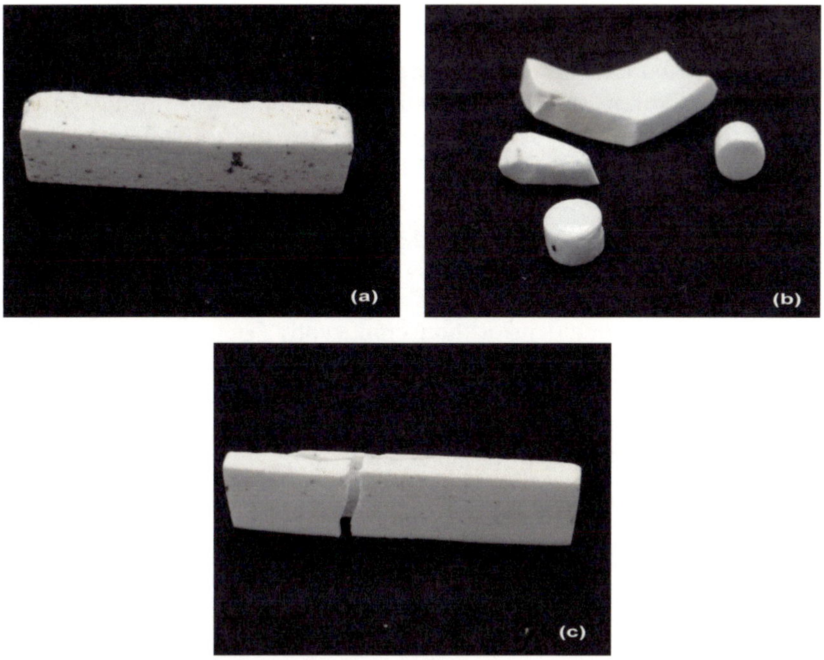

Fig. 9. (a-c): The effect of heating temperature and time duration on physical surface of sintered specimens (a) Open pores on the surface of MAS-G4 (b) Swelling of MAS material (c) brittle and broken of sintered slab.

Fig. 10. (a-c): Photograph of MAS-G4 glass specimen (a) green and sintered specimens (b) microscopic image and (c) SEM micrograph of polished sintered MAS-G4 glass specimen.

3.5 Particle size distribution

The particle size distribution curve of specimen MAS-G8 powder is shown in Fig. 11. This curve revealed that the particle size distribution is < 10μm with median particle size is around 5.3μm. Smaller particle size and particle size distribution in narrow range is essential for good sinterability. Experimental results proved the same. It is also found that specimen MAS-G8 sintered at temperature 1040°C provided the material of smaller and uniform particle distribution and density ~2.35 g/cm^3 (93% of theoretical density) having 3-4% through porosity. Because of larger surface area of fine powder, the progress of sintering process accelerates even at low temperatures. However, fine particles, higher the tendency to form agglomerates retarding densification significantly (Ting, & R. Y. Lin, 1995). During pressing, it was observed that if the initial charge was not pulverized for sufficiently long time after calcination, the sintered briquette had a lot of open pores on the surface. So, in order to minimize the porosity in the sintered product, quite long period was used for pulverization and finally 72 h time duration was selected.

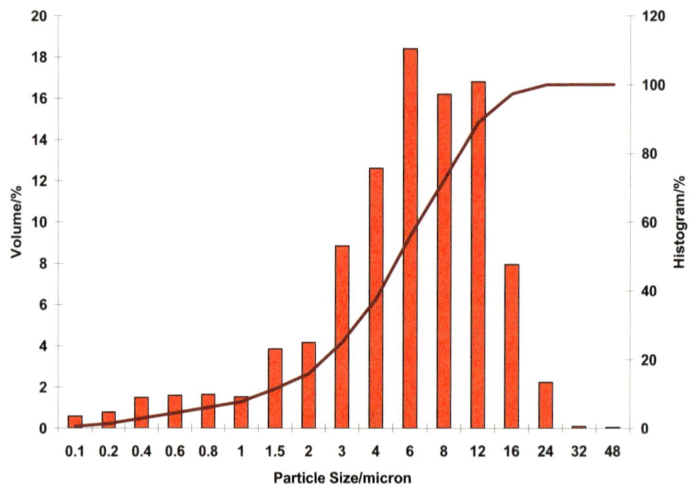

Fig. 11. Particle size distribution of magnesium aluminum silicate glass ceramic MAS-G8 powder.

3.6 Porosity

Porosity or void fraction is a measure of void spaces in a material, and is the ratio of the volume of all the pores in a material to the total volume. Radiography of specimen MAS-G4 and MAS-G8 was conducted in order to observe the porosity. Fig. 12 and Fig.13 show radiographic studies for the estimation of porosity. Porosity is the result of gas entrapment in the solidifying glass ceramic materials. Porosity can take many shapes on a radiograph but often appears as dark round or irregular spots or specks appearing singularly, in clusters, or in rows. Radiographic images as indicated in Fig. 12b and Fig.13b do not show any dark round or irregular spots or specks and do not predict any sign of porosity. It means that the specimens MAS-G8 and MAS-G4 have through porosity. However, some surface porosity was observed by SEM in both specimens MAS-G8 and MAS-G4 as shown in Fig. 14(a-b).

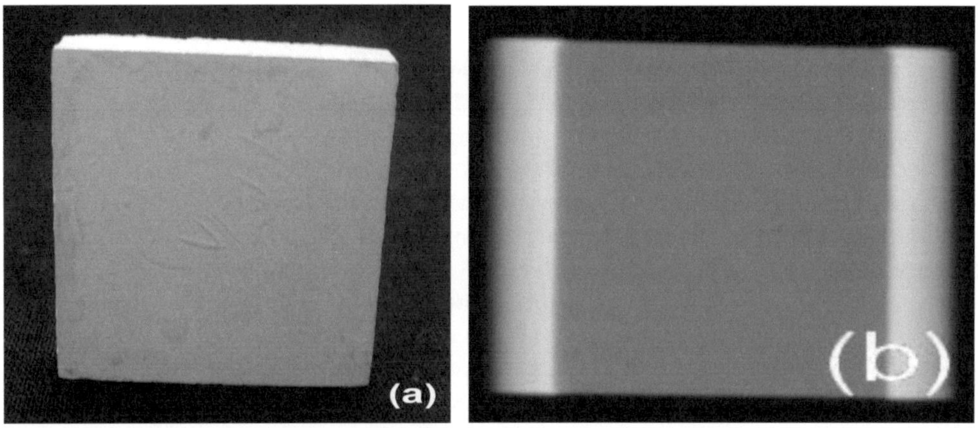

Fig. 12. (a-b): Radiographic studies of sintered specimens of magnesium aluminum silicate glass ceramic: (a) sintered MAS-G4 (b) Radiographic image of MAS-G4.

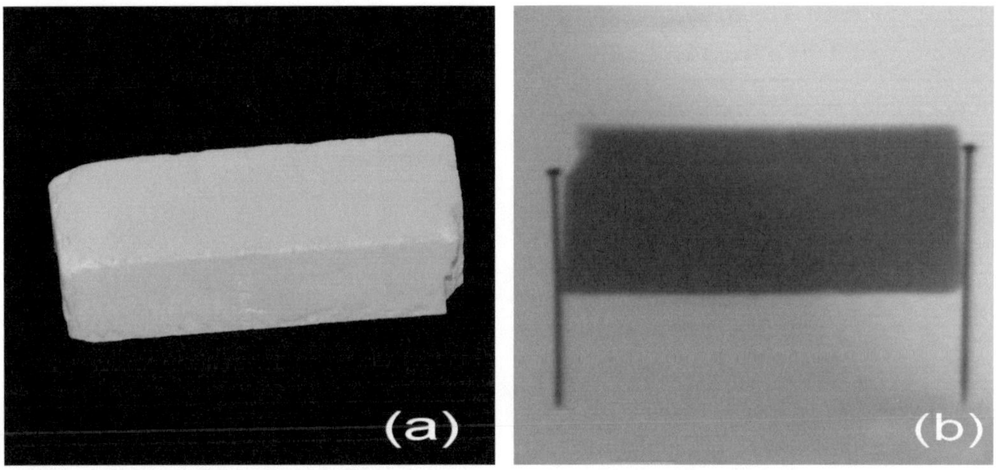

Fig. 13. (a-b): Radiographic studies of sintered specimen of magnesium aluminum silicate glass ceramic (a) sintered MAS-G8) (b) Radiographic image of MAS-G8.

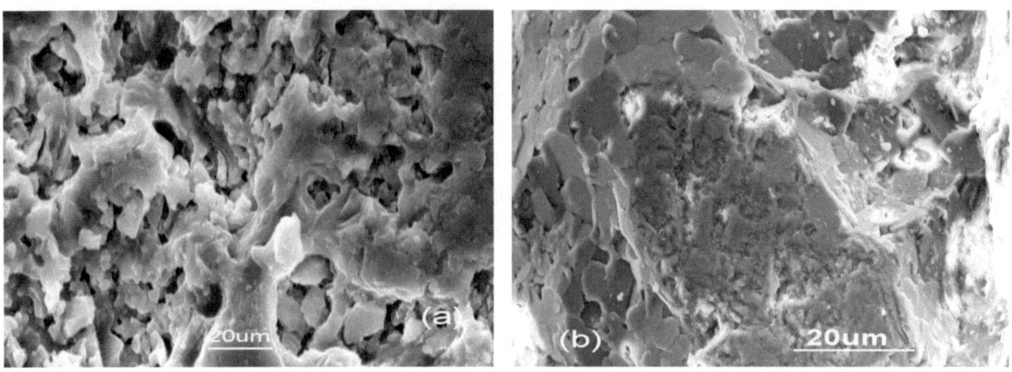

Fig. 14. (a-b): SEM image of magnesium aluminum silicate glass ceramic (a) MAS-G8 and (b) MAS-G4.

3.7 Resistance measurement

The impedance plane plots of MAS glass specimens sintered at 1040 and 1050°C at room temperature are shown in Fig.15 and Fig.16. The Fig.15b and Fig.16b show the typical impedance plane plots between real (Z/) and imaginary parts (Z//) of impedance. An arc-like behavior was noticed in both the plots and the intersection of the arc at low frequency (right hand side) gives the total resistance of the specimens , whereas, intersection at high frequency (the extension of arc on the left hand side) passes through origin. When MAS-G4 glass specimen heated to 1050°C, the intersection of arc on the right hand side shows a decrease in the magnitude of Z/, when sintering temperature is increased from 1040°C to 1450°C. This decrease can be explained as heating will cause grain growth which results decrease in impedance values of the specimen. The impedance values calculated from Fig.15b and Fig.16b, revealed high magnitude of resistance i.e., 1.28 x 10^9 and 2.72 x 10^9 ohms for MAS-G4 and MAS-G8 specimens respectively. Real part of permittivity (dielectric) shows the dispersion at low frequencies but at high frequency the dielectric value is least around 1.0 in both MAS glass ceramic specimens, which is according to literature (Hu et.al. 2001). It was observed that MAS-G8 glass specimen is more resistive and insulator in nature than the MAS-G4 and these result values confirmed the XRD, SEM and radiographic results of same sintered specimens.

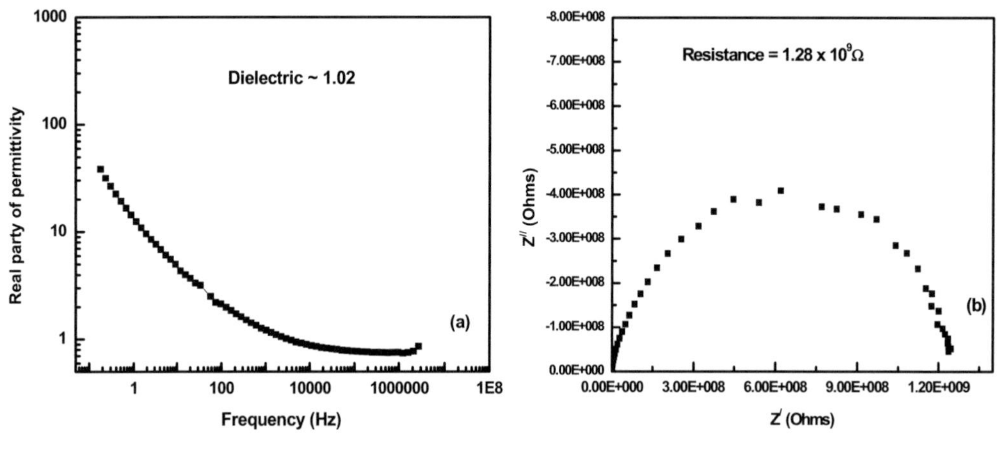

Fig. 15. Impedance analysis of MAS-G4 specimen sintered at 1050°C, (a) Variation of real part of impedance as a function of frequency in Hz (b) Impedance plane plot of MAS-G4 specimen at RT.

3.8 Effect of K_2CO_3 concentration

The physico-chemical characteristics and phase evolution during firing of MAS-K series ceramic bodies was investigated as a function of K_2CO_3 concentration and firing temperatures. The physico-chemical characteristics of sintered MAS glass specimens with variation of K_2CO_3 compositions expressed as MAS-K1- MAS-K10 and are presented in **Table 5**. Figure 15 shows the XRD patterns of MAS specimens MAS-K2-MAS-K5. The phase evolution examined by X-ray diffraction (XRD) revealed that leucite, and sillimanite are the main phases in the sintered bodies. Besides the above mentioned crystalline phases, other crystalline Mg_2SiO_4 and SiO_2 phases are also identified in Fig.16. In addition, glassy phase fluorophlogopite appears as result of different chemical reactions during firing in some sintered MAS specimens. The relative densities (after sintering at 1040-1060°C) of MAS glass specimens were found to be decreased (85-74%) with increase of stoichiometric compositions of K_2CO_3. These stoichiometric variations altered the density, porosity and other physical properties in agreement with microstructure analyzed by SEM already shown in Fig.14(a-b), where a high fraction of porosity was observed in MAS-G10 specimen.

3.9 Influence of cooling rate after high temperature sintering treatment

MAS glass ceramic specimen (MAS-K6) was examined through cooling and quenching after high temperature sintering treatment in order to observed the influence of cooling rate on the development of phases, morphology (grain boundaries) and their impact on physical properties by studying the process of crystallization. The MAS glass ceramic (MAS-K6) specimen was heated to 1040°C, held for 4h and then cooled. The heating and cooling rates were kept 5°C/min and 5-10°C/min respectively. XRD was used to detect the developed phases on different cooing time. Figure 15 shows the XRD patterns of MAS specimen MAS-K6 (fully cooled down). The XRD data revealed that fluorophlogopite, leucite, and sillimanite are the main phases developed in the sintered bodies along with some other crystalline phases i.e. Mg_2SiO_4 and SiO_2. The observed physical characteristics, morphology and phase evolution are given in **Table 6.** It was noticed that in addition, the glassy phase, fluorophlogopite appears as result of different chemical reactions during firing and cooling. The change in density and other physical appearance were also observed due to different cooling temperatures as shown in Fig.16 (a-d). When the MAS glass material was rapid quenched after sintering exhibit silica-rich glassy phases at grain boundaries with enrichment of sillimanite, Mg_2SiO_4 and less amount of fluorophlogopite, leucite phases, Fig.16 a & f and the surface was found to be porous due to the evaporation of absorbed water. When the MAS material specimen was slowly cooled in furnace, the silica glassy phases were only present along with enrichment of fluorophlogopite and leucite phases, Fig.16 b& c.

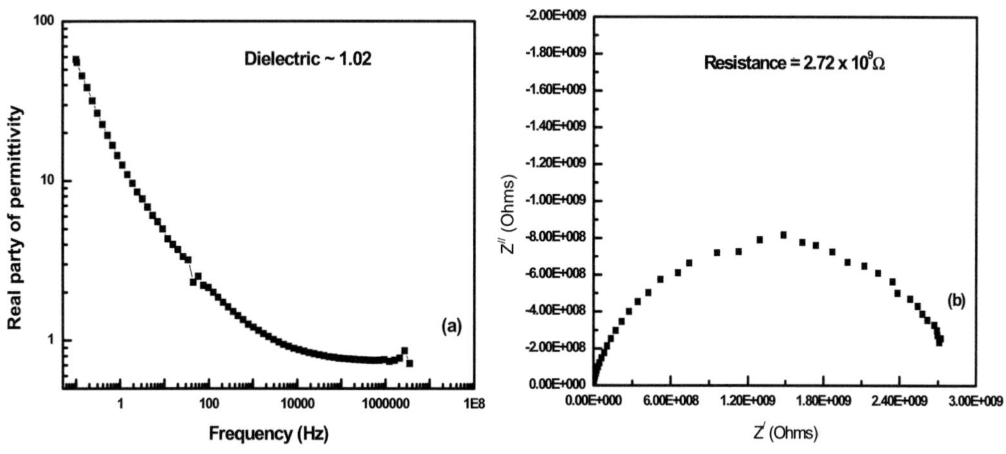

Fig. 16. Impedance analysis of MAS-G8 specimen sintered at 1040°C, (a) Variation of real part of impedance as a function of frequency in Hz (b) Impedance plane plot of MAS-G8 specimen at RT.

Specimen #	Chemical Composition (Wt %)						Firing Temperature °C	Sintered Density (g/cc)	XRD (%Phase)				Properties
	SiO_2	Al_2O_3	MgO	K_2CO_3	B_2O_3	MgF_2			F	L	S	O	
MAS-K1	43.8	17.6	16.5	12.3	5.8	4	1015	2.42	38	15	42	5	Brittle
MAS-K2	35.6	22.5	18.7	12.6	5.8	4.8	1020	2.32	34	25	29	12	Brittle and deformed
MAS-K3	47.2	12.6	16.9	14.7	5.9	4.5	1025	2.27	36	44	14	3	Brittle and swell up
MAS - K4	40.2	15.5	17.9	15.4	5.7	5.3	1030	2.34	42	12	37	9	Hard and Brittle
MAS-K5	36.4	24.7	13.5	11.4	3.6	10.4	1040	2.38	57	3	40	-	Mechinable
MAS-K6	45.3	12.4	17.1	13.7	6.8	4.7	1040	2.39	52	39	9	-	Mechinable
MAS-K7	45.5	16.8	15.9	12.6	4	5.2	1040	2.28	52	4	38	4	Mechinable
MAS-K8	37.5	20.2	17.7	12.8	5	6.8	1045	2.41	54	5	39	2	Mechinable
MAS-K9	37.5	20.2	17.7	12.8	5	6.8	1050	2.41	51	7	42	-	Mechinable
MAS-K10	38.5	16.4	19.4	11.6	7.5	6.6	1045	2.41	49	36	7	8	Hard and Brittle
MAS - K11	36.2	23.6	16.1	9.8	4	10.3	1070	2.37	46	5	40	9	Hard and Brittle

F= Fluorophlogopite L= Leucite (L) S = Sillimanite O= other phases like Mg_2SiO_4 and SiO_2

Table 5. Effect of K_2CO_3 concentration on sintered density, phase purity and physical properties of MAS glass ceramic materials.

3.10 Elemental contents

The elemental composition of prepared powder and sintered specimens of MAS-G8 glass ceramic was measured by electron probe micro analyzer (EPMA) attached with scanning electron microscope and these elemental contents were also identified by XRF. Fig. 17(a-b) shows the composition analysis of MAS-G8 specimen sintered at 1040°C estimated by EPMA and also analyzed on XRF in order to detect the contents Al, Mg, K and Si.

Physico-chemical properties	Sintering temperature =1040°C				
	Specimens collected after quenching (cooling rate 10°C/min).				
	1040	740	440	140	30
Density (g/cm^3)	2.05	2.19	2.34	2.31	2.28
Phase formation (%)	F= 25, L= 21 S= 48 O= 6	F= 42, L= 18 S= 31 O= 9	F= 47, L= 12 S= 37 O= 4	F= 50, L= 8 S= 37 O= 5	F= 52, L= 4 S= 38 O= 4
Characteristic	Swell up, hard and brittle	Broken , hard and brittle	Broken , hard and brittle	hard and to some extent Machinable	Machinable
Physical appearance (Shape)	Deformed, grey white color with more porosity	Deformed grey-white color with more porosity	Off-white and shining with less porosity	White with porosity (5-6%	White with porosity (5-6%)
Withdrawal of specimen conditions	After thermal treatment, specimen was taken out from the furnace and quenched in air until room temperature	After thermal treatment, specimen was remained in furnace in respective time interval. Cooling at furnace rate contemplates cutting off the electric current to the furnace and allowing the furnace to cool down for desire time and also until room temperature.			
Photographs Figure 16(a-f)	Figure 16f	Figure 16e	Figure 16d	Figure 16c	Figure 16b

F= Fluorophlogopite L= Leucite (L) S = Sillimanite O= other phases like Mg_2SiO_4 and SiO_2

Table 6. Influence of cooling rate on MAS glass ceramic specimen (MAS-K6) after high temperature sintering treatment.

3.11 Acid effect

The sintered specimens were tested for resistance to acids and bases when subjected to 5% hydrochloric acid, hydrofluoric acid at 95°C for 24h for acids and 6h for sodium hydroxide and sodium carbonate bases **Table 7.** The chemical resistance i.e. weight loss per unit area (mg/cm^2) of MAS-4b either acids or base was found significant and values were found similar with (Wawrziniak et al., 1980).

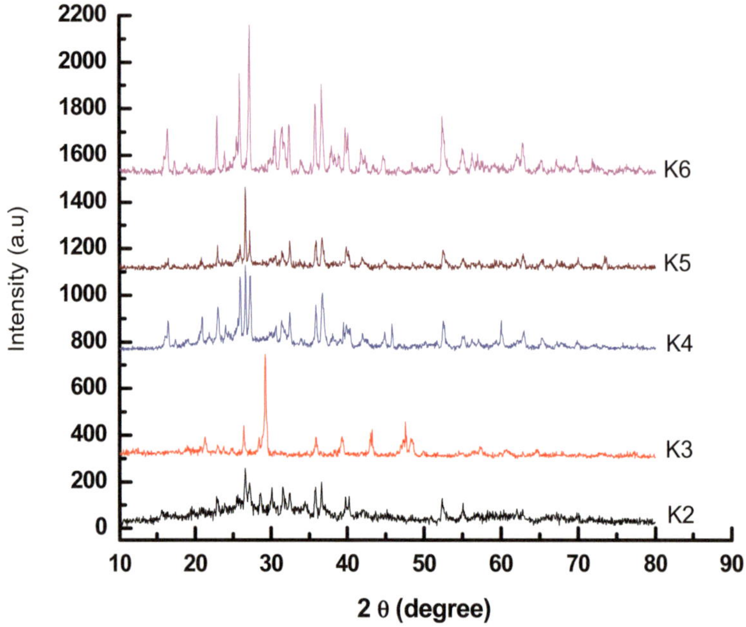

Fig. 17. XRD patterns of MAS glass ceramic specimens (MAS-K1-MASK5) as a function of variation of K_2CO_3 concentration and firing temperatures.

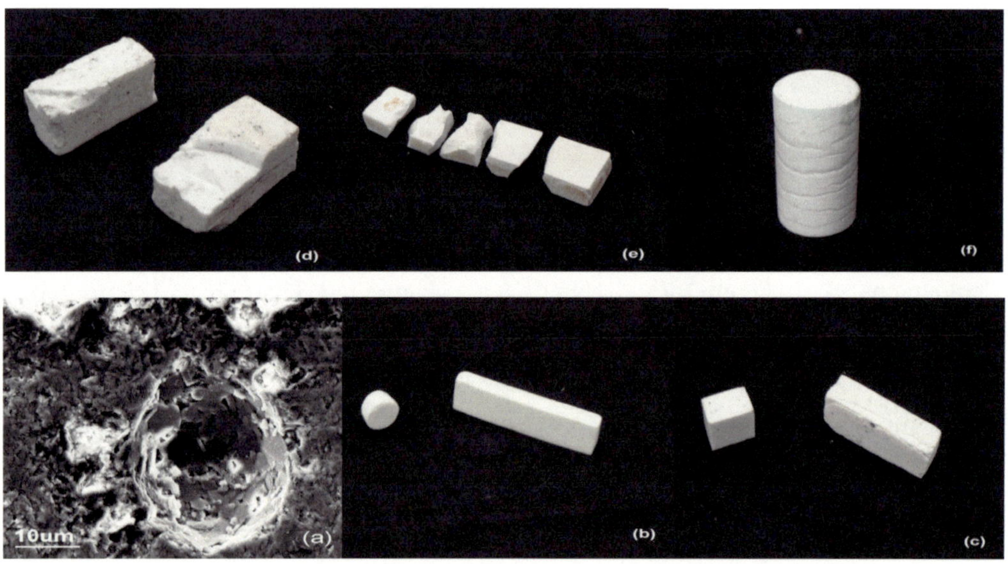

Fig. 18. (a-f): SEM image and photographs of MAS glass ceramic as a function of cooling rates after high sintering temperature (a) SEM image of MAS-K6 specimen sintered at 1040°C and cooled at 30°C (b) MAS-K6 specimen sintered at 1040°C and cooled at 30°C (c) sintered at 1040°C and cooled at 140°C (d) sintered at 1040°C and cooled at 440°C (e) sintered at 1040°C and cooled at 740°C (f) sintered at 1040°C and cooled at 1040°C.

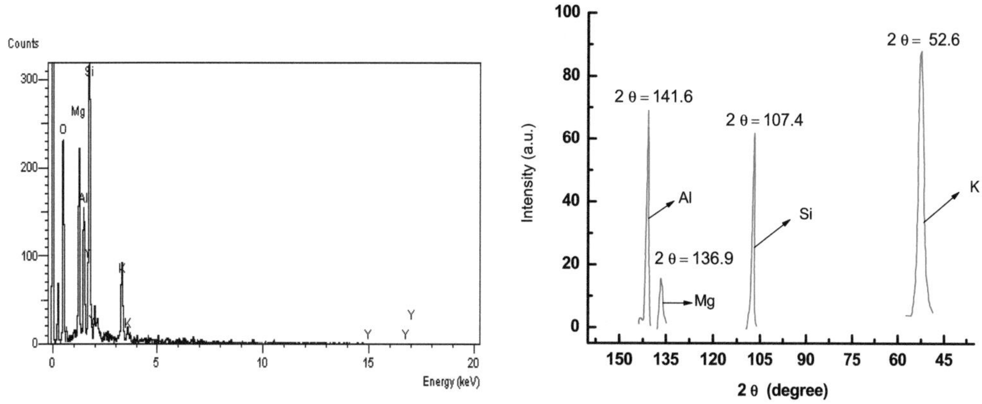

Fig. 19. (a-b): Chemical composition analysis of MAS-G8 (a) EPMA spectra (b) XRF patterns.

Specimen #	Resistance to Acid Wt-loss per unit area				Resistance to Base Wt-loss per unit area			
	Temperature °C	Time (h)	HCl (5%) (mg/cm²)	HF (5%) (mg/cm²)	Temperature °C	Time (h)	NaOH (5%) (mg/cm²)	Na_2CO_3 (5%) (mg/cm²)
MAS- G4	95	24	67	17.51	95	6	10.62	Nil
MAS- G6	95	24	87	156	95	6	9.52	2.13
MAS- G8	95	24	48	6	95	6	11.82	1.35

Table 7. Acids and bases effect on magnesium aluminum silicate glass ceramic specimens.

On the basis of experiments performed and capability for preparation of MAS glass ceramic materials using sintering route with 93% relative density contain 57% fluorophlogopite phase responsible for machinability. The high voltage insulators are required indifferent shapes and sizes so for these reasons some MAS glass ceramic materials have been developed by sintering route in different shape and size. The photographs of developed MAS glass ceramic items are as shown in Fig. 20.

Fig. 20. (a-c): MAS glass ceramic materials in different shape and size (a) showing solid slab, cylindrical rod, green crucible, disc (b) green and sintered slabs (c) sintered rod.

4. Conclusion

Magnesium aluminum silicate (MAS) glass ceramic was prepared successfully by sintering route and characterized in terms of both phase evolution during firing and microstructure at the optimum sintering temperature. The chemical composition of specimen MAS-G8 was found appropriate as compared to other compositions. The 9.14% weight loss was observed in specimen MAS-G8 after heating up to $900^{o}C$ by TG-DTA and found thermally stable above $900^{o}C$. MAS-G8 sintered at temperature $1040^{o}C$ provided smaller and uniform particle size distribution, 93% theoretical density. The phase evolution examined by XRD revealed that three main phase fluorophlogopite (F), sillimanite (S) and leucite (L) were present in the prepared magnesium aluminum silicate glass ceramic specimens as result of different chemical reactions during firing; however the specimen MAS-G8 consists of predominant fluorophlogopite glassy phase with uniform particle size distribution. Radiographic studies show the least (3-4%) through porosity whereas SEM image indicates some surface porosity.

5. Acknowledgement

The authors wish to acknowledge the contributions of M. Masood, M.M.R. Baig, Zahid Hussain, Aurangzeb, M. Hussain, Zahir Ahmad, Shahid Ayub for drawing 3D figures on AutoCAD and also providing assistance during the preparation of sintered specimens, measurement of density and XRD/XRF analysis.

6. References

Abo-Mosallam, H.A.; Salama, S.N. & Salman S.M.(2009). Formulation and characterization of glass-ceramics based on $Na_2Ca_2Si_3O_9$-$Ca_5(PO_4)_3F$-Mg_2SiO_4-system in relation to their biological activity, *J. Mater. Sci. Mater.* In Med., 20, 12, (2009) 2385-2394, ISSN: 1573-4838.

Bhargava, A.K. (2005). *Engineering Materials: Polymers Ceramics and Composites,* PHI Learning Pvt. Ltd., (2005), ISBN: 8120325834, 9788120325838

Baik, D.S.; No, K.S.; Chun, J.S.; Yoon Y.Y. & Cho, H.Y.(1995). Effect of the aspect ratio of mica crystals and crystallinity on the microhardness and machinability of mica glass-ceramics, *J. Mater. Process. Tech.*, 67, 1-3, (1997) 50-54, ISSN: 0924-0136

Barsoum, M. (1994). *Fundamental of Ceramics,* McGrew-Hill, Singapore, 1997, ISBN-0-07-005521-1

Boccaccini, A.R. (1997). Machinability and brittleness of glass-ceramics, *J. Mater. Process Technol.*, 65, (1997) 302, ISSN: 0924-0136

Bapna, M. S., & Mueller, H. (1996). *J. Biomaterials,* 17, (1996) 2045-2052, ISSN: 0142-9612

Bozadzhiev L.S.; Georgiev, G.T. & Bozadzhiev, R.L.(2011) *A Glass-Ceramic Material for Fixation of Radioactive Waste, Science of Sintering,* 43, (2011) 225-229, ISSN: 0350-820X

Callister, W.D. (2010). *Materials science and engineering: an introduction,* 8th edition, John Wiley & Sons, E-Book: Wiley Desktop Edition, (2007), ISBN: 978-0-470-93281-0

Carter, C.B. & Norton, M.G. (2007). *Ceramic materials: science and engineering,* Springer, (2007), ISBN: 978-0-387-46270-7

Cattell, M.J.; Chadwick, T.C.; Knowles, J.C.; Clarke, R.L. & Samarawickrama, D.Y. (2006) *The nucleation and crystallization of fine grained leucite glass-ceramics for dental applications,* Dent.Mater., 22,10, (2006) 925-33, ISSN: 0109-5641

Chowdhury, T. (2010). *Ceramic Insulator,* VDM Verlag, Dr. Mueller E.K., (2010), ISBN: 3639306414, 9783639306415

Chen, Q. Z.; Li, Y.; Jin, L. Y.; Quinn, J. M. & Komesaroff, P. A. , A new sol-gel process for producing Na_2O-containing bioactive glass ceramics, *Acta Biomaterialia* 6, 10, (2010) 4143-4153, ISSN: 1742-7061

Ciacchi, F.T.; Crane, K.M.; & Badwal, S.P.S. (1994). Evaluation of commercial zirconia powders for solid oxide fuel cells. *Solid St. Ionics,* 73, (1994) 49-61, ISSN: 0167-2738

David E. C. & Bruce K. Zoitos(1992). Corrosion of glass, ceramics, and ceramic superconductors: principles, testing, characterization, and applications, William Andrew, (1992), ISBN: 081551283X, 9780815512837

Doremus, R.H. (1994). *Glass Science,* 2nd Edition, Jhon Wiley & Sons, (1994), ISBN:0471891746, 9780471891741

Durrani S.K.; Akhtar, J.; Ahmad, M.; M.J. Moughal M.J. (2005), *J. Mater. Sci. Technol.*, 21, 4, (2005) 563-570, ISSN 1005-0302

Durrani, S.K.; Hussain, M.A.; Hussain, S.Z.; Akhtar, J.; Saeed, A.; Hussain, N. & Ahmed, N. (2010). Fabication of magnesium aluminum silicate glass ceramics by sintering route, *Materials Science-Poland,* 28, (2010) 459, ISSN 0137-1339

Emad M El-Meliegy (2004), Machinable spodumene-fluorophlogopite glass ceramic, *Ceram. Internat.*, 30, 6, (2004) 1059-1065. ISSN: 0272-8842

German, R.M. (1996). *Sintering Theory and Practice,* Wiley-Interscience, New York, NY, (1996), ISBN-13: 978-0471057864

Goswami, M.; Sarkar, A.; Mirza, T.; Shrikhande, V.K.; Sangeeta, Gurumurthy, K.R. & Kothiyal, G.P. (2002). Study of some thermal and mechanical properties of magnesium aluminium silicate glass ceramic, *Ceram. Internat.* 28, 6, (2002) 585-592, ISSN: 0272-8842

Hattori, T.; Shigemori, A.; Mohri, J.I.; Yoshimura, M. &, Somiya, s. (1982). *Fabrication of Nonadditive Mica Ceramics by Hot Isostatic Processing,* J. Amer. Ceram. Soc. Vol. 65, 9, (1982) c142, ISSN: 1551-2916

Hench, L.L. & Wilson, J. (1993). *An Introduction to bioceramics,* World Scientific, World Scientific Publishing Co Pte Ltd. (1993), ISBN: 9789810214005

Holkova, *Z.et al.,* (2003). *Ceramics- Silikáty,* Vol. 47, (2003) 13, ISSN: 08625468

Hölland, W. & Beall, G.H. (2002). *Glass-ceramic Technology,* The American Ceramic Society, Ohio, (2002), ISBN 1-57498-107-2

Hu.J. & Qim, H. (2001). J.Magn. Magn. Mater., *231,* (2001) L1, ISSN: 0304-8853

Hussain, S.Z.; Durrani, S.K.; Hussain, M.A.;.Akhtar, J.; Saeed, A.; Hussain, N. & Ahmed, N. (2010). Phase and Thermal Analysis of Magnesium Aluminum Silicate Glass Ceramic, *J. Pak. Mate. Soc. 4,* 1, (2010), ISSN:1994-6899

Izquierdo-Barba, I.; Salinas, J. & Vallet-Regi, M. (1999). *J.Biomedical research,* 47, (1999). 243-250, ISSN: 1549-3296.

McClune, W.F. (Ed.), Powder Diffraction File, Inorganic Phaess, International Centre for Diffraction Data, 1601 Park Lane, Swarthmore, PA 19081-2389, (1987).

Koshiro, I.; Mori, K.; Miyamoto, H.; & Nanjyo, F. (1995). *4th Symposium on SOFC In Japan edited by SOFC Society of Japan,* Electrochemical Society of Japan, Tokyo, (1995) 33-40, ISSN 1344-3542

Lei, B.; Chen, X.; Wang, Y.; Zhao, N.; Miao, G.; Li, Z. & Lin, C. Fabrication of porous,bioactive glass particles by one step sintering, *Mater.Lett.* 64, 21, (2010) 2293-2295, ISSN 0167-577X

Lewis M.H.(1989). *Glasses and Glass-ceramics,* Chapman & Hall the University of Michigan, USA., (1989), ISBN : 0412276909, 9780412276903

Manfredini, T.; Pellacani, G.C.; & Rincón, J. Ma. (1997). *Glass-Ceramics: Fundamentals and Applications,* Publisher Mucchi, Modena, (1997), ISBN: 887000287X, 9788870002874

Margha, F. H.; Abdel-Hameed, S.A.M.; Ghonim, N.A.E.; Ali, S.A.; Kato, S.; Satokawa, S. & Kojima, T. (2009). Crystallization behaviour and hardness of glass ceramics rich in nanocrystals of ZrO_2, *Ceram. Internat.* 35, I3, (2009) 1133-1137, ISSN: 0272-8842

Michael, B. (2007). Barium lanthanum silicate glass-ceramics, *US Patent 7189668,* 2007.

Miller, R. & Miller, M. R.(2002), *Electronics the Easy Way,* Barron's Educational Series, (2002), ISBN: 0764119818, 9780764119811

Minh, N.Q. & Takahashi, T (1995). *Science and Technology of Ceramic Fuel Cells,* Elsevier, Amsterdam, (1995), ISBN 13: 978-0-444-89568-4.

Mitchell, B.S.(2004) *An introduction to materials engineering and science for chemical and materials engineers,* Wiley-IEEE, (2004), ISBN: 0471436232, 9780471436232

Moghadam F.K. & Steveson, D. A. (1982). *J. Amer., Ceram. Soc.,* 65, (1982), 213-15, ISSN: 1551-2916

Montedo O. R. K.; Floriano, F.J.; Filho, J. O.; Angoletto E. & Bernardin A.M. (2009). Sintering behavior of LZSA glass-ceramics, *Mater. Res.* 12, (2009), 197-200, ISSN: 1516-1439

Mostaghaci, H. (1996). *Advanced ceramic materials: applications of advanced materials in a high-tech society* , Trans Tech Publications, (1996), ISBN : 0878497455, 9780878497454

Moulson A.J. & Hebert J.M. (1992). *ElectroCeramic Materials Properties Application,* Chapman & Hall, New York, NY 10001, USA, (1992), ISBN 0471 49748 7

Novaes de Oliveira A.P.; Manfredini T.; Leonelli C. & Peellacani, G. C.(1994). Physical Properties of Quenched Glasses in the Li_2O-ZrO_2-SiO_2 System *J. Amer. Ceram. Soc.,* 79, (1994) 1092-1094, ISSN: 1551-2916

Olevsky,E.A.; Bordia, R.(2009) *Advances in Sintering Science and Technology,Vol.211,Ceramic Transactions Series,* John Wiley & Sons, (2009), ISBN: 0470408499, 9780470408490

Pannhorst, W. (1997). Glass ceramics: State-of-the-art, *J. Non. Cryst, Solids.* 219 (1997) 198-204, ISSN 0022-3093

Partridge, G. (1994). An overview of glass-ceramics. Part. 1. Development and principal bulk applications. *Glass Technol.*, 35, (1994), 116-127, ISSN 0017-1050

Perdomol, F.; Avaca, L. A.; Aegerter, M. A. & Lima-Neto, P. De.(1998). *J. Mater. Sci. Lett.* 17, (1998) 295, ISSN 0261-8028

Radojic, L.J. & Nikolic, L.J. (1991). *Eur. Ceram. Soc.* 7, (1991) 11, ISSN: 0955-2219

Rahaman, M.N. (2003). *Ceramic Processing and Sintering,* 2nd Ed. Vol. 23. New York: Taylor & Francis, Inc., (2003), ISBN-13: 978-0824709884

Rajendran, V. (2004). *Materials Science,* Tata McGraw-Hill Education, (2004), p.40-44, ISBN: 0070583692, 9780070583696

Rice, R. & Rice, W. (2002) *Ceramic Fabrication Technology, Volume 20, Materials engineering* CRC Press, (2002), ISBN:0824745736, 9780824745738

Richerson, D.W. (1992). *Modern Ceramic Engineering: Properties, Processing and Use in Design,* 2nd Ed. New York: Marcel Dekker, (1992), ISBN-13: 978-0824786342

Rincón, J. Ma. (1992). Principles of Nucleation and Controlled Crystallization of Glasses. *Polym. Plast. Technol. Eng.*, 3, (1992) 309-357, ISSN: 0360-2559

Ring, T.A. (1996). *Fundamentals of ceramic powder processing and synthesis,* Academic Press, (1996), ISBN: 0125889305, 9780125889308

Roy, S. & Basu.B. (2006). *Microstructure Development in Machinable Mica Based Dental Glass Ceramics.* Trends in Biomaterials and Artificial Organs, Vol.20, No.(1), (2006) 90-100, ISSN 0971-1198

Shackelford, J. F. & Doremus R.H. (2008). *Ceramic and glass materials: structure, properties and processing,* Springer, (2008), ISBN 0387733612, 9780387733616

Shinroku, S. (1987). *Fine Ceramics,* Elsever Science Publishing, Co.,Vanderbilt Avenue, New York, USA. (1987), ISBN-13: 978-0444011930

Simmons, J.H.; Uhlmann D.R. & Beall G.H. (1982), *Nucleation and crystallization in glasses,* American Ceramic Society, 1982 Science, ISBN: 060800720X, 9780608007205

Smothers, .W (2009) *5th Annual Conference on Composites and Advanced Ceramic Materials: Ceramic Engineering and Science Proceedings, Volume 2, Issues 7-8,* Wiley & Sons John Wiley & Sons (2009) , ISNB:0470291516, 9780470291511

Steffestun, B. & Frischat, G.H. Alkali-Resistant Magnesium Aluminosilicate Oxynitride Glasses. J. Am. Ceram. Soc. 76 (1993), 699-704, ISSN: 1551 2916

Strnad, Z. (1986). *Glass Ceramic materials:: liquid phase separation, nucleation, and crystallization in glasses,* Elsevier, Amsterdam, (1986),ISBN: 0444995242, 9780444995247

Swab, J.J. (2009). *Advances in ceramic armor V: ceramic engineering and science proceedings.* I5, Vol.30, John Wiley and Sons, (2009), ISBN: 0470457554, 9780470457559

Ting, J. M. & Lin, R.Y. Effect of particle size distribution on sintering, *J. Mater. Sci.* 30, (1995) 2382-2389, ISSN 0022-2461

Valenzuela, R., (2005). *Magnetic Ceramics,* Cambridge University Press, (2005) ISBN: 0521018439, 9780521018432

Yuan, Q..L.; Zhang, P.; Gao, L.; Peng, H.; Ren, X. & Zhang, D.(2010). $MgO-Al_2O_3-SiO_2$ Glass-Ceramic Prepared by Sol-Gel Method, *Advan. Mater. Res.* 92, (2010) 131-137, ISSN: 1022-6680

Yu, Y., Wang, X., Cao, Y. & Hu, X., *Appl. Surf. Sci.,* 172, 3-4, (2001) 260-264, ISSN: 0169-4332

Wang. Y., (2008). *Introduction to ceramics,* McGraw-Hill Education (Asia), (2008), ISBN: 0071274979, 9780071274975

Wawrziniak,W. & Grunov, r. (1980). *Silikat technik,* 31, (1980) 238, ISSN: 0037-5233.

13

A Novel Approach to Develop Chalcogenide Glasses and Glass-Ceramics by Pulsed Current Electrical Sintering (PCES)

Gaelle Delaizir[1] and Laurent Calvez[2]
[1]*Groupe d'Etudes des Matériaux Hétérogènes (GEMH, ENSCI), Centre Européen de la Céramique, 12 rue Atlantis, Limoges,*
[2]*Equipe Verres et Céramiques, UMR CNRS Sciences Chimiques de Rennes, Université de Rennes 1, Rennes Cedex, France*

1. Introduction

Chalcogenide glasses have been in the last decades, of paramount interest for night vision devices because of their remarkable transparency in the two atmospheric windows (3-5µm and 8-12µm). Chalcogenide glasses tend to replace, at least partially, the expensive mono-crystalline Ge or polycrystalline ZnSe for infrared (IR) lenses (Zhang et al., 2003). The ease of processing due to their viscoplastic property and the lower cost of chalcogenide glasses compared to mono-crystalline Ge have made them one of the best candidates for lenses of optical thermal imaging cameras but are also very efficient for various fields of applications working in the second and third atmospheric windows. Thus, chalcogenide glasses are at the centre of active and passive applications such as night vision (Guillevic et al., 2009), generation of new infrared sources (Troles et al., 2010), electronic devices (Danto et al., 2010), chemical and biological sensors to detect CO_2 or tumors respectively (Wilhelm et al., 2007), etc. They are also promising materials for energy applications, such as solid electrolyte (Hayashi et al., 2001) or thermoelectric materials (Goncalves et al., 2010). These glasses that contain no oxygen in their composition are usually synthesized in vacuumed silica ampoules using the so-called melt-quenching technique. The low thermal conductivity of silica limits the cooling rate during quenching. This usually leads to heterogeneous composition in the case of unstable glass composition (that tends to crystallize) and thus, reduces the glassy domain as well as the available diameters of these glasses. In this chapter, we describe a novel approach to synthesize chalcogenide glass bulks with large diameters including unstable compositions. This technique combines either the mechanical alloying to get amorphous powder or the grinding of glass obtained from previous small diameter melt-quenching technique and the Pulsed Current Electrical Sintering (PCES) also known as Spark Plasma Sintering (SPS) (Hubert et al., submitted). This new technique allows both the sintering of amorphous powder and its shaping in one step in few minutes. This paves the way for a new set of glasses previously impossible to synthesize, especially for crystallization concerns. Indeed, the fast heating rates reached by SPS (Joule effect heating) prevents the glass from undesirable crystallization. Also, the SPS technique is efficient to make IR glass-ceramics from glass powder with reduced time of crystallization in comparison with conventional technique, i.e, heating a glass bulk in a conventional furnace (Delaizir et al., 2010). Two different compositions have been tested: $62.5GeS_2$-$12.5Sb_2S_3$-25CsCl (% mol) and $80GeSe_2$-$20Ga_2Se_3$ (% mol); the results are reported.

Finally, this chapter reports on an analysis of SPS consolidation of the latest chalcogenide glass composition little studied so far as well as the mechanisms of sintering and devitrification through the SPS technique.

2. Chalcogenide glasses and glass-ceramics

2.1 Background

Chalcogenide glasses are, by definition, glasses which contain at least one of the three elements: sulphur, selenium and tellurium and no oxygen. Depending on the application, other chemical elements such as arsenic, antimony, germanium, etc are added to improve mechanical and optical properties and also to increase the stability against devitrification. The most interesting property of these glasses is associated with their transparency in the mid and far infrared (IR).

Glass-ceramics are defined as polycrystalline solids prepared by the controlled crystallization of glasses with an appropriate thermal treatment. They can also be seen as composite materials made of a glassy matrix containing crystals as fillers [Mc Millan, 1979]. These kinds of materials may have exceptional properties that can be optimized according to the targeted applications. In the literature, Stookey from Corning is the first who discovered accidentally this new material, called glass-ceramics, in the middle of the 1950s. Glass-ceramics are preferred to glasses because of their better mechanical properties and are usually preferred to ceramics because of the ease of processing. Indeed, moulding is easier, faster and cheaper than solid state sintering which is the common way to produce crystalline ceramics.

2.2 Synthesis

2.2.1 Glass

2.2.1.1 Melt-quenching technique

The chemical elements constituting the glass, weighed in predetermined quantities are introduced in the set-up as shown in Fig. 1. They are put under secondary vacuum during few hours then the reactional tube is obtained by sealing the silica tube (Fig. 1).

The reactional tube is then heated in a rocking furnace at the very low rate of 2°C/min (to avoid the rapid formation of vapor and therefore the explosion of the tube) until the homogenization temperature which depends on the glass composition. The glassy composition stays few hours at this temperature. The tube containing the melt is quenched into water or air. The glass rod is then annealed in an annealing furnace previously heated at a temperature $T_{annealing}$ near the glass transition temperature Tg and then slowly cooled down to room temperature to release mechanical stress induced by the quenching. We therefore obtain a glass rod (Fig. 2). Usually to reach a high transmittance, raw elements such as sulphur or selenium are purified to avoid losses of transmission due to absorption bands (S-H, Se-H, O-H, etc)

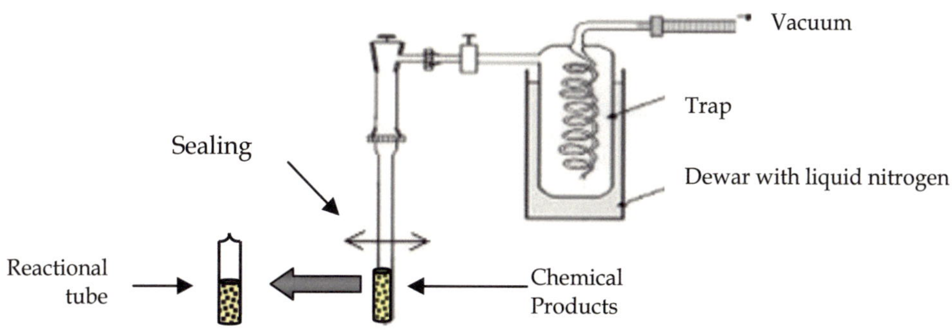

Fig. 1. Experimental set-up.

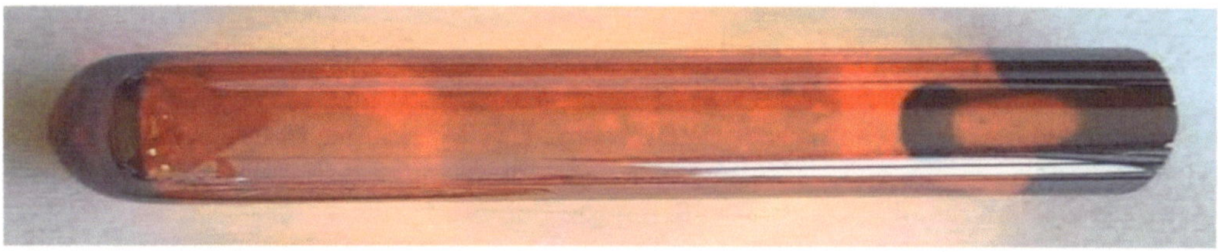

Fig. 2. Example of glass rod of the GeS_2-Sb_2S_3-CsCl composition.

2.2.1.2 Mechanical milling

High energy ball milling process has been demonstrated to be a promising method for the preparation of amorphous powder with fine microstructure. The amorphization reaction during mechanical milling is usually attributed to microstructural breakdown followed by the interdiffusion of elements (Johnson, 1986) or mechanically driven atomic mixing (Lund et al., 2004 & Delogu et al., 2005) among previously formed nanocrystalline multilayers. In mechanical milling experiments, the kinetic process of the amorphization reactions usually proceeds slowly and therefore a glass forming composition is determined only after milling for an extended time (usually >100 h) (Eckert et al., 1997, Schurak et al., 1999 & Choi et al., 2006) (Fig. 3). The ability to synthesize a dense material from amorphous powders obtained by mechanical-alloying has recently been demonstrated for many systems regarding metallic glasses (Yan et al., 2008, Patil et al., 2005 & Kim et al., 2005). Also, amorphous powders in the systems Li_2S-P_2S_5 or AgI-As_2Se_3 have been synthesized using the same method for solid electrolyte applications (Hayashi et al., 2001, Sekine et al., 2007 & Trevey et al., 2009). Few papers report the amorphization of metallic elements used in the manufacture of IR glasses; they are limited to the study of Ge-Se powder but none of these papers are relevant for the production of optical devices (Shirakawa et al., 2001 & Machado et al., 2005).

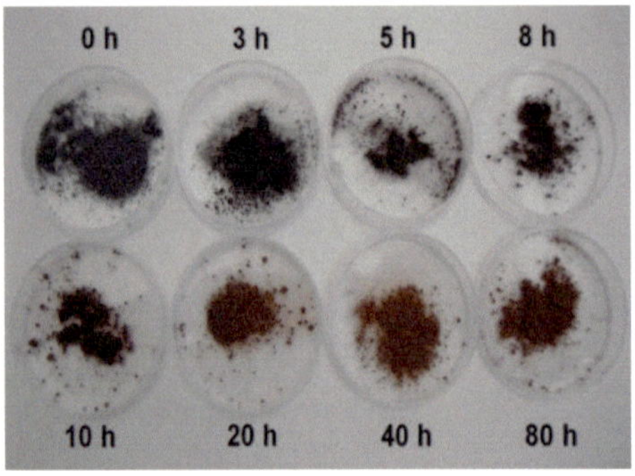

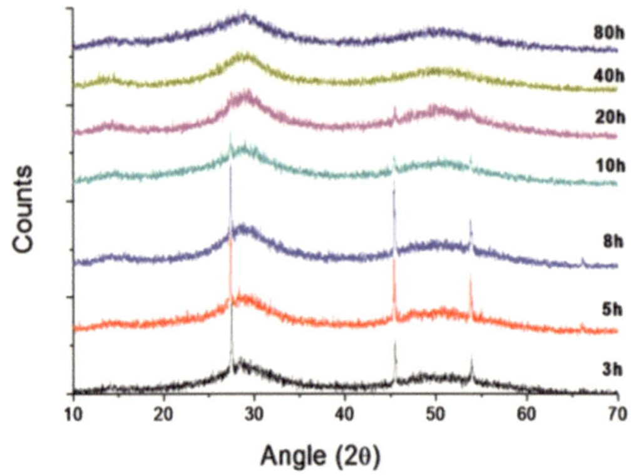

(a) (b)

Fig. 3. (a) Evolution of the 80 $GeSe_2$ - 20 Ga_2Se_3 powder color with the milling duration (b) XRD patterns of the corresponding powders obtained at different milling durations showing the progressive amorphization of the powder.

2.2.2 Glass-ceramics

Glass-ceramics synthesis, or the devitrification process of the glassy matrix, implies a two-step procedure: nucleation and crystal growth.

Nucleation represents the first step of glass devitrification. It consists in inducing germinate from which the growth can start. It is based on kinetic parameters. Nucleation may be homogeneous or heterogeneous. In homogeneous nucleation, the first tiny seeds are of the same constitution as the crystals which grow upon them, whereas, in the case of heterogeneous nucleation, the nuclei can be quite different chemically from the crystals which are deposited. Some substances enabling or hastening bulk nucleation can be added to glass composition; they are termed nucleating agent. We can distinguish two types of nucleating agents. Metallic nucleating agent such as Au, Cu, Pt, etc, are added to the glass in very small amounts (0.01 to 1% mass). The mechanisms of the effect of these nucleation agents in increasing the nucleation rate of the principal crystalline phase is quite complex but is based on heterogeneous nucleation. A second group of nucleating agents including TiO_2, ZrO_2, SnO_2, P_2O_5 (in the case of oxide glass-ceramics) or metallic sulphide can be added in greater amounts (mostly up to 20%) in order to get oxide glass-ceramics. They are part of the oxide glassy composition and they are found to be effective nucleating agents in specific initiation of bulk nucleation.

The so called TTT curve (Time, Transformation rate, Temperature) can predict the time needed to crystallize a fraction of glass at one given temperature. The advantage of the TTT curve (Fig. 4a) lies in the fact that it permits the determination of a critical point for which the time needed for crystallization is minimal and the temperature for instability is maximal. Avrami's equation permits to build TTT curve (Avrami, 1939).

$$x = 1 - e^{-f} \text{ with } f = (kt)^n \tag{1}$$

Where n is the Avrami exponent. This equation is valid if the nucleation is monotonous.

It is assumed that k, the rate constant, varies with time according to the Arrhenius law:

$$k = k_0 \times \exp(\frac{-E}{RT}) \quad (2)$$

where E is the activation energy and k_0, a frequency factor.

Without nucleation, crystal growth cannot happen and without growth, no crystals can appear. These processes must take place within a certain range of temperatures which is critical for the devitrification (Fig. 4b). Above the melting temperature, Tm, the liquid constitutes the stable phase. When the liquid cools down, the crystal growth can theoretically happen between T2 and T4. However, the initial nuclei needed before the crystal growth can theoretically happen between T1 and T2. The critical range is therefore between T2 and T3 (in the case of non addition of nucleating agents in the glass).

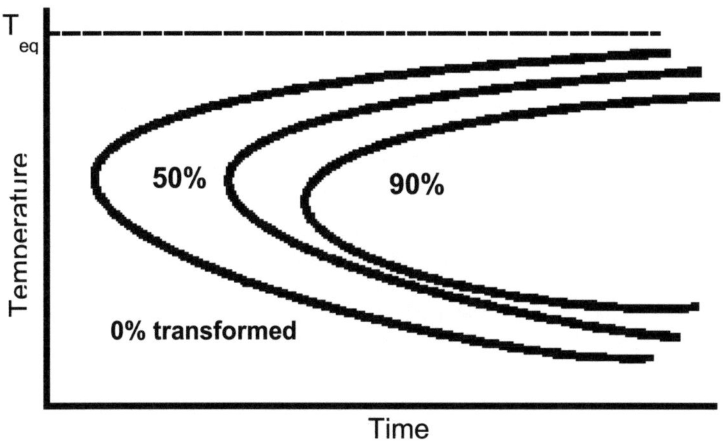

(a)

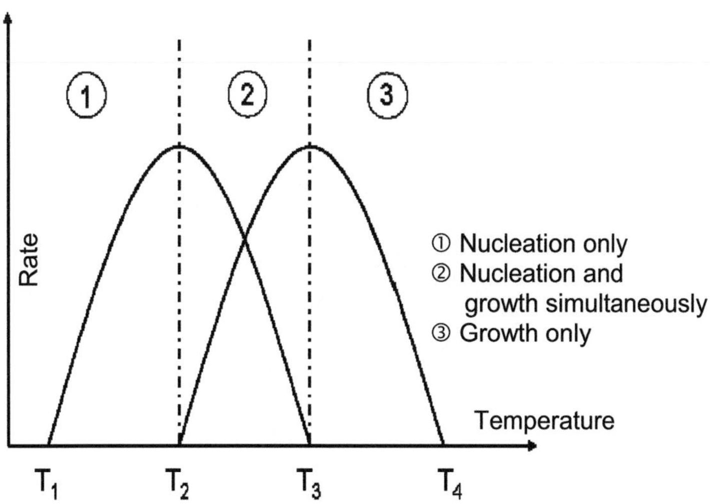

(b)

Fig. 4. (a) Example of TTT curve, (b) Nucleation and growth phenomena as a function of temperature.

The common treatment to obtain glass-ceramics is an appropriate thermal treatment from a base glass. The thermal treatment used in industry to get this material is a "two plateaus" treatment. It consists of heating the glassy matrix (base glass) at a temperature above the glass transition temperature, Tg in order to induce nuclei in the glass. The temperature is then increased to a second plateau to induce the growth of these nuclei. A second technique consists in a single plateau. The glass is heated at a temperature above Tg but below the crystallization temperature, Tc. This technique allows the nucleation phenomenon and avoids excessive growth.

Oxide glass-ceramics are by far the most studied glass-ceramics. They have been widely investigated since 1950 and the research associated to this area is now slowing down. Today, research is more focused on the nucleation and growth phenomena to have a better understanding. However chalcogenide glass-ceramics still remain of great interest because of their transparency in the infrared range associated to better mechanical properties. As previously mentioned, potential applications are infrared lenses for thermal camera.

Chalcogenide glass-ceramics transparent in the range 8-12 µm were first synthesized in 1973 by Mecholsky in the system 0.3 PbSe-0.7 $Ge_{1.5}As_{0.5}Se_3$ with a 60% crystalline fraction. He showed that the glass-ceramic modulus of rupture was increased to as much as twice that of the base glass and the Vickers hardness increased by 30% (Mecholsky et al., 1976).

Other researchers worked on systems such as As-Ge-Se-Sn (Cheng, 1982), Ga-Ge-Sb-Se (Ma et al., 2003) or Ge-Te-Se (Song et al., 1997) but the reproducibility of the glass-ceramics synthesis remained difficult.

First chalco-halide glass-ceramics, transparent in the far infrared (10 µm) was obtained in 2003 within the system GeS_2-Sb_2S_3-CsCl in the "Glass and Ceramic" laboratory in Rennes (France) (Zhang et al., 2004) (Fig. 5). The simultaneous presence of ionic and covalent compounds prevent from the rapid and uncontrollable crystallization. Three years later, glass-ceramics transparent until 14µm, covering the second and third atmospheric windows entirely, were synthesized in the system GeS_2-Ga_2Se_3-CsCl (Calvez et al., 2007).

(a) (b)

(c) (d)

Fig. 5. Glass composition 62.5GeS_2-12.5Sb_2S_3-25CsCl heated at 290°C for different crystallization times (a) No thermal treatment (b) 7h (c) 73h and (d) 144h.

2.3 Physical properties

2.3.1 Thermal properties

Thermal characteristics of a glass, such as glass transition temperature Tg and crystallization temperature Tc, are determined using Differential Scanning Calorimetry (DSC). Fig. 6 represents the thermogram heat flow versus temperature for one given glass undergoing crystallization phenomenon (exothermic peak). Glasses which are stable against devitrification do not present any crystallization peak.

The glass transition temperature, Tg, is the main characteristic of a glass. Before Tg, the viscosity is infinite (solid state), at Tg, the viscosity is equal to 10^{13} poises (10^{12} Pa.s) and after Tg, the viscosity decreases as the temperature increases, therefore, the material can be easily shaped. The crystallization phenomenon is characterized by the rearrangement of atoms in organized lattice due to the change of viscosity. Crystallization is at the origin of the loss of the viscoplastic properties as well as the optical transparency. The stability against devitrification is associated with the difference Tc-Tg. The higher is the difference between Tc and Tg and better is the stability against devitrification.

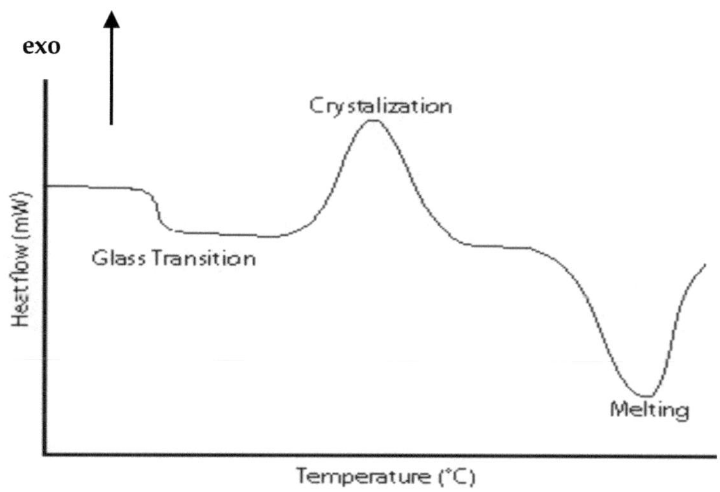

Fig. 6. Typical DSC curve for a glass.

2.3.2 Optical properties

Optical transmission of glasses is characterized by its optical window. At shorter wavelengths, the band gap limits the optical window while at longer wavelengths the optical window is limited by the multi-phonon absorption.

The band gap results from electronic transition inside the glass. Photons with sufficient energy are absorbed by exciting electrons across the forbidden band-gap. The electrons are excited from the top of the valence band to the bottom of the conduction band. In glasses, additional states exist just above the valence band and just below the conduction band. These states are present because the disorder creates localized electronic states. These localized states participate in the absorption process.

In chalcogenide glasses, absorption phenomena are due to the excitation of non bonding electrons of chalcogenide chemical element: S, Se or Te. Because non bonding electrons of selenium ($4s^2p^4$) or tellurium ($5s^2p^4$) are higher in energy than bonding electrons of sulphur ($3s^2p^4$), they are more excitable. Therefore, the band gap shifts from visible with sulphur-based glass to near infrared for selenium or tellurium-based glass.

The multi-phonon absorption at longer wavelengths deals with interaction between light and vibration modes of the chemical bonds inside the glass. The phonon energy, E, is directly linked to the atoms weight and is inversely proportional (Kittel, 1998). The large atomic mass of chalcogenide elements causes the phonon vibrations to have low energies. Materials with high phonon energies have multi-phonon absorption in the mid- and near-infrared region. This is especially true for materials with lightweight, strongly bounded atoms such as silica glasses and it limits their usefulness for infrared applications. Chalcogenide glasses typically have optical windows that extend into the far infrared beyond 12 µm (Klocek et al., 1987).

The transmission spectra (Fig. 7) illustrate the multi-phonon absorption for different types of glass (oxide, fluoride and chalcogenide glasses with S, Se and Te). The molecular weight (MW) is in the order: MW(O)<MW(F)<MW(S)<MW(Se)<MW(Te), therefore the multi-phonon absorption for chalcogenide glasses is shifted toward longer wavelengths compared to oxide or fluoride based-glass compositions.

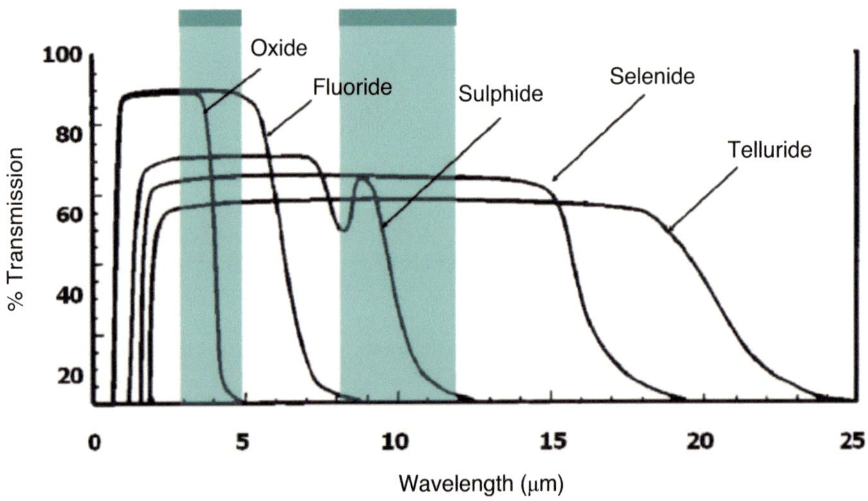

Fig. 7. Optical window of different types of glass.

2.3.3 Other properties

General observation regarding glass-ceramics compared to the derivative base glass shows an increase of toughness, stiffness and hardness as well as an interruption of the crack propagation due to crystals (Choi et al., 2003) (Fig. 8). Mechanical properties of glass-ceramics are influenced by several factors such as particle size and volume fraction of crystalline phase (which can be up to 90%), interfacial bond strength, differences in elastic modulus and thermal expansion between the glassy matrix and the crystals, etc.

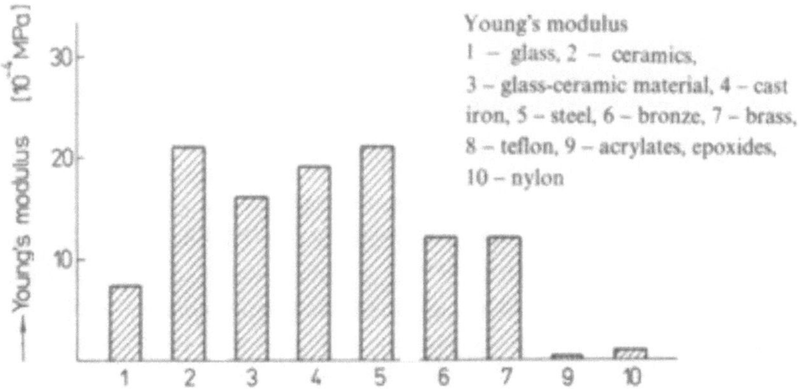

Fig. 8. Young's modulus for different materials (Strnad, 1986).

The increase of mechanical properties is maybe the most important advantage of glass-ceramics over glasses but glass-ceramics present other great advantages. Because of their adjustable coefficient of thermal expansion, glass-ceramics are resistant to thermal shock and permit the sealing to a variety of metals (mostly for oxide glass-ceramics). Then depending on the crystal size, glass-ceramics can be totally transparent or opaque. Oxide glass-ceramics are also used in electronic for their wide range of dielectric constants and are chosen instead of glasses for their lower dielectric losses. They are also corrosion resistant against weathering and a wide range of chemicals.

Electrical properties of chalcogenide glasses depend on their chemical composition. They can be seen as semiconductors because they possess a band gap energy (~2 eV) characteristic of semiconductor materials (l-3 eV). This band-gap (E_g) depends on the glass composition and is smaller for tellurium based-glass composition ($E_g(S)>E_g(Se)>E_g(Te)$) (Table 1). The electrical conductivity of semi-conductors at room temperature is in the range 10^{2}-10^{-9} Ω^{-1}.cm^{-1} (Kittel, 1998). The electrical conductivities of sulfur and selenium based-glass compositions are very low. They can therefore be seen as insulator in contrary to telluride based-glass composition.

	Eg (eV)	σ (Ω-1.cm-1)
As2S3	2.12	10-17
As2Se3	1.53	10-12
As2Te3	0.3	10-4

Table 1. Comparison of band-gap and electrical conductivity of sulphide, selenide and telluride based-glass composition (Elliott, 1991).

3. The Pulsed Current Electrical Sintering (PCES) technique

3.1 Outline

The Pulsed Current Electrical Sintering (PCES) also known as Spark Plasma Sintering (SPS), Field Assisted Sintering Technique (FAST) or Electric Current Activated Sintering (ECAS) is a powerful technique for powder consolidation. This technology started in the late 1920s when a sintering process using electrical energizing was introduced in the USA. In the meantime, in Japan, an ongoing research on the process of pulsed current activated sintering intensified and was patented in the 1960s [Inoue, 1962 & Inoue, 1966). This method seems to be characterized by technical and economical advantages over conventional sintering methods such as Hot Uniaxial Pressing (HUP), Hot Isostatic Pressing (HIP), etc. Indeed, the fast heating rate, the short holding times, the absence of sintering aids, the lower sintering temperatures, the improved properties of the process ceramic in comparison with other sintering methods, no need of cold compaction make the PCES a very competitive technique for industrial applications and for the field of materials research. As a proof, the number of research papers is exponential since the late 1990s when research lab worldwide started to be equipped with this technology [Orru et al., 2009]. Moreover, with the fast heating rates and short holding times, the processing of nano-powders was then possible to keep the nanostructure of the ceramics.

3.2 Process description

The description of the process is illustrated in Fig. 9. DC pulses allow the conducting die (graphite, WC, stainless steel) to heat by Joule effect while a uniaxial load can be applied. The PCES can therefore be seen as similar as Hot Uniaxial Pressing (HUP) technique with faster heating rate up to 1000°C/min. However, specific experimental sets-up have been designed in order to apply isostatic (Saito & Sawaoka, 1973) or quasi-isostatic (Song et al., 2004) pressure to the sample to be consolidated. The sintered powder can be either conductive or insulating. Depending on the electrical properties of the powder, the heating mechanisms are different (Anselmi-Tamburini et al., 2005). In the case of insulating powder, the heating mostly occurs through the container (die) while conducting powders are heated by Joule effect and by heat transfer from the container and electrodes. If the powder is conductive, an insulating die can be used. The mechanical load applied depends on the nature of the container. The load is usually limited to 120MPa in the case of graphite die but can be much more in the case of WC die (800MPa). However, graphite die can be heated up to 2000°C while WC can be heated up to 900°C depending on the applied load. Different working atmosphere can be applied during experiments: vacuum, neutral, reducing, etc.

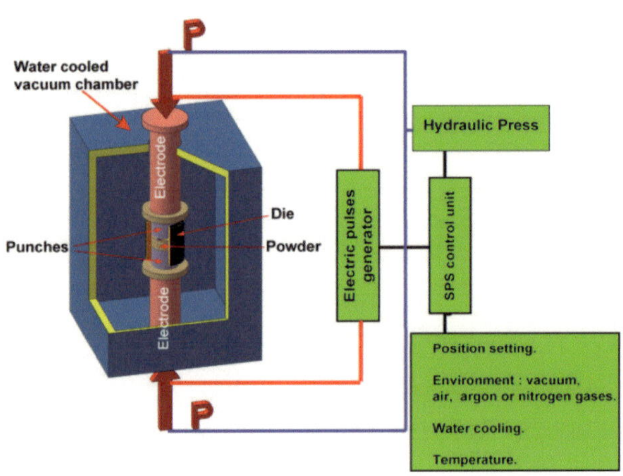

Fig. 9. Process description of PCES technique (Hungria et al., 2009).

This technique is also very interesting as it permits to sinter and shape in a single step a wide range of materials into conventional (cylinder, etc) and non conventional shapes (ball, etc) (Hungria et al., 2009). The diameter, depending on the machine and the sintering conditions, can range from 8mm to 100mm or up. It should be however mentioned that in PCES processes the problem of adequate electrical conductance of the powders and the achievement of homogenous temperature distribution is particularly acute, especially for large diameter samples. In fact, current and consequent temperature distributions within the sample are very sensitive to the homogeneity of density distribution. Moreover, large density spatial variation, especially at the beginning of current flow, may result in high local overheating or even melting (Orru et al., 2009). That is why, some researchers focused now on modeling the thermal gradient during experiment. Then, some graphite pollution can occur during SPS experiment that is detrimental for optical applications for example (Bernard-Granger et al., 2009). This can be solved by using inner tantalum (Ta) foil or boron nitride (BN) layer as graphite barrier diffusion.

3.3 State of the art of the materials investigated

A large number of sintered materials have been tested until now among which we can cite: metals and alloys, intermetallics, carbides, borides, nitrides, silicides, oxides, sialon, ceramic–metal and ceramic–intermetallic composites, ceramic–ceramic composites, hydroxyapatite-based materials, chalcogenides, polymer-based materials, functional graded materials, systems for joining, graphite/carbon-based materials including composites containing carbon-nanotubes, multi-materials including solid state batteries (Aboulaich et al., 2011) and other systems. A review of the materials tested can be found elsewhere Orru et al., 2009). The highlights are that higher densities can be achieved at lower sintering temperature in shorter sintering time and that smaller grain sizes ceramics are typically obtained with a consequent direct effect on nearly all properties investigated on the fabricated dense materials.

In the literature, the sintering of glasses by SPS mostly concerns conductor metallic glasses (Li et al., 2009; Shin et al., 2010 & Choi et al. 2007). These glasses which composition can include several chemical elements among Zr, Ti, Al, Cu, Pd, Mg, Pt, Ca, Fe, Ni, Co and sometimes rare earth are largely studied for their good mechanical properties. In this case, the SPS technique is very interesting since the heating rates are fast enough to avoid the crystallization phenomenon of these unstable glasses. These amorphous materials are most often obtained by mechanical alloying, melt-spinning or atomization. Recently, Nowak *et al.* (Nowak et al., 2011) and Perriere *et al.* (Perriere et al., 2011) studied the phenomenon during SPS treatment at the origin of the microstructure and densification of glasses. The devitrification in metallic glasses during SPS treatment has also been studied in order to obtain composite materials or glass-ceramics (Li et al., 2010). Except this study on metallic glasses, we can cite the research activity of Mayerhofer *et al.* on the sintering of amorphous silica nano-particles by SPS (Mayerhofer et al., 2008). The sintering of chalcogenide glasses and the *in situ* synthesis of glass-ceramics by SPS has been little studied to the best of our knowledge. We can identify a study on oxide glasses from the SiO_2-Al_2O_3-Li_2O:Er^{3+} system (Riello et al., 2006) and the ones we propose in this chapter on the GeS_2-Sb_2S_3-CsCl and $GeSe_2$-Ga_2Se_3 glassy systems (Hubert et al., submitted & Delaizir et al., 2010). These last studies clearly show the advantage of SPS for the shaping of glass and the gain of time concerning the synthesis of glass-ceramics in comparison with a conventional technique in a conventional furnace.

Recent results also suggest that the Spark Plasma Sintering is a new technique to achieve very fast solid state chemistry (Galy et al., 2008). This technique appears as a new synthesis technique which permits to decrease both the temperature and time reaction while mastering the particle size. Even though all the mechanisms are not well understood, it is generally agreed that an accelerated diffusion process due to the electrical discharge is at the origin of the fast reactivity by SPS (Galy et al., 2008).

The SPS or PCES still remains controversial with plasma formation or not, removal of oxides (breakdown of oxides films) and adsorbed gases from the particle surfaces with a resulting cleaning effect, high localized temperature at the contact area between particles, enhanced diffusion of materials at forming particle necks (Orru et al., 2009) since no evidence has been proven. Thermal gradients are also very discussed especially for large diameters. However research on SPS mechanisms to answer to these questions as well as modeling is of growing and fundamental interests.

4. The Pulsed Current Electrical Sintering (PCES) technique applied to the preparation of chalcogenide glasses and glass-ceramics

4.1 Principle

As previously mentioned, the common way to prepare chalcogenide glasses is the melt-quenching technique in silica tube under vacuum. Due to the low thermal conductivity of silica during quenching, this leads to heterogeneous glass sample when the composition is unstable ($Tc-Tg<100°C$). Usually the inner part of the glass is crystallized and the outer part amorphous. Therefore, this limits the preparation of glass samples to small diameters to ensure homogeneity of the glass. Glass-ceramics are obtained by heating a glass bulk with an appropriate thermal treatment in a furnace. According to targeted crystal fraction and crystals size, this can take up to hundreds of hours. It is well known that grained glass samples devitrify more easily than glass bulk samples and upon a process of amorphous state sintering and simultaneous or subsequent crystallization, glass-ceramic materials can be obtained (Gutzow et al., 1998).

Grained glass samples can be either obtained by mechanical milling or by grinding and sieving pre-existing glass bulk synthesized by the conventional melt-quenching technique.

The phenomenon at the origin of easier devitrification is mostly due to pre-existing structural defects and new surfaces that act as nucleation sites in powders obtained from mechanical milling process or grinding. Therefore, it should be possible to change the kinetic process in comparison with devitrification of common glass bulks.

Moreover, the SPS or PCES technique has fast heating rates that prevent the glass powder from uncontrollable ceramization. By combining glass powder obtained from mechanical milling or grinding of pre-existing glass bulk and SPS technique, the idea is to develop an easy way to produce glass/glass-ceramics bulks with large diameters even for unstable compositions that are of great interest for optics.

4.2 Glasses of investigation and results

So far two compositions have been tested: $62.5GeS_2$-$12.5Sb_2S_3$-25Cscl and $80GeSe_2$-$20Ga_2Se_3$ (mol %). However this technique is believed to be suitable for all glass compositions to get amorphous bulks as well as glass-ceramics. These compositions were chosen for their potential applications (Huber et al., submitted & Delaizir et al., 2010).

The glass composition $62.5GeS_2$-$12.5Sb_2S_3$-25Cscl has been obtained through a melt-quenching technique and the resulting bulk has been grinded and sieved (45µm) while the glass composition $80GeSe_2$-$20Ga_2Se_3$ has been obtained through an 80h mechanical milling process from high purity metallic elements Ge, Ga and Se (Fig. 3a). The size distribution, measured using laser diffusion technique shows low mean particle size (D50 = 3.72 µm). The thermal properties of these glasses are summarized in Table 2. Tc_1 corresponds to the crystallization of CsCl and $GeGa_4Se_8$ (or Ga_2Se_3) species respectively for the $62.5GeS_2$-$12.5Sb_2S_3$-25Cscl and $80GeSe_2$-$20Ga_2Se_3$ compositions. Tc_2 corresponds to the crystallization of $GeSe_2$.

	Tg (°C)	Tc_1 (°C)	Tc_2 (°C)
$62.5GeS_2$-$12.5Sb_2S_3$-25Cscl	260	380	-
$80GeSe_2$-$20Ga_2Se_3$	347	449	470

Table 2. Thermal properties of two chalcogenide glasses.

The powders were then inserted into a graphite die (Ø8, 20 or 35mm) with inner tantalum foil to prevent the glass from carbon contamination (Fig. 10). It is noteworthy that the glass powder is yellow due to the presence of sulphur in the composition. The corresponding sintered glass is red since the optical band gap is in the 600-700nm region. According to the glass composition, different SPS parameters were tested. The typical load ranges from 50 to 100MPa and the optimized sintering temperatures were found to be respectively 290°C and 390°C for the $62.5GeS_2$-$12.5Sb_2S_3$-25Cscl and $80GeSe_2$-$20Ga_2Se_3$ glass compositions. The dwell times at the sintering temperature range from 2 minutes to get amorphous bulk to 90 minutes to get glass-ceramics as shown by XRD patterns (Fig. 11).

Fig. 10. Photograph of the processed glass sample for the composition $62.5GeS_2$-$12.5Sb_2S_3$-25Cscl.

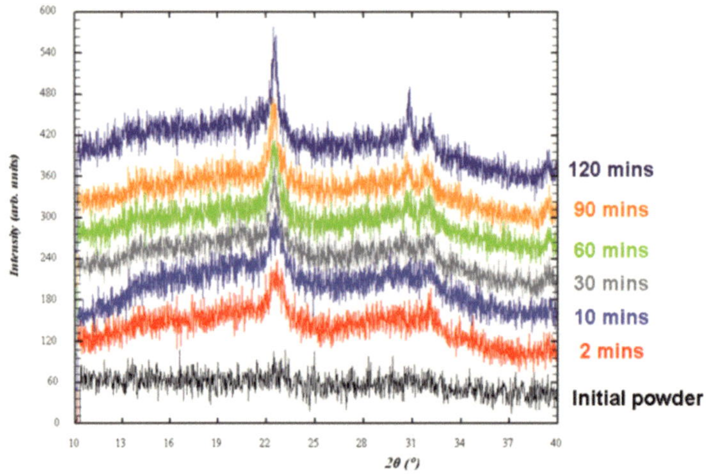

(a)

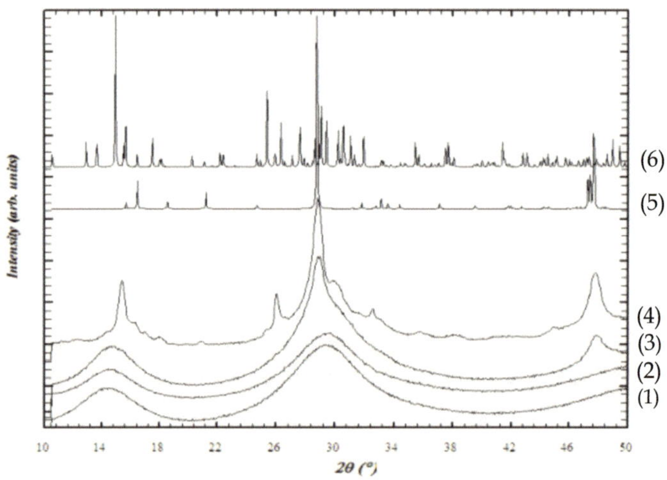

(b)

Fig. 11. X-ray diffraction patterns of (a) glass-ceramics obtained from the $62.5GeS_2$–$12.5Sb_2S_3$–25CsCl-based glass powder for different SPS treatment times at 290°C under 100 MPa, (b) glass-ceramics obtained from the $80GeSe_2$-$20Ga_2Se_3$ glass powder for different SPS treatment times at 390°C under 50 MPa, (1) initial $80GeSe_2$-$20Ga_2Se_3$ powder, (2) 2'SPS treatment time, (3) 30' SPS treatment time, (4) 60' SPS treatment time, (5) crystalline Ga_2Se_3, (6) crystalline $GeSe_2$.

The transmission of the processed glassy samples whatever their dimensions is good in the mid infrared range as observable in the picture taken by a thermal camera working in the third atmospheric window from 8 to 12μm (Figs. 12a and 12b). The absorption observed in this picture is firstly due to reflection on both faces because of the high refractive index of the $80GeSe_2$-$20Ga_2Se_3$ glass composition (n~2.41). Secondly, the presence of oxygen leading to the formation of Ge-O bonds inside the bulk glass induces phonon absorption and finally a slight diffusion of C graphite (papyex) or tantalum used as inner part of the die can also interfere (Fig. 12c). These all parameters can be improved by working under controlled atmosphere from the beginning to the end of the experiments and by pre-compacting the powder at room temperature. The sintering of large diameter is challenging in SPS experiment due to thermal gradient between the center and the periphery of the pellet. A micro-diffraction experiment has been carried out to investigate the amorphous character of the glass bulk along the diameter. Results show that no crystallization peaks are observed in the 36mm diameter glass and the glass is still amorphous (Fig. 12d). For recall, this glass composition doesn't exceed a diameter of 9mm using the conventional way of synthesis of chalcogenide glasses.

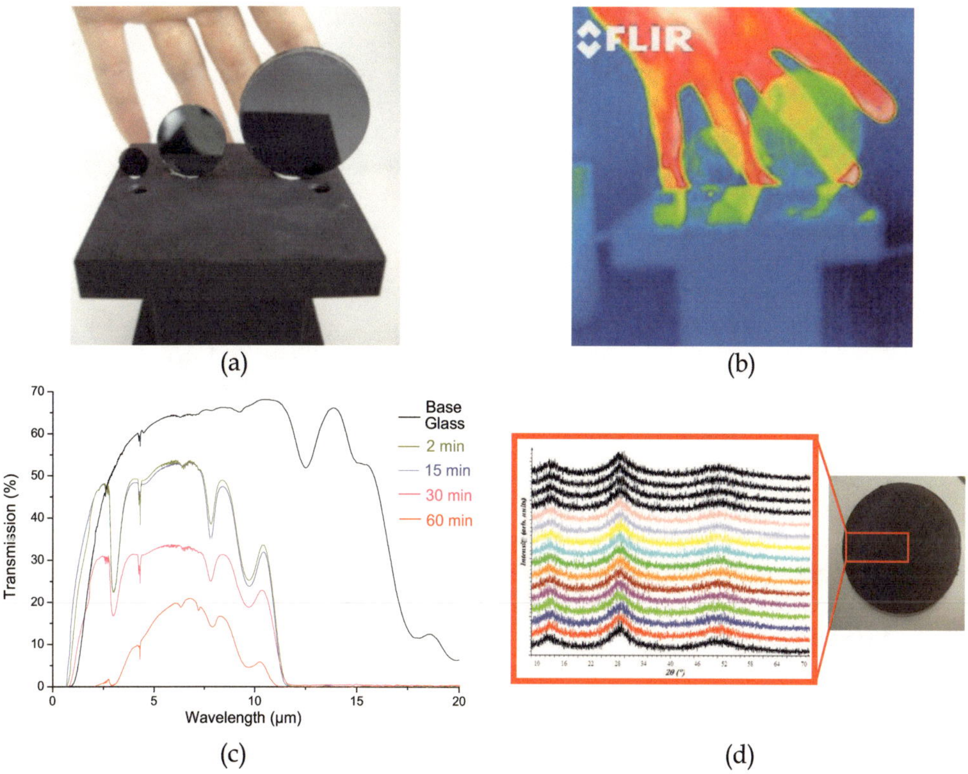

Fig. 12. (a) Photograph of the SPS processed $80GeSe_2$-$20Ga_2Se_3$ glasses (9, 20, 36mm diameter), (b) IR picture of the corresponding glasses observed with a 8-12μm thermal camera, (c) IR spectra of the base glass synthesized in sealed silica tube, finely grinded and sintered by SPS (390°C, 50MPa, 2'), glass and glass-ceramics made by mechanical milling and SPS at 390°C for 2, 15, 30 and 60 min under 50MPa, (d) X-Ray micro-diffraction patterns of the 36mm diameter sample from the center to the periphery.

Similar results are obtained with the $62.5GeS_2$-$12.5Sb_2S_3$-25Cscl glass composition that has been grinded from a bulk obtained by melt-quenching and sieved (Fig. 13). It is noteworthy that better optical transmissions are obtained with this technique in comparison with the sintering of powder obtained by mechanical milling. This supposes that some oxidation may occur during the long time mechanical milling process leading to lower optical properties.

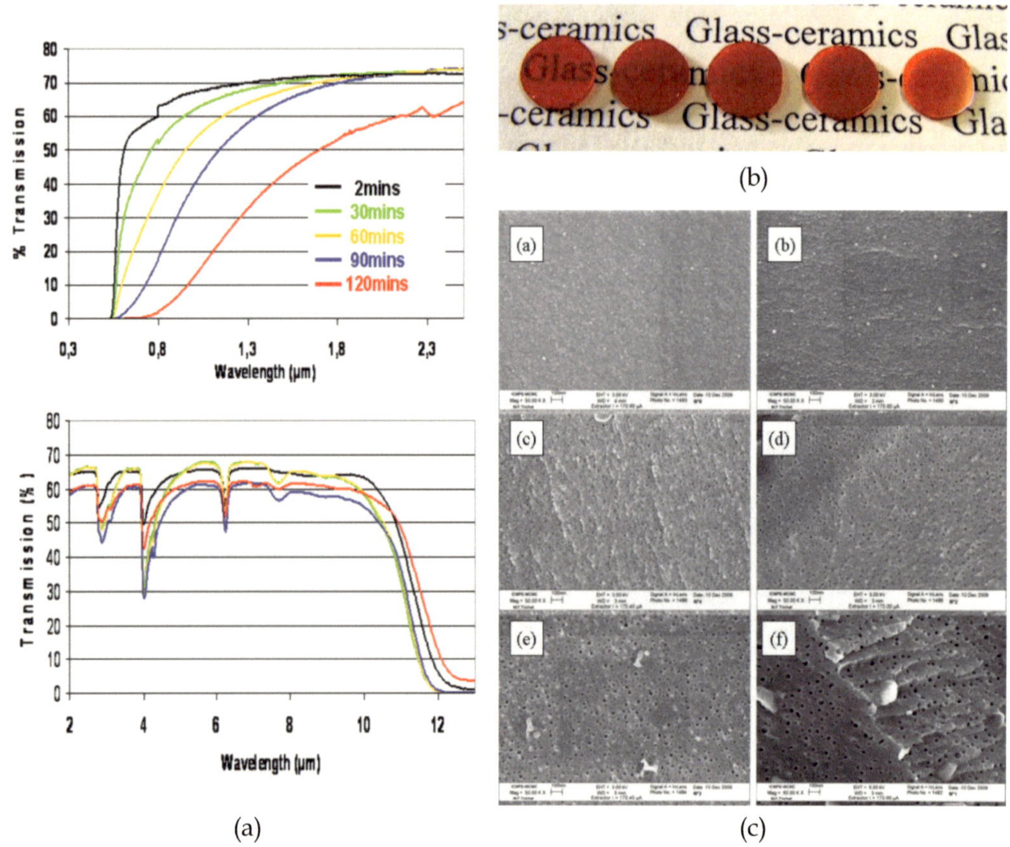

Fig. 13. (a) Transmission of different glass-ceramics obtained from the $62.5GeS_2$-$12.5Sb_2S_3$-25CsCl-based glass powder for SPS treatment times of 2 min, 30 min, 60 min, 90 min, and 120 min at 290°C under 100 MPa (thickness 1 mm) (b) Visual aspect of glass-ceramics samples for SPS treatment times of (a) 10 min, (b) 30 min, (c) 60 min, (d) 90 min, and (e) 120 min at 290°C under 100 MPa, (c) Observation of glass-ceramics under scanning electronic microscope for SPS treatment times of 2 min, 10 min, 30 min, 60 min, 90 min, and 120 min at 290°C under 100 MPa.

4.3 Mechanisms of sintering

A specific study on the mechanisms of sintering of chalcogenide glassy powder by SPS has been investigated on the special composition $80GeSe_2$-$20Ga_2Se_3$.

As previously mentioned, the Spark Plasma Sintering (SPS) technique is a non-conventional sintering technique based on the use of pulsed current. It is well known that the current distributions in the SPS die and through the sample is different in the case of a conductor and an insulator material (Anselmi-Tamburini et al., 2005). In the case of a non-electrically conducting sample such as alumina or the $80GeSe_2$-$20Ga_2Se_3$ glassy powder in our case, no Joule heating is expected through the sample and the heating of the sample is mostly due to heating of the die. This is in contrast with the case of conducting materials such as metals sample, where Joule heating starts immediately through the sample.

Due to the insulating property of the glass composition (Giridhar et al., 1990 & Raspopova et al., 1980), it is thus believed that heating mostly occurs through heating of carbon die. This is confirmed by Perriere et al. who suggest that (Perriere et al., 2011)

$$\frac{1}{I_{applied}} = \frac{1}{i_{die}} + \frac{1}{i_{powder}} \tag{3}$$

considering an equivalent electrical outline of the SPS apparatus. Therefore,

$$i_{powder} = I_{applied} \times \frac{1}{1+\frac{R_{powder}}{R_{die}}} \tag{4}$$

With i_{powder} and i_{die} the respectively electrical current passing through the chalcogenide glass powder and the die, $I_{applied}$ the macroscopic current intensity applied to the system,

$$R_{powder} = \frac{\rho_{powder} x l_{powder}}{S_{powder}} \quad (5), \; R_{die} = \frac{\rho_{die} x l_{die}}{S_{die}} \tag{6}$$

with ρ_{die} and ρ_{powder} the respectively electrical resistivity of the die and the powder, l_{die} and l_{powder} the length of the die and the powder crossed by the current, S_{die} and S_{powder} the respectively areas of the die or the powder crossed by the current.

Considering an electrical conductivity σ_{powder}=1×10^{-5} S.m^{-1} at 390°C (Giridhar et al., 1990), i.e ρ_{powder}= 1×10^{5} Ω.m and ρ_{die}= 20×10^{-6} Ω.m, l_{die}=30mm, l_{powder}=1.5mm, $Ø_{die}$=25mm and $Ø_{powder}$=8mm, we end with $i_{powder} \rightarrow 0$. The macroscopic current is therefore mostly driven by the graphite die. In the case of the 80$GeSe_2$-20Ga_2Se_3 amorphous powder, the SPS experiment can thus be seen as a Hot Uniaxial Pressing (HUP) experiment with faster heating rate.

The sintering of glass occurs by viscous flow, to reduce surface energy of a porous compact, through neck growth and densification involving deformation of initial particles (Rahamn, 2008). Different mechanisms leading to the production of glass-ceramics have been hypothesized. The first one would be the densification of glassy powder through viscous sintering followed by subsequent devitrification of the matrix. The second mechanism would imply densification and concurrent gradual crystallization of the matrix through the growth of neck between glassy particles. For one given glass composition, the corresponding crystalline phase has a considerably higher viscosity than the amorphous phase, so the sintering of polycrystalline material is more difficult than the amorphous phase. This suggests that the first mechanism described above, i.e achieving full density prior to any significant crystallization should be favored. However, in the specific case of metallic glasses, it is reported that concurrent mechanisms are involved during SPS treatment (Perriere et al., 2011).

Sintering of powdered glasses based on a viscous flow mechanism has been intensively studied by Frenkel, Scherer or Kingery and Berg (Frenkel, 1945; Kingery, 1955 & Scherer, 1977). Especially Murray et al. described the densification process of amorphous powder under load. The theoretical model based on viscous flow mechanism obeys the following relationship (Murray et al., 1954):

$$dD/_{dt} = 3P/4\eta(1-D) \quad \text{or} \quad \ln(1-D) = {-3P}/_{4\eta t} + \ln(1-D_i) \tag{7}$$

Where D is the relative density ($D=\rho/\rho_s$, ρ: the measured density and ρ_s, the theoretical density), Di the initial relative density, η the viscosity and t, the time. The slope of the ln (1-D) versus time t can be used to calculate the viscosity at one given temperature.

A formal sintering analysis, in order to help formulate hypotheses concerning the mechanism(s) controlling densification and devitrification has been performed. The influence of the dwell temperature (250°C, 280°C, 310°C, 330°C, 350°C, 370°C and 390°C) as well as the dwell time (2, 15 and 60 min) were studied while keeping the load constant (50MPa). These temperatures were chosen according to the sintering curve (Fig. 14) obtained from an SPS experiment (SPS parameters: 390°C, 50MPa, 2 minutes dwell time) where the full densification of glass is achieved without crystallization (D_{glass}=4.39). The heating rate is 100°C/min until 200°C and is then decreased to 30°C/min until the set temperature, i.e 390°C. The sintering of the glass is a two-step process. At 270°C, the shrinkage starts intensively. Viscous flow between particles leads to the formation of a neck. The growth of the necks occurs until 350°C which corresponds to the Tg of the glass (347°C). Then the mechanism of densification slows down and restarts around 360°C with lower amplitude. This second mechanism is believed to correspond to the closure of residual porosity measured from the Archimedean technique which is less than 10% at this stage.

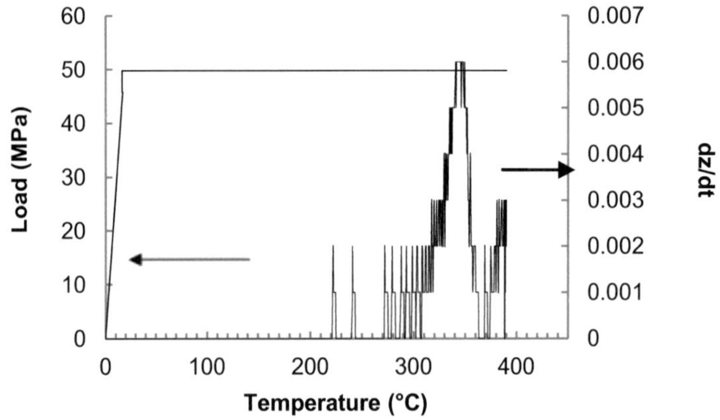

Fig. 14. Shrinkage rate (dz/dt) and load applied as a function of temperature during SPS experiment.

Fig. 15 shows micrographs obtained on the fracture surface of powder compacts heat treated at different temperatures from 250°C to 390°C for 2 minutes dwell time (load of 50MPa). At 290°C, far below the glass transition temperature Tg, the particles begin to soften and fuse together but a lot of inter-granular porosity is still present (27%). At 390°C, particles are totally fused together and no porosity is observed.

The relative densities, D, according to the dwell temperature and dwell time are reported in Figs. 16a and 16b. At 250°C, the compactness is only 70%. The residual porosity decreases gradually until 390°C where the full densification is obtained.

The influence of the dwell time at one given temperature was also studied. For all the dwell temperatures, the compactness regularly increases upon the increase in dwell time, for example from 90% (370°C, 2 min) to 96% (370°C, 30 min).

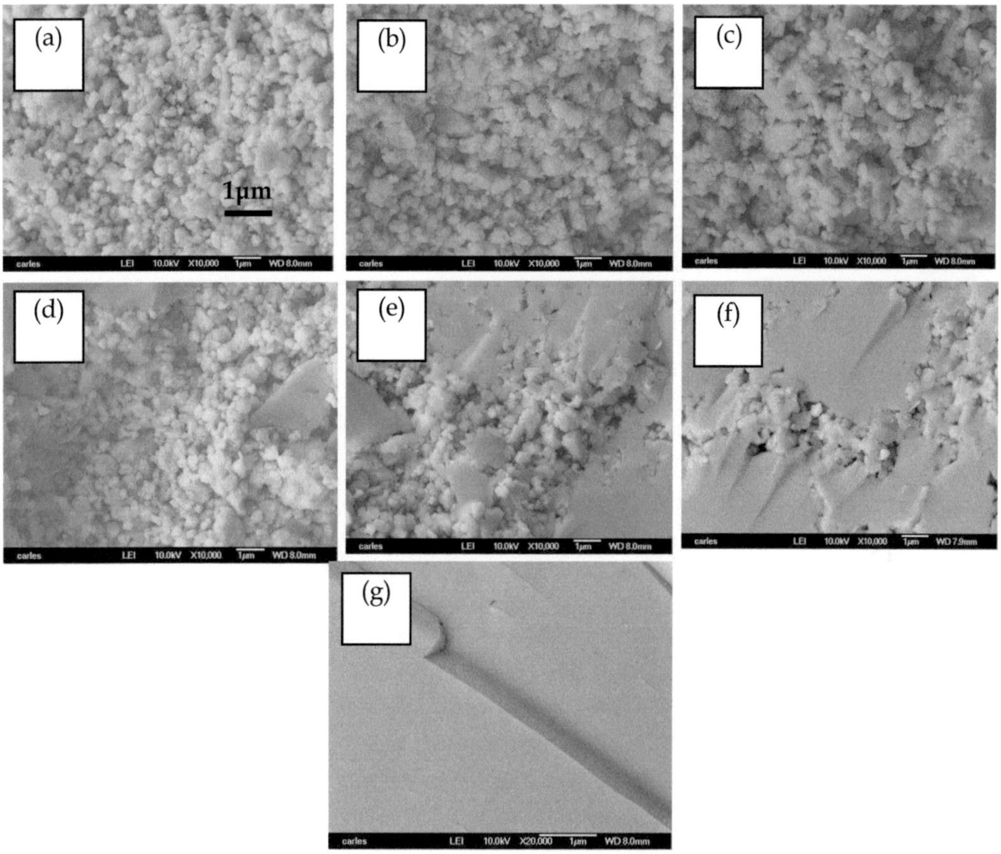

Fig. 15. FEC SEM photographs of $80GeSe_2$-$20Ga_2Se_3$ SPS glass compacts sintered at different dwell temperatures (50MPa, 2 minutes dwell time), (a) 250°C (b) 280°C (c) 310°C (d) 330°C (e) 350°C (f) 370°C (g) 390°C (same scale for all the photographs).

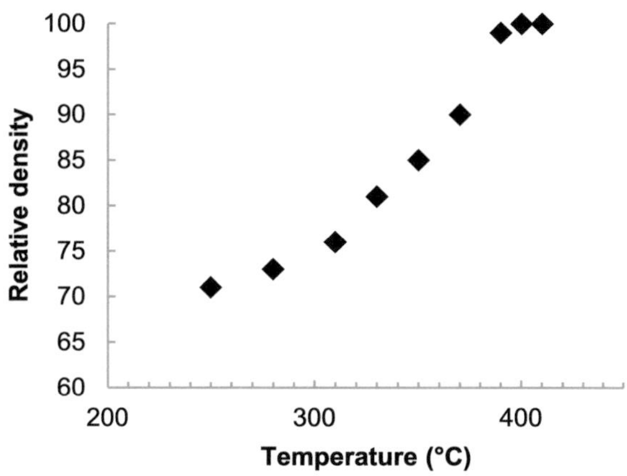

(a)

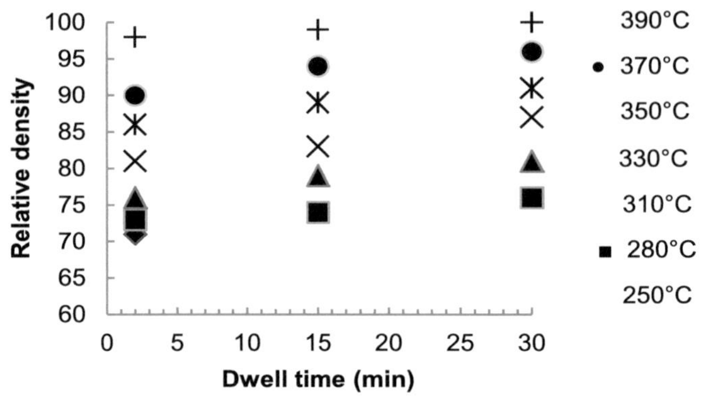

(b)

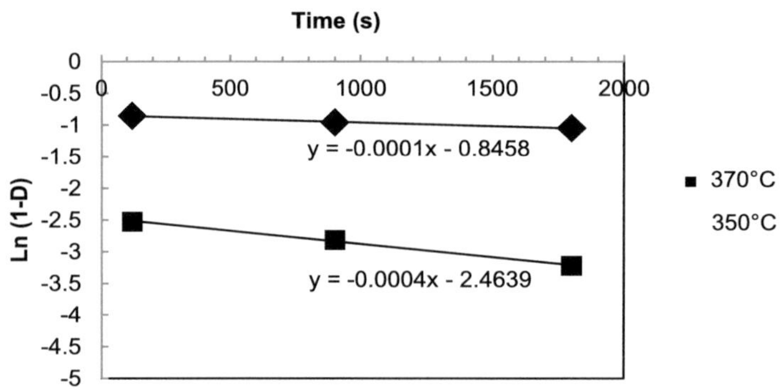

(c)

Fig. 16. (a) Influence of the dwell temperature on the relative density (50MPa, 2minutes dwell time), (b) Influence of the dwell time on the relative density for several dwell temperatures, (c) Plots of the ln (1-D) vs t curve for two dwell temperatures.

At 390°C for 2 minutes dwell time, the almost full densification of the glass is reached and the glass remains amorphous as shown by XRD patterns (Fig. 17). At this temperature only, the increase of dwell time induces partial devitrification of the glass as shown in Figs. 17 and 18. The crystals composition is Ga_2Se_3 (or $GaGe_4Se_8$) (Roze et al., 2008). Thus, there is clear evidence of densification mechanism prior to devitrification.

From eq. (7), we can deduce the viscosity for one given temperature (Fig. 16c). For temperatures close to the glass transition temperature, Tg, or higher temperatures, calculations give $\eta(350°C)\sim 2\times10^{11}$ Pa.s and $\eta(370°C)\sim 9\times10^{10}$ Pa.s (Fig. 16c). Considering a Newtonian behavior in this range of temperature (Roze et al., 2011), the dependence of its viscosity as a function of temperature can be described by a simple Arrhenius equation:

$$\eta = \eta_0 e^{E_\eta / RT} \quad (8)$$

where η is the viscosity, η_0 the pre-exponential factor, R the universal gas constant, T the temperature and E_η is the activation energy of viscous flow.

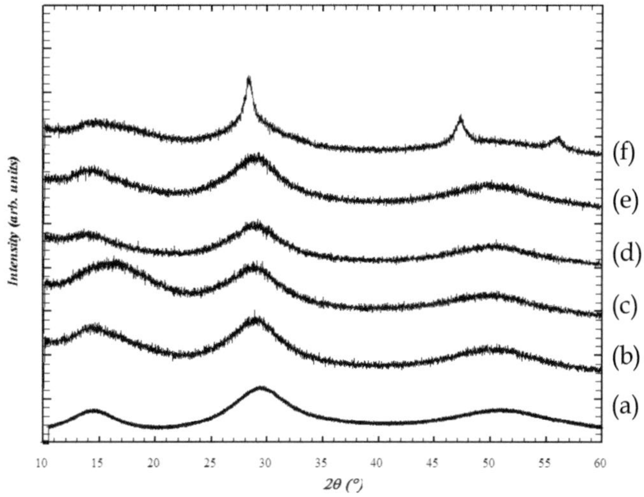

Fig. 17. XRD patterns of (a) $80GeSe_2$-$20Ga_2Se_3$ initial powder, (b) SPS glass sample (370°C, 50MPa, 15') (c) SPS glass sample (370°C, 50MPa, 30') (d) SPS glass sample (390°C, 50MPa, 2') (e) SPS glass sample (390°C, 50MPa, 15') (f) SPS glass sample (390°C, 50MPa, 30').

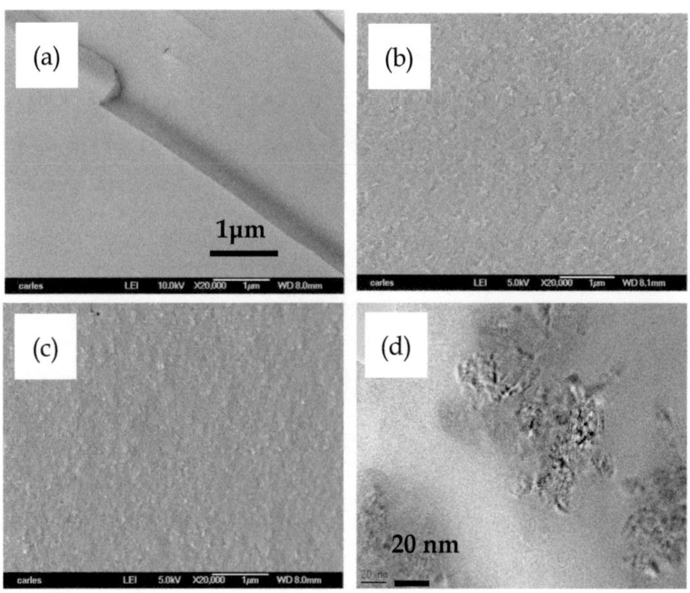

Fig. 18. FEG-SEM photographs of $80GeSe_2$-$20Ga_2Se_3$ SPS samples sintered at 390°C under 50MPa for different dwell times (a) 2 minutes, (b) 15 minutes, (c) 30 minutes, (d) TEM observation of glass-ceramic (c) obtained for SPS treatment time of 30 min (390°C, 50MPa) showing Ga_2Se_3 crystals.

We can thus deduce the activation energy of viscosity between Tg and Tc_1 from the curve ln η=f(1/T) which is found to be equal to ~500±100kJ.mol^{-1} (Fig. 19). This value has been compared to literature on 80$GeSe_2$-20Ga_2Se_3 synthesized by the common melt-quenching technique instead of mechanical alloying (Roze et al., 2011). From the data viscosity, similar activation energy is found.

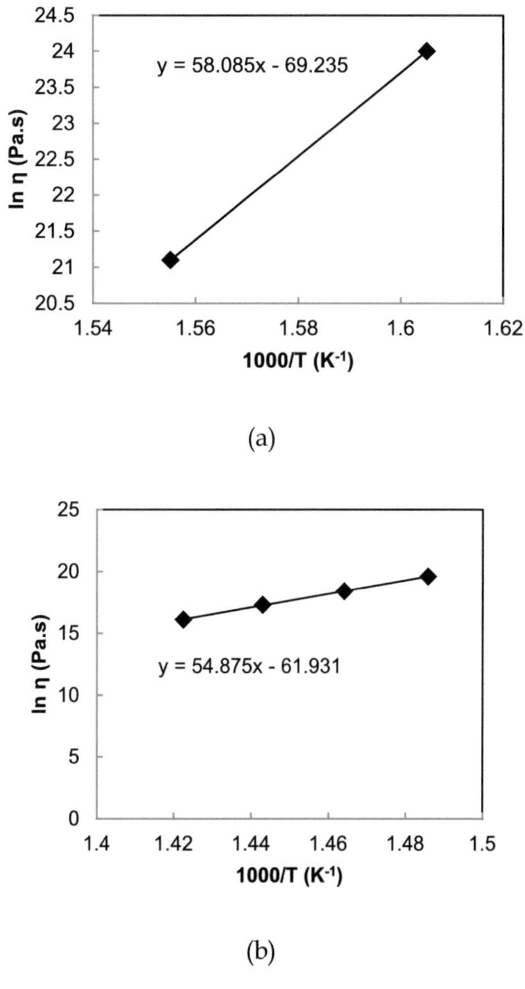

Fig. 19. Plots of the curve ln η= f(1/T) in the case of our study (a) and in [63] (b).

5. Conclusion

This chapter provides an overview of a new technique to synthesize chalcogenide glasses and glass-ceramics. This technique combines either the synthesis of glass powder by mechanical milling or the grinding and sieving of pre-existing bulk obtained by the conventional melt-quenching technique and the PCES technique (or SPS). Many chalcogenide glasses presenting real potential applications were set aside since the common synthesis in sealed silica tubes under vacuum did not permit to produce samples big enough for applications without crystallization and inhomogeneity concerns. It offers the possibility to make glasses or glass-ceramics of desired shape and size with sintering slightly higher than the glass transition temperature in few minutes. Indeed, the fast heating rates of PCES allow getting amorphous bulks in few minutes and partial devitrification starts after tens of minutes. Thus, this new process would lead to lower manufacturing costs of new chalcogenide glasses and glass-ceramics which could be adapted to numerous set-ups such as thermal imaging, infrared laser sources and detectors, thermoelectric devices, etc. Ultimately, this new synthetic route will lead to larger scale production of infrared optics, broadening the target market to a more general public.

The mechanisms of sintering of one given chalcogenide glass composition that has an electrical insulating property have been investigated. Results show that heating of the glass powder occurs mainly through the heating of the die due to the electrical insulating property of the glassy powder. It is also shown that densification of the glass powder occurs prior to the devitrification of the glass. The model of Murray et al. has been successfully applied to our glass and allows the determination of viscosity and activation energy for viscosity between Tg and Tc.

6. References

Aboulaich, A.; Bouchet, R.; Delaizir, G.; Seznec, V.; Tortet, L.; Rozier, P.; Morcrette, M.; Tarascon, J. M.; Viallet, V. & Dollé, M. (2011). A New Approach to Develop Safe All-Inorganic Monolithic Li-Ion Batteries. *Advanced Energy Materials*, Vol 2, pp. 179-183

Anselmi-Tamburini, U.; Gennari, S.; Garay, J. E. & Munir, Z. A. (2005). Fundamental investigations on the spark plasma sintering/synthesis process: II. Modeling of current and temperature distributions. *Materials Science and Engineering A*, Vol 394, pp. 139–148

Avrami, M. (1939). Kinetics of phase change. I, General theory. *Journal of Chemical Physics*, Vol 7, pp. 1103-1112

Bernard-Granger, G; Benameur, N.; Guizard, C. & Nygren, M. (2009). Influence of graphite contamination on the optical properties of transparent spinel obtained by spark plasma sintering. *Scripta Materialia*, Vol 60, pp. 164–167

Calvez, L.; Ma, H. L.; Lucas, J.; Zhang, X. H. (2007). Selenium-based glasses and glass-ceramics transmitting light from the visible to the far IR. *Advanced Materials*, Vol 19, pp. 129-132

Cheng, J. (1982). Properties and structure of the infrared-transmitting arsenic-germanium - selenium-tin glass-ceramic system. *Huadong Huagong Xueyuan Xuebao*, Vol 3, pp. 337-351

Choi, S. R.; Gyekenyesi, J. P. (2003). Results of mechanical testing for Pyroceram glassceramic. NASA/TM, (NASA/TM-2003-212487), pp. 1-84

Choi, P. P.; Kim, J. S.; Nguyen, O. T. H. ; Kwon, D. H. & Kwon, Y. S. (2006). Al-La-Ni-Fe Amorphous Alloys and Amorphous-crystalline Composites Produced by Mechanical Alloying. *Materials Science Forum*, Vol 510–511, pp. 290–293

Choi, P.; Kim, J. S; Nguyen, O. T. H.; Kwon, D. H.; Kwon, Y. S. & Kim, J. C. (2007). Al-La-Ni-Fe bulk metallic glasses produced by mechanical alloying and spark-plasma sintering. *Materials Science and Engineering: A*, Vol 449-451, pp. 1119-1122

Danto, S.; Ruff, Z.; Wang, Z.; Joannopoulos, J. D. & Fink, Y. (2010). Ovonic memory switching in multimaterials fibers. *Advanced Functional Materials*, Vol 21, pp. 1095-1101

Delaizir, G.; Dollé, M.; Rozier, P. & Zhang, X. H. (2010). Spark Plasma Sintering: An easy way to Make Infrared Transparent Glass–Ceramics. *Journal of the American Ceramic Society*, Vol 93, pp. 2495-2498

Delogu, F. & Cocco, G. (2005). Numerical simulations of structural modifications at a Ni-Zr sliding interface. *Physical Review B*, Vol 72 pp. 014124

Eckert, J.; Seidel, M.; Schlorke, N.; Kubler, A. & Schultz, L. (1997). Synthesis and Properties of Mechanically Alloyed and Nanocrystalline Materials. *Materials Science Forum*, Vol 235, pp. 23–28

Elliott, S.R. (1991). Chalcogenide glasses. *Materials Science and technology*, Vol 9, pp. 375

Frenkel, J. (1945). Viscous flow of crystalline bodies under the action of surface tension. *Journal of Physics*, Vol 9, pp. 385-391

Galy, J.; Dolle, M.; Hungria, T.; Rozier, P. & Monchoux, J. P. (2008). A new way to make solid state chemistry: Spark plasma synthesis of copper or silver vanadium oxide bronzes. *Solid State Sciences*, Vol 10, pp. 976–981

Giridhar, A. & Mahadevan, S. (1990). Chemical ordering in Ge-Ga-Se glasses. *Journal of Non-Crystalline Solids*, Vol 126, pp. 161-169

Goncalves, A. P; Lopes, E. B.; Rouleau, O. & Godart, C. (2010). Conducting glasses as new potential thermoelectric materials: the Cu–Ge–Te case. *Journal of Materials Chemistry*, Vol 20, pp. 1516–1521

Guillevic, E.; Zhang, X. H.; Pain, T.; Calvez, L.; Adam, J. L.; Lucas, J.; Guilloux-Viry, M.; Ollivier, S. & Gadret, G. (2009). Optimization of chalcogenide glass in the As-Se-S system for automotive applications. *Optical Materials*, Vol 31, pp. 1688-1692

Gutzow, I.; Pascova, R.; Karamanov, A. & Schmelzer, J. (1998). The kinetics of surface induced sinter crystallization and the formation of glass-ceramic materials. *Journal of Materials Sciences*, Vol 33, pp. 5265-5273

Hayashi, A.; Hama, S.; Morimoto, H.; Tatsumisago, M. & Minami, T. (2001). Preparation of $Li_2S–P_2S_5$ Amorphous Solid Electrolytes by Mechanical Milling. *Journal of the American Ceramic Society*, Vol 84, pp. 477-479

Hubert, M.; Delaizir, G.; Monnier, J.; Godart, C.; Ma, H. L.; Zhang, X. H. & Calvez, L. (2011) submitted to *Journal of Materials Chemistry*

Hungria, T.; Galy, J. & Castro, A. (2009). Spark Plasma Sintering as a Useful Technique to the Nanostructuration of Piezo-Ferroelectric Materials. *Advanced Engineering Materials*, Vol 11, pp. 615-631

Inoue, K., (1962) U.S. Patent No. 3241956

Inoue, K., (1966) U.S. Patent No. 3250892

Johnson, W. L. (1986). Thermodynamic and kinetic aspects of the crystal to glass transformation in metallic materials. *Progress in Materials Science*, Vol 30, pp. 81–134

Kingery, W. D. & Berg, M. (1955). Study of the Initial Stages of Sintering Solids by Viscous Flow Evaporation-Condensation, and Self-Diffusion. *Journal of Applied Physics*, Vol 26, pp. 1205-1212

Kittel, C. (1998). Physique de l'état solide. Dunod, Paris

Kim, T. S.; Lee, J. K.; Kim, H. J. & Bae, J. C. (2005). Consolidation of Cu47Ti34Zr11Ni8 bulk amorphous alloy powders. *Materials Science and Engineering: A*, Vol 402, pp. 228-233

Klocek, P.; Roth, M. & Rock, R. D. (1987). Chalcogenide glass optical fibers and image properties and applications. *Optical Engineering*, Vol 26, pp. 88-95

Li, Q.; Wang, G.; Song, X.; Fan, L.; Hu, W.; Xiao, F.; Yang, Q.; Ma, M.; Zhang, J. & Liu, R. (2009). $Ti_{50}Cu_{23}Ni_{20}Sn_7$ bulk metallic glasses prepared by mechanical alloying and spark-plasma sintering. *Journal of Materials Processing Technology*, Vol 209, pp. 3285-3288

Li, Y. Y.; Yang, C.; Qu, S. G.; Li, X. Q. & Chen, W. P. (2010). Nucleation and growth mechanism of crystalline phase for fabrication of ultrafine-grained $Ti_{66}Nb_{13}Cu_8Ni_{6.8}Al_{6.2}$ composites by spark plasma sintering and crystallization of amorphous phase. *Materials Science and Engineering: A*, Vol 528, pp. 486-493

Lund, A. C. & Schuh, C. A. (2004). Topological and chemical arrangement of binary alloys during severe deformation. *Journal of Applied Physics*, Vol 95, pp. 4815–4822

Ma, H. L.; Zhang, X. H. & Lucas, J. (2003). Infrared transmitting chalcogenide glass ceramics. *Journal of Non-Crystalline Solids*, Vol 317, pp. 270-274

Machado, K. D ; De Lima, J. C. ; Campos, C. E. M ; Gasperini, A. A. M. ; De Souza, S. M. ; Maurmann, C. E ; Grandi, T. A. & Pizani, P. S. (2005). Reverse Monte Carlo simulations and Raman scattering of an amorphous $GeSe_4$ alloy produced by mechanical alloying. *Solid State Communications, Vol* 133, pp. 411-416

Mayerhöfer, T. G; Shen, Z.; Leonova, E.; Edén, M.; Kriltz, A.; Popp, J. (2008). Consolidated silica glass from nanoparticles. *Journal of Solid State Chemistry*, Vol 181, pp. 2442-2447

Mc. Millan, P. W. (1979). Glass ceramics, Academic Press, second edition

Mecholsky, J. J.; Moynihan, C. T.; Macedo, P. B. & Srinivasan, G. R. (1976). Microstructure and properties of an infra-red transmitting chalcogenide glass-ceramic. *Journal of Materials Science*, Vol 11, pp. 1952-1960

Murray, P.; Rodgers, E. P. & Williams, A. E. (1954). Practical and theoretical aspects of the hot pressing of refractory oxides. *Transactions and Journal of the British Ceramic Society*, Vol 53, pp. 474-509

Nowak, S.; Perrière, L.; Dembinski, L.; Tusseau-Nenez, S. & Champion, Y. (2011). Approach of the spark plasma sintering mechanism in $Zr_{57}Cu_{20}Al_{10}Ni_8Ti_5$ metallic glass. *Journal of Alloys and Compounds*, Vol 509, pp. 1011-1019

Orru, R.; Licheri, R.; Mario Loccib, A.; Cincotti, A. & Cao, G. (2009). Consolidation/synthesis of materials by electric current activated/assisted sintering. *Materials Science and Engineering R*, Vol 63, pp. 127–287

Patil, U.; Hong, S. J. & Suryanarayana, C. (2005). An Unusual Phase Transformation during Mechanical Alloying of an Fe-based Bulk Metallic Glass Composition. *Journal of Alloys and Compounds*, Vol 389, pp. 121

Perriere, L.; Thai, M. T.; Tusseau-Nenez, S.; Bletry, M. & Champion, Y. (2011). Spark Plasma Sintering of a Zr-Based Metallic Glass. *Advanced Engineering Materials*, Vol 13, pp. 581-586

Rahamn, M. N. (2008). Sintering of ceramic. CRC Press edited by Taylor and Francis group, pp 47

Raspopova, E. M.; Kolomiets, B. T.; Shilo, V. P. & Smirnova, N. N. (1980). Electrical properties of Ga-Ge-Se chalcogenide glasses doped with REE. *Physica Status Solidi A*, Vol 59, pp. K235-K240

Riello, P.; Bucella, S.; Zamengo, L.; Anselmi-Tamburini, U.; Francini, R.; Pietrantoni, S. & Munir, Z. A. (2006). Erbium-doped LAS glass ceramics prepared by spark plasma sintering (SPS). *Journal of the European Ceramic Society*, Vol 26, pp. 3301-3306

Roze, M.; Calvez, L.; Ledemi, Y.; Ma, H. L.; Lucas, J. & Zhang, X. H. (2008). Optical and thermo-mechanical properties of new Ge-Ga-Se-AgI glasses. *Journal of Optoelectronics and Advanced Materials*, Vol 10, pp. 141-144

Roze, M.; Calvez, L.; Hubert, M.; Toupin, P.; Bureau, B.; Boussard-Pledel, C. & Zhang, X. H. (2011). Molded Glass–Ceramics for Infrared Applications. *International Journal of Applied Glass Science*, Vol 2, pp. 129–136

Saito, S. & Sawaoka, A. (1973). *Powder Metallurgy International*, Vol 5, pp. 70-75

Scherer, G. W. (1977). Sintering of Low Density Glasses: II. Experimental Study. *Journal of the American Ceramic Society*, Vol 60, pp. 239-243

Schurack, F.; Borner, I.; Eckert, J. & Schultz, L. (1999). Synthesis and Properties of Mechanically Alloyed and Ball Milled High Strength Amorphous or Quasicristalline Al-Alloys. *Materials Science Forum*, Vol 312–314, pp 49–54

Sekine, M.; Suzuki, Y.; Ueno, H.; Onodera, Y.; Usuki, T.; Nasu, T. & Wei, S. (2007). Appearance of fast ionic conduction in AgI-doped chalcogenide glass powders prepared by mechanical milling. *Journal of Non-Crystalline Solids*, Vol 353, pp. 2069-2073

Shin, S.; Kim, T. S. & Kang, S. K. (2010). The influence of spark plasma sintering temperature on the mechanical properties and corrosion resistance of $Zr_{65}Al_{10}Ni_{10}Cu_{15}$ metallic glass powder. *Intermetallics*, Vol 18, pp. 2005-2008

Shirakawa, Y.; Matsuda, T.; Tani, Y.; Shimosaka, A. & Hidaka, J. (2001). Amorphization of Ge-GeSe mixtures in mechanical alloying process. *Journal of Non-Crystalline Solids, Vol* 293-295, pp. 764-768

Song, S. M.; Choi, S. Y. & Lee, Y. K. (1997). Crystallization property effects in $Ge_{30}Se_{60}Te_{10}$. *Journal of Non-Crystalline Solids*, Vol 217, pp. 79-82

Song, Z.; Kishimoto, S. & Shinya, N. (2004). A Novel Pulse-Current-Assisted Sintering Method for Fabrication of Metallic Cellular Structures. *Advanced Engineering Materials*, Vol 6, pp. 211-214

Strnad, Z. (1986). Glass-Ceramic Materials. Publishers of Technical Literature, Prague, 185

Trevey, J.; Jang, J. S.; Jung, Y. S.; Stoldt, C. R. & Lee, S. H. (2009). Glass–ceramic $Li_2S–P_2S_5$ electrolytes prepared by a single step ball billing process and their application for all-solid-state lithium–ion batteries. *Electrochemistry Communications*, Vol 11, pp. 1830-1833

Troles, J.; Coulombier, Q. ; Canat, G. ; Duhant, M. ; Renard, W. ; Toupin, P. ; Calvez, L. ; Renversez, G. ; Smektala, F. ; El Amraoui, M. ; Adam, J. L. ; Chartier, T. ; Mechin, D. & Brilland, L. (2010). Low loss microstructured chalcogenide fibers for large non linear effects at 1995 nm. *Optics Express*, Vol 18, pp. 26647-26654

Wilhelm, A.; Boussard-Pledel, C.; Coulombier, Q.; Lucas, J.; Bureau, B. & Lucas, P. (2007). Development of Far-Infrared-Transmitting Te Based Glasses Suitable for Carbon Dioxide Detection and Space Optics. *Advanced Materials*, Vol 19, pp. 3796-3800

Yan, X.; Song, X.; Lu, N.; Li, E. & Zhang, J. (2008). A novel route for preparing binary Sm-Co bulk amorphous alloys. *Materials Letters*, Vol 62, pp. 2862-2864

Zhang, X. H.; Guimond, Y. & Bellec, Y. (2003). Production of complex chalcogenide glass optics by molding for thermal imaging. *Journal of Non-Crystalline Solids*, Vol 326-327, pp. 519-523

Zhang, X. H.; Ma, H. L. & Lucas, J. (2004). A new class of infrared transmitting glass-ceramic based on controlled nucleation and growth of alkali halide in a sulphide based glass matrix. *Journal of Non-Crystalline Solids*, Vol 337, pp. 130-135

Part 4

Impregnation of Doping Materials

14

Effect of TiO_2 Addition on the Sintering Process of Magnesium Oxide from Seawater

Vanja Martinac
University of Split / Faculty of Chemistry and Technology
Croatia

1. Introduction

Magnesium oxide is one of the most important materials used in the production of high-temperature-resistant ceramics. Due to its high refractory properties (MgO melts at (2823 ± 40) °C), MgO ceramic is non-toxic and chemically inert in basic environments et elevated temperatures, resistant to the effect of metal melts, acid gases, alkali slag, neutral salts, and react with carbon only above 1800 °C. Today, in large-scale technicall processes, magnesia (MgO) for refractories is produced from two sources: natural and synthetic. Magnesia from natural sources constitutes 82 % of the world's magnesia installed capacity. The dominant source is magnesite ($MgCO_3$) which occurs in both a macro and a cryptocrystalline forms. Less significant are dolomite ($CaCO_3 \cdot MgCO_3$), hydromagnesite ($3MgCO_3 \cdot Mg(OH)_2 \cdot 3H_2O$), brucite ($Mg(OH)_2$) and serpentine ($Mg_3(Si_2O_5)(OH)_4$). Synthetic materials are manufactured either from seawater or from magnesia rich brines. Magnesium oxide obtained from sea water is a high-quality refractory material, and its advantages lie not only in the huge reserves of seawater (1 m^3 contains 0.945 kg of magnesium), but in the higher purity of the sintered magnesium oxide (≥ 98 % MgO). The production of magnesium oxide from seawater is a well-know industrial process (Bocanegra-Bernal, 2008; Bonney, 1982; Gilpin & Heasman, 1977; Heasman, 1979; Maddan, 2001; Martinac, 1994; Petric & Petric, 1980, Rabadžhieva et al., 1997) and has been studied all over the world for a number of years. For most of the second half of the twentieth century, seawater provided almost 50 % of the magnesium produced in the western world, and today it still remains a major source of magnesium oxide in many countries. The process involves the extraction of dissolved magnesium, which has a concentration of around 1.3 g dm^{-3} in seawater (Brown et al., 1997), and 3 to 40 times this values for brines, and the reaction of magnesium salts (chloride and sulphate) with lime or dolomite lime to produce a magnesium hydroxide precipitate. The precipitate is washed and calcined to form caustic magnesia. The apparently simple chemistry of the process is unfortunately complicated in practice because seawater is not a pure solution of magnesium salts and dolomite or limestone, although abundant, are never found free of impurities. Boron is a particulary problematic impurity for the magnesia used as a high quality refractory material. Thus, boron can be a problem in refractory magnesia for specialized refractory applications where a high hot strength is required. Taking into consideration that B_2O_3 is common impurity in seawater derived magnesia, the aim of this study was to examine the possibility of adding TiO_2 in quantities of 1, 2 and 5 wt.-% for reducing the boron content in the product, i.e. sintered magnesium oxide obtained from seawater. The purpose of this paper was, first, to reduce the B_2O_3 content in magnesium oxide from seawater as much as possible in ensure a high-purity product, because the hot-strength properties of certain refractory products are significantly affected by their boron content, and, second to sinter the individual products and determine the properties of samples sintered depending on the precipitation method and the boron content in the magnesium oxide.

2. MgO from seawater

Processing of seawater magnesium involves precipitation of magnesium hydroxide in seawater reacting with an alkaline base, such as calcined dolomite or calcined limestone. If dolomite lime is used as precipitation agent, the chemical reaction is as follows:

$$2\,CaO{\cdot}MgO(s) + 2\,Mg^{2+} + SO_4^{2-} + 2\,Cl^- + 4\,H_2O = 4\,Mg(OH)_2(s) + CaSO_4(s) + Ca^{2+} + 2\,Cl^- \quad (1)$$

The composition of the dolomite lime (from the location Đipalo near the town of Sinj, Croatia) used for precipitating the magnesium hydroxide from seawater was as follows (wt.-%): MgO = 42.27%, CaO = 57.55 %, SiO_2 = 0.076%, Al_2O_3 = 0.042%, Fe_2O_3 = 0.064%, and the composition of the seawater (from the location at the promontory of the hill Marjan near the the Oceanographic Institute in Split, Croatia) was as follows: MgO = 2.423 g dm^{-3} and CaO = 0.604 g dm^{-3}.

Impurities from seawater and from precipitation agent get into the magnesium hydroxide precipitate, so that special attention has to be paid to precipitate purity, depending on the product application. Thus, seawater is pretreated by acidifying with H_2SO_4 to lower its pH from the normal 8.2 to 4.0, in order to remove bicarbonate (HCO_3^-) and carbonate (CO_3^{2-}) ions. The chemical reaction are:

$$Ca^{2+} + CO_3^{2-} + H^+ + HSO_4^- = Ca^{2+} + SO_4^{2-} + H_2O + CO_2(aq) \quad (2)$$

$$Ca^{2+} + 2\,HCO_3^- + H^+ + HSO_4^- = Ca^{2+} + SO_4^{2-} + 2\,H_2O + CO_2(aq) \quad (3)$$

The calcium sulphate formed remained in the solution. Seawater was then passed through the desorption tower packed with Rasching rings where it flowed downward against a rinsing stream of air. The liberated carbon dioxide (CO_2) gas was removed from falling water drops by the ascending airflow. In this way seawater derived lime contamination of the magnesia can be minimised. The flow rate of the induced air was 120 $dm^3\,h^{-1}$, and the volumetric flow rate of the seawater through the desorption tower was 6 $dm^3\,h^{-1}$. After the pretreatment of the seawatwer, a calculated amount of dolomite lime was added to precipitate the magnesium hydroxide. The magnesium oxide used was obtained from seawater by substoichiometric precipitation (where precipitation of magnesium hydroxide took place with 80% of the stoichiometric quantity of the dolomite lime) and by overstoichiometric precipitation (which took place with 120% of the stoichiometric quantity of the dolomite lime). The precipitation reaction lasted for 30 min; a magnetic stirrer was used. After magnesium hydroxide precipitation, settling took place. The sedimentation rate was increased by addition of the optimum amount of the anionic Flokal-B flocculent (polyacrilamide) (produced by Župa-Kruševac, Serbia). The precipitate obtained was then decanted and rinsed. The rinsing and decantation procedure was repeted five times with approximately 1 dm^3 of distilled water as rinsing agent. After that, the magnesium hydroxide precipitate was filtered through a number of funnels. The rinsing agent used with the $Mg(OH)_2$ precipitate on the filter paper was the same as the one used for rinsing by decantation. This procedure was also repeted five times, i.e. until rinsing was completely carried out. The magnesium hydroxide thus obtained was dried at 105 °C and then calcined at 950 °C for 5 h to form caustic magnesia. The boron content was determined potentiometrically. The variation coefficient for the potentiometric method employed in boron determination is ± 1% (Culkin, 1975). The results listed represent an average value of a number of measurements (an average of five analyses in each case). Table 1 shows the chemical composition of magnesium oxide obtained from seawater with regard to magnesium oxide, calcium oxide, and boron(III) oxide.

Sample	MgO / wt.-%	CaO / wt.-%	B_2O_3 / wt.-%
MgO (80% precipitation)	99.20	0.59	0.193
MgO (120% precipitation)	98.25	1.32	0.056

Table 1. Chemical composition of magnesium oxide obtained from seawater.

The substoichiometric precipitation of magnesium hydroxide from seawater is a very convenient precipitation method in the so-called «wet phase» (Petric & Petric, 1980), as it significantly increases the thickener capacity, i.e. the magnesium hydroxide settling rate which is the «bottleneck» of this technology. This is very important for the design of the thickener as its construction is the time-controlling factor in plants of this type. At precipitation of 80% the capacity of the thickener (calculated according to Kynch) increases by 71% in relation to complete precipitation (Martinac et al., 1997). Substoichiometric precipitation significantly increases the sedimentation rate of the magnesium hydroxide precipitate formed, due to the decreased thickeness of the double electrical layer around the magnesium hydroxide particle. A consequence of the increased adsorption of Mg^{2+} ions onto $Mg(OH)_2$ particles is a decrease in the zeta-potential. Therefore, substoichiometric precipitation increases the coagulation stability of the given $Mg(OH)_2$-seawater system. Also, one of the advantages of substoichiometric (80%) precipitation lies in the reduced quantity of concentrated HCl needed to neutralize waste seawater after sedimentation. This quantity amounts to only 1.1 g of concentrated HCl per kg of magnesium oxide, while it is 210.5 of concentrated HCl per kg of magnesium oxide with the overstoichiometric (120%) precipitation (Petric et al., 1991). In such a case, i.e. when this precipitation method is employed, the boron content adsorbed onto the magnesium hydroxide during the precipitation process is somewhat higher than during overstoichiometric precipitation, and should therefore be reduced. Boron oxide is a common impurity in magnesia obtained from seawater; it is capable of acting as a powerful fluxing agent for the calcium silicate phases which can be present in refractory grades of magnesia.

2.1 TiO_2 addition as sintering aid

The use of sintering aids - small additions of various compounds that enhance densification, or allow it to occur at a lower temperature during sintering - is quite common in the production of ceramic bodies. The most commonly used additives are oxides (Li_2O, Al_2O_3, Cr_2O_3, Fe_2O_3, SiO_2, TiO_2, ZrO_2 and V_2O_5) and some halides, such as LiF and LiCl. The effect of small additions of this compaunds on the sintering of magnesium oxide has been studied in detail (Chaudhuri, 1990, 1992, 1999; Ćosić et al., 1987; Lee, 1998; Lucion, 2004; Martinac et al., 1996; Petric et al., 1987, 1989, 1994, 1999) and has received wide attention. Additions of tetravalent Si, Ti and Zr enhance sintering. There is a general consensus regarding the way in which many of these additives operate, based on a mechanism where intergranular liquid phase are formed which can restrict grain growth, assist the grain-boundary sliding and accelerate mass transport during sintering. It has been established that the addition of TiO_2 greatly affect properties of magnesium oxide obtained from seawater; even a small addition of 0.5 wt.-% TiO_2 significantly increases product density at 1300 °C (Petric et.al., 1989). The densities amount to 94% of the theoretical density (ρ_t = 3.576 g cm^{-3}) for durationof isothermal heating 5 h. The addition of TiO_2 promotes low-temperature densification of magnesium oxide, proportional to the extend of solid solution formation and vacancy formation. In that case the sintering was intensified in the presence of the liquid phase in the MgO-TiO_2 system. It is evident that TiO_2 addition is more efficient at lower temperatures than at the higher ones. The effect of ultravalent ions (such as Ti^{4+}) in the periclase crystal structure creates lattice defects in the form of cation vacancies (Fig. 1) which promote material transport and sintering at relatively low temperatures. At higher temperatures, such as 1600 °C and 1700 °C, the effect of this aid is less prominent. We can assume the mass

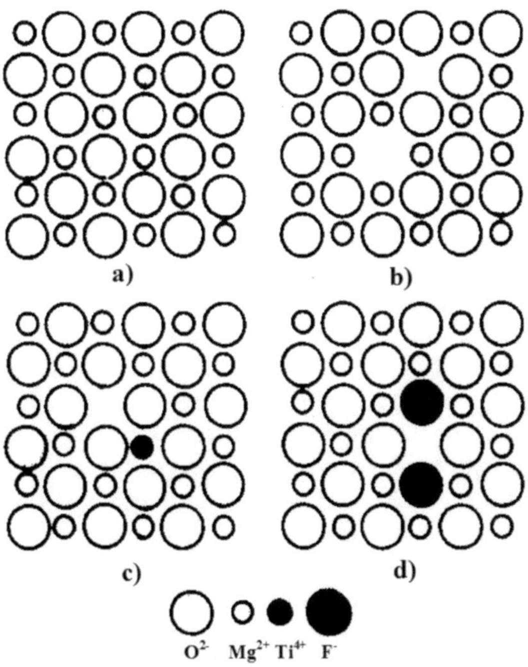

Fig. 1. Schematic representation of a small section of a periclase crystal (MgO), a) at low temperatures (intrinsic) and b) Schottky defect (anionic and cationic vacancies). The ions originally at the vacant lattice sites have been removed to the surface, c) The crystal has a Ti^{4+} ion that induces a cation vacancy, d) This crystalstal has a F^- ion inducing a cation vacancies.

transfer, as in the case with pure magnesium oxide to be determined by diffusion of O_2^- ions through the MgO lattice as the slower diffusion species. Higher temperatures improve mobility in elements forming the crystal lattice, due to which an interface is formed between particles of compact powder, porosity is eliminated and the whole system shrinks. The densities amount to 94-97% of the theoretical densities at 1600 °C, and 96-98% at 1700 °C, for duration of isothermal heating 1-5 h, and with 1, 2 and 5 wt.-% TiO_2 added. Data on apparent porosity in sintered samples point to a very low presence of open pores in the system. The pores present are mainly the closed ones. Accordingly, total porosity is almost indentical to closed porosity. An apparent porosity ranges from 0.15-0.10% at 1600 °C and 0.05-0.03% at 1700 °C, for soaking time 1-5 h for sintered magnesium oxide samples (80% precipitation) and 0.16-0.11% at 1600 °C and 0.04-0.01% at 1700 °C for sintering magnesium oxide samples (120% precipitation) under the same operating conditions (Petric et al., 1999). The low values obtained for the densification during isothermal heating in the samples examined indicate that a great part of densification process takes place during heating, i.e. before the maximum sintering temperature is reached.

The addition of TiO_2 also greatly affects the removal of boron from the sample into air, i.e., TiO_2 reduces the B_2O_3 content during isothermal sintering of magnesium oxide obtained from seawater (Martinac, 1994). The boron content of seawater presents a problem because the hot-strength properties of certain specialized magnesia refractory products are markedly affected by their boron content. Boron is present in seawater in part as the non-dissociated orthoborate acid H_3BO_3 and partly as the borate ion $H_2BO_3^-$. The concentration of the higher oxidation level ions HBO_3^{2-} and BO_3^{3-} is very low. The orthoborate acid is a weak acid with the following dissociation constants:

$$H_3BO_3 = H^+ + H_2BO_3^- \qquad K_1 = 5.8 \cdot 10^{-10} \tag{4}$$

$$H_2BO_3^- = H^+ + HBO_3^{2-} \qquad K_2 = 1.8 \cdot 10^{-13} \tag{5}$$

$$HBO_3^{2-} = H^+ + BO_3^{3-} \qquad K_3 = 1.6 \cdot 10^{-14} \tag{6}$$

By calculating the dissociation rate, one can establish the molal concentration of $H_2BO_3^-$, HBO_3^{2-}, and BO_3^{3-}, as well as the molal dissociation rate for every degree of dissociation of the orthoborate acid. For 80 % precipitation of magnesium hydroxide from seawater by dolomite lime, the pH value is 9.6 during reaction precipitation and settling of the precipitate formed. In that case the orthoborate acid dissociation in the first degree is 69.78 %, which contributes to a significant increase of the B_2O_3 content in the product, i.e. in magnesium oxide obtained from seawater (0.193 wt.-%). Under the conditions more favorable to coprecipitation, the boron contamination of the magnesium hydroxide can be as high as the equivalent of 0.5 parts B_2O_3 per 100 parts of magnesia. However, using specific reaction conditions as well as addition TiO_2, the boron contamination can be virtually eliminated. The addition of TiO_2 proved rather interesting since the content of B_2O_3 is reduced in a sintered samples by means of TiO_2. Mixtures of magnesium oxide were prepared in the above composition (Tab. 1), with 1, 2 and 5 wt.-% TiO_2, respectively. The dopant oxide used was an analytical reagent grade titania (TiO_2 p.a.), in rutile form, produced by Merck. The chemical analysis of TiO_2 p.a. is given in Tab.2.

	TiO_2 (99 %)
Water soluble matter	0.3 %
Chloride (Cl)	0.01 %
Sulphate (SO4)	0.05 %
Heavy metals (such as Pb)	0.001 %
Iron (Fe)	0.005 %
Arsenic (As)	0.0002 %

Table 2. Chemical analysis (wt.-%) of TiO_2 p.a. (Merck).

Samples were homogenized by manual stirring in ethanol absolute (C_2H_6O p.a.) for 30 min. After drying (at 80 °C) the mixture was crushed into fine powder and the powders were mixed well again. The mixtures were compacted by a cold-pressing process. The process was carried out in a hydraulic press at pressure of 625 MPa. The compacts were sintered at temperatures of 1300 °C and 1500 °C, with an isothermal heating duration of τ = 1, 3 and 5 h. The sintering at 1300 °C was carried out in an electric furnace. A gas furnace, made by a French firm, Mecker, (Type 553) with zirconium(IV) oxide lining, was used for sintering at 1500 °C. The furnace was heated by burning a mixture of propane-butane in the air, with oxygen added to achieve high temperature. It took approximately 2 h to reach the maximum temperature in the furnaces. In both cases, after sintering, the samples were left to cool in the furnace. Tabs. 3 and 4 show the results obtained for the effect of TiO_2 on the content of B_2O_3 in magnesium oxide samples after sintering at 1300 °C and 1500 °C, taking into account the method of obtaining magnesium hydroxide from seawater as well as the operating conditions listed. The results shown represent an average of a number of measurements. The standard deviation, σ, for MgO (80 % precipitation) was: $\sigma_{max} = 9.8 \cdot 10^{-3}$ and $\sigma_{min} = 4.4 \cdot 10^{-3}$. The standard deviation for MgO (120 % precipitation) was: $\sigma_{max} = 5.0 \cdot 10^{-3}$ and $\sigma_{min} = 1.5 \cdot 10^{-3}$.

t / °C	τ /h	B_2O_3 (wt.-%) in MgO without addition	B_2O_3 (wt.-%) in MgO + 1 wt.-% TiO_2	B_2O_3 (wt.-%) in MgO + 2 wt.-% TiO_2	B_2O_3 (wt.-%) in MgO + 5 wt.-% TiO_2
1300	1	0.1934	0.1395	0.0789	0.0652
	3	0.1655	0.1363	0.0752	0.0638
	5	0.1192	0.0852	0.0645	0.0587
1500	1	0.1265	0.0434	0.0396	0.0264
	3	0.0756	0.0184	0.0170	
	5	0.0689	0.0173	0.0159	0.0131

Table 3. Effect of TiO_2 on the B_2O_3 content in the sintered magnesium oxide samples (80 % precipitation) at t = 1300 °C, 1500 °C, τ = 1, 3, 5 h, p = 625 MPa.

t / °C	τ /h	B_2O_3 (wt.-%) in MgO without addition	B_2O_3 (wt.-%) in MgO + 1 wt.-% TiO_2	B_2O_3 (wt.-%) in MgO + 2 wt.-% TiO_2	B_2O_3 (wt.-%) in MgO + 5 wt.-% TiO_2
1300	1	0.0512	0.0428	0.0293	0.0165
	3	0.0459		0.0109	0.0086
	5	0.0376	0.0384	0.0096	0.0053
1500	1	0.0453	0.0431	0.0116	0.0062
	3	0.0400	0.0331	0.0100	0.0060
	5	0.0318	0.0204	0.0050	0.0035

Table 4. Effect of TiO_2 on the B_2O_3 content in the sintered magnesium oxide samples (120 % precipitation) at t = 1300 °C, 1500 °C, τ = 1, 3, 5 h, p = 625 MPa.

The experimental dana indicate that the TiO_2 addition together with the temperature and duration of isothermal heating significantly reduces the B_2O_3 content during sintering. Different behaviour patterns relative to the B_2O_3 content were noticed in magnesium oxide obtained by 80 % or by 120 % precipitation of magnesium hydroxide in seawater; this is due to different contents of CaO in those samples. It was noted that the presence of calcium oxide caused the retention of boron in the samples during sintering. With the magnesium oxide (120 % precipitation) the content of CaO = 1.32 wt.-% is significantly higher than with the magnesium oxide (80 % precipitation) where CaO = 0.59 wt.-%, i.e. there is a significantly larger quantity of CaO than in case of 80 % precipitation which favors the $Ca_2B_2O_5$ formation reaction. Namely, based on a previous paper (Petric et al., 1987) the presence of dicalcium borate ($Ca_2B_2O_5$) was proved in sintered samples by the method of X-ray diffraction, that is, it was established that B_2O_3 transforms into $Ca_2B_2O_5$ through the reaction with CaO. Also the studies (Chaudhuri et al., 1992, 1999; Ćosić et al., 1989; Čeh & Kolar, 1994) show that the method of X-ray diffraction and EDAX analysis indicate that in the sintering process the TiO_2 added reacts with CaO from the MgO-CaO solid solution and transforms into calcium titanate $CaTiO_3$. Therefore, TiO_2 binds a part of CaO in $CaTiO_3$ and thus reduces the CaO content which reacts with B_2O_5. So a smaller quantity of $Ca_2B_2O_5$ is formed which remains in the sintered samples while a greater part of B_2O_3 evaporates. This is the way in which the TiO_2 reduces the quantity of B_2O_3 in a sample. The higher the CaO content, the more B_2O_3 is retained in the sintered samples. With MgO (80% precipitation) already a small amount of TiO_2 (wt. = 1%) binds almost all of CaO present. With MgO (120 % precipitation) CaO is in excess and favors $Ca_2B_2O_5$ formation; in MgO (80 % precipitation) a greater part of B_2O_3 evaporates from the sample into the atmosphere. In the magnesium oxide (120 % precipitation) it can be seen that a higher quantity of TiO_2 (2 – 5 wt.-%) binds almost all of CaO and effects boron removal significantly. Therefore, the final content of B_2O_3 in the sintered samples depends both on the CaO and TiO_2 content. These two mutually dependent reactions of formation of $Ca_2B_2O_5$ and $CaTiO_3$ which cause B_2O_3 content reduction during sintering, are:

$$2CaO + B_2O_3 = Ca_2B_2O_5 \tag{7}$$

$$CaO + TiO_2 = CaTiO_3 \tag{8}$$

In order to examine the effect of TiO_2 on the reduction of the B_2O_3 content in samples sintered, experimental results on the fraction of evaporated boron and the degree of reaction CaO with TiO_2 has been examined relative to the temperature and the duration of isothermal sintering for magnesium oxide samples obtained from seawatwr by 80% and 120% precipitation, with addition of wt. = 1, 2 and 5% TiO_2 respectively, according to expressions used in the open system thermodynamics (De Groot & Mazur, 1984; Haase, 1990; Lavenda, 1993; Prigogine, 1968). A system of equations dealt with the open system thermodynamics has therefore been considered, and coefficients L_{11}, L_{12} and L_{22} that describe the mutual effect of two simultaneous irreversible processes examined, have been calculated based on an important theorem due to Onsanger. Generally, the phenomenological relationship may be written in the following form:

$$J_i = \Sigma L_{ij} X_j \tag{9}$$

For each force X, there is a corresponding conjugate primary flow J. These phenomena, and other like them, are called cross-effects. The coefficients L_{ij} (with $i \neq j$) are called phenomenological coefficients. For the system with two flows caused by two driving forces, i.e., with two simultaneous irreversible processes, phenomenological dependencies can be expressed in the following way:

$$J_1 = L_{11} X_1 + L_{12} X_2 \tag{10}$$

$$J_2 = L_{21} X_1 + L_{22} X_2 \tag{11}$$

where J_1 and J_2 denote flows and X_1 and X_2 denote the forces causing these flows. Coefficients L_{ij} (with $i \neq j$) describe the interference of the two irreversible processes i and j. There exists a so-called Onsanger reciprocity ratio between cross coefficients L_{ij} and L_{ji} which can be expressed by following equations:

$$L_{ij} = L_{ji} \quad (ij = 1, \dots n;\ i \neq j) \tag{12}$$

or

$$\left(\frac{\partial J_i}{\partial X_j}\right)_{X_i=0,\, i\neq j} = \left(\frac{\partial J_j}{\partial X_i}\right)_{X_j=0,\, j\neq i} \tag{13}$$

These Onsanger reciprocity relations state that when the flux, corresponding to the irreversible process i, is influenced by the force X_j of the irreversible process j, then the flux j is also influenced by the force X_i through the same interference coefficient L_{ij}. Equation (12) allows a reduction in the number of phenomenological coefficients, i.e., the interaction coefficients L_{12} and L_{21} are equal. The coefficients L_{ij} in the system of two equations, i.e., for n = 2, must satisfy the following conditions:

$$L_{11} \geq 0; \quad L_{22} \geq 0 \tag{14}$$

$$(L_{12} + L_{21})^2 \leq 4\, L_{11}L_{22} \tag{15}$$

Using again the Onsanger relation $L_{12} = L_{21}$, equation (15) now becomes

$$L_{11}\, L_{22} - L_{12}^2 \geq 0 \tag{16}$$

The conjugate coefficients (i.e. L_{11} and L_{22}) must be positive. Obviously, the crossed coefficients or interference coefficients (L_{12} and L_{21}) have no definite sign. They may be either positive or negative; their magnitude being limited only by equation (15). If the system of phenomenological Eqs. (10) and (11) is applied to the $Ca_2B_2O_5$ and $CaTiO_3$ formation reactions, which are interdependent, we assume the linear relations:

$$J_1 = L_{11}\, t' + L_{12}\, \tau \tag{17}$$

$$J_2 = L_{21}\, t' + L_{22}\, \tau \tag{18}$$

where J_1 is the percent of B_2O_3 removed during sintering, and calculated from experimental data on the B_2O_3 content in sintered samples and on the content B_2O_3 in calcined magnesium oxide, i.e. the sample before sintering, J_2 is the percent of CaO which reacted with TiO_2, τ is the duration of isothermal heating (h), and t' is the themperature at 10^{-2} (°C), i.e., $t' = t \cdot 10^{-2}$ (°C). From this we see that we may regard t' and τ as driving forces corresponding to the fluxes J_1 and J_2, respectively. Tabs. 5 and 6 present the values obtained for dependence of J_1 and J_2 on the temperature (t') and duratin of isothermal heating (τ) for sintered magnesium oxide samples (80 % and 120 % precipitation), with different quantities of sintering

1 wt.-% TiO_2							
	J_1				J_2		
t'/τ	1	3	5	t'/τ	1	3	5
13	27.87	-	55.95	13	61.19	-	76.46
15	77.56	90.49	91.05	15	85.54	93.25	93.54
2 wt.-% TiO_2							
	J_1				J_2		
t'/τ	1	3	5	t'/τ	1	3	5
13	59.20	61.12	66.65	13	77.57	78.56	81.45
15	79.52	91.21	91.78	15	87.57	93.62	93.89
5 wt.-% TiO_2							
	J_1				J_2		
t'/τ	1	3	5	t'/τ	1	3	5
13	66.29	67.01	69.70	13	81.26	81.62	83.04
15	86.35	-	93.23	15	91.10	-	94.66

Table 5. Dependence of J_1 and J_2 on themperature (t') and duration of isothermal heating (τ) for the sintered magnesium oxide samples (80 % precipitation) with different quantities of sintering aid.

aid, respecitvely. The coefficients L_{11}, L_{12} and L_{22} in eqs. (17) and (18) were calculated by a computer using combination of the mean values method with the least squares method. After calculating the coefficients, the equations for J_1 and J_2 for each percent of TiO_2 added, for the magnesium oxide (80 % precipitation) and the magnesium oxide (120 % precipitation) are shown in Tab. 7. Thus, the experimental data J_1, i.e. the percent of B_2O_3 «removed» during sintering process, and J_2, i.e. the percent of CaO which reacted with TiO_2, which also indirectly affects the content of B_2O_3 were used to calculate the coefficients L_{11}, L_{12} and L_{22}. The calculated phenomenological coefficients L_{11}, L_{12} and L_{22} describe simultaneneous irreversible processes (reactions) and provide an insight into the interdependence of both reactions.

1 wt.-% TiO_2							
	J_1				J_2		
t' / τ	1	3	5	t' / τ	1	3	5
13	23.81	-	31.69	13	52.52	-	52.52
15	23.31	41.16	63.70	15	93.06	94.26	95.76
2 wt.-% TiO_2							
	J_1				J_2		
t' / τ	1	3	5	t' / τ	1	3	5
13	47.85	80.64	83.02	13	95.32	95.57	97.73
15	96.84	97.01	97.64	15	79.36	82.21	91.10
5 wt.-% TiO_2							
	J_1				J_2		
t' / τ	1	3	5	t' / τ	1	3	5
13	70.64	84.75	90.50	13	96.89	97.85	98.24
15	89.02	89.23	93.75	15	97.47	97.50	97.81

Table 6. Dependence of J_1 and J_2 on themperature (t') and duration of isothermal heating (τ) for the sintered magnesium oxide samples (120 % precipitation) with different quantities of sintering aid.

For MgO (80 % precipitation)	For MgO (120 % precipitation)
For 1 wt.-% TiO_2	
$J_1 = 3.9411\ t' + 4.6189\ \tau$	$J_1 = 1.8155\ t' + 3.7793\ \tau$
$J_2 = 4.6189\ t' + 4.8140\ \tau$	$J_2 = 3.7793\ t' + 5.9798\ \tau$
For 2 wt.-% TiO_2	
$J_1 = 4.3421\ t' + 4.9044\ \tau$	$J_1 = 4.9090\ t' + 5.2364\ \tau$
$J_2 = 4.9044\ t' + 4.6067\ \tau$	$J_2 = 5.2364\ t' + 4.7514\ \tau$
For 5 wt.-% TiO_2	
$J_1 = 4.4856\ t' + 5.0239\ \tau$	$J_1 = 4.9594\ t' + 5.6129\ \tau$
$J_2 = 5.0239\ t' + 4.3569\ \tau$	$J_2 = 5.6129\ t' + 2.6914\ \tau$

Table 7. Equations for J_1 and J_2 with the calculated coefficients L_{11}, L_{12} and L_{22} for each percent of TiO_2 added, for the sintered magnesium oxide samples MgO (80 % precipitation) and MgO (120 % precipitation), respectively.

The coefficient values L_{11}, L_{12} and L_{22} calculated depend on the quantity of TiO_2 added. Therefore, the dependence of the coefficients value L_{11}, L_{12} and L_{22} on percent TiO_2 was calculated. The relationship between the phenomenological coefficients and the percent of TiO_2 added can be expressed by the following equation:

$$Y = A x^2 + B x + C \tag{19}$$

where Y is the phenomenological coefficients L_{11}, L_{12} and L_{22}, x is the percent of TiO_2 added and A, B and C are constants. The coefficients were calculated by the least squares method and are shown by the equations: For the sintered magnesium oxide samples (80 % precipitation):

$$L_{11} = - 0.0833 x^2 + 0.6659 x + 3.3635 \tag{20}$$

$$L_{12} = - 0.0614 x^2 + 0.4697 x + 4.2106 \tag{21}$$

$$L_{22} = 0.0310 x^2 - 0.3003 x + 5.0833 \tag{22}$$

For the sintered magnesium oxide samples (120 % precipitation):

$$L_{11} = - 0.7692 x^2 + 5.4010 x - 2.8164 \tag{23}$$

$$L_{12} = - 0.3293 x^2 + 2.4333 x + 1.6803 \tag{24}$$

$$L_{22} = 0.1354 x^2 - 1.6347 x + 7.4791 \tag{25}$$

where x is the percent of TiO_2. These equations describing dependence of L to x make it possible to calculate the coefficients L_{11}, L_{12} and L_{22} for other percentages of x in the range from 1 wt.-% to 5 wt.-% TiO_2. As CaO simultaneously reacts with both B_2O_3 and TiO_2 two described reactions of formation of dicalcium borate and calcium titanate are related, and it was of interest to calculate the coefficients for Eqs. (17) and (18), as well as their dependence on the percentage of TiO_2 added. The analysis provides the opportunity to determine which percentage of TiO_2 should be added to the sample once to CaO and B_2O_3 contents are known. Thermodynamical analysis of the magnesium oxide sintering process with varying quantities of added TiO_2 has made possible to predict mathematically, without experiments, the B_2O_3 content in samples sintered relative to the temperature and the duration of isothermal sintering, as well as on the properties of initial magnesium oxide samples. The method of describing a system by application of equations studied in the open system thermodynamics can be used in some other cases when similar laws are involved, i.e. when due to a motive force in a system, a flow of mass or energy occurs.

3. Conclusion

The effect of TiO_2 addition on the B_2O_3 content of sintered samples, i.e. on product properties, has been examined. The addition of TiO_2 reduces the B_2O_3 content in the isothermal sintering process, as it binds a part of CaO in calcium titanate, $CaTiO_3$, so that a greater part of B_2O_3 evaporates from the system during sintering. Depending on the CaO content of the sample, i.e, the method of obtaining magnesium hydroxide from seawater, it has been found that in magnesium oxide (80 % precipitation) a lower quantity of TiO_2 (1 wt.-%) binds almost all the CaO present (which has not reacted with B_2O_3). In the magnesium oxide (120 % precipitation) it takes 2 wt.-% TiO_2 to bind all the CaO present (which has not reacted), so that only a greater quantity (5 wt.-%) TiO_2 affects boron removal during sintering to a greater degree. The higher the CaO content, the more B_2O_3 is retained in the sintered samples. Two mutually dependent reactions of formation of $Ca_2B_2O_5$ and $CaTiO_3$ were analysed, and phenomenological coefficients calculated according to expresions used in the open system thermodynamics. Calculated phenomenological coefficients L_{11}, L_{12} and L_{22} describe the mutual interdependence of two simultaneous irreversible processes, based on an important theorem due to Onsanger. It is thus possible to calculate the quantity of boron (B_2O_3) removed during the sintering process, i.e. the quantity of B_2O_3 which remains in the sample sintered, for the area examined. Analogous consideration can be carried out for all the other cases when similar laws are involved, i.e. when mass or energy flows occur in the system due to a motive force.

4. Acknowledgment

The results shown arise from the research project «Activated sintering of magnesium oxide» which is financial supported by the Ministry of Science, Education and Sports of the Republic of Croatia.

5. References

Bocanegra-Bernal, M. H. (2008). Microstructural evolution during sintering in MgO Power Precipitated from Seawater under Induced Agglomeration conditions, *Powder Technology*, Vol.186, No.3, (September 2008), pp. (267-272), ISSN 032-5910

Bonney, O. V. (1982). Recovery of Magnesium as Magnesium Hydroxide from Seawater, US Pat. 43 149 85, 9 February 1982: *Chemical Abstract*, Vol.96, No.125549

Brown, E. et al. (1997). *Seawater: Its Composition, Properties and Behaviour*, 2nd Ed., Butterworth Heinemann in association with The Open University, ISBN 0 7506 3715 3, Walton Hall, Milton Keynes, MK7 6AA, England

Chaudhuri, M. N.; Kumar, A.; Bhadra, A. K. & Banerjee, G. (1990). Sintering and Grain Growth in Indian Magnesites Doped with Titanium Dioxide, *Interceramics*, Vol.39, No.4/5, (April 1990), pp. (26-30), ISSN 0020-5214

Chaudhuri, M. N.; Kumar, A.; Bhadra, A. K.; Banerjee, G. & Sarkar, S. L. (1992). Microstructure of Sintered Natural Indian Magnesites with Titania Addition, *American **Ceramic** Society **Bulletin***, Vol.71, No.3, (March 1992), pp. (345-348), ISSN 0002-7812

Chaudhuri, M. N.; Banerjee, G.; Kumar, A. & Sarkar, S. L. (1999). Secondary Phases in Natural Magnesite Sintered with Addition of Titania, *Journal of Materials Science*, Vol.34, No.23, (December 1999), pp. (5821-5825),ISSN 0022-2461

Culkin, F. (1975). The Major Constituents of Seawater, In: *Chemical Oceanography*, Vol.1, J. P. Riley & G. Skirrow, (Eds.), pp. (136-151), Academic Press, ISBN: 0125887019/0-12-588701-9, London

Čeh, M. & Kolar, D. (1994), Solubility of CaO in $CaTiO_3$, *Journal of Materials Science*, Vol.29, Issue 23, (January 1994),pp. (6295-6300), ISSN 0022-2461

Ćosić, M.; Pavlovski, B. & Tkalčec, E. (1989), Activated Sintering of Magnesium Oxide Derived from Serpentine, *Science of Sintering*, Vol.21, No.3, (September 1989), pp. (161-174), ISSN 0350-820X

De Groot, S. R. & Mazur, P. (1984). *Non-equilibrium Thermodynamics*, Dover Publications, ISBN 0 486 64741 2, New York

Gilpin, W. C. & Heasman, N. (1977). Recovery of Magnesium Compaunds from Seawater, *Chemistry and Industry*, Vol.16, No.6, (July 1977) pp. (567-572), ISSN 0009-3068

Haase, R. (1990), *Thermodynamics of irreversible processes*, Dover Publications, ISBN 0486663566, New York

Heasman, N. (1979). New Developments in Seawater Derived Magnesia, *Gas Wärme International*, Vol.28, No.6-7,(June 1997), pp. (329-397), ISSN 0020-9384

Lavenda, B. H. (1993). *Thermodynamics of Irreversible Processes*, Dover Publications, ISBN 0486675769, New York

Lee, Y. B.; Park, H. C. & OH, K. D. (1998). Sintering and Microstructure Development in the System $MgO-TiO_2$, *Journal of Materials Science*, Vol.33, No.16, (August 1998), pp. (4321-4325), ISSN 0022-2461

Lucion, T.; Duvigneaud, P. H.; Laudet, A.; Stenger, J. F. & Gueguen, E. (2004). Effect of TiO_2 Additions on the Densification of MgO and MgO-CaO Mixtures, *Key Engineering Materials*, Vols.264-268, Issue I, pp. (209-212), on line avaible since 2004/May/15 at http://www.scientific.net

Maddan, O. Lee (2001). Apparatus and Method for Producing Magnesium from Seawater, US Pat. 6 267 854B1, 31 July 2001: *Chemical Abstract,* Vol. 134, No.286989M

Martinac, V. (1994). A Study of Isothermal Sintering of Magnesium Oxide, PhD Thesis, University of Split, Faculty of Chemistry and Technology, Split, 1994

Martinac, V., Labor, M. & Petric, N. (1996). Effect of TiO_2, SiO_2 and Al_2O_3 on Properties of Sintered Magnesium Oxide from Seawater, *Materials Chemistry and Physics*, Vol.46, Issue 1, (October 1996), pp. (23-30),ISSN 0254-0584

Martinac, V.; Labor, M.; Petric, N. & Arbunić, N. (1997). Sedimentation of Magnesium Hydroxide in Seawater and its Effect on Plant Capacity, *Indian Journal of Marine Science*, Vol.26, No.4, (December 1997), pp. (335-340), ISSN 0379-5136

Petric, B. & Petric, N. (1980). Investigations of the Rate of Sedimentation of Magnesium Hydroxide Obtained from Seawater, *Industrial and Engineering Chemistry Process Design and Development*, Vo.19, No.3, (July 1980), pp. (329-335), ISSN 0196-4305

Petric, N.; Petric, B.; Tkalčec, E.; Martinac, V.; Bogdanić, N.; Mirošević-Anzulović, M. (1987), Effect of Additives onSintering of Magnesium oxide Obtained from Seawater, *Science of Sintering*, Vol.19, No.2, (May 1987),pp. 81-87., ISSN 0350-820X

Petric, N.; Petric, B.; Martinac, V. & Mirošević-Anzulović, M. (1989), A Study of Isothermal Sintering and Properties of Magnesium Oxide from Seawater, In: *Science of Sintering: New Directions for Materials Processing and Microstructural Control*, D. P. Uskoković, N. Palmour III & R. M. Spring, (Eds.), pp. (565-572), Plenum Press, ISBN 0-306-43528-4, New York & London

Petric, N.; Petric, B. & Martinac, V. (1991). Examination of Boron Content and Properties of Magnesium Oxide Obtained from Seawater, *Journal of Chemical Technology & Biotechnology*, Vol.52, No.4, (June 1991), pp. (519-526), ISSN 0268-2575

Petric, N.; Petric, B.; Martinac, V.; Labor, M. & Mirošević-Anzulović, M. (1994). Effect of TiO_2 on Properties of Magnesium Oxide Obtained from Seawater, *Journal of Materials Science*, Vol.29, Issue 24, (January 1994), pp. (6548-65590), ISSN 0022-2461

Petric, N.; Martinac, V.; Labor, M. & Mirošević-Anzulović, M. (1999). Activated Sintering of Magnesium Oxide from Seawater, *Chemistry Engineering & Technology*, Vol.22, No.5, (May 1999), pp. (451-456), ISSN 0930-7516

Prigogine, I. (1968). *Introduction to the Thermodynamics of IrreversibleProcesses*, 3rd Ed., Wiley, ISBN 0470699280, London

Rabadžhieva, D.; Ivanova, K.; Balarev Hr. & Trendafelov, D. (1997). Polučenie hidroksida magnija iz ostatočnoi raplji pri dobljiče soli iz morskoi vodlji, *Žurnal Priklačnoji Himii*, Vol.70, No.3, (Mart 1997), pp. (375-380), ISSN 1070-4272

15

The Role of Sintering in the Synthesis of Luminescence Phosphors

Arunachalam Lakshmanan
Saveetha Engineering College,
Thandalam, Chennai,
India

1. Introduction

The phenomena of calcination, roasting and sintering are closely related and often used intermittently. *Calcination* is the process of subjecting a substance to the action of heat, but without melting or fusion, for the purpose of causing some change in its physical or chemical constitution. The objects of calcination are usually: (1) to drive off water, present as absorbed moisture, as "water of crystallization," or as "water of constitution"; (2) to drive off carbon dioxide, sulphur dioxide, or other volatile constituent; (3) to oxidize a part or the whole of the substance. The process of calcination derives its name from the Latin *calcinare* (to burn lime) due to its most common application, the decomposition of calcium carbonate (limestone) to calcium oxide (lime) and carbon dioxide in order to produce cement. In *roasting*, the minerals impose heartburn, which is used to drive out volatile components whereas in *sintering*, small pieces of ore or powder are heated to make bonding. *Sintering* is a method for making objects from powder through agglomeration by heating the material in a furnace to 80-90% of its melting point until its particles adhere to each other. It is known as solid state sintering. The clay particles sinter even before they actually begin to melt into a glassy state (vitrification). The production of powder metal components can be summarized in three steps; powder preparation, compaction and sintering.

Sintering is traditionally used for manufacturing ceramic objects, and has also found uses in such fields as powder metallurgy and synthesis of impurity doped luminescence phosphors. The source of power for solid-state processes is the change in free or chemical potential energy between the neck and the surface of the particle. This energy creates a transfer of material though the fastest means possible; if transfer were to take place from the particle volume or the grain boundary between particles then there would be particle reduction and pore destruction. The pore elimination occurs faster for a trial with many pores of uniform size and higher porosity where the boundary diffusion distance is smaller. Control of temperature is very important to the sintering the process, since grain-boundary diffusion and volume diffusion rely heavily upon temperature, the size and distribution of particles of the material, the materials composition, and often the sintering environment to be controlled. Through diffusion and other mass transport mechanisms, material from the particles is carried to the necks **(Fig.1)**, allowing them to grow as the particle bonding enters the *intermediate stage*. The intermediate stage of bonding is characterized by the pores

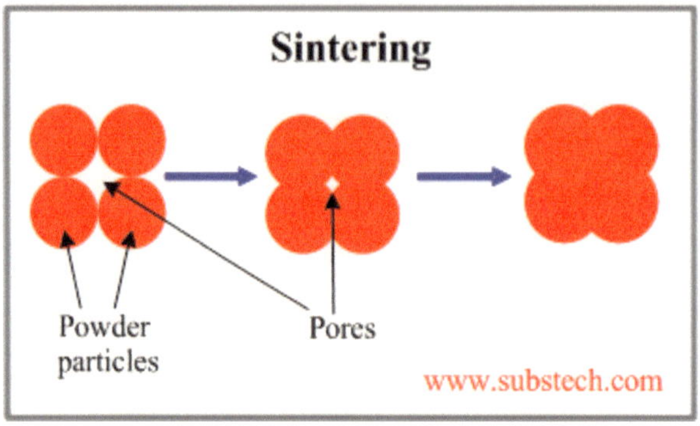

(a)

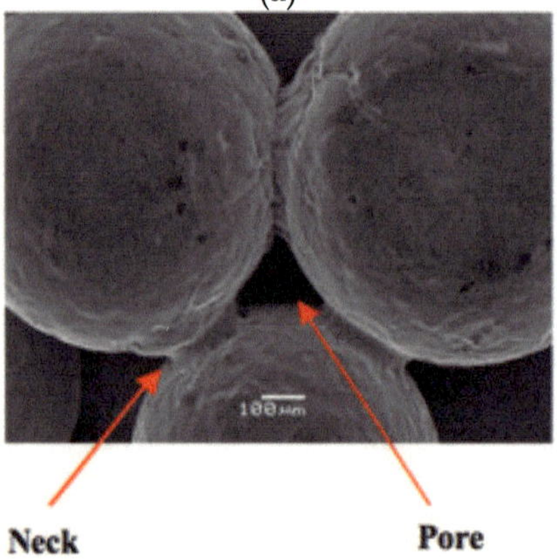

(b)

Fig. 1. (a) Process of sintering (b) The initial stage of the bonding occurs as small "necks" form between the particles.

beginning to round. As the mass transport continues, the pores will become even more rounded and some will appear to be isolated away from the grain boundaries of the particles. This is referred to as the *final stage* of bonding. The final step of the sintering process is to cool the bonded compact to a temperature at which it can be handled. This cooling is performed in an atmosphere that is no longer required to chemically react with the compact. The atmosphere in this stage of the process aids in the transport of the heat away from the compact and minimizes the re-oxidation of the compact during cooling. There are two types of sintering: with pressure (also known as hot pressing), and without pressure. Pressureless sintering is possible with graded metal-ceramic composites, with a nanoparticle sintering aid and bulk molding technology.

Luminescence phosphors owe their practical importance to their property of absorbing incident energy and converting it into visible radiations. This phenomenon, known as luminescence, is driven by electronic processes in the material due to the presence of trapping levels created by the presence of impurity atoms or lattice defects. *Solid-state diffusion* (SSD) reaction is the most popular method used in the synthesis of commercial luminescence phosphors as it is easily reproducible and amenable to large scale production. The products obtained yield a high luminescence efficiency. However, SSD has some disadvantages, such as (1) process complexity and energy-consuming (firing at high temperature, repetitive heat treatment, milling, and sieving), (2) inhomogeneous mixing and contamination by impurities, (3) product with irregularly shaped and aggregated particles unsuitable for screen brightness and high resolution. Deagglomeration of sintered phosphor chunks is quite cumbersome involving pulverizing, milling, sieving etc. As a result, many attempts have been carried out to find alternative methods for the preparation of phosphors. Superior display performance requires improvement in phosphors particle characteristics such as grain morphology and particle size on the luminescent intensity, efficiency, and resolution. Powders with optimal properties are obtained by different methods such as *chemical precipitation, the sol-gel, solution combustion, plasma chemical, hydrothermal, spray pyrolysis, microwave* etc. However, in most cases, high temperature (although lower than those used in SSD) sintering of samples prepared by these methods was often found to be essential as it increased their luminescence efficiencies due to improved crystallization and optimal incorporation of dopants in the host crystals.

2. Effects of sintering fluxes on morphology

$BaMgAl_{10}O_{17}:Eu^{2+}$ (BAM:Eu) phosphor is an important blue-emitting phosphor and has found widespread applications in plasma display panels (PDPS) and fluorescent lamps. The BAM phosphor powders synthesized with individual flux materials, such as AlF_3, NH_4F, LiF, and so on, have been found to exhibit different morphologies. Flux materials are usually compounds of alkali- or alkaline earth metals with lower melting temperatures than that of the host. In this study, $BaMgAl_{10}O_{17}:Eu^{2+}$ phosphor was prepared with fluxes by spray drying and post-treatment processes. The phosphor prepared with combination of KF and H_3BO_3 resulted in fairly uniform hexagonal plate-like morphology **(Fig. 2),** and the morphology as well as the plate size is actually in between those obtained by each of these fluxes. However, the phosphor prepared with the combination of KF and NaCl gives particles showing two distinct morphologies, including thin hexagonal plates and rounded particles **(Fig. 3).** These morphologies had appeared to be a mixture of the products

Fig. 2. SEM photographs of the BAM : Eu phosphor prepared with KF and H_3BO_3 **[1].**

Fig. 3. SEM photographs of the BAM : Eu phosphor prepared with KF and NaCl **[1].**

resulting from each individual of the two fluxes. These powders show different photoluminescence (PL) intensities. To sum up, by selecting distinct individual or binary fluxes, the morphologies, particle size, and the PL intensities of BAM phosphor can be controlled. Beside, both larger crystal size and appropriate aspect ratio play a crucial role on enhancing the luminescence of BAM phosphor.

3. Role of flux in calcination temperature

$CaAl_2O_4:Eu^{3+},R^+$ (R=Li, Na, K) phosphors were initially prepared by mixing stoichiometric amounts of $CaCO_3$, Al_2O_3 (A.R.), Eu_2O_3 (99.99%), and with or without one of Li_2CO_3, Na_2CO_3 or K_2CO_3 (A.R.) flux using solid state reaction technique at high temperature. Then a certain quantity of flux H_3BO_3 were added. The quantity of the flux H_3BO_3 is very crucial and dictates the calcination and reduction temperatures. The X-ray diffraction patterns of $CaAl_2O_4:Eu^{3+}, Li^+$ sample (Eu^{3+} and Li^+ were 3 mol.%) calcined at 1000, 1100, 1200 and 1300 °C for 4 h are shown in **Fig. 4**. After calcined at 1000 and 1100 °C, the precipitated precursors showed some characteristic peaks of Al_2O_3 and CaO besides the characteristic peaks of $CaAl_2O_4$. When the temperature was increased to 1200 °C, only the $CaAl_2O_4$ phase was detected (JCPDS card No. 23-1036), and no other products or starting materials were observed. The high intensity of the peaks reveals the high crystallinity of the synthesized

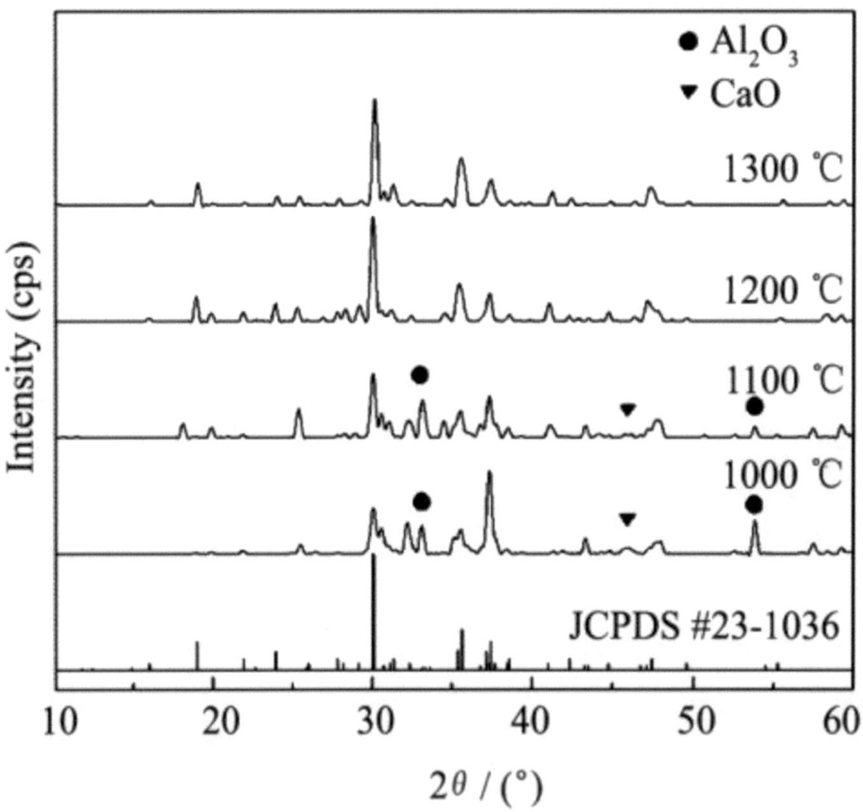

Fig. 4. XRD patterns of $CaAl_2O_4:Eu^{3+},Li^+$ phosphors sintered at different temperatures **[2]**.

powders. The sintering temperature was optimized to be 1200 °C. The luminescence intensity of $CaAl_2O_4:Eu^{3+}$ was significantly enhanced by co-doping with alkali metal ions, probably due to the charge compensation. Furthermore, the emission intensities were gradually enhanced when the radius of R^+ became smaller from K^+ to Li^+ ion. It was probably due to the difference of ionic radii which would give rise to the diversity of sub-lattice structure around the luminescent center ions. This fundamental work might be important in developing new luminescent devices applicable for tricolor lamps, light emitting diodes and other fields.

In Y_2O_3:Eu sintered at 700-1200^{0}C in air, Li_2CO_3 flux was found to: (i) enhance the crystalline growth, ii) improved the grain size slightly morphology from a plate like structure to spherical shape, and iii) improved significantly its PL sensitivity. The optimal red PL was achieved when the $Y_2O_3:Eu^{3+},Li^+$ phosphor was synthesized using 11 mol% Eu_2O_3 and 70 mol% Li_2CO_3 and sintered at 1,200°C for 5 h.

In $ZnWO_4$, the maximum PL intensity was obtained when the sintering temperature was 1,100°C. A significant decrease in PL intensity was measured when the phosphor was sintered at 1,200°C. This decrease was attributed to a change in the crystallinity of the phosphor, in which (020) ZnWO4 was the dominant crystalline phase. Empirically, the change in crystallinity alters the emission mechanisms of the phosphor. The growth of larger phosphor grains was another reason for the decrease in luminescence. Furthermore, the PL spectrum was broadened when the sintering temperature increased. Apparently, oxygen vacancies were involved in the phosphor crystal, and the bluish-green emission was related to electron transitions from the energy levels of the ionized oxygen vacancies to the phosphor valance band. The concentration of oxygen vacancies usually increases with an increase in sintering temperature and a broadened emission is thus observed.

Significantly, on UV illumination, a white-light phosphor could be achieved if the bluish-green $ZnWO_4$ and red $Y_2O_3:Eu^{3+},Li^+$ phosphors were blended.

4. Molten salt sintering

Combined co-precipitation with the molten salt method, a new technology for preparation of $Y_2O_3:Eu^{3+}$ and $YAG:Ce^{3+}$ phosphors was proposed with the controlled size and higher luminescent intensity. With rare earths oxide as raw materials, the molten salt method was compared with solid phase method. Some main principles for the selection of molten salt system were, i) the melting point should lower the temperature of phosphor preparation, ii) the difference of boiling point and melting point should be as wide as possible, and iii) the molten salt must not be hazardous to luminescent intensity. The best multiple molten salt system for $Y_2O_3:Eu^{3+}$ and $YAG:Ce^{3+}$ were $NaCl+S+Na_2CO_3$ and $Na_2SO_4+BaF_2$, respectively **[2]**. Molten salt sintering improved the crystal degree and configuration of phosphors, resulting in higher luminescent intensity. Using YCl_3 and $EuCl_3$ as raw material, the preparation of $Y_2O_3:Eu^{3+}$ precursor was investigated concerning some factors, such as temperature, complexing agent, precipitation agent and the dripping mode. The size of precursor was the smallest at pH=7 and the complexing agent could control the release velocity of rare ion effectively. With citric acid as a complexing agent, the size of precursor and sintering sample was the smallest and the luminescent intensity of sintering sample was the highest. Probably, the citric acid could complex effectively the earth ionic and buffer the pH during the precipitation process in the presence of ammonia and therefore enhanced the precursor density and activity. For the preparation of $Y_{3-x}Ce_xAl_5O_{12}$ (YAG:Ce^{3+}) precursor by co-precipitation, the optimal process condition were: the concentration of salt was as low as about 0.05M, precipitation agent was NH_4HCO_3, pH=8, temperature=90℃, and adverse dripping mode was preferred. Because the precursor was a sol mixture of $Y_2(CO_3)$,nH_2O and $NH_4AlO(OH)HCO_3$, it was easy to agglomerate after drying. It was found that the agglomeration problem could be solved by adding active carbon before precipitation. For active carbon, its numerous capillary frameworks might disconnect the sol effectively and its incomplete combustion in sintering was helpful for the deoxidization from Ce^{4+} to Ce^{3+}. Through adjusting components of multiple molten salt and mole ratio of the molten salt to precursor, the optimal sintering conditions for preparation of Y_2O_3:Eu^{3+} and YAG:Ce^{3+} were obtained. The samples were sphere-like particles whose average size were 1~3μm and 3~5μm and luminescent intensity were 11% and 8% better than commercial phosphor respectively. The results for different sintering temperature indicated that molten salt could reduce activation energy of phosphor like a kind of catalyst, leading to lower sintering temperature than solid phase method. The formation of sphere-like particles might be owing to the surface tension difference between liquid molten salt and phosphor, and the existence of double layer insured the dispersion of particles. The liquid molten salt provided the stable high temperature field and liquid environment and promoted the crystal degree, resulted in the increased luminescent intensity. In addition, the mole ratio of Y:Al:Ce was investigated for increasing luminescent intensity of YAG:Ce^{3+}. Luminescent intensity of sample was enhanced evidently when the mole ratio of Y:Al:Ce was reduced from 2.94:5:0.06 to 2.90:5:0.06 in $Y_{3-x}Ce_xAl_5O_{12}$. A little lack of Y in crystal lattice might help to increase luminescent intensity, which coincided with the theory of radiation from crystal lattice defect. The structure of Y_2O_3:Eu^{3+} and YAG:Ce^{3+} was body-centered cubic structure and yttrium aluminum garnet structure respectively showing that the molten salt did not enter into the crystal lattice of phosphor. Compared with the traditional solid phase method, the new technology can obtain the controlled size and higher luminescent intensity phosphor through only one sintering process, avoiding comminuting process required in solid phase method. In addition, it is a new energy-saving process with lower sintering temperature and has a potential application in preparation of phosphor with excellent performance.

5. Thermal stability against sintering and crystal structure

The thermal stability of $BaAl_2Si_2O_8$:Eu^{2+} (BAS:Eu^{2+}) phosphor used in PDP was found to depend on its polymorph property - hexagonal and monoclinic crystal structure. The monoclinic BAS:Eu^{2+} when baked at 500 °C in air for 30 min, showed the same PL intensity as the fresh one, whereas the baked hexagonal one lost its PL intensity significantly **(Fig.5)**. Electron spin resonance studies on Eu^{2+} and Rietveld refinement showed that the difference of thermal stability between hexagonal and monoclinic BAS:Eu^{2+} could be ascribed to both the crystal structure of host materials and the average inter-atomic distances between the Eu^{2+} ion and oxygen which plays the key role of shield for Eu^{2+} ions against an oxidation atmosphere.

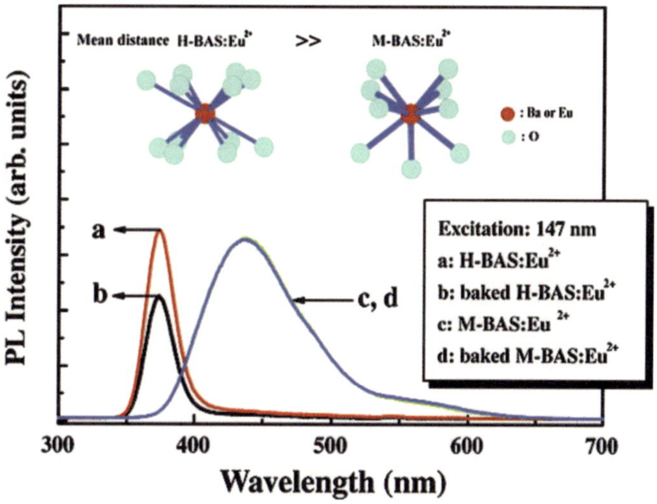

Fig. 5. Emission spectra of $BaAl_2Si_2O_8:Eu^{2+}$ ($BAS:Eu^{2+}$) phosphor under VUV (147 nm) excitation **[3]**.

6. Effect of sintering on photo and thermo-luminescence in $CaSO_4$: Dy phosphor

Sieving *before* the high-temperature sintering treatment has successfully eliminated particle agglomeration during subsequent sintering, and has further enhanced its thermoluminescence (TL) sensitivity to γ-rays. The reduction in TL sensitivity of higher sized grains observed following the procedure of sieving *after* sintering has also more or less vanished. Maximum TL sensitivity is seen after sintering around 700°C, whereas maximum PL sensitivity is seen after sintering around 325°C. While the observed increase in TL sensitivity (by 30%) with increasing sintering temperature in the range 325-700°C is explained on the basis of diffusion of Dy^{3+} ions from the surface to the whole volume of the grains (0-75 μm), the drastic decrease (by a factor of 3) in PL sensitivity with increasing sintering temperature is explained on the basis of change in the Dy^{3+} environment on the grain surface perhaps due to oxygen incorporation **(Fig.6)**. Washing with water and acetone, which affect mainly the surface traps, enhances the PL sensitivity of $CaSO_4$:Dy slightly; however, it does not influence TL sensitivity very significantly. Grinding reduces PL in general, but no such trend was noticed in TSL which supports the conclusion that PL originates mainly from surface traps since grinding affects mainly the grain surface. However, the sharp reduction in TL and PL sensitivities observed at 400°C indicates that an unusual process takes place near that sintering temperature. TG-DTA (thermogravimetric and differential thermal analysis) data indicate that dysprosium sulphate dopant in $CaSO_4$ which are hydrated at RT become anhydrous at 400^{0}C and the water molecules released possibly damage the crystal lattice which get restored at higher sintering temperatures. As a result, $CaSO_4$:Dy annealed at 400^{0}C show a slightly reduced TL while those annealed at 300^{0}C or at 700^{0}C do not show any such reduction. The water molecules released at 400^{0}C possibly displace Dy from its lattice site causing a reduction in TL. No such reduction in TLis observed on annealing at 300^{0}C since the water molecules do not get dislodged from the lattice. Annealing at 700^{0}C possibly restore the Dy in its original lattice sites and hence restore the TL sensitivity.

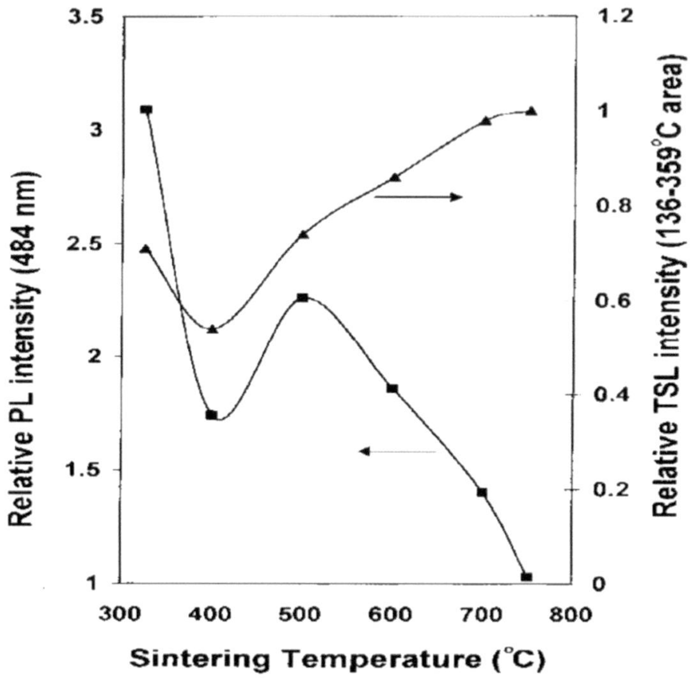

Fig. 6. Dependence of the Thermostimulated luminescence (TSL, 136–359°C area) as well as 484 nm PL emission (λ_{exi} = 350 nm) intensities of unwashed as prepared samples (<75μm grain size and sieving carried out before sintering) of $CaSO_4$:Dy on the sintering temperature **[4]**.

7. Precipitation and sintering – TG/DTA studies

Eu^{2+} activated long lasting $Sr_4Al_{14}O_{25}$ nano sized phosphor synthesized by precipitation method is a revealing study. $Al(NO_3)_3 \cdot 9H_2O$, $Sr(NO_3)_2$, $(NH_4)_2CO_3$, $Eu(NO_3)_3$ and $Dy(NO_3)_3$, all in analytical purity, were the starting materials. The $(NH_4)_2CO_3$ solution was added in droplets to produce a white precursor. After drying at 120°C for 24 h, the final luminescent powders were obtained by calcinating the dried precursor at different temperature from 1200 to 1300 °C in a reducing environment of 5%H_2 + 95%N_2. Thermogravimetric (TG) and differential thermal analysis (DTA) studies revealed that the endothermic peaks A and B in DSC (differential scanning calorimetry) curve can be assigned to the dehydration of $Al(OH)_3$ and the other two endothermic peaks C and D can be assigned to the decomposition of $SrCO_3$, since the dehydration usually occurs at a relatively low-temperature. The TG curve also shows a good accordance with the DSC result. It can be seen from the TG curve that the weight loss (WL) after the dehydration is 20.6%, very close to the theoretical value 22.5%. And the weight loss after the decomposition of the carbonate is 10.2%, which is also very close to the theoretical value 10.5%.It is obvious that the TG curve reveals no weight loss after 1100°C, indicating that the two exothermic peaks E and F may result from the chemical reaction between the calcined oxides at high-temperature. XRD data reveal the formation of the orthorhombic aluminate could take several steps, and the low-temperature products after the dehydration and decomposition, say Al_2O_3 and SrO, will react spontaneously to formthe monoclinic $SrAl_2O_4$ around 1200°C. When the calcination temperature increased to

1250°C, some peaks of new phases, $SrAl_{12}O_{19}$ and $Sr_4Al_{14}O_{25}$, are identified from the XRD pattern, showing the complex of the phase transition in this temperature region. However, with the calcination temperature increasing to 1300°C, almost all the diffraction peaks can be indexed to the orthorhombic $Sr_4Al_{14}O_{25}$ phase when referring to PDF 74-1810, implying that $SrAl_2O_4$ and $SrAl_{12}O_{19}$ can be viewed as the intermediate phases during the process of $Sr_4Al_{14}O_{25}:Eu^{2+},Dy^{3+}$ preparation. $Sr_4Al_{14}O_{25}:Eu^{2+},Dy^{3+}$ phosphor exhibited better afterglow property than the $SrAl_2O_4:Eu^{2+},Dy^{3+}$ phosphor due to a deeper trap level and a higher trap concentration formed in the host material. When compared to the powder obtained in conventional method, the nano sized powders revealed a blue shift in emission spectrum due to the decrease in grain size **(Figs.7-10)**.

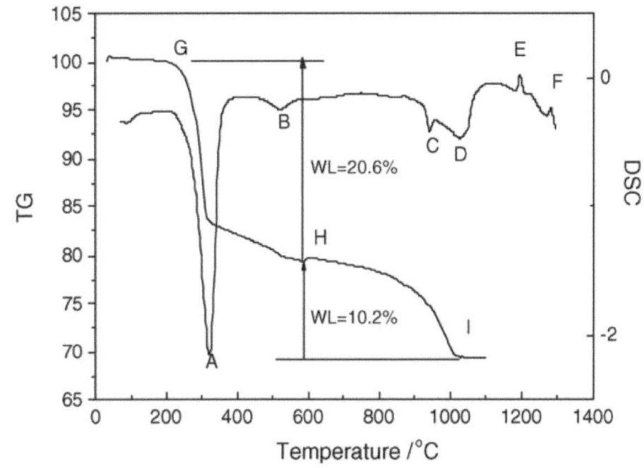

Fig. 7. DSC and TG profiles of precursors, showing the dehydration/decomposition reaction process during calcinations. Note- **DTA** detects any change in all categories of materials; **DSC** determines the temperature and heat of transformation **[5].**

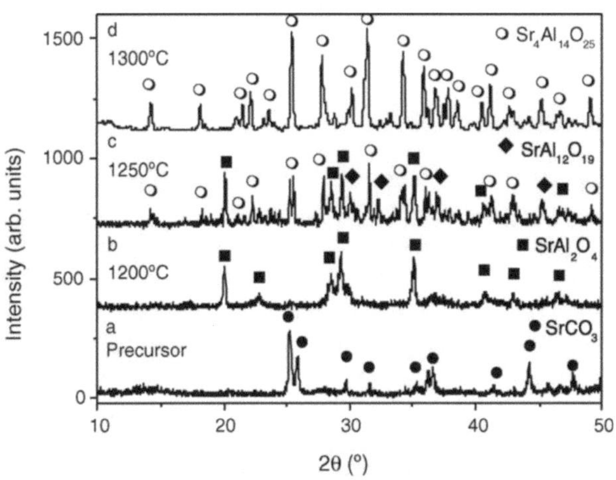

Fig. 8. XRD patterns of the precursors calcined at different temperature: (a) raw powder, (b) 1200 °C, (c) 1250°C and (d) 1300 °C, indicating the phase transformation during the calcinations **[5].**

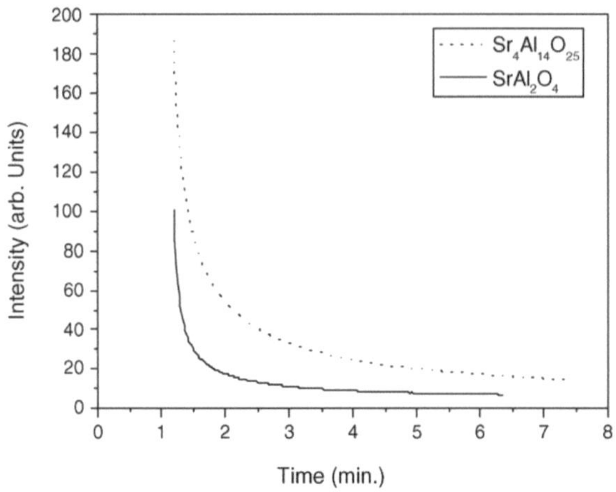

Fig. 9. Afterglow property of the nano scaled phosphors **[5].**

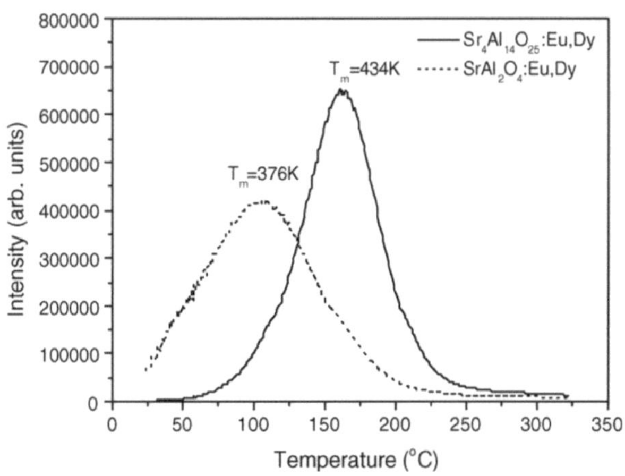

Fig. 10. TL curve for nano scaled $SrAl_2O_4$:Eu,Dy and $Sr_4Al_{14}O_{25}$:Eu,Dy phosphors **[5].**

8. Pelletization and sintering

The change in TL properties due to compression and subsequent heating are well studied in LiF TL dosimetry phosphor. Changes in TL at contacts and in the vicinity of ores in carbonate host rocks and near faults have been studied. In the history of formation of terrestrial planets, the collision of solid bodies is central at an early stage of their evolution. Analysis of shock processes in natural minerals play an important role in geology. Influence of thermal history, deformation and stresses associated with major tectonic dislocations, meteoritic craters and underground nuclear explosions on the TL of several natural minerals have been discussed earlier. Shock-induced TL in Oligoclase, Quartz and Calcite have been studied in this regard. An enhancement in TL intensity and shifting of the glow peaks to relatively higher temperatures in pelletized kyanite samples when compared to single crystals which were attributed to the particle nature of the phosphor and/or pressure-induced defects. A shift in the temperature of glow peaks of γ- and X-ray irradiated quartz with pressure was attributed to an electron trap that is getting shallower with increasing pressure.

During the preparation of sintered pellets for applications in radiation dosimetry, Mg-doped LiF phosphors compressed with a pressure of 100 atm (10.13MPa) exhibited a significant glow curve change and a TL sensitivity decrease by a factor of about 10–16. Phosphors finely crushed in a mortar did not exhibit such a behavior. These changes are reportedly caused due to lattice deformations generated by the static pressure. The deformations quench the original TL peak near 200⁰C but produced new traps giving rise to peaks between 230 and 400⁰C. Pressure applied to the irradiated phosphor empty the filled traps and create new traps. However, the original glow curve shape and TL sensitivity were restored when the LiF phosphor was heated to a temperature higher than 200⁰C even for a shorter period. The sintered pellets were found reusable after a 400⁰C, 1h+ 80⁰C, 4 h annealing treatment. Subsequent investigations revealed that the pressure-induced defomation disappears fully only when the annealing temperatures are above 350°C; annealing temperatures higher than 100°C cause further complicated variations in the glow curve shape of LiF-based phosphors. Plastic deformation of LiF:Mg,Ti followed by irradiation decreased the TL intensities, with 200⁰C peak decreasing more drastically than 100⁰C peak. The latter reported that it is more likely that new TL traps are created by dislocation intersections which then compete with the previously existing traps. Plastic deformation after exposure to ionizing radiation increase the F band (250 nm) absorption in LiF:Mg,Ti, while decreasing the absorptions at 310 and 380 nm which are related to TL traps. The rate of change with deformation of the three bands indicates that F-centres are created when 310-and 380-nm absorption centres are destroyed. These results were interpreted in terms of the removal of F-centres from the defect complexes causing the 310- and 380-nm absorption bands.

Studies by the author **[6]** showed pressure-induced changes in the luminescent properties of gypsum ($CaSO_4.2H_2O$), calcite ($CaCO_3$), $CaSO_4$:Dy and anhydrite ($CaSO_4$). A systematic reduction in the TL as well as PL (in the case of $CaSO_4$:Dy) sensitivity with the pressure applied is seen with the former three materials **(Fig.11).** However, an exactly reverse trend,

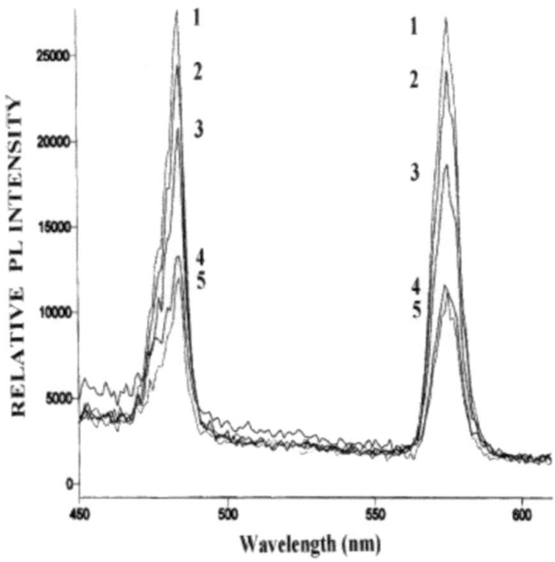

Fig. 11. Dependence of PL emission spectra (λ_{exi} = 350 nm) of unirradiated $CaSO_4$:Dy on the pressure applied (in MPa), 1:0, 2:0.7, 3:1.05, 4:1.23, 5:1.4.**[6].**

i.e., a systematic increase in the TL sensitivity with the pressure was witnessed in the case of anhydrite. While the application of pressure (7MPa) reduced the TL sensitivity of $CaSO_4$:Dy to 43% of its initial value, a 700°C, 1 h anneal could restore its value to only 51%. This shows that the pressure-induced changes are more or less permanent, unlike the case reported with LiF. No change in the XRD data was seen in all these samples on the application of pressure (7MPa) which shows that the changes observed in TL and PL should be attributed to the damage (in the cases of calcite, gypsum and $CaSO_4$:Dy) / creation (in the case of anhydrite) of traps / luminescent centres rather than to the damage to their crystal structure. Pressure applied before or after the irradiation produced no difference in $CaSO_4$:Dy. However, in γ-ray exposed anhydrite and gypsum, pressure-induced changes in the TL sensitivity were quite complex though no change in its glow curve structure was seen. Results showed that in anhydrite, there is a competition between two opposing pressure-induced phenomena—one is radiation damage to the filled traps and the other is the increased luminescence efficiency.

9. Activator diffusion during sintering

$CaSO_4$ can be pelletized only with suitable additives. However, Dy^{3+} could be incorporated in $CaSO_4$ lattice only by recrystallization in H_2SO_4 medium. A recent study **[7]** has shown that Li coactivator could be successfully introduced in $CaSO_4$:Dy/Tm at a concentration of 0.06% during a subsequent step of cold pressing at RT and sintering of the pellets at 700°C. A number of alternative lithium compounds have been used in this role and they include Li2CO3, LiCl, LiF, Li2B4O7 and Li2SO4. Addition of lithium significantly shifted the major TL peak of $CaSO_4$:Dy/Tm from 220°C down to 120°C. The fact that similar results were obtained earlier with Li coactivator added during initial crystal growth show that certain impurities can easily diffuse into host crystal during pelletization followed by sintering. In this case Dy^{3+} incorporation in Ca^{2+} sites creates cation vacancies into which Li^+ co-dopant

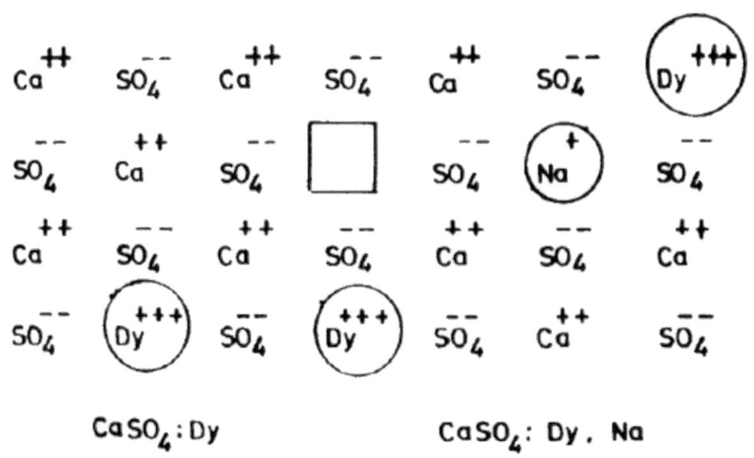

Fig. 12. Pictorial representation of $CaSO_4$:Dy and $CaSO_4$:Dy,Na. Trivalent Dy could be incorporated into $CaSO_4$ lattice only by recrystallization in H_2SO_4 medium. Monovalent co-dopants such as Na^+ or Li^+ could, however, be incorporated into CaSO4:Dy lattice by cold pressing followed by sintering [8].

could easily diffuse in. Such a technique could offer an easy recipe for the preparation of doped samples. Unlike the case with Li co-dopant, Dy^{3+} or Tm^{3+} could not be doped by this (pelletization followed by sintering) technique. This indicates that incorporation of trivalent dopants require greater activation energy than monovalent co-dopants. This is understandable since trivalent dopants in a divalent host require activation energy for the creation of cation vacancies **(Fig.12).**

10. Effect of atmosphere during sintering

Phosphors tend to get oxidised during sintering in air. For example $CaSO_4$ gets oxidized to CaO at temperatures above 800^{0}C in air. As a result the luminescence efficiency of $CaSO_4$:Dy gets reduced at high sintering temperatures. In addition there are certain activators such as Mn, Cu etc which tend to get oxidised during sintering in air. Certain other activators such Eu and Ce tend to exhibit dual valence state. Quite often during sintering, reducing atmosphere is essential to prevent Eu^{2+} and Ce^{3+} from getting oxidized to Eu^{3+} and Ce^{4+}, respectively. While phosphors containing Eu^{2+} activator (eg., BAM:Eu^{2+}) gives intense blue emission, those containing Eu^{3+} activator (eg., Y_2O_3:Eu^{3+}) gives intense red emission. So depending on the phosphor and activator, sintering should be carried out either in a reducing atmosphere or in air. Reducing atmospheres are usually obtained with H_2/N_2 mixture. Alternately, the phosphors to be sintered are covered with carbon powder in closed crucibles so as to create reducing CO atmosphere when burnt with limited oxygen.

CaS:Eu^{2+} red-emitting phosphors particles, were prepared by the precipitation method with calcium acetate and Na_2S as starting materials, followed by sintering in the atmosphere over the mixture of sulfur powder, Na_2CO_3, and carbon-containing compounds such as tartaric acid, citric avid, glucose, and cane sugar. CaS:Eu^{2+} particles without additive show inhomogeneous, rough and aggregation with the size of 75–125 nm, but the spherical particles with mean size of about 110 nm were obtained by adding carbon-containing compounds **(Fig.13).** Compared with phosphor without additive, the addition of carbon-containing materials induced a remarkable increase of PL, in the order of cane sugar, glucose, citric acid, and tartaric acid. This enhancement is due to the improvement of crystallinity, particle morphology and size distribution of the samples by adding carbon-containing additive.

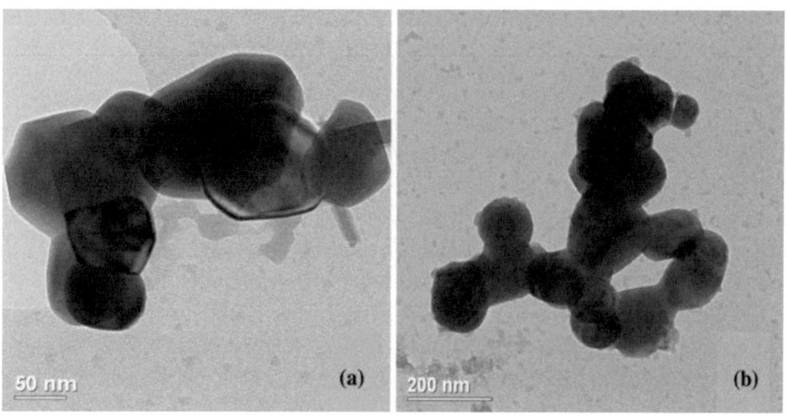

Fig. 13. Transmission electron (TEM) micrographs of CaS:Eu^{2+} obtained by the precipitation method without additive (a) and with cane sugar additive (b) **[9].**

Yttria-stabilized zirconia (YSZ) is a zirconium-oxide based ceramic, in which the particular crystal structure of zirconium oxide is made stable at room temperature by an addition of yttrium oxide. These oxides are commonly called "zirconia" (ZrO_2) and "yttria" (Y_2O_3), hence the name. The addition of yttria to pure zirconia replaces some of the Zr^{4+} ions in the zirconia lattice with Y^{3+} ions. This produces oxygen vacancies, as three O^{2-} ions replace four O^{2-} ions. It also permits yttrium stabilized zirconia to conduct O^{2-} ions, provided there is sufficient vacancy site mobility, a property that increases with temperature. This ability to conduct O^{2-} ions makes yttria-stabilized zirconia well suited to use in solid oxide fuel cells, although it requires that they operate at high enough temperatures.

11. Effect of calcination temperature

Rare-earth ions can easily replace yttrium ions because their properties are similar. All peaks in XRD pattern of $YInGe_2O_7$ doped with 5 mol.% Eu^{3+} and calcined at various temperatures from 1100^{0}C to 1400^{0}C in air for 10 h could be attributed to the monoclinic $YInGe_2O_7$ phase. Trivalent europium ions (94.7 pm) were introduced to substitute trivalent yttrium ions (90 pm) in the (Y, Eu) $InGe_2O_7$ system. The variations are almost the same for Eu^{3+} and Y^{3+} ion radii, both easily forming a solid solution. Additionally, there were no charge compensation issues when Eu^{3+} ions substituted Y^{3+} ions in the $YInGe_2O_7$ lattice because both have the same valence. The full-width at half-maximum (FWHM) of these peaks seemed to decrease and the crystallinity of $YInGe_2O_7{:}Eu^{3+}$ became better with an increase of the calcination temperature to 1200^{0}C. From Fig.14 it is observed that the emission intensity increases with calcination temperature, with a maximum value at1200^{0}C.

This was caused by the reduction of nonradiative recombination effects, i.e., quenching sites and surface defects trapped by increased crystallinity and decreased defects in the crystal and is in agreement with the results of the XRD analysis, as optimum crystallinity was presented at 1200°C. Higher calcination temperature enhances atomic mobility and causes grain growth, resulting in better crystallinity. When calcination temperature increased further, the emission intensity decreased significantly as shown in **Fig.14.** When the calcination temperature was higher than 1200^{0}C, the second phase of In_2O_3 (JCPD no. 06-0416) was observed in the XRD pattern. The amount of the second phase increased with calcination temperature. The second phase might be produced because the increased calcination temperature leads to a nonstoichiometric system because the melting point of In_2O_3 (850^{0}C) is lower than that of Y_2O_3 (2410^{0}C) and GeO_2 (1086^0 C).

For the preparation of $ZnWO_4$ phosphor, ZnO powder used as the source material was mixed with WO_3 (99.9%) at different concentrations (10–60 mol%). The mixed powders were blended with deionized water then milled for 24 h. Subsequently, the solution was dried in an oven and sintered at 700–1,200°C for 1–8 h. Finally, the synthesized powders were ground, and the $ZnWO_4$ phosphor was prepared. The crystallization of the $ZnWO_4$ phosphor was improved by increasing the sintering temperature from 800°C to 1,200°C as seen in **Fig.15.** Meanwhile, the decreased FWHMs in the XRD patterns show that the grain size of the phosphor increased with the sintering temperature. Improvement in crystallization is normally accompanied by an increase in phosphor particle size. The particle sizes of $ZnWO_4$ phosphor were approximately 16.8, 17.9, 20.2, 19.6, 19.6, and 26.5 nm for sintering temperatures increasing in steps of 100°C from 700°C to 1,200°C, respectively. SEM result is consistent with the results of XRD analysis. Significant changes in grain shape and size were observed with an increase in the sintering time. Optimal crystallization was realized in the case of the $ZnWO_4$ phosphor synthesized using 50 mol% WO_3 at 1,100°C for 3 h. The maximum emission intensity was achieved when the phosphor exhibited optimal crystallization.

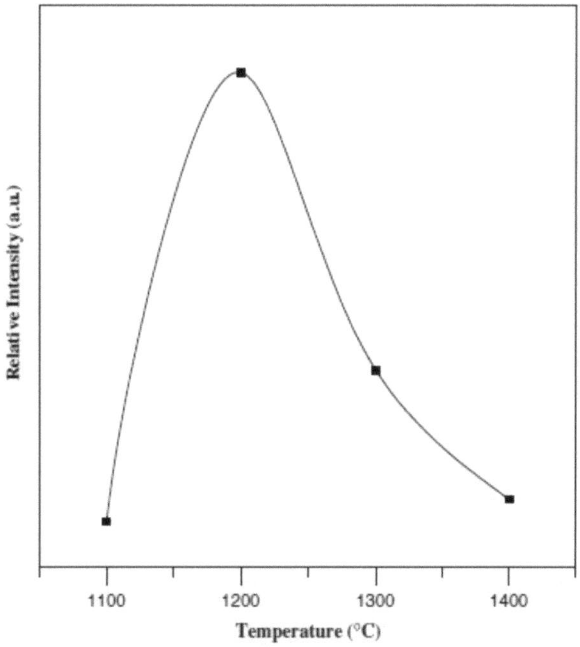

Fig. 14. The relative emission intensity versus the calcined temperature of $YInGe_2O_7$:5 mole% Eu^{3+} under an excitation of 393 nm. The signals were detected at 611 nm **[10]**.

12. Phase change during sintering

Luminescence properties (both PL and TL) of Tricalcium phosphate (TCP) are very sensitive to its crystal phase (α/β). TCP crystals, in both α and β phases, were synthesized through two different routes, viz. wet precipitation and high temperature solid state reaction **[12]**. The doping was done during the synthesis using suitable compounds of Dy and Eu. In the wet precipitation method used, the wet reaction is carried out using calcium nitrate and diammonium hydrogen phosphate in an ammoniated solution. The precipitation of tricalcium phosphate occurs through the chemical reaction $3Ca(NO_3)_2 + 2(NH_4)_2HPO_4 + 2NH_4OH \rightarrow Ca_3(PO_4)_2 + 6NH_4NO_3 + 2H_2O$. the supernatant liquid was decanted to collect the precipitate. It was then centrifuged thrice using distilled water and finely filtered. The filtrate was dried at 100°C in a hot air oven overnight and then calcined at 300°C in a muffle furnace for 3 h to remove any traces of other compounds. The calcined material was ground to form fine powder and graded using standard sieves. It was then sintered at high temperatures for 2 h in a programmable furnace to obtain the required phase (900°C for β-TCP and 1300°C for α-TCP). Various samples were prepared using dysprosium and europium as dopant. The doping was done by adding oxides of the dopant elements (dysprosium and europium) dissolved in minimum quantity of dilute nitric acid. The solid state synthesis of tricalcium phosphate was done through a high temperature firing of the

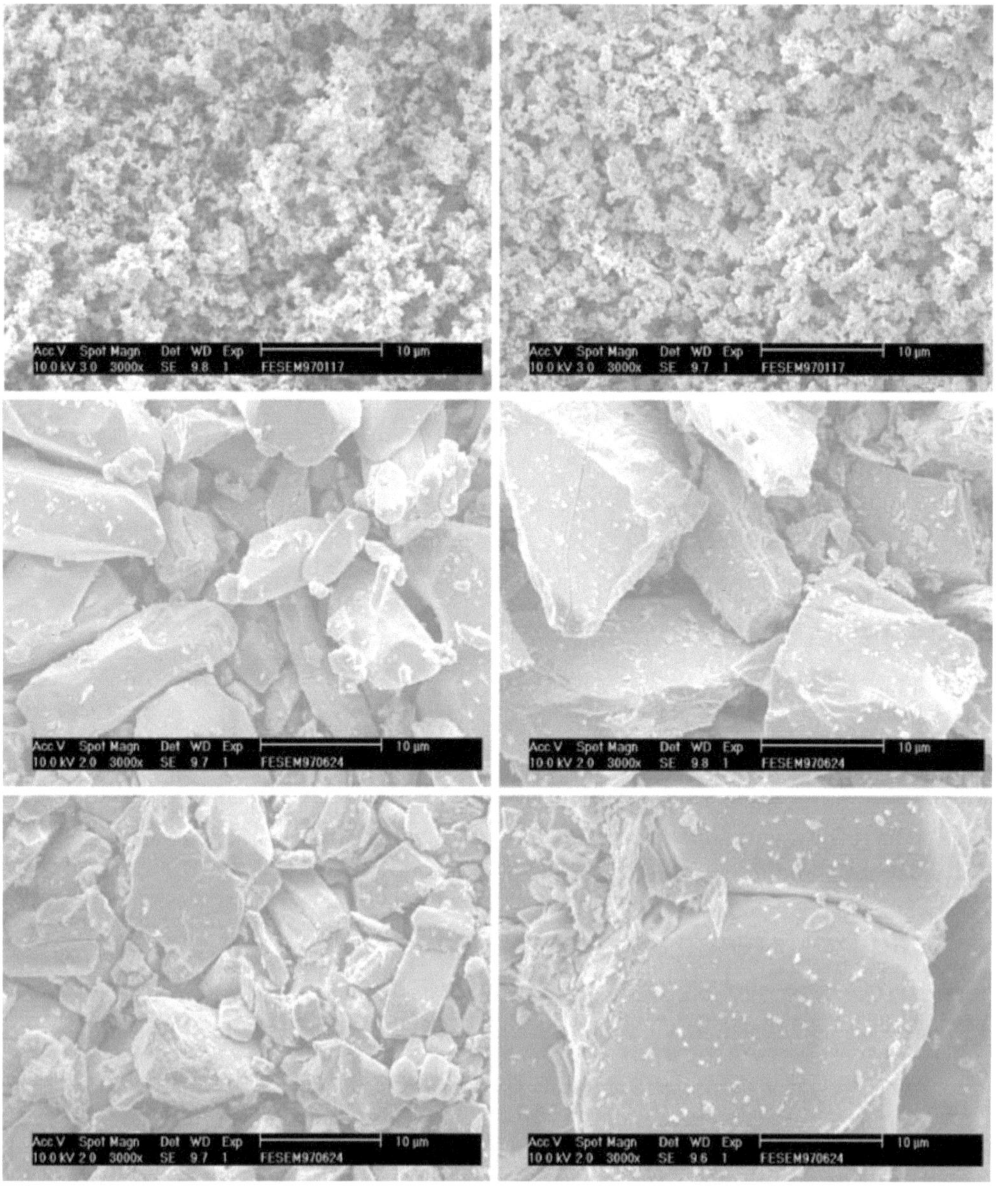

Fig. 15. Scanning electron microscope images of the $ZnWO_4$ phosphor prepared under different synthesis conditions; WO_3 concentrations at (a) 10 and (b) 50 mol% and sintered at 700°C for 3 h; WO_3 concentration at 50 wt% and sintered at (c) 1,100°C and (d) 1,200°C for 3 h; WO_3 concentration at 50 wt% and sintered at 1,100°C for (e) 1 and (f) 8 h **[11].**

powder mixture of calcium oxide (CaO) and dicalcium phosphate ($CaHPO_4$). The reaction governing the process is $CaO + 2CaHPO_4 \rightarrow Ca_3(PO_4)_2 + H_2O$. The dopant, in powder form, in appropriate amount was added to this, again mixed thoroughly for 1 h, and then transferred to a porcelain crucible. The powder was heated at about 300°C, then mixed for 1 h, and annealed at 1100°C for 2 h. The resulting compound was crushed to powder from and graded using standard sieves. The samples prepared by wet precipitation annealed at 900°C) and by solid state sintering (annealed at 1100°C) techniques gave the spectra which match perfectly with that of β-TCP (Whitlockite mineral phase, JCPDS File No. 09-0169). The XRD pattern of sample prepared by wet precipitation method and annealed at 1300°C matched with that of α-TCP (JCPDS file number 09-0348). Though the TL efficiency of α-TCP is more compared to β-TCP, the former is less suited for TL applications as its peak temperature is on the lower side, which indicates high fading rate. An interesting outcome of the TL studies on doped tricalcium phosphate is that the phase transition from β to α during annealing could easily be identified through emission parameters. Dy is found to be an efficient dopant in β- TCP matrix compared to europium for maximum TL efficiency.

Aluminum oxide, commonly referred to as alumina, possesses strong ionic interatomic bonding giving rise to its desirable material characteristics. It can exist in several crystalline phases which all revert to the most stable hexagonal alpha phase at elevated temperatures. This is the phase of particular interest for structural applications. Alpha phase alumina is the strongest and stiffest of the oxide ceramics. Its high hardness, excellent dielectric properties, refractoriness and good thermal properties make it the material of choice for a wide range of applications. In a recent study, the α-alumina samples were prepared using a commercial ultra-pure γ-alumina powder obtained by thermal decomposition of ammonium alum ($NH_4Al(SO_4)_2$, $12H_2O$). The study of Zr, Th and Ca doping is performed through the impregnation of the γ-alumina powder by an alcoholic solution of zirconium or calcium chloride or of thorium nitrate **[13].** A further thermal treatment allows decomposition of the solvent and the salts (drying at 100°C for 24 h) and diffusion of the doping species inside the host material (600°C for 24 h). The doped γ-alumina is then transformed into doped α-alumina by a calcination at 1450°C for 2 h under a pure gas flow (O_2 or Ar+2% H_2). Th^{4+} and Zr^{4+} give rise to the same peaks on the glow curves, but the intensity is higher for Th^{4+}. For both dopants, the doping promotes shrinkage in the case of very low concentrations, in fact, until the cation integrates into the host material.

Carbon doped α-Al_2O_3 is a well known optically stimulated luminescence material used in radiation dosimetry **[14].** α-Al_2O_3 doped either with Tb^{3+} or Tm^{3+} has been prepared by combustion synthesis techniques for TL ionizing radiation dosimetry applications. In this method, the reactants (aluminum nitrate, urea and terbium or thulium nitrate) are ignited in a muffle furnace at temperatures as low as 500 °C. This synthesis route is an alternative technique to the conventional fabrication methods of materials based on α-Al_2O_3 (Czochralsky, Vernuil), where high melting temperatures and reducing atmospheres are required. After combustion, the samples were annealed at temperatures ranging from 1000 to 1400 °C for 4 h in order to obtain the pure α-phase structure and were then irradiated with a Co-60 gamma radiation source. The annealed samples present a well defined TL glow peak with a maximum at approximately 200 °C and linear TL response in the dose range 0.5–5 Gy. It was observed that a 0.1 mol% concentration of Tb^{3+} or Tm^{3+} and annealing at 1400 °C optimize the TL sensitivity. The highest sensitivity was found for Tm^{3+} doped samples which were approximately 25 times more sensitive than Tb^{3+} doped samples. These results strongly suggest that combustion synthesis is a suitable technique to prepare doped aluminum oxide material and that Tm^{3+} doped α-Al_2O_3 is a potential material for TL radiation dosimetry **[15].**

13. Sintering enhances thin film electroluminescence

$(Zn_xCd_{1-x})S$ phosphors are of considerable interest because of their use as window material in solar cells. While powder phosphors were prepared by heating at 900ºC in argon atmosphere, thin films (15-25 μm) were obtained by painting uniformly the mixture of phosphor powder along with 30% of $CdCl_2$ and 65% propylene glycol on SnO_2 substrate, drying in air at 120ºC for 2h and then sintering at 625ºC in N_2 atmosphere for 30 min. $CdCl_2$ is used as a sintering aid in the preparation of thin films of $(Zn_xCd_{1-x})S$. During sintering Zn is gradually replaced by Cd through the reaction $ZnS + CdCl_2 \rightarrow ZnCl_2 + CdS$. This is confirmed by the red shift in the optical absorption edge of thin film when compared to that of powder phosphor. In sintered films new centres are created which increases the luminescence efficiency and emission occurs in green, yellow-orange and red regions as against only yellow-orange emission in (Zn,Cd)S:Mn,Sm phosphors. While the 525 nm emission band was interpreted as due to free to bound transition from conduction band and acceptor band, the 625 nm emission band was attributed to the radiative recombination of electrons from the donor and acceptor levels. Excitation from Sm^{3+} ion can be transferred to donor-acceptor pair which leads to enhanced emission at 625 nm. In addition, new traps (donor levels) are created in ZnS type phosphors containing Cl^- as co-activator when Zn was gradually replaced by Cd. Moreover, during sintering, Zn vacancies (V_{Zn}) are created which act as deep acceptors.

14. Sintering in vacuum

While annealing Zr_2O in air at 1300ºC was accompanied with pronounced phase changes, phase composition remained unchanged even with prolonged holding at 1300ºC in vacuum. Vacuum sintering of powders with complex morphology at a high temperature thus allowed obtaining a dense ceramic based on Zr_2O_3 and Y_2O_3 with no monoclinic phase, despite the strong recrystallization of grains of the tetragonal phase.

15. Combustion synthesis

Self-propagating synthesis (SHS), also known as combustion synthesis, is a process that involves a reaction, which is sufficiently exothermic to sustain itself. This has led to a low cost, energhetically efficient method for the development of advanced materials. **Solution combustion synthesis (SCS)** is a promising method to prepare high-purity, small-sized and spherical particle phosphors because the starting raw materials are homogeneously mixed in liquid phases, and the high temperature generated instantly by exothermic reaction can volatilize low boiling point impurities leading to purer products. In addition, SCS results in products with narrow particle distribution because of the decrease in reaction time (a few seconds during the combustion reaction) **[16].** The mechanism of nanoparticle forming in SCS is shown in **Fig.16.** When heated rapidly at 550º C, the solution containing stoichiometric amount of redox mixture boils, dehydrate, followed by decomposition generating combustible gases. The volatile combustible gases ignite and burn with a flame. The large amount of escaping gases dissipates heat and thereby prevents the material from sintering and thus provides conditions for formation of nanocrystalline phase. Also, as the gases escape they leave voluminous, foaming and crystalline fine powder occupying the entire volume of the container and have no chance of forming agglomerations unlike in the other conventional processes. Therefore, in combustion synthesis, instantaneous and *in situ* very high temperature, combined with release of large volume of volatiles from liquid mixture results in production of nanoparticles. The photographs in **Figs.17a and b** show the blue luminescence emission of as prepared (a) $BaMgAl_{11}O_{17}:Eu^{2+}$ and (b) $BaMgAl_{11}O_{17}:Mn^{2+}$ by solution combustion synthesis under UV illumination.

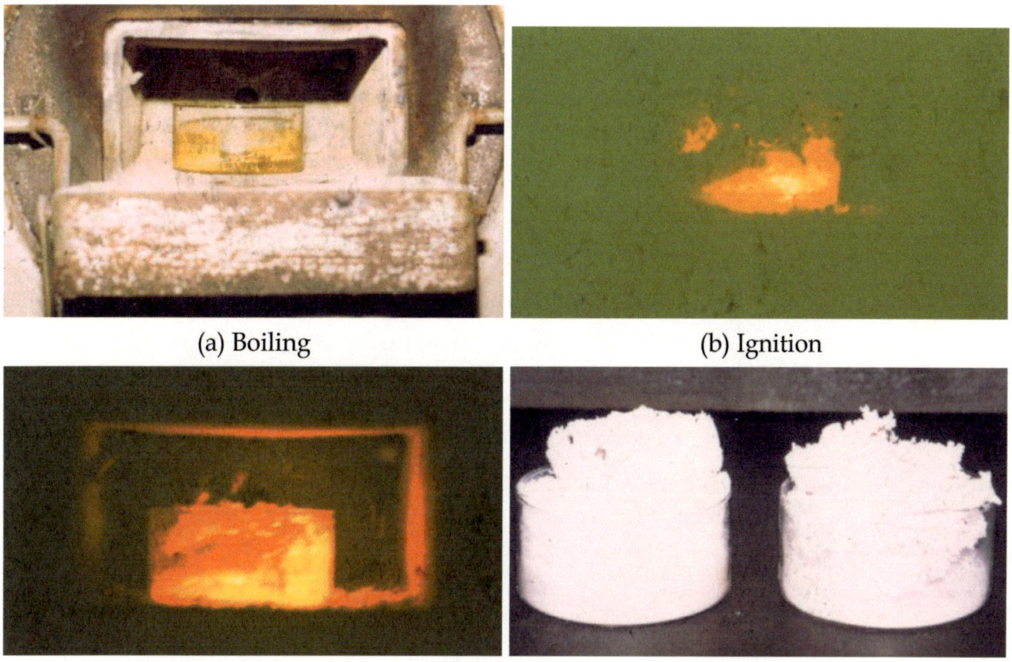

(a) Boiling (b) Ignition

(c) Propagation (d) Final Product

Fig. 16. Various steps in solution combustion process.

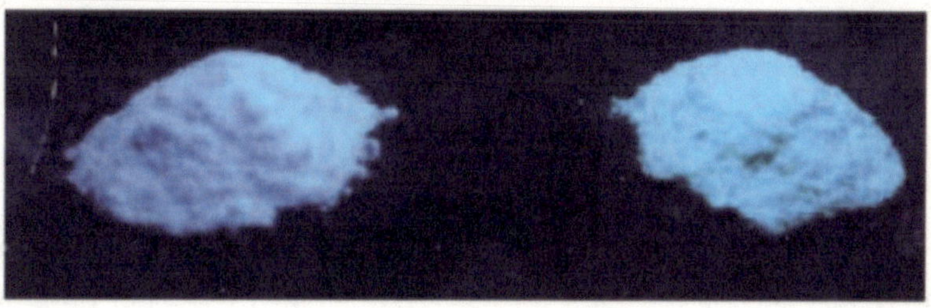

(a) (b)

Fig. 17. Photographs of as prepared (a) $BaMgAl_{11}O_{17}:Eu^{2+}$ and (b) $BaMgAl_{11}O_{17}:Mn^{2+}$ by solutiuon combustion synthesis show their blue luminescence emission under UV illumination **[17]**.

Several luminescence phosphors have been successfully prepared by SCS technique. For instance, Ce-doped yttrium aluminum garnet (YAG, $Y_3Al_5O_{12}$) phosphor powders were synthesized using the combustion method. The luminescence, formation process, and structure of the phosphor powders were investigated by X-ray diffraction (XRD) analysis, scanning electron microscopy (SEM), and photoluminescence (PL) spectroscopy. The XRD patterns show that the YAG phase can be produced with this method by sintering at 1000 °C for 2 h. This temperature is much lower than that required to synthesize the YAG phase using the conventional solid-state reaction method. No intermediate phases such as yttrium aluminum perovskite (YAP; $YAlO_3$) or yttrium aluminum monoclinic (YAM; $Y_4Al_2O_9$) were observed as a result of the sintering process. The powders were found to absorb excitation energies in the range of 410–510 nm. Furthermore, the crystalline YAG:Ce powder phosphor produced broad emission peaks in the range of 480–600 nm with the maximum intensity at 528 nm **[18]. Fig. 18** shows the TEM picture of BAM:Eu nano particles prepared by solution combustion synthesis [15].

Fig. 18. TEM picture of BAM:Eu nano particles prepared by solution combustion synthesis [15].

16. Sol-gel method

The sol–gel process refers broadly to the room temperature solution routes for preparing oxide materials. The solutions of precursors are reacted to form the irreversible gels that dry and shrink to rigid oxide glasses and powders. Traditionally, phosphor films are generally deposited by using the sputtering technique and only in recent decades, the sol–gel method has been considered as a low cost alternative approach for the preparation of novel nano-structured materials including luminescent powders and films. The sol-gel route presents a lot of advantages: low-temperature synthesis, possible formation of powders with uniform grain morphology and achievement of homogeneous multicomponent films. An advantage in using the nano-crystalline phosphors is that non-radiative transition could significantly be controlled with a decrease in the crystal grain size. The preparation of the Zn_2SiO_4 precursor followed a simple route and was completed at room temperature and room humidity: firstly, $Zn(Ac)_2$ was dissolved in a mixture solution of de-ionized water and ethanol; secondly, stoichiometric TEOS was added into the above solution with agitation for homogenization; thirdly, $Eu(Ac)_3$ and $Tb(Ac)_3$ were added into the precursor in the amount of 10 mol% ($Eu:Zn_2SiO_4$). Finally, a proper amount of 0.1M HCl was applied as catalyst for the hydrolysis of TEOS. The obtained precursor was transparent and clear, and was stable for several months if sealed at room temperature. Transparent gel can be obtained by leaving the precursor in air for about 2 or 3 days. Dry gel was prepared by baking the gel at 120°C for 5 min. Powder phosphors were prepared by sintering the dry gel for 30 min at different temperatures from 650 to 850°C. XRD data revealed that good crystallization of the powders could be obtained at about 850°C, which is about 450°C lower than that of the conventional solid-state reaction method **(Fig.19).** Phase analysis indicates that the obtained product has a willemite structure (rhombohedral). At the same time, a little amount of triclinic Zn_2SiO_4 was also detected. The TEM observation of the powders indicates that the particles are in the range 40–100 nm in diameter and have a good crystallinity.

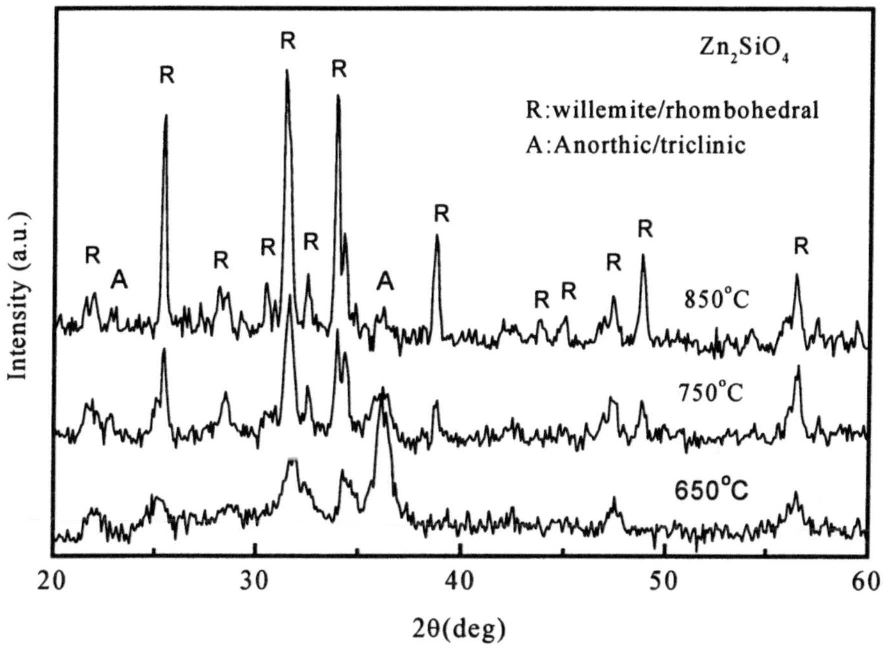

Fig. 19. XRD patterns of sol–gel derived Zn_2SiO_4 powders [19].

TG and DTA analysis in sol-gel route - In order to study the thermal decomposition of undoped xerogels, thermal gravimetry analysis (TGA) combined with infrared (IR) spectroscopy were used. TG and DTA (differential thermal analysis) results show cinetics of emitted gases during TG analysis, determined through the intensity evolution of the infrared main characteristic bands of each species. The total weight loss remains between -35 and -40% whatever the Tb^{3+} concentration is and occurs in three steps as shown in **Fig.20.** The first one (25–200°C) corresponds to the *departure of adsorbed moisture and chemically bonded alcohol molecules,* in agreement with **Fig. 21.** This stage leads to a wide *endothermic peak.* The second and main weight loss (−19%) lying from 200 to 600°C is associated with an *exothermic phenomenon.* This stage can be *assigned to the pyrolysis of the organic parts* of the alkoxide groups and likely to more strongly adsorbed alcohol molecules. It is in good agreement with IR measurements which show that most of the residual organic groups are removed before

600°C. Above 600°C, only a weak weight loss takes place, ascribed to the withdrawal of the last residual alkoxy groups embedded in YAG matrix. Furthermore, *a sharp exothermic peak at about 894°C indicates the onset of YAG crystallization.* This is consistent with the XRD results

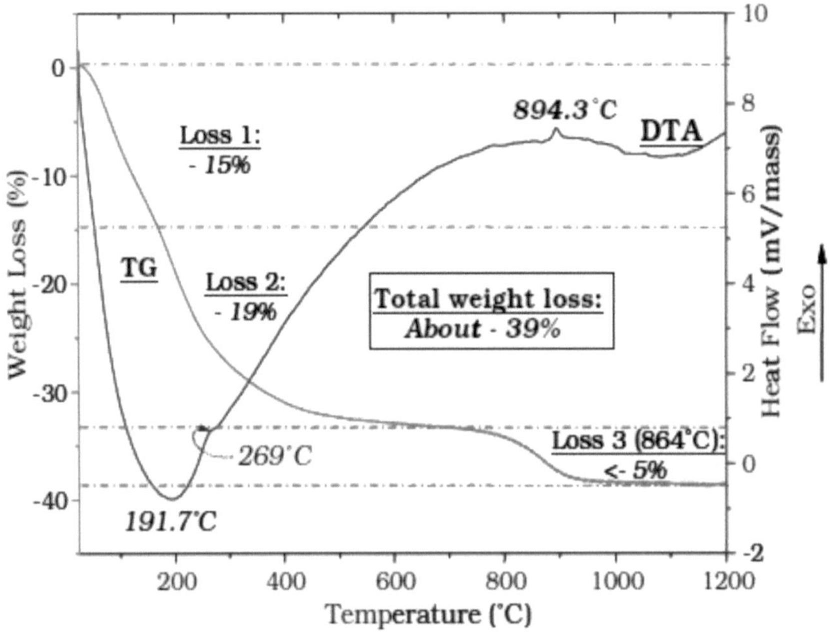

Fig. 20. TG and DTA profiles obtained fromYAG precursor gel **[20].**

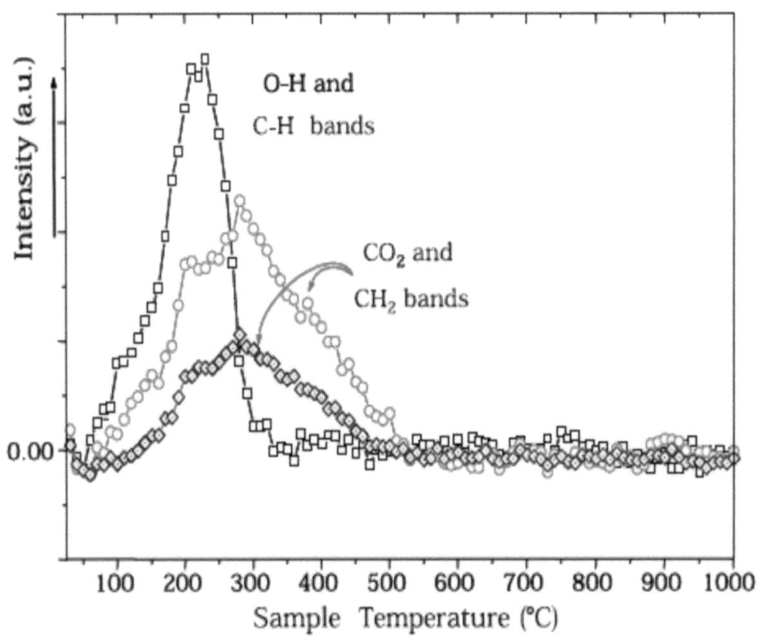

Fig. 21. Cinetics of gases emitted during TG analysis **[20].**

which showed that YAG sample begins to crystallize between 850 and 900°C. No significant weight loss appears thereafter, assuming the formation of the final product $Y_3Al_5O_{12}$.

17. Infrared spectroscopy

In the xerogel FTIR spectrum of of xerogel and undoped YAG sintered for 4 h at 1100°C, the presence of the characteristic organic bands is noticed. The broad band ranging from 2600 cm^{-1} to 3800 cm^{-1} is ascribed to −CH3 and −OH stretching in the isopropanol and alkoxy groups. This latter also involves −OH stretching resulting from water added in excess for hydrolysis and adsorbed from air moisture. A peak of weak intensity at about 2350 cm^{-1} is attributed to the adsorbed carbon dioxide from the atmosphere. The bands lying from 1200 to 1600 cm^{-1} are attributed to C-O and -CH_3 stretches bonds of organic groups whereas the peaks at about 1638 cm^{-1} is likely due to water added in excess to hydrolyze the heterometallic isopropoxide sol. Moreover, the several broad bands observed within the 400–850 cm^{-1} region of the IR spectrum correspond to M-O bonds (M=Y or Al) vibrations in YAG lattice. After sintering, specific bands related to the solvent and alkoxy groups significantly decrease or disappear. The carbon dioxide peak remains. Specific Al-O and Y-O vibrations peaks below 800 cm^{-1} are clearly identified as YAG ones.

18. Scanning electron microscopy

It is known that luminescent properties of a phosphor depend on its particles shape and size. On this account, the morphology of Tb^{3+} activated YAG samples has been studied by scanning electron microscopy (SEM). SEM micrographs of YAG:Tb^{3+}(5%) powder recorded at 800× magnification are shown in Fig.22. From the first picture related to the as-prepared xerogel **(Fig.22a),** it can be seen irregular size blocks. Voids and pores are also observed. The micrograph of heat-treated sample **(Fig.22b)** exhibits a denser network with fewer voids and narrow size distribution. For the two samples, the largest particles can reach several tens of micrometers.

Fig. 22. SEM images recorded at 800× magnification from $Y_3A_{15}O_{12}$:Tb^{3+}(5%) unheated (a) and annealed at 1100°C for 4 h (b) **[20].**

19. NMR

$Y_3Al_5O_{12}$ generally adopts a cubic garnet structure with lattice parameter of 12 Å (space group Ia¯3d). Its structure consists of a network where aluminum atoms reside both in octahedral and tetrahedral interstices whereas yttrium atoms occupy dodecahedral sites. Since the chemical shift of ^{27}Al NMR is sensitive to the local coordination, this technique is largely applied for checking the different phases and coordination states of Al centers in aluminates. As a result, MAS NMR ^{27}Al study has been undertaken on YAG samples in order to apprehend what kind of interatomic movements occur during sol-gel process. octahedral AlO_6 sites resonate between 15 and 30 ppm, the much less common AlO_5 sites between 40 and 25 ppm and tetrahedral AlO_4 between 80 and 50 ppm. On this account, the 2.6 ppm major resonance in the spectrum of the uncalcined xerogel corresponds to six-coordinate aluminum. Besides, two other distinct peaks located at about 35.4 and 61 ppm can be assigned to five and four-fold coordinated aluminum atoms. It can be noticed that the resonance at 35.4 ppm can also be consistent with a tetrahedral site distorted due to the presence of oxygen defects. ^{27}Al NMR spectral features of the sintered YAG powder reveal three signals at 0.426, 23.6 and 62.7 ppm, which can be imputed to the three types of coordination. The octahedral band has shrinked into the sharp signal at 0.426 ppm. Furthermore, since it has been demonstrated that Al atoms can only occupy tetrahedral and octahedral sites in crystallized YAG, we can deduce that the 35.4 and 23.6 ppm signals respectively in xerogel and crystallized powder correspond to distorted tetrahedral sites and not to five-fold coordinated one. Besides, by integrating peak areas, the ^{27}Al tetrahedral/octahedral ratio has been determined. Contrary to other studies, it remains the same after calcination, even if the resonance relative to "distorted tetrahedral" sites has weakened in favour of tetrahedral one **[20].**

20. Co-precipitation and sintering

The author has developed a simplified co-precipitation technique for the synthesis of $CaSO_4$:Dy TLD phosphor, which circumvent the cumbersome procedure, used so far, namely, evaporation of (highly corrosive) concentrated H_2SO_4 by recrysttalization. High TL sensitivity, uniform microcrystalline morphology, lower grain size (see SEM pictures shown below left) suitable for manufacturing dosimeters in solid form, better glow curve structure, lesser glow peak shift and better linearity and simplified preparation technique make the new phosphor a better alternative **(Figs.23 & 24).** The new recipe of $CaSO_4$:Dy based on co-precipitation technique is not only economical but also compatible for large scale production. Sintering of the co-precipitated phosphor at 700^{0}C in air increased its TL sensitivity by more than a factor of 2 (see TL glow curves shown below right) due to improved crystallinity and diffusion of Dy^{3+} ions from the surface to the whole volume of the grains as stated earlier.

In this experiment, initially, $CaSO_4.2H_2O$ and Dy_2O_3 salts were dissolved in hot concentrated H_2SO_4 acid and during slow dilution - water was added drop wise into the above hot solution - $CaSO_4$:Dy was found to precipitate. As per conventional solution chemistry, only $CaSO_4$ should precipitate since $CaSO_4$ does not dissolve in dilute H_2SO_4 acid. The Dy^{3+} ions should remain dissolved in the acid-water solution due to the high solubility of $Dy_2(SO_4)_3$ in water and in H_2SO_4 acid. The formation $CaSO_4$:Dy, as confirmed by TL and PL studies, however indicated that a considerable amount of Dy ions present in the solution actually diffuse into the $CaSO_4$ host lattice during precipitation. This is considered as an amazing result which confirms that there is a competition between the solubility of Dy ions in $CaSO_4$ lattice and their solubility in acid solution. At low Dy concentrations (~ 0.1 mol%) normally used in TL phosphors, nearly 90% of Dy gets into the $CaSO_4$ lattice. Only at high Dy concentrations, the percentage of Dy getting into $CaSO_4$ lattice goes down.

21. Combustion and pyrolysis

A number of display phosphors has been recently synthesized in the author's laboratory using combustion and pyrolysis technique. In the pyrolysis technique, the constituent chemicals decompose and fuse by the action of heat at relatively low external temperatures ($\leq 1000^0C$) in air with adequate ventilation so that the gaseous products released during decomposition reactions escape and the fusion process is complete, Limitations of conventional solid state method are inhomogeneity of the product, formation of large particles with low surface area and hence mechanical particle size reduction is required, which introduces impurity and defects and presence of defects, which are harmful to luminescence. The problem of inhomogeneity could be mitigated by the use of non-conventional methods (wet-chemical) which include solution combustion. Combustion is an exothermic reaction and occurs with the evolution of heat and light. This method was accidentally discovered in 1988 in Prof. Patil's laboratory in India. The first synthesis of Eu^{3+} doped $LnBO_3$ (Ln=La, Gd and Y) borate phosphors by combustion method was made by his group. The emission spectrum of $LaBO_3:Eu^{3+}$ consisted of two bands at 615 and 595 nm and these bands were attributed to $^5D_0 \rightarrow ^7F_2$ and $^5D_0 \rightarrow ^7F_1$ transition of 9-coordinated Eu^{3+} ions, respectively. But, there were three bands at 625, 610 and 595 nm observed for $GdBO_3:Eu^{3+}$ and $YBO_3:Eu^{3+}$ phosphors. The band at 595 nm was attributed to magnetic dipole $^5D_0 \rightarrow ^7F_1$ transition of Eu^{3+}, whereas the bands at 625 nm and 610 nm were attributed to electric dipole $^5D_0 \rightarrow ^7F_2$ transition for 12 and 8 coordinated Eu^{3+} ions, respectively. Since electric dipole transition, $^5D_0 \rightarrow ^7F_2$ depends upon the structure, two bands were observed for differently coordinated Eu^{3+} in $GdBO_3:Eu^{3+}$ and $YBO_3:Eu^{3+}$.

$(Y,Gd)BO_3:Eu^{3+}$ phosphor used in PDP displays was prepared earlier by combustion method using amino acetic acid as the combustion agent and then sintered at 1000 °C for 30 min. However, details of the recipe used and the PL sensitivity comparisons with commercial phosphor were not reported. In author's lab, raw materials used for the synthesis of $(Y,Gd)BO_3:Eu^{3+}$ and $YBO_3:Eu^{3+}$ phosphors using combustion technique were Y_2O_3, Gd_2O_3, H_3BO_3 and Eu_2O_3. NH_4NO_3 and urea (CH_4N_2O) were used as oxidizer (O) and fuel (F) respectively. The O/F ratio was kept at unity. After wet mixing in a porcelain crucible, combustion was carried out at 600 °C for 15 min in a muffle furnace with adequate ventilation so that the gaseous products released during combustion escape and the combustion process is complete. Combustion of the borate materials with urea fuel resulted in smoldering without flame. In contrast, with Oxalyl dihydrazide (ODH – $C_2H_6N_4O_2$) fuel, the combustion was reported to be flaming and the flame temperature measured using optical pyrometer was about 1400±100 °C. ODH process showed very sharp powder XRD pattern whereas urea process exhibited very broad powder XRD peaks. Therefore, by changing the fuel one could control particulate properties. However, ODH is nearly 50 times more expensive than urea and hence we used only urea as the fuel in the combustion process. The material obtained after combustion was agglomerated and since it was not a hard material, a mild grinding in an agate mortar and pestle was found sufficient to remove the agglomeration which was then transferred to an alumina crucible before sintering at 1000 °C for 2 h in air in the furnace. The material obtained was once again hand ground mildly to obtain grain sizes below 30 μm. The grain morphology and their size distribution are being studied. The increase in emission intensity of the combustion synthesized red phosphors on calcination at high temperatures has been attributed to improved crystallinity. For PL intensity comparisons, $(Y,Gd)BO_3:Eu^{3+}$ PDP phosphor obtained from LG chemicals (Korea) was used.

Fig. 23. SEM photographs of $CaSO_4$:Dy TL phosphor grains obtained by (A) co-precipitation (in as-grown condition) and (B) conventional recrystallization (after grinding) **[21].**

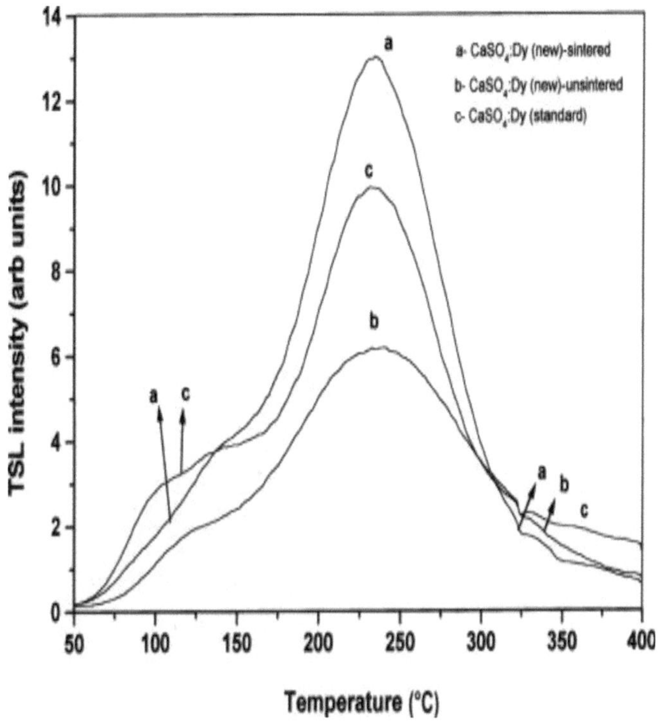

Fig. 24. Comparison of TL glow curves of $CaSO_4$:Dy phosphor obtained by recrystallization (c) and co-precipitation (a and b) The TL sensitivity of sample b increases by more than a factor of 2 on sintering at 700⁰C for 1h **[21].**

The PL intensities of 592 nm emission peak on 393 nm excitation of $(Y_{0.54}Gd_{046})_x$ $(BO_3)_y$:Eu^{3+} (5 mol%) were found to vary with y/x molar ratio. Optimal PL efficiency is obtained at the y/x molar ratio of 1.38. XRD data reveal that at this molar ratio, interfering phases such as $(Y_{0.95},Eu_{0.05})_2O_3$ and $(Y,Eu)_3BO_6$ are negligible and H_3BO_3 had been fully used to build $(Y,Gd,Eu)BO_3$ crystal resulting in the maximized luminescence efficiency **(Fig.25).** A 12-fold increase in its luminescence efficiency was seen with an increase in Eu concentration from 0.5 to 8.4 mol%. No saturation or activation quenching of the PL efficiency was seen till the highest Eu concentration (8.4 mol%) level studied. PL efficiencies at still higher Eu concentrations are being studied. Interestingly, at 8.4 mol% Eu concentration, the PL sensitivity of $(Y_{0.54}Gd_{046})_x$ $(BO_3)_y$:Eu^{3+} is 40% higher than that of the commercial $(Y,Gd)BO_3$:Eu^{3+} PDP phosphor. It is concluded that $(Y_{0.54}Gd_{046})_x$ $(BO_3)_y$:Eu^{3+}, at the y/x molar ratio of 1.38, developed in this work is a potential candidate for application as a red phosphor for nUV LED.

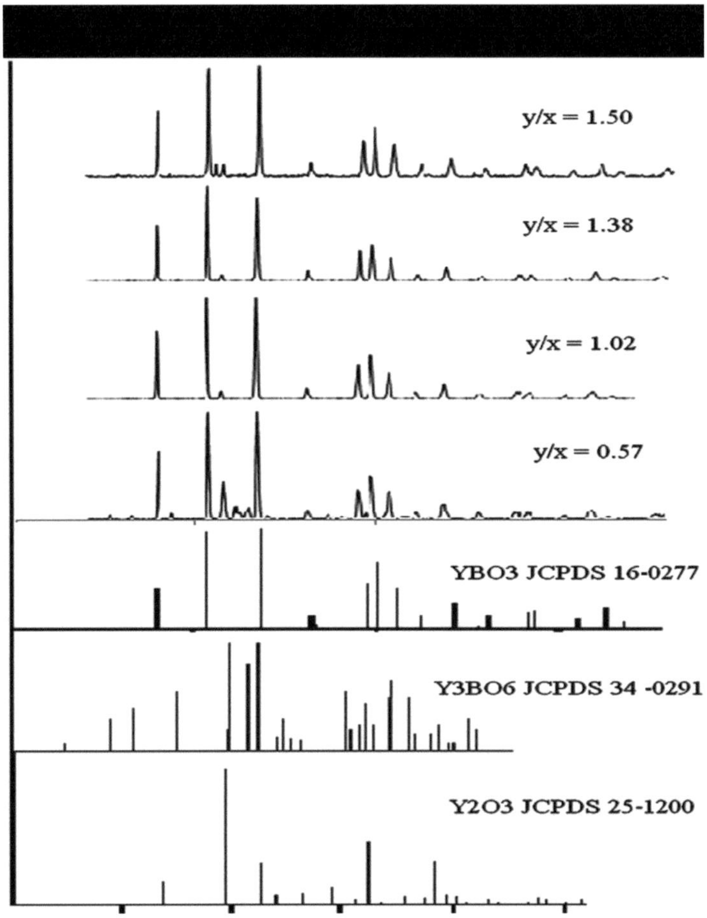

Fig. 25. XRD spectra of $(Y_{0.54}Gd_{046})_x$ $(BO_3)_y$:Eu^{3+} (5 mol%) sintered at 1000^0C as a function of BO_3 concentration. y/x ratio: 0.57, 1.02, 1.38 and 1.50. The standard XRD spectra of YBO_3 (JCPDS 16-0277), Y_3BO_6 (JCPDS 34-0291) and Y_2O_3 (JCPDS 25-1200) are also shown **[22].**

Development of high sensitive $(Y,Gd)BO_3$:Eu^{3+} phosphor also assumes significance since it is widely employed as red phosphor for plasma display panels (PDP) due to the BO_3 mediated energy absorption in the 130–170 nm VUV region giving rise to 593 nm emission. However the color purity of all YBO_3 based phosphors developed needs improvement since the 593 nm orange emission corresponding to the $^5D_0$–$^7F_1$ magnetic dipole transition is greater than the red components from electric dipole transitions in the 610–680 nm region. Studies carried out in this direction by others have succeeded only partially so far.**Fig. 26** shows luminescence from some of the phosphors prepared by pyrolysis.

Fig. 26. Luminescence from some of the phosphors prepared by pyrolysis in author's lab are shown below under UV and 450 nm LED illumination. From left to right: Zn_2SiO_4:Mn, YAG:Ce LED and Y_2O_3:Eu^{3+}.

22. Microwave synthesis

Firing (sintering) is a necessary step for solid-state phosphor synthesis since activation energy must be supplied for the activators to go into the crystal structure of the host material. By solid-state reaction, conventional synthesis of phosphors takes many hours even with flux. Microwave processing is a relatively new technique and characterized by substantially accelerated reaction kinetics in the material systems if properly chosen. Using microwave processing, various phosphors have been synthesized. The high efficiency phosphors developed for field emission displays, plasma displays, and white light emitting diodes (LED) tend to be degraded by the operating environment and/or the devices' manufacturing conditions. Transparent oxide coating on phosphor powders is an effective approach to protecting phosphors from ageing, The coating can isolate the phosphor from the environment and could protect both the phosphor and the device. Microwave synthesis of some important phosphors and microwave processing of some phosphors coated with oxides are reported. In one study, microwave synthesis was carried out in a 2.45 GHz, 6 kW capability multimode microwave furnace with atmosphere controlled, although, only about 1 kW or less power was needed in the experiments. The powder sample (30-70 g each) was loaded in an alumina crucible, which was then placed in a properly insulated package with SiC as microwave susceptor. The insulation package was made of porous FiberFrax Duraboard 3000. The temperature of the sample was monitored with an optical pyrometer. The temperature was controlled by adjusting input power. During the microwave processing, the sample was rotating horizontally about the axis. The samples were microwave heated up and held at the designed temperatures for typically 10-20 min. The microwave-synthesized products were characterized for particle size, brightness, phase composition, morphology, luminescence emission, and color coordinates. Optimization of the parameters is required to achieve desired properties **[23].**

23. Spray pyrolysis

The aerosol processing, for example spray pyrolysis, is Known as a promising technique, especially to prepare the multi-component oxides like the $CaMgSi_2O_6$:Eu (CMS) phosphor. **Fig. 27** shows a simplified schematic diagram displaying the whole preparation steps of CMS phosphor by the spray pyrolysis. An ultrasonic aerosol generator with six vibrators (1.7 MHz) was used to atomize the precursor solution and the produced droplets were carried by the air (45 L min^{-1}) into a hot furnace with a quartz tube (length: 1200mm and inner diameter: 50 mm) at 900 °C. The formed particles were collected by a Teflon bag filter and posttreated at the temperature of 1000–1250°C for 5 h under a reducing atmosphere (5% H_2/N_2 mixture gas) for crystallization and activation of divalent europium. To test the thermal degradation, the prepared CMS and commercial BAM samples were mixed with commercially available binder paste (Daejoo corporation) which is practically used for the formation of phosphor films in PDPs. Next, the phosphor paste, wherein the fraction of powder was 50 wt.%, was fired at 500 °C for 30 min in the air environment. The crystal phase of the prepared CMS particles and PL were then analyzed.X-ray storage phosphors based on BaFBr:Eu are also made using this technique.

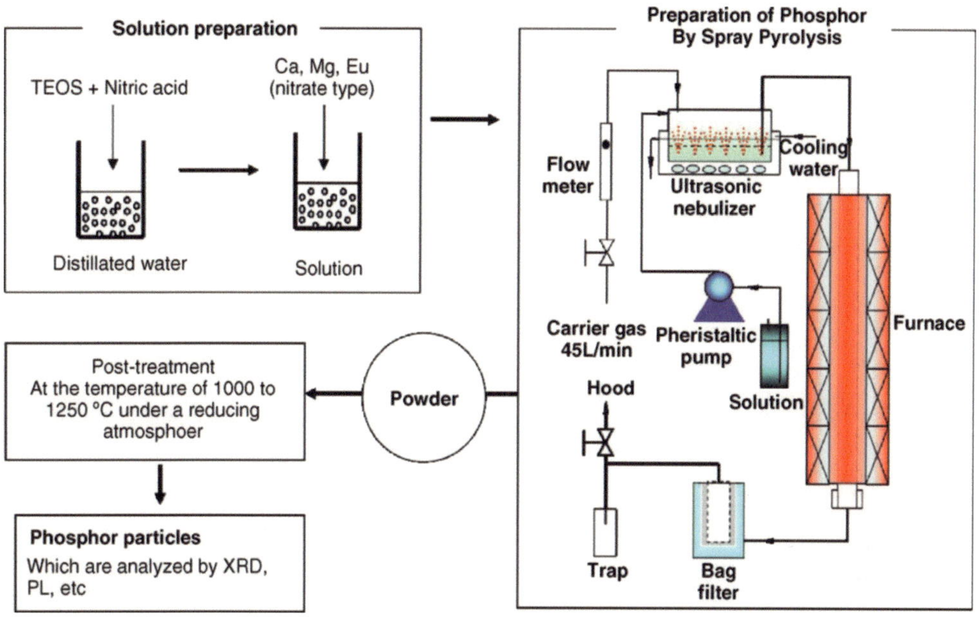

Fig. 27. Schematic diagram displaying the whole preparation steps of $CaMgSi_2O_6$:Eu blue phosphor particles by the spray pyrolysis **[24].**

24. Hydrothermal synthesis

The hydrothermal (HT) method, which uses autogenous pressure developed at temperatures above the boiling point of water, has been used especially in the synthesis of various ceramic oxide powders. The advantage of the hydrothermal method is that ceramic materials can be synthesized at relatively low temperatures (100-300°C) without milling or calcinations [11]. The particle size and shape can also be controlled by various processing variables such as temperature, pH, and the addition of surfactants ormineralizers. The reaction is controlled by dissolution/precipitation of reactants in an aqueous medium. Therefore, the above processing variables are thought to have a significant influence on the dissolution or precipitation behavior, even though the exact roles or effects are not fully understood and differ in various systems.

For the hydrothermal synthesis, stoichiometric amounts of Gd_2O_3 and Eu_2O_3 were dissolved in 100 ml of distilled water acidified by the addition of nitric acid. After the complete dissolution of these oxides, NH_4OH solution was added drop wise until the pH of the solution reached 9. White precipitates were instantaneously formed and so-obtained precipitates were washed several times with distilled water by centrifugation. The washed precipitates were mixed with 50 ml solution of H_3BO_3 and distilled water. The amount of H_3BO_3 in the solution was adjusted so as to have an appropriatemole ratio with respect to Gd and Eu to give $GdBO_3$:Eu^{3+}. NH_4OH was used to adjust pH of the precursor solution to 7–10. After vigorous stirring, the precursor solution was placed in a Teflon-lined stainless steel autoclave with a volume of 100 ml. The solution was heated at 200-240°C for 3–10 h" and cooled to room temperature. The resulting powders were filtered and washed several times with distilled water by centrifugation, and finally suspended in ethanol then dried at 60°C for 5 h. An outline of the experimental procedure is shown in **Fig. 28.** For comparison, the bulk $GdBO_3:Eu^{3+}$ samples are prepared by a solid-state (SS) reaction at 500 and 1100^{0}C each for 2 h, respectively, from the mixture of Gd_2O_3, Eu_2O_3, and H_3BO_3 (20 wt.% excess).A high pH and excess H_3BO_3 promote the growth of $GdBO_3$ in a preferred orientation i.e., (2 0 0).

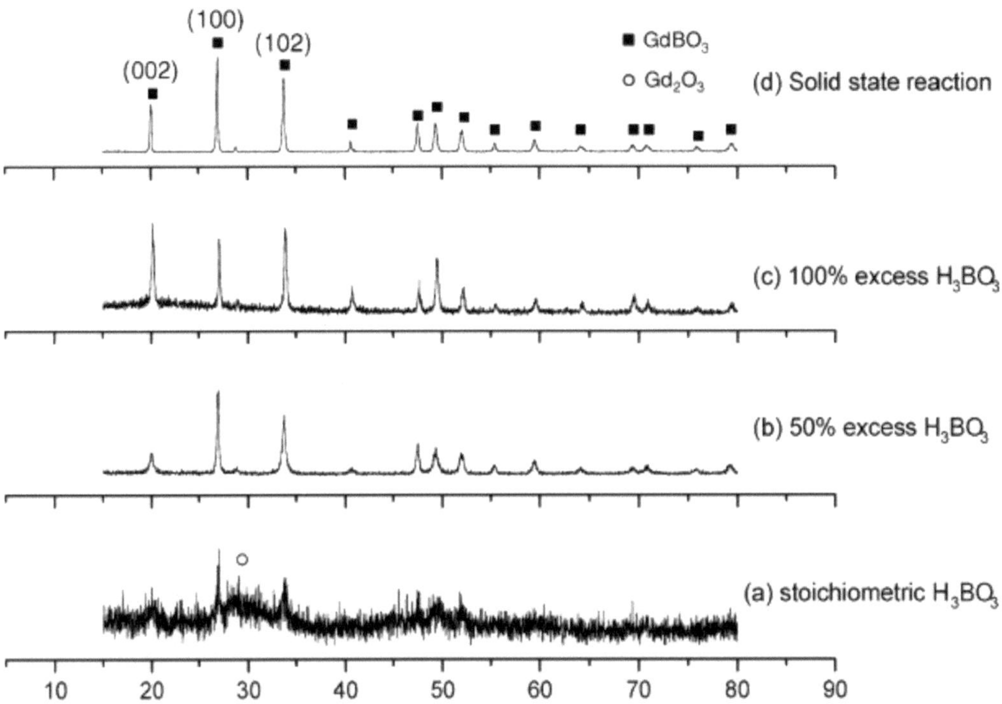

Fig. 28. XRD patterns of $Gd_{0.9}BO_3:Eu_{0.1}^{3+}$ powders synthesized with (a) stoichiometric, (b) excess amount of H_3BO_3, (c) by hydrothermal method (at 200^{0}C, 10h, pH8) and (d) by solid state reaction [25].

25. References

[1] Chun-Chieh Lin, Wei-Ren Liu, and Nae-Lih Wu. Effects of Sintering Fluxes on Structure and Luminescence of BaMgAl1O17:Eu2+ Phosphor. Abstract #2345, 218th ECS Meeting, © 2010 The Electrochemical Society

[2] L. Guanghuan , L. Tao , S. Yanhua, G. Guimei, X. Jijing, AN Baichao, G. Shucai and H. Guangyan. Preparation and luminescent properties of CaA12O4:Eu3+,R+ (R=Li, Na, K) Phosphors. J. Rare Earths, Vol. 28, No. 1, Feb. 2010, p. 22

[3] W. B. Im, Y.I. Kim and D. Y. Jeon. Thermal Stability Study of BaAl2Si2O8:Eu2+ Phosphor Using Its Polymorphism for Plasma Display Panel Application.Chem. Mater., 2006, 18 (5), pp 1190–1195. DOI: 10.1021/cm051894v

[4] A R Lakshmanan, M T Jose, V Ponnusamy and P R Vivek kumar. Luminescence in CaSO4:Dy Phosphor - Dependence on Grain Agglomeration, Sintering Temperature, Sieving and Washing. J.Phys.D:Appl. Phys. 35, 386 (2002)

[5] C. Chang, Z. Yuan and D. Mao. Eu2+ activated long persistent strontium aluminate nano scaled phosphor prepared by precipitation method. Journal of Alloys and Compounds 415 (2006) 220–224

[6] A.R. Lakshmanan. Effect of pressure on the luminescence properties of gypsum, anhydrite, calcite and Dy doped CaSO4. Rad. Meas. 39, 235-240 (2005)

[7] Y. Wang, N. Can, P. Townsend. Influence of Li dopants on thermoluminescence spectra of CaSO4: with Dy or Tm. Luminescence in Rare-Earth Doped CaSO4 Phosphors. Rad. Meas. (in press)

[8] A.R. Lakshmanan. Photoluminescence and thermostimulated luminescence processes in rare-earth-doped CaSO4 phosphors. Prog. Mat. Science 44, 1-187 (1999)

[9] N.H. He and Y. Zhu. Improvement of morphology and luminescence of CaS:Eu2+ red emitting phosphor particles via carbon containing additive strategy. J. Mater. Sci. 43, 1515-1519 (2008)

[10] Y.S. Chang.The Effects of Heat Treatment on the Crystallinity and Luminescence Properties of YInGe2O7 Doped with Eu3+ Ions. Journal of Electronic Materials, Vol. 37, No. 7 1024-28 (2008)

[11] Su-Hua Yang & Fu-Shou Tsai & Jia-Xing Chen. Characterizations of white-light ZnWO4 phosphor prepared by blending complementary phosphor. J Solid State Electrochem. 14:937–943 (2010)

[12] K Madhukumar, H.K. Varma, M. Komath, T.S. Elias, V.Padmanabhan and C.M.K. Nair. Photoluminescence and Thermoluminescence properties of tricalcium phosphate phosphors doped with dysprosium and europium. Bull. Mater. Sci., Vol. 30, No. 5, October 2007, pp. 527–534.

[13] F. ferey, P. Grosseau, B. Guilhot, P. Iacconi and M. Benabdesselam. Thermoluminescence and sintering of high purity α- alumina doped by Zr, Th and Ca. Solid State Ionics, 2001, 141-2, 567-74.

[14] V. Schembri and B.J. Heijmen. Optically stimulated luminescence (OSL) of carbon-doped aluminum oxide (Al2O3:C) for film dosimetry in radiotherapy. Med. Phys.34(6) 2113-8 (2007)

[15] V.S.M. Barros, W.M. de Azevedo, H.J. Khoury, M.E.A. Andrade and P. Linhares Filho. Thermoluminescence study of aluminum oxide doped with terbium and thulium. Radiation Measurements. Volume 45, Issues 3-6, March-July 2010, Pages 435-437

[16] S. Ekambaram, K.C. Patil and M.J. Maaza. Synthesis of lamp phosphors: facile combustion approach, J. Alloys Comp. 393, 81 (2005)

[17] K.C. Patil, M.S. Hegde, T. Rattan and S.T. Aruna. Chemistry of nanocrystalline oxide materials:combustion synthesis, properties and applications. Advanced Ceramics. Pub. World Scientific, 2008, ISBN 9812793143

[18] S.K. Lee, H.H. Yoon, S.J. Park, K.H. Kim and H.W Choi. Photoluminescence Characteristics of Y3Al5O12:Ce3+ Phosphors Synthesized Using the Combustion Method with Various Reagents. Jpn. J. Appl. Phys. 46 (2007) pp. 7983-7986.

[19] H.X. Zhang, S. Buddhudu_, C.H. Kam, Y. Zhou, Y. L. Lam, K.S. Wong, B.S. Ooi, S.L. Ng, W.X. Que. Luminescence of Eu3+ and Tb3+ doped Zn2SiO4 nanometer powder phosphors. Materials Chemistry and Physics 68 (2001) 31–35

[20] A. Potdevin, G. Chadeyron, D. Boyer and R. Mahiou. Sol-gel elaboration and characterization of YAG:Tb3+ powdered phosphors. J Mater. Sci 41 (2006) 2201–2209

[21] A.R. Lakshmanan, M.T. Jose, O. Annalakshmi. Synthesis of high sensitive CaSO4:Dy TLD Phosphor by co-precipitation Technique. Radiat. Prot. Dosim. pp. 1-9 (2008) doi: 10, 1093/rpd/ncn 215.

[22] A,Lakshmanan, R.S. Bhaskar, Preema C. Thomas, R. Satheesh Kumar, V. Siva Kumar and M.T. Jose. A red phosphor for nUV LED based on (Y,Gd)BO3 :Eu3+. Mat. Letters 64, 1809-1812, 2010

[23] Kyeong Youl Jung, Hyun Woo Lee, Yun Chan Kang, Seung Bin Park,§ and Young Suk YangLuminescent Properties of (Ba,Sr)MgAl10O17:Mn,Eu Green Phosphor Prepared by Spray Pyrolysis under VUV ExcitationChem. Mater., 2005, 17 (10), pp 2729–2734 DOI: 10.1021/cm050074f

[24] Taehyung Kim, Shinhoo Kang. Hydrothermal synthesis and photoluminescence properties of nano-crystalline GdBO3:Eu3+ phosphor. Materials Research Bulletin 40 (2005) 1945–1954

16

Sintering and Characterization of Rare Earth Doped Bismuth Titanate Ceramics Prepared by Soft Combustion Synthesis

Umar Al-Amani Azlan, Warapong Krengvirat, Ahmad Fauzi Mohd Noor, Khairunisak Abd. Razak and Srimala Sreekantan
Universiti Sains Malaysia, USM, Pulau Pinang
Malaysia

1. Introduction

Nowadays, ferroelectric ceramics and thin films have attracted much attention for various studies which are generally used for numerous potential applications in ferroelectric random access memory (FRAM), in microelectronic mechanical system (MEMS), non-linear optical devices, surface acoustic wave devices, tunable capacitors, sensing applications or pyroelectric detectors (Besland et al., 2006, Yang et al., 2008). The main focus to develop the ferroelectric thin films has started in 1980s (Besland et al., 2006). Several methods were initially used for deposition of thin films such as conventional dipping, sputtering and spin coating techniques. Typically, the deposition of thin films will be more complex to obtain a good layer that consists of several ferroelectric compounds on substrate. Furthermore, high precision deposition technique is essential to control the desired thickness and surface layer of ferroelectric compounds. As far as our concern, the up to date technique such as physical vapor deposition (PVD), RF sputtering, chemical vapor deposition (CVD) and metal-organic chemical vapor deposition have been frequently used in many studies to obtain a better ferroelectric thin films condition. Nevertheless, a major concern on expensive equipment and experience user limit this technique in many studies. In order to develop the ferroelectric materials, the preparation in the form of bulk ceramics has been extensively studied. Up to now, several methods, including solid state reaction, hydrothermal synthesis, mechanical activation technique, sol-gel method, co-precipitation method were used for the preparation of bulk ceramics. Recently, the soft combustion synthesis is used as alternative route since it offers several beneficial points to the processing element and the properties of ceramics (Yan ,Razak, 2010). Bismuth titanate, $Bi_4Ti_3O_{12}$ or BTO has received a lot of attention as dielectric and ferroelectric materials. Many studies have been conducted in various processing route to improve the microstructure that has a significance effect on dielectric and ferroelectric properties (Hardy et al., 2004, Pookmanee et al., 2004, Zhi-hui et al., 2010). In addition, a modification on basic compound is essential to enhance those properties. Since BTO is also sought as a potential material for dielectric application, in this chapter the effect of Sm^{3+} and Pr^{3+} doped-BTO was prepared and characterized by soft combustion technique. In order to investigate a possible application as wireless dielectric antenna, the dielectric study at different frequencies was carried out.

2. Bismuth titanate and other properties

Considerable attention has recently been paid to bismuth layer-structured ferroelectric (BLSF) as ferroelectric materials instead of unfriendly lead (Pb)-based ferroelectrics because of its excellent fatigue resistance and Pb-free chemical composition (AlguerÃ³ et al., 2006, Subbarao, 1961, Xue et al., 2009, Yang et al., 2003). The general formula is given by $(Bi_2O_2)^{2+}$ $(A_{m-1}B_mO_{3m-1})^{2-}$ where A = Bi^{3+}, Pb^{2+} Sr^{2+}, Ba^{2+}, etc. and B = Nb^{5+}, Ta^{5+}, Ti^{4+}, etc. m = 1, 2, 3, 4, 5, etc. $(Bi_2O_2)^{2+}$ is the bismuth oxide layer and $(A_{m-1}B_mO_{3m+1})^{2-}$ is the pseudo perovskite layer (Armstrong ,Newnham, 1972, Newnham et al., 1971, Yan et al., 2006). BLSF is expected to have various numbers of pseudo perovskite blocks in unit cells. BLSF, bismuth titanate, $Bi_4Ti_3O_{12}$ (m = 3) or BTO has three pseudo perovskite blocks in half-unit cells. In simple words, its structure can be described as formed by three unit cells of $(Bi_2Ti_3O_{10})^{2-}$ with perovskite like structure interleaved with $(Bi_2O_2)^{2+}$ layers (Ng et al., 2002). BTO is an attractive material that has low processing temperature (700-750°C) than other BLSF (e.g. $SrBi_2Ta_2O_9$) and strong anisotropy of the spontaneous polarization (P_s) along the *a*-axis (~ 50 μC/cm²) and *c*-axis (~ 4 μC/cm²) (Wang et al., 1999, Zhi-hui et al., 2010). However, the low remanent polarization (P_r = 5 μC/cm²), low fatigue resistance and high dielectric loss of BTO would limits its application in FRAM applications. The reduction in remanent polarization and fatigue with high dielectric losses become more serious issues due to defects in perovskite structure whereby the Bi ions volatile during sintering process and create the Bi vacancies accompanied by oxygen vacancies. Nevertheless, there are advantages on BTO whereby it has high Curie temperature at 675°C and high dielectric permittivity (~ 200), making this material for other possible applications such as capacitors, antennas, sensors and piezoelectric (Golda et al., 2011).

3. Rare-earth doping and other properties

Recent studies revealed that ion substitution in perovskite BTO structure is an efficient technique for improving the drawbacks of BTO as ferroelectric ceramics (Cui ,Hu, 2009, Santos et al., 2009, SimÃµes et al., 2008). It was reported that the fatigue free films with excellent ferroelectric properties are obtained by substitution of Bi-site ions in BTO films for La^{3+} ions using a pulsed laser deposition (Park et al., 1999). This ion substitution resulted in large remanent polarization value over 20μC/cm², which was considerably higher than that of BTO films. Furthermore, it has good fatigue resistance, low leakage current at 10^{-7} A/cm² at 5V and low processing temperature approximately in the range of 650 to 700°C. Recently, it has been reported that the substitution of Nd^{3+} in BTO thin films was more effective for improving the ferroelectric properties than La substitution (Kim ,Kim, 2005). This result can be explained by the fact that the substitution of Bi^{3+} by rare-earth ions with a smaller ionic radius for the Bi^{3+} site is effective in improving the ferroelectric properties. In this case, the ionic radius of Nd^{3+} is much smaller than those of Bi^{3+} and La^{3+}. According to both studies, it is necessary to find out more rare-earth elements with smaller ionic radius in other to enhance the ferroelectric properties of BTO ceramics. Besides that, a small amount of rare-earth elements is important to tailor the microstructures of BTO. Recently, it was reported that Nd doping into BTO ceramics act as a grain-growth inhibitor whereby a remarkable decrease in the grain size with fine and homogeneous microstructure (Kan et al., 2008). It is well known that typical BTO powders are attributed to high anisotropic grains, in which the ferroelectric properties are grain orientation dependent. Thus, it can be said that the homogeneity in microstructure is strongly influenced by rare-earth content. In addition, the corresponding microstructure can produce better ferroelectric properties. The summary of doping studies in BTO and their properties are listed in **Table 1**. Based on this summary, the selection of processing route is important to determine the grain orientation and microstructure as well as dielectric and ferroelectric properties. Besides that, the doping studies can improve the dielectric and ferroelectric properties of BTO.

Rare-earth	Process	XRD	Microstructure	Dopant content	ε_r	tan δ	P_r	E_c
La^{3+} (47)	Polymeric precursor method			0 0.25 0.5 0.75	86 104 112 148 at 1 MHz	0.008 0.0011 0.0068 0.0018	15.1 20.2 20.2 20.3	1.45 1.09 0.66 0.99
La^{3+} (38)	Sol-gel	Randomly oriented at 700°C for 30 minute		0 0.85			12.5 18.6	2.8 2.8
La^{3+} (18)	Conventional solid state reaction			0 0.5 0.75 1 1.5 2	130 145 147 186 168 101 at 100 kHz	0.022 0.023 0.024 0.022 0.017 0.014		
Nd^{3+} (50)	Pulsed laser deposition	Highly c-axis oriented		0 0.85	103 177 at 500 kHz	0.063 0.068	4.5 8.6	142 88
Nd^{3+} (24)	Solid state reaction			0.75 0.80 0.85 0.90	174 143 158 179	0.0603 0.0055 0.0044 0.0070	 11.1 	
Nd^{3+} (52)	Solid state reaction		Plate-like grains	0 0.25 0.5 0.75 1	108 128 132 148 160	0.0078 0.0045 0.0044 0.0040 0.0039	5.2 5.5 9.3 11.1 9.8	27 35.2 45.2 42 34
Nd^{3+} (53)	Metal organic solution decompositio n	Highly c-axis oriented grains	Plate-like grains	0.5 0.75	172 156		17 19	150 98
Sm^{3+} (56)	Solid state reaction	Randomly oriented	Plate-like grains	0.8	270	0.003	16	70

Table 1. Summary of doping studies in BTO and their properties.

4. Processing route for preparation the BTO and rare-earth doping

As is known, synthesis process plays a crucial role to determine the microstructure of the ceramics as well as the control purity and stoichiometry. Different synthesis methods have been developed for the production of perovskite powders, like solid state reaction, sol-gel technique, hydrothermal synthesis, co-precipitation and combustion synthesis (Hardy et al., 2004, Kim, 2006, Kojima et al., 2009, Macedo et al., 2004, Pookmanee ,Phanichphant, 2009). It was reported that the ferroelectric properties of BTO can also be improved with a specific control in terms of microstructure, chemical homogeneity and its purity (Lu et al., 2005, Yang et al., 2008). Nevertheless, there are several merits and drawbacks of each synthesis process in order to control the ferroelectric domains through the preferred microstructure as well as crystal structure. Thus, the description of each synthesis process is discussed in the following subsection.

4.1 Conventional solid state reaction

The conventional solid state reaction is mostly used for preparation of bulk ceramics. It is an endothermic reaction used to produce simple oxide from carbonates, hydroxides and other metal salts. Such conventional reaction often results in high agglomeration and compositional inhomogeneity of powders because of high calcination temperature and repeated grinding. As a result, the sinterability of ceramics is fairly low subsequently a higher sintering temperature is required to enhance their properties. Subbarao (Subbarao, 1961, Subbarao, 1962) prepared the BTO ceramics using the solid state reaction and sintered from 1000 to 1250°C to achieve the theoretical density of about 80 %. In some cases, the sintering condition with longer soaking time is needed to enhance other properties. Watcharapasorn et al. (Watcharapasorn et al., 2010) studied the grain growth behavior of BTO ceramics using different sintering conditions. It was reported that the sintering of ceramics for longer time could render a material with more isotropic microstructure with reduced preferred orientation. Nevertheless, the increase in relative density (91 - 94 %) was very small with increasing soaking time.

4.2 Mechanical activation technique

Mechanical activation technique was initially derived from mechanical alloying for synthesizing alloys and intermetallics. The corresponding technique is a common part of the powder preparation route in the field of ceramics where high-energy ball milling has become a conventional method for producing nanocrystalline materials. This technique uses low-cost and widely available oxides as starting materials and skips the calcination step at an intermediate temperature, leading to a simplified process (StojanoviÄ‡ et al., 2008). Furthermore, the mechanically derived powders have higher sinterability than those powders synthesized by the conventional ball milling. Kong et al. (Kong et al., 2001) obtained the large P_r (24 μC/cm^2) and low E_c (11 kV/cm) for BTO ceramics with better density of 98 % after low temperature sintering at 850°C for the powder derived from mechanical activation technique. Stojanovic et al. (Stojanovic et al., 2006, Stojanovic et al., 2006) reported that the BTO powder can be directly synthesized using high impact milling for about 3 to 12 hours and then sintered at 1000°C for 2 h. Han et al. (Han ,Ko, 2009) stated the formation of BTO phase is highly dependent on the processing parameters particularly the impact energy or milling intensity. Zdujic et al. (ZdujiÄ‡ et al., 2006) reported that a mixture of α-Bi_2O_3 transformed to $Bi_2O_2CO_3$ at a milling intensity of ~ 0.49 W/g, which in turn was converted directly into a nanocrystalline BTO phase when the intensity was increased to ~ 2.68 W/g. Thus, it can be concluded that the parameter of mechanical milling has a significance result on the particle morphology and sinterability.

4.3 Sol-gel synthesis

The most popular wet chemical technique like sol-gel synthesis is widely used since it offers excellent uniformity over a large area, easy composition control, short fabrication time, as well as a low temperature process at comparatively a low cost (Du et al., 2007). This technique can be used to prepare the samples in the form of bulk ceramics and thin films. Several factors that need to be considered in a sol-gel synthesis are solvent, precursors, catalyst, pH, additives and mechanical agitation (Du et al., 2007, Guo et al., 2007, Ke et al., 2010). These factors greatly influence the powder size and other properties. Du et al. (Du et al., 2007, Du et al., 2008) reported that a highly stable and homogeneous BTO powders was produced at calcination temperature as low as 550°C, which is fairly low in wet chemical technique.

4.4 Hydrothermal synthesis

Another wet chemical technique is known as hydrothermal synthesis. In hydrothermal synthesis, the reaction mixture is heated above the boiling point of water in an autoclave or other closed system and the sample is exposed to steam at high pressures (Pookmanee et al., 2004, Shi et al., 2000, Yang et al., 2003). In addition, the parameter of Teflon-lined vessel such as temperature and reaction time are mainly important factor to determine the phase structure and particle morphology (Pookmanee et al., 2004). It was also reported that the hydrothermally powder was significantly influenced by different mineralizer KOH content and molar ratio of Bi/Ti (Shi et al., 2000). Recently, Xie et al. (Xie et al., 2007) reported that the concentration of KOH, reaction time and temperature had a significant effect on the phase composition and morphology of the resultant single crystals. Many authors reported that hydrothermal synthesis has several advantages including narrow particle size distribution, highly purity with fine powder, and low degree of agglomeration. In processing stand point, the hydrothermal synthesis is able to synthesize powder at a much lower temperature compared to other methods. Nevertheless, the synthesis in an aqueous environment causes water to be incorporated into the powder, thus causing deterioration in the electrical properties (Yan ,Razak, 2010).

4.5 Co-precipitation method

In order to prepare the controlled morphology, narrow particle size distribution, high purity and high degree of crystallinity as well as possible reduction in sintering temperature, the co-precipitation method might be a promising route instead of other wet chemical route. Precipitation is the formation of a solid product or powder from a liquid solution which initiated by either changing the solution temperature, pressure, pH or using a chemical precipitation agent so as to exceed the solubility limit of the desired species (Pookmanee, Phanichphant, 2009, Thongtem ,Thongtem, 2004). In general, co-precipitation reaction relies on dissolving the metal salts, commonly metal chlorides, nitrates and hydroxides followed by a rapid pH change to form precipitate. The precipitate must be thoroughly washed to get rid of the impurities from the solutions prior to calcination. It was reported that the well-dispersed particles of about 10 nm began to form a BTO phase at 470°C. The phase formation was complete after a 550°C for 30 minute heat treatment. It was finally sintered at 750°C for 1 hour to achieve a sample of high density of 97.2% (Kan et al., 2002).

4.6 Soft combustion synthesis

The synthesis of BTO powders using combustion reactions, which provides good compositional control, is an alternative synthesis method which worth pursuing. The combustion synthesis enables synthesis at low temperatures and the products obtained are in a finely divided state with large surface areas. Furthermore, the nature of combustion synthesis is characterized by simple experimental set-up, short reaction time between the preparation of the reactants and the availability of the final product and less in external energy consumption (Aruna ,Mukasyan, 2008, Patil et al., 2002). Typically, the mixture of reactants consists of metal nitrate and a suitable organic fuel such as urea, glycine and citric acid. Additionally, the temperature is essential to boil the mixture until the ignition and self-sustaining reaction takes off. The large amount of gases formed can result in the appearance of a flame, which can reach temperatures in excess of 1000°C. In some cases, the external source like simple calcination is necessary to accomplish the synthesis of the appropriate phase. This is because the energy released from the exothermic reaction between the nitrate and the fuel is usually ignited at a temperature much lower than the actual phase formation. Thus, the single phase formation is not ease to produce. Recently, our group had performed a modification on soft combustion synthesis, whereby nitrate salts, Bismuth (Bi) and organic Titanium (IV) isopropoxide (Ti) were dissolved into 2-methaoxyethanol and acetylacetone. In addition, the organic fuel was not used in this work. To introduce the doping content, the Sm^{3+} and Pr^{3+} from nitrate salts were also used. The observation of the soft combustion will be discussed in the following section.

5. Observation during the combustion process

Figure 1 shows the actual condition before and after combustion on BTO and Pr^{3+} doping. The Bi-Ti precursor was observed in clear-yellowish solution (**Figure 1a**) whereas the Pr^{3+} precursor was found in clear-greenish solution (**Figure 1b**). The Sm^{3+} precursor was observed in transparent solution (not shown here). The Bi-Ti precursor was then stirred at 40°C for 2 hours and the colour of the solution changed slightly milky-yellowish as shown in **Figure 1c**. The solution was then continuously evaporated and temperature maintained at ~90°C. The colour of the solution changed. It was found that higher Pr^{3+} doping tends to prolong hydrolysis process. After that, the temperature increased rapidly to ~120°C and the solution completely evaporated, resulting dark-yellowish gel (**Figure 1e**). The gel started to induced ignition at ~150°C. Metal nitrates were decomposed to form metal oxides and nitrogen. The compound was eventually acting as an oxidizer to continue the combustion synthesis, which was accompanied by the releasing of voluminous gases. At the end of this stage, flaming occurred and resulted foamy-like structure as shown in **Figure 1f**. The flaming temperature was found to be approximate 230°C. In order to remove the carbon content in as-combusted powders, the calcination is necessary to enhance the degree of crystallinity with high purity of BTO content.

Fig. 1. Evolution of (a) Bi-Ti precursor (b) Pr precursor (c) Bi-Ti precursor stirred at 40°C for 2 hour (d) Bi-Ti precursor heated at 90°C (e) viscous gel and (f) as-combusted powder.

6. Powder characterization

6.1 Effect of calcination temperature on BTO

X-ray diffraction (XRD) was conducted on the as-combusted powders which calcined at different temperatures and the result is shown in **Figure 2**. The study on different calcination temperature is essential in this work. The main reason is to determine the optimum calcination temperature, which will be used for the following characterization in particular dielectric study. As seen in **Figure 2**, the BTO phase was observed in as-combusted powder. The main peak corresponding to BTO was found at ~29°. However, the presence other intermediate phases such as BTO7.7 and BTO2 were also identified and marked in XRD pattern. At calcination temperature of 600°C, there was a tremendous increase at the main peak (~29°). This inedicates that the presence more BTO phase was observed after calcination process. In addition, the peaks corresponding to intermediate phases were decreased. Further increase calcination temperature, the intermediate phases were gradually eliminated. The intermediate phases completely disappeared at 750 and 800°C. In other words, the BTO phase was successfully formed as single phase at temperature as low as 750°C. It also suggest that the optimum calcination temperature for BTO is 750°C. This temperature is probably lower than other processing route such as conventional solid state reaction and some other wet chemical synthesis (Kan et al., 2003, Pookmanee, 2008). Moreover, as the calcination temperature was increased, the XRD peaks were sharper and the stable phase BTO powders with higher crystallinity could be obtained. **Table 2** presents the variation of lattice parameters and crystallite sizes of BTO calcined at different temperatures. As can be seen in **Table 2**, the corresponding lattice *a*-, *b*-, and *c*-parameters as well as volume cell units, V were observed to exhibit in various values as a result of calcination temperature. In addition, the crystal structure of calcined powder was confirmed to belong orthrhombic, in which it is a typical structure for BTO (Hervoches ,Lightfoot, 1999, Kim ,Jeon, 2004). Besides that, the crystallite size of calcined powder was increased with increasing calcination temperature. The increase in a such way was observed in many studies (Hou et al., 2010, Pookmanee ,Phanichphant, 2009, Umar Al-Amani et al., 2010).

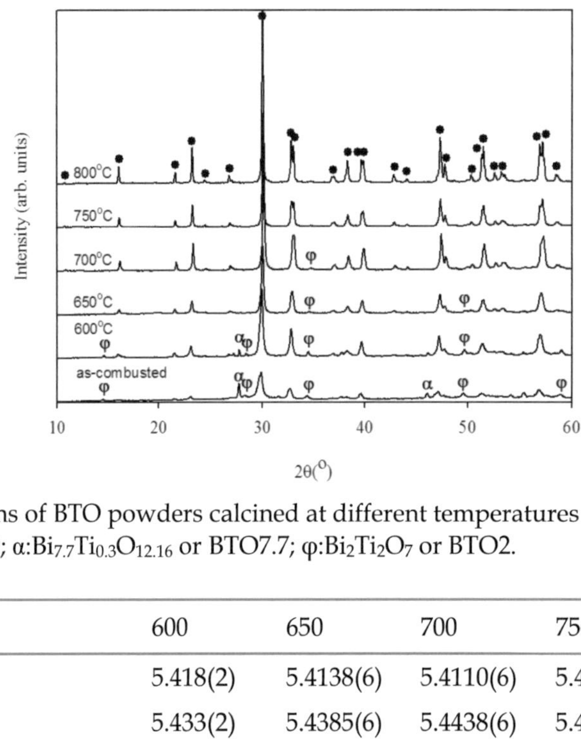

Fig. 2. XRD patterns of BTO powders calcined at different temperatures for 3 hour. •:$Bi_4Ti_3O_{12}$ or BTO; α:$Bi_{7.7}Ti_{0.3}O_{12.16}$ or BTO7.7; φ:$Bi_2Ti_2O_7$ or BTO2.

Calcination, °C	600	650	700	750	800
a/ Å	5.418(2)	5.4138(6)	5.4110(6)	5.4093(2)	5.4066(2)
b/ Å	5.433(2)	5.4385(6)	5.4438(6)	5.4418(2)	5.4429(2)
c/ Å	32.765(6)	32.810(3)	32.853(4)	32.830(1)	32.817(1)
V/Å^3	964.4709	966.0297	967.7028	966.4148	965.7186
Crystallite Size/ nm	24.87	37.19	50.54	71.34	98.07

Table 2. Lattice parameters and crystallite sizes of BTO.

6.2 Effect of Sm^{3+} and Pr^{3+} doping on crystal structure and lattice parameter

The introduction of Sm^{3+} and Pr^{3+} to form $Bi_{4-x}Sm_xTi_3O_{12}$ (BSmT) and $Bi_{4-x}Pr_xTi_3O_{12}$ (BPrT) was expected to have a major changes on calcination temperature, crystal structure, lattice parameter and crystallite size. Based on previous studies, the ionic radii of Sm^{3+} and Pr^{3+} were reported around 0.108 nm and 0.113 nm, respectively, whereas the ionic radii of Bi^{3+} was about 0.117 nm (Garg et al., 2005, Hu et al., 2005). In general, the substitution of Sm^{3+} or Pr^{3+} for Bi^{3+} with larger difference in ionic radii size would lead to larger structure distortion of BTO lattice

(Hu et al., 2005). In addition, the increase in dopant contents would also result in difference in structure distortion (Kan et al., 2004, Kan et al., 2008). In order to determine the effect of Sm^{3+} and Pr^{3+} doping on crystal structure and lattice parameter, the calcined powder for respective contents were analyzed by XRD. Interestingly, the calcination temperature was successfully reduced from 750°C (for BTO) to 650°C (for BSmT and BPrT). According to **Figure 3**, the XRD patterns of BSmT and BPrT powders calcined at 650°C for 3 hour with different Sm^{3+} and Pr^{3+} contents were presented. Regardless of Sm^{3+} and Pr^{3+} contents, the formation of a single phase BTO was observed in **Figure 3(a)** and **Figure 3(b)** for both doping powders. This indicates that the perovskite phase was fully formed in calcined powders and all of them have a bismuth-layered structure. In comparison to the BTO calcined powder at 650°C for 3 hour (see **Figure 2**), the intermediate phases such as BTO2 and BTO7.7 were eliminated as a result of Sm^{3+} and Pr^{3+} doping. This result indicates that the Sm^{3+} and Pr^{3+} ions in the BSmT and BPrT, respectively, are incorporated into the pseudo-perovskite structure, substituting for the Bi^{3+} ions. Besides that, the lattice parameters and crystallite size of BSmT and BPrT were greatly influenced by Sm^{3+} and Pr^{3+} doping, as shown in **Table 3** and **Table 4**, respectively. Based on both tables, the *a*-parameter rapidly approached the *b*-parameter with increasing Sm^{3+} and Pr^{3+} content. The closed value of the *a*- and *b*-parameters was obtained at 1.0 mole of Sm^{3+} and Pr^{3+}, corresponding to the increment in the symmetry of the crystal structure. This also suggested that the orthorhombic structure was formed when doping content was equivalent to 0.25, 0.5 and 0.75 whereas the tetragonal structure was formed doping content was equivalent to 1.0. Further observation shows that the *c*-parameter slightly changed with increasing Sm^{3+} and Pr^{3+} content. It was reported that the variation of *c*-parameter is attributed to the rotation of the TiO_6 octahedron and the reduction in the oxygen deficient (Yoneda et al., 2006). It is evident that Sm^{3+} and Pr^{3+} had substituted in $(Bi_2Ti_3O_{10})^{2-}$ perovskite-type layers and O vacancies in the TiO_6 octahedron; these would eventually result in the shrinkage of unit cell. The crystallite size of BTO was found to be 37.19 nm and decreased continuously to 24.69 nm and 24.56 nm with 1.0 of Sm^{3+} and Pr^{3+}, respectively. It is attributed to the reduction of space charge density and oxygen vacancies with increasing Pr^{3+} content, which act as grain growth inhibiter. This is supported by the finding of Xiang et al. (Xiang et al., 2006). Based on XRD studies, it can conclude that the optimum calcination temperature for BTO was determined at 750°C, whereas the BSmT and BPrT were around 650°C.

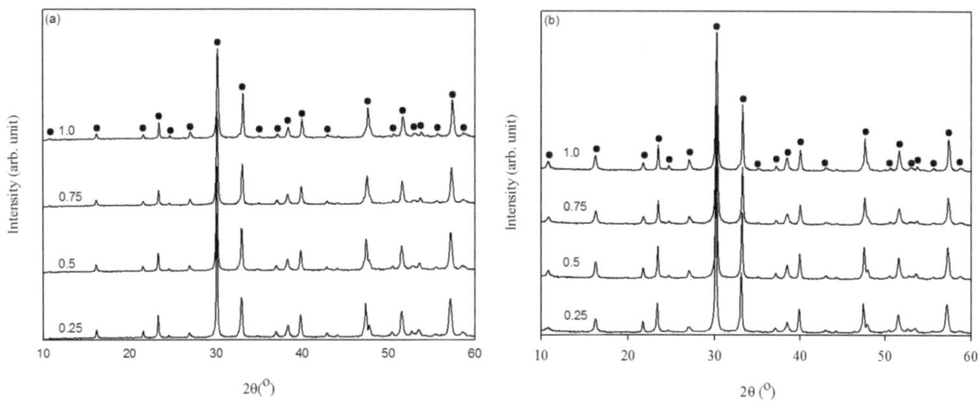

Fig. 3. XRD patterns of (a) BSmT and (b) BPrT powders calcined at 650°C for 3 hour with different Sm^{3+} and Pr^{3+} contents. •:$Bi_4Ti_3O_{12}$ or BTO.

Sm^{3+} content	0.25	0.5	0.75	1.0
a/ Å	5.4076(3)	5.4018(4)	5.3952(5)	5.3910(1)
b/ Å	5.4283(3)	5.4155(4)	5.4059(5)	5.3940(1)
c/ Å	32.798(2)	32.803(2)	32.800(2)	32.777(2)
V/$Å^3$	962.74	959.6153	956.6382	953.0192
Crystallite size/nm	54.89	48.47	45.27	24.69

Table 3. Lattice parameters and crystallite sizes of BSmT.

Pr^{3+} content	0.25	0.5	0.75	1.0
a/ Å	5.4210(9)	5.4157(8)	5.4099(1)	5.4067(8)
b/ Å	5.4306(9)	5.4213(8)	5.4161(1)	5.4054(8)
c/ Å	32.783(3)	32.810(2)	32.816(3)	32.817(3)
V/$Å^3$	965.0938	963.3097	961.527	959.0734
Crystallite size/ nm	36.44	32.86	28.55	24.56

Table 4. Lattice parameters and crystallite sizes of BPrT.

6.3 Grain morphology of BSmT and BPrT powders

To gain an insight into the formation of BTO prepared using different doping content, the calcined powders were monitored by taking field emission scanning electron microscopy (FESEM) micrographs. **Figure 4** shows the morphology of BTO, BSmT and BPrT powders. In order to observe the increase in particle size of BTO, the morphology at 650°C and 750°C were displayed in **Figure 4a** and **Figure 4b**, respectively. It was found that the particle size is relatively expanded with increasing temperature. It was also determined that the particle size in range of 0.1 - 0.2 µm and 0.3 - 0.5 µm were found at 650°C and 750°C, respectively. It clearly observed that plate-like morphology was formed at 750°C instead of 650°C. The formation of such morphology was observed in many studies in which the plate-like structure with highly anisotropic properties is one of typical shape for pure BTO (Chen et al., 2006). In addition, the variation in particle size is also attributed to a greater distortion of perovskite-layer along *ab*-plane (particle length) as compared with *c*-axis (particle thickness). The morphology of BSmT and BPrT powders with different doping contents were observed and depicted in **Figure 4c - 4f**. In this section, the selected micrographs for each dopant with doping content of 0.25 and 1.0 were presented. The selection of the minimum and maximum doping contents is necessary to determine the variation size and shape of resultant particles. It was found that the particle size decreased with increasing doping content, corresponding to the lower diffusivity of both doping content compared to Bi^{3+}, resulting to the suppression of the grain growth (Goh et al., 2009). It was determined that the particle size in range of 0.2 - 0.4 µm and 0.1 - 0.2 µm were observed when Sm^{3+} contents were equivalent to 0.25 and 1.0, respectively. Meanwhile, the average particle size in range of 0.1 - 0.2 µm and 0.05 - 0.1 µm were found when Pr^{3+} contents were equivalent to 0.25 and 1.0, respectively. It is also noticed that the size of plate-like particle decreased relatively with increasing Sm^{3+} and Pr^{3+}, corresponding to the greater relaxation in the perovskite-layer. In order to see the difference of the particle size between Sm^{3+} and Pr^{3+}, the doping content was fixed at 0.25. As can be seen in **Figure 4c** and **Figure 4e** for Sm^{3+} and Pr^{3+}, respectively, the particle size of Sm^{3+} doping was found to substantially larger than Pr^{3+} doping. This might be attributed to the difference in ionic radii which also resulted in different diffusivity.

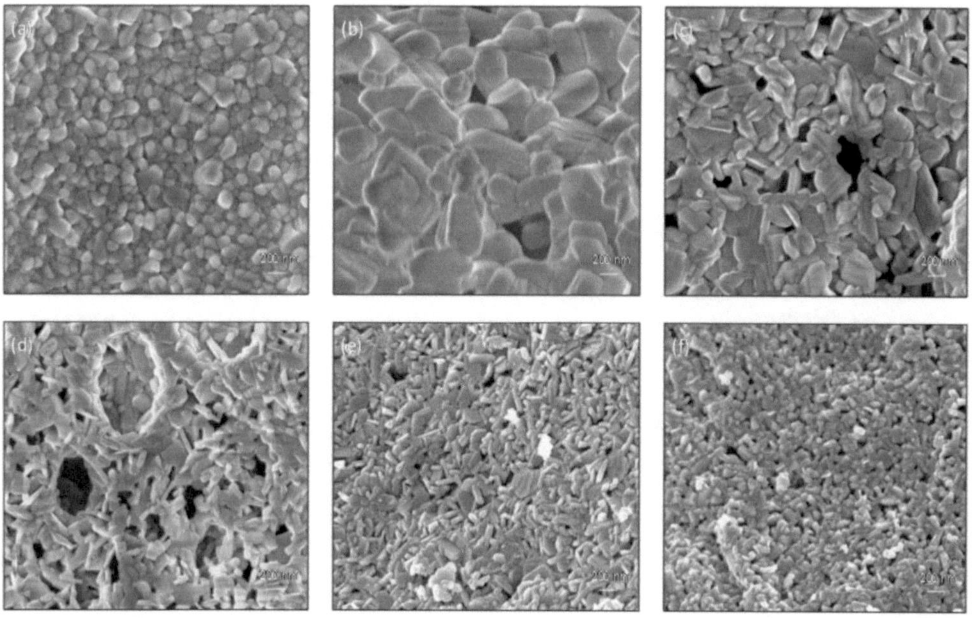

Fig. 4. Morphologies of (a) BTO:650°C, (b) BTO:750°C, (c) BSmT:0.25, (d) BSmT:1.0, (e) BPrT:0.25 and (f) BPrT:1.0.

6.4 Comparison of lattice vibration BTO and Sm^{3+} doping

In order to enhance the understanding of the doping effect from the structural point of view, Raman scattering study is a very useful tool for investigating the lattice vibrational modes, which can provide details of lattice vibrations changes. **Figure 5** shows the Raman spectra of BTO and BSmT powders at room temperature from 100 to 2000 cm^{-1}. Theoretically, the Raman selection rules allow 24 Raman active modes for orthorhombic BTO. (Kojima, 2000, Kojima ,Shimada, 1996). However, as shown in **Figure 5a**, the Raman spectrum of BTO less than 9 active modes were observed which is partially due to the possible overlap of the same symmetry vibrations or the weak features of some Raman bands (Liang et al., 2009). As can be seen in **Figure 5a**, the Raman modes at 193, 228, 267, 330, 353, 537, 563, 614 and 850 cm^{-1} were observed in BTO. All the Raman modes are also characterized as the vibrational modes of BTO which can be classified as internal modes of TiO_6 octahedra. According to Kojima et al. (Kojima ,Shimada, 1996), the internal modes of TiO_6 octahedra appear above 200 cm^{-1}. The mode at 850 cm^{-1} is attributed to the symmetric Ti – O stretching vibration of atom inside the TiO_6 octahedron whereas the mode at 614 cm^{-1} corresponds to the symmetry one. The two modes at 537 and 563 cm^{-1} correspond to the opposing excursions of the external apical oxygen (O) atoms of the $Ti0_6$ octahedron. The 228 and 267 cm^{-1} modes are ascribed to the O – Ti – O bending vibration. Although the mode at 228 cm^{-1} is Raman inactive according to the *Oh* symmetry of TiO_6, it is often observed because of the distortion of octahedron. The mode at 330 cm^{-1} was from a combination of the stretching and bending vibrations of the TiO_6 octahedron. In addition, the formation of BTO with orthorhombic structure is identified by the splitting mode at 193 and 228 cm^{-1}, and 537 and 563cm^{-1}. Nevertheless, the Raman modes of the lower wavenumber at 116 cm^{-1} was not found in this spectrum, to show the vibrations between Bi and O atoms. However, the reason of missing mode is still not clear. The effect of Sm^{3+} doping on the structure change of BTO on the basis of the Raman modes is presented in **Figure 5b**. It was clearly observed that the peak intensity decreased with increasing Sm^{3+} contents from 0.25 to 0.75. It is believed to be associated with strong interactions between the ionic bonds; corresponding to the smaller ionic radius of Sm^{3+} (0.108 nm) compared with Bi^{3+} (0.117 nm). With the increase of Sm content, the distortion structure would be more and the grain size would be smaller. This finding is in line with the XRD pattern and FESEM micrograph. It was reported that the duplet peaks observed in Raman spectra tend to merge into one mode when the Bi^{3+} is substituted by rare-earth elements (Wu et al., 2005). Similar observation was discovered in BST. In the present work, the duplet peaks at 228-267 cm^{-1}, 330-353 cm^{-1} and 537-563 cm^{-1} were found to merged into a single peak at 264 cm^{-1}, 322 cm^{-1} and 542 cm^{-1} for x= 0.25 and 264 cm^{-1}, 330 cm^{-1} and 553 cm^{-1} for x= 0.75.

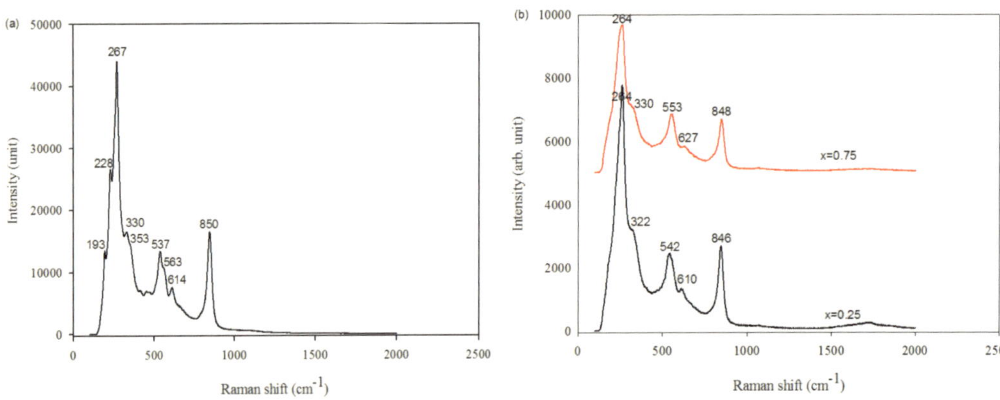

Fig. 5. Raman spectra of (a) BTO and (b) BSmT at different Sm^{3+} content.

7. Bulk ceramic characterization

7.1 Effect of sintering temperature of BTO

Figure 6 shows the XRD patterns of BTO ceramics sintered at different temperature for 3 hour. As can be seen from this figure, the BTO was formed with random oriented grains in which the strongest peak was found at (117) instead of (00l). Besides that, the increase in calcination temperature also implies the improvement of crystallinity and the enhancement of crystallite size. This can be explained by the width of the diffraction lines, which decreased, whilst the intensity increased. The crystallite sizes for ceramics sintered at 900, 1000 and 1100oC were calculated to be approximately 104.66, 126.52 and 130.22 nm, respectively.

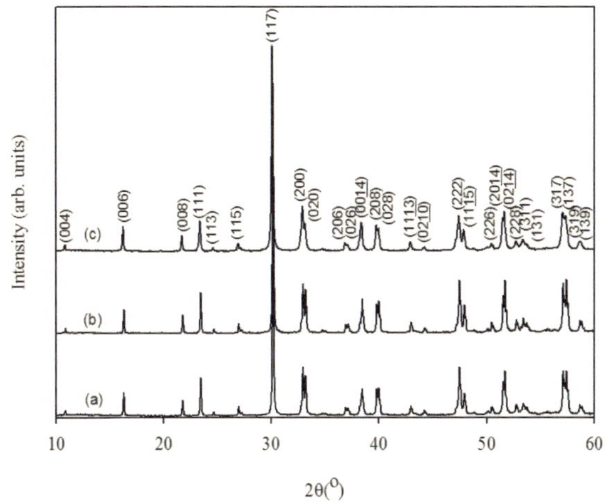

Fig. 6. XRD patterns of BTO ceramics sintered at different temperatures for 3 hour: (a) 900°C, (b) 1000°C and (c) 1100°C.

7.2 Effect of Sm^{3+} and Pr^{3+} doping on grain orientation

The preferred grain orientation was obviously affected by rare-earth doping in BTO particularly after sintering. Kannan et al. (Kannan et al., 2006) reported the reflections corresponding to (00*l*) plane along the *c*-axis was observed with increasing Nd^{3+} content from 0 to 0.25. It can give us another hint that preferred grain orientation would be different when rare-earth was introduced into BTO lattice. As discussed in **Figure 6**, the pure BTO was formed with no preferred grain orientation. In **Figure 7**, the XRD patterns of BST and BPT ceramics sintered at 1100°C for 3 hour were presented. As seen from this figure, the Sm^{3+} and Pr^{3+} doping show a highly *c*-axis oriented growth with increasing Sm^{3+} and Pr^{3+} contents. The XRD peak corresponding to (00*l*) plane was clearly observed with higher intensity as compared to (117) plane. To simplify the discussion, the peak at (00$\underline{14}$) and (117) are taken into consideration to determine the degree of *c*-axis orientation, α_c by Lotgering factor (Yang et al., 2008):

$$\alpha_c = I(00\underline{14})/[I(00\underline{14}) + I(117)]$$

The values of α_c calculated for the Sm^{3+} and Pr^{3+} doping with various contents are listed in **Table 5**. It was found that the degree of *c*-axis orientation increased with increasing Sm^{3+} and Pr^{3+} content. This indicates that the doping content has a significance result on grain orientation.

Sm^{3+} content	α_c (%)	Pr^{3+} content	α_c (%)
0.25	41.40	0.25	48.34
0.5	58.12	0.5	86.61
0.75	58.89	0.75	96.78
1.0	86.17	1.0	98.79

Table 5. Lotgering factor of the degree of *c*-axis orientation.

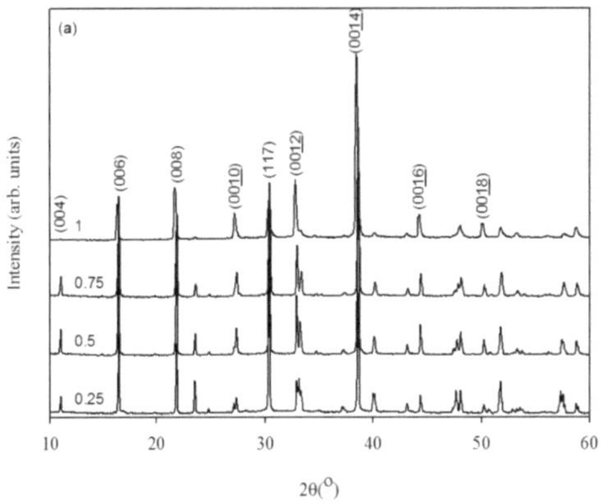

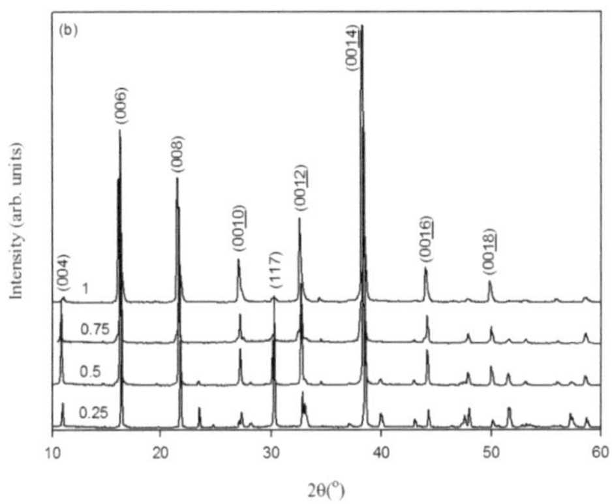

Fig. 7. XRD patterns of (a) BSmT and (b) BPrT ceramics sintered at 1100°C for 3 hour.

7.3 Microstructure of bulk ceramics after sintering

It is more interesting to observe the microstructure of bulk ceramics after sintering, as shown in **Figure 8**. Prior to view the micrograph using field emission scanning electron microscopy (FESEM), the surface of ceramics were polished on SiC papers grit 1000 followed by finer grit 2000. The polished ceramics were placed into ultrasonic for 10 minute to remove contaminants. The polished ceramics were thermally etched with temperature of 100°C lower than the sintering temperature for 30 minute. As can be seen in **Figure 8a**, the microstructure of the BTO ceramic shows a random arrangement of elongated-like grains, several of which are highly elongated. On the other hand, the microstructure of the BSmT and BPrT ceramics show a random arrangement of plate-like grains, as observed in **Figure 8b-8e**. It was also noticed that the average grain size relatively decrease with increasing Sm^{3+} and Pr^{3+} contents, indicating a strong influence of doping concentration which resulted in less amount of Bi^{3+} in BTO. The micrographs also revealed that the plate-like grains were not homogeneously distributed when Sm^{3+} and Pr^{3+} are equivalent to 0.25. Nevertheless, homogeneous microstructures with small grain size were found from the BSmT and BPrT ceramics with 1.0. The resultant micrographs was mainly attributed to a greater suppresion of the Bi^{3+} volatility by substitution of low diffusivity of Sm^{3+} and Pr^{3+}, which eventually inhibits the grain growth.

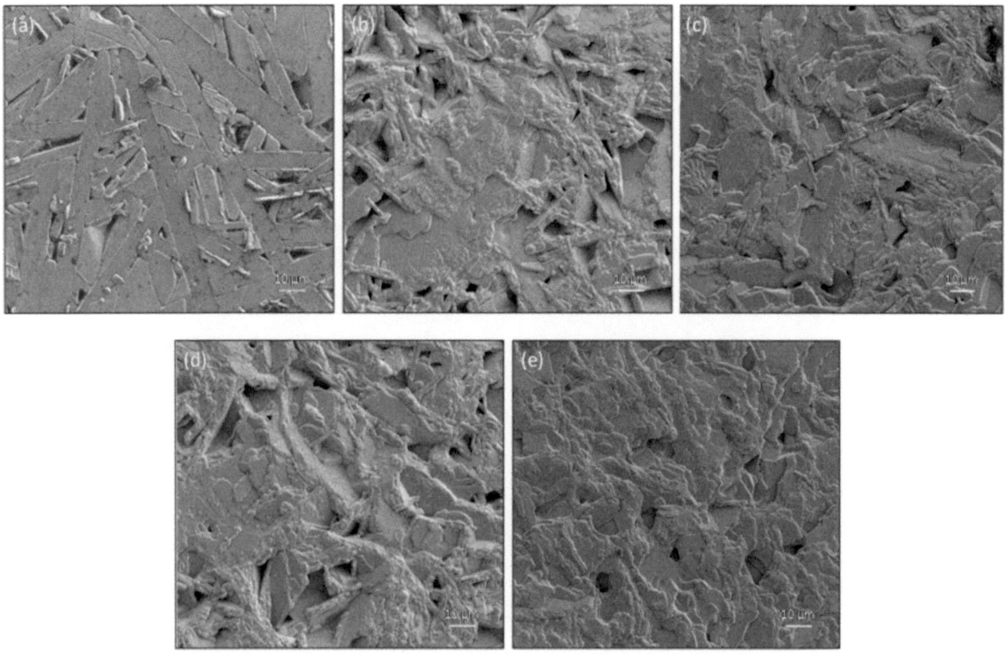

Fig. 8. FESEM micrograph of (a) BTO, (b) BSmT:025, (c) BSmT:1.0, (d) BPrT:025 and (e) BPrT:1.0, sintered at 1100°C for 3 hour.

7.4 Effect of sintering temperature and Sm^{3+}, Pr^{3+} content on relative density

The effect of sintering temperature on relative density of BTO ceramics was studied and depicted in **Figure 9a**. With increasing temperature, the relative density of ceramics was also increased up to 93% at 1100°C. This indicates that the densification behavior of BTO ceramics is temperature dependent. Thus, the sintering temperature at 1100°C was used for densification process with doping content. **Figure 9b** and **Figure 9c** show the relative density of Sm^{3+} and Pr^{3+} doping, respectively. As seen, the densities of BTO with doping content are different from one to another. It was determined that the densities was in range of 92 - 95%, indicating that a slight improvement as compared to pure BTO. The increase in relative density is associated to the decrease in Bi-loss during sintering resulting from the substitution effect by Sm^{3+} and Pr^{3+}. It can be said the small difference in relative density is another indicator to show the improvement of densification behavior in a such way.

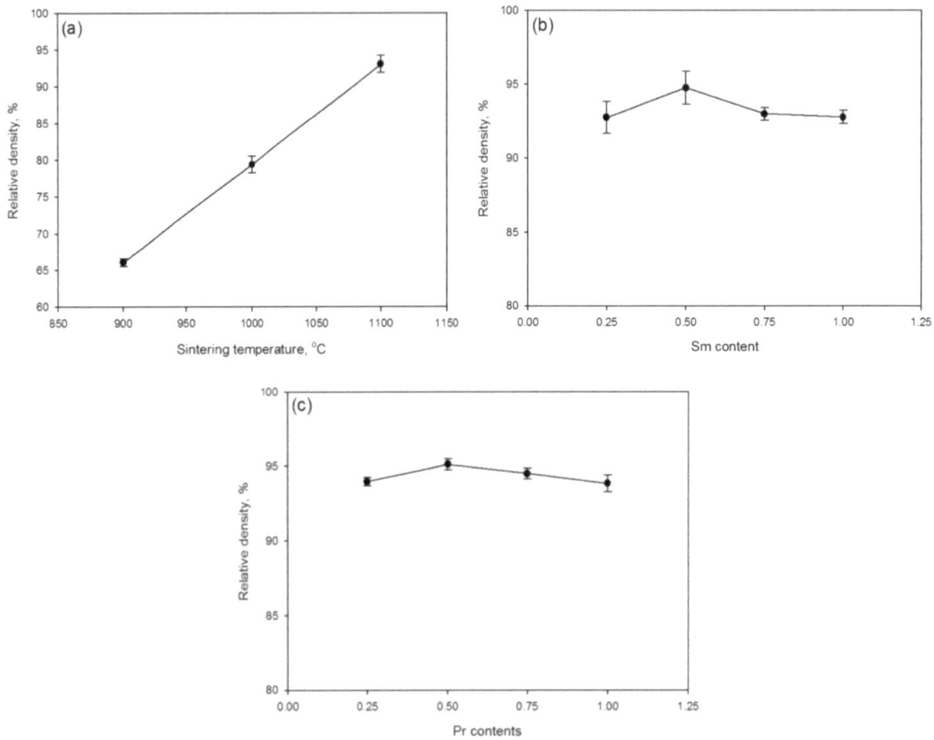

Fig. 9. Relative density of (a) BTO at different sintering temperatures, (b) BSmT with different Sm^{3+} contents and (c) BPrT with different Pr^{3+} contents; sintered at 1100°C.

7.5 Effect of Sm^{3+} and Pr^{3+} doping on dielectric properties

The effect of Sm^{3+} and Pr^{3+} contents in BSmT and BPrT on dielectric properties were studied. In this study, the measurement of dielectric constant, ε_r and dielectric loss, tan δ were performed at 1 kHz and at room temperature, 25°C and the results were presented in **Figure 10**. The variation of ε_r and tan δ were clearly observed in **Figure 10a** and **Figure 10b** for the BSmT and BPrT ceramics, respectively. The ε_r of the BSmT ceramics were in the range between 78 and 95, whereas the ε_r of the BPrT ceramics were in the range between 75 an 105. This indicates that the ε_r of the BPrT ceramics was slightly larger than the BSmT ceramics. This is associated to the larger ionic radii of Pr^{3+} than that of Sm^{3+}. This result can be explained by a shift of TiO^6 octahedra in a layered structure due to the substitution of larger ionic radii than that of Bi^{3+}. It was also noticed that the tan δ abruptly decreased with increasing Sm^{3+} and Pr^{3+} contents from 0.25 to 0.5. The decrement in the tan δ was attributed to a better electric flux caused by the reduction of grain imperfection. It was also supported by the increase in relative density, in which the ceramics appeared to be dense. Above 0.5, the tan δ were almost consistent with small difference in its value, corresponding to the reduction of the defects such as bismuth and oxygen vacancies. In order to see possible application as dielectric antenna, the dielectric study at different frequencies was performed. It was reported that the development of wireless technologies application requires very stringent criteria for dielectric ceramics materials. Typically, the dielectric ceramic materials must have a high dielectric constant, low dielectric loss and a a thermally stable resonant frequency coefficient (LazareviÄ‡ et al., 2005). Thus, this study is essential to determine the potential application of such a field for the BTO, BSmT and BPrT ceramics.

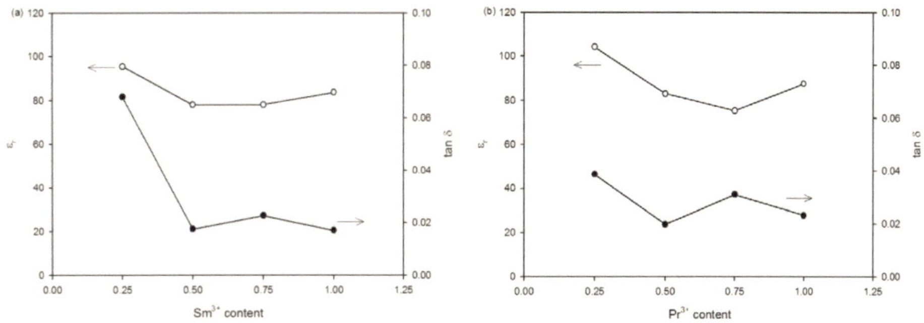

Fig. 10. Dielectric constant, ε_r and dielectric loss, tan δ at different doping contents: (a) BSmT and (b) BPrT.

7.6 Effect of various frequencies on the dielectric properties of the BTO, BSmT and BPrT ceramics

As reported in previous studies, the dielectric constant, ε_r and dielectric loss, tan δ were strongly dependent on frequency (Rachna et al., 2010, SimÃµes et al., 2008). In this work, the dielectric poperties were measured at different frequency ranges from 1 MHz to 1 GHz. As it can be seen from **Figure 11**, the ε_r of the BTO ceramic shows very obvious dispersion with

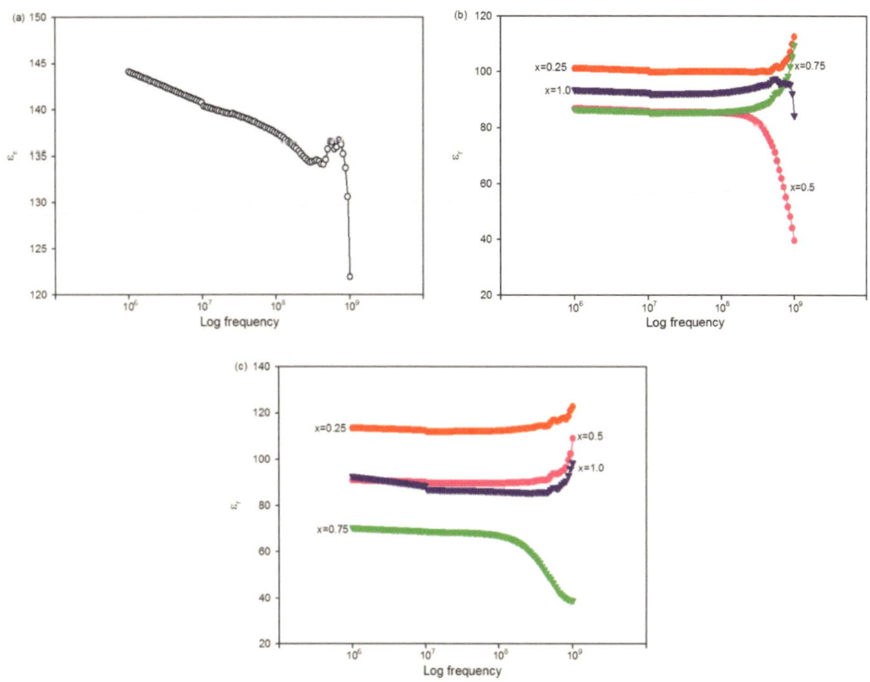

Fig. 11. Dielectric constant, ε_r of (a) BTO, (b) BSmT and (c) BPrT measured at high frequency range at room temperature.

frequency, indicating that the corresponding ceramic possess high defect concentration such as bismuth and oxygen vacancies. On the other hand, the ε_r of the BSmT and BPrT ceramics show very little dispersion from 1 MHz to 100 MHz. However, the ε_r shows very obvious dispersion above 100 MHz. This indicates the ε_r will be more complex at higher frequency range between 100 MHz and 1 GHz. The dielectric loss, tan δ at different frequencies was depicted in **Figure 12**. It was found that the tan δ of was slowly increased from 1 MHz to 10 MHz and abruptly increased from 10 MHz to 1 GHz, as shown in **Figure 12a**. In addition, the presence of relaxation peak in the tan δ was observed, as shown in inset **Figure 12a**. This indicates that the relaxation peak was observed approximately 700 MHz. It can be said that the increase trend in tan δ was also found in the BSmT and BPrT ceramics, as shown in **Figure 12b** and **Figure 12c**, respectively. Furthermore, the relaxation peaks in the tan δ was also observed around 700 MHz, which is almost comparable to BTO ceramic. The dielectric loss relaxation peak phenomenon can be explained by the Debye-like model for relaxation effects. The dielectric loss peak is maximal at the resonant frequency, which is the reciprocal of the relaxation time (Sulaiman et al., 2010). The dielectric loss relaxation may be generated by several possible factors such as surface roughness, distribution grain sizes and many more (Sulaiman et al., 2010, Wang et al., 2010). From the FESEM micrographs (see **Figure 8c to Figure 8e**) revealed that BSmT and BPrT with 1.0 have homogeneous distribution of grain sizes compared to 0.25.

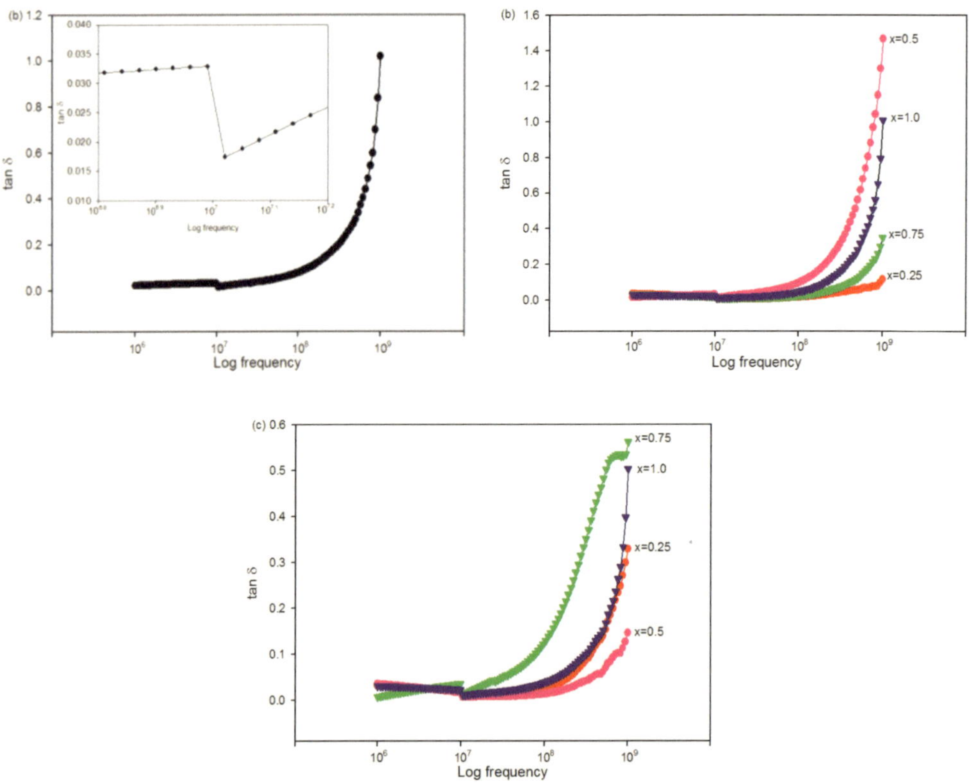

Fig. 12. Dielectric loss, tan δ of (a) BTO, inset showing the Debye relaxation effect, (b) BSmT and (c) BPrT measured at high frequency range at room temperature.

Therefore, the BSmT and BPrT with 1.0 gave a lower dielectric loss at relaxation frequency due to the above reason. Besides that, the improved ε_r with little dispersion and small variation in tan δ (or almost constant) can also suggest that the BSmT and BPrT ceramics are possible to be applied for wireless dielectric antenna applications instead of the BTO ceramic. However, a detail study is necessary to focus on return loss with a specific design of ceramic.

8. Conclusion

Based on this work, the rare-earth doping i.e. Sm^{3+} and Pr^{3+} had successfully improved the processing and properties of pure BTO ceramics. The calcination temperature was greatly reduced from 750°C to 650°C in order to form a single phase structure. The particle size of plate-like structure decreased continuously with increasing Sm^{3+} and Pr^{3+} content. The peak intensity and peak width in Raman spectrum were apparently low and broaden with increasing Sm^{3+} content. The Lotgering factor showed the enhancement in degree of *c*-axis orientation. The microstructure of the Sm^{3+} and Pr^{3+} doping showed a random arrangement of plate-like grains in which the grain size was relatively decrease at higher doping content. A great in densification behavior was also observed with Sm^{3+} and Pr^{3+} doping which resulted in the relative density of about 92-95% at 1100°C. The dielectric constant, ε_r of the BPrT ceramics was slightly larger than the BSmT ceramics, which can be explained in terms of larger ionic radii of Pr^{3+} than that of Sm^{3+}. The dielectric loss, tan δ of the BSmT and BPrT ceramics were greatly improved when dopant content above 0.5. For frequency study, the the ε_r of the BSmT and BPrT ceramics show very little dispersion from 1 MHz to 100 MHz instead of above 100 MHz. The relaxation peak in tan δ was observed approximately 700 MHz for all ceramics with different dopant contents. Based of frequency study, the BSmT and BPrT can be used as potential wireless dielectric antenna applications.

9. Acknowledgment

The authors appreciate the technical support from the School of Materials and Mineral Resources Engineering, Universiti Sains Malaysia. This research was supported by the USM Short term grant 6035276, USM-RU grant 1001/PBahan/8042018 and 811069.

10. References

Algueró, M., Ferrer, P., Vila, E., Iglesias, J. E. & Castro, A., Journal of the American Ceramic Society, 89 (2006) 3340.

Armstrong, R. A. & Newnham, R. E., Materials Research Bulletin, 7 (1972) 1025.

Aruna, S. T. & Mukasyan, A. S., Current Opinion in Solid State and Materials Science, 12 (2008) 44.

Besland, M. P., Djani-Ait Aissa, H., Barroy, P. R. J., Lafane, S., Tessier, P. Y., Angleraud, B., Richard-Plouet, M., Brohan, L. & Djouadi, M. A., Thin Solid Films, 495 (2006) 86.

Chen, W., Kinemuchi, Y., Watari, K., Tamura, T. & Miwa, K., Journal of the American Ceramic Society, 89 (2006) 490.

Cui, L. & Hu, Y. J., Physica B: Condensed Matter, 404 (2009) 150.
Du, X., Xu, Y., Ma, H., Wang, J. & Li, X., Journal of the American Ceramic Society, 90 (2007) 1382.
Du, X., Xu, Y., Ma, H., Wang, J. & Li, X., Journal of the American Ceramic Society, 91 (2008) 2079.
Garg, A., Hu, X. & Barber, Z. H., Ferroelectrics, 328 (2005) 93.
Goh, P. Y., Razak, K. A. & Sreekantan, S., Journal of Alloys and Compounds, 475 (2009) 758.
Golda, R. A., Marikani, A. & Padiyan, D. P., Ceramics International, (2011)
Guo, D. Y., Li, M. Y., Liu, J., Fu, L., Wang, J., Yu, B. F. & Yang, B., Materials Science and Engineering B: Solid-State Materials for Advanced Technology, 142 (2007) 135.
Han, K. & Ko, T., Journal of Alloys and Compounds, 473 (2009) 490.
Hardy, A., Mondelaers, D., Van Bael, M. K., Mullens, J., Van Poucke, L. C., Vanhoyland, G. & D'Haen, J., Journal of the European Ceramic Society, 24 (2004) 905.
Hervoches, C. H. & Lightfoot, P., Chemistry of Materials, 11 (1999) 3359.
Hou, J., Kumar, R., Qu, Y. & Krsmanovic, D., Journal of Nanoparticle Research, 12 (2010) 563.
Hu, X., Garg, A. & Barber, Z. H., Thin Solid Films, 484 (2005) 188.
Kan, Y. M., Zhang, G. J., Wang, P. L. & Cheng, Y. B., Journal of the European Ceramic Society, 28 (2008) 1641.
Kan, Y., Jin, X., Wang, P., Li, Y., Cheng, Y.-B. & Yan, D., Materials Research Bulletin, 38 (2003) 567.
Kan, Y., Jin, X., Zhang, G., Wang, P., Cheng, Y. B. & Yan, D., Journal of Materials Chemistry, 14 (2004) 3566.
Kan, Y., Wang, P., Li, Y., Cheng, Y. B. & Yan, D., Materials Letters, 56 (2002) 910.
Kannan, C. V., Cheng, Z. X., Kimura, H., Shimamura, K., Miyazaki, A. & Kitamura, K., Journal of Crystal Growth, 292 (2006) 485.
Ke, H., Wang, W., Chen, L., Xu, J., Jia, D., Lu, Z. & Zhou, Y., Journal of Sol-Gel Science and Technology, 53 (2010) 135.
Kim, J. S. & Kim, S. S., Applied Physics A: Materials Science and Processing, 81 (2005) 1427.
Kim, J. S., Integrated Ferroelectrics, 79 (2006) 139.
Kim, Y. I. & Jeon, M. K., Materials Letters, 58 (2004) 1889.
Kojima, S. & Shimada, S., Physica B: Condensed Matter, 219-220 (1996) 617.
Kojima, S., Ferroelectrics, 239 (2000) 55.
Kojima, T., Yoshida, I., Uekawa, N. & Kakegawa, K., Journal of the European Ceramic Society, 29 (2009) 431.
Kong, L. B., Ma, J., Zhu, W. & Tan, O. K., Materials Letters, 51 (2001) 108.
Lazarevic, Z. Stojanovic, B. D. & Varela J. A., Science of Sintering, 37 (2005) 199.
Liang, K., Qi, Y. & Lu, C., Journal of Raman Spectroscopy, 40 (2009) 2088.
Lu, C. J., Qiao, Y., Qi, Y. J., Chen, X. Q. & Zhu, J. S., Applied Physics Letters, 87 (2005) 1.
Macedo, Z. S., Ferrari, C. R. & Hernandes, A. C., Powder Technology, 139 (2004) 175.
Newnham, R. E., Wolfe, R. W. & Dorrian, J. F., Materials Research Bulletin, 6 (1971) 1029.
Ng, S. H., Xue, J. & Wang, J., Journal of the American Ceramic Society, 85 (2002) 2660.
Park, B. H., Kang, B. S., Bu, S. D., Noh, T. W., Lee, J. & Jo, W., Nature, 401 (1999) 682.
Patil, K. C., Aruna, S. T. & Mimani, T., Current Opinion in Solid State and Materials Science, 6 (2002) 507.

Pookmanee, P. & Phanichphant, S., Journal of Ceramic Processing Research, 10 (2009) 448.
Pookmanee, P., Journal of Ceramic Processing Research, 9 (2008) 30.
Pookmanee, P., Uriwilast, P. & Phanichpant, S., Ceramics International, 30 (2004) 1913.
Rachna, S., Bhattacharyya, S. & Gupta, S. M., Materials Science and Engineering B: Solid-State Materials for Advanced Technology, 175 (2010) 207.
Santos, V. B., M'Peko, J. C., Mir, M., Mastelaro, V. R. & Hernandes, A. C., Journal of the European Ceramic Society, 29 (2009) 751.
Shi, Y., Cao, C. & Feng, S., Materials Letters, 46 (2000) 270.
Simões, A. Z., Stojanovic, B. D., Ramirez, M. A., Cavalheiro, A. A., Longo, E. & Varela, J. A., Ceramics International, 34 (2008) 257.
Stojanovic, B. D., Paiva-Santos, C. O., Cilense, M., Jovalekic, C. & Lazarevic, Z. Z., Materials Research Bulletin, 43 (2008) 1743.
Stojanovic, B. D., Paiva-Santos, C. O., Jovalekic, C., Simoes, A. Z., Filho, F. M., Lazarevic, Z. & Varela, J. A., Materials Chemistry and Physics, 96 (2006) 471.
Stojanovic, B. D., Simoes, A. Z., Paiva-Santos, C. O., Quinelato, C., Longo, E. & Varela, J. A., Ceramics International, 32 (2006) 707.
Subbarao, E. C., Journal of Physics and Chemistry of Solids, 23 (1962) 665.
Subbarao, E. C., Physical Review, 122 (1961) 804.
Sulaiman, M. A., Hutagalung, S. D., Ain, M. F. & Ahmad, Z. A., Journal of Alloys and Compounds, 493 (2010) 486.
Thongtem, T. & Thongtem, S., Ceramics International, 30 (2004) 1463.
Umar Al-Amani, A., Srimala, S., Ahmad Fauzi, M. N. & Khairunisak, A. R., Malaysian Journal of Microscopy, 6 (2010) 69.
Wang, C. M., Lin, S. Y., Kao, K. S., Chen, Y. C. & Weng, S. C., Journal of Alloys and Compounds, 491 (2010) 423.
Wang, X. S., Zhang, Y. J., Zhang, L. Y. & Yao, X., Applied Physics A: Materials Science and Processing, 68 (1999) 547.
Watcharapasorn, A., Siriprapa P. & Jiansirisomboon S., Journal of the European Ceramic Society, 30 (2010) 87.
Wu, D., Deng, Y., Mak, C. L., Wong, K. H., Li A. D., Zhang M. S. & Ming N. B., Applied Physics A: Materials Science and Processing, 80 (2005) 607.
Xiang, P. H., Kinemuchi, Y. & Watari, K., Materials Letters, 60 (2006) 2837.
Xie, L., Ma, J., Wu, P., Tian, H., Zhao, Z., Zhou, J., Hu, Y., Wang, Y., Tao, J. & Zhu, X., Materials Research Bulletin, 42 (2007) 389.
Xue, K. H., Celinska, J. & Paz De Araujo, C. A., Applied Physics Letters, 95 (2009)
Yan, H., Zhang, H., Zhang, Z., Ubic, R. & Reece, M. J., Journal of the European Ceramic Society, 26 (2006) 2785.
Yan, N. C. & Razak, K. A., Journal of Alloys and Compounds, 509 (2011) 942–947
Yang, B., Zhang, D. M., Zhou, B., Huang, L. H., Zheng, C. D., Wu, Y. Y., Guo, D. Y. & Yu, J., Journal of Crystal Growth, 310 (2008) 4511.
Yang, Q., Li, Y., Yin, Q., Wang, P. & Cheng, Y. B., Journal of the European Ceramic Society, 23 (2003) 161.
Yoneda, Y., Kohara, S. & Mizuki, J., Japanese Journal of Applied Physics, Part 1: Regular Papers and Short Notes and Review Papers, 45 (2006) 7556.

Zdujić, M., Jovalekić, Č., Poleti, D., Veljković, I. & Karanović, L., Journal of Non-Crystalline Solids, 352 (2006) 3058.

Zhi-hui, C., Jun-fu, Q., Cheng, L., Jian-ning, D. & Yuan-yuan, Z., Ceramics International, 36 (2010) 241.

17

High Pressure Sintering of WC-10Co Doped with Rare-Earth Elements

C.M.F.G. Marques, G.S. Bobrovnitchii and J.N.F Holanda
Northern Fluminense State University - UENF, Laboratory of Advanced Materials - LAMAV, Campos dos Goytacazes, RJ, Brazil

1. Introduction

The cemented carbides are usually known as the hard metals. They are typically two phase composite materials, in which a hard refractory carbide phase is bonded by an iron-group metal, usually metallic cobalt (Thümmler e Oberacker, 1993). This is due to its ability to wetting the carbide particles, which allows the agglomeration of particles, giving the material toughness. The tungsten carbide (WC) due to its metallic nature presents an excellent thermal conductivity. Thus, the cemented carbide is a typical example of composite material.

The properties of cemented carbides are related to a combination of the properties of individual constituents. Thus, the attractive properties of such materials are related to the carbide hardness combined with toughness of the binder metal. When compared to other systems, WC-Co exhibits the best combination of mechanical strength, wear resistance and good toughness (Schwarzkopf and Kieffer, 1986). Considering this fact, WC-Co is an unique system in this field of materials.

The demands of the industrial and economical development worldwide at the beginning of the twentieth century resulted in the development of the cemented carbide (Williams, 1998). Since that time the cemented carbide is a class of materials of great technological importance, whose main applications are in machining operations and as tooling for metal forming or cutting. In addition, the cemented carbides can also be applied in drawing dies, drawing, stamping and metal forming, mining and wear-resistant pieces. Thus, these materials are applied in those activities that require high abrasion resistance, hardness and impact resistance, such as heavy machining and drilling.

The properties of the cemented carbides can be modified from changes in the proportions of the materials used in its manufacture. There are several cemented carbide compositions in which the cobalt phase can vary from 3 to 30 wt.%. However, the most usual composition and of high interest in this chapter is that based on 90 wt.% tungsten carbide (WC) and 10 wt.% cobalt (Co) (German, 1994).

The cemented carbide is conventionally produced by powder metallurgy processing via liquid phase sintering. The elemental powders are comminuted, mixed, compacted, and sintered under various conditions. The processes commonly used are pressing/sintering, hot pressing, and sinter/hot isostatic pressing (HIP) (North et al., 1991; White, 1998). Recently, an unusual process for the production of cemented carbide using a high pressure technique has been investigated (Rodrigues et al., 2006). On the other hand, the effect of the rare-earth elements addition on the sintering behavior of cemented carbide has also been investigated. In particular, the beneficial effects of the rare-earth elements on the mechanical properties, microstructure and cutting performance have been extensively investigated (Li et al., 1986; Yao et al., 1987; Liang et al., 1989; Shan, 1990; Luo, 1991; Liu et al., 1992; Chenguang, 1992; Pan, 1993; Yang et al., 1993; Deng, 1993; Li et al, 1993; He et al., 1994; Li et al, 1994; Yan et al., 1995; Cheng and Yu, 1995; Yuan et al., 1995; Li, 1996; He, 1996; Ji et al., 1996; Xu et al., 2001; Gomes, 2004). However, the production of cemented carbide doped with rare-earth elements using a high pressure technique is still to be investigated.

This chapter focuses on the possibility of production of WC-10wt.%Co doped with rare-earth elements under high pressure and high temperature (HPHT) conditions. Emphasis is given on the effects of the rare-earth additions on the densification behavior and physical and mechanical properties of the end product.

2. Preparation and testing of the cemented carbide

2.1 Materials and mixture preparation

The materials used in powder form were: tungsten carbide (WC), cobalt (Co), cerium oxide (CeO_2), lanthanum oxide (La_2O_3), calcite ($CaCO_3$), and graphite.

The tungsten carbide powder of high purity used was provided by Wolfram Bergbau-Und-GmbH, whose main characteristics are presented in Table 1. The cobalt powder of high purity was provided by Vetec Química Fina, whose grain size is < 400 mesh. The rare-earth oxides (> 99 %) were provided by Vetec Química Fina. The calcite and graphite were used in assembling the reactive cell in the high-pressure experiments. The graphite powder of high purity was provided by Nacional Grafite.

Characteristic	Range
Total C, wt.%	6.15
Free C, wt.%	0.04
Density, g/cm^3	15.65
Mean particle size, μm	1.07
Specific surface (BET), m^2/g	0.42

Table 1. Characteristics of the WC powder.

Several WC10wt.%Co mixtures containing up to 2 wt.% of La_2O_3 and CeO_2 of the cobalt phase were prepared (Table 2). The samples of dry powders were mixed and homogenized in a laboratory mixer during 1 h.

Samples	WC10wt.%Co	Rare-earth	
		La_2O_3	CeO_2
AL1	100.0	-	-
AL2	99.5	0.5	-
AL3	99.0	1.0	-
AL4	98.5	1.5	-
AL5	98.0	2.0	-
AL6	99.5	-	0.5
AL7	99.0	-	1.0
AL8	98.5	-	1.5
AL9	98.0	-	2.0

Table 2. Composition of the cemented carbides doped with rare-earths.

2.2 High pressure sintering of the cemented carbide

The mixed powders initially were submitted to a single action die-compaction step using a hydraulic press (Danpresse, model DC 20) in a cylindrical 7 mm diameter steel die without lubrificants at 800 MPa. The compact dimensions were 7 mm in diameter and 7 mm in height. After uniaxial compaction, the consolidated samples were pressed under high pressure and high temperature (HPHT).

The high pressure compaction needs an apparatus that can stably generate a pressure of more than 1 GPa. A toroidal type high pressure device was used in the experiments. Fig. 1 shows a typical drawing of the high-pressure device (Ramalho, 1998). This device allows the compaction tests in cylindrical powder samples until 7 mm of diameter and 9 mm of height. The high-pressure tests were carried out in a special hydraulic press (Ryazantyashpressmash, DO 138B model). The tests were carried out as follows: the pre-compacted mixed powder sample was placed without slackness into central aperture of the deformable capsule (Fig. 1). The deformable capsule is made of a calcite based powder compact (95 % $CaCO_3$ + 5 % (SiO_2, Al_2O_3, Fe_2O_3)). The calcite based material is a solid pressure medium well known in the field of ultrahigh pressure technology (Vianna et al., 2001). For effective heat conduction in the sample, the capsule was covered with two caps composed of 50 % of graphite and 50 % of calcite. A high pressure anvil made of hard metal was used as mould, where the capsule was fixed in its concavity, and then sealed with the outer side of the mould. Finally, the complete device was axially placed in the mobile table of the press. When the force is applied, the capsule is deformed with concomitant formation of a gasket and the generation of a high pressure. A high pressure of 5.5 GPa was applied to the samples. The high pressure into the compression chamber was transmitted to the cemented carbide powder samples by the capsule. The high pressure level reached was maintained for 30 s, and then the temperature was raised to 1400 °C. These conditions of high pressure and high temperature were maintained for a time of 40 s.

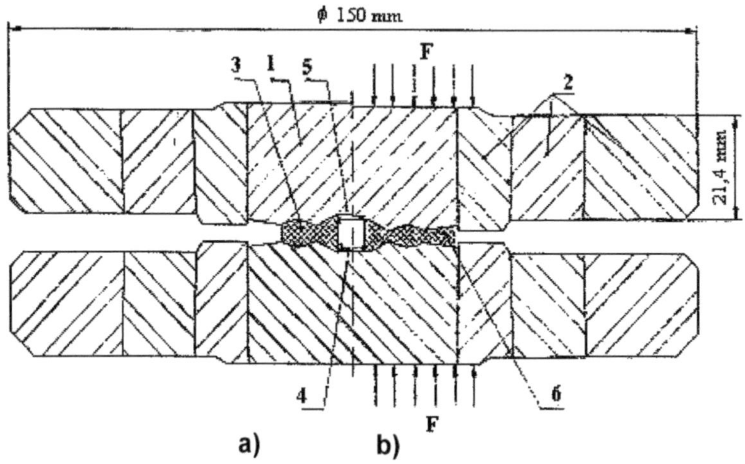

Fig. 1. Anvil type high pressure device with toroidal concavity for press of 630 tonnes: a) the high pressure device without applying pressure; and b) the high pressure device with application of pressure. 1) Anvil of WC; 2) multi-rings; 3) deformable capsule; 4) sample of WC10wt.%Co; 5) disk of protection; and 6) gasket (Ramalho, 1998).

2.3 Characterization of the cemented carbide pellets

After high pressure sintering, the cylindrical pellets (6 mm in diameter and 4 mm in height) were removed from the capsule. The surface in contact with the graphite was clean and was later made a grinding surface resulting in parallel tops and cylindrical side.

For metallographic analysis, the sample was embedded in thermosetting polymer resin and then was made polishing in diamond paste. The metallographic analysis was made using an optical microscope (Zeiss, model Neophot-32) that contains an image capture system with CCD camera attached.

Scanning electron microscopy operating at 10 kV (model SSX 550, Shimadzu) was used to examine the gold-coated fracture surfaces of the sintered cemented carbide pellets via secondary electron images (SEI). EDS also was used to identify the constituent elements of the sintered samples.

The crystalline phases after sintering were investigated between $2\theta = 20^{o}$ and $2\theta = 90^{o}$ via X-ray diffraction analysis (Shimadzu, model XRD 7000) with Cu-Kα radiation (40 kV, 40 mA). The phases were identified from peak positions and intensities using reference data from the JCPDS handbook.

The following properties have been determined: apparent density, relative density, coercive force, mechanical strength, elasticity modulus, microhardness, and wear resistance. Archimedes method of immersion in water was used to determine the apparent density. For this purpose, the cemented carbide pellets were weighed at dry state (M1), then boiled in water for 2 h, cooled, and weighed again a second time in water (M2). The pellets were weighed again at the saturated wet state (M3). The apparent density (ρ_a) of pellets was determined according to:

$$\rho_a = M1/M3 - M2 \quad (1)$$

The relative density (ρ_r) was determined according to the following expression:

$$\rho_r = \rho_a / \rho_t \quad (2)$$

in which ρ_a is the apparent density (g/cm^3) and ρ_t is the theoretical density of the WC10wt.%Co (14.53 g/cm^3).

The coercive force of the pellets was determined using a coercive force meter, which create a magnetic field. The test of coercive force was performed as follow. Initially, the coercive force meter was reset. The pellets were placed in the polarized magnetization devices. It was then made to read the display of the coercive force.

In this work the mechanical strength of the pellets was evaluated through axial compression strength due to the size of the pellets obtained. This means that the values of mechanical strength of the samples obtained in this work are for comparison only among themselves. The axial compressive strength (σ_c) was determined using an universal testing machine (EMIC, model DL - 10000) at a loading rate of 0.5 mm/min according to

$$\sigma_c = 4P / \pi D^2 \quad (3)$$

in which P is the load at rupture and D the specimen diameter.

The axial compressive elasticity modulus (E_{CA}) was determined using the stress-strain curve according to

$$E_{CA} = \sigma_{AC} / \varepsilon \quad (4)$$

in which σ_{AC} is the axial compressive tension and ε is the relative deformation.

The Vickers microhardness tests were performed using a microhardness apparatus coupled to an optical microscope according to

$$HV = 0.189 P / d^2 \quad (5)$$

in which P is the applied load (kgf) and d is the average length of the impression diagonal (mm).

The abrasion wear tests were performed using an abrasion meter (AROTEC, model AROPOL E) with maximum speed of 620 rpm and disk of carborundun. The following procedure was adopted: i) the sample is weighed before the wear test; ii) the sample was fixed in a chuck property for the test; iii) the disk is rotated and applied a vertical load on the sample fixed; iv) the sample was kept fixed in a straight line for 10 min; and v) the sample is weighed after the test for determining the mass loss. Thus, the wear resistance of the sintered pellets was determined according to

$$\Delta M = m_i - m_f / m_i \times 100 \quad (6)$$

in which m_i is the initial mass (g) of the pellets and m_f is the final mass (g) obtained after the wear test.

3. Results and discussion

3.1 Structural characterization

XRD patterns of the WC10wt.%Co samples are presented in Fig. 2. As expected, the results showed that at room temperature (Fig. 2a) the mixed powder is composed of the crystalline phases WC and Co. The low intensity of the peaks of cobalt may be related to Cu-Kα radiation used. When the mixed powder is sintered under high pressure and high temperature (5.5 GPa and 1400 °C), as shown in Fig. 2b, characteristic peaks of WC, Co and Co_3W_3C are identified. In addition, a small amount of the y phase was also identified.

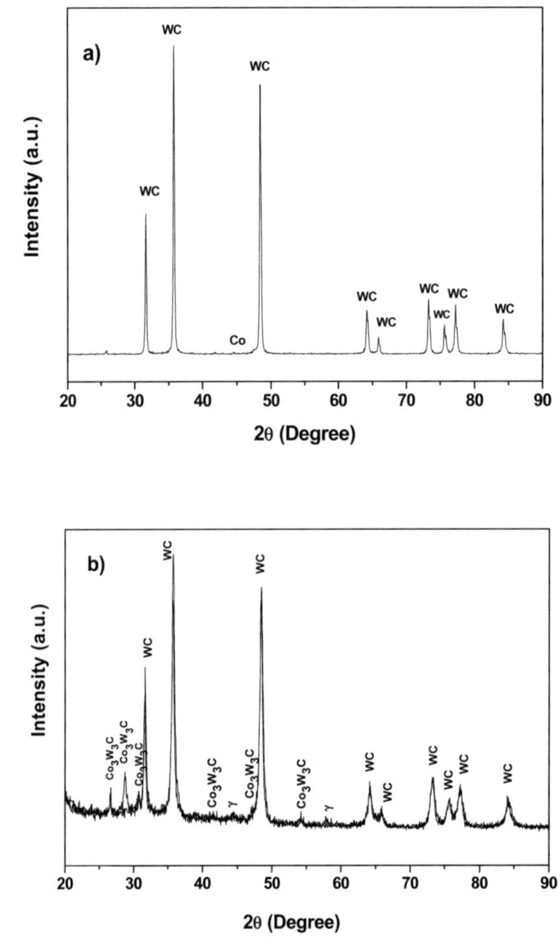

Fig. 2. XRD patterns of the WC10wt.%Co: a) before sintering; and b) after sintering.

The XRD analysis showed the formation of intermediate phases such as Co_3W_3C and y. These phases formed are undesired because they are harmful to the mechanical properties of cemented carbide. The appearance of these phases during sintering is consistent with the binary diagram of WC-Co (Silva, 1996). It is possible that these phases can be related to the following factors: i) sensitivity of the WC to the loss of carbon as consequence of their low formation energy; ii) sintering atmosphere with characteristic oxidant; and iii) the presence in the starting powders of compounds that react with the carbide, consumining the carbon. Another important aspect that must be considered is that the HPHT sintering process is very fast. This can take the carbon to react with the adsorbed oxygen also quickly, resulting in decrease or loss of carbon of the WC. The consequence is not having a good dissolution of free carbon in the liquid phase, so new phases can occur by diffusion.

Figures 3 and 4 show the X-ray diffraction patterns of the cemented carbide samples doped com rare-earth elements.

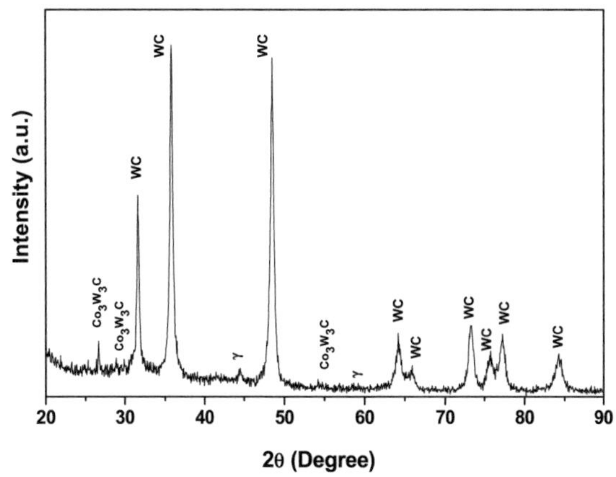

Fig. 3. XRD pattern of the WC10wt.%Co doped with 2 % of La_2O_3 sintered under HPHT.

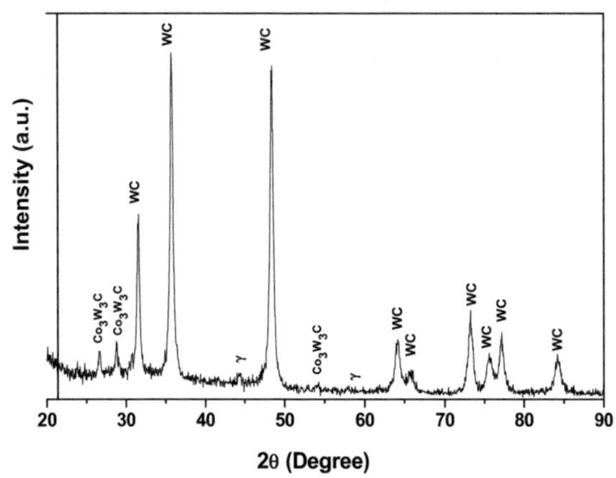

Fig. 4. XRD pattern of the WC10wt.%Co doped with 2 % of CeO_2 sintered under HPHT.

As can be observed, the samples doped with rare-earth elements (La_2O_3 (Fig. 3) and CeO_2 (Fig. 4)) had a phase composition very similar to that of the rare-earth free samples. It is noticed that only small differences in the peak intensities occurred. This means that the formation of these intermediate phases is not associated with the rare-earth elements incorporated in the cemented carbide studied.

3.2 Densification

The apparent density of the pellets of WC10wt.%Co doped with rare-earth elements sintered under HPHT is shown in Fig. 5. It can be observed that the samples containing rare-earth elements showed higher apparent density. The highest values were obtained for samples containing 2 % La_2O_3 (14.00 g/cm³) and 0.5 % CeO_2 (13.53 g/cm³).

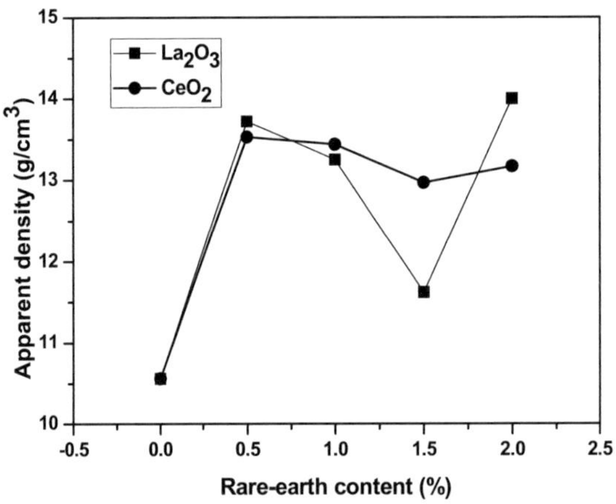

Fig. 5. Apparent density of the cemented carbides sintered under HPHT.

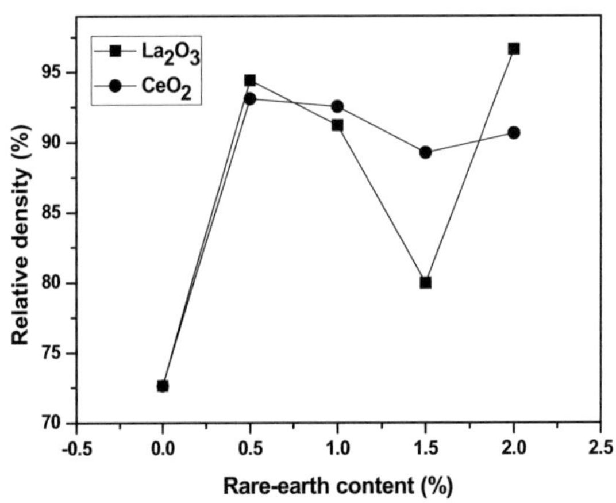

Fig. 6. Relative density of the cemented carbides sintered under HPHT.

The densification behavior of the cemented carbide as a function of the content of rare-earth elements is shown in Fig. 6. The value of theoretical density for the cemented carbide used (WC10wt.%Co) was 14.54 g/cm3.

According to the literature (Arbilla et al., 1996), the manufacturing of cemented carbide (WC10wt.%Co) is basically the liquid phase sintering of powder mixture of tungsten carbide and cobalt at a temperature of approximately 1400 °C. At this temperature, the cobalt becomes liquid and diffuses into the structure and, after cooling, the desired properties are obtained. In this work, however, an unusual process was used for the production of cemented carbide doped with rare-earth elements.

The results in Fig. 5 indicated that the WC10wt.%Co-free rare-earth elements showed a low value of apparent density (10.56 g/cm^3), reaching a densification of about 73 % (Fig. 6). Whereas the samples studied were subjected to the HPHT process, it was expected a higher value. Furthermore, this value is lower than in the cemented carbide densification obtained in industrial routines (Gomes et al., 2004).

The sintering when performed via the HPHT process tends to promote greater adhesion between the particles and higher densification in the material. However, the actions of certain factors must be taken into account because they can lead to some variations such as (Rodrigues et al., 2006; Bobrovnitchii, 2001): i) the presence of gradients of temperature and pressure in the compression chamber of the DAP may result in residual stresses and cracks in the samples; and ii) the rapid reduction of pressure and temperature can cause defects in the samples.

It can also be seen in Figure 6 that the samples containing rare-earth element showed a high densification. In particular, the samples with higher degree of densification were those containing 2 % of La_2O_3 (96.63 %) and 0.5 % of CeO_2 (93.10 %) in relation the cobalt phase. Importantly, all cemented carbide samples obtained had the same procedure of manufacture. This fact confirms that the rare-earth elements incorporated into the WC10wt.%Co contribute effectively to the high densification of the pellets sintered under HPHT conditions. This result is consistent with the literature (Xu et al., 2001; Gomes, 2004).

According to the literature (Ji et al., 1996), the addition of rare-earth element leads to decreased porosity of the cemented carbide. As the cemented carbide is a fragile PM material, porosity always causes great tension, which is a source of fracture. There is a direct relationship between mechanical strength and porosity of the cemented carbide, in which fewer pores and smaller pore size results in higher mechanical strength. During the sintering cycle, gaseous impurities, for example, oxygen and sulfur in cemented carbide is partially released in gaseous form. If they are not partially removed such defects as pores will take shape. Since the form of sulfide and oxide compounds of rare earths, which are stable at high temperatures, gaseous impurities are reduced and thus reduce the possible formation of pores. However, reducing the temperature required for the liquid phase by the addition of rare-earth element implies that under the same conditions of sintering, the sintering time by liquid phase has been extended, which leads to significant displacement of material, filling of pores with the liquid phase and decrease in number and size of pores. This fact is combinated with the effect of high pressure and high temperature that promote the filling of empty spaces by the binder metal.

3.3 Coercive force

Figures 7 and 8 present the values of coercive force (Hc) of the cemented carbide produced via HPHT as a function of the rare-earth elements added. The results show that the addition of rare-earth elements in the structure of the cemented carbide influences the coercive field.

This magnetic property is very sensitive to microstructure and chemical composition of the cemented carbide.

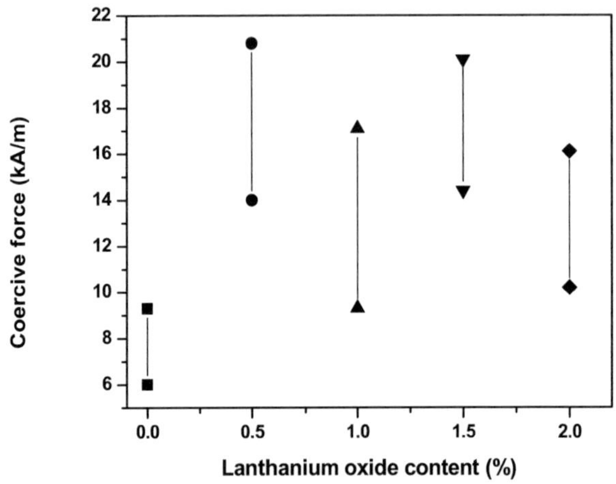

Fig. 7. Coercive force of the cemented carbides doped with lanthanum oxide sintered under HPHT.

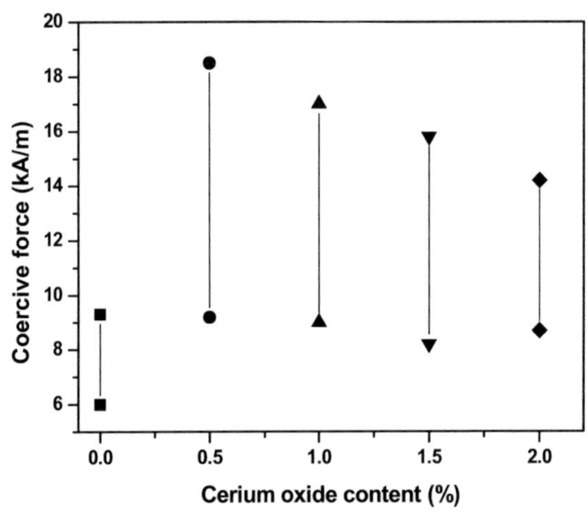

Fig. 8. Coercive force of the cemented carbides doped with cerium oxide sintered under HPHT.

As observed in Figs 7 and 8 there is variability in the values obtained for the coercive force. However, this variability is also observed in the literature (Brookes, 1998) for the cemented carbide. The coercive force values obtained in this work are also within the range of Hc for the cemented carbide (WC10wt.%Co).

The measures of coercive force of the sintered pellets indicated an increase in the coercive field of the WC10wt.%Co doped with rare-earth elements. This increase in the coercive field may have been influenced by the WC grain size resulting from the refinement of the microstructure caused by the rare-earth elements. It is reported in the literature (Xu et al., 2001) that these elements inhibit the grain growth of WC thereby making thinner. The values of Hc increase with the refinement of microstructure, because this makes it difficult to magnetization of the material in the direction of the applied field. Thus, it is necessary to generate higher coercive force to magnetize the WC10wt.%Co. The high value of Hc indicates that the cemented carbide studied corresponded to the class of magnetically hard materials, the application it is intended for light machining. The results also showed that the lanthanum oxide (La_2O_3) was the most effective in increasing the coercive field of the WC10wt.%Co. The sample with 0.5 %of lanthanum oxide reached the highest value of Hc (14.0 - 20.8 kA/m). In the case of addition of cerium oxide (CeO_2), the sample with 0.5 % of CeO_2 reached the highest value of Hc (9.2 – 18.5 kA/m).

3.4 Axial compressive strength

The behavior of the axial compressive strength as a function of the rare-earth content of the pellets of cemented carbides sintered under HPHT is shown in Fig. 9.

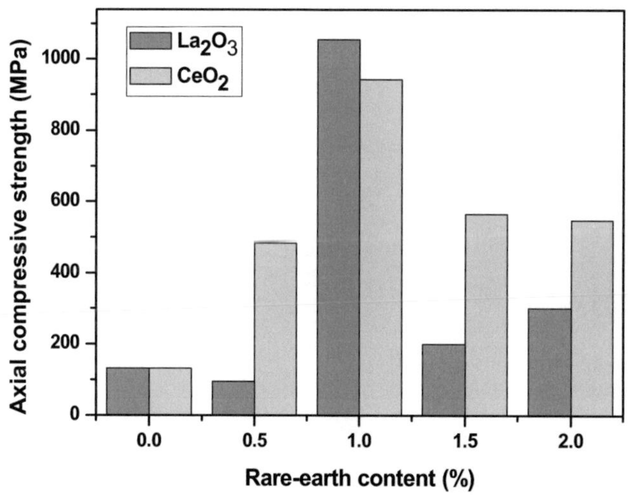

Fig. 9. Axial compressive strength of the cemented carbides sintered under HPHT.

The effect of the addition of rare-earth elements was to increase the compressive strength of the pieces of WC10wt.%Co. This is mainly associated with increased densification of the cemented carbide containing rare-earth during the sintering process. In addition, the rare-earth additive can inhibit the martensitic phase transformation of Co phase (ε-Co), which weakens the material resulting in the increase of α-Co (CFC) that contributes directly to increased mechanical strength (Xu et al., 2001).

The results of Figure 9 also showed different behavior for the axial compressive strength, depending on the rare-earth added. The highest values of mechanical strength of the cemented carbide doped with rare-earth elements were obtained for the samples containing 1 % of La_2O_3 (1055 MPa) and 1 % of CeO_2 (943 MPa). In addition, the complex variation observed for the mechanical strength may be associated with the fact that some samples did not show regular cylinder geometry (Fig. 10).

Fig. 10. Appearance of the pellets of WC10wt.%Co obtained via HPHT.

3.5 Axial compressive elasticity modulus

The axial compressive elasticity modulus of the cemented carbides obtained from the tension-deformation curve resulting from axial compression tests is shown in Fig. 11.

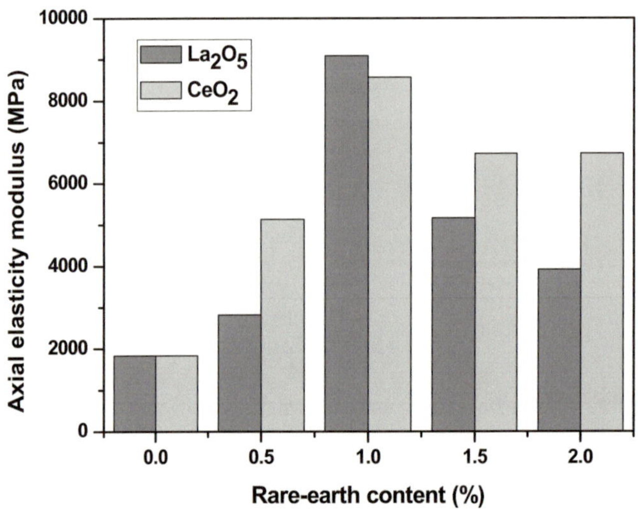

Fig. 11. Axial elasticity modulus of the cemented carbides sintered under HPHT.

The cemented carbides are characterized by a high modulus of elasticity. In a way similar to the axial compressive strength, it is observed that the axial compressive elasticity modulus increased with the incorporation of rare-earth oxides. This is mainly due to higher densification of the cemented carbide doped with these additives.

The results of Figure 11 also showed that the highest values of axial compressive elasticity modulus of the cemented carbide doped with rare-earth elements were obtained for the samples containing 1 % of La_2O_3 (9095 MPa) and 1 % of CeO_2 (8573 MPa).

3.6 Microhardenss

Figure 12 shows the values of microhardness for the pellets of WC10wt.%Co doped with rare-earth elements. The results show that the microhardness increases with the addition of rare-earth elements. This increase may be due to increased densification with the rare earth added. On the other hand, the complex variation in microhardness values may have been influenced by the geometric irregularity at the top and bottom of the samples caused by the sintering process as previously described.

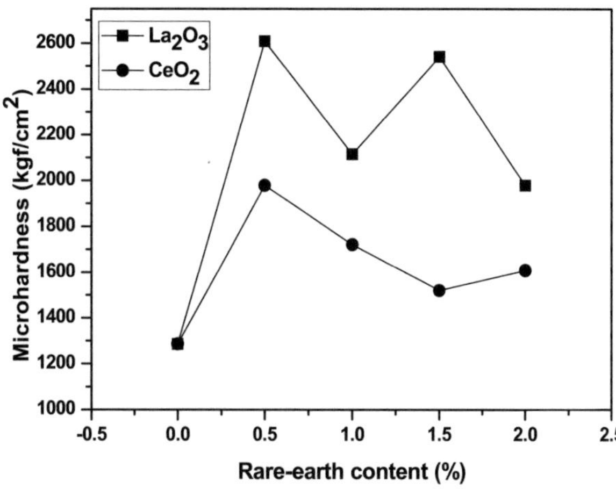

Fig. 12. Microhardness of the cemented carbides sintered under HPHT.

It can be seen in Fig. 12 an increase in microhardness when 0.5 % of La_2O_3 is added. A decrease in microhardness occurs with the addition of 1 % of La_2O_3. With the addition of 1.5 % of La_2O_3, the microhardness increases again and back to decrease with 2 % of La_2O_3. For cemented carbide containing cerium oxide, there is an increase with the addition of 0.5 % and a subsequent decrease between 1 to 1.5 %. An increase occurs again with the addition of 2 % of CeO_2. It can also be observed that the highest values of microhardness of the cemented carbide doped with rare-earth elements were obtained for the samples containing 0.5 % of La_2O_3 (2609 HV) and 0.5 % of CeO_2 (1979 HV).

3.7 Wear resistance

Table 3 presents the values of wear resistance for the pellets of cemented carbide doped with rare-earth elements obtained under HPHT conditions. The results indicated that all samples of cemented carbide containing rare-earth elements showed less mass loss compared to the reference sample (WC10wt.%Co). This was expected, since the decrease in grain size caused by the rare-earth elements and also the positive influence of high pressure leads to increased microhardness and, consequently, to reduce the wear.

Samples	Wear resistance (%)
AL1	7.7
AL2	3.7
AL3	0.2
AL4	0.1
AL5	0.1
AL6	0.1
AL7	1.7
AL8	0.2
AL9	0.3

Table 3. Wear resistance of the cemented carbides sintered under HPHT.

The results in Table 3 also show that the best samples in terms of wear resistance are those with 1.0 to 2.0 % of La_2O_3 and 0.5, 1.5 and 2.0 % of CeO_2.

3.8 Microstructural analysis

Optical micrograph of the sample of WC10wt.%Co free of rare-earth sintered under HPHT is shown in Fig. 13. One can observe a non-homogeneous distribution of WC by the structure, indicating a possible change of the sintering conditions. On the other hand, there is a good distribution of cobalt for samples containing rare-earth elements (Figs. 14 e 15). This means better mixing or influence of the rare-earth about the process.

SEM micrographs of the fractured surfaces of the samples of WC10wt.%Co with and without rare-earth additives are presented in Figs. 16 - 18. It may be noted that the appearance of the fractured surface is typical of brittle fracture for all samples.

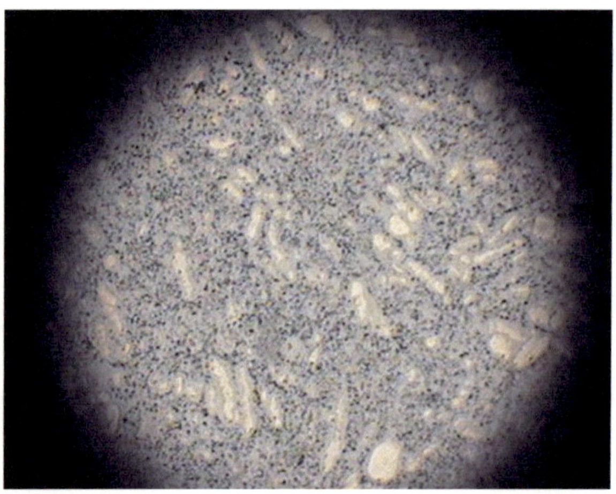

Fig. 13. Optical micrograph of the AL1 sample sintered under HPHT.

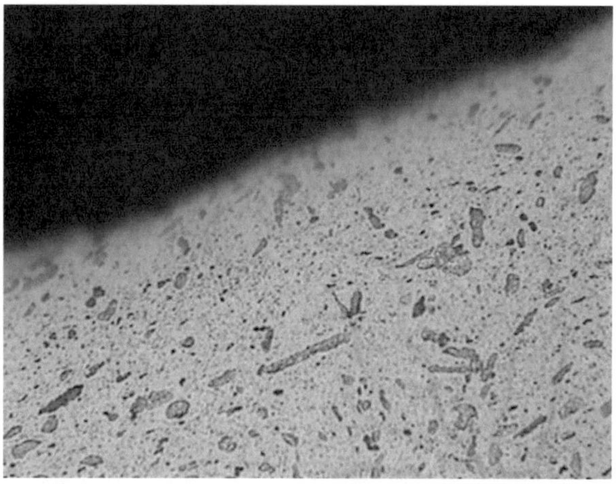

Fig. 14. Optical micrograph of the AL5 sample sintered under HPHT.

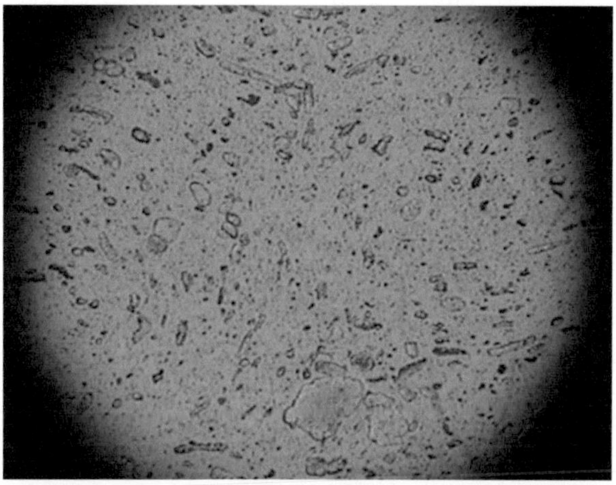

Fig. 15. Optical micrograph of the AL6 sample sintered under HPHT.

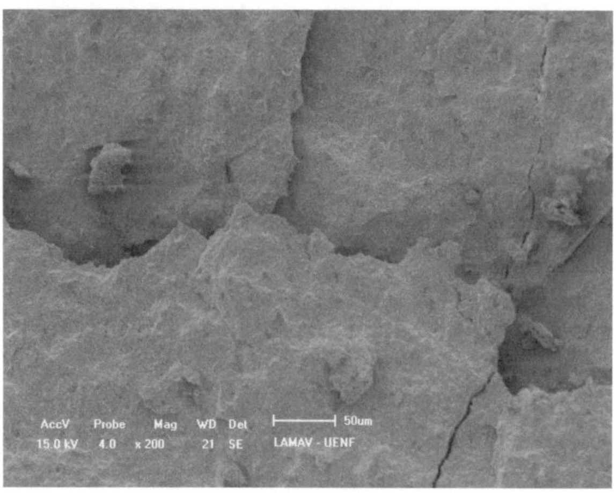

Fig. 16. Fracture surface of the of the AL1 sample sintered under HPHT.

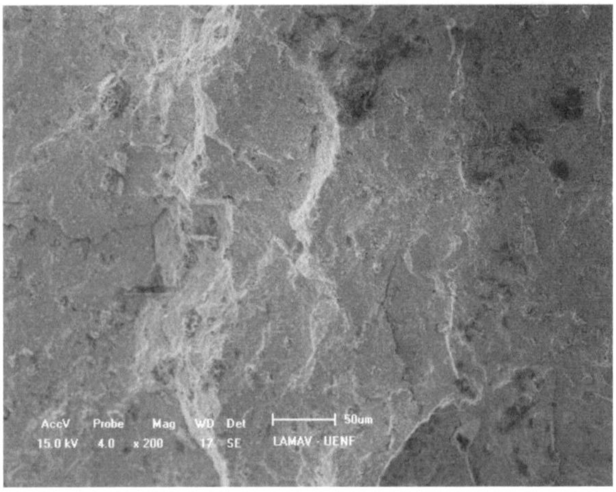

Fig. 17. Fracture surface of the of the AL5 sample sintered under HPHT.

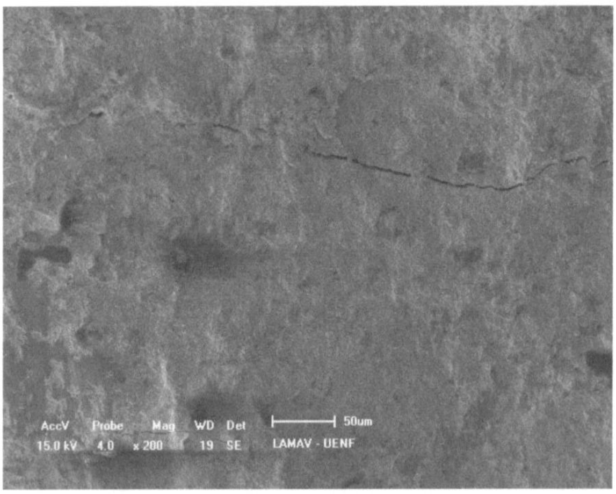

Fig. 18. Fracture surface of the of the AL7 sample sintered under HPHT.

Line spectrums for the samples of cemented carbides determined using EDS are shown in Figs. 19-21. The following constituent elements of the cemented carbides sintered under HPHT were detected: W, C, Co, La, and Ce.

The sintered microstructures of the AL5 (2 % La_2O_3) and AL7 (1 % CeO_2) samples obtained via SEI/SEM are shown in Figs. 22-23, respectively. One can clearly observe the formation of cobalt lakes surrounding the tungsten carbide grains and pore possible. This may be related to different factors: i) mixing process inefficient; and ii) HPHT processing. According to North et al. (1992), high pressure applied continuously during the heating cycle can provide in some regions the formation of cobalt lakes, which persist at high temperatures even with some structural rearrangement of the WC.

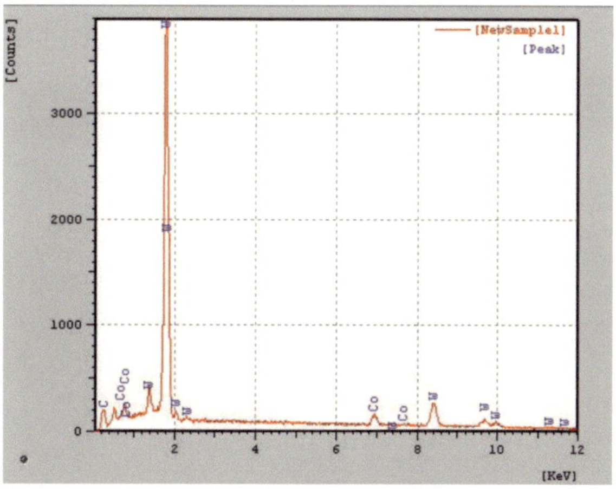

Fig. 19. EDS spectrum for the AL1 sample sintered under HPHT.

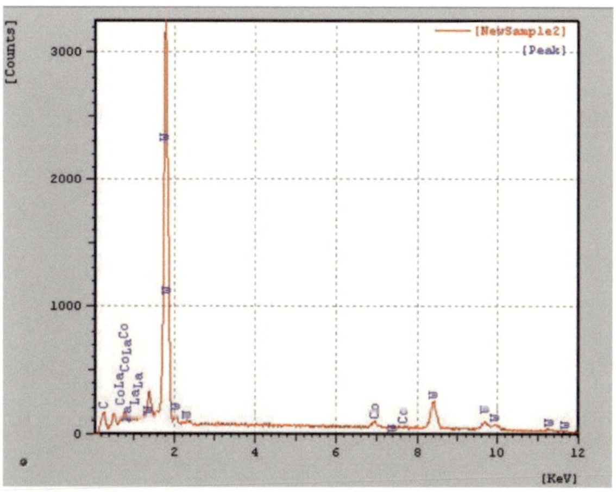

Fig. 20. EDS spectrum for the AL2 sample sintered under HPHT.

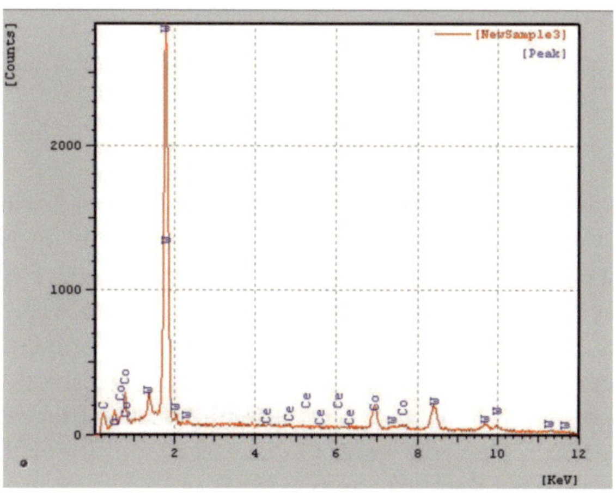

Fig. 21. EDS spectrum for the AL7 sample sintered under HPHT.

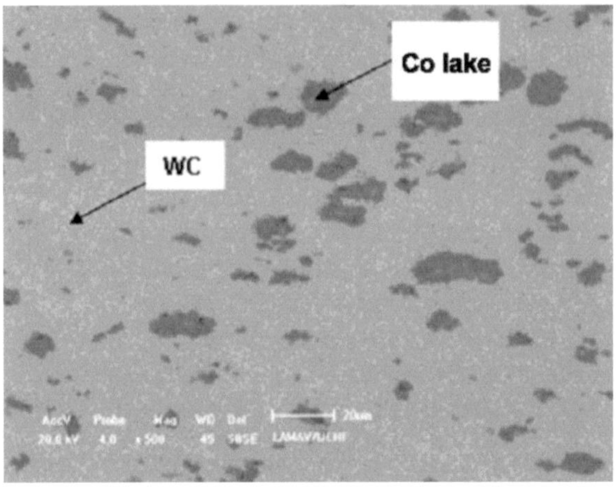

Fig. 22. SEM micrograph of the AL5 sample sintered under HPHT.

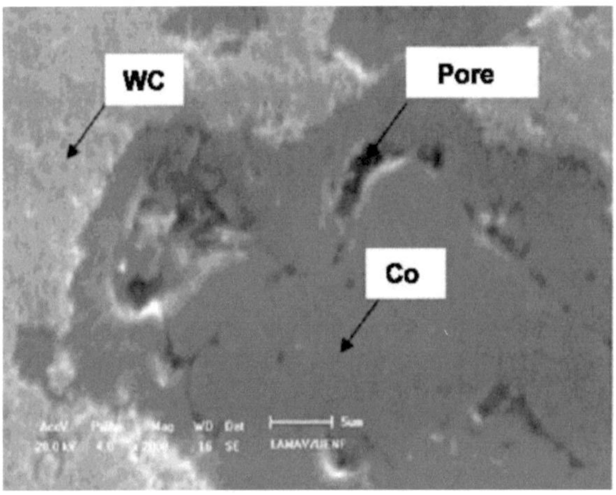

Fig. 23. SEM micrograph of the AL7 sample sintered under HPHT.

4. Conclusion

In this chapter the cemented carbide (WC10wt.%Co) powder doped with different rare-earth elements (La_2O_3 and CeO_2) sintered under HPHT conditions was investigated. The method used can be an alternative for WC10wt.%Co dense pellets processing. High final densities near theoretical density have been obtained. It was also found that the incorporation of rare-earth elements positively influenced the physical and mechanical properties of the cemented carbide. These properties include increased densification, increase in coercive field, axial compressive strength, axial elasticity, microhardness and wear resistance. This behavior is due to the addition of rare-earth elements leads to decreased porosity of the cemented carbide. Moreover, it was observed that the lanthanum oxide (La_2O_3) was more effective in improving the properties of the cemented carbide investigated.

5. Acknowledgements

The authors acknowledge the FAPERJ and CNPq for supporting this work.

6. References

Arbilla, G.; Corrêa, S.M. & Carvalho, M.S. (1996). *Ciência Hoje*, Vol. 21,No. 122

Bobrovnitchii, G.S. (2001). Tendências do Desenvolvimento das Tecnologias de produção dos Materiais Superduros no Brasil, *Proceedings of the 56th ABM Congress*, Belo Horizonte, Brazil

Brooks, A.J.K. (1998). *Hard Metals and Hard Materials*, Third Ediiton, East Barnet Hertfordshire, United Kingdom

Cheng, J. & Yu, Q. (1995). Wear Resistance of Rare-Earth Cemented Carbide Tools, *Mech Eng*, Vol. 4, pp. 9-10.

Chenguang, L. (1992). Composition Morphology and Distribution of RE Element Containing Phases in Cemented Carbide, *International Journal of Refractory Metals and Hard Materials*, Vol. 11, pp. 295-302

Deng, Q. (1993). Progress in the Study on Strengthening Mechanisms of Rare-Earth Cemented Carbide, *Rare Metals*, Vol. 17, pp. 442-454

German, R.M. (1994). *Powder Metallurgy Science*, MPIF, 2nd Edition, New Jersey, USA

Gomes, C.M.F. (2004). Estudo da Influência da Adição de Terras-Raras na Sinterização e Propriedades Mecânicas de Carbeto Cementado, *Master Thesis*, UENF, Campos dos Goytacazes, Brazil

He, C. (1996). Study on the Cutting Behavior of Rare-Earth Cemented Carbide, *Cemented Carbide*, Vol. 13, pp. 23-26

He, C.; Wang, Y. & Ma, F. (1994). The Researching of Application of Hard Metal by Adding RE Element Cemented Carbide, *Cemented Carbide*, Vol. 11, pp. 129-133

Ji, X.; Jiang-gao, Y. & Xing-hua, G. (1996). Application of Rare-Earth Elements in Cemented Carbide Inserts, Drawing Dies and Mining Tools, *Materials Science and Engineering A*, Vol. 209, pp. 287-293

Li, B. (1996). Effects of Adding Method of Rare-Earth on the Physical and Mechanical Properties and Performances of Cemented Carbide, *Cemented Carbide*, Vol. 13, pp. 15-18

Li, G.; Yan, L. & Li, Z. (1986). Applications of Rare Earth Elements in Cemented Carbide, *Powder Metallurgy Technology*, Vol. 4, pp. 25-29

Li,G.; Yan, L.; Zhang, H. & Li, F. (1994). Research on Rare-Earth Cemented Carbide YG6R, *Powder Metallurgy Technology*, Vol. 12, pp. 206-209

Li, G.; Yan, L. & Zou, H. (1993). Distribution and Existent Morphology of Rare-Earths in Cemented Carbide, *Powder Metallurgy Technology*, Vol. 11, pp. 83-89

Liang, P.; Su, H. & Jiang, N. (1989). Development of Cemented Carbide with Rare-Earth Elements, *Cemented Carbide*, Vol. 6, pp. 31-36

Liu, N.; Hu, Z. & Kui, K. (1992). The Microstructure, Mechanical Properties and Development of Cemented Carbide with Rare-Earth Elements, *Cemented Carbide*, Vol. 10, pp. 50-53

Luo, Z. (1991). Research on the Effect of Rare-Earth Elements on the Properties of Cemented Carbide, *Cemented Carbide*, Vol. 8, pp. 12-19

North, B.; Greenfield, M.S.; McCoy, W.C. & Vankirk, J.S. (1992). Effect of Pressure Sinter and Post-Treatment on the Strength of Cemented Carbides, *Advances in Powder Metallurgy and Particulate Materials*, Vol. 8, pp. 111-123

North, B.; Pfouts, W.R. & Greenfield, M.S. (1991). Pressure Sinter and HIP on Cemented Crabides, *Metal Powder Report*, pp. 40-45

Pan, Q. (1993). Effects of Rare-Earth Oxide on the Properties of WC-Co Cemented Carbide, Rare Met Mater Eng, Vol. 22, pp. 35-38

Ramalho, A.M. (1998). A Influência da Construção e dos Materiais sobre o Estado de Tensões do Cintamento dos Dispositivos de Alta Pressão do Tipo Bigorna Usados Na Produção de Materiais Superduros, *Master Thesis*, UENF, Campos dos Goytacazes, Brazil

Rodrigues, M.F.; Bobrovnitchii, G.S.; Quintanilha, R.; Cândido, R.; Silva, G. & Figueira, M. (2006). Sinterização da Liga WC/10Co por Altas Pressões, Revista Matéria, Vol. 11, No. 3, pp. 174-180

Schwarzkopf, P. & Kieffer, R. (1986). *Cemented Carbides*, MacMillan, New York, USA

Shan, S. (1990). Review of the Effect of Rare-Earth Element on the Cemented Carbides, *Rare Met Cemented Carbide*, Vol. 4, pp. 39-42

Silva, A.G.P. (1996). Study on Sintering and Grain Growth of WC-Based Hard Metals, Dissertation, Technischen Universität Wien, Wien, Austria

Thümmler, F. & Oberacker, R. (1993). *Introduction to Powder Metallurgy*, The Institute of Materials, Cambridge, Great Britain

White, C. (1998). History of Powder Metallurgy, In: *Handbook of Powder Metal Technologies and Applications*, APMI, New York, USA

Willians, B. (1998). The History of Hard Metal, *World Tooling*, No. 3, pp. 4-7

Vianna, W.S.; Bobrovnitchii, G.S. & Monteiro, S.N. (2001). Study of the Variation of the Electric Current During the Synthesis of Diamond Powders, *Journal of Super Hard Materials*, Vol. 5, No. 133, pp. 9-13

Xu, C.; Ai, X. & Huang, C. (2001). Research and Development of Rare-Earth Cemented Carbides, *International Journal of Refractory Metals and Hard Materials*, Vol. 19, pp. 159-168

Yan, K.; Meng, T. & Yu, J. (1995). The Effect on the Mechanical Properties and Cutting Performances of Cemented Inserted with Rare-Earth Metal Elements, Journal of the Southwest Jiaotong University, Vol. 30, pp. 46-51

Yang, J.; Xiong, J. & Liu, J. (1993). Overview on the Present Research Situation of Cemented Carbide with Rare-Earths, *Cemented Carbide*, Vol. 10, pp. 46-49

Yao, X.; Huang, P. & Lu, H. (1987). Effects of Rare-Earth Oxide on the Martensitic Transformation of Cobalt and Mechanical Properties of Cemented Carbide, *Powder Metallurgy Technology*, Vol. 5, pp. 200-204

Yuan, Y.; Bai, Y. & Feng, H. (1995). Microstructure and Properties of the Cemented Crabide with Rare-Earth Element, *Powder Metallurgy Technology*, Vol. 13, pp. 88-91

Part 5

Conventional Sintering

18

Mechanisms of Microstructure Control in Conventional Sintering

Adriana Scoton Antonio Chinelatto[1],
Elíria Maria de Jesus Agnolon Pallone[2],
Ana Maria de Souza[1], Milena Kowalczuk Manosso[1],
Adilson Luiz Chinelatto[1] and Roberto Tomasi[3]
[1]*Department of Materials Engineering - State University of Ponta Grossa*
[2]*Department of Basic Sciences – FZEA - São Paulo University*
[3]*Department of Materials Engineering - Federal University of São Carlos*
Brazil

1. Introduction

The manner and mechanisms involved on the sintering process are essential investigation to achieve the required microstructure and final properties in solids. During the conventional sintering of a compacted powder, densification and grain growth occur simultaneously through atomic diffusion mechanisms. Many researchers have been working on reducing the grain size below 1 µm aiming to improve some properties, such as strength, toughness and wear resistance in ceramics (Greer, 1998; Inoue & Masumoto, 1993; Morris, 1998). In order to obtain ultra-fine ceramic microstructures, nanocrystalline powders can be used. Although the sinterability of nanoparticles is superior to that of fine particles due to the higher sintering stress, densification of these powders is often accompanied by grain growth (Suryanarayana, 1995).

Hot pressing sintering (He & Ma, 2000; Porat et al., 1996), spark plasma sintering (Gao et al., 2000; Chakravarty et al., 2008) or pulse electric current sintering (Zhou et al., 2004) are typical techniques employed to produce nanostructured ceramics. However, many of these techniques are not economically viable depending on the use of the final product. Thus, conventional pressureless sintering is still a more attractive sintering method to produce ceramic products, mainly due to its simplicity and cost compared to other methods. In the conventional pressureless sintering, a controlled grain size with high densification could be achieved by adequate control procedures of the heating curve – herein defined as the maximization of the final density with minimum grain growth.

One hypothesis to the heating curve control can be achieved by improving the narrowing of grain size distribution in a pre-densification sintering stage followed by a final densification stage namely at a maximum densification rate temperature (Chu et al., 1991; Lin & DeJonghe, 1997a, 1997b). In a thermodynamics point-of-view, another hypothesis is regarded to control the heating schedule at temperature ranging the active grain boundary diffusion. Note, however, that the grain boundary migration is sufficiently sluggish and the densification could occur without grain growth. The aforementioned hypothesis was proposed by Chen and Wang (Chen & Wang, 2000) and has been successfully applied to different types of materials. A second phase can be added to preserve fine grains. In this case, grain boundary inhibition can be due to the pinning effect, which is associated with particles locations at grain boundaries or triple junctions (Chaim e al. 1998; Trombini et al., 2007). This drag pinning effect associated with heating curve control can be more effective to suppress the grain growth.

2. Nanostructure materials

The size control of microstructural elements has been always considered as one of the most important factors in control of several properties in the development of new materials or design new microstructures. As a historical example, it can be mentioned that the grain refining of metallic materials, which results in increased mechanical strength, tenacity, occurrence of superplastic, etc. New materials with sub-micron grain size have been developed recently as commercial materials and the latest generation of this development is the nanostructured materials (Inoue & Masumoto, 1993).

Nanostructured materials (also called nanocrystalline materials, nanophasics materials or nanometer-sized crystalline solids) are known to have properties or combinations of properties, which may be new or even superior to conventional materials (Greer, 1998). Nanostructured materials can be defined as a system containing at least one microstructural nano characteristic (with sizes ranging up to 100-150nm). Due the extremely small dimensions, a large volume fractions atoms located in grain boundaries (Morris, 1998), which gives them a unique combination of composition and microstructure (Suryanarayana, 1995). Generally, these materials exhibit high strength and hardness, increased diffusivity, improved ductility and toughness, reduced elastic modulus, lower thermal conductivity when compared to larger grain size materials (~10μm) (Suryanarayana, 1995).

Since nanocrystalline materials contain a large fraction of atoms in grain boundaries, many of these interfaces provide high density of short diffusion paths. Therefore, it is expected that these materials show increased diffusivity compared to polycrystalline materials of the same composition and conventional particle size (of the order of microns) (He & Ma, 2000). The consequence of such increased diffusivity is increased sinterability of nanometric powders, which causes decrease in sintering temperature of these powders when compared to the same material with conventional particle size (Porat et al., 1996).

The driving force for sintering or "sintering stress" of nanocrystalline ceramics with pores in the range of 5nm is about 400MPa (considering γ about $1Jm^{-2}$), while for conventional ceramics with pores around 1 μm it is 2 MPa. Thus, a nanocrystalline ceramic must have a great thermodynamic driving force for retraction, which must densify extremely well even under unfavorable kinetic conditions such as low temperatures (Suryanarayana, 1995).

The interest in this nanostructured materials area has grown due to the availability of nanocrystalline ceramic powders. These nanocrystalline powders can be synthesized using different techniques, but its consolidation into dense ceramics without significant grain growth is still a challenge.

Compaction and sintering of ultra fine and/or nanoscale powder have a positive set of aspects over behavior during processing and final properties of products; however, there are also several processing difficulties. Main positive aspects include: increased reactivity between reagents and solid particles and between particles and the gas phase, which are important processes in synthesis; increased sintering rate and particularly lowering of sintering temperature, which can reduced by half the material's melting point (Hahn, 1993; Mayo, 1996).

On the other hand, also due to the large surface area and large excess of free energy nanometric powder systems, there are many detrimental aspects to the processing and obtaining the refined and homogeneous microstructures. Some of these aspects are: a very strong tendency to agglomeration of primary particles of nanometric powders; difficulties of mixing and homogenization of compression due to the strong attraction between particles; demand for greater sintering atmosphere control, not only due to the higher reactivity, but also the possibility of formation of thermodynamically unstable phases and appearance of strong effect of adsorbed gases on the surface (Allen et al., 1996; Averback et al., 1992).

Many studies (Chen & Chen, 1996, 1997) on nanometric size particles have shown reduction of sintering temperature. Hahn et al. (Hanh, 1990), studying the sintering of nanometric TiO_2 (12nm), Y_2O_3 (4 nm) and ZrO_2 (8nm), found lower sintering temperatures than those conventional. The sintering of TiO_2 occurred at 1000°C while conventional TiO_2 sintering requires temperatures above 1400°C. The same pattern of reduced sintering temperature was observed for Y_2O_3 and ZrO_2. In spite of the proven decrease in sintering temperature of nanometric powders, its densification is often accompanied by a large grain growth, causing lost of their nanocrystalline ceramics characteristics.

2.1 Effect of heating curve in the sintering

Production of polycrystalline ceramics with high density and small grain size have been studied for several processing routes. Among these routes may be cited: colloidal processes of powder with controlled particle size distribution (Sigmund & Bergström, 2000; Lim et al., 1997), sintering under pressure (He & Ma, 2000; Weibel et al.,1997), use of additives incorporated into a second phase or in solid solution (Novkov, 2006; Erkalfa et al., 1996), spark plasma sintering (Gao et al., 2000; Chakravarty et al., 2008, Bernard-Granger & Guizard, 2007), pulse electric current sintering (Zhou et al., 2004), etc. Usually these methods have several limitations on usage, in addition to requiring more complex and expensive equipment. Thus, sintering without pressure is even a more desirable sintering method to produce ceramic products, mainly due to its simplicity and cost when compared to other methods.

In pressureless sintering, beyond the control of powders' characteristics, control of the sintering process has a major effect on final material's density and microstructure. This method is often unable to prepare dense ceramics with ultrafine grain size, once way the final sintering stage, both densification and grain growth occur by the same diffusion mechanisms (Mazahery et al., 2009).

Heating curve control to manipulate the microstructure during sintering is a route that has been studied and offers advantages such as simplicity and economy. The rate-controlled sintering (Brook, 1982, German, 1996) is one of the ways in which the relationship between densification rate and grain growth rate is determined to identify the sintering temperature at which densification rate is maximized (Chu et al., 1991). Ragulya and Skoroklod (Ragulya & Skoroklod, 1995) studied the rate-controlled sintering of ultra fine nickel powders obtaining sintered samples with high densities (~ 99% TD) and grain size smaller than 100nm. Based on their results, they stated that rate-controlled sintering is a possible route for obtaining dense materials with nanocrystalline structure.

A direct consequence of the rate-controlled sintering method is fast firing (Harmer & Brook, 1981), which can produce dense materials with small grain size, minimizing the time of exposure at temperatures where grain growth is fast compared with densification (Chu et al., 1991). This is possible because, generally, coalescence mechanisms (eg, surface diffusion and vapor transport) prevail over densification mechanisms (eg, volumetric diffusion by grain boundary diffusion) at low temperatures. In this case, shorter times at lower temperatures reduce growth, so that the driving force for densification is not decreased significantly (Lin & DeJonghe, 1997). In case of alumina (Harmer et al., 1979) for example, the activation energy for densification is greater than that for grain growth, and high sintering temperatures the most suitable (Harmer & Brook, 1981).

Kim and Kim (Kim & Kim, 1993) studying the effect of heating rate on shrinkage of pores in yttria-zirconia doped, stated that growth of pores is also inhibited by the fast firing process, helping thus to increase the densification. Searcy (Beruto et al., 1989; Searcy, 1987) suggested that the beneficial effects attributed to the fast firing may be due in part to temperature gradients developed in the sample during heating.

Rate-controlled sintering is more efficient for non-agglomerated powders, in which the microstructure develops relatively homogeneous. However, benefits of these techniques have proved less effective for agglomerated systems. The difficulty of obtaining homogeneous green microstructures using ultra-fine powders, owing to their high degree of agglomeration leads to inhomogeneous, low densification rate and limited final density (Rosen & Bowen, 1988; Inada et al. 1990; Dynys & Hallonen, 1984).

Recently, the availability of many different production routes for ultrafine and nanosized ceramic powders have led research to focus increasingly on the processing of these types of powders. Transformation processes that occur at low temperatures have been observed and studied particularly before or at the beginning of the densification stages of sintering. These processes, which have been reported for coarsening and particle repacking (Chen & Chen, 1996, 1997), affect the subsequent sintering stages. When these processes are controlled, it is possible to obtain dense and fine microstructures. One way to control these processes is optimizing the material's heating curve by pre-treating it at low temperatures.

De Jonghe et al. (Chu et al., 1991; Lin & DeJonghe, 1997a, 1997b) found that pre-heat treatment (50 to 100 hours) at low temperatures (800 ºC), in which little or no densification occurs, can improve densification and microstructure of a high purity alumina with and without MgO addition. A consequence of these pre-treatments was reducing the densification rate in the initial stages of sintering. However, benefits of evolving a more homogeneous microstructure are evidenced in the final stages of sintering, allowing a final microstructure refinement. According to DeJonghe et al. (Chu et al., 1991; Lin & DeJonghe, 1997a, 1997b), the pre-treatment leads to more compaction due to the strong increase in the neck formation among particles, promotes the elimination of fine particles, probably through the ripening process of Ostvald and produces a narrower distribution in pore size. These factors make decrease the density fluctuation during sintering, thus favoring the achievement of more uniform microstructures. The best microstructural homogeneity, both in relation to pores and particles, retards the closing of the pore network, so that pores remain open until higher densities inhibiting grain growth more effectively (Lin & DeJonghe, 1997b).

Kim and Kishi (Kim & Kishi, 1996) observed the effect of pre-treatments on strength and subcritical crack growth in alumina. Alumina sintered by hot pressing had 400-500MPa resistance while alumina subjected to a pre-treatment (1000 to 1200 °C for 10 hours) increased their resistance to 750MPa. They concluded that the fracture toughness of grain boundary is increased with pre-treatment and toughness of grain boundary reduces the rate of subcritical crack growth sintering resulting in increased strength of the material. Sato and Carry (Sato & Carry, 1995) studied the effect of particle size and pre-treatment on ultra-fine alumina and found that the pre-treatment delays the start of abnormal grain growth, creating a more uniform microstructure before densification. Chinelatto et al. (Chinelatto et al., 2008) studied the influence of heating curve on the sintering of alumina subjected to high-energy milling and observed that the isothermal treatments at a temperature below the beginning of linear shrinkage cause the fine particles to disappear, narrowing the final grain size distribution.

A new sintering process in two steps was proposed by Chen (Chen & Wang, 2000). The author showed the possibility of obtaining fully dense bodies and sizes of nanosized grains in sintering without applying pressure. This rapid sintering technique inhibits grain growth that occurs in the final stages of sintering and consists of a heating curve in which the ceramic body is subjected to a rapid peak in temperature followed by cooling to the sintering level. Thus, there is densification of the material without the characteristic grain growth. Suppression of grain growth in the final stage of sintering was achieved by exploiting the difference between the kinetics of diffusion in the grain boundary and controlled grain boundary migration rate. Chen and colleagues used the technique of two-step sintering nanosized powders of Y_2O_3 (Wang et al., 2006a), $BaTiO_3$ ferrites and Ni-Cu-Zn (Wang et al., 2006b). Other studies are reported in the literature using the two-step sintering to post nanometric TiO_2 (Mazaheri, 2008a), yttria stabilized zirconia (Mazaheri, 2008b), zirconia (Tartaj, 2009), abrasive alumina with additions of MgO-CaO-SiO_2 (Li et al., 2008), alumina-zirconia (Wang et. al., 2009) among others.

According to Chen and Wang (Chen & Wang, 2000; Wang et al., 2006a), in a temperature range called kinetic window, occurs by grain boundary or volumetric diffusion while the grain boundary movement is restricted, so that the densification occurs without, however, growth occurs grain. The sintering temperature in this region results in elimination of residual porosity without grain growth at final stage of work. Suppression of grain growth but not densification is consistent with a network of grain boundaries anchored by triple junction points, which have higher activation energy for migration than grain boundaries (Wang et al., 2006a).

The choice of temperature for both steps is essential for successful sintering. If densities greater than a critical value are reached in the first heating stage, the density of triple junctions decreases, so the effect of the triple points drag mechanism is reduced and the grain growth control is injured in the final sintering stages. On the other hand, if densities are lower than certain critical value, it is not possible to achieve material's densification in the second sintering stage (Chen & Wang, 2000; Hesabi et al., 2009).

Zhou et al. (Zhou et al., 2003) showed that triple junction at large grain sizes is not significant since its volume fraction is negligible compared with the total interface fraction. It is believed that occurs when the passage to the second stage of sintering, the energy in the triple junction during the whole period of time, remains constant. If there is increase in temperature it may be due to increased energy of the system, so there may be a greater mobility of the triple junction in comparison with the grain boundary, so the contour can move freely without any difficulty, indicating a common growth grain. At low temperature, the triple junctions make difficult the movement of grain boundaries not allowing grain growth occurrence (Czubayko et al., 1998).

Nanometric and sub-micrometric alumina powders (Hesabi et al., 2009; Li & Ye, 2006) were also sintered in two steps. Ye and Li (Li & Ye, 2006) found that it is necessary that nanosized alumina powders reach 85% theoretical density in the first stage of sintering, so it can be fully densified at the second level, while Bodisova (Bodisova et al., 2007) showed that density should not be less than 92% theoretical density to achieve full densification without grain growth in the second level for post sub-micron alumina.

2.2 Addition of particles of a second phase

A strategy used to achieve nanometric grain sizes is through addition of solutes or particles of a second phase in single-phase ceramics, which reduce the grain boundary mobility or fix the grain boundary, respectively (Novkov, 2006). This strategy has been used successfully by many researchers. Chaim et al. (Chaim et al., 1998) added 4 wt% trivalent oxides (Y, La, Bi) and tetravalent oxides (Ce, Th) in nanocrystalline zirconia powder and found that Y_2O_3, CeO_2 and ThO_2 inhibit grain growth during sintering. According to Mayo (Mayo, 1996), Hahn et al. added to a powder Y_2O_3 nanocrystalline TiO_2 to limit grain growth. Part of Y_2O_3 dissolved in the regions of grain boundaries and partly reacted with TiO_2 to form a second phase in grain boundaries. These two effects have limited the growth of grains so that the Y_2O_3 sintered without applying pressure reached 90% density with 50nm grain size and Y_2O_3; when adding TiO_2, sintered under the same conditions, it showed 30nm grain size with 99% density.

Recent studies have shown that grain growth inhibition during sintering, which favors increase in mechanical properties of the nanocomposite, occurs by adding small amounts of nanosized zirconia inclusions in a ceramic body of alumina matrix. Grain growth inhibition has also been observed with nanometric inclusions of silicon carbide. However, densification during sintering is difficult by the presence of zirconia in alumina. Other problems were reported in the literature: tendency to particles agglomeration and difficulty to dispersion of nanosized particles of zirconia in alumina matrix, particularly for mechanical mixing methods (Sakka & Hiraga, 1999; Susuki, 2001).

Trombini et al. (Trombini et al., 2007) dispersed powder of alumina and zirconia separately, which allowed them to obtain a complete and homogeneous dispersion of nanosized particles of zirconia in alumina matrix. The Spark Plasma Sintering (SPS) could be used to obtain samples with densities close to theoretical density with very homogeneous microstructure and grain size similar to the initial particle size of powder with at 1300 °C sintering temperature. Pierri et al. (Pierri et al., 2005) observed that the presence of small amounts of zirconia (1 vol%) was sufficient to cause an grain growth inhibition of alumina, allowing the sintering process without application of pressure that results in higher final densities and increased mechanical strength and wear resistance.

3. Experimental procedure

Initially, the alumina powder was processed to remove the hard agglomerates. The following procedure was used: powder was dispersed in isopropyl alcohol with 0.2 w% of PABA (4-aminobenzoic acid) and 0.5 w% of oleic acid. The suspension was submitted to a ball mill during 10 h, using zirconia balls (ball/powder in mass ratio of 2:1) in a polypropylene vial. Suspension was dried at 75°Cand then pulverized and sieved.

For dispersion of zirconia nanometric powder in the alumina powder a ZrO_2 suspension was prepared through traditional balls milling (ZrO_2 balls with 5mm diameter) using 0.5 wt% of deflocculant PABA (4-aminobenzoic acid) in alcoholic medium with a balls/powder mass ratio of 4:1. After 12 hours milling, suspension was separated through the milling and reserved. Simultaneously, Al_2O_3 suspension in alcoholic medium was prepared with 0.2 wt% PABA with a balls/powder ratio of 5:1 for 1 hour in balls mill. 5 vol% ZrO_2 previously prepared were added to this suspension under agitation. Then, final suspension was mixed in conventional balls mill for 22 hours. Finally, 0.5w% oleic acid was added to the suspension and mixed for 2 more hours. The obtained mixtures were dried at room temperature under flowing air.

Prior to sintering experiments, samples of pure alumina were uniaxially pressed under 80 MPa into cylindrical compacts (ø=10 mm, and height of about 5 mm) and isostatically cold-pressed under 200 MPa. Samples were heat-treated at 600°C in air for 1 h to eliminate organic materials. Green density of samples was about 59% of the theoretical density (%TD).

4. Results and discussion

Conventional sintering experiments were carried out at temperature between 900 and 1500°C in air with 2 h dwell time. This sample also was sintered in a Netzsch - DIL 402C dilatometer at 15°C/min constant heating rate in air atmosphere. Based on these results, steps for the sintering were defined. The sintering process was performed in electric furnace (Model Lindberg) in the presence of $MoSi_2$ heating elements in air atmosphere.

In addition to thermal analysis by dilatometry, sintered samples were further characterized by the apparent density taken the Archimedes method as reference, grain size measurements using an image analysis program, and the microstructure was analyzed by scanning electronic microscopy (SEM).

Figure 1 shows the linear shrinkage rate versus temperature during heating in dilatometer at15°C/min constant heating rate. Sintering shrinkage started at 1030°C and maximum shrinkage rate occurred at 1345°C. In figure 1, two different areas can be defined: the first area beginning between 900°C and 1000°C and until approximately 1080°C and refers to the temperature range before sample shrinkage beginning. As shrinkage is directly related to the densification of ceramic body during the sintering process, sintered samples did not begin the densification at temperatures lower than 1080°C having a rearrangement process, coarsening of the particles and appearance of contact points among particles. The second area can be defined as the one where shrinkage occurs, from approximately 1080˚C to 1500ºC. In this area, shrinkage rate reaches the maximum value at approximately 1350°C.

The sintering temperature effect on the densification and grain growth of compacts sintered at temperature ranging between 900 and 1500ºC for 1 hour is shown in figure 2. No significant densification was observed below 1030ºC confirming the dilatometric results (figure 1). Densification was accelerated at the temperature between 1100ºC and 1350ºC, without, however, presenting great grain growth. At higher temperature, densification was minimal but the grain growth was fast. Final grain size of the nearly fully dense structure was higher than 1800 nm. While relative density increased from 95% to 99.2% with increase in temperature from 1300ºC to 1500ºC, the average grain size became coarser from 480 nm to 1800 nm; in other words, there was more than 250% increase in grain size.

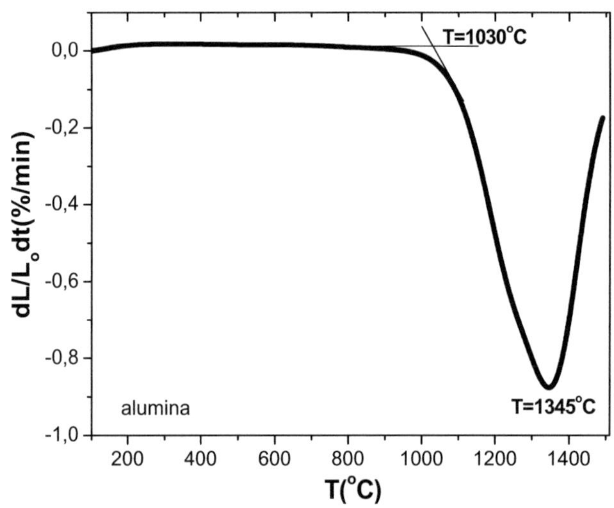

Fig. 1. Linear shrinkage rate versus temperature during heating in dilatometer at 15°C/min constant heating rate.

It has reported that dispersed open pores can pin grain boundaries and hinder grain-boundary migration in the second stage of sintering, for which the grain growth is suppressed (German, 1996). In contrast, a very sharp ascending of grain size is observable in the final sintering stage (relative density above 90% TD); however, there is remarkable increase in density. It has been confirmed that open pores referring to the intermediate stage of sintering collapse to form the closed ones after the final stage starts. Such a collapse results in a substantial decrease in pore pinning, which triggers the accelerated grain growth.

Considering the results of sintering experiments (figures 1 and 2) and to suppress the accelerated grain growth at the final sintering stage, two different sintering heating curves were applied to produce densification of Al_2O_3 compacts. These experiments were carried out using 15ºC/min heating rate.

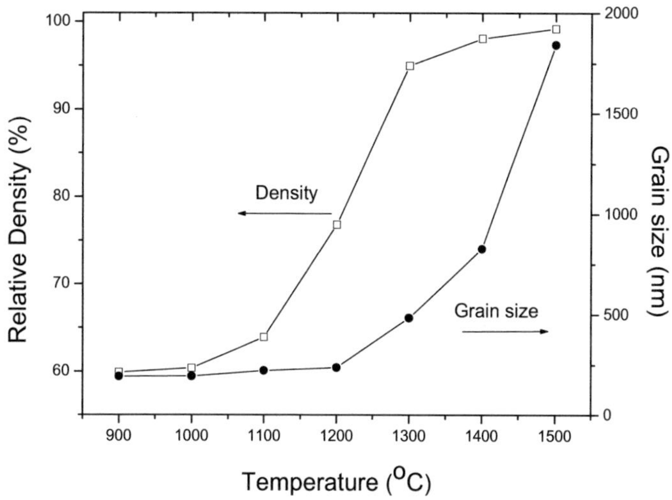

Fig. 2. Density and grain size of Al_2O_3 compacts after sintering at various temperatures for 1 hour.

In the first sintering heating curve one hypothesis was assumed: the maximization of final density with minimum grain growth could be achieved by improving the narrowing of grain size distribution at a pre-densification sintering stage and producing the final densification at a maximum densification rate. To confirm this hypothesis, a temperature below the onset of the densification process was chosen. This effect can be observed in samples with the first step at 1050°C. Samples produced by the first step at 1050°C followed by a second step at 1500°C showed significantly smaller final grain sizes as shown in figure 3 (Chinelatto et al., 2010).

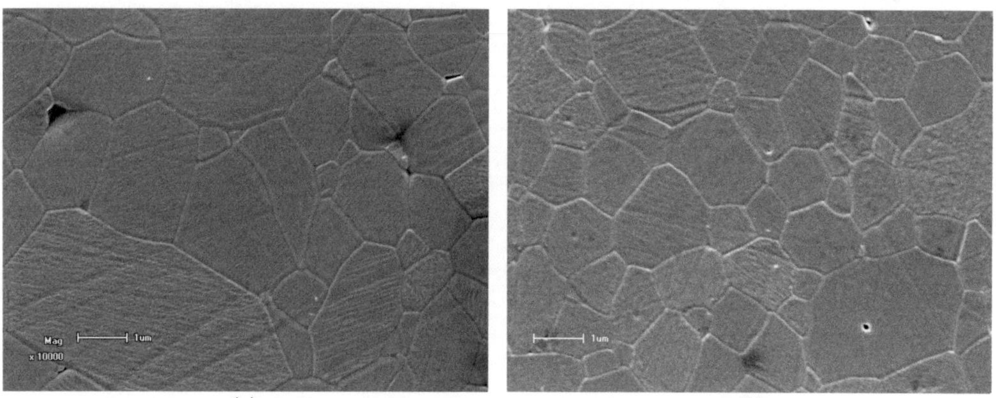

(a) (b)

Fig. 3. SEM micrographs of alumina samples after two-step sintering: a) T=1500°C/2 h; and b) T=1050°C/2 h and T=1500°C/2 h.

Figure 4 shows micrographs of surface fracture of alumina compacts, one heated at 1050°C and cooled immediately upon reaching that temperature; and the other heated to the same temperature and kept at such temperature for 2 hours. The heat treatment made finest particles to disappear and slightly coarsened the coarsest particles, decreasing the specific surface area and slightly increasing the mean grain size as indicated in Table 1. De Jonghe et al. (Lin & DeJonghe, 1997a, 1997b) suggested that during the first step, coarsening of the microstructure by surface diffusion, vapor transport, or some combination of these mechanisms produces a more uniform microstructure by an Ostwald ripening process. The evolution to a more homogeneous microstructure can be expected from the trend of porous system to evolve towards a quasi-steady state structure. Such steady-state structural distributions are generally significantly narrower than that usually produced in a powder compact (Chu et al., 1991).

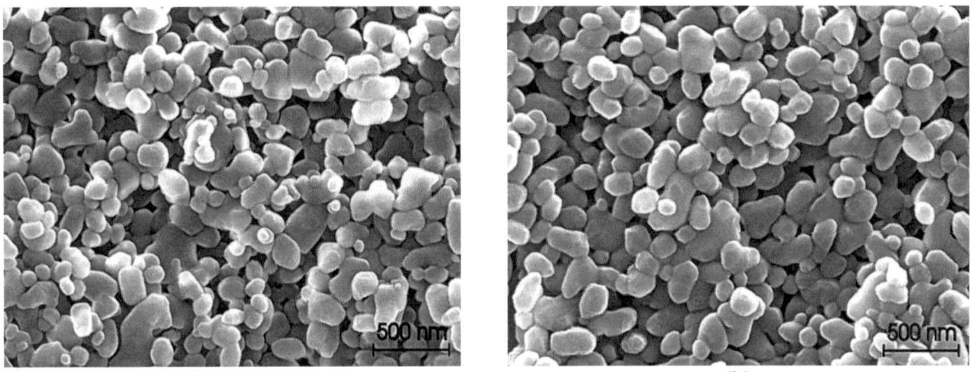

(a) (b)

Fig. 4. SEM fracture surfaces of Al_2O_3 compacts: (a) heated at 1050°C and cooled immediately e (b) heated at 1050°C for 2 hours.

	1050°C	1050°C/2 hours
Superficial area (m^2/g)	12.3	10.5
Mean grain size (nm)	119 ± 33	133 ± 28

Table 1. **Superficial area and mean grain size of particles.**

Other heating curves were developed applying sintering curves coherent with the temperature ranges in which the two processes, i.e., narrowing grain size distribution and final densification, were expected to occur. The following conditions were defined for the sintering heating curves: the first step for alumina was at 1050°C and 1000°C and the second step was at the maximum sintering temperature of 1350°C. Table 2 describes the sintering conditions and findings regarding density and average grain size of samples produced in the two-step sintering experiments.

Changes in the relative density and mean grain size with the holding time obtained are shown in Figure 5. Increased holding time results in increase of relative density and decrease in mean grain size.

Sintering procedure	Relative Density (%TD)	Mean Grain Size (nm)
TSS1 - T_1=1000°C/3h and T_2=1350°C/3h	93.8	797.4
TSS2 - T_1=1000°C/6h and T_2=1350°C/3h	94.2	763.5
TSS3 - T_1=1000°C/9h and T_2=1350°C/3h	94.6	683.5
TSS4 - T_1=1050°C/3h and T_2=1350°C/3h	93.9	717.1
TSS5- T_1=1050°C/6h and T_2=1350°C/3h	94.1	685.1
TSS6 - T_1=1050°C/9h and T_2=1350°C/3h	94.2	659.3

Table 2. **Sintering procedure and results of relative density (%TD) and mean grain size of alumina samples.**

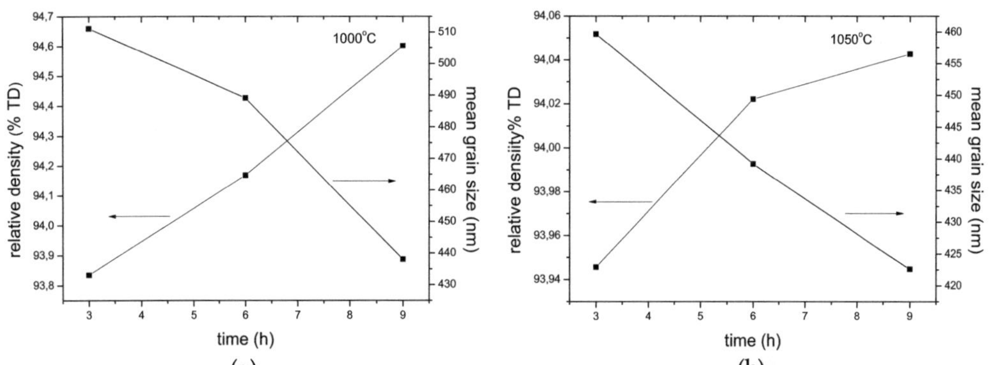

Fig. 5. Relative density and mean grain size of alumina compacts sintered versus holding time at: (a) 1000°C and (b) 1050°C.

According to Lin and Dejonghe (Lin & DeJonghe, 1997a, 1997b), with the steps at low temperature, the onset of densification is delayed due to the elimination of the finest particles (and smallest pores associated with them) during the first step. The local densification associated with the finest particles in the conventional sintering is significantly reduced in compacts subjected to the first step. Thus, removal of the finest particles due to the first step reduces the differential densification and formation of densest regions in the early sintering stages. This fact causes reduction in density fluctuations in the compact and promotes a more homogeneous final microstructure.

The other two-step sintering is based on works of Chen and Wang (Chen &Wang, 2000), in which samples are first heated to a higher temperature to achieve intermediate density, and then cooled down and kept at lower temperature until they are dense. A pre-requisite for successful densification during the second step of sintering is that pores become subcritical and unstable against shrinkage.

Chen and Wang (Chen &Wang, 2000) have explained that to achieve densification without grain growth, grain-boundary diffusion needs to remain active, while the grain-boundary migration must to be suppressed. A mechanism to inhibit grain-boundary movement is a triple-point (junction) drag. Consequently, to prevent accelerated grain growth, it is essential to decrease grain-boundary mobility. The grain growth entails a competition between grain-boundary mobility and junction mobility. Once the latter becomes less at low temperatures in which junctions are rather motionless, the mentioned drag would occur. Therefore, the grain growth is prohibited. Network mobility follows the grain-boundary mobility at high temperatures. At low temperatures, junction mobility dominates. Below the temperature at which the two rates become equal, junction mobility is essentially reduced despite the considerable grain-boundary diffusion.

Figure 2 shows that grain growth is most intense at temperatures above 1400ºC. Since samples conditions after the first stage affect the second stage of sintering, grain growth resulting from heating in the first stage must be avoided. Thus, the temperature chosen for the first stage of sintering was 1400 °C.

Figure 6 shows the behavior of relative density (%TD) versus temperature for sintering at constant heating rate of alumina. Density of alumina when temperature reaches 1400°C is 81% TD. The relative density during sintering was determined from green density (dv) and measured shrinkage (ΔL/Lo), using the approximate Eq. (1), assuming that deformation is isotropic and all axial strain is devoted to specimen's densification (Ray, 1985).

$$d_i = \frac{d_v}{(1+\frac{\Delta L}{L_o})^3} \tag{1}$$

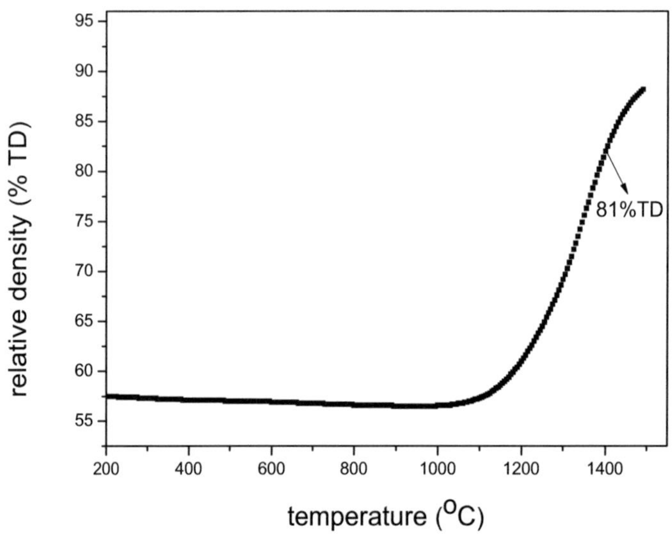

Fig. 6. Variation of relative density (% TD) versus temperature for alumina sintered at 15°C/min until the temperature of 1500 °C.

SEM micrograph of alumina when it reaches 1400ºC in the first step of sintering is showed in Figure 7. The mean grain size of alumina in this condition is about 330 nm.

Fig. 7. SEM micrograph of alumina sintered at 1400°C.

To choose the temperature for the second step T2, it is necessary to choose a temperature in which volume diffusion or grain boundary diffusion operate while the grain boundary movement is restricted (Mazaheri et al., 2008). Therefore, the second step temperatures were 1260°C and 1300°C. Sintering conditions and results of density relative and mean grain size are presented in table 3.

Sintering procedure	Relative density (%TD)	Mean grain size (nm)
TSS7 - T_1=1400°C and T_2=1260°C/3h	91.0	518.8
TSS8 - T_1=1400°C and T_2= 1260°C/6h	92.1	579.2
TSS9 - T_1=1400°C and T_2= 1260°C/9h	93.1	647.8
TSS10 - T_1=1400°C and T_2= 1300°C/3h	95.9	668.5
TSS11 - T_1=1400°C and T_2= 1300°C/6h	96.5	692.5
TSS-12 - T_1=1400°C and T_2= 1300°C/9h	96.6	718.7

Table 3. Sintering procedure and results of relative density (%TD) and mean grain size of alumina samples.

According to the results of the second step of TSS10, TSS11 and TSS12, holding the samples at 1300°C resulted in accentuate densification. Diffusive mechanisms that seem to be time dependent are therefore active at this stage. Grain-boundary diffusion and volumetric diffusion are possibly responsible for the shrinkage of the samples. On the other hand, TSS7, TSS8 and TSS9 do not lead to a dense structure, showing the inactivity of the grain-boundary diffusion at 1260°C. Considering all these facts, one can infer that 1300°C is the minimum temperature after which the grain boundary diffusion mechanism dominates.

Due to the relatively low temperature of the second stage (1260°C), densification stops before reaching a fully dense sample. A similar trend has also been reported for the two-step sintering behavior of Y_2O_3 (Wang and Chen, 2006) and ZnO (Mazaheri et. al., 2008) confirming that the reason for exhaustion in the second stage of densification is attributed to low temperature which retards grain-boundary diffusion as the sintering mechanism.

It can be seen in figure 8 that density variation results in increased grain size of the sample, showing that this condition is not yet the ideal to control the grain size in two-step sintering.

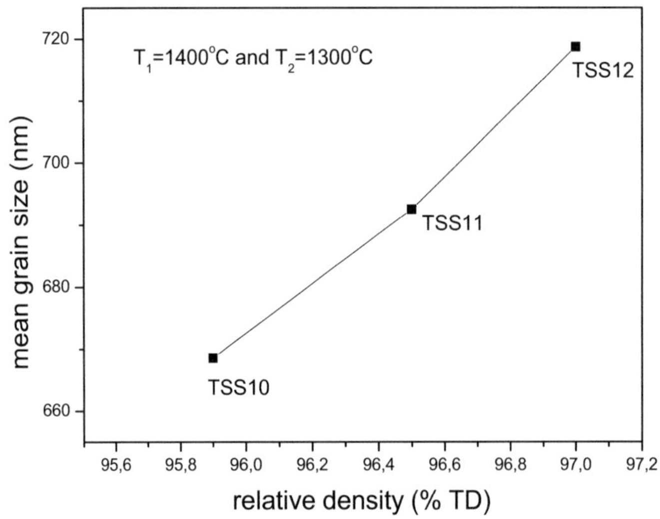

Fig. 8. Grain size/relative density trajectory obtained by two-step sintering T_1=1400ºC and T_2=1300ºC.

On the other hand, comparing the two-step sintering with conventional sintering, it is observed that the two-step sintering is efficient to control the grain growth. Figure 9 shows the micrographs of alumina sintered at 1500ºC for 2 hours and sintered at TSS3 and TSS12 conditions.

The heating curve control, through using steps of sintering, associated with control of grain size by addition of nanometric zirconia inclusions is also control the microstrucuture in conventional sintering.

Figure 10 shows the linear shrinkage rate as function of temperature for alumina-5%vol zirconia at 15ºC/min constant heating rate and 1500ºC. The presence of zirconia particles increases the maximum densification rate temperature; for alumina this temperature is 1350ºC (figure 1) and for alumina-zirconia the temperature is increased for 1440ºC. The temperature at the beginning of shrinkage process is also altered from 1030 to 1210ºC with the addition of zirconia particles. Zirconia inclusions hinder the movement of grain boundary, reducing the densification rate and grain growth (Hori et al., 1985; Liu et al., 1998; Stearns & Harmer, 1996). In the figure 11 (a) and (b), that shows the micrographs of samples of alumina-zirconia and alumina, respectively, sintered at 1500ºC for 2h, the influence of zirconia on microstructure evolution is noted through observing the grain growth behavior. The addition of nanometric zirconia is very efficient to promote a controlled grain growth. The inhibitive trend is due to the pinning effect which is associated with locations of small zirconia particles at grain boundaries or triple junctions of alumina.

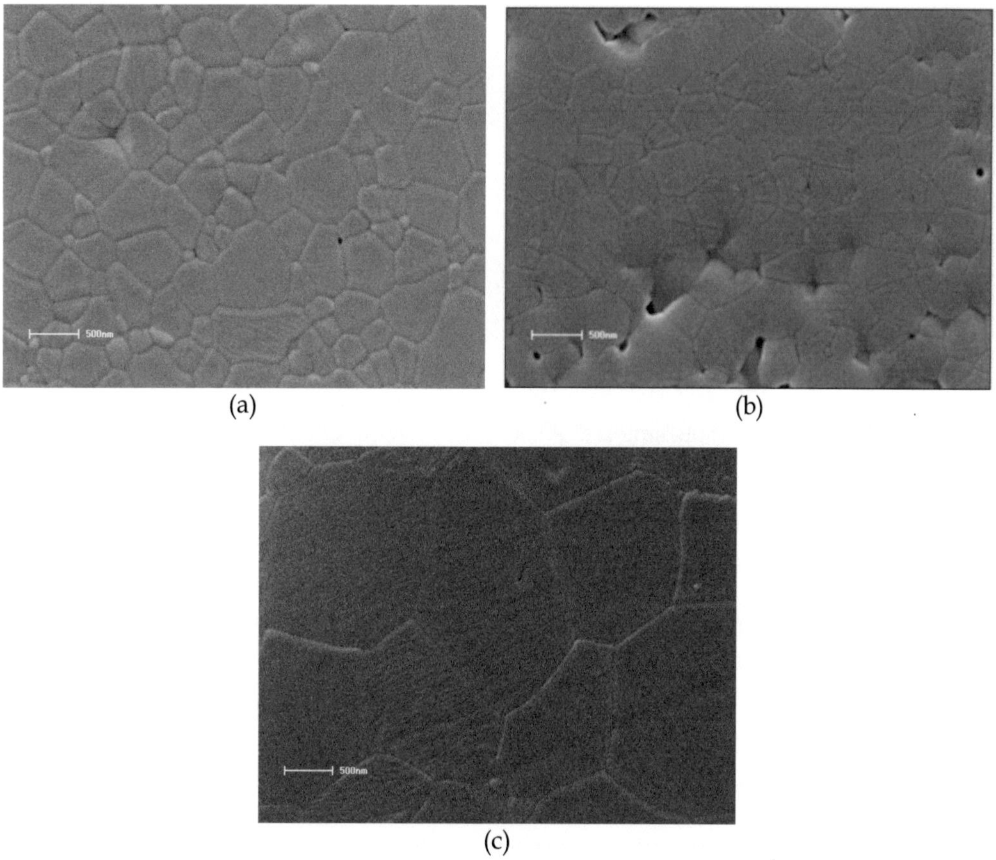

Fig. 9. SEM micrographs (a) TSS3; (b) TSS12 and (c) CS.

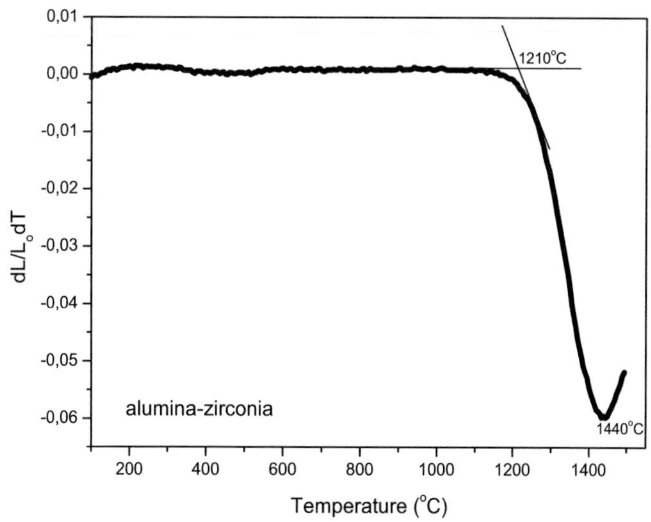

Fig. 10. Linear shrinkage rate versus temperature during heating in dilatometer at 15°C/min constant heating rate for alumina-zirconia compacts.

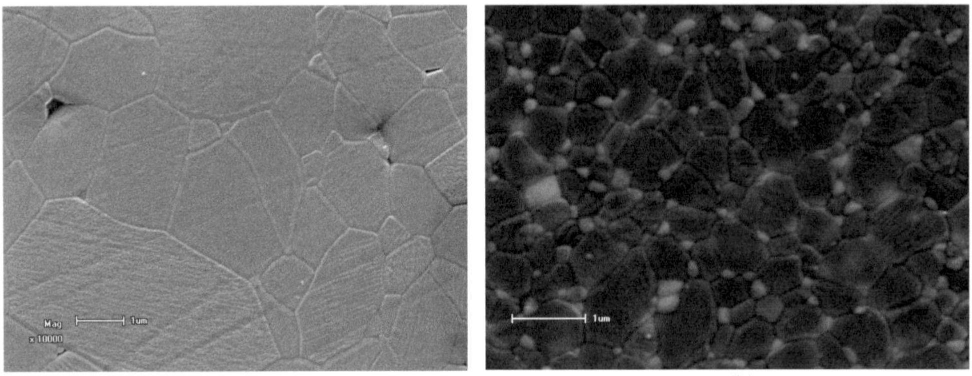

(a) (b)

Fig. 11. SEM images of sample sintered at 1500ºC for 2 h: (a) alumina and (b) alumina-zirconia.

The heating curve control, combined with the presence of nanoparticles inclusions can further optimize the microstructure control. Table 4 shows the sintering procedure and results of relative density (%TD) and mean grain size for alumina-zirconia samples. Results for TSS13 and TSS14 conditions show that the two-step sintering promoted reduction of the mean grain size compared to the conventional sintering (CS1) (Manosso et al., 2010).

Sintering procedure	**Relative density (%TD)**	**Mean grain size (nm)**
CS - T=1500ºC/2h	99.0	550
TSS13 - T_1=1460ºC/h and T_2=1350ºC/3h	97.8	330
TSS14 - T_1=1300ºC/2h and T_2=1460ºC/2h	99.7	410

Table 4. Sintering procedure and results of relative density (%TD) and mean grain size of alumina-zirconia samples.

The microstructure of the sample heated at 1460ºC and cooled immediately and the sample sintered under TSS13 conditions are showed in Figure 12. It can be noticed an initial densification and 77%DT relative density for this sample. When the sample was heated at 1460ºC and cooled down to 1350ºC (TSS13 condition) the sample could be densified without grain growth (see table 2). It means that, 77% DT reached density in the first step at high temperature for this sample can be considered the critical density. In spite of the smaller grain size presented by TSS1 condition, its relative density was lower than densities of TSS14 and CS conditions. It suggests that the time of soaking in the second step can still be prolonged. Many studies (Tarjat & Trajat, 2009; Mazaheri et al., 2008) have been demonstrated that long times in the second steps allowed the total densification without grain growth.

In the TSS14 condition, pre-densification sintering stage at 1300ºC for 2 hours was effective in grain growth control and final densification. Figure 13 (a), (b) and (c) presents the microstructure of alumina-zirconia sintered under CS1, TSS13 and TSS14 conditions, respectively. These micrographs confirmed that the two-step sintering used have been efficient to the sintering process control. It can be observed that the sample conventionally sintered presents larger grain size. Finally, it was observed that the step of sintering with addition of inclusions is also efficient in grain growth.

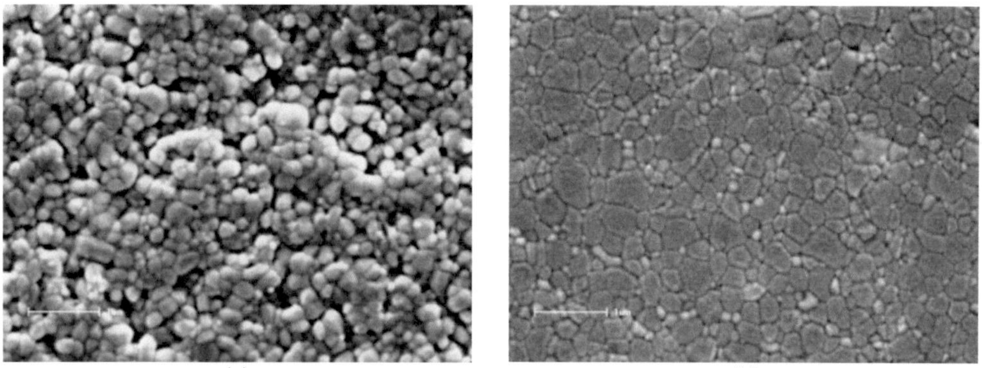

(a) (b)

Fig. 12. SEM micrographs of sintered samples: (a) T_1=1460°C and cooled and (b) TSS13 condition.

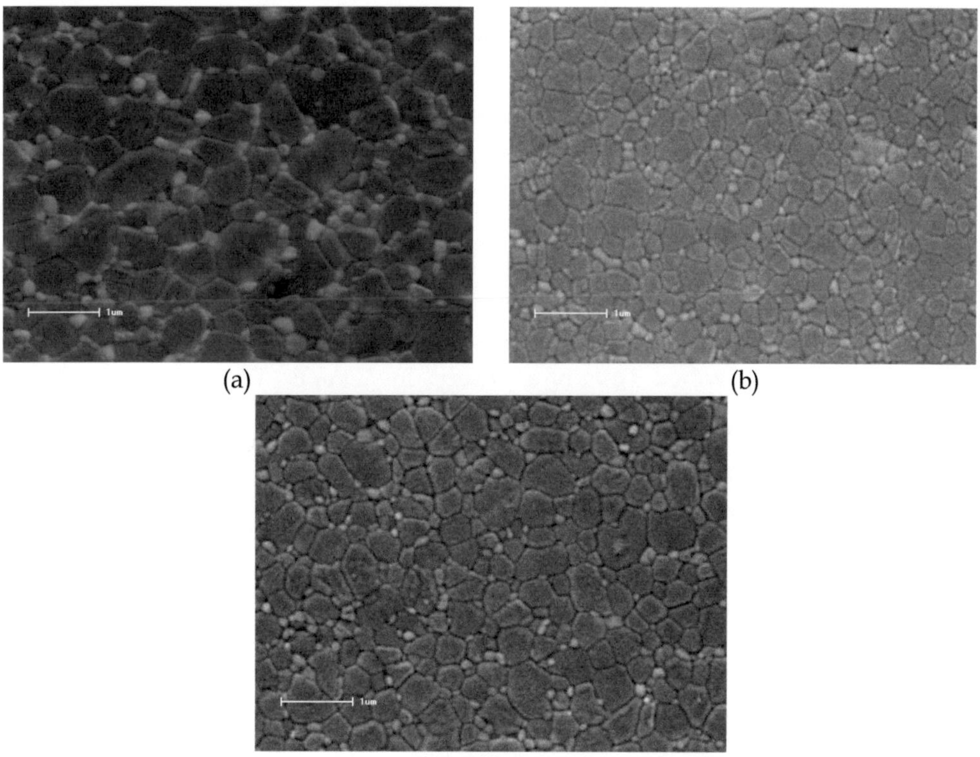

(a) (b)

(c)

Fig. 13. SEM image of the alumina-zirconia sintered under conditions: (a) CS; (b) TSS13; (c) TSS14.

5. Conclusion

The introduction of isothermal treatments in the heating curve at temperatures below the beginning of an accentuated shrinkage process influenced the development of the final microstructure, promoting a microstructural refinement of particles compacts.

The heating curve control is a simple and efficient method to control ceramic microstructure, although it is difficult to achieve an optimum condition for accessing a successful regime. The main characteristics of the heating curve control are: nanostructured ceramics can be obtained with nearly full densities; it is not necessary sophisticated and unavailable equipment, like those used for spark plasma sinterig and hot isostatic pressing; and and it is possible to achieve fully dense structures at lower temperatures.

The heating curve control, combined with the presence of nanoparticles inclusions can further optimize the microstructure control. Fine grains in the sintering induce a pinning effect on grain boundary migration and the degree of grain growth during sintering is effectively reduced.

6. References

Allen, A. J.; Krueger, S.; Skandan, G.; Long, G. L.; Hahn, H.; Kerch, H. M.; Parker, J. C. & Ali, M. N. (1996). Microstructural Evolution during the Sintering of Nanostructured Ceramic Oxides *J. Am. Ceram. Soc.*, Vol. 79, No. 5, (May 1996), pp. (1201-1212), ISSN 0002-7820.

Averback, R. S.; Höfler, H. J.; Hahn, H. & Logas, J. C. (1992). Sintering and Grain Growth in Nanocrystalline Ceramics, *Nano Mat.*, Vol. 1, No. 1, (March-April 1992) ,pp. (173-178), ISSN 0965-9773.

Bernard-Granger, G. & Guizard, C. (2007). Spark Plasma Sintering of a Commercially Available Granulated Zirconia Powder: I. Sintering Path and Hypotheses about the Mechanism(s) Controlling Densification, *Act. Mater.*, Vol. 55, No. 10, (June 2007), pp. (3493-3504), ISSN 1359-6454.

Beruto, D.; Botter, R. & Searcy, A. W. (1989). The Influence of Thermal Clycling on Densification Further: Tests of a Theory, In: Ceramic Transaction, v.1 Ceramic Powder Science IIB. Ed by Fuller, E. R. Jr.; Husner, H & Messing, G.L., pp. (911-918), Am. Ceram. Soc. Inc., ISBN: 0916094316 Westervillw, OH.

Bodisova, K.; Sajgalik, P.; Galusek, D. & Svancare, P. (2007). Two-Stage Sintering of Alumina with Submicrometer Grain Size, *J. Am. Ceram. Soc.*, Vol. 90, No. 1, (January 2007), pp. (330-332), ISSN 0002-7820.

Brook, R.J. (1982). Fabrication Principles for the Production of Ceramics with Superior Mechanical Properties, *Proc. Br. Ceram. Soc*, Vol. 32, pp. (7-24), ISSN 0524-5141.

Chaim, R.; Basat, G. & Kats-Demyanets, A. (1998). Effect of Oxide Additives on Grain Growth During Sintering of Nanocrystalline Zirconia Alloys, *Mater. Let.*, Vol. 35, No. 3-4 , (May 1998), pp. (245-250), ISSN 0167-577X.

Chakravarty, D.; Bysakh, S.; Muraleedharan, K.; Rao, T. N. & Sundaresan, R. (2008). Spark Plasma Sintering of Magnesia-Doped Alumina with High Hardness and Fracture Toughness, *J. Am. Ceram. Soc.*, Vol. 91, No. 1, (January 2008), pp. 203-208 ISSN 0002-7820.

Chen, P. L. & Chen, I. W. (1996). Sintering of Fine Oxide Powders: I, Microstructural Evolution, *J. Am. Ceram. Soc.*, Vol. 79, No. 12, (December 1996), pp. (3129-3141), ISSN 0002-7820.

Chen, P. L. & Chen, I. W. (1997). Sintering of Fine Oxide Powders: II, Sintering Mechanism, *J. Am. Ceram. Soc.*, Vol. 80, No. 3, (March 1997), pp. (637-645), ISSN 0002-7820.

Chen, I. W. & Wang, X. H. (2000). Sintering dense nanocrystalline ceramics without final-stage grain growth, *Nature*, Vol. 404, (March 2000), pp. (168-171), ISSN 0028-0826.

Chinelatto, A. S. A.; Manosso, M. K.; Pallone, E. M. J. A.; Souza, A. M. & Chinelatto, A.L. (2010). Effect of the Two-Step Sintering in the Microstructure of Ultrafine Alumina, *Adv. Sci.Tech.*, Vol. 62, (October 2010), pp. (221-226), ISSN 1662-0356.

Chinelatto,A. S. A.; Pallone, E. M. J. A.; Trombini, V. & Tomasi, R. (2008). Influence of Heating Curve on the Sintering of Alumina Subjected to High-Energy Milling. *Ceram. Int.* Vol. 34, No. 8, (December 2008), pp. (2121–2127), ISSN 0272-8842.

Chu, M.Y.; DeJonghe, L.C.; Lin, M.K.F. & Lin, F.J.T. (1991). Precoarsing to Improve Microstructure and Sintering of Powder Compacts, *J. Am. Ceram. Soc.*, Vol. 74, No 11, (November 1991), pp. (2902-2911), ISSN 0002-7820.

Czubayko, U.; Sursaeva, V.G.; Gottstein, G. & Shvindlerman, L. S. (1998). Influence of Triple Junctions on Grain Boundary Motion, *Act. Mater.* Vol. 46, No. 16, (October 1998), pp. (5863-5871), ISSN 1359-6454.

Dynys, F. W. & Hallonen, T.W. (1984). Influence of Aggregates on Sintering, *J. Am. Ceram. Soc.*, Vol. 67, No. 9, (September 1984), pp. (596-601), ISSN 0002-7820.

Erkalfa, H.; Misirk, Z. & Baykara, T. (1996). Effect of Additives on the Densification and Microstructural Development of Low-Grade Alumina Powders, *J. Mat. Proc. Tec.*, Vol. 62, No. 1-3, (November 1996), pp.(108-115), INSS 0924-0136.

Gao, L.; Hong, J. S.; Miyamoto, H & Torre, D. D. L.(2000). Bending Strength and Microstructure of Al_2O_3 Ceramics Densified by Spark Plasma Sintering, *J. Eur. Ceram. Soc.*, Vol. 20, No. 12, (November 2000), pp. (2149-2152) ISSN 0955-2219.

German, R. M. (1996). *Sintering: Theory and Practice*. Ed. John Wiley & Sons, ISBN 978-0471057864, New York.

Greer, A. L. (1998). Nanostrucuture Materials - from Fundamentals to Applications. *Mat. Sci. Forum*, Vol. 269-272, pp. (3 -10), ISSN 0255-5476.

Hahn, H. (1993). Microstrucutre and Properties of Nanostructured Oxides, *Nano. Mat.*, Vol. 2, No. 3, (May-June 1993), pp. (251-265), ISSN 0965-9773.

Hahn, H.; Logas, J & Averback, R. S. (1990). Sintering Characteristics of Nanocrytalline TiO_2, *J. Mater. Res.*, Vol. 5, No. 3, (May 1990), pp. (609-614), 1990, ISSN 0884-2914.

Harmer, M.P. & Brook, R.J. (1981). Fast Firing Microstructural Benefits, *J. Br. Ceram. Soc.*, Vol. 80, No. 5, pp. (147-48), ISSN 0307-7357.

Harmer, M. P.; Roberts, E. W. & Brook, R. J. (1979). Rapid Sintering of Pure and Doped α-Al_2O_3, *J. Br. Ceram. Soc.*, Vol. 78, No. 1, pp. (22-25), ISSN 0307-7357.

He, Z. & Ma, J. (2000). Grain Growth Rate Constant of Hot-Pressed Alumina Ceramics, *Materials Letters*, Vol. 44, No.1, (May 2000), pp.14-18, ISSN 0167-577X.

Hesabi, Z. R.; Haghighatzadeh, M. ; Mazaheri, M. ; Galusek, D. & Sadrnezhaad, S. K. (2009). Suppression of Grain Growth in Sub-Micrometer Alumina via Two-Step Sintering Method, *J. Eur. Ceram. Soc.*, Vol. 29, No. 8, (May 2009), pp. (1371-1377), ISSN 0955-2219.

Hori, S.; Kurita, R.; Yoshimura, M. & Somiya, S. (1985). Suppressed grain growth in final-stage sintering of Al2O3 with dispersed ZrO2 particles, *J. Mat. Sci. Let.*, Vol. 4, No. 9, (September 1985), pp. (1067-1070), ISSN 0261-8028.

Inada, S.; Kimura, T. & Yamaguchi, T. (1990). Effect of Green Compact Structure on the Sintering of Alumina, *Ceram. Int.*, Vol. 16, No. 6, pp. (369-373), ISSN 0272-8842.

Inoue, A. & Masumoto, T. (1993). Nanocrystalline Alloys Produced by Crystallization of Amorphous Alloys, In: *Current Topics in Amorphous Materials: Physics an Technology*, Ed. by Y.Sakurai, Y.Hamakawa, Y.Masumoto, K.Shirae and K. Suzuki, pp. (177-184), Elsevier Science Ltd, ISBN 9780-444815767, Amsterdam.

Kim, B. N. & Kiski, T. (1996). Strengthening Mechanism of Alumina Ceramics Prepared by Precoarsening Treatments, *Mat. Sci. Eng. A*, Vol. 215, No. 1-2, (September 1996), pp. (18-25). ISSN 0921-5093.

Li, J. & Ye, Y. (2006). Densification And Grain Growth of Al2O3 Nanoceramics During Pressureless Sintering, *J. Am. Ceram. Soc.*, Vol. 89, No. 1, (January 2006), pp. (139-143), ISSN 0002-7820.

Li, Z.; Li, Z; Zhang, A. & Zhu, Y. (2008). Two-Step Sintering Behavior of Sol-Gel Derived Nanocrystalline Corundum Abrasive with MgO-CaO-SiO_2 Additions, *J. Sol. Gel Sci. Technol.*, Vol. 48, No. 3, (December 2008), pp. (283-288), ISSN 0928-0707.

Liao, S. C.; Chen, Y, J; Kear, B. H. & Mayo, W. E. (1998). High Pressure/Low Temperature Sintering of Nanocrystalline Alumina, *Nano Mater.*, Vol. 10, No. 6, (August 1998), pp. (1063-1079), ISSN 0965-9773.

Lim, L. C.; Wong, P. M. & Ma, J. (1997). Colloidal Processing of Sub-Micron Alumina Powder Compacts, *J. Mat. Proc. Tec.*, Vol. 67, No. 1-3, (May 1997), pp. (137–142), INSS 0924-0136.

Lin, F. J. T. & DeJonghe, L. C. (1997a). Initial Coarsening and Microstructural Evolution of Fast-Fired and MgO-Doped Alumina, *J. Am. Ceram. Soc.*, Vol. 80, No. 11, (November 1997), pp. (2891-2896), ISSN 0002-7820.

Lin, F. J. T. & DeJonghe, L. C. (1997b). Microstructure Refinement of Sintered Alumina by Two-Step Sintering Technique, *J. Am. Ceram. Soc.*, Vol. 80, No. 10, (October 1997), pp. (2269-2277), ISSN 0002-7820.

Liu, G. L.; Qiu, H.; Todd, R.; Brook, R. J. & Guo, J. K. (1998). Processing and Mechanical Behavior of Al2O3/ZrO2 Nanocomposites, *Mat. Res. Bull.*, Vol. 33, No. 2, (February 1998), pp. (281-288), ISSN 0025-5408.

Manosso, M. K.; Pallone, E. M. J. A.; Souza, A. M.; Chinelatto, A.L. & Chinelatto, A. S. A. (2010). Two-Steps Sintering of Alumina-Zirconia Ceramics, *Mat. Sci. For.*, Vol. 660-661, (October 2010), pp. (819-825), ISSN 1662-9752.

Mayo, M. J. (1996). Processing of Nanocrystalline Ceramics from Ultrafine Particles, *Int. Mat. Rev.*, Vol. 41, No. 3, (January 1996), pp. (85-115), ISSN 0950-6608.

Mazaheri, M; Hesabi, Z. R. & Sadrnezhaad, S. K. (2008). Two-Step Sintering of Titania Nanoceramics Assisted by Anatase-to-Rutile Phase Transformation, *Scr. Mat.*, Vol. 59, No. 2, (July 2008), pp. (139-142), ISSN 1359-6462.

Mazaheri, M.; Simchi, A. & Golestani-Fardi, F. (2008). Densification and Grain Growth of Nanocrystalline 3Y-TZP During Two-Steps Sintering, *J. Eur. Ceram. Soc.*, Vol. 28, No. 15, (November 2008), pp. (2933-2939), ISSN 0955-2219.

Mazaheri, M.; Zahedi, A. M. & Sadrnezhaad, S. K. (2008). Two-Step Sintering of Nanocrystalline ZnO Compacts: Effect of Temperature on Densification and Grain Growth, *J. Am. Ceram. Soc.*, Vol. 91, No. 1, (January 2001), pp.(56-63), ISSN 0002-7820.

Mazaheri, M.; Zahedi, A. M..; Haghighatzadeh, M. & Sadrnezhaad, S.K. (2009). Sintering of Titania Nanoceramic: Densification and Grain Growth, *Ceram. Int.*, Vol. 35, No. 2, (March 2009), pp. (685-691), ISSN 0272-8842.

Morris, D. G. (1998).What Have We Learned about Nanoscale Materials? The Past and Future. *Mat. Sci. Forum*, Vol. 268-272, pp. (11-14), ISSN 0255-5476.

Novikov, V. Y. (2006). Grain Growth Controlled by Mobile Particles on Grain Boundaries, *Scr. Mat.*, Vol. 55, No. 3, (August 2006), pp.(243-246), ISSN 1359-6462.

Pierri, J. J.; Maestrelli, S. C.; Pallone, E. M. J. A. & Tomasi, R. (2005). Dispersão de Nanopartículas de ZrO_2 Visando Produção de Nanocompósitos de ZrO_2 em Matriz de Al_2O_3, *Cerâmica,* Vol. 51, No. 317, (Jan-Mar 2005), pp. (08-12), ISSN 0366-6913.

Porat, R.; Berger, S. & Rosen, A. (1996). Dilatometric Study of the Sintering Mechanism of Nanocrystalline Cemented Carbides, *Nano Mater.*, Vol. 7, No. 4, (May-June 1996) pp. (429-436), ISSN 0965-9773.

Ragulya, A.V. & Skorokhod, V.V. (1995). Rate-Controlled Sintering of Ultrafine Nickel Powder, *Nano. Mat.*, Vol. 5, No. 7-8, (September-December 1995), pp. (835-843), ISSN 0965-9773.

Rosen, A. & Bowen, H.K. (1988). Influence of Various Consolidation Techniques on the Green Microstructure and Sintering Behavior of Alumina Powders, *J. Am. Ceram. Soc.*, Vol. 71, No. 11, (November 1988), pp. (970-977), ISSN 0002-7820.

Sakka, Y. & Hiraga, K. (1999). Preparation Methods and Superplastic Properties of Fine-Grained Zirconia and Alumina Based Ceramics, *Nippon Kagaku Kaishi*, Vol. 8, pp. (497–508), ISSN 0369-4577.

Sigmund, W.M.; Bell, N.S. & Bergström, L. (2000). Novel Powder-Processing Methods for Advanced Ceramics, *J. Am. Ceram. Soc.*, Vol. 83, No. 7, (July 2000), pp. 1557–1574, ISSN 0002-7820.

Suzuki, T.; Sakka, Y.; Nakano, K. & Hiraga, K. (2001). Effect of Ultrasonication on the Microstructure and Tensile Elongation of Zirconia-Dispersed Alumina Ceramics Prepared by Colloidal Processing, *J. Am. Ceram. Soc.*, Vol. 84, No. 9, (September 2001), pp.(2132–2134), ISSN 0002-7820.

Sato, E. & Carry, C. (1995). Effect of Powder Granulometry and Pré-Treatment on Sintering Behavior of Submicron-Grained α-Alumina, *J. Eur. Ceram. Soc.*, Vol. 15, No. 1, pp. (9-16), ISSN 0955-2219.

Searcy, A. W. (1987). Theory for Sintering in Temperature Gradients: Role of Long-Range Mass Transport, *J. Am. Ceram. Soc.*, Vol. 70, No. 3, (March 1987), pp. (C61-C62), ISSN 0002-7820.

Stearns, L. & Harmer, M. P. (1996). Particle-Inhibited Grain Growth in Al_2O_3-SiC: I, Experimental Results, *J. Am. Ceram. Soc.*, Vol. 79, No. 12, (November 1996), pp. (3013-3019), ISSN 0002-7820.

Suryanarayana, C. (1995). Nanocrystalline Materials , *Int. Mat. Reviews*, Vol. 40, No. 2, pp. (41-64), ISSN 0950-6608.

Tartaj, J. & Tartaj, P. (2009). Two-Stage Sintering of Nanosize Pure Zirconia, *J. Am. Ceram. Soc.*, Vol. 92, No. Supplement s1, (January 2009), pp. (S103-S106). ISSN 0002-7820.

Trombini, V.; Pallone, E. M. J. A.; Munir, Z. A. & Tomasi, R. (2007). Spark Plasma Sintering (SPS) de Nanocompósitos de Al_2O_3-ZrO_2, *Cerâmica*, Vol. 53, No. 325 , (Jan-Mar 2007), pp. (62-67), ISSN 0366-6913.

Wang, C. J.; Huang, C. Y. & Wu, Y. C. (2009). Two-step Sintering of Fine Alumina–Zirconia Ceramics, *Ceram. Int.*, Vol. 35, No. 4, (May 2009), pp. (1467-1472), ISSN 0272-8842.

Wang, X. H.; Chen, P. L. & Chen, I. W. (2006a). Two-Step Sintering of Ceramics with Constant Grain-Size, I. Y_2O_3, *J. Am. Ceram. Soc.*, Vol. 89, No 2, (February 2006), pp. (431-437), ISSN 0002-7820.

Wang, X. H.; Deng, X. Y.; Bai, H. I.; Zhou, H.; Qu, W. G.; Li, L.T. & Chen, I.W. (2006b). Two-Step Sintering of Ceramics with Constant Grain-Size, II.BaTiO3 and Ni-Cu-Zn Ferrite, *J. Am. Ceram. Soc.*, Vol. 89, No. 2, (February 2006), pp.(438-443), ISSN 0002-7820.

Weibel, A.; Bouchet, R.; Denoyel, R. & Knauth, P. (2007). Hot Pressing of Nanocrystalline TiO_2 (Anatase) Ceramics with Controlled Microstructure, *J. Eur. Ceram. Soc.*, Vol. 27, No. 7, pp. (2641-2646), ISSN 0955-2219.

Zhou, Y.; Erb, U.; Aust, K. T. & Palumbo, G. (2003). The Effects of Triple Junctions and Grain Boundaries on Hardness and Young Modulus in Nanostructured Ni-P, *Scr. Mater.*, Vol.48, No. 6, (March 2003), pp. (825-830), ISSN 1359-6462.

Zhou, Y.; Hirao, K.; Yamauchi, Y. & Kanzaki, S. (2004). Densification and Grain Growth in Pulse Electric Current Sintering of Alumina, *J. Eur. Ceram. Soc.* Vol. 24, No. 12, pp. (3465-3470), ISSN 0955-2219.

19

Two-Step Sintering Applied to Ceramics

Gislâine Bezerra Pinto Ferreira, José Ferreira da Silva Jr, Rubens Maribondo do Nascimento, Uílame Umbelino Gomes and Antonio Eduardo Martinelli
Federal University of Rio Grande do Norte
Brazil

1. Introduction

During the process of sintering of ceramics, it is necessary to apply high temperature owing the high melting point of the raw materials. In general, a ceramist, wishing to produce a material with particular properties, must identify the required microstructure and then design processing conditions that will produce this required microstructure (Lutgard et al., 2003). One of the options to adapt the microstructure is a technique called two step sintering (TSS), this technique has been applied to the sintering of ceramic oxides to achieve full density without grain growth in final stage of sintering without loss densification (Chen & Wand, 2000). The two-step sintering process consists in to heat a ceramic body to a peak temperature (T1) to achieve an intermediate density and then the temperature is reduced to a dwell temperature (T2), which is held till full density is achieved. To succeed in two-step sintering, a sufficiently high relative density (70% or greater) needs to be achieved at T1 (Chen & Wang, 2000 & Chen, 2000). Once this critical density is reached, a lower temperature, T2, used for the isothermal hold will be sufficient to achieve full density. Difference between kinetics of grain boundary diffusion and grain boundary migration is used to obtain almost full dense, nanostructured ceramics. During the last stage of the sintering occur the grain growth in materials, this implicate in final properties, like mechanical resistance, density, ionic and electrical conductivity and others (Robert et al., 2003). The two-step sintering has been applied in many mateirals with the main goal of avoiding the grain growth in final stage of sintering, the results show the TSS is a technique efficient for it. Some application for two-step sintering are materials which need high density and small grain size, for example electrolytes of solid oxide fuel cell, as ceramics based in Y_2O_3 and CeO_2, both with and without doppant (Wang et al., 2006; Wright, 2008 & Lapa, 2009). Others examples in which TSS are used also as nanostructural fosterite (Fathi, et al., 2009), alumina-zirconia ceramics (Wang et al., 2008), TiBaO3 and Ni-Cu-Zn Ferrite (Wang et al., 2006), ZnO (Shahraki et al., 2010). In this cases, the researchs are getting the relative density higher than 97% and the size grains in level sub-micrometer.

2. Mechanisms of two-step sintering

During the process of the TSS the first step needs high temperature enough to achieve the critical diameter spherical (d_c) of the core to become the crystallization, in this step the relative density (ρ) need be the same or higher than 75% of the teorical density to obtain unstable pores and the sintering the material be kept. In the second step is necessary the temperature to keep the densification until the end of the sintering, but avoiding the grain growth. An important reason to have an understanding of two-step sintering is the possibility to increase the heat rate of sintering, to avoid the grain growth and to obtain a material with improved mechanical, thermal, electrical and optical properties in the materials (Chen & Wang, 2000). The absence of grain growth in second-step sintering has important implications for kinetics. Grain coarsening creates a powerful dynamic that constantly refreshes the microstructure. Statistically, only one-eighth of all grains survive every time the size of the grains doubles. This evolution can be a source of enhanced kinetics. Even without grain growth, enhanced kinetics has also been suspected in cases when microstructure evolution is otherwise robust; for example, in fine-grain superplasticity (McFadden et al., 1999; Wakai et al., 1990 & Chen & Xue, 1990). Because second-step sintering proceeds in a ´frozen' microstructure, it should have slower kinetics. Yet the slower kinetics is sufficient for reaching full density, while providing the benefit of suppressing grain growth. The diffusion kinetics is quantified in the frozen microstructure by measuring the densification rates in the second step, and comparing them with the prediction based on Herring's dimensional argument (Herring, 1950) for normalized densification rate (dr/rdt, where r is relative density and t is time):

$$\frac{d\rho}{\rho dt} = F(\rho)\left(\frac{\gamma\Omega}{GkT}\right)\left(\frac{\delta D}{G^3}\right) \tag{1}$$

Here γ is surface energy, Ω is atomic volume, G is grain size, d is grain-boundary thickness, and D is grain-boundary diffusivity. In the above, $\gamma\Omega/GkT$ may be viewed as the normalized driving force, and $\delta D/G^3$ is the standard kinetic factor that enters the strain-rate equation for grain-boundary processes such as sintering, diffusional creep and superplasticity. The remaining dimensionless prefactor on the right-hand side, F, is independent of the grain size as such, but could depend on other aspects of the microstructure such as density and pore distribution (Wei & Wang, 2000).

Grain boundaries in ceramics have been extensively investigated in recent years with the intent to understand their structures and mechanical/electrical properties. Grain boundaries are also important for kinetic phenomena, such as sintering, grain growth, diffusional creep and superplasticity. Their importance increases as the grain size decreases, since the ratio of grain boundary to the grain interior is inversely proportional to the grain size. In addition, this ratio also dictates that there is a large capillary pressure (and its variation) in fine grain materials. For a typical grain boundary energy (and surface energy) in ceramics of 1 J/m^2, the capillary pressure is of the order of 2000 MPa at a grain size of 1 nm, 200 MPa at 10 nm, and 20MPa at 100 nm. These pressures are rather significant and may cause additional kinetic effects at intermediate and high temperatures. In general, the dominant kinetic paths in submicron powders and ceramic bodies are surfaces and grain boundaries, the latter becoming increasingly important as the relative density reaches toward 100%. "Clean" experiments on grain boundary kinetics without the "contamination" of surface effects can be undertaken provided full density is first achieved. Common ceramic firing processes, however, always induce rapid grain growth when the relative density exceeds 85%, because of the breakdown of pore channels at three grain junctions and the resulting reduction of the pore drag on grain boundary migration. Nevertheless, the combination of good powder processing, fine starting powders, and low sintering temperatures may help to achieve submicron grain sizes in fully dense bodies. Such ceramics are suitable for kinetic studies of grain boundaries (Chen, 2000). The feasibility of densification without grain growth relies on the suppression of grain-boundary migration while keeping grain boundary diffusion active. Two-step sintering can be used to achieve a relative density of 98% by exploiting the ''kinetic window' that separates grain-boundary diffusion and grain-boundary migration. When conditions for two-step sintering fall below the ''kinetic window,'' a density _ 96% cannot be achieved even if a starting density of 70% is achieved at T1, as grain growth may still be suppressed but densification will be exhausted. Above the ''kinetic window,'' grain growth is likely to occur (Wright, 2008). The suppression of the final-stage grain growth is achieved by exploiting the difference in kinetics between grain boundary diffusion and grain-boundary migration. Such a process should facilitate the cost-effective preparation of other nanocrystalline materials for practical applications. To succeed in two-step sintering, a suficiently high starting density should be obtained during the first step. When the density is above 70%, porosimetry data have shown that all pores in Y_2O_3 become subcritical and unstable against shrinkage (which occurs by capillary action). These pores can be called as long as grain-boundary diffusion allows it, even if the particle network is frozen as it clearly is in the second step (Chen & Wang, 2000). From the thermodynamics aspect, at a temperature range where grain boundary diffusion is active, but grain boundary migration is sufficiently sluggish, densification would continue without any significant grain growth. On basis of this idea was developed to suppress the accelerated grain growth at the final stage of sintering by triple junctions. To take the advantage of boundary dragging by triple junctions, a critical density at first should be achieved where sufficient triple junctions exist throughout the body as pins. Then with decreasing the sintering temperature to a critical degree, the grain growth would be stopped by triple junctions while densification may not be impaired. In doing so, samples have to be exposed to prolonged isothermal heating at the second (low temperature) step. As in a TSS regime, the triple junctions are going to prohibit grain growth, while unstable pores can shrink with low temperature annealing, seemingly the source of different densities lies in the pore size and distribution which needs to be further investigated. Certainly, formation of inhomogeneous porosity due to the increased tendency of nanopowder to form agglomerates complicates the situation. To solve this problem, one can use larger particles with lower agglomeration degree and shape green bodies with advanced methods to obtain a more homogenous structure. Under this condition, one can expect successful TSS at lower temperatures (Hesabi et al., 2008).

Sophisticated firing profiles are an alternative to compositional effects, to obtain dense ceramics with proper microstructure. Two-step sintering profiles, including optimized combinations of peak and dwell sintering temperatures, produced nanostructured materials with high densification at reasonably low temperatures, due to different grain growth and densification kinetics (Chen & Wang, 2000). Was observed that samples processed by two-step sintering show best results the density, around 94% of theoretical density and sub-micrometer grain size (Lapa, 2009).

3. Grain boundary kinetics during intermediate and final stage sintering

Sintering data are often used to infer the rate-controlling mechanism following the scaling analysis of Herring (Herring, 1950 & Herring, 1951).

$$\frac{d\rho}{dt} = \left(\frac{f(\rho)}{kTR^m}\right) D_0\, exp(-Q/kT) \qquad (2)$$

In the above, ρ is the relative density, m is either 3 for lattice diffusion or 4 for grain boundary diffusion, D_0 and Q refer to the pre-factor and activation energy of either lattice diffusion or grain boundary diffusion, and f is a constant that is dependent on the pore/grain geometry of the sintering body. Over a range of relative density, from 60% to 90%, some model calculations suggest that f is relatively constant (Coble, 1965 & Swinkels and Ashby, 1981). Thus, Eq. (2) can be used to deduce the diffusivity and the rate controlling mechanism if the densification rate and the grain size are known. In practice, plotting Log $(Td\rho=dt)R^m$ against $1/T$ usually yields a straight line regardless whether m is chosen as 3 or 4. This is due to the relatively poor resolution of such plotting method, the unavoidable scatter of the data, and the uncertainty of the value of $f(\rho)$. Therefore, it is usually not possible to definitively state that the sintering mechanism is via grain boundary diffusion or lattice diffusion based on the scaling analysis alone. On the other hand, the inferred values of diffusivities, and especially those of the activation energy, are usually quite different depending on whether m is chosen as 3 or 4. Thus, when independent diffusivity data are available, e.g., from grain boundary mobility measurements, a self-consistency check may be applied to infer whether the lattice or grain boundary mechanism applies. Using this method, is possible to conclude that later stage sintering of submicron CeO_2 and Y_2O_3 powders is controlled by grain boundary diffusion (Chen & Chen, 1997). As cited above, to achieve densification without grain growth, is necessary to first fire the ceramic at a higher temperature ($T1$) to a relative density of 75% or more, then sinter it at a lower temperature ($T2$) for an extended time to reach full density. This schedule is different from the conventional practice for sintering ceramics, in which the temperature always increases or, at least is held constant at the highest temperature, until densification is complete. The temperature $T2$ required for the second step decreased with the increasing grain size. However, if $T2$ is too low, then sintering proceeds for a while and then becomes exhausted. On the other hand, if $T2$ is too high, grain growth still occurs in the second step. Chen 2000, observed that the same shape in different Y_2O_3 but the medium temperature is shifted depending on the solute added. The results imply that grain boundaries previously stabilized at a higher temperature are difficult to migrate at a lower temperature even though they may still provide fast diffusion paths. The presence of impurity or solute segregation is not essential for this observation, since the same observation was found in pure Y_2O_3 as well as Y_2O_3 doped with both diffusion enhancing and diffusion-suppressing solutes. Therefore, the suppression of migration is not due to solute pinning. This suggests, for the first time that the mechanism for grain boundary migration is not grain boundary diffusion even in a pure substance. If the activation energy of the additional step is higher than that of grain boundary diffusion, it could explain why grain boundary migration is inhibited at low temperatures but not at high temperature. The most likely candidates for such step are movement of nodal points or nodal lines on the grain boundary, such as four-grain junctions, pore-grain boundary junctions, or three-grain junctions. The structures of these nodal points and lines may be special and they could become stabilized by prior high temperature treatment, rendering them difficult to alter to accommodate the subsequent movement of migrating grain boundary at low temperatures. Empirically, this may be modeled by assigning mobility to the nodal point (line), whose ratio to grain boundary mobility decreases with increasing temperature.

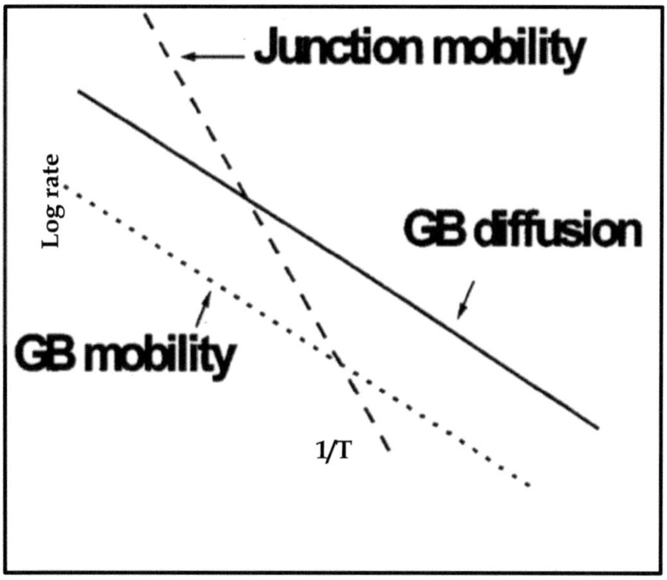

Fig. 1. Schematic Arrhenius plot for grain boundary diffusion, mobility of pore/grain-boundary junction or four-grain junction, and intrinsic mobility of grain boundary (without extrinsic drag due to nodal points/lines) (Source: Gary J. Wright, 2008).

Simple arguments then show that below a certain temperature, equilibrium grain boundary migration does not obtain since the boundaries are effectively pinned by the nodal points (lines). As mentioned above, enhanced grain boundary migration is often observed in superplastic deformation of fine grain oxides. The grain growth in this case is found to be controlled by the plastic strain. Indeed the ratio of grain size is essentially of the same order as the ratio of specimen dimensions before and after deformation (Chen & Xue, 1990). This may be regarded as opposite to the suppression of grain boundary migration described above. It is likely that in both cases, the dynamics of the nodal line/point are important. In superplasticity, the dynamics are enhanced to facilitate grain boundary migration. In low temperature sintering, the dynamics are inhibited to suppress grain boundary migration. A better knowledge of the structures of the grain boundary nodal points and lines, in both equilibrium configurations and in dynamic configurations would be required for a full understanding of the grain boundary kinetics. One interesting observation though is that a parallel effect of solute is seen in all three cases: normal grain growth, dynamic grain growth, and sintering without grain growth. For example, solutes that enhance normal grain growth also cause faster dynamic grain growth, and solutes that suppress normal grain growth likewise show a higher temperature *T2* in the kinetic window for sintering without grain growth. Thus, while the kinetics of the nodal point/line may be distinct from that of grain boundary diffusion, they may not be entirely independent of each other. Recent studies of high-purity zinc have shown that grain-boundary migration can be severely hampered by the slow mobility of grain junctions at lower temperatures, the latter having a higher activation energy (Czubayko, et al., 1998). It is possible that a similar process, in which grain junctions as well as grain boundary/pore junctions impede grain-boundary migration, may here explain the apparent suppression of grain growth at lower temperatures. Interface kinetics in very fine grain polycrystals is sometimes limited due to difficulties in maintaining sources and sinks to accommodate point defects 21±23. This leads to a threshold energy or stress, of the order of 2g/G. For a grain size of 100 nm, this amount to 20MPa, which is rather substantial compared to capillary pressure and could be the cause for the suppression. This effect should diminish at larger grain sizes, allowing the kinetic window to extend to lower temperatures. Therefore, in exploiting the difference in the kinetics of grain-boundary diffusion and grain-boundary migration to achieve densification without growth at lower temperatures, it is still advisable to utilize dopants to `tune' the overall kinetics.

4. Influence of triple junctions on grain boundary motion

According Czubayco et al., during the formation of granular structure of a polycrystalline material, both grain boundaries and triple junctions influence in characteristics. In the past, just the of grain boundaries motion were studied, while the influence of triple junctions on grain boundary has not attention necessary. In the last years, the velocity of the junction motion, the shape of the intersecting grain boundaries must be measured. Moreover, the steady-state motion of a grain boundary system with a triple junction is only possible in a very narrow range of geometrical boundary configurations. It is usually assumed that triple junctions do not influence the motion of the adjoining grain boundaries and that their role is reduced to control the thermodynamic equilibrium angles at the junction during the boundary motion. A specific mobility of triple junctions was first introduced by Shvindlerman and co-workers (Galina et al., 1987), who considered the steady-state motion of a grain boundary system with a triple junction. The geometry of the used boundary system is shown in Fig. 2. The boundaries of this system are perpendicular to the plane of the diagram, and far from the triple junction they run parallel to one another and to the x-axis.

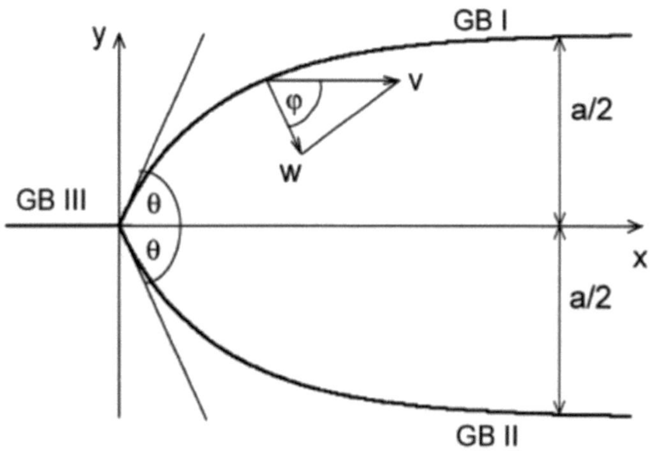

Fig. 2. Geometry of the grain boundary system with triple junction in the course of steady state motion (From: Czubayko et al., 1998).

The three boundaries of the system are considered identical, in particular their surface tension σ and their mobility m^{GB}. Furthermore, it is assumed that m^{GB} and σ are independent of the inclination of the grain boundaries. These assumptions define the problem to be symmetric with regard to the x-axis. With these simplifications some very important features of the motion of this system can be established. (a) A steady-state motion of the whole system is possible indeed. (b) The dimensionless criterion Λ:

$$\Lambda = \frac{m^{TJ}a}{m^{GB}} = \frac{2\theta}{2\cos\theta - 1} \quad (3)$$

describes the drag influence of the triple junction on the motion of the entire boundary system. For Λ >>1 the junction does not drag the motion of the boundary system, and the angle θ tends to the equilibrium value $\pi/3$. In such a case the velocity v of the motion of the entire boundary system is independent of the mobility of the triple junction and is determined by the grain boundary mobility and the acting driving force:

$$v = \frac{2\pi m^{GB}\sigma}{3a} \quad (4)$$

In contrast, for Λ<<1 the motion of the system is controlled by the motion of the triple junction and the angle θ tends to zero. The velocity depends only on the triple junction mobility and the grain boundary surface tension σ:

$$v = \sigma m^{TJ} \quad (5)$$

Owing to the fact that there are no measurements and no data of triple junction mobility, we cannot even estimate whether the ratio m^{TJ}/m^{GB} is finite. On the other hand, in the course of triple junction motion the straight grain boundary (Fig. 2, GB III) has to be extended. The velocity of its formation is unknown, but the kinetics of it should depend on the structure and properties of the generated grain boundary. Insofar as the rate of formation of this boundary can be interpreted as the velocity of the triple junction, which is proportional to its mobility according to equation (5). In the following a boundary system as shown in Fig. 2 with two identical curved boundaries (GB I and II) and a di€erent straight boundary (GB III) will be considered. The respective surface tensions and the mobilities of the boundaries are:

$$\sigma 1 = \sigma 2 \equiv \sigma \neq \sigma 3, m_1^{GB} = m_2^{GB} \equiv m^{GB} \neq m_3^{GB} \quad (6)$$

In this case the shape of a steadily moving boundary system can be expressed by the equation

$$\frac{d^2y}{dx^2} = -\frac{v}{m^{GB}\sigma}\frac{dy}{dx}\left[1 + \left(\frac{dy}{dx}\right)^2\right] \quad (7)$$

with the boundary conditions

$$y(0) = 0, y(\infty) = \frac{a}{2}, y'(0) = \tan\theta \quad (8)$$

as obvious from Fig. 2. Equations (6)-(8) completely define the problem. The shape of the stationary moving grain boundaries GB I and II (Fig. 1) is given by:

$$y(x) = \xi arccos\left(e^{-\frac{x}{\xi}+C_1}\right) + C_2,\ \xi = \frac{a}{2\theta} \quad (9)$$

$$C_1 = \frac{1}{2}ln(\sin\theta)^2, C_2 = \xi\left(\frac{\pi}{2} - \theta\right) \quad (10)$$

The steady-state velocity of GB I and II is

$$v^{GB} = \frac{2\theta m^{GB}\sigma}{a} \tag{11}$$

The velocity of the triple junction v^{TJ} can be expressed as (Soraes, et al., 1941), (Galina, et al., 1987), (Fradkov, et al., 1988):

$$v^{TJ} = m^{TJ}\sum \sigma_i \vec{\tau_i} \tag{12}$$

where every $\vec{\tau_i}$ is the unit vector normal to the triple line and aligned with the plane of the adjacent boundary. If the angles at the triple junction are in equilibrium, the driving force is equal to zero and for a finite triple junction mobility the velocity v^{TJ} should vanish as well. Consequently, for a finite m^{TJ}, the motion of the triple junction disturbs the equilibrium of the angles and, as a result, drags the motion of the boundaries. For the situation given in Fig. 2.

$$v^{TJ} = m^{TJ}(2\sigma\cos\theta - \sigma_3) \tag{13}$$

In the case of steady-state motion of the entire boundary system the velocity of the triple junction equals the velocity of the grain boundaries. Therefore, the steady-state value of the angle θ is determined by equations (11) and (12):

$$\frac{2\theta}{2\cos\theta - \frac{\sigma_3}{\sigma}} = \frac{m^{TJ}a}{m^{GB}} = \Lambda \tag{14}$$

The dimensionless criterion Λ reflects the drag influence of the triple junction on the migration of the system. One can distinguish two limiting cases:

$\Lambda \to 0$: In this case the angle θ tends to zero, i.e. the motion of the entire boundary system is governed by the mobility of the triple junction and the corresponding driving force. For the limit $\theta = 0^\circ$ the velocity of the system is given with equation (13) by

$$v = m^{TJ}(2\sigma - \sigma_3) \tag{15}$$

$\Lambda \to \infty$: In this case the angle θ tends to the value of thermodynamic equilibrium:

$$\theta = arccos\left(\frac{\sigma_3}{2\sigma}\right) = \theta_{eq.} \tag{16}$$

The motion of the system is independent of the triple junction mobility and is governed only by the grain boundary mobility and the corresponding driving force. The velocity of the boundary system in this case with equations (11) and (16) is given by:

$$v = \frac{2\theta_{eq} m_{GB\sigma}}{a} \tag{17}$$

The two states of motion of the entire grain boundary system can be distinguished experimentally for a known ratio $\sigma 3/\sigma$ by measuring the contact angle θ.

5. Influence of external shear stresses on grain boundary migration

A method to activate and investigate the migration of planar, symmetrical tilt boundaries is influenced by external shear stress. It is shown that low- as well as high-angle boundaries could be moved by this shear stress. From the activation parameters for grain boundary migration, the transition from low- to high-angle boundaries can be determined. The migration kinetics were compared with results on curved boundaries, and it was shown that the kinetics of stress induced motion were different from the migration kinetics of curvature driven boundaries. Washburn, et al. 1952 and Li, et al., 1953 investigated planar low-angle boundaries in Zn under the influence of an external shear stress and observed the motion with polarized light in an optical microscope. Symmetrical low angle tilt boundaries consist of periodic arrangements of a single sets of edge dislocations. An external shear stress perpendicular to the boundary plane will cause a force on each dislocation and in summary a driving force on the boundary. The samples were exposed to a shear stress ranging from 10^{-1} to 10^{-3}MPa. In aluminum (purity 99.999%) the yield stress is 15–20MPa, hence the applied shear stress is definitely in the elastic range. High angle symmetrical tilt boundaries also can be formally described as an arrangement of a single set of edge dislocations except that the dislocation cores overlap and the identity of the dislocations gets lost in the relaxed boundary structure. I showed that irrespective of the magnitude of the angle of rotation, grain boundaries can be moved under the action of the applied shear stress. The transition from low- to high-angle grain boundaries is revealed by a conspicuous step of the activation enthalpy at a misorientation angle of 13.6°. This holds for low angle as well as for high angle symmetrical tilt boundaries. For the curvature driven grain boundaries our results are in good agreement with previous experimental data [14] and one can see a strong dependency of the activation enthalpy on the misorientation angle, i.e. on the grain boundary structure. There is also a clear difference between the activation enthalpies for the stress induced motion of the planar high angle grain boundaries and the curvature driven migration of the curved high angle grain boundaries. Obviously, a dislocation in a high angle grain boundary does not relax completely its strain field and correspondingly, a biased elastic energy density induced by an applied shear stress will induce a force on all dislocations that comprise the grain boundary. The results prove that grain boundaries can be driven by an applied shear stress irrespective whether low- or high-angle boundaries. Obviously, the motion of the grain boundary is caused by the movement of the dislocations, which compose the grain boundary. The motion of an edge dislocation in a FCC crystal in reaction to an applied shear stress ought to be purely mechanical and not thermally activated. Obviously, the observed grain boundary motion is a thermally activated process controlled by diffusion. To understand this, one has to recognize first that grain boundary motion is a drift motion since it experiences a driving force that is smaller compared with thermal energy. Moreover, real boundaries are never perfect symmetrical tilt boundaries but always contain structural dislocations of other Burgers vectors. These dislocations have to be displaced by nonconservative motion to make the entire boundary migrate. The climb process requires diffusion, which can only be volume diffusion for low angle grain boundaries but grain boundary diffusion for high angle grain boundaries according to the observed activation enthalpies. The different behavior of curvature driven grain boundaries is not due to the curvature of the boundaries rather than due to a different effect of the respective driving force. While an applied shear stress couples with the dislocation content of the boundary in a curved grain boundary each individual atom experiences a drift pressure to move in order to reduce curvature.

6. Shape of the moving grain boundaries

The principal parameter which controls the motion of a grain boundary is the grain boundary mobility. In practically all relevant cases the motion of a straight grain boundary is the exception rather than the rule. That is why the shape of a moving grain boundary is of interest and it will be shown that the grain boundary shape is a source of new interesting and useful findings of grain boundary motion, for the interaction of a moving grain boundary with mobile particles, in particular. The experimentally derived shape of grain boundary 'quarter-loop' in Al-bicrystals of different purity was compared with theoretical calculations in the Lücke-Detert approximation. The shape of a moving GB quarter-loop was determined analytically under the assumption of uniform GB properties and quasi-two-dimensionality (Verhasselt, et al., 1999):

$$y(x) = \{-(b_F - b_L)arccos\left(\frac{\sin\Theta}{e^{x^*/b_F}}\right) + \frac{a}{2} - b_L\frac{\pi}{2} + b_F arccos\left(e^{b_F ln(\sin\Theta) - x/b_F}\right)$$

$$0 \le x \le x^* \frac{a}{2} - b_L\frac{\pi}{2} + b_L arccos\left(e^{b_L \ln(\sin\Theta) - x^*((b_L/b_F)-1) - x/b_L}\right) \quad (18)$$

$$x \ge x^*$$

where b_L is:

$$b_L = \frac{b_F\left(arc\ cos\left(\sin\Theta/e^{x^*/b_F} + \Theta - (\pi/2) - a/2\right)\right)}{arc\ cos\left(\sin\Theta/e^{x^*/b_F}\right) - (\pi/2)} b \quad (19)$$

The parameters in Eq. (18) are the width of the shrinking grain $a/2$, the angle Θ of the grain boundary with the free surfaces in the triple junction, the critical point x^*, and b_F and b_L: $b_L = m_L\sigma/V$; $b_F = m_F\sigma/V$, where m_L and m_F are the GB mobilities for 'loaded' and 'free' GB, respectively, σ is GB surface tension, V is a velocity of a quarter-loop. The first two parameters can be measured directly in the experiment. The latter two have to be chosen in an approximate way to fit the experimentally derived grain boundary shape. The point x^* is the point of intersection 'free' and 'loaded' segments of the GB. The value m_L/m_F is a measure for how drastic the change between the 'free' and the 'loaded' part in the point of intersection will be. The investigation proves that the influence of the impurity atoms on grain boundary properties and behavior is rather strong even in very pure materials. As mentioned above, the shape of a moving grain boundary is a new source of information on grain boundary migration. One example is given in Fig. 4, where the value of the critical distance x^*, normalized by the driving force (in terms of the quarter-loop width a) is plotted versus the impurity content. In accordance with the Lücke–Detert theory the critical velocity v^* (and rigidly bound to it the position of the critical point x^* on the quarter-loop) is determined by the balance between the maximum force of interaction of the impurity atoms with the boundary and the force, which is imposed by the energy dissipation caused by boundary motion across the matrix. The difference of the impurity drag for grain boundaries in samples with different amount of impurities is caused by the adsorption of impurities at the grain boundary. According to theory, the velocity should decrease proportionally to the inverse of the concentration of adsorbed atoms. Therefore x^* should increase with decreasing impurity content, as observed qualitatively (Fig. 4 and Fig. 5). However, a linear relation between the inverse of the impurity concentration and v^*, i.e. x^*, is not observed over the whole concentration range, which indicates a more complicated interaction of adsorbed atoms with the grain boundary. In such a case, x^*/a should increase more strongly with decreasing impurity content than it does linearly. This tendency is indeed observed (Fig. 4 and Fig. 5).

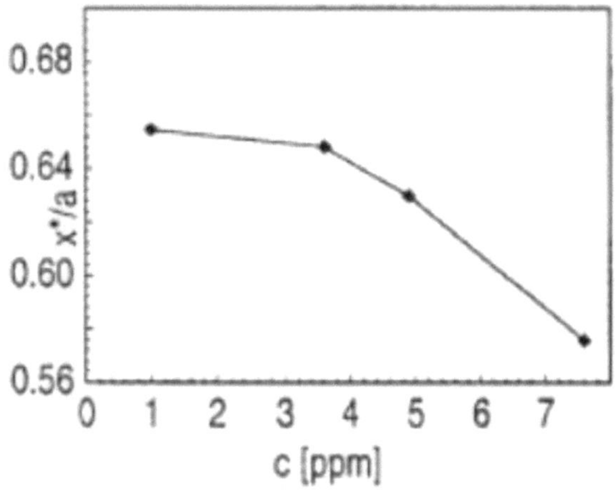

Fig. 4. Dependence of critical point x^*/a on impurity content; (b) reciprocal impurity content (From: Shvindlerman, Gottstein, 2001).

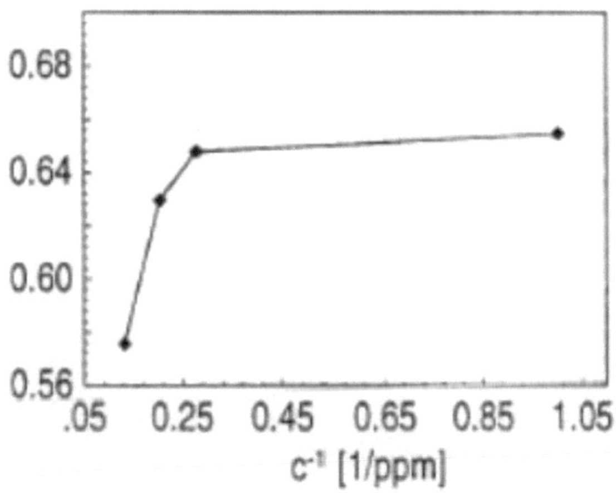

Fig. 5. Dependence of critical point x^*/a on reciprocal impurity content (From: Shvindlerman, Gottstein, 2001).

7. Dragging effect of tripe junction on grain boundary motion

In spite of the fact that a line (or column) of intersection of three boundaries constitutes a system with specific thermodynamic properties was realized more than 100 years ago (by Gibbs), the kinetic properties of this subject, in particular the mobility of triple junctions, and their influence on grain growth and relevant processes were ignored up to now. Although the number of triple junctions in polycrystals is comparable in magnitude with the number of boundaries, all peculiarities in the behavior of polycrystals during grain growth were solely attributed to the motion of grain boundaries so far. It was tacitly assumed in theoretical approaches, computer simulations and interpretation of experiment results that triple junctions do not disturb grain boundary motion and that their role in grain growth is reduced to preserve the thermodynamically prescribed equilibrium angles at the lines (or the points for 2-D systems) where boundaries meet. The most prominent example of how this assumption determines the fundamental concepts of grain structure evolution gives the Von Neumann–Mullins relation (Neumann, 1952 and Mullins, 1956). No doubt this relation forms the basis for practically all theoretical and experimental investigations as well as computer simulations of microstructure evolution in 2-D polycrystals in the course of grain growth. This relation is based on three essential assumptions, namely, (i) all grain boundaries possess equal mobilities and surface tensions, irrespective of their misorientation and crystallographic orientation of the boundaries; (ii) the mobility of a grain boundary is independent of its velocity; (iii) the third assumption relates directly to the triple junctions, namely, they do not affect grain boundary motion; therefore, the contact angles at triple junctions are in equilibrium and, due to the first assumption, are equal to 120°. As it was shown in (Neumann, 1952 and Mullins, 1956), for 2-D grain, the rate of change of the grain area S can be expressed by:

$$\frac{dS}{dt} = -A_b \oint d\varphi \tag{20}$$

where $A_b = m_b\sigma$; m_b being the grain boundary mobility, σ is the grain boundary surface tension. If the grain were bordered by a smooth line, the integral in Eq. (20) would equal 2π. However, owing to the discontinuous angular change at every triple junction, the angular interval $\Delta\varphi = \pi/3$ is subtracted from the total value of 2π for each triple junction. Consequently:

$$\frac{dS}{dt} = -A_b\left(2\pi - \frac{n\pi}{3}\right) = \frac{A_b\pi}{3}(n-6) \tag{21}$$

where n is the number of triple junctions for each respective grain, i.e. the topological class of the grain.

8. Conclusions

Many researchers have used the two-step sintering as a design process to obtain samples with a microstcture without grain growth in final stage of sintering. Same exemples that we can cite are:

- Chen, I.W. & Wang, X.H. (2000) obtained samples of the Y_2O_3 with a grain size of 60nm can be prepared by a simple two-step sintering method, at temperatures of about 1,000°C without applied pressure. The suppression of the final-stage grain growth is achieved by exploiting the difference in kinetics between grain boundary diffusion and grain-boundary migration. Such a process should facilitate the cost-effective preparation of other nanocrystalline materials for practical applications.
- 2: Lapa, et al., (2009) prepared samples of the yttrium and gadolinium-doped ceria-based electrolytes (20 at% dopant cation) with and without small Ga_2O_3-additions (0.5 mol%). The average grain sizes in the range 150–250 nm and densifications up to about 94% were found dependent on the sintering profile and presence of Ga. The grain boundary arcs in the impedance spectra increased significantly with Ga-doping, cancelling the apparently positive role of Ga on bulk transport, evidenced mostly in the case of yttrium-doped materials.
- 3: Wang, et al. (2006) used two-step sintering to sinter $BaTiO_3$ and Ni–Cu–Zn ferrite ceramics to high density with unprecedentedly fine grain size, by suppressing grain

growth in the final stage of densification. Dense $BaTiO_3$ ceramics with a grain size of 35 nm undergo distortions from cubic to various low-temperature ferroelectric structures. Dense fine grain Ni–Cu–Zn ferrite ceramics have the same saturation magnetization as their coarse grain counterparts.

9. Acknowledgment

The authors wish to thank PRH-ANP, CAPES, LMCME-UFRN, Materials Laboratory-UFRN and NEPGN-UFRN.

10. References

Arzt, E., Ashby, M. F. & Verrall, R. A. (1993) Interface-controlled diffusional creep. Acta Metall. 31, 1977± 1989.

Cannon, R. M., Rhodes, W. H. & Heuer, A. H. (1980) Plastic deformation of fine-grained alumina: I. interface-controlled diffusional creep. J. Am. Ceram. Soc. 63, 48-53

Chen, I.W. (1993). Mobility control of ceramic grain boundaries and interfaces, Materials Science and Engineering. 166. 51-58.

Chen, I.W. (2000). Grain boundary kinectics in oxide ceramics with the cubic fluorite crystal structure and its derivates. Interface Science. Vol. 8. Pp. 147-156.

Chen, I.W. & Wang, X.H. (2000). Sintering dense nanocrystaline ceramics without final-stage grain growth. Nature. Vol. 404.

Chen, I.W. & Xue, L.A. (1990). Development of superplastic structure ceramics. Journal of American Ceramics Society. Vol. 73.

Chen, P.L. & Chen, I.W. (1997). Journal of American Ceramics Society.

Coble, R.L. (1965). Intermediate-stage sintering : Modification and correction of a lattice-diffusion model. Vol. 36.

Czubayko, L., Sursaeva, V. G., Gottstein, G. & Shvindlerman, L. S., (1998) Influence of triple junctions on grain boundary motion. Acta Mater. 46, 5863±5871.

Galina, A. V., Fradkov, V. E. and Shvindlerman, L. S., (1987) Physics Metals Metallogr., 63, 165.

Herring, C. (1950). Effect of change of scale on sintering phenomena. Journal Applied to Physics. Vol. 21.

Herring, C. (1951). The physics of powder metallurgy. McGraw-Hill. New York. Pp. 143.

Hesabi, Z.R. ; Haghighatzadeh, M. ; Mazaheri, M. ; Galusek, D.S.K. & Sadrnezhaad. (2008) Suppression of grain growth in sub-micrometer alumina via two-step sintering method. Journal of the European Ceramic Society.

Jonghe, L.C. ; Rahaman, M.N. (2003). Sinterig ceramics. Handbook of Advanced Ceramics.

Land, T. A., Martin, T. L., Potapenko, S., Palmore, G. T. & De Yoreo, J. J. (1999) Recovery of surfaces from impurity poisoning during crystal growth. Nature 399, 442±445.

Lapa, C.M. ; Souza, D.P. ; Figueiredo, F.M.L. & Marques, F.M.B. (2009). Electrical and microstructural characterization of two-step sintered ceria based electrolytes. Journal of Power Sources. Vol. 187. Pp. 204-208.

Lapa, C.M. ; Souza, D.P. ; Figueiredo, F.M.L. & Marques, F.M.B. (2009). Two-step sintering ceria-based electrolytes. International Journal of Hydrogen Energy. Pp. 1-5.

Li, C.H., Edwards, E.H., Washburn, J., Parker, E.R., (1954) Recent observations on the motion of small angle dislocation boundaries Acta Met. 2, 322–333.

Mullins, W.W.,(1956), Two-dimensional motion of idealized grain boundaries. J. Appl. Phys. 27, 900–904.

Fathi, M.H.; Kharaziha, M. (2009). Two-step sintering of dense, nanostructural forsterite. Materials Letters - 63, Pp. 1455–1458.

Fradkov, V. E. and Shvindlerman, L. S., (1988) Structure and Properties of Interfaces in Metals. Nauka, Moscow, p. 213.

McFadden, S.X. ; Mishra, R.S. ; Valiev, R.Z. ; Zhilyaev, A.P. & Mukherjee, A.K. (1999). Low-temperature superplasticity in nanostructured nickel and metal alloys. Nature. Vol. 396. Pp. 684-686.

Robert, C.L. ; Ansart, F. ; Degolet, C.L. ; Gaudon, M. & Rousset, A. (2003). Dense yttria stabilized zirconia: sintering and microstructure Ceramics International. Vol. 29. Pp. 151-158.

Shahraki, M.M.; Shojaee, S.A.; Sani, M.A.D.; Nemati, A. & Safaee, I. (2010). Two-step sintering of ZnO varistors. Solid State Ionics.

Shvindlerman, L.S., Gottstein, G., (2001), Grain boundary and triple junction migration Materials Science and Engineering, A302, 141–150.

Soraes, A., Ferro, A. C. and Fortes, M. A., (1941) Scripta metallurgycal., 1985, 19.

Swinkels, F.B. & Ashby M.F. (1981). Acta Metallurgycal. Vol. 29. Pp. 259.

Verhasselt, J.C., Gottstein, G., Molodov, D.A., Shvindlerman, L.S., (1999) Acta Mater. 47, 887–892.

Von Neumann, J., (1952) in: Metal Interfaces, American Society for Testing Materials, Cleveland, OH, P. 108.

Wakai, F. et al. (1990) A superplastic covalent crystal composite. Nature. Vol. 344.

Wang, J.C.; Huang, C.Y.; Wu, Y.C. (2008) Two-step sintering of fine alumina-zirconia ceramics. Ceramics international.

Wang, X.H. ; Chen, P.L. & Chen, I.W. (2006). Two-step sintering of ceramics with constant grain-size, I Y_2O_3. Journal American Ceramics Society. Vol. 89. Pp. 431-437.

Wang, X.H.; Deng, X.Y.; Bai, H.L.; Zhou H.; Qu, W.G. & Li, L.T. (2006). Two-step sintering of ceramics with constant grain-size, II BaTiO3 and Ni-Cu-Zn ferrite. Journal American Ceramics Society.Vol. 89. 438-443.

Washburn, J., Parker,E.R., J. (1952), Journal of Metals. 4, 1076–1078.

Winning, M., Gottstein,G., Shvindlerman, L.S., in: T. Sakai, G. Suzuki (Eds.), (1999), Recrystallization and Related Phenomena, The Japan Institute of Metals, pp. 451–456.

Wright, G.J. (2008). Constrained sintering of yttria-stabilized zirconia electrolytes: The influence of two-step sintering profiles on microstructure and gas permeance. International Journal Applied to Ceramic Technologies. Vol. 5. Pp. 589-596.

20

$Ba_{1-X}Sr_XTiO_3$ Ceramics Synthesized by an Alternative Solid-State Reaction Route

R.A. Vargas-Ortíz, F.J. Espinoza-Beltrán and J. Muñoz-Saldaña
Centro de Investigación y de Estudios Avanzados del IPN,
Unidad Querétaro, Libramiento Norponiente No. 2000,
Fracc. Real de Juriquilla, CP Querétaro, Qro.
México

1. Introduction

All materials respond to stimulus, whether it be an electric field, mechanical stress, heat or light. The manner and degree to which they respond varies and is often what determines which material is selected for a given application. On the most basic level, elastic materials deform in response to mechanical stress and return to their original form when the load is removed. Other materials conduct electricity in response to an applied voltage. Both of these are well-known phenomena, and materials with such behaviors are sometimes called "trivial". On the other hand are pyroelectric and piezoelectric materials, which generate an electric field with a stimulus of heat or mechanical stress, respectively (unexpected phenomenon) and are called "smart" or "functional" materials. Ferroelectric materials are materials that exhibit piezoelectricity and pyroelectricity, as well as the phenomenon which gives them their name (ferroelectricity).

Due to their unique properties, ferroelectric materials are widely used in all areas of electronics and microelectronics, such as cellular phones, computers, cars, airplanes and satellites [KENJI, BUCHANAN]. They have a high discharge dielectric constant (ε) [SHEPARD, RADHESHYAM ,ZHIN], which allows them to be used in high permitivity dielectric devices. Their pyroelectric behaviour is used in heat sensors [PADMAJA, YOO, WHATMORE], and their piezoelectricity is applied in devices like resonadores, sonars, horns, and actuators [GURURAJA, YAMASHITA 1997, YAMASHITA 1998, CHEN]. A combination of their properites are applied in electro-optical devices such as controlable diffraction grids, waveguides, etc. [HEIHACHI, HAMMER, BLOMQVIST]. They are also used in dynamic random access memory (DRAM) [KINGON 2000, KINGON 2006, KOTECKI] and non-volatile memory (NVRAM) [MASUI, KOHLSTEDT].

Many "novel" materials known today were developed many decades ago. Ferroelectric materials are no exception, having been discovered more than seven decades ago. Valasek reported the first ferroelectric material, Rochelle salt (potassium or sodium tetrahydrate tartatre, $KNaC_4H_4O_6 \cdot 4H_2O$) in 1921 [VALASEK]. Subsequently, potassium dihydrogen phosphate (KH_2PO_4) was identified by Busch and Scherrer in 1935 [BUSCH], and barium titanate ($BaTiO_3$ or BT) was noted for its unusual dielectric properties by Wainer and

Salomon in 1942-43 [WAINER]. The discovery of ferroelectricity in ceramics from the BaO-TiO_2 system was extremely important, as it was the first ferroelastic made from simple oxide materials. Since the discovery of $BaTiO_3$, several other oxide-based ferroelectric materials have been developed, such as strontium titanate ($SrTiO_3$ or ST), lead zirconate titanate (PZT), lead titanate ($PbTiO_3$, PT), lithium tantalate ($LiTaO_3$) phosphate, and potassium titanyl ($KTiOPO_4$) to name a few [MESCHKE, KUGEL, HIDAKA, GOPALAN, ROSENMAN]. The study of BT-based ceramics with stoichiometric compositions different from pure BT has become one of the most important subjects of ferroelectrics in recent years. Particularly, substitution of Sr^{2+} ions in place of Ba^{2+} ions into BT leads to a solid solution, barium strontium titanate (BSTx or $Ba_{(1-X)}Sr_XTiO_3$, where $0 \leq X \leq 1$). Ferroelectric materials have been synthesized by various techniques, the most commonly used today being the technical or sol-gel process for the production of powders or thin films [BOLAND, PARK, ZHU, KAMALASANAN]. Another technique used to obtain powders is hydrothermal process [XU, RAZAK, VOLD, CHENG]. Finally there is the conventional route solid-state reaction of mixed oxides [VITTAYAKORN, IANCULESCU, CHAISAN] to obtain powders and solid ceramics. The interest of processing highly dense BSTx ceramics is that the Curie temperature and thus the dielectric properties can be tuned using the chemical variations between $SrTiO_3$ and $BaTiO_3$ [BERBECARU, YUN].

This chapter provides the description of an alternative solid-state reaction route based on high energy ball milling and subsequent sintering for the synthesis and densification of BSTx bulk ceramics. It provides a more direct route than the conventional route of mixed oxides. In addition to presenting structural characterization and results of electrical measurements (dielectric constant versus temperature curves and ferroelectric hysteresis loops), a novel technique known as contact resonance piezoresponse force microscopy (CR-PFM) is applied in the detection and characterization of ferroelectric domains in the BSTx samples.

2. Experimental procedures

Titanium oxide (TiO_2 - 99.9% purity, anatase phase, Aldrich), barium carbonate ($BaCO_3$ - 99.9% purity, Aldrich) and strontium carbonate ($SrCO_3$ - 99.9% purity, Aldrich) powders were weighed according to the stoichiometric proportion of Equation 1. The mixture was ball-milled in a high energy vibratory mill (SPEX 8000) for 6 h using a nylamid vial with 10 toughened zirconia balls (10 mm diameter) and a ball-to-powder weight-ratio of 10:1. The milled powders were uniaxially pressed in a stainless steel cylindrical die (10 mm inner diameter) using 1 GPa pressure. The green compacts were placed on top of TiO_2 substrates inside alumina crucibles (to avoid the reaction of barium oxide with the crucibles) and reactively sintered using a two-step heating program (Figure 1). The compacts are heated from room temperature up to 1273 K (reaction temperature) at 5 K/min and held there for 1 h. The temperature is then increased to between 1523 and 1573 K (sintering temperature) at 3.0 K/min and held there for 2 h. Finally they are cooled to room temperature at 3.0 K/min. The resulting BSTx sintered ceramics with the perovskite ABO_3 structure complied with the formula $Ba_{(1-X)}Sr_XTiO_3$ where X = 0, 0.1, 0.2 … 1 and were named accordingly, e.g. BST3 refers to the X = 0.3 composition. The starting powders, milled powders and sintered ceramics were characterized by x-ray diffraction (XRD) using a Rigaku Dmax-2100 diffractometer equipped with Co Kα radiation; scanning electron microscopy (SEM) using a

Philips XL30 ESEM; transmission electron microscopy (TEM) using a JEOL 2010; micro-Raman scattering using a DILOR unit; differential scanning calorimetry (DSC) using a Mettler Toledo; and thermogravimetric analysis (TGA) using a SDTA851 Mettler Toledo. The bulk density of the sintered ceramics was determined using the Archimedes method.

$$(1\text{-}X)BaCO_3 + XSrCO_3 + TiO_2 \rightarrow Ba_{(1-X)}Sr_XTiO_3 + CO_2\,(g) \tag{1}$$

This fabrication method is a modification of the conventional route (solid state reaction) for the manufacture of $Ba_{(1-X)}Sr_XTiO_3$ ceramics. It requires less processing steps (Figure 2), is more straightforward than other chemical processes and, most important, leads to the fabrication of high-density ceramics with very low porosity.

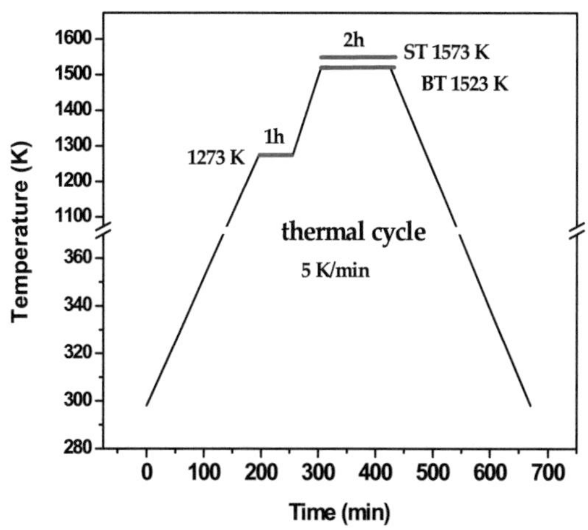

Fig. 1. Heat treatment used for the manufacture of BSTx.

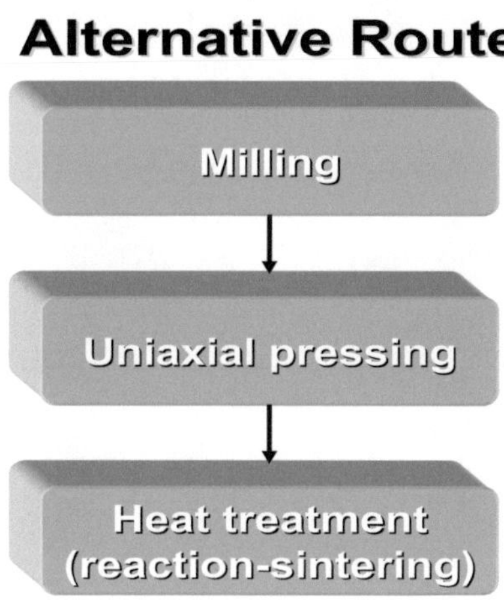

Fig. 2. Stages of the proposed alternative route for the manufacture of BSTx.

It must be noted that the high energy milling process used allows for homogenization and particle size reduction of the starting powders. It is a more efficient process than conventional attrition, planetarium or automatic-agate milling. The milled powders were compacted using much higher pressure (~1.0 GPa) than applied using the conventional approach (~0.1 GPa). This resulted in uniform green compacts which can conduct an homogeneous chemical reaction. Finally, the thermal treatment induces simultaneous reaction and sintering (reaction-sintering, Figure 1) as opposed to the conventional manufacturing process where the starting powders are milled, thermally-treated to react, a second milling process is performed, and then the twice-milled powders are pressed into compacts. Finally, a second thermal treatment (sintering) densifies the compacts.

3. Results and discussion

3.1 Fabrication of BSTx powders

The high energy milling process significantly affects the size and shape of the starting powders, as seen in the SEM micrographs of Figure 3. Figure 4 shows the TEM micrographs of the starting powders after milling for 6 h. A final particle size of less than 50 nm was attained for both extreme concentrations (only $BaCO_3$ and TiO_2, and only $SrCO_3$ and TiO_2).

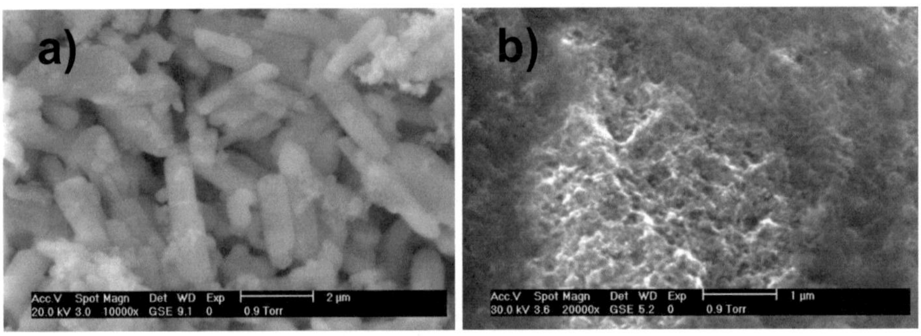

Fig. 3. SEM image of starting powders (a) milled for 6 hours, and (b) before milling.

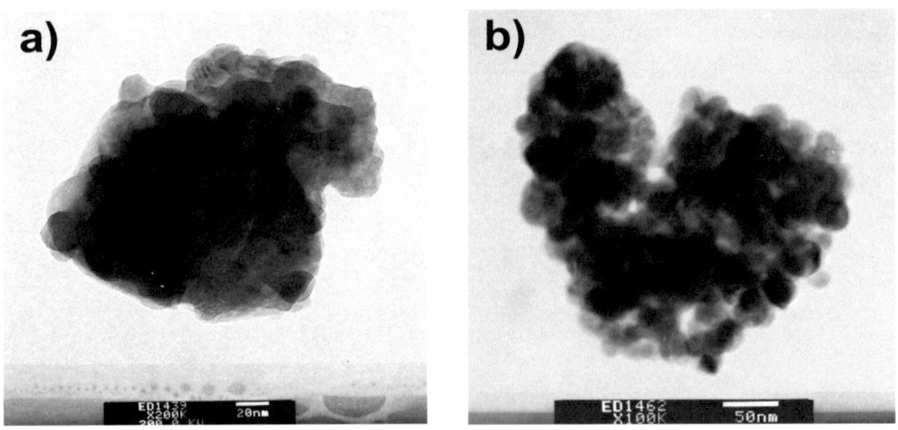

Fig. 4. TEM micrographs of (1-X)$BaCO_3$ + X$SrCO_3$+ TiO_2 powder milled for 6 hours, with (a) X = 0.0 and (b) X = 1.0.

The x-ray diffraction patterns for the milled $BaCO_3$, TiO_2 and $SrCO_3$ powder mixtures are presented in Figure 5. The main $BaCO_3$ peak, located at approximately $2\theta = 28°$ is present up to X = 0.9, and the small peak at approximately 30° belongs to TiO_2. As expected, increasing the $SrCO_3$ content increases the height of the $SrCO_3$ peak located at approximately 30°. There is a degree of amorphization due to the creation of defects in the crystal structure. Figure 6 presents the thermogravimetric analyses (TGA) curves for the same collection of samples as analyzed by x-ray diffraction (Figure 5). All the curves are similar in their key characteristics. The weight loss behavior for a single sample, with X = 0.65, is shown in Figure 8. There are four stages of weight loss centered at 403, 773, 973 and 1273 K. In the first stage, from room temperature (RT) up to 403 K, a weight reduction of approximately 1.3% occurs due to evaporation of water from the material surface. In the second stage, from 403 to 773 K, the weight loss of about 5.2% is related to the loss of chemically bound water in the form of OH groups from the BaOH that was formed when BaO combined with water during milling [BALÁZ]. This loss typically occurs between 473 and 873 K [ASIAIE]. The phenomenon of water loss has also been reported to occur in other carbonates between 593 and 723 K [DING]. The third stage, between 773 and 973 K, is not related to any structural modification of the BSTx samples. According to the literature, the decomposition of strontium and barium carbonates to form CO_2 occurs at higher temperatures, between 1023 and 1273 K [JUDD, L'VOV, MAITRA]. However, in this case, the generation of CO_2 begins as early as at 833 K and runs up to 1273 K. The x-ray difffraction patterns of Figure 7 show the appearance of a peak ($2\theta \approx 36°$) at 873 K, which corresponds to the formation of the perovskite structure of $BaTiO_3/SrTiO_3$. Therefore, the weight loss from 773 to 1323 K can be considered a single stage that varies with Sr content. It is directly related to the CO_2 excess from the carbonates used as starting powders (see Equation 1). It is the difference between the weight of the $(1-X)BaCO_3 + XSrCO_3 + TiO_2$ starting powders and that of the resulting $Ba_{(1-X)}Sr_XTiO_3$. Table 1 lists the total weight loss, the loss in the different stages, and the weight loss expected from CO_2 liberation. For the group of samples as a whole, an approximate 5% weight loss was observed between RT and 773 K, and the weight loss from 773 to 1323 K corresponds closely to the stoichiometric CO_2 loss.

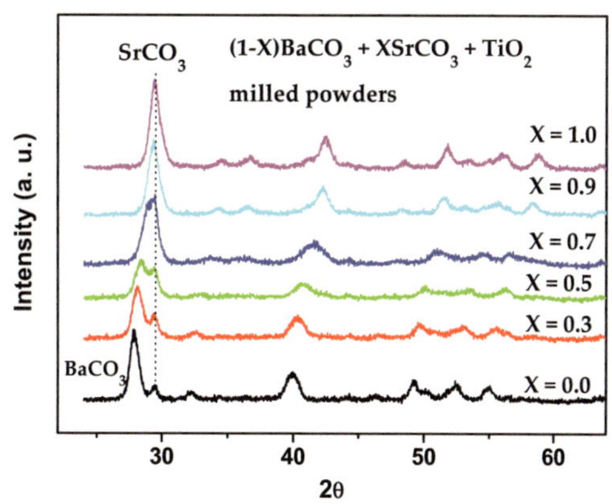

Fig. 5. X-ray diffraction patterns for $(1-X)BaCO_3 + XSrCO_3 + TiO_2$ milled powders.

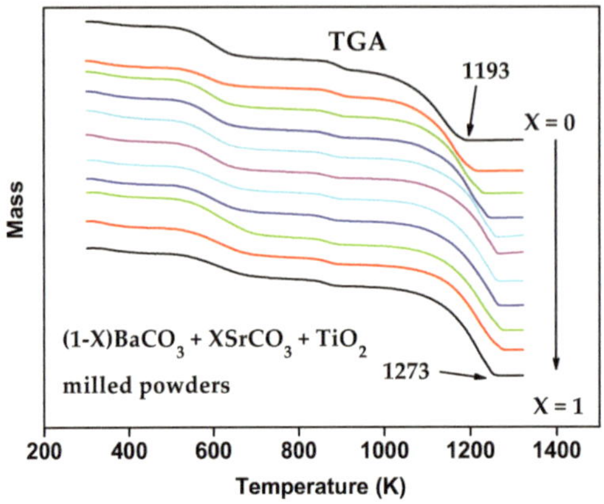

Fig. 6. Thermogravimetric curves of (1-X)$BaCO_3$ + X$SrCO_3$ + TiO_2 milled powders.

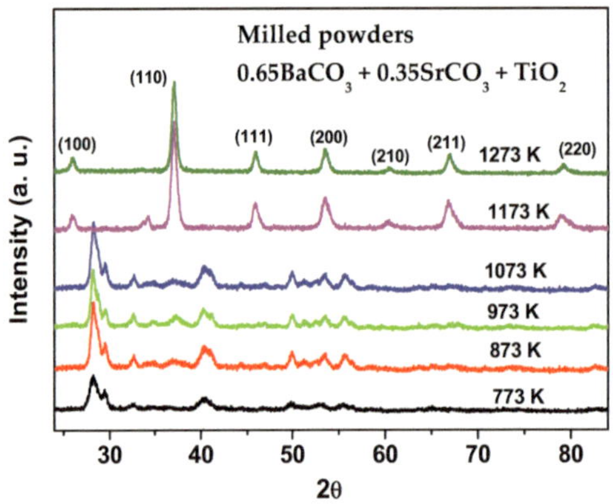

Fig. 7. XRD patterns of 0.65$BaCO_3$, 0.35$SrCO_3$ and TiO_2 powders milled for 6 hours and subsequently heat treated at different temperatures.

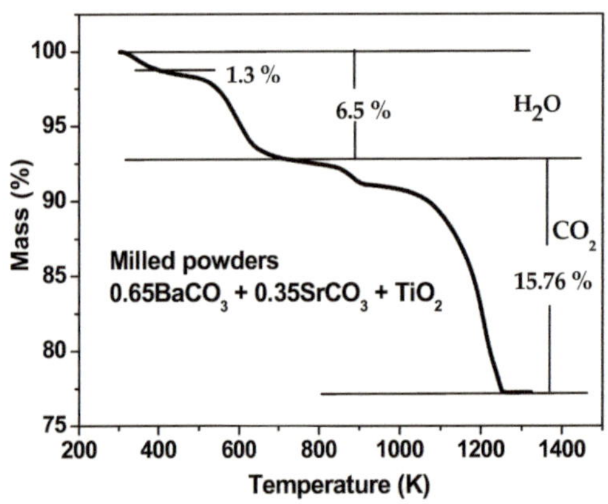

Fig. 8. Thermogravimetric analysis of 0.65$BaCO_3$, 0.35$SrCO_3$ and TiO_2 powders milled for 6 hours.

The thermogravimetric plots also show a correlation between the Sr content and the temperature at final weight loss (zero slope section). For example, the BST0 sample stops losing weight at around 1193 K but the BST10 (with highest Sr content) at around 1273 K. The temperature at which no more weight loss is observed marks the end of the reaction, as corroborated by the XRD measurements (Figure 7). The thermogravimetric curves can therefore be used as guides to establish the reaction temperature in the BSTx system.

Sample ID	% Total weight loss	% Weight loss from RT to 773 K	% Weight loss from 773 to 1323 K	% Weight loss CO_2 stoichiometric
BST0	20.43	5.56	14.87	15.87
BST1	19.38	4.12	15.26	16.16
BST2	21.02	5.48	15.54	16.47
BST3	21.59	5.61	15.98	16.78
BST4	21.52	5.94	15.58	17.10
BST5	21.39	5.80	15.59	17.44
BST6	21.76	4.85	16.91	17.79
BST7	22.48	5.19	17.29	18.15
BST8	24.29	5.76	18.53	18.53
BST9	23.46	5.42	18.04	18.93
BST10	22.87	4.52	18.35	19.34

Table 1. Thermogravimetric analysis of $(1-X)BaCO_3 + XSrCO_3 + TiO_2$ powder after high energy milling for 6 hours.

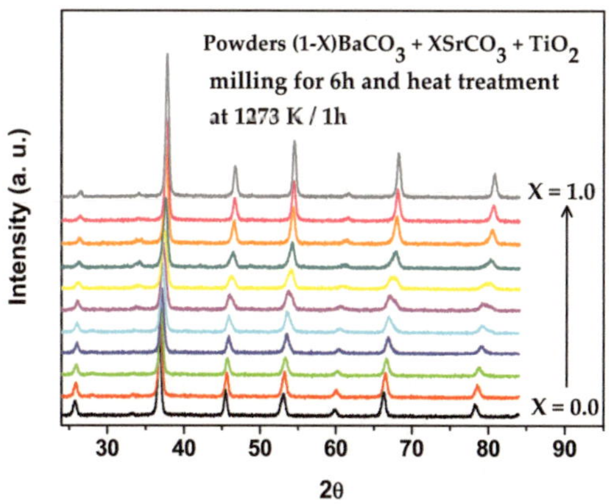

Fig. 9. X-ray diffraction patterns of $(1-X)BaCO_3 + XSrCO_3 + TiO_2$ powder milled and heat treated at 1273 K for 1 h. The crystal structure is perovskite-type ABO_3.

The diffraction patterns for the range of samples after milling and thermal treatment at 1273 K for 1 hour are shown in Figure 9. All the samples present the perovskite-type structure ABO_3, indicating the reaction was successfully completed. The formation temperature of the $BaTiO_3$ phase coincides with that reported by L. B. Kong [KONG], where the rutile phase of titanium oxide (TiO_2) was used instead of TiO_2 anatase phase. The peaks shift to higher angle as the Sr content increases. For example, the most intense $BaTiO_3$ peak in Figure 10 shifts from 36.79° for the $BaTiO_3$ (BST0) to 37.77° for $SrTiO_3$ (BST10). This shift corresponds to a reduction in unit cell size, consistent with the difference between the ionic radii of Ba^{2+} (1.34 Å) and Sr^{2+} (1.12 Å). The ABO_3 perovskite structure refers to the relative position of the Ba^{2+}, Sr^{2+} and Ti^{4+} ions with respect to oxygen (O^{2-}). The structure can present as different phases, depending on the material and temperature. For the BSTx system it may be cubic or tetragonal at room temperature, depending on the strontium content. In this case, we observed the cubic phase after 1 hour of heat treatment at 1273 K (Figure 9). The relationship between the cubic and tetragonal phase and the volume of the unit cell will be discussed in more detail when disucssing compaction.

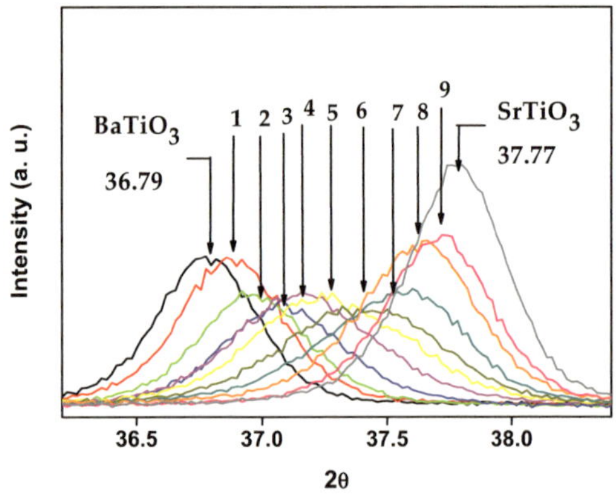

Fig. 10. Expanded view of the primary peak in Figure 9, showing the shift in peak position with increasing Sr content from X = 0 ($BaTiO_3$) to X = 1 ($SrTiO_3$).

3.2 Sintering

During conventional manufacture (solid-state reaction) the starting oxide powders are first milled and calcined. The reacted powders (having the desired stoichiometry) are milled once again to reduce the particle size. They are then compacted into disks or other shapes (green ceramic) and finally, thermally treated to sinter the compact. At the sintering temperature, the ceramic particles coalesce with each other to form grains, the material shrinks and the pores are eliminated [KINGERY]. The alternative manufacturing route of the present work is more straightforward. The starting powders are milled only once to homogeneize and reduce the particle size before being compacted into disks or other shapes. The green ceramics are thermally treated at the reaction temperature (1273 K) to obtain the perovskite-type structure ABO_3. A second heating step is subsequently applied to sinter the ceramic (Figure 1). In this way, dense ceramics (>90% of the theoretical density) are obtained. The sintering temperatures of the samples with intermediate stoichiometry varied between 1523 K for pure $BaTiO_3$ and 1573 K for pure $SrTiO_3$. The temperatures were empirically determined. Such reduced sintering temperatures can only be applied to powders with particle size smaller than 50 nm, which is uniquely attained in our alternative route, thanks to the high energy milling. Whenever the $BaTiO_3$ samples were sintered at temperatures higher than 1523 K, severe strain was observed as the melting temperature of the material (1898 K [PRADEEP]) was approached. Ceramic shrinkage, or pore reduction, is directly related to the initial particle size, as shown in Equation 2 [KINGERY]:

$$d - d_0 = (2k)^{1/2} t^{1/2} \tag{2}$$

where *d* and d_0 are the initial and final grain diameters, respectively, *k* is a constant, and *t* is the processing time. Sintering temperatures for the conventional route of ceramic preparation are reported to be between 1623 and 1703 K [ZHONG, WODECKADUS, TERANISHI]. The x-ray diffraction patterns of the thermally treated BSTx ceramics (Figure 11) clearly show the perovskite-type structure. As in the case of the powders, the peaks shift to higher angle as the Sr content increases. From Figure 12 it can be seen that the strongest peak shifts linearly (~0.1°) for each 0.1 increase in X (Sr content), and thus can be used to determine the sample composition (stoichiometry). Figure 13 plots the peak position as a function of the Sr content. All the thermally treated BSTx ceramics have the correct stoichiometry and present the perovskite-type structure ABO_3. However, the structure can be either tetragonal or cubic at room temperature depending on the processing conditions. For the $BaTiO_3$ the relative position of the (002) and (200) peaks determine the tetragonality of the phase, i.e., the ratio of the lattice parameters c/a. Such peaks appear at approximately 2θ = 53°.

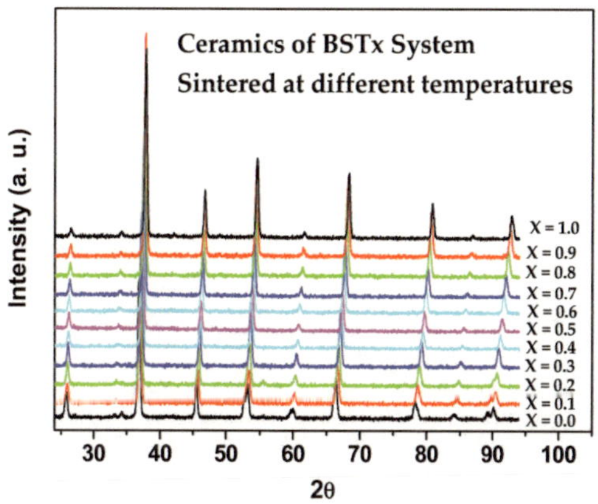

Fig. 11. X-ray diffraction patterns of BSTx ceramics system sintered from 1523 to 1573 K for 2h.

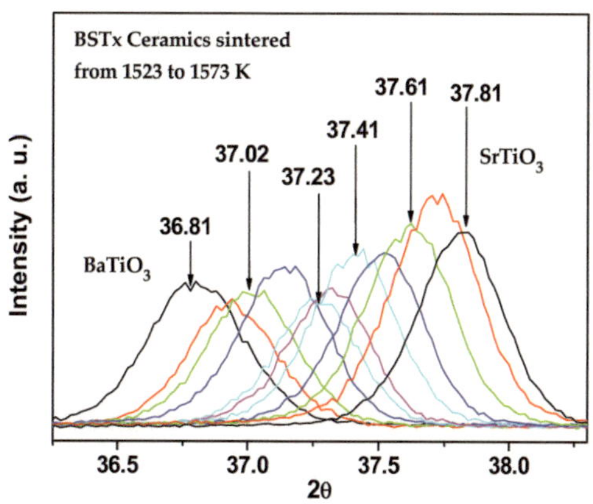

Fig. 12. Main XRD peaks of BSTx ceramic system sintered from 1523 to 1573 K of the figure 11.

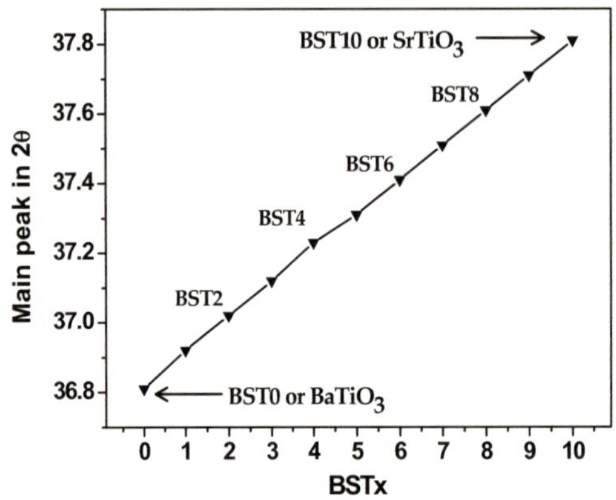

Fig. 13. Main diffraction peak position (Figure 12) as a function of stoichiometric composition.

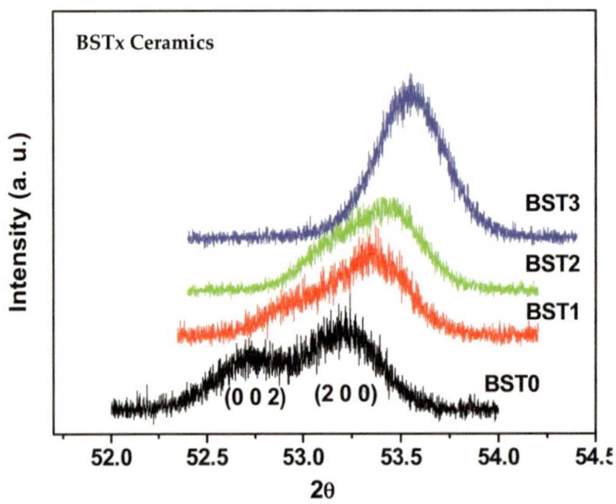

Fig. 14. Separation (or lack therof) of the (002) and (200) reflections of sintered ceramics BST0, BST1, BST2 and BST3 providing a measure of the c/a ration.

The phase is cubic and paralectric if the c/a ratio = 1. In such a case there is only a single peak near 2θ = 53°. If c/a > 1 the phase is tetragonal and ferroelectric. In such case there is a double peak. Figure 14 shows the BST0, BST1 and BST2 samples have a double peak, while sample BST3 apparently has only a single peak. Rietveld refinement (using Maud Program) was performed on the diffraction patterns of Figure 11 in order to determine the unit cell lattice parameters. The results, presented in Figure 15, show a gradual decrease in lattice parameters with increasing Sr content. This is due to the substitution of strontium Sr^{2+} ions (with ionic radius of 1.12 Å) for barium Ba^{2+} ions (with ionic radius of 1.34 Å). The tetragonality (c/a) decreases also from 1.008 (BST0) to 1.0014 (BST3) while for BST4 to BST10 it has value of 1. Differential scanning calorimetry (DSC), Raman spectroscopy, as well as the ferroelectric and dielectric measurements confirmed this phase transition, and are presented below, along with the cubic-to-tetragonal phase transition temperature (Curie temperature).

Ceramic Sample	Position in 2θ	Maximun intensity of the peak	Peak width	Lattice parameter of unit cell	Tetragonality c/a
BST0	52.694	704	0.485	a = 3.9953	1.0089
	53.227	1265	0.543	c = 4.0311	
BST1	52.942	388	0.520	a = 3.9844	1.0067
	53.368	1190	0.487	c = 4.0114	
BST2	53.124	490	0.446	a = 3.9785	1.0048
	53.459	1185	0.477	c = 3.9979	
BST3	53.457	1327	0.335	a = 3.9684	1.0014
	53.611	1764	0.360	c = 3.9737	
BST4	53.725	10	0.146	a = 3.9611	1

Table 2. Unit cell parameters and fitting parameters of peaks around 53° in 2θ.

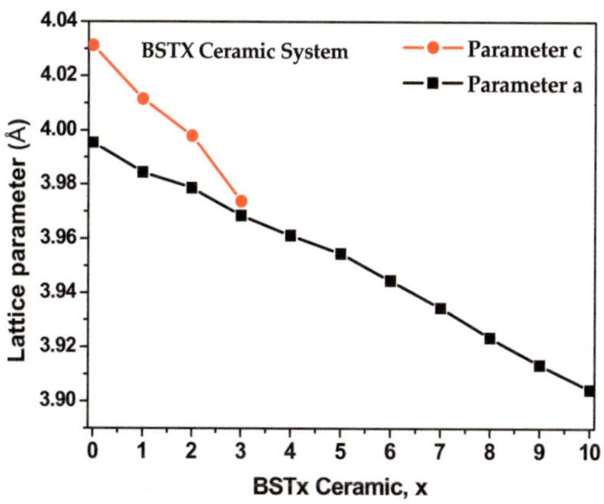

Fig. 15. Lattice parameter as a function of Sr content.

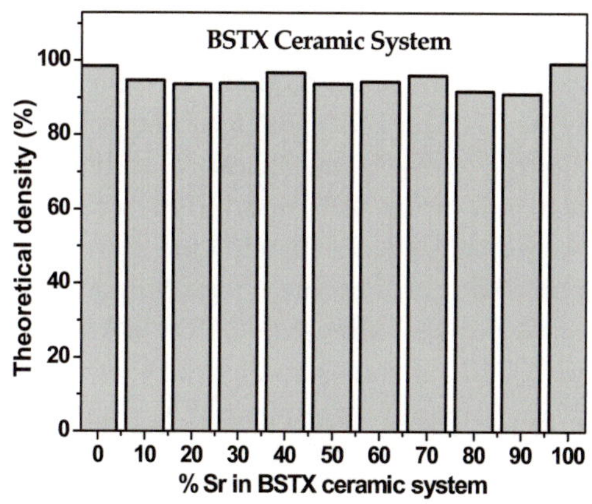

Fig. 16. Percentage of theoretical density of sintered compacts as a function of Sr content.

3.3 Density

With the actual volume of the unit cell and considering the number of barium, strontium, titanium and oxygen atoms composing the ABO_3 unit cell, the theoretical density can be calculated using:

$$\text{Theoretical Density} = \frac{\text{Unit cell mass}}{\text{Unit cell volume}} \tag{3}$$

The unit cell volume ($a^2 \times c$) was calculated using the results of the Rietveld analysis. The mass of the unit cell is calculated considering 3 oxygen, 1 titanium, (x) strontium and (1-x) barium atoms. The bulk density of the samples was determined by the Archimedes method. Figure 16 plots the bulk density as percentage of the theoretical density. All samples have bulk densities higher than 90% of theoretical, proving that the alternative fabrication route can attain a high densification of the ceramics. The density caluclation and measurment results are presented in Table 3.

BSTx Sample	Unit cell volume ($Å^3$)	Theoretical density ($g\ m^{-3}$)	Bulk density ($g\ m^{-3}$)	% Theoretical density
BST0	64.349	6.016	5.928	98.526
BST1	63.684	5.950	5.630	94.632
BST2	63.283	5.857	5.486	93.661
BST3	62.675	5.782	5.430	93.910
BST4	62.152	5.698	5.513	96.758
BST5	61.839	5.593	5.246	93.798
BST6	61.377	5.501	5.191	94.364
BST7	60.905	5.408	5.191	95.986
BST8	60.393	5.317	4.883	91.838
BST9	59.934	5.220	4.755	91.080
BST10	59.514	5.118	5.078	99.217

Table 3. Theoretical and bulk (measured) Densities of BSTx ceramic system.

SEM micrographs of transversely fractured sections from the BST0, BST4, BST8 and BST10 samples are shown in Figure 17. The BST0 ($BaTiO_3$) and BST10 ($SrTiO_3$) ceramics have a uniform compact morphology, present no cracks and have low porosity. The grain boundaries are not observable. Samples BST0 and BST10 have a bulk density of 98.52% and 99.21% of the theoretical, respectively. Even with the lower-than-conventional sintering temperatures, it is possible that the observed morphology resulted from liquid phase formation due to the small particle size [KINGERY, BARSOUM, BARRY]. Grain size distributions for BST0 and BST10 are shown in Figure 18. For the BST0 sample, grain sizes between 1 and 3.5 μm (2 μm average) were measured, while for the BST10 sample, sizes ranged ranged from 1 μm to 2.6 μm (1.4 μm average). The BST4 and BST8 samples, consisting of a barium-strontium solid solution $(Ba,Sr)TiO_3$, have a distinctly smaller grain size, with distributions of 0.2 to 1.3 μm (650 nm average) and 0.3 to 0.9 μm (550 nm average), respectively. Generally speaking, the conventional route for fabrication of mixed oxide ceramics leads to grains larger than 1 μm, in most cases on the order of several microns [LIN, LIOU, LU]. Smaller grains are only attained with nanopowders and special sintering processes such as spark plasma sintering (SPS) [DENG] or hot pressing [XIAO]. Other methods for inhibiting grain growth include the incorporation of 1 wt% of Na, Mn or Mg ions [LIOU].

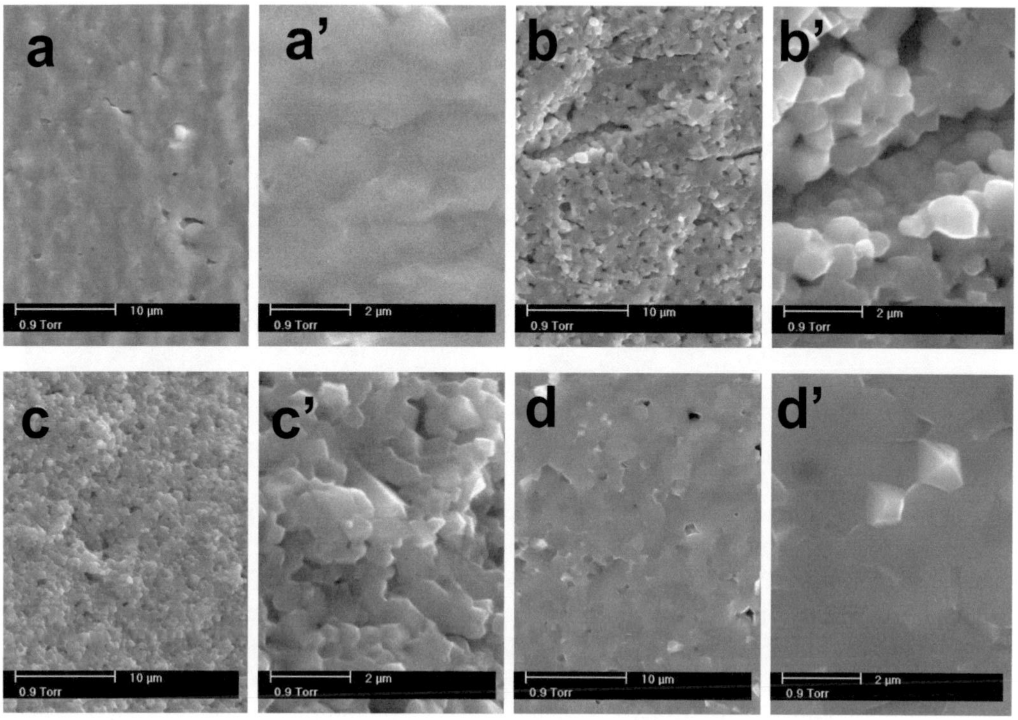

Fig. 17. SEM micrographs of transversely fractured BST0, BST4, BST8 and BST10 samples at two magnifications: (a) BST0 sintered at 1523 K, (b) BST4 sintered at 1543 K, (c) BST8 sintered at 1573 K, (d) BST10 sintered at 1573 K.

3.4 Curie temperature (Tc) via differential scanning calorimetry

DSC measurements were conducted from 203 to 423 K in a nitrogen atmosphere with a heating rate of 20°C/min. Figure 19 presents the DSC curves for samples BST0 to BST3. The cubic to tetragonal phase transition (Curie temperature, Tc) is an endothermic event. Tc decreases with increasing content of Sr in the samples, i.e., as the Sr^{2+} ions replace Ba^{2+} ions. This behavior was previously reported by Rupprecht and Bell [RUPPRECHET].

The linear dependence between the Tc and at.% Sr is described by Equation 4, the result of fitting the experimental data (Figure 20):

$$Tc = 128.4871 - 31.469*X \tag{4}$$

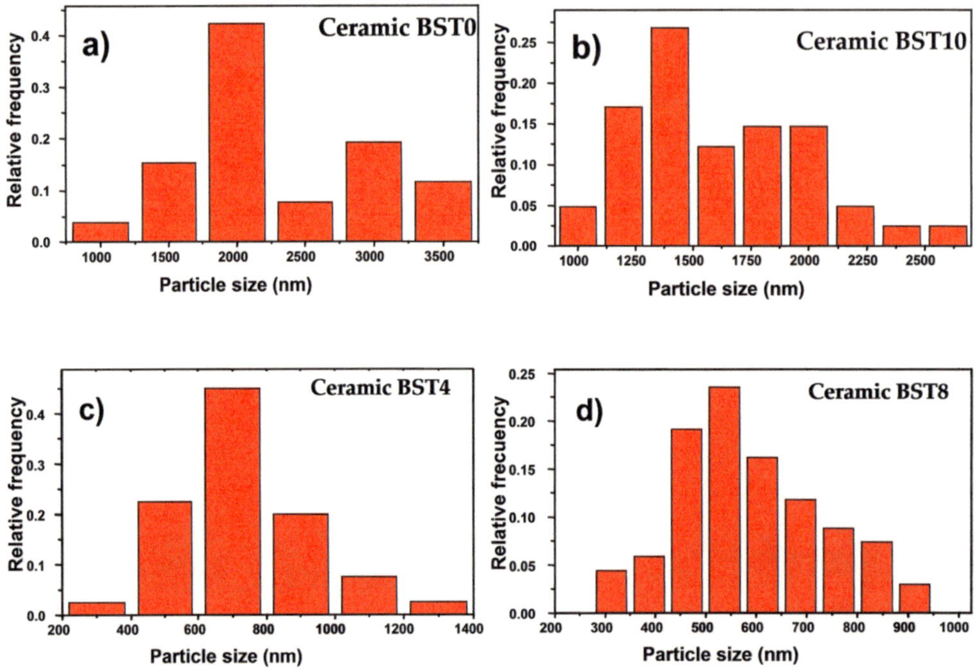

Fig. 18. Grain size distribution of (a) BST0, (b) BST10, (c) BST4 and (d) BST8 ceramic samples.

where Tc is the Curie temperature and x is the Sr content in at.%. Table 4 presents the Curie temperatures of samples BST0 to BST3 determined by Equation 4. The equation was extrapolated to the composition of samples BST4 to BST10. The determined Curie temperatures resulted not so different from those reported in the literature, for example, the BST35 system (35 mol% of Sr) has an approximate Tc of 292 K [ALI] compared to 291.35 K determined with Equation 4. The BST3 system was reported to have a Tc of 306~307 K [PITICESCU], compared to 307.08 K calculated with Equation 4. The $Ba_{60}Sr_{40}TiO_3$ system (BST4) has a Tc of 272 K (274.15 K) [FETEIRA], compared to our calculation of 276.6 K.

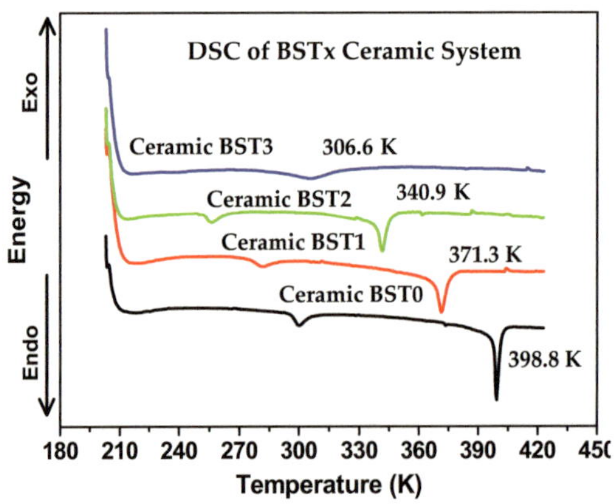

Fig. 19. Differential scanning calorimetry curves of the BST0, BST1, BST2 and BST3 samples.

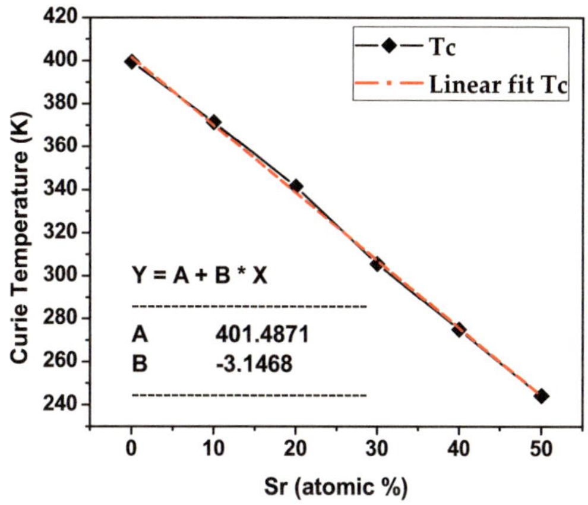

Fig. 20. Linear fit of the Curie temperature (Tc) from the DSC curves.

Sample ID	Calculated Tc (K)	Experimental Tc (K)
BST0	401.48	399.80
BST1	370.02	371.36
BST2	338.55	340.96
BST3	307.08	306.62
BST4	275.61	273.0
BST5	244.15	241.65
BST6*	212.68	
BST7*	181.21	
BST8*	149.74	
BST9*	118.27	
BST10*	86.80	

Table 4. Experimental Curie temperatures (via DSC), and those calculated using Equation 4. Tc for samples BST6 – BST10 were not experimentally determined due to exceeding the temperature range of the differential scanning calorimeter.

3.5 Curie temperature (Tc) via Raman spectroscopy

Figure 21 shows the Raman scattering spectrum (radiation wavelength = 514.5 nm) for $BaTiO_3$ (BST0) with perovskite-type structure ABO_3 and tetragonal phase at room temperature. Raman active phonons for the P4/mmm tetragonal symmetry are represented by 3A1 + B1 + 4E. Long-range electrostatic forces induce a splitting in the transverse and longitudinal phonons, resulting in a split of the Raman active phonons represented by 3 [A1 (TO) + A1 (LO)] + B1 + 4 [E (TO) + E (LO)] [SHIRATORI]. Raman shift bands are reported at 250, 520 and 720 cm^{-1} with a sharp peak at around 306 cm^{-1} [DIDOMENICO, ROUSSEAU, BASKARA].

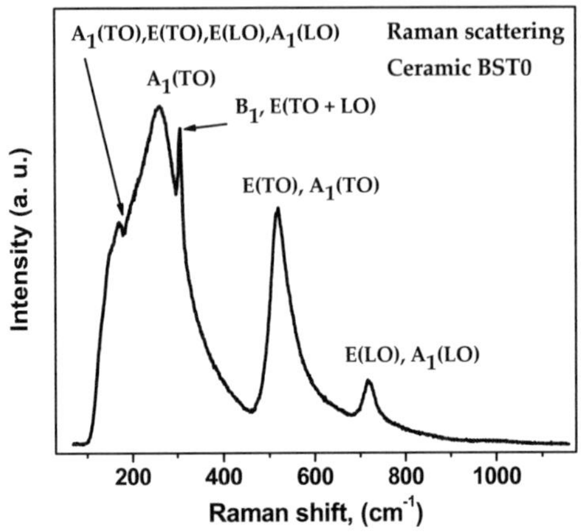

Fig. 21. Raman scattering spectra for BST0.

The shoulder at around 180 cm^{-1} in bulk $BaTiO_3$ is attributed to the coupling of the three disharmonic phonons A1 (TO) [VENKATESWARAN, FREY]. Figure 23 shows the Raman scattering spectra of the BST0 sample at different temperatures. The 250, 520 and 720 cm^{-1} bands as well as the 306 cm^{-1} peak decrease gradually as the temperature increases. At 403 K, the sharp 306 cm^{-1} peak disappears, indicating the transition from cubic to tetragonal phase (Tc). The transition temperature was previously reported by C. H. Perry [PERRY]. Thus, the sharp peak around 306 cm^{-1} indicates whether the BSTx system is in the tetragonal or cubic phase. The relative intensity of the 306 cm^{-1} peak as a function of temperature for BST0 is presented in Figure 22. Figures 24 and 25 show the temperature-dependent Raman scattering spectra for BST1 and BST2, respectively. The cubic to tetragonal phase transition is observed in the range from 363 to 373 K for BST1 and 333 to 343 K for BST2. Both ranges match those determined by DSC. Figure 26 shows the Raman scattering spectra for BST1 to BST5 at room temperature. Samples BST4 and BST5 do not present the sharp peak at 306 cm^{-1}, i.e., they have a stable cubic phase and a tetragonality of 1 (c/a = 1). These results are consistent the XRD and DSC results presented earlier.

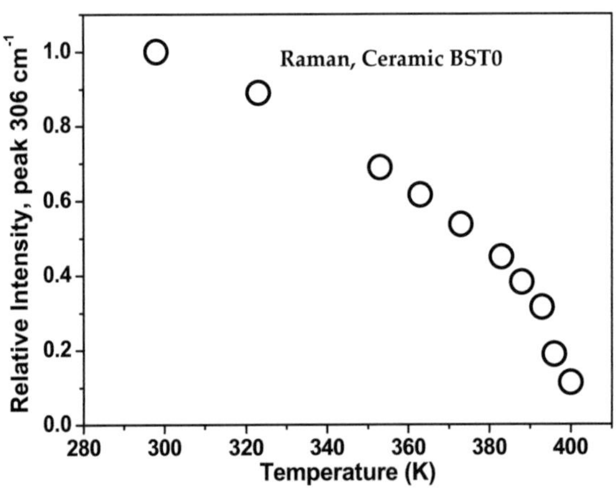

Fig. 22. Realtive peak intensity for the 306 cm^{-1} Raman reflection (from Figure 21) as a function of temperature for BST0.

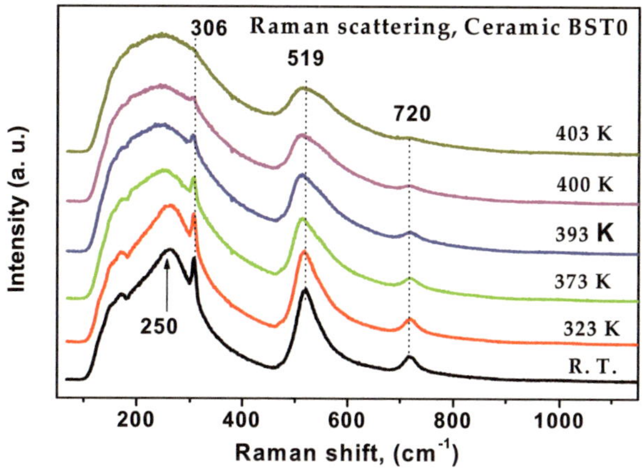

Fig. 23. Raman scattering spectra at different temperatures for BST0.

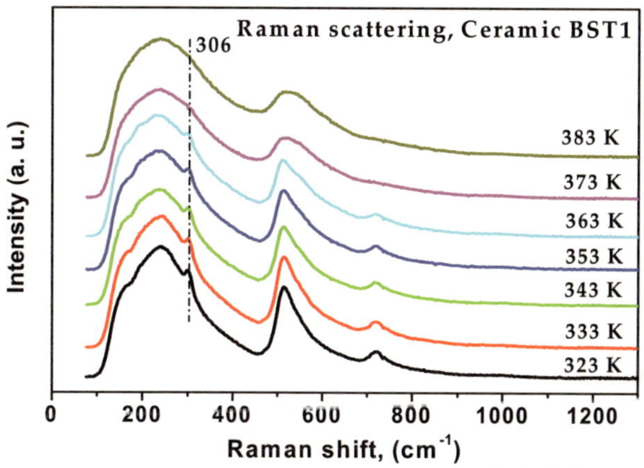

Fig. 24. Raman scattering spectra at different temperatures for BST1.

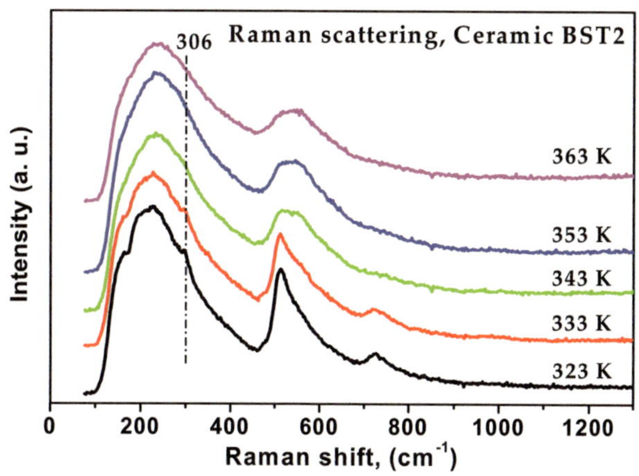

Fig. 25. Raman scattering spectra at different temperatures for BST2.

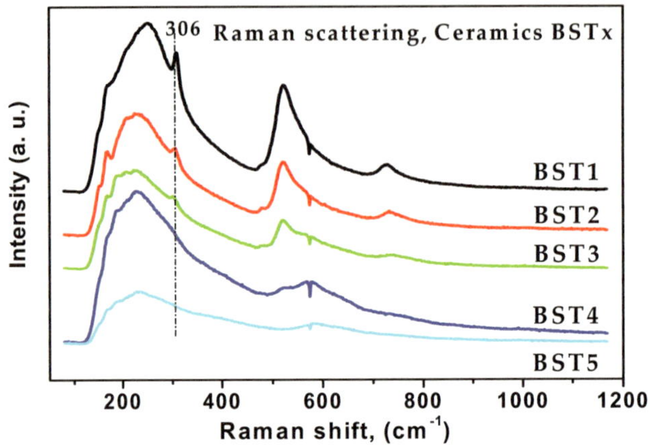

Fig. 26. Raman scattering spectra at room temperature for BST1, BST2, BST3, BST4 and BST5.

3.6 Dielectric and ferroelectric properties

3.6.1 Dielectric constant

When a polycrystalline ferroelectric ceramic is cooled below its Curie temperature, some of its properties undergo strong changes. For example, the dielectric constant (ε) shows a maximum at Tc for BST0, BST1 and BST3 at 0.1, 1.0, and 100 kHz (Figure 27), a typical ferroelectric behavior of perovskite-type materials [MILLAR, STANFORD]. A widening of the peak at Tc has been reported to occur with decreasing grain size [KINOSHITA, SAKABE]. The dielectric constant has a magnitude larger than 1000 within most of the measured temperature interval, and it decreases at higher frequencies. This effect has been observed in $BaTiO_3$ ceramics containing Zr [DEB] and in those containing Na and Bi (BTNx) [GAO]. The dielectric constant has polar and ionic parts, therefore, the dielectric dispersion can be attributed to the dipoles ceasing to contribute to the dielectric constant as the frequency increases [MERZ, 1954]. The dielectric relaxation effect occurs at frequencies where the electric dipoles can no longer follow the oscillation of the applied electric field. The relaxation frequency can be determined by a drop in the real part of ε and a maximum in the imaginary part [RAVEZ]. Although there is dielectric relaxation in the BSTx samples, the material does not present typical reflexor behavior. That is, there is no change in the position of Tc when the frequency changes. Another

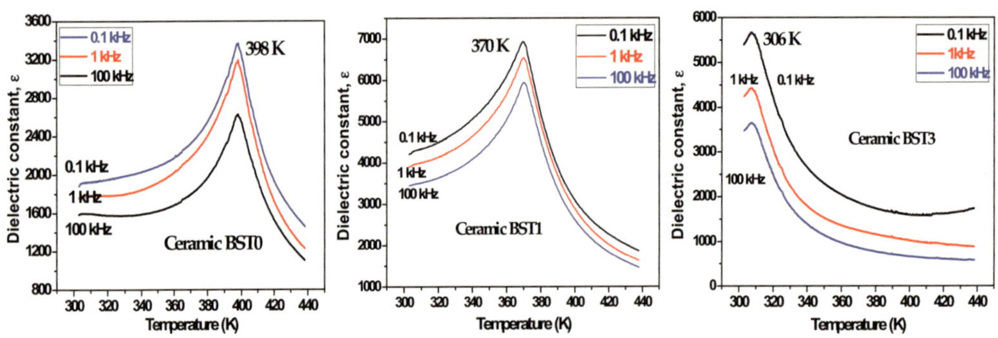

Fig. 27. Dielectric constant curves for (a) BST0, (b) BST1 and (c) BST3.

ferroelectric material with such a behavior is $Pb(Mg_{1/3}Nb_{2/3})O_3$ [KOVALA]. The determined Tc's agree well with those obtained by DSC and Raman. The dielectric constant peak values are 3.179, 6.540 and 4.432 for BST0, BST1 and BST3 ceramics respectively, 2 to 3 orders of magnitude larger than those of other materials conventionally used in capacitors or CMOS (complementary metal oxide semiconductor) devices [ROBERTO, WILK].

All of the BSTx samples showed a maximum in the dielectric constant at Tc. Above this temperature, the dielectric behavior obeys the Curie-Weiss law and has the form [BURFOOT]:

$$\varepsilon = \frac{C}{T - T_0} \tag{5}$$

where ε is the dielectric constant (material permittivity), C is the Curie-Weiss constant, T is the temperature of the material and T_0 is the Curie-Weiss temperature.

3.6.2 Ferroelectric hysteresis loops

When an external electric field $\boldsymbol{E}$ is applied to a dielectric material, it produces a $\boldsymbol{P}$ vs. $\boldsymbol{E}$ curve. In the case of ferroelectric materials there is a delay in the $\boldsymbol{P}$ response to the $\boldsymbol{E}$ stimulus, i.e., hysteresis. A freshly manufactured ferroelectric has a zero spontaneous net polarization ($\boldsymbol{P_s=0}$). When an external electric field is applied, nucleation and growth of the ferroelectric domains occur [MERZ, 1954].

The shape of the ferroelectric curve P = f(E) depends on both time and temperature. In the present work, ferroelectric measurements were performed at room temperature (~298 K) at a fixed frequency of 100 Hz using a comercial Sawyer-Tower circuit (ferroelectric RADIANT Test System) [SAWYER and TOWER]. Figure 28 presents the hysteresis loops for BST0, BST1, BST2 and BST3. The **P-E** urves exhibit the typical behavior of polycrystalline ferroelectric ceramics [MERZ, 1953]. The remanent polarization (P_r) is low compared to that of $BaTiO_3$ single crystals (>20 μC/cm^2 [SRIVASTAVA]), but higher than that of nanocrystalline $BaTiO_3$ (P_r < 1 μC/cm^2 [BUSCAGLIA]). It is comparable to that of $BaTiO_3$ ceramics with grain sizes of around 1 μm [TAKEUCHI].

3.7 Piezoresponse Force Microscopy (PFM)

3.7.1 Ferroelectric domain observation

Ferroelectric materials are composed of ferroelectric domains, which can be observed by polarized light microscopy [RUPPRECHT, ARLT], scanning electron microscopy [CHOU, ROUSSEAU] and transmission electron microscopy [FREY, GANPULE]. To be detected by these techniques, some sort of chemical attack is necessary to reveal the ferroelectric domains, as they demonstrate a preferential rate of erosion [FETEIRA, LAURELL]. Furthermore, these techniques do not directly indicate the direction of polarization (direction of the domain); they only discriminate one domain from another. Piezoresponse force microscopy (PFM) does not suffer from these shortcomings [RABE], allowing visualization of ferroelectric domains with sizes on the order of 1 μm [SAURENBACH, WITTBORN], accurate detection of the polarization direction [ENG 1998, CHO] and reconstruction of the three-dimensional orientation of the domains [ENG, 1999]. Figure 29 is a diagram of oriented domains in ferroelectric grains. Figure 29 (a shows a domain up and one down (antiparallel ↑↓), separated by domain walls (gray stripe). The adjacent domains are oriented in opposite directions (180°). If these domains are observed from above, we would see that the direction of the domains are perpendicular to the surface, either pointing down into the sample (crosses) or up out of the sample (circles)., Both types are described as out-of-plane (OOP) (Figure 30). Figure 29 (b shows domains oriented at 90° (↑→) relative to one another. If these domains are observed from above, we would see domains in the plane of or parallel to the surface. These dominains are described as in-plane (IP) (Figure 30).

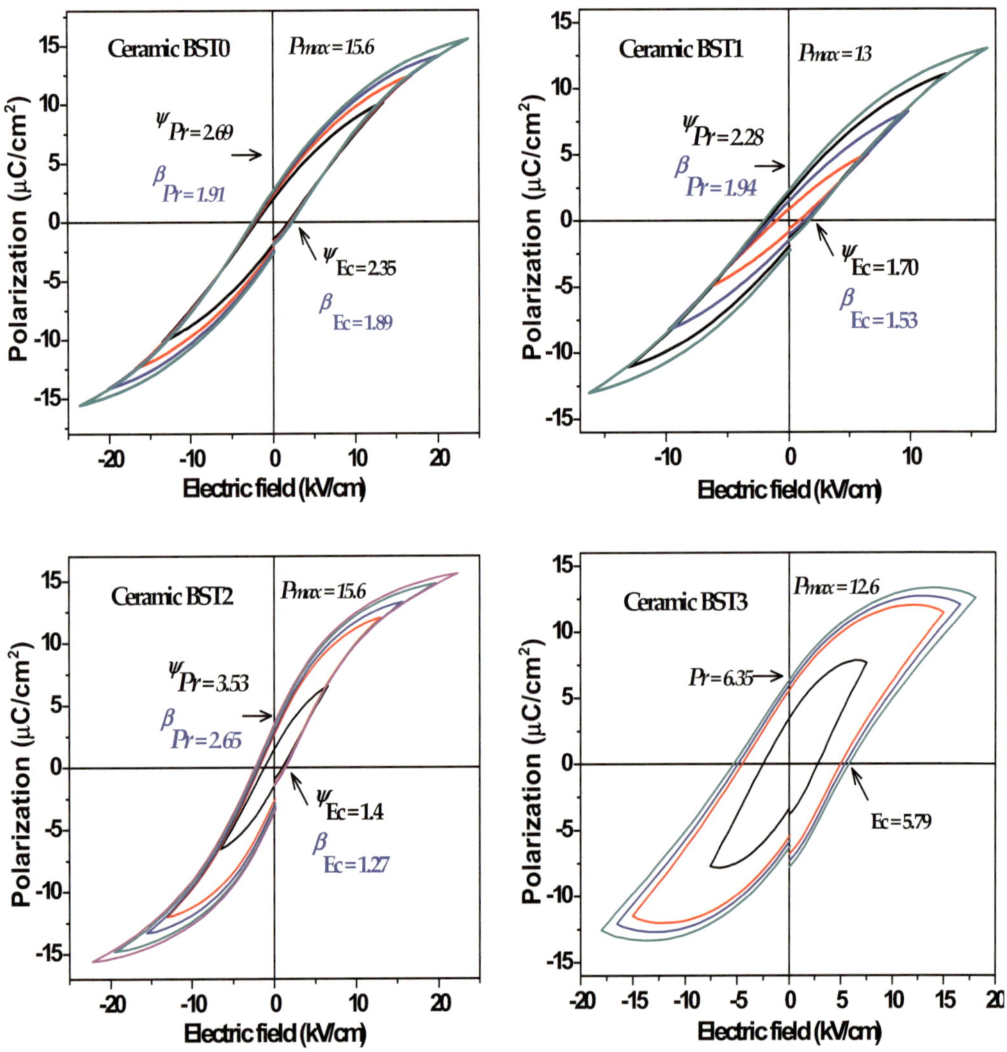

Fig. 28. P-E loops obtained for BST0, BST1, BST2 and BST3.

It isimportant to note that Figure 29 represents the ideal case of monocrystal domains ideally oriented, i.e., the direction of the observed polarization vectors are orthogonal. For the real case of a polycrystalline ceramic, the direction of the polarization vector would be random, i.e., a domain can point in any direction in 3D space. Therefore, the signals from piezoresponse measurements in OOP or IP are the projections of these random polarization vectors.

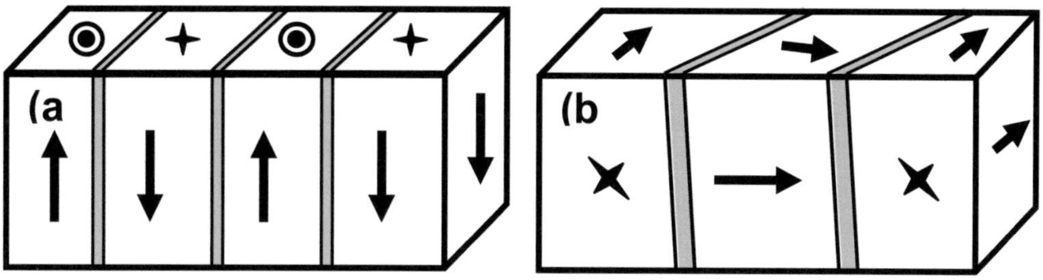

Fig. 29. Ferroelectric domains with differents orientations of the polarization vector. (a out-of-plane (OOP) and (b in-plane (IP).

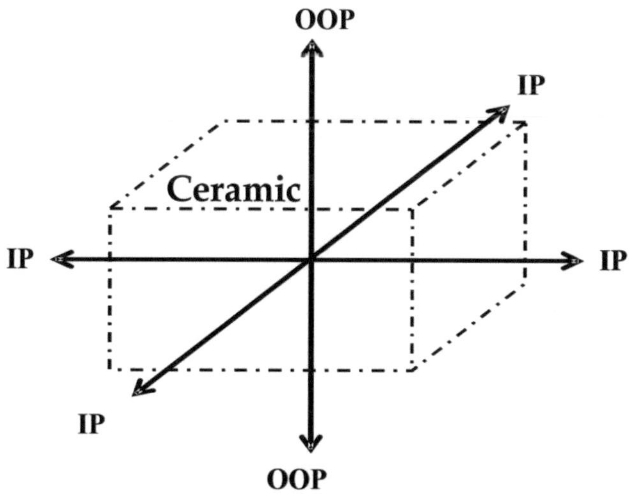

Fig. 30. Standard terminology for polarization vector orientations relative to the observed surface in piezoresponse force microscopy (PFM). IP refers to in-plane, and OOP is out-of-plane.

3.7.2 Contact Resonance-Enhanced Piezoresponse Force Microscopy (CR-PFM)

In addition to the conventional PFM technique there is another mode called contact resonance-enhanced PFM or CR-PFM [HARNAGEA]. R-PFM is based on the same principles of operation of as PFM, the only difference being the value of the frequency of the alternating voltage applied to the tip (V_{AC}). For contact resonance frequencies (CR-PFM), the strain amplitude response of the material can be more than one or two orders of magnitude higher than the amplitudes recorded by conventional PFM (quality factor Q) [HARNAGEA]. In other words, R-PFM is more sensitive than PFM and can be applied to materials whose piezoelectric constants (d_{ij}) are very small. Table 5 lists some materials and their piezoelectric constant (coefficient) .

Material	Piezoelectric coefficient, d_{33} (pm/V)
$Sr_{0.61}Ba_{0.39}Nb_2O_6$	200
$BaTiO_3$	190
PZT4	291
PZT5a	373
Quartz	3

Table 5. Piezoelectric constants of different materials.

The piezoelectric constant (d_{33}) indicates how much the material will be deformed (in picometers) for each applied volt (V) applied. or Fmaterials with a low piezoelectric constant, it would be necessary to amplify the piezoresponse signal by applying a high voltage, risking a change in piezoelectric response. In such cases, R-PFM is an alternative to observe ferroelectric domains while still using a low voltage.

3.7.3 Results of R-PFM for BSTx samples

R-PFM measurements for the BSTx samples were made using the modified conventional method on an atomic force microscope (AFM, Veeco di DimensionTM 3100), usig conductive tips of Cr/Pt (Budget Sensors Tap I300E) with a force constant k = 40 N/m, and a free-resonance frequency was 300 kHz. The samples were polished starting with No. 500 sandpaper down to 0.3 μm alumina. The samples were not attacked by any chemical or mechanochemical process, insuring that topography did not contribute to the piezoresponse signal. AFM images of the contact resonance mode piezoresponse (CR-PFM) for BST0, with a grain size of 1-2 μm, are shown Figure 31. The three images, taken of the same area and at the same time, show (a) the topography, (b) the OOP piezoresponse amplitude, and (c) the OOP piezoresponse phase. Measurements were performed over a 5 x 5 μm area, using an applied voltage was 2V and a contact resonance frequency of 1350 kHz. The amplitude and phase piezoresponse images present different characteristics. In the Figure 31b, the piezoresponse amplitude image allows us to visualize domains that leave or enter of the surface (OOP), where the amplitude *A* can be the same for antiparallel dominions (↑↓). Different regions are distinguished by their gray tones, delimited by black contours, which correspond to the ferroelectric domain walls. The domain walls are not observed in the topography, as that would have required etching (chemical attack) [FETEIRA]. Regions that appear as similar shades of gray in the piezoresponse amplitude image (Figure 31b) appear as contrasting bright and dark in the piezoresponse phase image (Figure 31c). This indicates a phase shift of 180° from one domain to another, but with the same piezoresponse amplitude.

These results show that the R-PFM images of amplitude and phase are not influenced by the topography of the samples. In contrast, standard measurements of piezoresponse force microscopy (PFM) were performed at 20 kHz with an alternating voltage (V_{ac}) of 2-15 V without obtaining a piezoelectric response. A piezoresponse was obtained with contact resonance frequencies (~1350 kHz), with a quality factor Q betwen 50 and 100. Taking into account both the quality factor Q and the piezoresponse signal reported about 190 pm/V and to 419 pm/V [SHAO] for the case of $BaTiO_3$, we can estimate a small piezoelectric constant of our material is of the order of units or tens of pm/V.

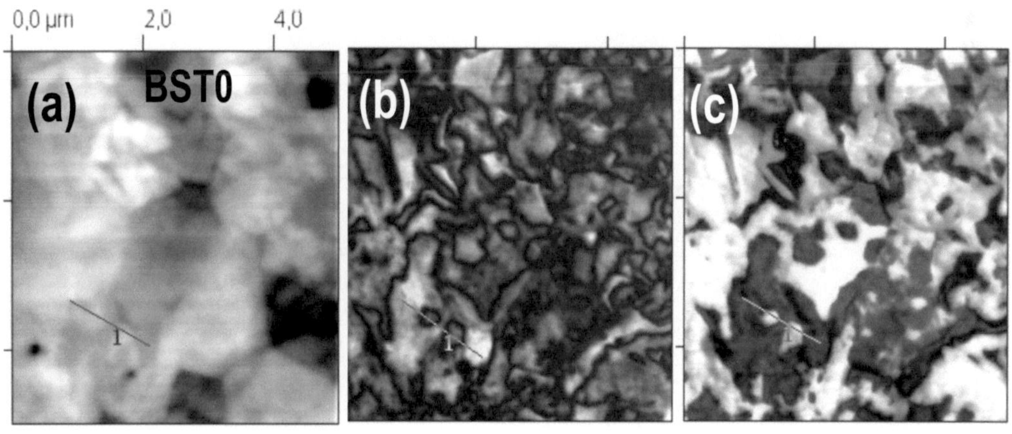

Fig. 31. R-PFM of BST0 at a frequency of 1.350 MHz and an excitation voltage of 2 V. (a) topography, (b) piezorepsonse amplitude, and (c) piezoresponse phase.

4. Conclusion

The mixed oxides modified route (solid-state reaction) is a direct alternative to obtain highly densified BSTx ceramics. In this route, the high-energy ball milling and the applied heat treatment allowed the preparation of nanometric powders (less than 200nm) with Perovskite-type structure ABO_3. The Curie temperature of the BSTx ceramics was unambiguosly determined as a function of temperature by several techniques: Raman spectroscopy, differential scaning calorimetry, and measurements of dielectric constant. This temperature was succesfully tuned from 87 K to 400 K by varying the Sr/Ba ratio, as expected. However, a shifting in the orthorhombic to tetragonal phase transition was observed in the sintered ceramics. For instance, the OTPT for the BST0-BST2 samples was shifted down 17-27 K with respect to the literature. The origin of this shifting is probably residual stresses associated to the fine-grained microstructure of the sintered samples.

The group of BST0, BST1, BST2 and BST3 ceramics present P(E) curves with ferroelectric behavior at room temperature. The other cases present paraelectric behavior. Moreover, the BST0, BST1and BST2 samples present rather low maximum and remanent polarization and coercive fields. BSTx ceramics (x = 0, 1, 2, 3) displayed piezoelectric response in the contact resonance piezoresponse force microscopy mode (CR-PFM). The polycrystalline BSTx ceramics showed ferroelectric domains with sizes several times smaller than the grains.

5. Acknowledgment

This work was partially supported by CONACYT, Mexico. The authors would like to thank Michael Boldrick Ph. D. and Rodrigo Mayen Mondragón Ph. D. for their help regarding the english translation. Besides, we thank Pedro García J., J. Eleazar Urbina A., M. Adelaido Hernández L., Francisco Rodríguez M., Agustín Galindo S., Rivelino Flores F., Alfredo Muñoz S., Ma. del Carmen Delgado C. and Eduardo Larios for their technical aid.

6. References

Ali, N. J. & S. J. Milne, "Comparison of powder synthesis routes for fabricating $(Ba_{0.65}Sr_{0.35})TiO_3$ ceramics", *J. Mater. Res.*, 21 (2006) 1390-1398.

Arlt, G. & P. Saxo, "Domain configuration and equilibrium size of domain in $BaTiO_3$ ceramics", *J. Appl. Phys.*, 5 (1980) 4956-4960.

Asiaie, R.; Weidong Zhu, Sheikh A. Akbar, & Prabir K. Dutta, "Characterization of submicron particles of tetragonal $BaTiO_3$", *Chem. Mater.*, 8 (1996) 226-234.

Baláz, P. & B. Plesingerova, "Thermal properties of mechanochemically pretreated precursors of $BaTiO_3$ synthesis", *J. Therm. Anal. Cal.*, 59 (2000) 1017-1021.

Barry C. (2007). "Ceramic Material Science and Engineering", Springer, USA.

Barsoum M. W. (2003). "FUNDAMENTALS OF CERAMICS", McGraw, USA.

Baskara, N.; Anil Hule, Chutan Bhongale, Ramaswamy Murugan & Hua Chang, "Phase transition studies of ceramic $BaTiO_3$ using thermo-Raman and dielectric constant measurement", *J. Appl. Phys.*, 91 (2002) 10038-10043.

Berbecaru, C.; H. V. Alexandru, C. Porosnicu, A. Velea, A. Ioachim, L. Nedelcu & M. Toacsan, "Ceramic materials $Ba_{(1-x)}Sr_xTiO_3$ for electronics – Synthesis and characterization", Thin *Solid Films*, 2008, 516, (22), 8210–8214.

Blomqvist, M.; Sergey Khartsev & Alex Grishin "Electrooptic ferroelectric $Na_{0.5}K_{0.5}NbO_3$ films", *IEEE Photonics Technology Letters*, 17 (2005) 1638-1640.

Boland, S. W.; Suresh C. Pillai, Weing-Duo Yang & Sossina M. Haile, "Preparation of $(Pb,Ba)TiO_3$ powders and highly oriented thin films by a sol-gel process", *J. Mater. Res.*, 19 (2004) 1492-1498.

Burfoot. Jack C. "*Ferroelectrics*", D. Van Nostrand Company LTD, London.

Buscaglia, M. T.; Massimo Viviani, Vincenzo Buscaglia, Liliana Mitoseriu, Andrea Testino, Paolo Nanni, Zhe Zhao, Mats Nygren, Catalin Harnagea, Daniele Piazza, & Carmen Galassi, "High dielectric constant and frozen macroscopic polarization in dense nanocrystalline $BaTiO_3$ ceramics", *Phys. Rev. B*, 73 (2006) 064114.

Busch, G. & P. Scherrer, *Naturwiss*, 23 (1935) 737.

Chaisan W.; S. Ananta & T. Tunkasiri, "Synthesis of barium titanate-lead zirconate solid solution by a modified mixed-oxide synthetic route", *Curr. Appl. Phys.*, 4 (2004) 182-185.

Chen, Chin-Yen & Hur-Lon Lin, "Piezoelectric properties of $Pb(Mn_{1/3}Nb_{2/3})O_3$-$PbZrO_3$-$PbTiO_3$ ceramics with sintering aid of 2CaO-FeO_2 compound", *Ceram. Int.*, 30 (2004) 2075-2079.

Cheng, H.; Jiming Ma & Zhenguo Zhao, "Hydrothermal synthesis of PbO-TiO_2 solid solution", *Chem. Mater.*, 6 (1994) 1033-1040.

Cho, Y.; Satoshi Kazuta & Kaori Matsuura, "Scanning nonlinear dielectric with nanometer resolution", *Appl. Phys. Lett.*, 75 (1999) 2833-2834.

Chou, Jung-Fang, Ming-Hong Lin & Hong-Yang Lu, "Ferroelectric domains in pressureless-sintered barium titanate", *Acta Mater.*, 48(2000) 3569.

Deb, K. K.; M. D. Hill & J. F. Nelly, "Pyroelectric characteristics of modified barium titanate ceramics", *J. Mater. Res.*, 7 (1992) 3296-3304.

Deng, X.; Xiaohui Wang, Hai Wen, Liangliang Chen, Lei Chen, & Longtu Li, "Ferroelectric properties of nanocrystalline barium titanate ceramics", *Appl. Phys. Lett.*, 88 (2006) 252905.

DiDomenico, M.; Jr., S. P. S. Porto, S.H. Wemple & R. P. Barman, "Raman spectrum of single-domain $BaTiO_3$", *Phys. Rev.*, 174 (1968) 522-530.

Ding, Z.; R. L. Frost & J. T. Kloprogge, "Thermal Activation of Cooper Carbonate", *J. Mater. Sci. Lett.*, 21 (2002) 981-983.

Eng, L. M., H.; –J. Güntherodt, G. A. Schneider, U. Köpe & J. Muñoz Saldaña, "Nanoscale reconstruction of surface crystallography from three-dimensional polarization distribution in ferroelectric barium-titanate ceramics", *Appl. Phys. Lett.*, 74 (1999) 233-235.

Eng, L. M.; H. –J. Güntherodt, G. Rosenman, A. Skilar, M. Oron, M Katz and D. Eger, "Nondestructive imaging & characterization of ferroelectric domains in periodically poled cristal", *J. Appl. Phys.*, 83 (1998) 5973-5977.

Feteira, A.; Derek C. Sinclair, Ian M. Reaney, Yoshitaka Somiya, & Michael T. Lanagan, "$BaTiO_3$-Based Ceramics for Tunable Microwave Applications", *J. Am. Ceram. Soc.*, 87 (2004) 1082-1087.

Frey, M. H. & D. A. Payne, "Grain-size effect on structure and phase transformations for barium titanate ", *Phys. Rev. B*, 54 (1996) 3158-3167.

Frey, M. H. & D. A. Payne, "Grain-size effect on structure and phase transformations for barium titanate", *Phys. Rev. B*, 54 (1996) 3158-3168.

Ganpule, C. S., V. Nagarajan, B. K. Hill, A. L. Roytburd, E. D. Williams & R. Armes, "Imaging three-dimensional polarization in epitaxial polydomain ferroelectric thin films", J. *Appl. Phys.*, 91 (2002) 1477-1481.

Gao, L.; Yanqiu Huang, Yan Hu & Hongyan Du, "Dielectric and ferroelectric properties of (1-x)$BaTiO_3$-x$Bi_{0.5}Na_{0.5}TiO_3$ ceramics", *Ceram. Int.*, 33 (2007) 1041-1046.

Gopalan, V. & Terence E. Mitchell, "Wall velocities, switching times, and the stabilization mechanism of 180° domains in congruent $LiTaO_3$ crystals", *J. Appl. Phys.*, 83 (1998) 941-954.

Gururaja T. R; Walter A. Schulze, Leslie E. Cross, Robert E. Newnham, Bertram A. Auld & Yuzhong J. Wang, "Piezoelectric Composite Materials for Ultrasonic Transducer Applications", *IEEE*, 32 (1985) 481-498.

Hammer, J. M. "Digital electro-optic grating deflector and modulator", *Appl. Phys. Lett.*, 18 (1971) 147-149.

Harnagea, C.; Alain Pignolet, Marin Alexe & Dietrich Hesse, "Higher-Order Electrochemical Response of Thin Films by Contact Resonant Piezoresponse Force Microscopy", *IEEE*, 53 (2006) 2309-2321.

Hidaka T.; T. Maruyama, M Saitoh & N. Mikoshiba, "Formation and observation of 50 nm polarized domains in $PbZr_{1-x}Ti_xO_3$ thin film using scanning probe microscope", *Appl. Phys. Lett.*, 68 (1996) 2358-2359.

Ianculescu, A.; A. Brăileanu & Georgeta Voicu, "Synthesis, microstructure and dielectric properties of antimony-doped strontium titanate ceramics", *J. Eur. Ceram. Soc.*, 27 (2007) 1123-1127.

Judd, M. D. & M. I. Pope, "Energy of activation for the decomposition of the alkaline-earth carbonates from thermogravimetric data", *J. Therm. Anal.*, 4 (1972) 31-38.

Kamalasanan, M. N.; N. Deepal Kumar & Subhas Chandra, "Structural and microstructural evolution of barium titanate thin films deposited by the sol-gel process", *J. Appl. Phys.*, 76 (1994) 4603-4609.

Kingery. (1960). "Introduction to Ceramics", Second Ed., John Wiley & Sons, USA.

Kingon, A.; "Is the ultimate memory in sight?", *Nat. Mater.*, 5 (2006) 251-252.

Kingon, Angus I,; Jon-Paul Maria & S. K. Streiffer, "Alternative dielectrics to silicon dioxide for memory and logic devices", *Nature*, 406 (2000) 1032-1038.

Kinoshita Kyoichi & Akihiko Yamaji, "Grain-size Effects On Dielectric Properties in Barium Titanate Ceramics", *J. Appl. Phys.*, 47 (1976) 371-373.

Kohlstedt, H.; Y. Mustafa, A. Gerber, A. Petraru, M. Fitsilis, R. Meyer, U. Böttger & R Waser, "Current status and chellenges of ferroelectric memory devices", *Microelectronic Engineering*, 80 (2005) 296-304.

Kong, L. B.; J. Ma, H. Huang, R.F. Zhang & W.X. Que, "Barium titanate derived from mechanochemically actived powders" *J. Alloys Compd.*, 337 (2002) 226-230.

Kotecki, D. E.; "(Ba,Sr)TiO_3 dielectrics for future stacked capacitors DRAM", *IBM J. Res. Develop.*, 43 (1999) 367-380.

Kovala, V.; Carlos Alemany, Jaroslav Briancin, Helena Brunckovà & Karol Saksl, "Effect of PMN modification on structure and electrical response of xPMN-(1-x)PZT ceramic systems", *J. Eur. Ceram. Soc.*, 23 (2003) 1157-1166.

Kugel, V. D.; G Rosenman & D. Shur, "Electron emission from $LiNbO_3$ crystals with domains of inverted polarization", *J. Phys. D: Appl. Phys.*, 28 (1995) 2360-2364.

L'vov, Boris V. & Valery L. Ugolkov, "Peculiarities of $CaCO_3$, $SrCO_3$ and $BaCO_3$ decomposition in CO_2 as a proof of their primary dissociative evaporation", *Thermochim. Acta*, 410 (2004) 47-55.

Laurell, F.; M. G. Roelofs, W. Bindloss, H. Hsiung, A Suna and J. D. Bierlein, "Detection of ferroelectric domain reversal in $KTiOPO_4$", *J. Appl. Phys.*, 71 (1992) 4664-4670.

Lin, Ming-Hong; Jung-Fang Chou & Hong-Yang Lu, "Grain-Growth Inhibition in Na2O-Doped TiO_2-Excess Barium Titanate Ceramic", *J. Am. Ceram. Soc.*, 83 (2000) 2155-2162.

Liou, Jih_Wei & d Bi-Shiou Chiou, "Effect of Direct-Current Biasing on the Dielectric Properties of Barium Strontium Titanate", *J. Am. Ceram. Soc.*, 80 (1997) 3093-3099.

Liou, Jing-Kai; Ming-Hong Lin & Hong-Yang Lu, "Crystallographic Facetting in Sintered Barium Titanate", *J. Am. Ceram. Soc.*, 85 (2002) 2931-2937.

Lu, Chung-Hsin; Wei-Hsing Tuan & Buh-Kuan Fang, "Effects of Pre-sintering Heat Treatment on the Microstructure of Barium Titanate", *J. Mater. Sci. Lett.*, 15 (1996) 43-45.

Masui, S.; Shunsuke Fueki, Koichi Masuntani, Amane Inoue, Toshiyuki Teramoto, Tetsuo Suzuki & Shoichiro Kawashima, "The Application of FeRAM to Future Information Technology World", *Topics Appl. Phys.*, 93 (2004) 271-284.

Merz Walter J. "Double Hysteresis Loop of $BaTiO_3$ at the Curie Point", *Phys. Rev.*, 91 (1953) 513-514.

Merz, Walter J. "Domain formation and Domain Wall Motions in Ferroelectric $BaTiO_3$ Single Crystal", *Phys. Rev.*, 95 (1954) 690-698.

Meschke, F.; A. Kolleck & G. A. Schneider, "R-curve behaviour of $BaTiO_3$ due to stress-induced ferroelastic domain switching", *J. Eur. Ceram. Soc.*, 17 (1997) 1143-1149.

Millar C. A. "Hysteresis Loss and Dielectric Constant in Barium Titanate", *Brit. J. Appl. Phys.*, 18 (1967) 1689-1697.

Padmaja, G.; Ashok K. Batra, James R. Curie, Mohan D. Aggarwal, Mohammad A. Alim & Ravindra B. Lal "Pyroelectric ceramics for infrared detection applications", *Mater. Lett.*, 60 (2006) 1937-1942.

Park, Yong-I, "Effect of composition on ferroelectric properties of sol-gel derived lead bismuth titanate ($PbBi_4Ti_4O_{15}$) thi films", *J. Mater. Sci.*, 36 (2001) 1261-1269.

Perry, C. H. & D. B. Hall, "Temperature Dependence of the Raman Spectrum of the $BaTiO_3$", *Phys. Rev. Lett.*, 15 (1965) 700-702

Piticescu, R. M.; P. Vilarnho, L. M. Popescu, R. R. Piticescu, "Hydrothermal synthesis of perovskite based materials for microelectronic applications", *J. Optoelectron. Adv. Mater.*, 8 (2006) 543-547.

Pradeep, P. P.; Subhash H. & Risbud, "Low-temperature synthesis and processing of electronic materials in the $BaO-TiO_2$ system", *J. Mater. Sci.*, 25 (1990) 1169-1183.

Rabe, U.; M. Kopycinska, S. Hirsekorn, J. Muñoz Saldaña, G. A. Schneider & W. Arnold, "High-resolution characterization of piezoelectric ceramics by ultrasonic scanning force microscopy techniques", *J. Phys. D: Appl. Phys.*, 35 (2002) 2621-2635.

Radheshyam, R. & Seema Sharma, "Structural and dielectric propierties of Sb-doped PLZT ceramics", *Ceram. Int.*, 30 (2004) 1295-1299.

Ravez, J. "Ferroelectricity in Solid State Chemistry", *Chem.*, 3 (2000) 267-283.

Razak, K. A.; A. Asadov & W. Gao, "Properties of BST prepared by high temperature hydrothermal process", *Ceram. Int.*, 33 (2007) 1495-1502.

Relva, B. C. (2004). "Ceramic Materials for Electronics", Marcel Dekker, USA.

Roberto, J. "High Dielectric Constant Oxides", *Eur. Phys. J. Appl. Phys.*, 28 (2004) 265–291.

Rosenman, G.; A. Skliar & I. Lareah, "Observation of ferroelectric domain structures by secondary-electron microscopy in as-grown $KTiOPO_4$ crystals", *Phys. Rev. B*, 54 (1996) 6222-6226.

Rousseau, D. L. & S. P. S. Porto, "Auger-like Resonant Interference in Raman Scattering From One and Two-Phonon States of $BaTiO_3$", *Phys. Rev. Lett.*, 20 (1968) 1354-1357.

Rupprecht, G. & R. O. Bell, "Microwave Losses in Strontium Titanate above the Phase Transition", *Phys. Rev.*, 125 (1962) 1915-1920.

S. Maitra, N. Chakrabarty & J. Pramanik, "Decomposition kinetics of alkaline heart carbonates by integral approximation method", *Ceramica*, 54 (2008) 268-272.

Sakabe, Y.; N. Wada & Y. Hamaji, "Grain Size Effects on Dielectric Properties and Crystal Structure of Fine-grained $BaTiO_3$ Ceramics", *J. Korean Phys. Soc.*, 32 (1998) 260-264.

Sato, H. & Kohji Toda, "An Application of $Pb(Zr,Ti)O_3$ Ceramic to Opto-Electronic Devices", *Appl. Phys.*, 13 (1977) 25-28.

Saurenbach, F. & B. D. Terris, "Imaging of ferroelectric domain walls by force microscopy", *Appl. Phys. Lett.*, 56 (1990) 1703-1705.

Sawyer, B. & C. H. Tower, "Salt rochelle as a dielectric", *Phys. Rev.*, 35 (1930) 269-273.

Shao, S.; Jialiang Zhang, Zong Zhang, Peng Zheng, Minglei Zhao, Jichao Li1 & Chunlei Wang, "High piezoelectric properties and domain configuration in $BaTiO_3$ ceramics obtained through the solid-state reaction route", *J. Phys. D: Appl. Phys.*, 41 (2008) 125408 (5pp).

Shepard, R. "Dielectric and Piezoelectric Properties of Barium Titanate", *Phys. Rev.*, 71 (1947) 890-895.

Shiratori, Y.; C. Pithan, J. Dornseiffer & R. Waser, "Raman scattering studies on nanocrystalline $BaTiO_3$ Part II - consolidated polycrystalline ceramics", *J. Raman Spectrosc.*, (2007).

Srivastava, N. & G.J. Weng, "The influence of a compressive stress on the nonlinear response of ferroelectric crystals", *Int. J. Plas.*, 23 (2007) 1860-1873.

Stanford, A. L. "Dielectric resonance in Ferroelectric Titanates in the Microwave Region", *Phys. Rev.*, 124 (1961) 408-410.

Takashi Teranishi, Takuya Oshina, Hiroaki Takeda & Takaaki Tsurimi, "Polarization behavior in diffuse phase transition of $Ba_xSr_{1-x}TiO_3$ ceramics", *J. Appl. Phys.*, 105 (2009) 054111.

Takeuchi, T.; Claudio Capiglia, NaliniBalakrishnan, Yasuo Takeda & Hiroyuki Kageyama, "Preparation of fine-grained $BaTiO_3$ ceramics by spark plasma sintering", *J. Mater. Res.*, 17 (2002) 575-581.

Uchino, Kenji. (2000). "Ferroelectric Devices", Marcel Dekker, USA.

Valasek, J. "Piezo-Electric And Allied Phenomena In Rochelle Salt", *Phys. Rev.*, 17 (1921) 475-481.

Venkateswaran, UD.; Naik VM & Naik R, "High-pressure Raman studies of polycrystalline $BaTiO_3$ ", *Phys. Rev. B*, 58 (1998) 14256.

Vittayakorn, N.; Theerachai Bongkarn & Gobwute Rujijanagul, "Phase transition, mechanical, dielectric and piezoelectric properties of Perovskite $(Pb_{1-x}Ba_x)ZrO_3$ ceramics", *Physica B*, 387 (2007) 415-420.

Vold, R. E.; R. Biederman, G. A. Rossettu JR., & A. Sacco JR., "Hydrothermal synthesis of lead doped barium titanate", *J. Mater. Sci.*, 36 (2001) 2019-2026.

Wainer, E. & A. N. Salomon, Titanium Alloy Manufacturing Company Elect. Rep. 8 (1942), 9 and 10 (1943).

Whatmore, R. W.; Qi Zhang, Christopher P. Shaw, Robert A. Dorey & Jeffery R. Alcock, "Pyroelectric ceramics and thin films for applications in uncooled infra-red sensor arrays", *Phys. Scr. T.*, 129 (2007) 6-11.

Wilk, G. D.; R. M. Wallace & J. M. Anthony, "High-k Gate Dielectrics: Current Status and Materials Properties Considerations", *Appl. Phys. Rev.*, 89 (2001) 5243-5275.

Wittborn. J.; C. Canalias, K. V. Rao, R. Clemens, H. Karlsson & F. Laurell, "Nanoscale imaging of domain and domain walls in periodically poled ferroelectrics using atomic force microscopy", *Appl. Phys. Lett.*, 80 (2002) 1622-1624.

Wodecka-Dus, B.; A. Lisinska-Cekaj. T. Orkisz, M. Adamczyk, K. Osinska, L. Kozielski & D. Czekaj, "The sol-gel synthesis of barium strontium titanate ceramics", *Mater. Sci. Poland*, 25 (2007) 791-799.

Xiao, C. J.; W. W. Zhang, Z. H. Chi, F. Y. Li, S. M. Feng, C. Q. Jin, X. H. Wang, L. T. Li, & R. Z. Chen, "Ferroelectric $BaTiO_3$ nanoceramics prepared by a three-step high-pressure sintering method", *Phys. Stat. Sol.*, 204 (2007) 874-880.

Xu, H. & Lian Gao, "Hydrothermal sinthesis of high-purity $BaTiO_3$ powders: control of powder and size, sintering density, and dielectric properties", *Mater. Lett.*, 58 (2004) 1582-1586.

Yamashita, Kouchi Harada, Yasuharu Hosono, Shinya Natsume & Noboru Ichinose, "Effects of B-site Ions on the Electrochemical Coupling Factors of $(Pb(B'B'')O_3$-$PbTiO_3$ Piezoelectric Materials", *Jpn. J. Appl. Phys.*, 37 (1998) 5288-5291.

Yamashita, Y.; Yasuharu Hosono & Noboru Ichinose, "Phase Stability, Dielectrcis and Peizoelectric Properties of the $Pb(Sc_{1/2}Nb_{1/2})O_3$-$Pb(Zn_{1/3}Nb_{2/3})O_3$-$PbTiO_3$", *Jpn. J. Appl. Phys.*, 36 (1997) 1141-1145.

Yoo, J. H.; W. Gao & K. H. Yoon, "Pyroelectric and dielectric bolometer properties of Sr modified $BaTiO_3$ ceramics", *J. Mater. Sci.*, 34 (1999) 5361-5369.

Yun, S. N.; X. L. Wang & D. L. Xu, "Influence of processing parameters on the structure and properties of barium strontium titanate ceramics", *Mater. Res. Bull.*, 2008, 43, (8–9), 1989–1995.

Zhi, Y.; Ruyan Guo, A.S. Bhalla, "Dielectric polarization and strain behavior of $Ba(Ti_{0.92}Zr_{0.08})O_3$ single crystal", *Mater. Lett.*, 57 (2002) 349-354.

Zhong, Z. & Patrick K. Gallagher, "Combustion synthesis and characterization of $BaTiO_3$", *J. Mater. Res.*, 10 (1995) 945-952.

Zhu, M.; Lei Hou, Yudong Hou, Jingbing Liu, Hao Wang & Hui Yan, "Lead-free $(K_{0.5}Bi_{0.5})TiO_3$ powders and ceramics prepared by sol-gel method", *Mater. Chem. Phys.*, 99 (2006) 329-332.

Part 6

Superconducting Ceramics

21

Sintering Process and Its Mechanism of MgB_2 Superconductors

Zongqing Ma and Yongchang Liu
Tianjin Key Lab of Composite and Functional Materials,
School of Materials Science & Engineering,
Tianjin University, Tianjin,
P R China

1. Introduction

1.1 The phase formation mechanism of MgB_2 during sintering

The superconductivity at 39 K discovered in MgB_2 among simple binary chemical composition attracted much interest in its fabrication techniques and practical applications [1]. MgB_2 superconductor exhibits many impressive properties such as highest critical temperature amongst intermetallic superconductors which means low cooling costs, impressive grain boundary transparency to the flow of current which leads to greater critical current density [2-4], comparatively large coherence length which allows a better Josephson junction fabrication, low material cost which will lead to a cheaper superconductor technology, simple crystal structure, etc. Hence, MgB_2 superconductors, especially the MgB_2 wires and coils, have the outstanding potential to be integrated into diverse commercial applications, such as, magnetic resonance imaging (MRI) [5, 6] , fault current limiters (FCL), Josephson junctions and SQUID [7, 8, 9], transformers, motors, generators, adiabatic demagnetization refrigerators, magnetic separators, magnetic levitation applications, energy storage, and high energy physical applications. But the MgB_2 itself is mechanically hard and brittle and therefore not amenable to drawing into the desired wire and tape geometry. Thus, the powder-in-tube (PIT) technique that was used to make the Y-Ba-Cu-O oxide superconductor has been employed in the fabrication of MgB_2 wires and tapes these years [10-14]. So far, in-situ sintering, including the in-situ PIT, from the mixture of magnesium and boron is the major method to fabricate MgB_2 superconductors (bulks, wires and tapes). The corresponding sintering parameters have a significant influence on the superconducting properties of MgB_2. Thus it is necessary to investigate the sintering mechanism of MgB_2 superconductors.

The reaction process and MgB_2 phase formation mechanism during the sintering have been studied by different methods, such as differential thermal analysis (DTA) [15-21], *in-situ* XRD measurement [22-25], *in-situ* resistance measurement [26, 27] and temperature dependent magnetization (M–T) measurements [28].

1.2 Sintering of Mg-B precursor powders over a wide temperature range

It can be seen from the DTA data of the Mg + $2B_{amorphous}$ precursor composition shown in Fig. 1 that the first exothermic peak occurs in the temperature range below 650 ^{o}C (the melting point of Mg) for samples heated at either 20 °C /min or 5 °C /min [29]. Previous studies have suggested different origins of this peak; some speculate that it is due to the reaction between Mg and impurity B_2O_3 in the original B powder [30, 31], whereas others suggest that it is associated with the solid-solid reaction between Mg and B [16, 18]. In general, there is consensus about the origin of the second and third DTA peaks, which are due to melting of Mg and the liquid-solid reaction between and B, respectively.

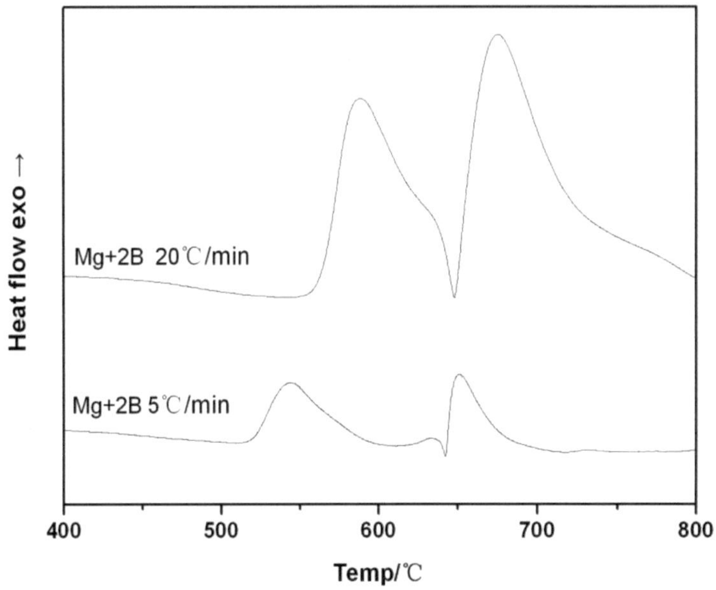

Fig. 1. The measured thermal analysis curves during the sintering of Mg + 2B sample with heating rates of 20 °C·min^{-1} and 5 °C·min^{-1}[29].

With aim of clarifying the origin of first exothermic peak in the DTA curves, the phase evolution of Mg + $2B_{amorphous}$ system was detected by in-situ X-ray diffractometer (XRD) during the sintering up to 750 °C and the measured results are shown in Fig. 2. It is found that obvious MgB_2 phase peaks can be recognized only after 550 °C. In fact, all the measured results of in-situ resistance, *in-situ* XRD and the temperature dependent magnetization during sintering of a mixed powder of Mg : B = 1 : 2 indicate that the MgB_2 phase begins to form before the Mg melting [22, 24, 25, 27, 28]. In this case, the exothermic peak in the DTA curves before the Mg melting occurs, and should be attributed to the solid-solid reaction between Mg and B. The phase formation of MgB_2 during the sintering process, therefore, proceeds via solid-solid reaction, Mg melting and, finally, liquid-solid reaction.

Observing the in-situ XRD patterns in Fig. 2 carefully, a lot of Mg is still present as primary phase in the Mg+2B sample sintered at 600 °C. The result implies that the MgB_2 phase formed at the solid-solid reaction stage is limited due to the low atomic diffusion rate in the solid state. On the other hand, also as shown in Fig. 2, MgB_2 phase forms on a much larger scale and becomes primary phase immediately following completion of the melting of Mg when the sintering temperature reaches 650 °C. As a result, in order to obtain complete MgB_2 phase rapidly, the MgB_2 superconductors were generally synthesized by sintering at high temperature in the past decade.

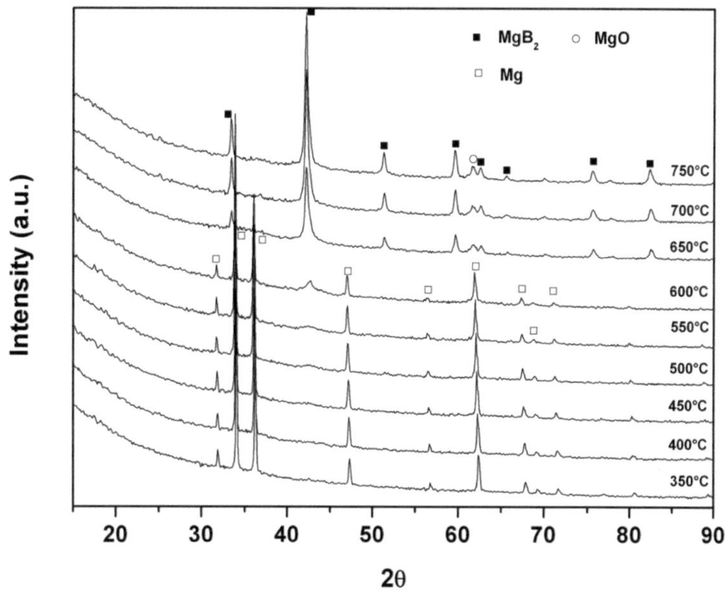

Fig. 2. The in-situ XRD patterns of (a) Cu-doped MgB_2 sample and (b) undoped MgB_2 sample.

1.3 Sintering process and mechanism of MgB_2 superconductors at high temperature

Since most of MgB_2 superconductors are prepared by sintering at high temperature, it makes sense to investigate their phase formation process and sintering mechanism at high sintering temperature. At high temperature, the liquid-solid reaction between Mg and B is activated following completion of the melting of Mg. The MgB_2 phase formation mechanism at this stage should be very different to that at the solid-solid reaction stage due to the presence of the Mg melt.

A large number of small MgB_2 grains exist in the bulk material after the solid-solid reaction, together with residual Mg and B particles. When the sintering temperature is above 650 °C, residual Mg melted and the flowing liquid phase (Mg) increased the diffusion rate of atom and enlarged the contact area of reactants, which leads to a strong and complete reaction between residual Mg and B.

According to our previous study [17], this solid-liquid reaction stage follows Ostwald ripening mechanism and includes three important processes [32]:

i. *rearrangement of particles*. The molten Mg helps individual particles to slip, spin and reassemble;
ii. *solid-liquid reaction*. The residual B particles are entrapped by the molten Mg, which promotes a strong instantaneous contact reaction;
iii. *solution-reprecipitation and grain growth*. Small MgB_2 grains generally have a higher solubility in the liquid phase than larger grains [33] and will dissolve first in the Mg melt to yield an over-saturated solution. This will lead to the precipitation of MgB_2 on the surface of existing MgB_2 grains, which will lead to further grain growth, as shown in Fig. 3.

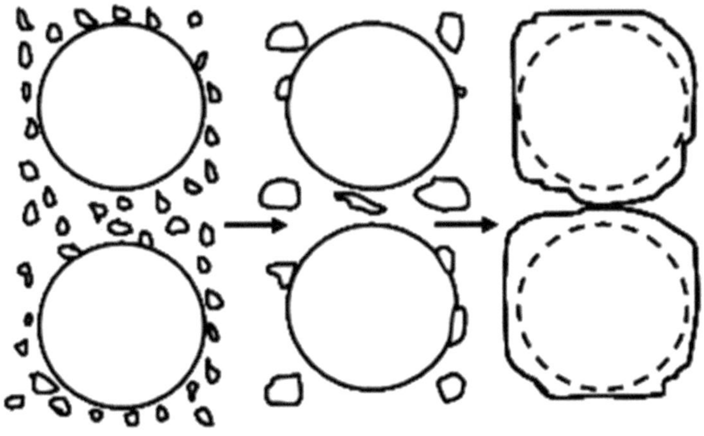

Fig. 3. Schematic illustration of the solution-reprecipitiation and growth process of grains during the liquid-solid stage [17].

According to above discussion, when the sintering temperature rise to 750 ^{o}C, the ending point of solid-liquid reaction in the DTA curve (see Fig. 1), complete MgB_2 phase can be obtained (see Fig. 2) and most of MgB_2 grains tend to be regular hexagon in the sample (see Fig. 4, the SEM image of sample sintered at 750 ^{o}C).

It is very difficult to calculate the kinetic parameters exactly from the DTA analysis data due to the overlap between the Mg melting and liquid-solid reaction thermal peaks. Hence, only limited kinetic information calculated from the *in-situ* X-ray diffraction measurement has been reported for the liquid-solid reaction between Mg and B. DeFouw *et al.* Ref. [23] reports that the liquid-solid reaction between Mg and B under isothermal conditions can be described by diffusion-controlled models of a reacting sphere with kinetics characterized by diffusion coefficients that increase with temperature from 2×10^{17} to 3×10^{16} s^{-1}, with associated activation energies of 123 ~ 143 $kJ\cdot mol^{-1}$. However, a very high heating rate was used in these studies to prevent the reaction between Mg and B occurring below a certain temperature (above the melting point of Mg). As a result, the sintering environment might be quite different from that in traditional sintering methods.

Previous studies on the phase formation mechanism of MgB_2 during liquid-solid sintering between Mg and B are deficient, and further investigation is necessary in addition to advanced test methods.

1.4 Conventional solid-state sintering of MgB_2 superconductors at low temperature

Though high-temperature sintering is the most popular method of synthesizing MgB_2 superconductors till now, the high volatility and tendency of Mg to oxidize at high temperature pose significant challenges to the fabrication of MgB_2 superconductors that exhibit excellent superconductive properties since these processes tend to generate voids and MgO impurities during in-situ sintering. Thus, recent studies have addressed the low-temperature preparation of MgB_2 superconductors in an attempt to reduce the oxidation and volatility of Mg.

As discussed above, the formation of the MgB_2 phase begins at a temperature below the melting point of Mg, which offers the prospect of sintering MgB_2 superconductors at relatively low temperature (i.e. below the melting point of Mg) in an attempt to avoid problems associated with the strong volatility and oxidation of Mg at high temperature. Rogado *et al.* [34] initially fabricated superconducting bulk MgB_2 samples by conventional solid state sintering at a temperature as low as 550 °C (see Fig. 5). This process required a sintering time of 16 hours to form the complete bulk MgB_2 phase, and the samples exhibited inferior superconducting properties than those sintered at high temperature. However, the result indicates that it is possible to fabricate MgB_2 superconductors at low temperature and this work resulted in increased attempts world-wide to develop a low-temperature sintering process for both undoped and doped MgB_2 bulk superconductors.

Fig. 4. SEM image of MgB_2 sample sintered at 750 °C.

Yamamoto *et al.* [35] found that MgB_2 bulk superconductors prepared by solid state sintering at 600°C for 240 h exhibited improved critical current densities at 20 K (see Fig. 11). This study confirmed the potential of the low temperature sintering technique for the fabrication of bulk MgB_2 superconductors. It also established that poor crystallinity is found to enhance H_{c2}, H_{irr} and J_c in MgB_2 at high fields, whereas strong grain connectivity, reduced MgO impurity content and a smaller grain size increases J_c at low fields.

MgB_2 wires and tapes with improved H_{irr} and J_c can also be prepared by low temperature sintering by an *in-situ* PIT technique. Goldacker *et al.* [36] reports the synthesis of thin, steel-reinforced MgB_2 wires with very high transport current densities at only 640 °C. These authors suggest that the low-temperature annealing could lead to a fine grain structure and a superconducting percolation path with very high associated critical current density. Moreover, the observation of a dramatically-reduced reaction layer between the filament and sheath in their samples is very promising for the production of filaments with small diameters in mono- and multifilamentary wire.

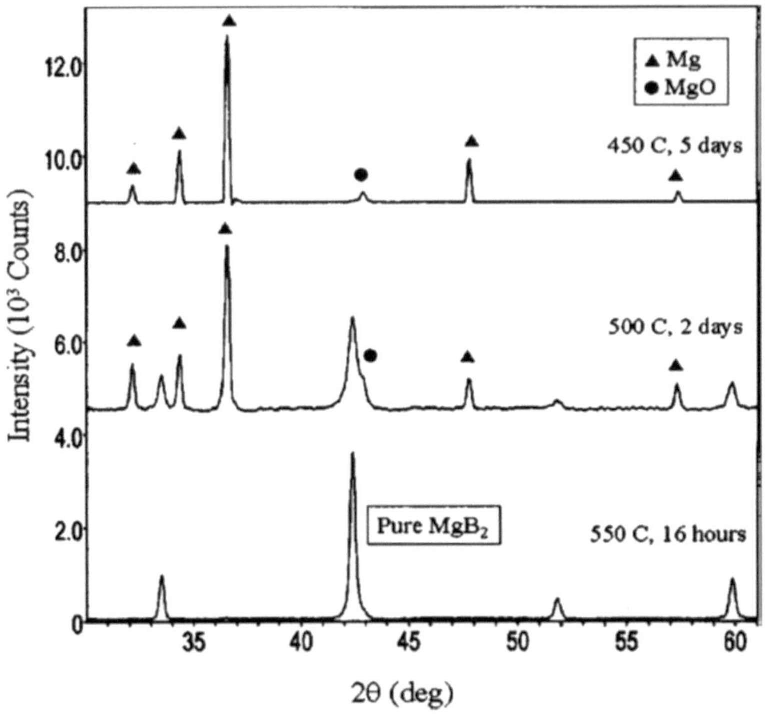

Fig. 5. Powder x-ray diffraction patterns of MgB_2 samples prepared using different heating conditions. Markers are placed above the peaks corresponding to Mg and MgO impurities. The unmarked peaks correspond to MgB_2 (from Ref. [34]).

Recently, a new process, called two-step heat-treatment, has been developed to fabricate undoped MgB_2 bulk superconductors [37]. In this process, short high-temperature sintering at 1100 °C is followed by low-temperature annealing. Samples prepared by this method exhibit, uniquely, well-connected small grains with a high level of disorder in the MgB_2 phase, which yields an in-field J_c of nearly one order of magnitude higher than for the samples prepared by single-step sintering at high or low temperature. However, the applicability of the two-step heat-treatment to the fabrication of MgB_2 wires has yet to be investigated.

To summarize, these MgB_2 superconductors synthesized at low temperature were generally cleaner and denser than the same samples sintered at high temperature due to the reduced volatility and oxidation of Mg, which could improve the connectivity between MgB_2 grains. Moreover, sintering at low temperature can also obtain the refined MgB_2 grains, which obviously strengthens the grain-boundary pinning. Both of factors are bound to result in the improvement of critical current density in the low-temperature sintered samples compared to the typical high-temperature sintered samples. From this point of view, the low-temperature synthesis might be the most promise and effective method in obtaining the higher J_c in MgB_2 superconductors. Hence, it makes sense to clarify the phase formation mechanism of MgB_2 during the low-temperature sintering.

1.5 Sintering kinetics of MgB_2 superconductors at low temperature

Analysis of the kinetics of the sintering process can be performed based on the DTA data using different computational methods. Yan *et al.* [18] calculated the activation energy of the solid-solid reaction at low sintering temperature using the Ozawa–Flynn–Wall and Kissinger as 58.2 and 72.8 $kJ \cdot mol^{-1}$, respectively. The value of the pre-exponential factor calculated using the Kissinger method is 2.0×10^{15} s^{-1}. They also report that the activation energy increases parabolic as the reaction progresses [18]. However, the study by Shi *et al.* [19] draws different conclusions to those of Yan et al. These authors use a new kinetic analysis (based on a variant on the Flynn-Wall-Ozawa method) under non-isothermal conditions and suggest that the solid-solid reaction between Mg and B powders follows an instantaneous nuclei growth (Avrami–Erofeev equation, n = 2) mechanism. The values of E decrease from 175.4 to 160.4 kJ mol^{-1} with the increase of the conversion degrees (α) from 0.1 to 0.8 in this model. However, the activation energy (E) increases to 222.7 kJ mol^{-1} [19] again as the conversion degree reaches 0.9.

On the other hand, the solid-solid reaction between Mg and B exothermal peak is partly overlapped with the Mg melting endothermic peak in the DTA curves. This phenomenon makes it difficult to calculate the kinetics parameters exactly from the thermal analysis data and also could be the reason why the previous results are different from different research groups. It is necessary to further investigate the phase formation mechanism of MgB_2 during the low-temperature sintering combined with advanced test methods.

In our recent work [38], in-situ X-ray diffraction technique is used to measure the degree of reaction between Mg and B as a function of time at several certain temperatures below Mg melting, respectively. Based on these isothermal data, the kinetics analysis of MgB_2 phase formation during the low-temperature sintering is carried out.

Bulk samples of MgB_2 were prepared by a solid-state sintering method using amorphous boron powder (99% purity, 25μm in size), magnesium powder (99.5% purity, 100 μm in size). Several reaction temperatures in the range of 550~600 °C, below the melting point of Mg, were chosen as the isothermal holding temperatures. Then the samples were fast-heated to the chosen temperature with a rate of 50 °C/min in order to prevent significant reaction between Mg and B before arriving at the isothermal annealing temperature. The x-ray diffraction measurement started as soon as the sample temperature reached the certain isothermal temperature and it will detect the sample every 15 min till the reaction is over. The weight fraction of synthesized MgB_2 which corresponds to the degree of reaction was calculated from the XRD data of sample obtained after different soaking time according to the External Standard Method.

Fig. 6 illustrates the typical X-ray diffraction patterns of the Mg-B sample isothermally annealed at 575 °C for different periods. One can see that no organized MgB_2 peak can be observed when the temperature just reaches 575 °C. It implies that the reaction between Mg and B did not occur during the rapid heating to the final isothermal holding temperature. As the holding time prolonging, the MgB_2 phase appears and increases gradually while the Mg phase decreases. However, the increase in the intensity of MgB_2 peaks becomes very slow and even stops when it reaches a certain degree despite of longer holding time (ie. longer than 480 min). Similar behavior is also found in the isothermal annealing experiments at 550 °C and 600 °C (the XRD patterns in not shown here).

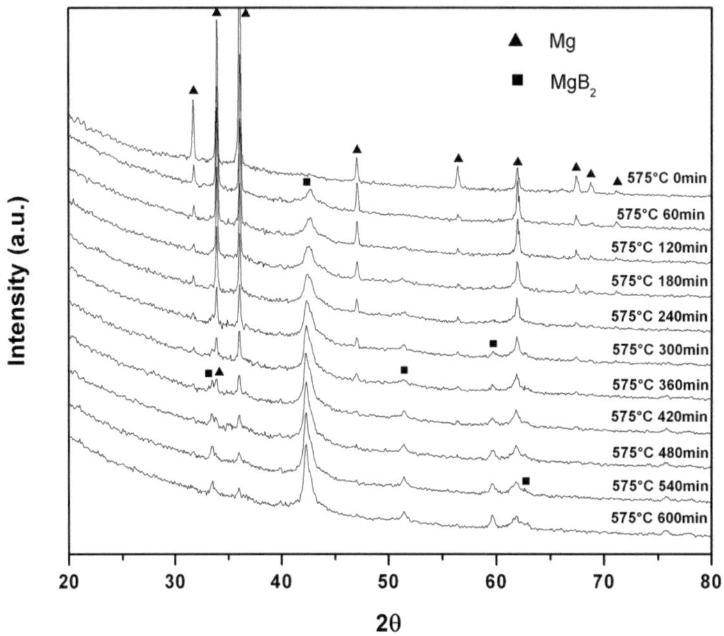

Fig. 6. Typical X-ray diffraction patterns of the Mg-B powder specimen isothermally annealed at 575 °C for different periods [38].

According to the XRD data, the weight fraction of the synthesized MgB_2 phase, which is considered as the degree of reaction, is calculated using the External Standard Method. Plots of degree of reaction vs. time are given in Fig. 7 for isothermal sintering at 550 °C, 575 °C

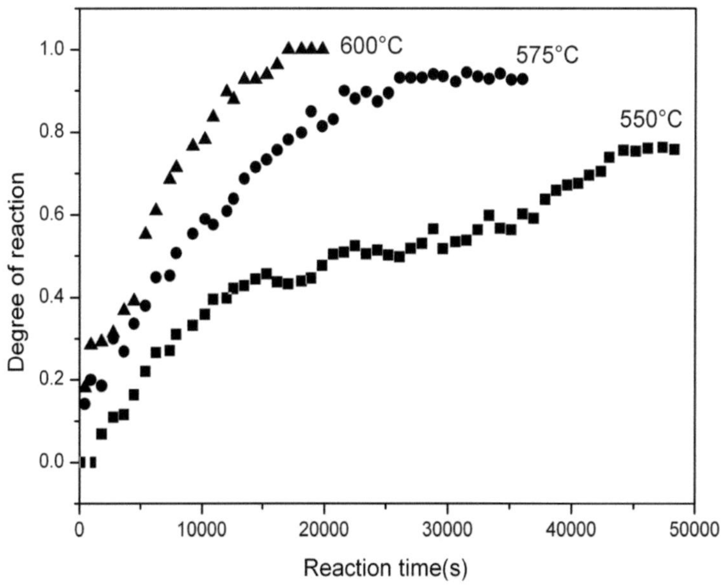

Fig. 7. Plots of degree of reaction vs. time isothermally-annealed Mg-B powder specimens at 550 °C, 575 °C and 600 °C, respectively [38].

and 600 °C, respectively. At 550 °C and 575 °C, the reaction seems stop even though the degree of reaction is still below 100% and there is residual Mg and B in the sample (see Fig. 6). It means that at the final stage of isothermal heating in present work, the reaction rate is so slow that it is difficult to observe the increase in the degree of reaction. At each isothermal annealing temperature, it can be also found that the reaction rate becomes slower and slower with the reaction time increasing and the degree of reaction seems unchanged at last.

Based on these isothermal data, kinetics analysis of the MgB_2 phase formation during the low-temperature sintering is carried out. Early kinetics studies employed the currently-accepted kinetic equation:

$$\frac{d\alpha}{dt} = k(T)f(\alpha) \tag{1}$$

Where *t* represents time, *a* is the degree of reaction, *T* is the temperature, k*(T)* is the temperature-dependent rate constant and *f(a)* is a function that represents the reaction model. *k(T)* can be described as:

$$k(T) = A\exp(-\frac{E}{RT}) \tag{2}$$

Where *A* is the pre-exponential factor, *E* is the activation energy and *R* is the gas constant.

Integrating Eq. (1), it comes:

$$g(\alpha) \equiv \int_0^{\alpha} [f(\alpha)]^{-1} d\alpha = k(T)t \tag{3}$$

Where *g(a)* is the integral form of *f(a)*.

After substitution for *k(T)* and rearranging, it yields:

$$\ln(t) = \frac{E}{RT} + \ln\left[\frac{g(\alpha)}{A}\right] \tag{4}$$

It needs different reaction time to reach certain degree of reaction at different isothermal holding temperatures. According to Eq. (4), at certain degree of reaction, one can plot the *lnt~1/T* and then do a linear fit. Following this way, the activation energy (*E*) can be obtained without referring to the reaction modes. Table 1 illustrates the reaction time and activation energy at different degree of reaction of the Mg-B sample. It is recognized that the activation energy (*E*) firstly decreases when *a* changes from 0.20 to 0.40 and then increases again when *a* changes from 0.50 to 0.80. It is indicated that the reaction between Mg and B at low temperature is not controlled by only one kinetics reaction model. At different stage of reaction, the kinetics model is varied.

In order to determine the kinetic modes of the reaction between Mg and B during the low-temperature sintering, Model fitting method is performed. Following this method, The determination of the *g(a)* term is achieved by fitting various reaction models to experimental data. As described below in Eq. (5), the relationship between *g(a)* and *t* should be linear.

$$g(\alpha) = A\exp(-\frac{E}{RT})t \tag{5}$$

Set of alternate reaction models is linearly-fitted to the experimental data obtained from the is-situ X-ray diffraction results at 575 °C and then obtained results are collected in Table 2. The coefficient r of the contracting cylinder, contracting sphere and one-dimensional diffusion models are the better ones. According the Eq. (5), the intercept t during the linear fitting should be zero. But the intercept of contracting cylinder model is 0.0821, which is too high compared to 0. And the corresponding value of r is also not as good as the case in the Contracting sphere model. Hence, the contracting cylinder model is ignored here. On the other hand, the coefficient r of the contracting sphere is better than that of the one-dimensional diffusion model. But the intercept of one-dimensional diffusion model is more near to 0 than that of the contracting sphere model.

Degree of reaction (a)	Reaction time (t) at 550 °C (s)	Reaction time (t) at 575 °C (s)	Reaction time (t) at 600 °C (s)	Activation energy (E) (kJ/mol)
0.20	5200	1500	520	275.39
0.30	6700	3600	2250	130.58
0.40	11700	5700	4500	114.75
0.50	21000	8100	5100	169.71
0.60	36000	11700	6150	181.71
0.70	42300	14100	7800	202.63
0.80	46800	18900	10260	211.74

Table 1. The reaction time (t) and activation energy (E) at different degree of reaction (α) during the low-temperature sintering of the Mg-B powders [38].

Reaction model	$G(a)$	Slope (b)	r-square	Intercept (a)
One-dimensional diffusion	α^2	3.49×10^{-5}	0.9851	-0.0165
Two-dimensional diffusion	$[1-(1-\alpha)^{1/2}]^{1/2}$	2.14×10^{-5}	0.9568	0.3411
Mampel (first order)	$-\ln(1-\alpha)$	9.20×10^{-5}	0.9720	-0.0060
Avrami-Erofeev	$[-\ln(1-\alpha)]^{1/2}$	4.53×10^{-5}	0.9830	0.4432
Avrami-Erofeev	$[-\ln(1-\alpha)]^{1/3}$	3.06×10^{-5}	0.9730	0.6088
Avrami-Erofeev	$[-\ln(1-\alpha)]^{1/4}$	2.32×10^{-5}	0.9648	0.6977
Contracting cylinder	$1-(1-\alpha)^{1/2}$	2.54×10^{-5}	0.9855	0.0821
Contracting sphere	$1-(1-\alpha)^{1/3}$	2.05×10^{-5}	0.9869	0.0417
Power law	$\alpha^{3/2}$	3.39×10^{-5}	0.9779	0.0711
Power law	$\alpha^{1/2}$	2.09×10^{-5}	0.9042	0.4934
Power law	$\alpha^{1/3}$	1.57×10^{-5}	0.8821	0.6276
Power law	$\alpha^{1/4}$	1.25×10^{-5}	0.8699	0.7061

Table 2. Linear fitting results of the experimental data obtained from the in-situ X-ray diffraction results at 575 °C by adopting alternate reaction models [38].

According to Eq. (4), the intercept of plot *lnt* ~*1/T* ($\ln(\frac{g(\alpha)}{A})$), can be obtained from the linear fitting. At certain degree of reaction, if the *g(a)* is given, the value of *A* can be calculated. Then $k(T) = A\exp(-\frac{E}{RT})$ can also be determined at certain temperatures. The calculated value of *k*(*T*) should be consistent with the linear fitted slope of plot *g*(*a*)~*t* if the given *g*(*a*) is the valid reaction model.

Following this method, we verify these two models discussed above at 575°C. At the initial stage (*a*=0.20) of reaction, for contracting sphere model ($1-(1-\alpha)^{1/3}$), the calculated *k*(*T*) is 2.09×10^{-5}, which is comparable to the corresponding value of slope given in Table 2. In the case of one-dimensional diffusion model (α^2), the calculated *k*(*T*) is 1.19×10^{-5}, which is much smaller than the corresponding value of the slope presented in Table 2. At the middle stage ((*a*=0.50) of reaction, the calculated value of *k*(*T*) from the contracting cylinder model is 2.18×10^{-5}, which is still comparable to the corresponding value of slope. For the one-dimensional diffusion model, the calculated *k*(*T*) is 2.64×10^{-5}, which is smaller than the corresponding value given in Table 2. However, at the final stage (*a*=0.80), the calculated *k*(*T*) from the one-dimensional diffusion model is 3.62×10^{-5}, which is more consistent with the corresponding value presented in Table 2 than that in the case of the contracting cylinder model (the calculated value is 2.33×10^{-5}).

Hence, one can say that the reaction between Mg and B during low-temperature sintering is firstly mainly controlled by the contacting sphere model, which is a kind of the phase boundary reaction mechanism. As the reaction prolongs, the one-dimensional diffusion model, one kind of diffusion-limited mechanism, gradually becomes dominant.

In our previous study [39], we have investigated the MgB_2 phase formation process during the sintering based on the phase identification and microstructure observation. It is found that the sintering 'necks' between individual Mg and B particles occurs at the first stage of the sintering. Then the solution active regions form at the sintering necks. With the isothermal heating prolonging, a few Mg and B atoms in the solution active regions can be activated and react with each other. The activated atoms are limited at this initial stage and thus the reaction rate is slow and mainly determined by the phase boundary reaction mechanism. Meanwhile, an MgB_2 phase layer is gradually formed at the sintering necks between Mg and B particles and the Mg atoms have to diffuse through the whole MgB_2 phase layer to reach the reaction interface, , as shown in Fig. 8. As the reaction prolonging, the synthesized MgB_2 layer becomes thicker and thicker, and it should be more and more difficult for Mg atoms to reach the reaction interface through diffusion. Finally, the reaction rate is controlled by the diffusion-limited mechanism. As a result, the reaction rate becomes slower and slower and it takes a very long time to be totally completed due to the slow diffusion rate of Mg. The corresponding activation energy is also decreased firstly and then increased again during the whole reaction process, as discussed above. It is also the reason why the residual Mg is still present even after holding for 10 h at 575 °C.

Based on above analysis and compared with previous studies [18, 19] in which they propose the reaction is controlled by single mechanism, the varied mechanisms are more valid and more consistent with the actual sintering process. The value of activation energy in present work is also comparable to the calculated value in Shi et al.'s study [19] whereas the models are different from theirs. But in their study, the activation energy is calculated using a model-free method, just as in our work.

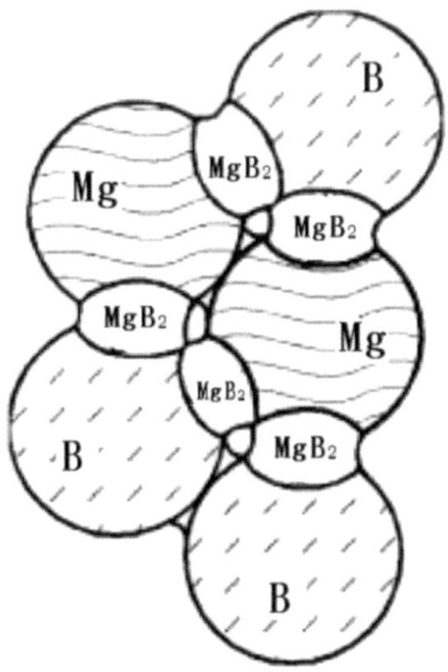

Fig. 8. Schematic illustration of solid-solid reaction between Mg and B particles based on the inter-diffusion mechanism [39].

Based on above discussion, it is concluded that the reaction between Mg and B during the low-temperature sintering is controlled by varied mechanisms. At initial stage, the reaction rate is mainly determined by the phase boundary reaction mechanism. As the reaction prolonging and the synthesized MgB_2 layer increasing, the diffusion-limited mechanism gradually becomes dominant. The corresponding activation energy is also decreased firstly and then increased again.

2. Accelerated sintering of MgB_2 with different metal and alloy additions at low temperature

According to the above sintering mechanism, the reaction between magnesium and boron at low temperature took a very long time to form the complete MgB_2 phase as the result of the low diffusion rate of Mg atom at solid state. In order to rapidly synthesize the complete MgB_2 phase through the low-temperature sintering, improving the diffusion efficiency of Mg atoms to the reaction interface is the key point, especially at the final stage of the sintering.

2.1 The influence of different metals and alloys on the sintering process and superconductive properties of MgB_2

In order to accelerate the diffusion rate of Mg atoms and thus improve sintering efficiency of MgB_2 at low temperature, different metals and alloys dopants were added into Mg-2B sintering system by world-wide research groups.

Shimoyama *et al.* [40] found that a small amount of silver addition decreases dramatically the reaction temperature of magnesium and boron in the formation of bulk MgB_2 without degradation of either the critical temperature or the critical current density. Although the added silver forms an Ag-Mg alloy after the heat treatment, these impurity particles exist mainly at the edge of voids in the sample microstructure and therefore do not provide a significant additional restriction to the effective current path. Accordingly MgB_2 bulks with excellent J_c properties have been fabricated at a temperature as low as 550 °C with the 3 at.% Ag doping. The sintering time of doped samples is also reduced significantly compared to that required for undoped samples fabricated by low temperature sintering. This effectively widens the processing window for the development of practical, low-cost MgB_2 superconductors by reaction at low temperature [40].

Hishinuma *et al.* [41] synthesized Mg_2Cu-doped MgB_2 wires with improved J_c by sintering at low temperature for 10 h. They found that the formation of the MgB_2 phase is improved due directly to the lower melting point of Mg_2Cu (568 °C) than Mg (650 °C), which can promote the diffusion of Mg in the partial liquid (the MgB_2 phase forms by the diffusion reaction between released Mg from Mg_2Cu and amorphous B powder [41]). The J_c of sample prepared in this way can be improved further in Mg_2Cu-doped MgB_2 wires by sintering at lower temperature (450 °C) for longer time (more than 100 h). The maximum core J_c value of these samples was found to be over 100, 000 A cm^{-2} at 4.2 K in a magnetic field of 5 T for a tape sintered for 200 h. Bulk MgB_2 has been fabricated successfully in other studies by Cu-doping and sintering at 575 °C for only 5 h [42]. Thermal analysis indicates that the Mg-Cu liquid forms through the eutectic reaction between Mg and Cu at about 485 °C, which leads to the accelerated formation of MgB_2 phase at low temperature. The SEM images of the sintered Cu-doped samples are shown in Fig. 9. It is observed that the undoped sample is porous and consists of small irregular MgB_2 particles and large regular Mg particles which are in poor connection with each other (see Fig. 9a). On the other hand, the MgB_2 particles of the doped sample become larger and more regular accompanying with the increasing amount of Cu addition. The doped samples also become denser with the amount of Cu addition increasing for the reason that the MgB_2 particles are in better connection with each other and give birth to less voids (see Figs. 9b-9d). Especially, as shown in the Fig. 9d, the MgB_2 grains in the $(Mg_{1.1}B_2)_{0.9}Cu_{0.1}$ sample exhibit platy structure with a typical hexagonal shape [42]. The high J_c in MgB_2 samples doped with Cu is attributed mainly to the grain boundary pinning mechanism that results from the formation of small MgB_2 grains during low temperature sintering. As with the Ag-doped samples, the concentration of Mg-Cu alloy in these samples tends to form at the edge of voids in the microstructure and does not degrade significantly the connectivity between MgB_2 grains, which contributes directly to enhanced J_c. Recently, the addition of Sn to the precursor powder has also been observed to assist the formation of the MgB_2 phase during low temperature sintering, and bulk Sn-doped MgB_2 prepared at 600 °C for 5 h exhibit good values of J_c [43].

Interestingly, although Ag and Sn addition can form a local eutectic liquid with the Mg precursor at lower temperature than the addition of Cu, the Cu has been found to play a more effective role in the improvement of MgB_2 phase formation than Ag and Sn at low temperature. The Cu-doped samples take significantly less time to form the primary MgB_2 phase than those containing similar concentrations of Ag or Sn at a similar sintering temperature [40, 42, 43]. Grivel *et al.* [44] observed a similar phenomenon in a study of the effects of both Cu and Ag on the kinetics of MgB_2 phase formation. The addition of 3 at.% Cu or Ag to a precursor mixture consisting of Mg and B powders results in a significant increase of MgB_2 phase formation kinetics in the temperature range below the melting point of Mg. The MgB_2 phase forms more rapidly in the precursors containing Cu than those containing Ag. These authors suggest that this behavior might be related to the lower solubility limit of Cu in solid Mg, compared to the case of the Mg-Ag system [44].

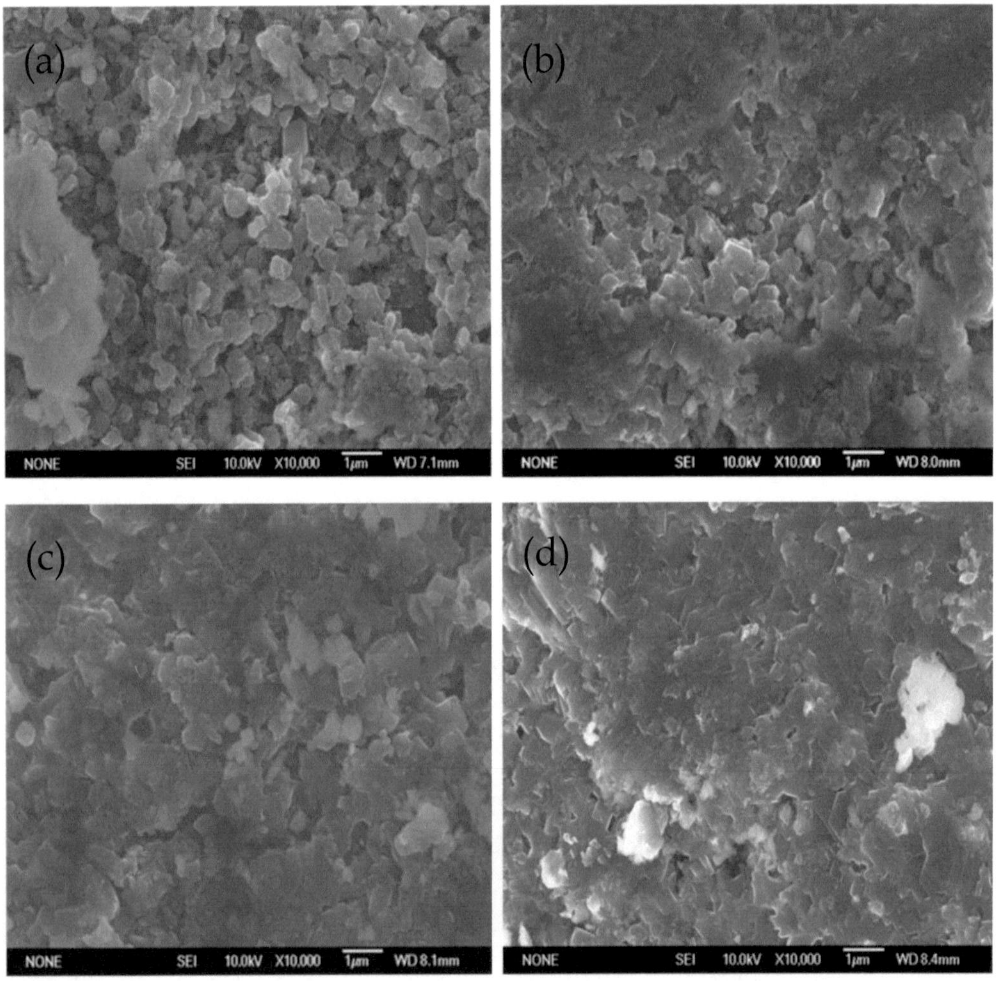

Fig. 9. SEM images of the microstructures of the $(Mg_{1.1}B_2)_{1-x}Cu_x$ samples sintered at 575 °C for 5 h with (a) x = 0.0, (b) x = 0.03, (c) x = 0.05 and (d) x = 0.10, respectively [42].

Based on the above discussion, the assisted sintering of MgB_2 with different metal or metal alloy additions at low temperature is convenient from a practical processing point of view and also reduces the processing time. In addition, these additives tend to be cheap and yield MgB_2 samples with improved J_c. Therefore, this technique appears to be the most promising way of preparing practical, low-cost MgB_2 superconductors at low temperature, compared to the use of different Mg-based precursors and the ball milling pretreatment of precursor powders.

2.2 The mechanism of metal-accelerated sintering at low temperature

The addition of minor metals or metal alloys additions represents the most convenient, effective and inexpensive way of preparing MgB_2 with excellent superconducting properties at low sintering temperatures. As a result, the accelerated sintering mechanism apparent in the effective processing of these samples should be clarified. The accelerated sintering mechanism of precursors containing Cu, for example, has been studied in detail using thermal analysis and activated sintering theory [44, 45, 46].

As discussed in section 1.1, thermal analysis of the sintering process of undoped MgB_2 reveals three peaks corresponding to solid-solid reaction, Mg melting and solid-liquid reaction (see Fig. 10). A similar process was observed in samples of composition $(Mg_{1.1}B_2)_{1-x}Cu_x$ with x = 0.01, 0.03, 0.05 and 0.10, except that the on-set temperature of each the three peaks decreased gradually with increasing amount of Cu addition. It should be noted that an apparent endothermic peak appears at about 485 °C, which is just below the first exothermic peak in the thermal analysis curves of the $(Mg_{1.1}B_2)_{0.90}Cu_{0.10}$ samples. By reference to the binary Mg-Cu phase diagram (see Fig. 11), it is apparent that the Mg-Cu liquid initially forms locally through the eutectic reaction during the sintering process of Mg-Cu-B system, resulting in the appearance of this apparent endothermic peak. The local formation of Mg-Cu liquid in the $(Mg_{1.1}B_2)_{1-x}Cu_x$ samples with x = 0.01, 0.03 and 0.05 is limited for the small amount of Cu added to the precursor, which results in an endothermic signal that is too small to be detected by the thermal analysis measurement in this temperature range [46].

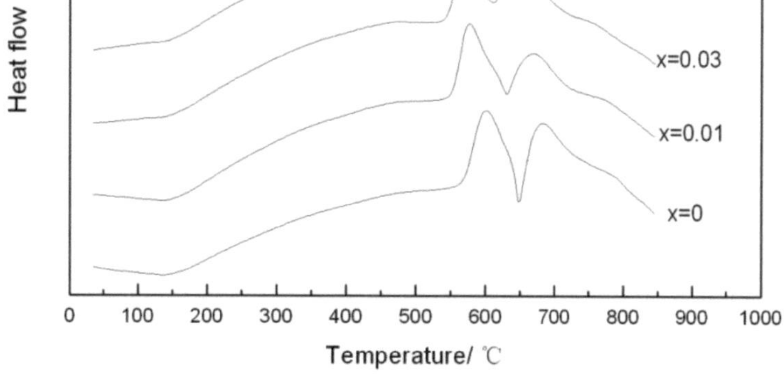

Fig. 10. Measured thermal analysis curves during the sintering of $(Mg_{1.1}B_2)_{1-x}Cu_x$ (with x = 0, 0.01, 0.03, 0.05 and 0.10) samples with an applied heating rate of 20 °C / min [46].

This raises the question of how the presence of a local Mg-Cu liquid accelerates the subsequent formation of MgB_2 phase. (in previous studies [45, 46], the accelerated sintering mechanism of MgB_2 with Cu addition is attributed to the activated sintering).

It is known that activated sintering of MgB_2 by chemical doping facilitates either a lower sintering temperature or a shorter sintering time. German *et al.* [45] have proposed the three criteria for activated sintering systems of solubility, segregation and diffusion as follows:

i. *Solubility* The additive A must have a high solubility for the base B, while the base B must have a low solubility for the additive A so that the additive can wet the base particles and then exhibit a favorable effect on diffusion.
ii. *Segregation* During the sintering, the additive must remain segregated at the inter-particle interfaces to remain effective during the entire sintering process.
iii. *Diffusion* The diffusivity of the base metal B in the additive layer must be higher than the self-diffusivity of the base metal B.

Accordingly, an ideal phase diagram for the activated sintering system can be constructed (see Fig. 12). The formation of the MgB_2 phase in the Mg-Cu-B system, is controlled mainly by the diffusion rate of Mg atoms. Only the effect of Cu addition on this diffusion rate is considered to be significant, and the effect of Cu addition on the B atoms can be neglected. Hence, the Mg-Cu-B system can be simplified as an Mg-Cu system for the analysis of the influence of Cu addition on the sintering process. As shown in the Mg-Cu phase diagram (see Fig. 11), Cu addition dramatically decreases the liquidus and the solidus, which implies that the local Mg-Cu liquid can segregate to the interface of Mg particles and therefore meets the *Segregation* criterion in German's study [45]. The high solubility of Cu for Mg (see Figs 11 and 12) enables the Mg-Cu liquid to wet the Mg particles and support the diffusion transport mechanism. As a result, this meets the *solubility* criterion. Generally, the atomic diffusivity in the liquid state is larger than that in the solid state, so the Mg-Cu liquid also meets the *Diffusion* criterion. Collectively, these observations suggest theoretically that the local Mg-Cu liquid meets all the criteria for the diffusion of Mg to B. Cu, therefore, can serve as the activated sintering addition and accelerate the formation of the MgB_2 phase. The conclusion can also be verified by the microstructure observation of the Cu-doped sample sintered at low temperature, as shown in Fig. 13 [46]. It was clear that Cu was concentrated at local region by the edge of voids while Mg was preferentially distributed inside of particles far away from the voids. The result indicated that the Mg-Cu alloys corresponding to the local Cu-Mg liquid during the sintering process mainly concentrated at the edge of voids. Since the void results from the diffusion of Mg atom into B during sintering as mentioned previously [21], the concentration of Mg-Cu alloys at the edge of the voids implied that the Mg-Cu liquid generated and segregated to the interface between Mg particles and B particles at the initial stage of the sintering process and then provided a high transport for the diffusion of Mg into B. After a period of sintering time, Mg was run out and voids formed at the former place of the Mg.

Other metals or metal alloys must first form local liquids with Mg before the formation of MgB_2 phase if they are to serve as activate additives during low temperature sintering. Whether or not these local liquids promote the formation of the MgB_2 phase and activate the sintering mechanism, should be verified by considering the criteria described above (for the case of Cu addition), the ideal phase diagram for the activated sintering system and the binary phase diagram of the added element and Mg. In the section 2.1, the addition of Cu was demonstrated to be more effective than Sn or Ag in accelerating the formation of the

MgB_2 phase at low temperature, even though the latter can form liquid at much lower temperature. Inspection of the relevant binary phase diagram of Sn and Ag with Mg (not shown here) indicates that the solubility limit of Mg in Sn is much lower than that in Cu. Hence, it is more difficult for the Sn-based local liquid to wet the Mg particles and promote the diffusion of Mg according to the *Solubility* criterion. As a result, the activated sintering of MgB_2 with Sn addition is much less efficient than Cu addition. On the other hand, the solubility limit of Mg in Ag is higher than that in Cu and the Ag-based local liquid should wet the Mg particles more easily and accelerate the diffusion of Mg more effectively. However, the solubility limit of Ag in solid Mg is also much higher than for the case of Cu, which means that the amount of local Ag-based liquid present will decrease due to the solution of Ag in the Mg solid. The effect of this is to lower the activated sintering efficiency compared to that obtained for a similar concentration of Cu addition.

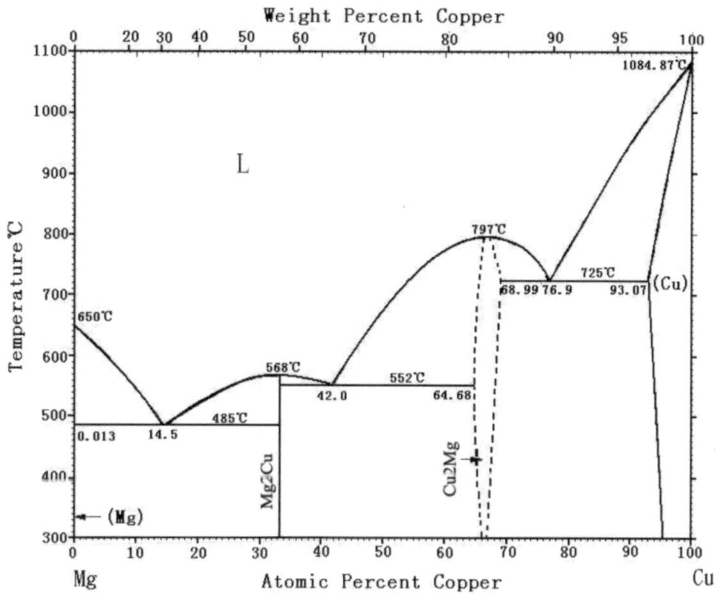

Fig. 11. Binary Mg-Cu phase diagram [48].

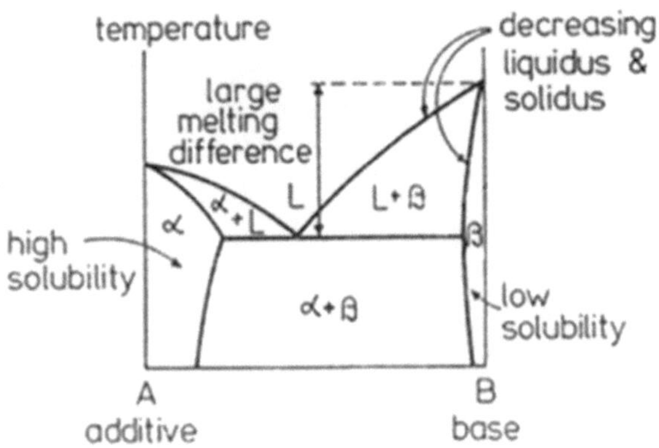

Fig. 12. An ideal phase diagram for the activated sintering system [45].

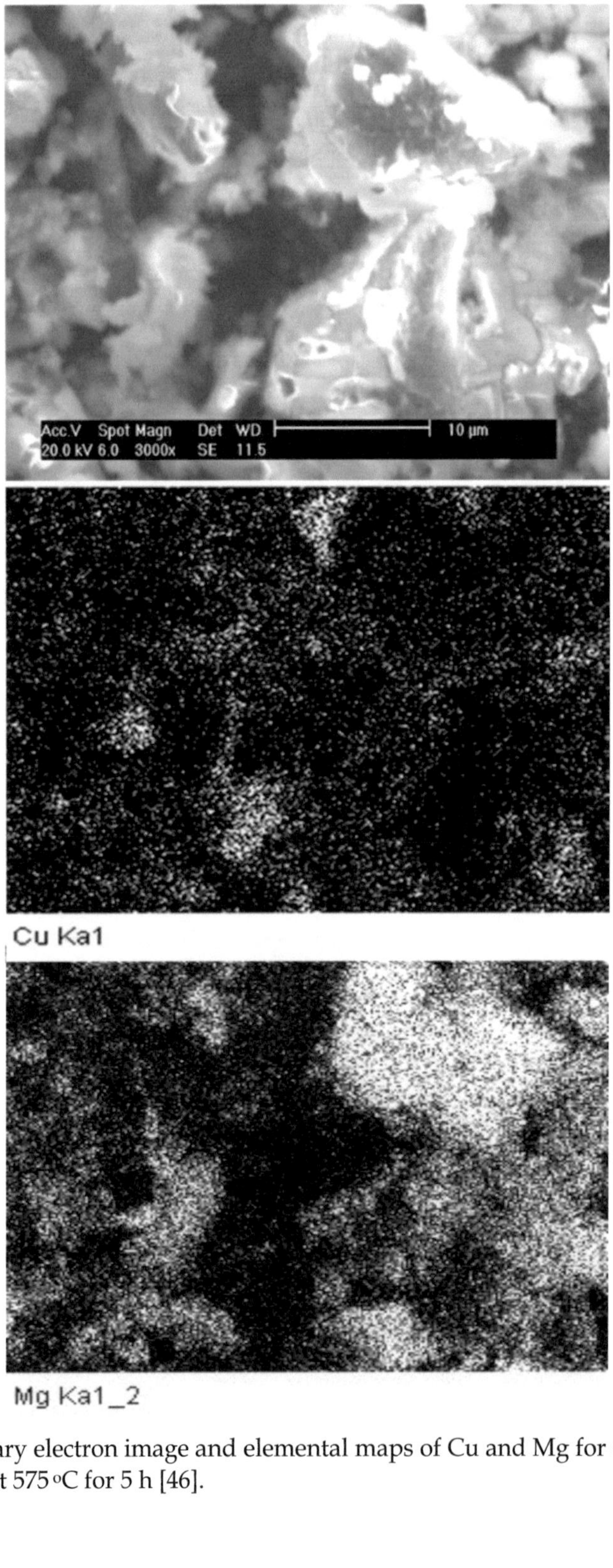

Fig. 13. A secondary electron image and elemental maps of Cu and Mg for $(Mg_{1.1}B_2)_{0.8}Cu_{0.2}$ sample sintered at 575 ^{o}C for 5 h [46].

In addition, following these criteria and inspection of the appropriate binary phase diagrams of the additive metal elements with Mg, effective activated addition for the low temperature sintering of MgB_2 could be achieved potentially using lots of candidate metals. Accordingly, Cu was finally determined as effective activator for improving sintering efficiency of MgB_2 in our work.

3. The effect of Cu activator on the microstructure and superconductive properties of MgB_2 prepared by sintering

Cu addition can improve the sintering efficiency of MgB_2 and thus selected as sintering activator. However, whether Cu activator optimizes the microstructure and superconductive properties of MgB_2? To answer this question, the effect of Cu activator on the microstructure of MgB_2 sintered at both low temperature and high temperature were investigated in detail.

3.1 Effect of Cu activator on the reduction of MgO impurity in MgB_2 sintered at high temperature

MgO is always present as the inevitable impurity phase during the sintering process of MgB_2 for the reason that Mg is very reactive with oxygen, which can be supplied by the gaseous O impurity in the protective Ar gas and the oxide impurity (such as B_2O_3 impurity in the B powders) in the starting materials. The presence of MgO impurity may be of great importance and yields a significant effect on the superconductive properties of MgB_2 superconductor. Although the MgO nanoinclusions within MgB_2 grains could serve as strong flux pinning centers when their size were comparable to the coherent length of MgB_2 (.approximately 6~7 nm), the presence of excess MgO phases or largesized MgO particles at the grain boundaries could result in the degradation of grain connectivity [49, 50]. Hence, it is essentially important to control the amount of MgO impurity during the sintering of MgB_2 samples.

In our previous work [51], based on the investigation of the effect of minor Cu addition on the phase formation of MgB_2, it is found that the minor Cu addition (<3 at %) could apparently reduce the amount of MgO impurity in the prepared MgB_2 samples, which provided a new route to govern the oxidation of Mg during the in-situ sintering of MgB_2 samples by altering the amount of Cu addition.

Figure 14 shows the X-ray diffraction patterns of the $(Mg_{1.1}B_2)_{1-x}Cu_x$ (x = 0.0, 0.01, 0.03 and 0.10) samples sintered at 850°C for 30min. It can be seen that all the sintered samples contain MgB_2 as the main phase. In the undoped samples, the MgO peaks are easily recognized, which suggests that some Mg was oxidized during the sintering process and thus MgO was the main impurity in the sintered samples. On the other hand, in the diffracted patterns of the Cu-doped samples, all the MgO phase peaks become weaker and even some peaks identified as MgO phase disappear with the amount of Cu addition increasing. This trend can be observed more clearly from the Fig. 15, which shows the most intense peak (the peak of (200) crystal plane) of MgO in the X-ray diffraction patterns of the sintered $(Mg_{1.1}B_2)_{1-x}Cu_x$ (x = 0.0, 0.01 and 0.03) samples. The results suggest that the minor Cu addition can depress the oxidation of Mg apparently during the in-situ sintering of MgB_2 samples.

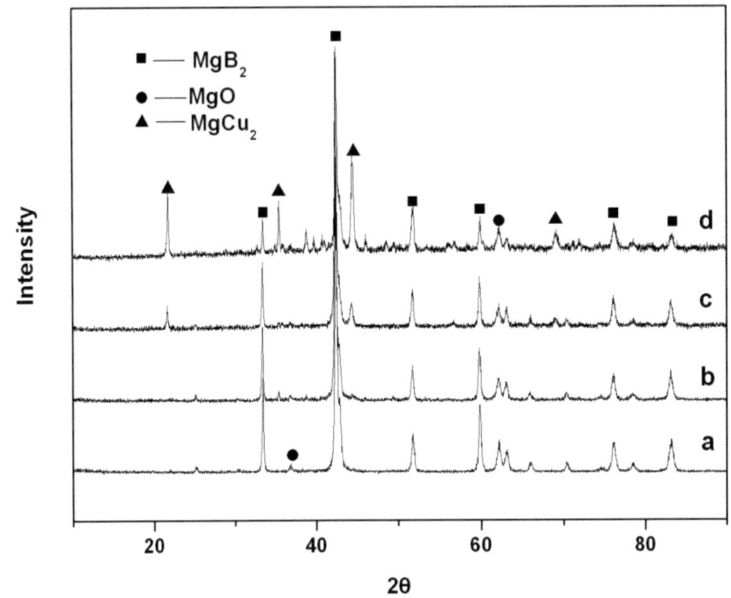

Fig. 14. X-ray diffraction patterns of the prepared $(Mg_{1.1}B_2)_{1-x}Cu_x$ samples sintered at 850°C for 30min with (a) x = 0, (b) x = 0.01, (c) x = 0.03, (d) x=0.10, respectively [51].

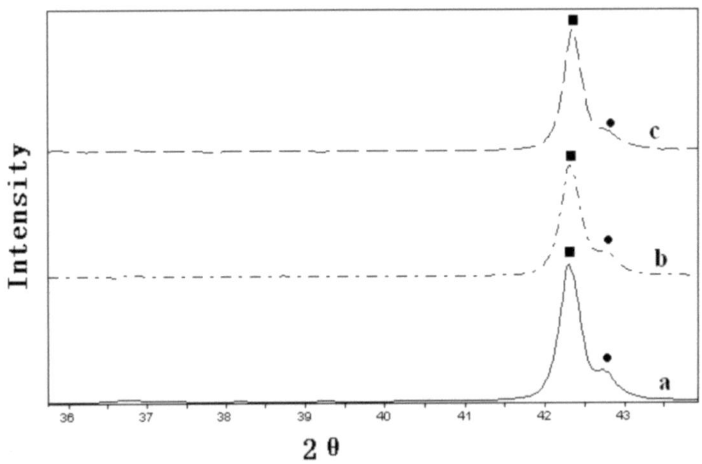

Fig. 15. The enlarged X-ray diffraction patterns around (200) of MgO impurity in the $(Mg_{1.1}B_2)_{1-x}Cu_x$ samples sintered at 850°C for 30min with (a) x = 0, (b) x = 0.01, (c) x = 0.03, respectively [51].

The weight fraction of MgO was calculated from the X-ray diffraction patterns according to the External Standard Method. Fig. 16 shows the weight fraction of MgO versus the amount of Cu addition in the sintered samples. From the figure, it is found that the weight fraction of MgO in the undoped MgB_2 sample is about 16.5%, which is comparable with the previous study [52]. On the other hand, the weight fraction of MgO decreases obviously from 16.5% to 12.5% with the amount of Cu addition increasing from 0.0 to 0.03. However, when the amount of Cu addition increases from 0.03 to 0.10, the weight fraction of MgO almost remains unchanged (see Fig. 16) while the $MgCu_2$ phase increases significantly (see Fig. 14). The excess Cu addition in the $(Mg_{1.1}B_2)_{0.9}Cu_{0.1}$ sample has no significant effect on the further decrease of MgO impurity. Besides, the excess $MgCu_2$ phase in the $(Mg_{1.1}B_2)_{0.9}Cu_{0.1}$ sample (see Fig. 1) may also depress the superconductivity properties of MgB_2 phase. Hence, we conclude that the x=0.03 Cu addition has the best effect on the decrease of MgO impurity during the in-situ sintering of MgB_2 samples.

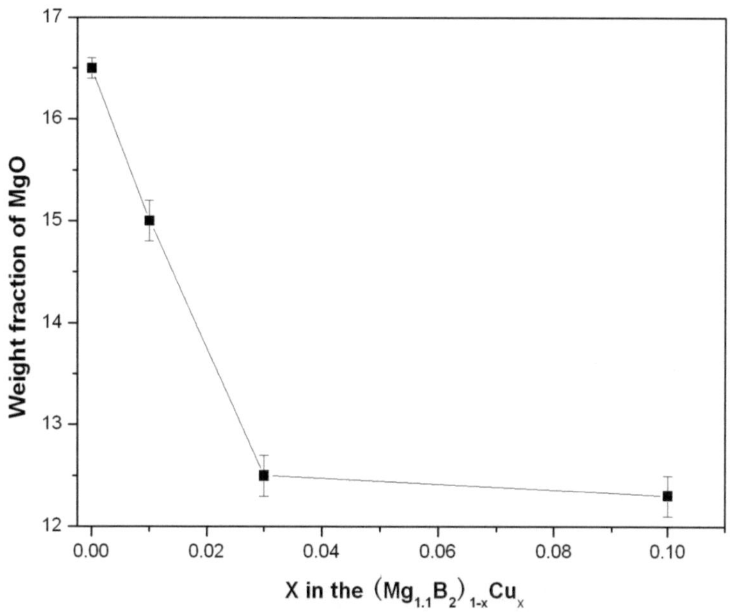

Fig. 16. The weight fraction of MgO versus the amount of Cu addition in the Cu-doped samples sintered at 850 °C for 30 min [51].

The scanning electron microscopy images of the sintered $(Mg_{1.1}B_2)_{1-x}Cu_x$ samples are shown in the Fig. 17. There is an MgO layer on the partial surface of MgB_2 grains in the undoped sample, as shown in Fig. 17a. The MgO layer consists of short MgO whiskers, which is similar with the MgO morphology observed in the study on the oxidation of MgB_2 [53]. On the other hand, in the Cu-doped samples, the amount of MgO impurity decreases with the increasing amount of Cu addition and at the same time the morphology of MgO transits from whiskers to nanoparticles (see Fig. 17b and Fig. 17c). The result of SEM images is consistent with the XRD pattern and they both reveal that the addition of minor Cu can apparently decrease the MgO impurity in the MgB_2 samples. From Fig. 17, it is also found that the MgB_2 grains become larger and more regular accompanying with the increasing amount of Cu addition, which indicates that the Cu addition can also promote the growth of MgB_2 grains at the same time with decreasing the MgO impurity in the prepared MgB_2 samples.

Fig. 17. Scanning electron microscopy images of the $(Mg_{1.1}B_2)_{1-x}Cu_x$ samples sintered at 850 °C for 30 min with (a) x = 0, (b) x = 0.01 and (c) x = 0.03, respectively. The MgO whiskers and nanoparticles are indicated by the black circles and arrows in the figures [51].

In order to investigate the effect of the decreasing MgO impurity induced by the minor Cu addition on the superconductive properties of MgB_2 samples, the corresponding T_c temperatures of all sintered samples were measured. Fig. 18 illustrates the temperature dependence of resistivity for the $(Mg_{1.1}B_2)_{1-x}Cu_x$ (with x = 0, 0.01 and 0.03) samples sintered at 850 °C for 30min. As shown in it, the undoped sample exhibits a slight suppression in the value of T_c compared to the typical pure MgB_2 samples, which can be attributed to the limit of MgB_2 intergranular connection caused by the excessive MgO impurity at the grain boundaries. However, in the Cu-doped samples, the values of T_c are over 38K and slightly increase from 38.1 K to 38.6 K with the increasing amount of Cu addition from x = 0.01 to x = 0.03, which is comparable to the pure MgB_2 samples (39 K). The observation could be explained by the decrease of MgO impurity and the growth of MgB_2 grains resulting from the minor Cu addition.

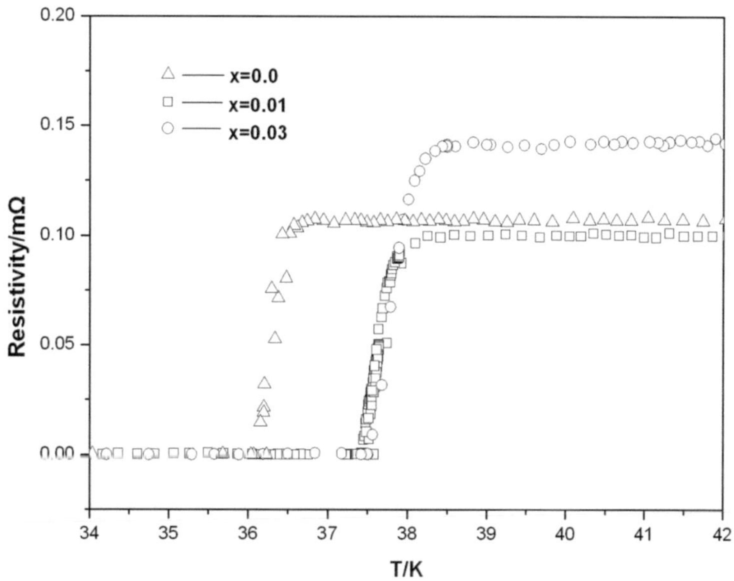

Fig. 18. The temperature dependence of resistivity for the $(Mg_{1.1}B_2)_{1-x}Cu_x$ (x = 0, 0.01 and 0.03) samples sintered at 850 °C for 30min [51].

In summary, the minor Cu addition can decrease the amount of MgO impurity and thus significantly improve the superconductive properties of MgB_2 bulks. However, how the Cu addition avoids the oxidation of Mg during the sintering process of MgB_2 is still to be answered.

It has been indicated that during the sintering process of Mg-Cu-B system, the Mg-Cu liquid locally formed firstly in the presence of Cu through an eutectic reaction. The local Mg-Cu liquid appearing at such low temperature could dissolve some Mg and wrap the neighboring Mg particles, which partly avoided Mg contacting with the gaseous O existing in the interspace of the pressed samples and the oxide impurity (such as B_2O_3 in the B powders) in the starting materials. Hence, the oxidation of Mg during the low-temperature (below the Mg melting point) sintering stage resulting from the gaseous O existing in the interspace and the oxide impurity in the starting materials can be depressed by the presence of local Mg-Cu liquid significantly.

When the temperature was above the Mg melting point (about 650 °C) during the sintering process, the unreacted Mg after the solid reaction stage would melt and volatilize severely as a result of the high vapor pressure of Mg liquid. The gaseous Mg mixed with the protective Ar gas and could react with the O_2 impurity in the protective Ar gas at such high temperature, which resulted in the increasing amount of MgO impurity deposited in the undoped MgB_2 samples after cooling to room temperature [54], as shown in Fig. 17a. On the other hand, the Cu addition could lower the melting point of Mg at the same time of decreasing the vapor pressure of Mg liquid at high temperature [55]. The decrease of the vapor pressure of Mg liquid led by the Cu addition could reduce the amount of the gaseous Mg from the volatilization of Mg, which thus decrease the amount of MgO impurity resulting from the oxidation of gaseous Mg in the doped samples (as shown in Fig. 17b and 17c).

3.2 The synthesis of lamellar MgB_2 crystalline by Cu activated sintering at low temperature

The microstructure of MgB_2 synthesized by Cu activated sintering at low temperature was also investigated in detail [56]. The SEM images of both sintered Cu-doped sample and undoped sample are illustrated in Fig. 19. Lamellar MgB_2 grains with typical hexagonal shape were observed in the Cu-doped sample sintered at low temperature (see Fig. 19a, denoted by the black arrows). There are few impurities between lamellar MgB_2 grains in the MgB_2 region and the Mg-Cu impurities mainly distribute in the region near the lamellar MgB_2 grains, as shown in Fig. 19a. One can also find that all of the MgB_2 grains in the lamellar structure almost share the same orientation except only a few of them. On the other hand, the MgB_2 grains in the undoped sample sintered at high temperature are nearly in the same size as those in the Cu-doped samples and most of them also exhibit typical hexagonal shape. But their orientation is random, which is the typical characteristic of MgB_2 grains sintered in the traditional solid-state sintering (see Fig. 19b). Hence, the MgB_2 grains in the lamellar structure seem to be in better connectivity with each other and there are also fewer voids between them when compared to the MgB_2 grains in the random orientation, as shown in Fig. 19.

It is proposed that Mg atoms could easily diffuse into B through the path of local Mg-Cu liquid and then react with B forming MgB_2 at the interface between Mg-Cu liquid and B particles. Since the local Mg-Cu liquid only serves as the path for the diffusion of Mg into B and does not react with B until all the Mg is run out, it is always present and provide the constant liquid environment for the nucleation and growth of MgB_2 grains. As we all known, there are two main mechanism forming the lamellar crystalline in the liquid sintering environment, two dimensional nucleation and screw dislocation nucleation. The concentration gradient is the driving force for both mechanisms during the isothermal liquid sintering. The surface of the grains in the lamellar crystalline formed following the two dimensional nucleation are generally more smooth and regular than the screw dislocation [57]. On the other hand, the two dimensional nucleation also needs higher supersaturation than the screw dislocation [58]. In present case, the surface of MgB_2 grains in the lamellar crystalline is smooth and regular and no obvious dislocations and impurities are observed.

Moreover, the supersaturation of MgB_2 in the Mg-Cu liquid must be high enough for the two dimensional nucleation due to the comparable less amount of Mg-Cu liquid (the amount of Cu addition is only 8 at%). Hence, it is proposed that the formation mechanism of the present lamellar MgB_2 grains is attributed to the two dimensional nucleation. The energy barrier of the two dimensional nucleation could be defined as follows:

$$(C / Ce)_{crit} = \exp(\pi h \Omega r^2 / 65 k^2 T^2) \tag{6}$$

$(C / Ce)_{crit}$ is the critical supersaturation, h is the step height, Ω is the atomic volume, r is the surface energy of crystal, k is the Boltzmann constant and T is the sintering temperature. It can be seen that the energy barrier of the two dimensional nucleation mainly depends

Fig. 19. The SEM images of sintered samples with (a) the Cu-doped sample sintered at 575 °C for 5h and (b) the undoped sample sintered at 750 °C for 1h [56].

on the step height and the surface energy of MgB_2 crystal. On the other hand, only when the MgB_2 supersaturation in the Mg-Cu liquid is higher than the critical supersaturation, the nucleation can start on the surface of B and form the new step. After that, the MgB_2 grains easily grow at this step and form a crystal layer. Accordingly, the schematic of the formation mechanism of the lamellar MgB_2 crytalline is shown in the Fig. 20 with (a) the initial stage, (b) the nucleation and growth stage and (c) the final stage. At the initial stage, the Mg that diffused to the surface of B through the Mg-Cu liquid would react with B as below: Mg + 2B = MgB_2, which can result in the concentration gradient of Mg in the interface (see Fig. 20a). As a result, a lot of Mg could diffuse into the interface and react with B forming MgB_2. Most of produced MgB_2 is dissolved in the Mg-Cu liquid and some MgB_2 will be physically absorbed on the surface of B. When the MgB_2 supersaturation is higher than the critical value for the nucleation, these absorbed MgB_2 will form two dimensional nuclei and produce a new step through the thermodynamic fluctuation. And then the dissolved MgB_2 will deposit on this step and the MgB_2 grains could rapidly grow on this step and form the crystal layer (see Fig. 20b). The new nuclei will continuously forming on the surface of MgB_2 crystal layer and then a new crystal layer will form again. As a result, the lamellar MgB_2 grains are obtained (see Fig. 20c).

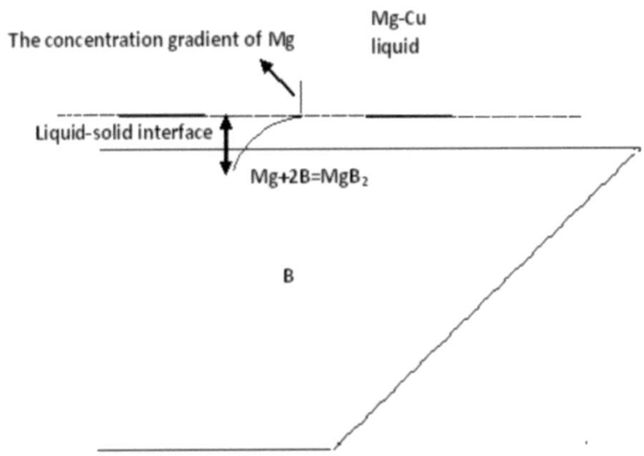

(a)

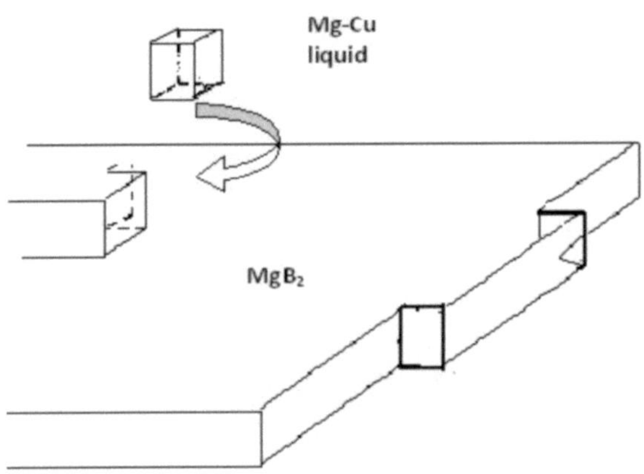

(b)

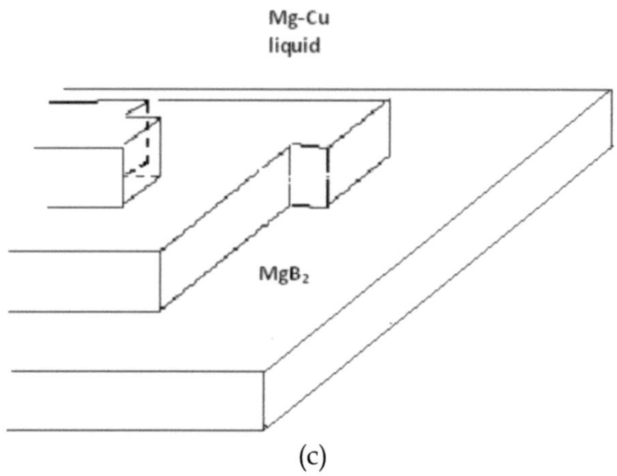

Fig. 20. The schematic of the formation mechanism of the lamellar MgB_2 crytalline with (a) the initial stage, (b) the nucleation and growth stage and (c) the final stage [56].

Fig. 21 shows the temperature dependence of resistivity of the Cu-doped sample and undoped sample. The resistivity of the Cu-doped samples is much lower than the undoped sample in the measured temperature region from 300K to 40K. The resistivity of MgB_2 sample should be increased with the addition of Cu, as reported previously [59]. Since the Mg-Cu alloys mainly concentrate around the voids and do not degrade the grain connectivity of MgB_2 phase in present sample, the resistivity is ought to keep unchanged and should not be lower than the undoped sample. Hence, the abnormal low resistivity must be attributed to the lamellar structure of MgB_2 grains.

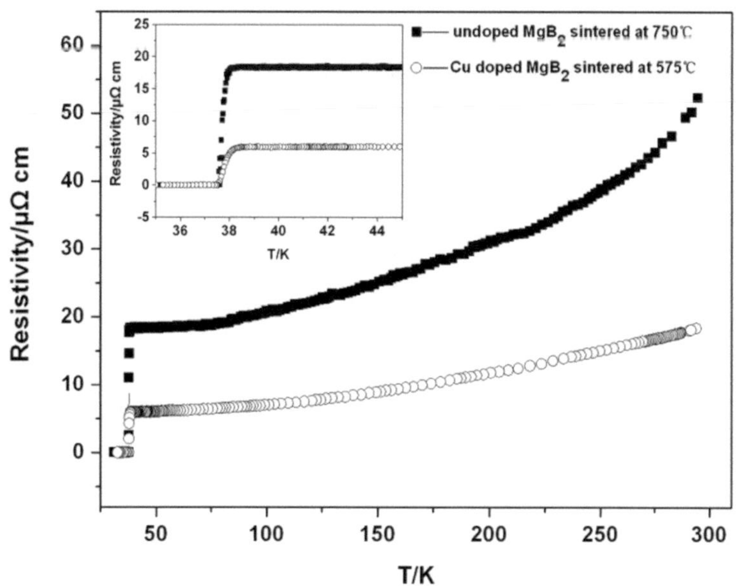

Fig. 21. The temperature dependence of resistivity of the Cu-doped sample and undoped sample [56].

According to the Rowell connectivity analysis, the calculated active cross-sectional area fraction (A_F) represents the connectivity factor between adjacent grains [60]. Here the A_F is estimated as:

$$A_F = \Delta \rho_{ideal} / (\rho_{300K} - \rho_{40K}) \quad (7)$$

$$\Delta \rho_{ideal} = \rho_{ideal(300K)} - \rho_{ideal(40K)} \quad (8)$$

Where ρ_{ideal} is the resistivity of a reference crystal and ρ_T is our measured resistivity at temperature T.

According to the previous studies [13, 60, 61], here the $\Delta \rho_{ideal}$ is 7.3 μΩcm. The results are listed in the Table 3. The value of A_F in the present undoped sample is comparable to that of samples sintered under the similar condition in previous reports [60]. The Cu-doped sample exhibits very excellent A_F, more than two times higher than that of the undoped samples. To further analyze the intergrain connectivity, the residual resistivity ratio (RRR) defined by ρ_{300K} / ρ_{40K} was also estimated, as shown in Table 3. All of the above results indicate that the lamellar MgB_2 grains possess much better grain connectivity than the typical morphology of MgB_2 grains. It should be pointed out that the onset of the transition temperature ($T_{c(onset)}$) of the Cu-doped sample is higher than that of undoped samples, which might be due to better connectivity and higher crystallinity of the lamellar grains (see Fig. 20a and Fig. 20b). However, the width of transition (ΔT_c) of the Cu-doped sample becomes a little wider than that of undoped sample. The transition broadening can be caused by the small grain size, inhomogeneity, impurities and so on. In present case, the lamellar MgB_2 grains shared the same orientation and lead to the intrinsic inhomogeneity in the doped sample, which could be the main factor broadening the transition width.

samples	ΔT_c (K)	$T_{c(onset)}$ (K)	ρ_{40K} (μΩcm)	ρ_{300K} (μΩcm)	RRR	A_P
Undoped MgB_2	0.25	38.0	18.388	51.449	2.780	0.221
Cu-doped MgB_2	0.40	38.3	5.980	18.355	3.069	0.590

Table 3 The transition temperature ($T_{c(onset)}$), width of transition (ΔT_c), measured resistivity values, residual resistivity ratio (RRR) and active cross-sectional area fraction (A_F) for Cu-doped sample and undoped sample, respectively.

In summary, the lamellar MgB_2 grains can be obtained by Cu-activated sintering at low temperature. This lamellar MgB_2 grains possess much better grain connectivity than the typical morphology of MgB_2 grains synthesized by the traditional solid-state sintering. Together with the proper methods increasing the pinning, the present lamellar MgB_2 grains might result in the further improvement of J_c.

4. References

[1] J. Nagamatsu, N. Nakagawa, T. Muranaka, Y. Zentani and J. Akimitsu: Nature, 2001 410 63-64.

[2] D.C. Larbalestier, L.D. Cooley, M.O. Rikel, A.A. Polyanskii, J. Jiang, S. Patnaik, X.Y. Cai, D.M. Feldmann, A. Gurevich, A.A. Squitieri, M.T. Naus, C.B. Eom, E.E. Hellstrom,

R.J. Cava, K.A. Regan, N. Rogado, M.A. Hayward, T. He, J.S. Slusky, P. Khalifah, K. Inumaru and M. Haas: Nature, 2001 410 186-189.
[3] K. Kawano, J.S. Abell, M. Kambara, N. Hari Babu and D.A. Cardwell: Appl. Phys. Lett., 2001 79 2216.
[4] V. Cambel, J. Fedor, D. Gregusova, P. Kovac and I. Husek: Supercond. Sci. Technol., 2005 18 417.
[5] V. Braccini, D. Nardelli, R. Penco and G. Grasso: Physica C, 2007 456 209-217.
[6] A. Serquis, L. Civale, J.Y. Coulter, D.L. Hammon, X.Z. Liao, Y.T. Zhu, D.E. Peterson, F.M. Mueller, V.F. Nesterenko and S.S. Indrakanti: Supercond. Sci. Technol., 2004 17 L35-L37.
[7] D. Mijatovic, A. Brinkman, D. Veldhuis, H. Hilgenkamp, H. Rogalla, G. Rijnders, D.H.A. Blank, A. V. Pogrebnyakov, J.M. Redwing, S.Y. Xu, Q. Li and X.X. Xi: Appl. Phys. Lett, 2005 87 192505.
[8] S.A. Cybart, K. Chen, Y. Cui, Q. Li, X.X. Xi and R.C. Dynes: Appl. Phys. Lett., 2006 88 012509.
[9] K. Chen, Y. Cui, Q. Li, X.X. Xi, S.A. Cybart, R.C. Dynes, X. Weng, E.C. Dickey and J.M. Redwing: Appl. Phys. Lett., 2006 88 222511.
[10] S. Jin, H. Mavoori, C. Bower and R.B. van Dover: Nature, 2001 411 563-565.
[11] S. Jin, R.C. Sherwood, R.B. Vandover, T.H. Tiefel and D.W. Johnson: Appl. Phys. Lett., 1987 51 203-204.
[12] H. Kumakura, A. Matsumoto, H. Fujii and K. Togano: Appl. Phys. Lett., 2001 79 2435-2437.
[13] P.C. Canfield, D.K. Finnemore, S.L. Bud'ko, J.E. Ostenson, G. Lapertot, C.E. Cunningham and C. Petrovic: Phys. Rev. Lett., 2001 86 2423-2426.
[14] A. Serquis, L. Civale, D.L. Hammon, J.Y. Coulter, X.Z. Liao, Y.T. Zhu, D.E. Peterson and F.M. Mueller: Appl. Phys. Lett., 2003 82 1754-1756.
[15] P.C. Canfield, D.K. Finnemore, S.L. Bud'ko, J.E. Ostenson, G. Lapertot, C.E. Cunningham and C. Petrovic: Phys. Rev. Lett., 2001 86 2423-2426.
[16] A.N. Baranov, V.L. Solozhenko, C. Lathe, V.Z. Turkevich and Y.W. Park: Supercond. Sci. Technol., 2003 16 1147-1151.
[17] J.C. Grivel, R. Pinholt, N.H. Andersen, P. Kovac, I. Husek and J. Homeyer: Supercond. Sci. Technol., 2006 19 96-101.
[18] Q.R. Feng, X. Wang, X.Y.Wang and G.C. Xiong: Solid State Commun., 2002 122 459.
[19] Q.R. Feng, X. Chen, Y.H. Wang, X. Wang, G.C. Xiong and Z.X. Gao: Physica C, 2003 386 653.
[20] Q.R. Feng, C. Chen, J. Xu, L.W. Kong, X. Chen, Y.Z. Wang, Y. Zhang and Z.X. Gao: Physica C, 2003 411 41-46.
[21] W. Goldacker, S.I. Schlachter, B. Obst and M. Eisterer: Supercond. Sci. Technol., 2004 17 S490.
[22] M. Bhatia, M.D. Sumption, S. Bohnenstiehl, S.A. Dregia, E.W. Collings, M. Tomsic and M. Rindfleisch: IEEE Trans Appl. Suprcond., 2007 17 2750-2753.
[23] G.H.S. Price: Int. Met., 1938 62 143.
[24] W.Z. Ostwald: Phys. Chem., 1900 34 495.
[25] W. Goldacker, S.I. Schlachter, B. Obst, B. Liu, J. Reiner and S. Zimmer: Supercond. Sci. Technol., 2004 17 S363–S368.
[26] M. Maeda, Y. Zhao, S.X. Dou, Y. Nakayama, T. Kawakami, H. Kobayashi and Y. Kubota: Supercond. Sci. Technol., 2008 21 032004.
[27] S. Soltanian, X.L. Wang, J. Horvat, S.X. Dou, M.D. Sumption, M. Bhatia, E.W. Collings, P. Munroe and M. Tomsic: Supercond. Sci. Technol., 2005 18 658–666.

[28] W.K. Yeoh, J. Horvat, J.H. Kim, X. Xu and S.X. Dou: Appl. Phys. Lett., 2007 90 122502.
[29] Y.C. Liu, Q.Z. Shi, Q. Zhao and Z.Q. Ma: J. Mater. Sci: Mater. Electron., 2007 18 855.
[30] M.S.A. Hossain, J.H. Kim, X. Xu, X.L. Wang, M. Rindfleisch, M. Tomic, M.D. Sumption, E. WCollings and S.X. Dou: Supercond. Sci. Technol., 2007 20 L51–L54.
[31] J.H. Kim, X. Xu, M.S.A. Hossain, D.Q. Shi, Y. Zhao, X.L. Wang, S.X. Dou, S. Choi and T. Kiyoshi: Appl. Phys. Lett., 2008 92 042506.
[32] G.H.S. Price: Int. Met., 1938 62 143.
[33] W.Z. Ostwald: Phys. Chem., 1900 34 495.
[34] N. Rogado, M.A. Hayward, K.A. Regan, Y. Wang, N.P. Ong, H.W. Zandbergen, J.M. Rowell and R.J. Cava: J. Appl. Phys., 2002 91 274-277.
[35] A.Yamamoto, J. Shimoyama, S. Ueda, Y. Katsura, S. Horii and K. Kishio: Supercond. Sci. Technol., 2005 18 116-121.
[36] W. Goldacker, S.I. Schlachter, B. Obst, B. Liu, J. Reiner and S. Zimmer: Supercond. Sci. Technol. 2004 17 S363–S368.
[37] M. Maeda, Y. Zhao, S.X. Dou, Y. Nakayama, T. Kawakami, H. Kobayashi and Y. Kubota: Supercond. Sci. Technol., 2008 21 032004.
[38] Z.Q. Ma and Y.C. Liu: Mater. Chem. Phys., 2011 126 114-117.
[39] Q.Z. Shi, Y.C. Liu, Q. Zhao and Z.Q. Ma: J. Alloys Compd., 2008 458 553-557.
[40] J. Shimoyama, K. Hanafusa, A. Yamamoto, Y. Katsura, S. Horii, K. Kishio and H. Kumakura: Supercond. Sci. Technol., 2007 20 307-311.
[41] Y. Hishinuma, A. Kikuchi, Y. Iijima, Y. Yoshida, T. Takeuchi and A. Nishimura: IEEE Trans. Appl. Supercond., 2007 17 2798-2801.
[42] Z.Q. Ma, Y.C. Liu, Q.Z. Shi, Q. Zhao and Z.M. Gao: Supercond. Sci. Technol., 2008 21 065004.
[43] Z. Q. Ma, H. Jiang and Y. C. Liu: Supercond. Sci. Technol., 2010 23 025005.
[44] J.C. Grivel, A. Abrahamsen and J. Bednarcik: Supercond. Sci. Technol., 2008 21 035006.
[45] C.M. Kipphut and R.M. German: Sci. sintering, 1988 20 31-41.
[46] Z.Q. Ma, Y.C. Liu, Q.Z. Shi, Q. Zhao and Z.M. Gao: Mater. Res. Bull., 2009 44 531-537.
[47] D.K. Finnemore, J.E. Ostenson, S.L. Bud'ko, G. Lapertot and P.C. Canfield: Phys. Rev. Lett., 2001 86 2420-2422.
[48] The Materials Information Society, Binary Alloy Phase Diagram, 2nd edn plus updates, Metals Park, OH: ASM International (1996).
[49] J.H. Kim, S.X. Dou, D.Q. Shi, et al: Supercond. Sci. Technol., 2007 20 1026-1031.
[50] C.H. Jiang, H. Hatakeyama, H. Kumakura: Physica C, 2005 423 45-50.
[51] Z. Q. Ma, Y.C. Liu, Q.Z. Shi, Q. Zhao and Z.M. Gao: J. Alloys Compd. 2009 471 105-108
[52] J.H. Kim, S.X. Dou, J.L. Wang, et al: Supercond. Sci. Technol., 2007 20 448-451.
[53] D. Yang, H. Sun, H. Lu, et al: Supercond. Sci. Technol., 2003, 16 576-581
[54] X.Z. Liao, A.C. Serquis, Y.T. Zhu, et al: Appl. Phys. Lett., 2002 80 4398-4400.
[55] V. Ganesan, H. Feufel, F. Sommer, et al: Metall. Mater. Trans. B, 1998 29 807-813.
[56] Z. Q. Ma, Y.C. Liu and Z.M. Gao: Scripta Mater. 2010 63 399-402
[57] A. E. Nielsen: *J. Crystal Growth,* 1984 67 289-310.
[58] L. M. Fabietti and R. Trivedi: *Metall. Trans. A,* 1991 22 1249-1258
[59] K. Singh, R. Mohan, N. Kaur, N. K. Gaur, M. Dixit, V. Shelke and R. K. Singh: *Physica C,* 2006 450 124–128.
[60] J. M. Rowell, S. Y. Xu, X. H. Zeng, A. V. Pogrebnyakov, Q. Li, X. X. Xi, J. M. Redwing, W. Tian and X. Pan: *Appl Phys Lett,* 2003 83 102-104.
[61] R. H. T. Wilke, S. L. Bud'ko, P. C. Canfield, D. K. Finnemore, R. J. Suplinskas and S. T. Hannahs: *Physica C,* 2005 424 1-16.

Part 7

Dielectrics and Opto-Electronic Materials

22

Reactive Sintering of Aluminum Titanate

Irene Barrios de Arenas
Instituto Universitario de Tecnología
"Dr Federico Rivero Palacio"
Venezuela

1. Introduction

The high thermal shock resistance due to the negligible thermal expansion coefficient, additional to its low thermal conductivity and good chemical resistance makes the aluminum titanate (Al_2TiO_5) a suitable material for different technological applications. It is a ceramic material consisting of a mixture of alumina (Al_2O_3) and titania (TiO_2) forming solid solution with stoichiometric proportion of the components: $Al_2O_3 \cdot TiO_2$ or Al_2TiO_5. It is prepared by heating a mixture of alumina and titania at temperature above 1350°C, in air atmosphere. Pure Aluminum Titanate is unstable at temperatures above 750°C, when the solid solution decomposes, following a eutectoid reaction, into two separate phases Al_2O_3 and TiO_2. For this reason Aluminum Titanate ceramics are doped usually with MgO, SiO_2 and ZrO_2 in order to stabilize the solid solution structure.

Unfortunately, the expansion crystal structure anisotropy that promotes the low thermal expansion coefficients provokes microcracking, as a result of anisotropy of thermal expansion along the three primary axes of the crystal lattice (a single crystal of Aluminum Titanate expands along two axes and contracts on the third one when heated), therefore a low mechanical strength and, on the other hand, the low thermal stability below 1280°C restricts its technical use.

Aluminum Titanate ceramic materials have many technological applications, among others, as thermal insulation liner, soot particulate filter in diesel engines, spacing rings of catalytic converters, in the foundry crucibles, launders, nozzles, riser tubes, pouring spouts and thermocouples for non-ferrous metallurgy and master moulds glass industries.

2. Fundamentals of low thermal expansion coefficient aluminum titanate (Al_2TiO_5) ceramics

There are two important features to achieve a very low thermal expansion coefficient, in crystalline ceramic structures highly anisotropic. The first aspect involves unit cell crystalline chemistry. The coefficients of thermal expansion of the crystal axes are controlled to develop solid solutions, in an attempt to ensure that the sum of the coefficients of the principal axes gives zero. In the case of polycrystalline ceramic materials, the volumetric thermal expansion coefficient is related to the sum of unit cells coefficients of thermal expansion. In orthorhombic crystal structures as that of the pseudobrookita, material object of this work, the relationship is:

$$\beta_v = \alpha_a + \alpha_b + \alpha_c \tag{1}$$

Where β_v = Volumetric thermal expansion coefficient
α_i = Thermal expansion coefficients of principal crystal axes

As the anisotropic crystalline structures have principal axes with positive and negative expansion coefficients, it is necessary to examine the thermal expansion coefficients of all the members of an isostructural family and chemically design a solid solution whose α_i addition is close to zero. Bayer (1971; 1973), studied the unit cell, of the pseudobrookita structure. Provided that the sum of the thermal expansion coefficients of the principal axes (α_i) add zero, it occurs an inevitable combination of positive and negative values. This condition leads to, very high (at GPa levels), micromechanical stresses at grain boundaries, during cooling from the temperatures of ceramic processing. The development of these internal stresses, promotes the breakdown of the grain boundaries, which causes a decrease in the structural integrity of the polycrystalline ceramic body. However, the existence of this microcracking depends on the microstructural grain size. Kuszyk and Bradt (1973) noted that the rigidity of the ceramic body decreased as increasing grain size, determining a critical grain size. Once determined this size, is simply necessary a process production control to achieve a compromise between the microcracking and the required structural mechanical resistance. Another possibility is to produce a material with large grain size and extensive microcracking with low mechanical resistance but where the main interest is the low thermal expansion (Hasselman, 1977; Stingl, 1986; Sheppard 1988; Huber, 1988). However, several researchers (Buessem, 1966; Cleveland, 1977; 1978) have suggested that the presence of the extensive internal microcracking, contributes to an increase in the resistance to fracture of these polycrystalline ceramics highly anisotropic, activating mechanisms such as: shielding, branching or cracks deviation. Experimentally, this hypothesis has not been demonstrated, so it is a concept that must be handled carefully.

3. Aluminum titanate (Al_2TiO_5) ceramics

The Al_2TiO_5 is a one mole Al_2O_3 and one mole TiO_2 compound. This material is conventionally prepared by reactive sintering of Al_2O_3 and TiO_2 powders. Its interest as polycrystalline ceramic material arose from the work of Bachmann (1948), who found that the thermal expansion of aluminum titanate, in the studied temperature range, could be lower than that of the vitreous silica. This material has interesting features for applications such as thermal insulator and can also withstand strong thermal gradients. This aluminum titanate emerged as a promising ceramics for technological applications; summarizing its most important physical properties, in Table 1.

The material presents two major problems: the thermodynamics instability of the Al_2TiO_5 below 1280 °C and its poor mechanical resistance related to an extensive microcracking which is, in turn, responsible for the low thermal expansion. Decomposition can be controlled or at least delayed, with oxides additions, such as MgO (Ishitsuka and col., 1987;

Wohlfrom and col., 1990) and Fe_2O_3 (Tilloca, G., 1991, Brown et al., 1994), which forms solid solutions between the Al_2TiO_5 and the isoestructurals $MgTi_2O_5$ and Fe_2TiO_5. The mechanical strength can be increased with good results preparing composite materials such as: Al_2TiO_5 - Mulita (Morishima and col. 1987), Al_2TiO_5 - Mulita - ZrO_2 (Wohlfrom et al., 1990).

Property	Al_2TiO_5	Reference
Density (g/cm^3)	3.702	Holcombe (1973)
Thermal Expansion Coefficient Average (x10^{-6} °C^{-1}) $\alpha_{a20\text{-}520} - \alpha_{a20\text{-}1000}$ $\alpha_{b20\text{-}520} - \alpha_{b20\text{-}1000}$ $\alpha_{c20\text{-}520} - \alpha_{c20\text{-}1000}$	-2.9 – -3 10.3 – 11.8 20.1 – 21.8	Wohlfromm (1990)
Thermal Expansion Coefficient Average (x10^{-6} °C^{-1}) Crystallographic $\alpha_{20\text{-}520} - \alpha_{20\text{-}1000}$ Macroscopic $\alpha_{20\text{-}1000}$ $\alpha_{20\text{-}1000}$ Anisotropy $\Delta\alpha_{20\text{-}520} - \Delta\alpha_{20\text{-}1000}$	9.2 – 10.2 1.0 – 1.5 1.5 – 1.7 23 –24.8	Stingl (1986) Milosevski (1995)
Melting Point (°C)	1860	Lang (1952)
Elastic Modulus E(GPa)	12 – 18 10 – 20 13 – 15	Stingl (1986) Cleveland (1978) Milosevski (1997)
Hardness, Hv (GPa)	5	Wohlfromm (1990)
Bending Strenght, σ (MPa)	4 – 20 25 – 40	Milosevski (1995)
Thermal shock resistance (Wm^{-1})	500	Stingl (1986)
Thermal Conductivity, k(W/mK)	1.5 -2.5	Stingl (1986) Milosevski (1997)

Table 1. Aluminum Titanate Physical Properties.

3.1 Equilibrium diagram

Lang et al. (1952) studied the Al_2O_3 -TiO_2 equilibrium diagram (Fig. 1), finding the existence of two allotropic forms of aluminum titanate: α- Al_2TiO_5, a high temperature phase, stable between 1820°C and the melting point at 1860±10°C and β-Al_2TiO_5, a low temperature phase stable from room temperature up to ≈ 750°C and from 1300°C up to inversion temperature 1820°C (at intermediate values, it has instability and decomposes to Al_2O_3 + TiO_2). The transformation between both phases is spontaneous and reversible; it was found that it is almost impossible to obtain α-Al_2TiO_5 at room temperature, being necessary cooling speeds greater than 800 K/h.)

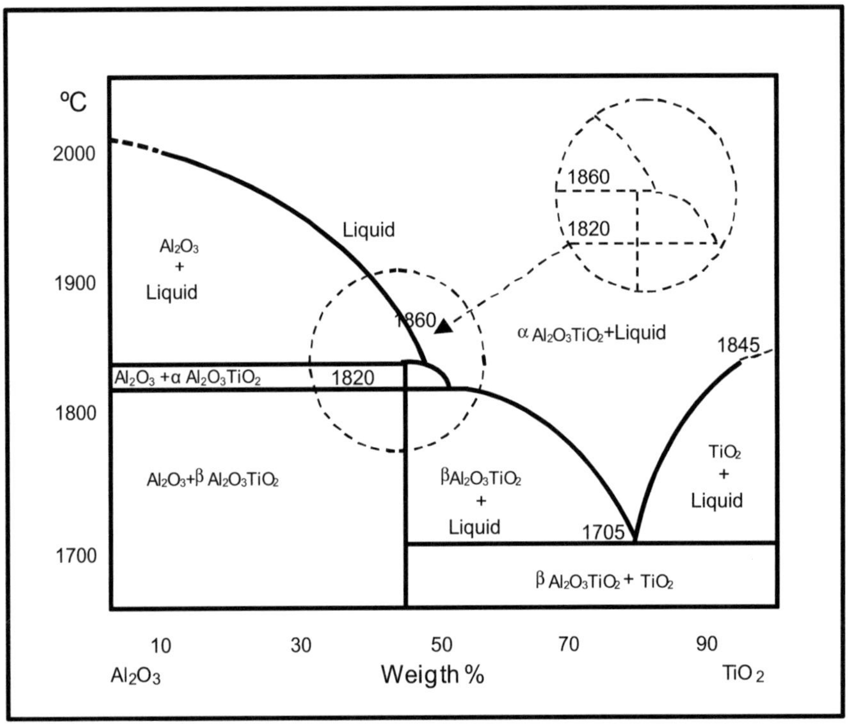

Fig. 1. Al_2O_3 -TiO_2 Equilibrium Diagram by Lang (1952).

Evidence suggests a congruent transformation of α- Al_2TiO_5 at 1860°C, but the possibility of an incongruent fusion or the existence of a solid solution between Al_2O_3 and Al_2TiO_5 could not be studied, due to the difficulties of obtaining accurate data, because of the high viscosity of liquid formed.

A second point of importance obtained from this research was the suggestion of an instability region for the β- aluminum titanate between 750°C and 130°C. This phenomenon has been confirmed by subsequent research (Fonseca & Baptista 2003). Lang et al., (1952) concluded that opposing formation and decomposition processes occurred in dynamic equilibrium, promoting decomposition at a certain temperature range, i.e. the β- Al_2TiO_5 phase is stable above 1300°C, below this temperature aluminum titanate undergoes an eutectoid transformation according to the reaction:

$$Al_2TiO_5 \Leftrightarrow Al_2O_3 + TiO_2 \quad (2)$$

3.1.1 Effect of oxygen partial pressure in aluminum titanate stability

The composition and mechanical properties of aluminum titanate are strongly influenced by the partial pressure of oxygen from the surrounding atmosphere. It is well known that the valence of the Ti cation in titanium oxides depends on the partial pressure of oxygen (Jürgen et al. 1996). At high oxygen pressures, Ti is tetravalent producing TiO_2. Due to entropic reasons, at low oxygen pressures (such as for example in air) there is a small fraction of Ti^{+3}, its amount depending on the temperature. By decreasing the partial pressure of oxygen furthermore, the Ti^{+3}/Ti^{+4} relationship increases continuously producing Ti_nO_{2n-1} Magneli phases and other sub-oxides such as Ti_4O_7, Ti_3O_5 and Ti_2O_3. The aluminum titanate phase shows similar behavior (Jürgen et al. 1996). Under high O_2 pressure the aluminum titanate is almost a stoichiometric composition phase Al_2TiO_5. Decreasing the oxygen partial pressure, due to the increase in the Ti^{+3}/Ti^{+4} ratios, it occurs a gradual interchange of Al^{+3} by Ti^{+3} in the aluminum titanate structure. This exchange results in the stoichiometric Al_2TiO_5 decomposition, to the reduced form of aluminum titanate, Al_2O_3 and oxygen according to the reaction:

$$(3\text{-}2z)[(Al^{+3})_2(Ti^{+4})_1(O^{-2})_5] = (Al_z^{+3}Ti_{1-z})_2(Ti^{+4})(O^{-2})_5+3(1\text{-}z)Al_2O_3+\tfrac{1}{2}(1\text{-}z)O_2 \qquad (3)$$

Decreasing the oxygen potential ($z \rightarrow 0$), the decomposition reaction eventually produces Ti_3O_5 titanium oxide. The degree of decomposition from Al_2TiO_5 to Ti_3O_5 can be related to a continuous change of the aluminum titanate parameter network c, (Asbrink et al., 1967).

Considering the solubility between the Al_2TiO_5 and the Ti_3O_5 under various oxygen pressures, the Al_2TiO_5 was described with a subnet model $(Al^{+3}, Ti^{+3})_2(Ti^{+4})_1(O^{-2})_5$, this model was derived from the Al_2TiO_5 orthorhombic structure (Epicer et al. 1991), taking into account the mutual exchange of trivalent cations in one subnet, while in the other subnet occupied by Ti^{+4} and O^{-2} species there is no influence.

Subsequently, Freudenberg (1987), brings together all the data obtained and proposed a modified diagram (Fig. 2). Where the only stable compound in the Al_2O_3 – TiO_2 system is considered to be, the β- Al_2TiO_5 phase; this compound decomposes above 1280 ± 1°C (Kato et al. 1980).

The Al_2TiO_5 divides the system in two sub-systems Al_2O_3 - Al_2TiO_5 and Al_2TiO_5 - TiO_2 with eutectics at titania 38.5 and 80 weight percent respectively (Fig. 2).

It is important to point out the remarkable solubility difference between corundum and titania, the Al_2O_3 in TiO_2 is ≈ one order in magnitude higher than the TiO_2 in Al_2O_3, and the Al_2O_3 and TiO_2 solubility are practically null in Al_2TiO_5; so it is considered aluminum titanate as a stoichiometric compound. However this claim is only correct for oxidizing atmospheres.

The Al_2O_3 has a 2.5% molar maximum solubility in TiO_2 (1.97 ± 0.18 in weight) at 1726°C, (Slepetys, 1969), whereas the solubility of the latter in the alumina is almost non-existent 0.35% molar between 1300 and 1700°C. While both oxides solubility in the Al_2TiO_5 is completely null (Golberg, 1968).

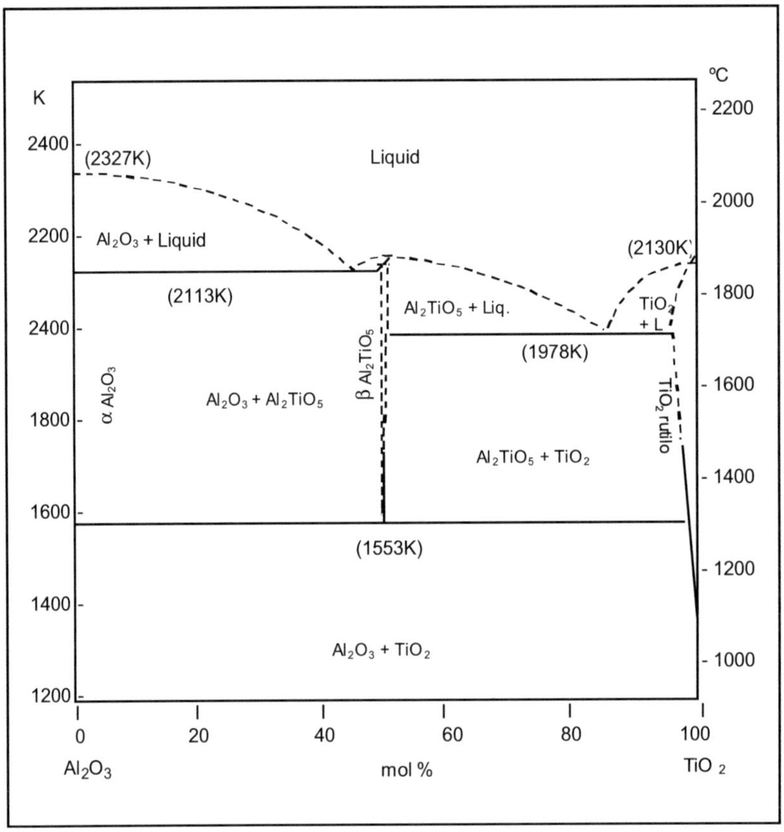

Fig. 2. Al_2O_3 -TiO_2 Equilibrium Diagram calculated in air, from experimental review by Freudenberg (1987).

4. Additives

The thermal instability of aluminum titanate and its low mechanical resistance are the main reasons for the additives use, taking into account these will influence the production process and the final product properties. An important characteristic for all additives is that they do not decrease significantly aluminum titanate thermomechanical properties. Small additions (≤ 5% by weight) are usually added with the aim of forming aluminum titanate solid solutions.

As was mentioned before, the aluminum titanate is formed above and decomposes below the equilibrium temperature 1280°C (Kato et al., 1980), with a free energy of formation given by:

$$\Delta G^o \ Al_2TiO_5 = \Delta H^o - \Delta S^o T \tag{4}$$

$$\Delta G^o \ Al_2TiO_5 = 17000 - 10.95T \tag{5}$$

The endothermic reaction is possible due to the entropy (ΔS°) positive contribution. So as other pseudobrookitas, Al_2TiO_5 can be stabilized entropically (Navrotsky 1975), with certain contributions to cation disorder (Morosin et al., 1972). It is conceivable that the positive effect of entropy can be reinforced with additional entropy in terms of mixing by the formation of aluminum titanate solid solutions. It has been determined empirically that solid solutions containing Fe^{+3} and Mg^{+2}, provide a lower decomposition temperature, i.e. increasing stability. On the other hand, solid solutions with Cr^{+3} promote a greater temperature of decomposition, i.e., reducing stability (Woermann 1985).

Jung et al. (1993), studied the replacement of Ti^{+4} by Ge^{+4} and Al^{+3} by Ga^{+3} and Ge solid solutions combined also with additions of MgO and Fe_2O_3, finding that the stabilizing effect of the additions decreased in the following order: Fe^{+3}, $Mg^{+2} > Ge^{+2} > Ga^{+3}$, corroborating data found in previous research that Fe^{+3}, Mg^{+2} are the best stabilizers so far.

Additions such as Fe_2O_3, MgO or SiO_2 were studied, the first two promoting structures of the pseudobrookites type Fe_2TiO_5 and $MgTi_2O_5$ giving complete solid solutions with Al_2TiO_5 (Brown 1994; Buscaglia et al., 1994; 1995; 1997). The SiO_2 has limited solubility (Ishitsuka 1987), however additions up to 3 weight percent produce a slight increase in the mechanical resistance, due to small amounts of liquid phase that densify the material but, larger amounts cause excessive growth of the grain that is detrimental to the mechanical resistance, (Thomas et al., 1989).

Liu et al., (1996), studied the thermal stability of Al_2TiO_5 with Fe_2TiO_5 and $MgTi_2O_5$ additions finding that material with Fe^{+3} additions did not show any significant mechanical properties decomposition or degradation and the material with Mg^{+2} annealed to 1000 - 1100°C showed an Al_2O_3 and TiO_2 breakdown.

5. Experimental procedure

The raw materials used were reactive grade: Al_2O_3 (D_{50}=0.60μm), TiO_2 (D_{50}=0.88μm), V_2O_5 (D_{50}=0.60μm), MnO (D_{50}=0.60μm), ferrosilicon (D_{50}=0.69μm), $FeTiO_3$ (D_{50}=0.82μm), and, alumina ball milled 98.5%$FeTiO_3$-1.2%SiO_2 purified mineral (D_{50}=0.88μm).

Two (2) equimolar mixtures of Al_2O_3 and TiO_2 (56wt%Al_2O_3 - 44wt%TiO_2) were homogeneously mixed with 3, 6 and 9 wt% of each additive using alumina jars and balls, during 6 hours. No binder has been added to the aqueous media powder mixture and it was dried out at 120°C for 24 hours. The material was crushed in an alumina mortar prior to the manufacture of samples by uniaxial die compaction at 300 MPa. Green bodies were reactive sintered at 1450°C, in air for 3 h. Heating was programmed at 5°C/min. whereas cooling at 15°C/min, in order to avoid eutectoid transformation: $Al_2TiO_5 \rightarrow Al_2O_3 + TiO_2$ (Kolomietsev et al.,1981).

X-ray diffraction (XRD) analysis has been performed on powders from crushed sintered samples, with grains below 30 μm suitable to obtain rigid specimens. The quantification of Al_2TiO_5 formed was determined by the internal standard method, through direct determination based on the methodology of Klug and Alexander (1954). In this study, the diffraction signals used were: Al_2TiO_5 (023), Al_2O_3 (104) and TiO_2 Rutile (110), which are representative of the three components of interest in the studied samples.

Sintered sample surfaces were carefully ceramographically prepared to minimize damage and, in some cases it was needed to chemically etch in ambient 15%HF solution for 60 s, to reveal grain boundaries. The microstructure characterization was carried out using compositional back scattered electron images (BSEI) from scanning electron microscopy (SEM) and energy-dispersive X-ray spectroscopy (EDS). Evaluation of grain size and phases present has been performed by image analysis.

In order to quantify the stabilization of Al_2TiO_5, sintered samples previously thermal treated at 1100°C for 100 h. were Si internal standard XRD analyzed, as in the as-sintered condition.

To determine the type of Fe ion in solution, it was used Mössbauer Spectroscopy with the isotope iron ^{57}Fe, in the samples with addition of ilmenite and ferrosilicon. The source used was ^{57}Co, the Mösssbauer transition is 14.41 keV, with the excited level of nuclear spin I = 3/2 and fundamental level I = 1/2. The extent of the isomeric shift provides information on the valence of the atom to which belong the core, as the electronic layers and therefore the density of electrons in the nucleus, are sensitive to chemical bonding.

Thermal expansion analysis in the temperature range of 25 to 1000°C and 1450°C, at 5°C/min heating and cooling ramps, has been performed on selected samples with good stabilization behavior.

6. Discussions of results

6.1 Additives selection

The additives selection is based on the cation radius, which must be related to the aluminum titanate cations i.e., Al^{+3} and Ti^{+4}, in order to be able to replace them in the solid solution to be formed. The thermal expansion can be related to the degree of crystalline distortion which is known to increase with the difference between the radii of the cations.

Consequently, the octahedral distortion is much greater in Al_2TiO_5 than in Fe_2TiO_5, due to the small radius of Al^{+3} ions, which facilitate a tendency to the tetrahedral coordination (Bayer, G., 1973). To avoid extreme distortion, the replacing selected cations must have a radii at least close to the Al^{+3}=0.54 Å; i.e., V^{+5}=0.59 Å; Mn^{+4} =0.76 Å; Si^{+4} =0.41 Å; Fe^{+3}=0.67 and Ti^{+4}=0.76. Hence, all additives used in this work fulfill the requirement. On the other hand, the aluminum titanate is formed by an equimolar reaction between Al_2O_3 and TiO_2. However, due to entropic reasons, in low O_2 pressure conditions, as in air for example, there is a small fraction of Ti^{+3}, its quantity being a function of the temperature. Hence, the reaction of transformation can be expressed in terms of the intermediate reaction of the titanium oxide Ti_3O_5 in the following manner:

$$\alpha\ Al_2O_3 + 1/3\ \mathbf{Ti_3O_5} + 1/6\ O_2 \Leftrightarrow Al_2TiO_5 \quad (6)$$

The $\mathbf{Ti_3O_5}$ can be seen in terms of $Ti^{+3}\ Ti^{+4}\ O_5^{-2}$ for which there is the possibility of forming "*limited solid solutions*" by cationic substitution as:

$$(1\text{-}x)Al_2TiO_5 + x\mathbf{Ti_3O_5} \Leftrightarrow (Al^{+3}{}_{1-x}Ti^{+3}{}_x)_2Ti^{+4}O_5^{-2} \quad (7)$$

Studying the affinity diagram of multivalent oxides of the transition metals with O_2, it was observed that below Ti, the order of decreasing affinity with O_2 is V, Mn and Fe (Fig.3.).

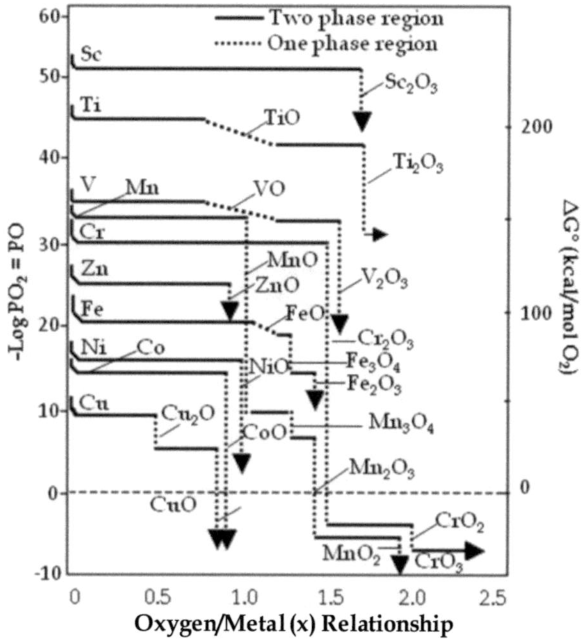

Fig. 3. Oxygen partial pressure (PO_2) vs. Oxygen/Metal (x en MO_x) for the 3d metals.

6.2 Structure analysis

The XRD results showed, for aluminum titanate without additive (Fig. 4.), that the selected temperature and time are sufficient for a near 100% Al_2TiO_5 reaction of formation, as the most important peaks correspond to this compound with a minimum of Al_2O_3 and TiO_2 remnants. It is important to point out that in all XRD, are represented the PDF values for all the constituents expected in each case, although they are not present.

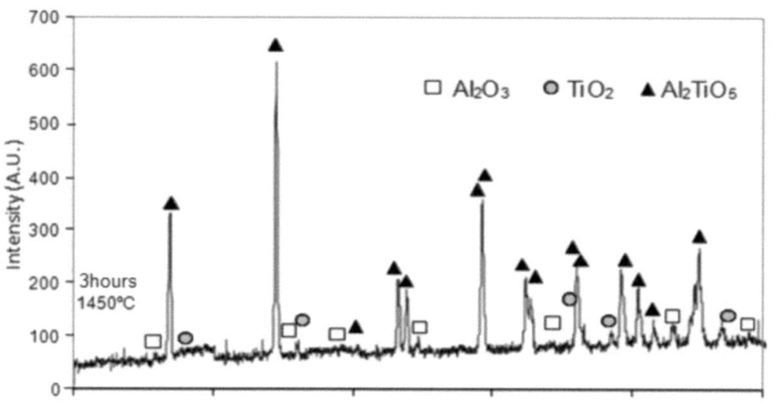

Fig. 4. X Ray Diffraction of equimolar mixture Al_2O_3 and TiO_2 without addition, sintered at 1450°C for 3 hours.

In the samples with V_2O_5, the formation of Al_2TiO_5 decreases as addition contents increase due to an intergranular liquid phase formed, identified by SEM-EDX, that inhibits the reaction between the main constituents (Fig. 5.)

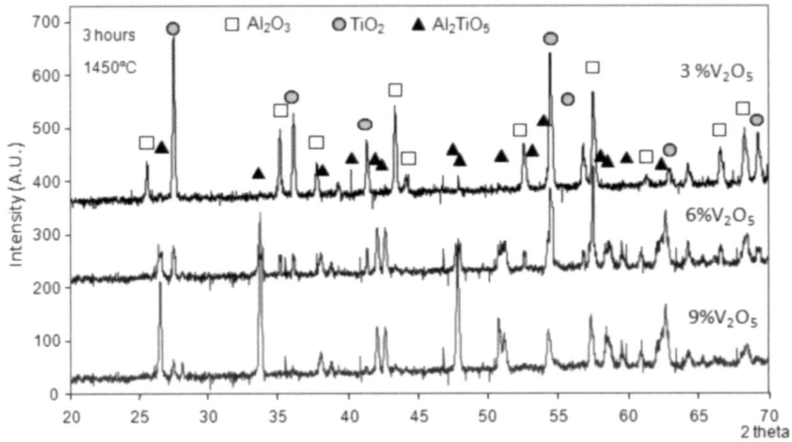

Fig. 5. X Ray Diffraction of equimolar mixture Al_2O_3 and TiO_2 with V_2O_5: 3, 6 and 9 wt% addition, sintered at 1450°C for 3 hours.

Regarding the MnO addition, it promotes Al_2TiO_5 formation with contents, as depicted by the aluminum titanate principal signals which increase in intensity whereas those of Al_2O_3 and TiO_2 are suppressed. $MnTiO_3$ appears as product of TiO_2 and MnO eutectic reaction. (Fig. 6.).

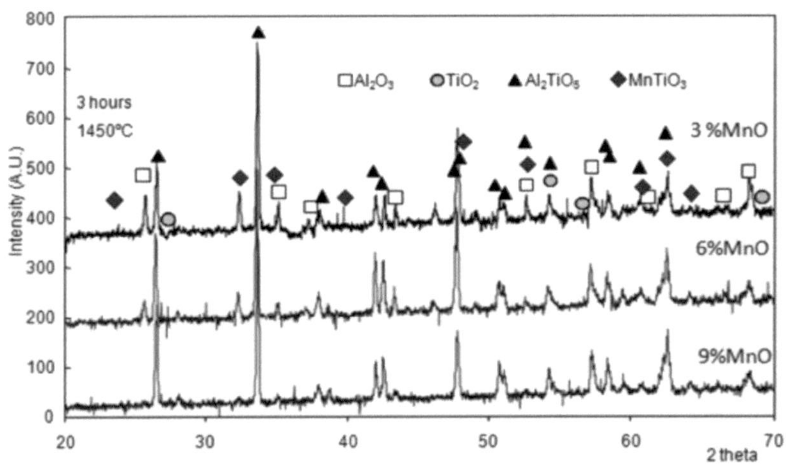

Fig. 6. X Ray Diffraction of equimolar mixture Al_2O_3 and TiO_2 with MnO: 3, 6 y 9 wt% addition, sintered at 1450°C for 3 hours.

In the case of ferrosilicon added compositions (Fig. 7), the Al_2TiO_5 formation reaction occurs but not complete and, the main signals of Al_2O_3 and TiO_2 are slightly shifted, corresponding to Al_2SiO_5 and $Al_4Ti_2SiO_{12}$ being the latter, a product of a ternary eutectic transformation.

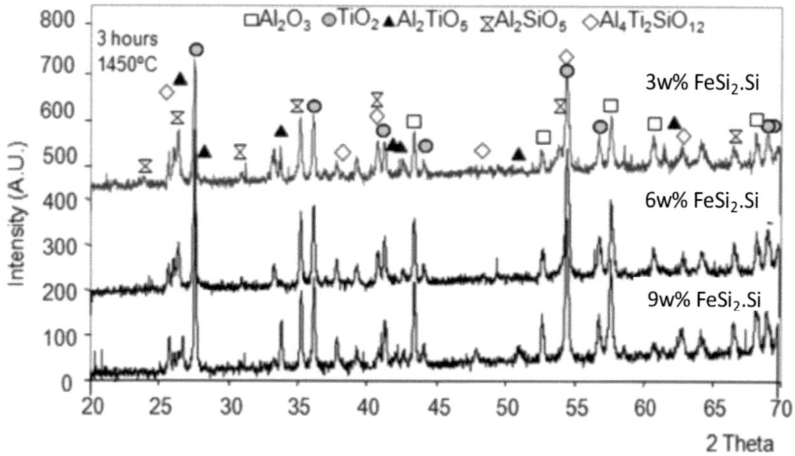

Fig. 7. X Ray Diffraction of equimolar mixture Al_2O_3 and TiO_2 with industrial $FeSi_2$.Si: 3, 6 and 9 wt% addition, sintered at 1450°C for 3 hours.

Both pure and concentrated mineral ilmenite ($FeTiO_3$) (Figs.8 and 9),promoted the formation of Al_2TiO_5 in all compositions studied. Phases such as Fe_2O_3, TiO_2, or Fe_2TiO_5 product of decomposition and reaction of the $FeTiO_3$, due to the oxidizing atmosphere, were not detected.

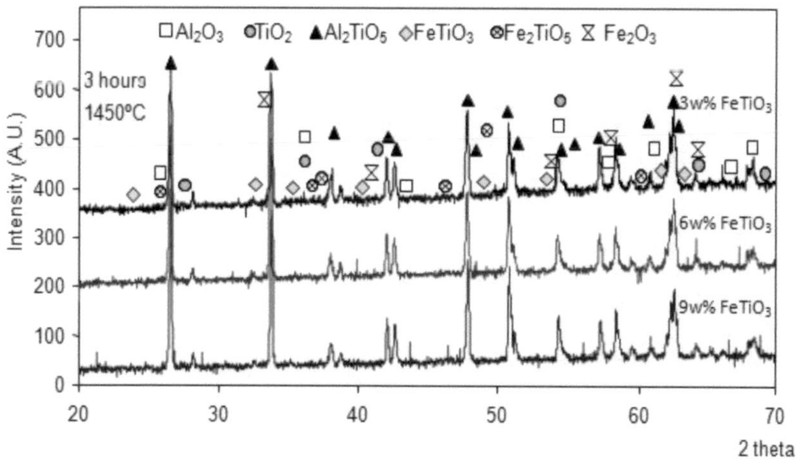

Fig. 8. X Ray Diffraction of equimolar mixture Al_2O_3 and TiO_2 with pure $FeTiO_3$: 3, 6 y 9 wt% addition, sintered at 1450°C for 3 hours.

Other remark is that expected phases, product of the reaction of contaminant SiO_2 with the parent Al_2O_3 and TiO_2, in the samples containing mineral did not show in the XRD spectra. Nevertheless, the most important reflections correspond to the Al_2TiO_5 corroborating the beneficial effect of this additive in its formation (Fig. 9).

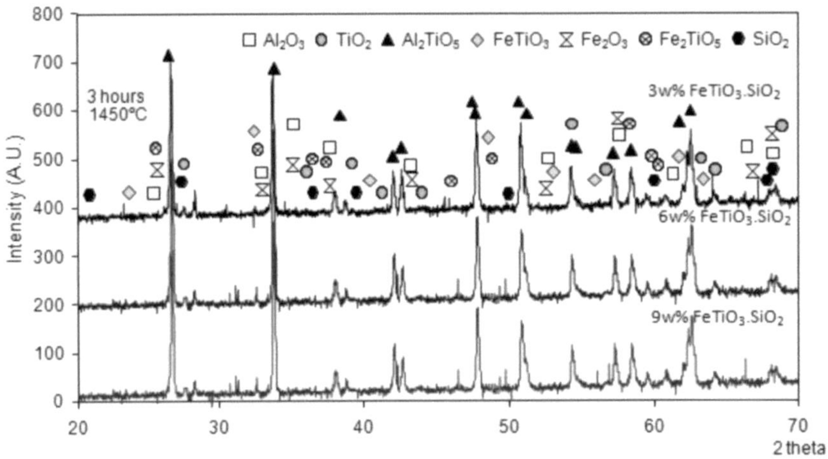

Fig. 9. X Ray Diffraction of equimolar mixture Al_2O_3 and TiO_2 with concentrated placer ilmenite ($FeTiO_3.SiO_2$): 3, 6 and 9 wt% addition, sintered at 1450°C for 3 hours.

6.2.1 Al_2TiO_5 formation phase quantification

The quantification of Al_2TiO_5 formed, was determined by the internal standard method; the achieved results are showed in Table 2.

AT+Additive (%)	%	%TiO_2 unreacted	% Al_2O_3 unreacted	%Al_2TiO_5 formed
V_2O_5	3	19.2	24.3	56.5
	6	20.1	25.5	54.4
	9	22.6	28.7	48.7
MnO	3	13.1	16.6	70.4
	6	11.2	14.1	74.7
	9	9.4	11.9	78.7
$FeTiO_3$ .SiO_2 (mineral)	3	5.5	5.8	88,7
	6	5.1	6.4	88.5
	9	5.3	628	88.5
$FeTiO_3$ (pure)	3	1.9	2.7	95,0
	6	2.3	2.8	94,8
	9	2.1	2.0	96,0
$FeSi_2$	3	21.3	27.1	51.7
	6	19.8	25.1	55.1
	9	16.9	21.5	61.6

Table 2. Al_2TiO_5 % phase formation by sintering at 1450°C/3hours.

It can be seen that the best results are obtained firstly for the pure ilmenite, secondly the mineral ilmenite, then the MnO, vanadium oxide and ferrosilicon additions respectively.

6.3 Microstructure analysis

The composition without addition (Fig. 10), shows the characteristic microstructure of the aluminum titanate: a porous and microcracked Al_2TiO_5 matrix phase and the presence of unreacted Al_2O_3 and TiO_2, due to the formation reaction kinetics, which is a process leaded by nucleation and growth of Al_2TiO_5 grains and finally the diffusion of the reactants remnants through the matrix, this is controlled for a very slow reacting species diffusion, as it was found by: Wohlfromm et al., (1991).

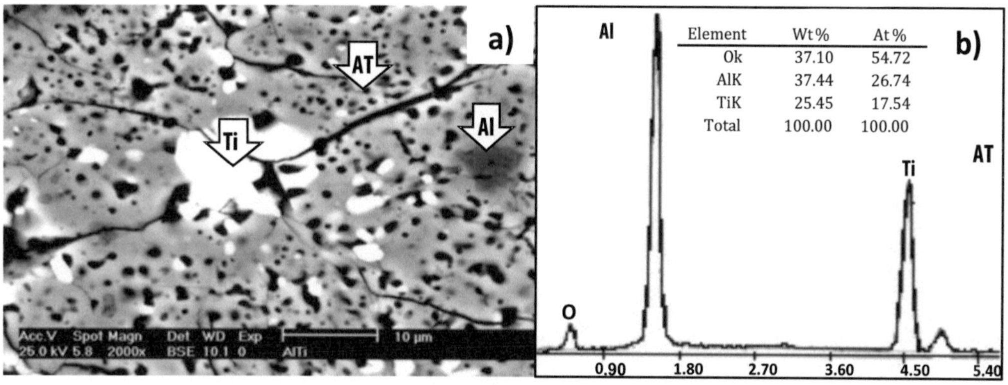

Fig. 10. a)BSE microstructure of Al_2O_3 and TiO_2 without addition, sintered at 1450°C for 3 hours, b) EDS of the matrix with an exact atomic relationship: 25 at% Al, 12.5 at% Ti and 62.5 at% O. (AT: Aluminum titanate; Ti: Titania; Al: Alumina).

The addition of the low melting point V_2O_5 (678°C) is evidenced in the microstructure with the presence of an abundant glassy intergranular phase, which constitutes a physical barrier between Al_2O_3 and TiO_2, retarding the Al_2TiO_5 formation (Fig. 11a).

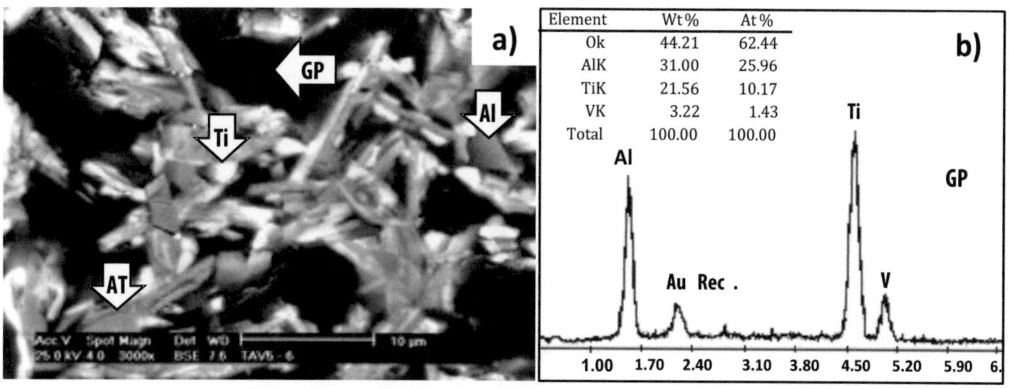

Fig. 11. a)BSE microstructure detail of Al_2O_3 and TiO_2 with V_2O_5: 6 wt% addition, sintered at 1450°C for 3 hours. b) EDS of the intergranular glassy phase (GP), appearing due to V_2O_5 low melting point. (AT: Aluminum titanate; Ti: Titania; Al: Alumina).

Althought localized EDS analysis on the glassy phase was carried out, the Al and Ti values obtained are due to the larger electron beam action volume compared to phase size (Fig. 11b).

The microstructure of MnO added samples shows extensive Al_2TiO_5 phase formation, with a minor presence of liquid phase, product of the two eutectic reactions, at 1290°C and 1330°C, between MnO and TiO_2. However, opposite to the V_2O_5 added samples, the reacting species diffusion and Al_2TiO_5 formation is accelerated with the MnO contents and, unreacted TiO_2 is absent in the microstructure, due to the secondary reactions. EDS analysis identified the intergranular eutectic phase as $2MnO.TiO_2$ and $MnO.TiO_2$ (Fig. 12). Microstructure grain size decreased with MnO contents in the sintered bodies.

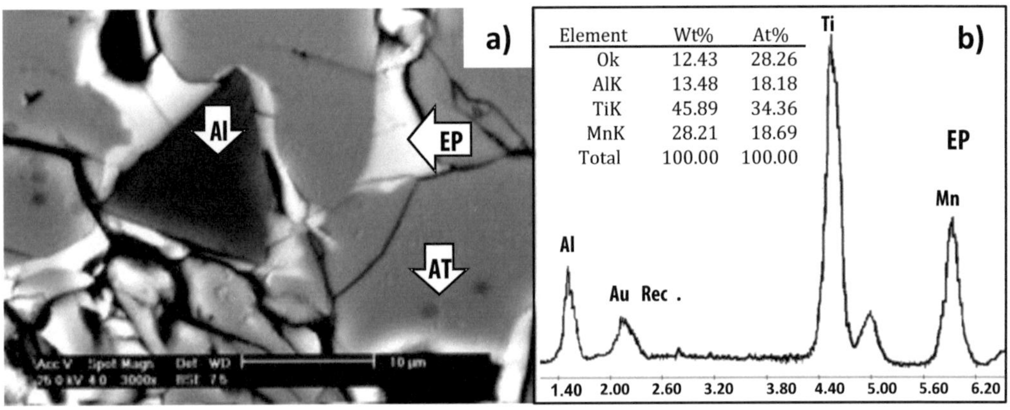

Fig. 12. a)BSE microstructural detail of Al_2O_3 and TiO_2 with MnO: 6 wt% addition, sintered at 1450°C for 3 hours. b) EDS of the intergranular $MnTiO_3$ eutectic phase (EP). (AT: Aluminum titanate; Al: Alumina).

The $FeSi_2.Si$ modified composition has a different microstructure to that obtained with other additives (Fig.13a-d).

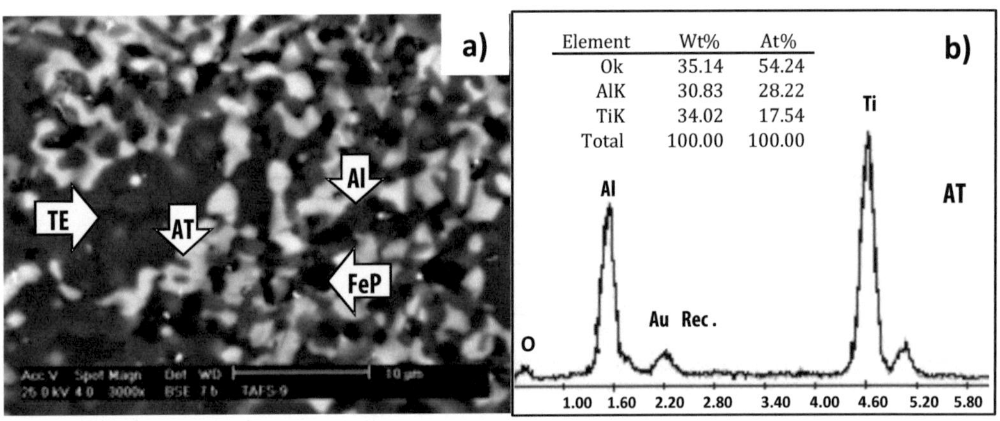

Fig. 13. a) BSE microstructural detail of Al_2O_3 and TiO_2 with $FeSi_2.Si$: 6 wt% addition, sintered at 1450°C for 3 hours. b)EDS of the Al2TiO5 matrix.

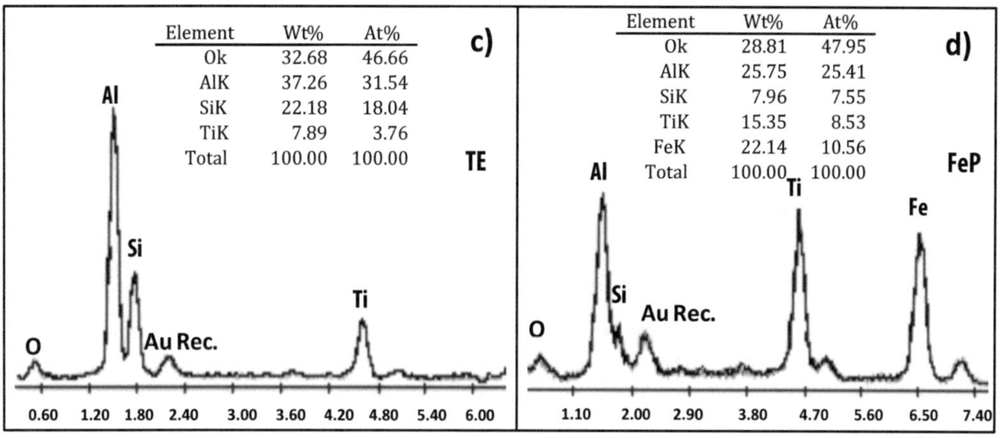

Fig. 13. c) EDS of ternary eutectic (TE) and d) EDS of the intergranular Fe rich phase (FeP).

A fraction of free Al_2O_3 remains as an intragranular phase, the TiO_2 reacts completely and besides the Al_2TiO_5 matrix phase a ternary eutectic reaction grainy phase is formed between the Al_2O_3-SiO_2-TiO_2 (as low as it is undetected by RXD), but increasing its quantity with the additive and, also a fourth intergranular phase rich in Fe is depicted (Fig.13d) (Arenas et al. 2011).

Pure and concentrated ilmenite ($FeTiO_3$), additions have a beneficial effect on grain growth control (Fig. 14 and 15). The SiO_2 left in the purified mineral promoted the formation of an intergranular liquid phase which could not be detected by XRD. Microstructures are practically free of unreacted original phases.

Fig. 14. a) BSE microstructural x1000 and a detail X3000, of Al_2O_3 and TiO_2 with pure $FeTiO_3$: 6 wt% addition, sintered at 1450°C for 3 hours. Notice the grain growth control.

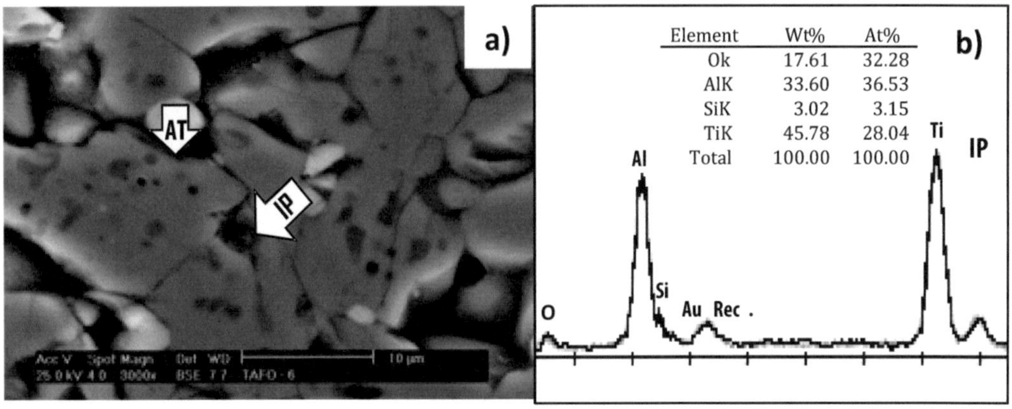

Fig. 15. a) BSE microstructural detail X3000, of Al_2O_3 and TiO_2 with with concentrated placer ilmenite ($FeTiO_3$. SiO_2): 6 wt% addition, sintered at 1450°C for 3 hours. b) EDS of intergranular phase due to SiO_2 presence.

6.4 Thermal stability

In order to determine the compositions stability, XRD analyses were performed on samples heat treated at 1100°C for 100 hours. The temperature selection is based on industrial applications working conditions and the maximum temperature for decomposition to occur.

The samples without additives and those with V_2O_5 and MnO showed a complete decomposition after heat treatment, as only the diffraction peaks of Al_2O_3 and TiO_2 were showed. (Fig. 16).

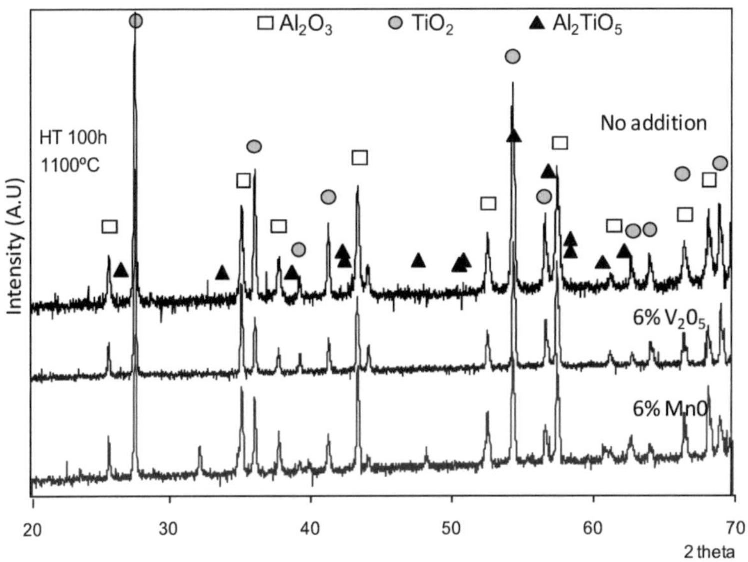

Fig. 16. X Ray Diffraction Al_2TiO_5 without addition, with 6wt%V_2O_5 and 6wt% MnO, heat treated for 100hours at 1100°C.

For $FeSi_2$.Si added samples (Fig.17), although the presence of oxidizing atmosphere, leads to the oxidation of Si and subsequent formation of ternary liquid phase between Al_2O_3, TiO_2 and SiO_2 which promotes a good densification, it has a minimal beneficial effect on stabilization. The presence of Al_2TiO_5 diffraction peaks is small if compared with that of Al_2O_3 and TiO_2, product of decomposition. It might be explained as, that only a fraction of the Fe ions from the $FeSi_2$ react and substitute the Al^{+3} ions, stabilizing the material.

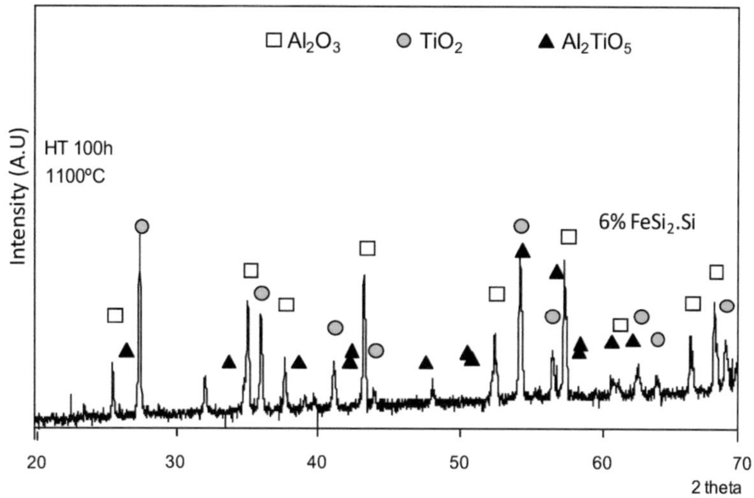

Fig. 17. X Ray Diffraction of Al_2TiO_5 with $FeSi_2$.Si 6 wt% addition, heat treated for 100hours at 1100°C.

The addition of pure $FeTiO_3$ (ilmenite) clearly shows an increase in the aluminum titanate stabilization (Fig. 18).

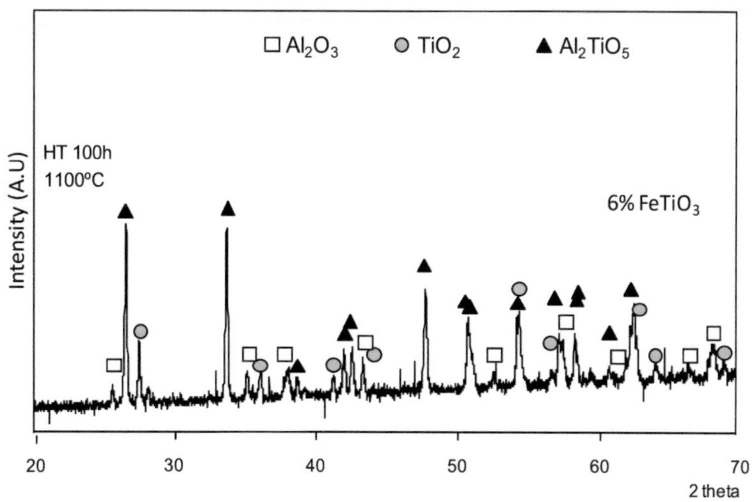

Fig. 18. X Ray Diffraction of Al_2TiO_5 with $FeTiO_3$ 6wt% addition, heat treated for 100hours at 1100°C.

This behavior agrees with the expected solid solution formation between $FeTiO_3$ and Al_2TiO_5, as depicted in the calorimetric studies (DSC). These experiments show the decomposition of ilmenite in air atmosphere to Fe_2O_3 and TiO_2 with the formation of Fe_2TiO_5 (Suresh et al.,1991), followed by Al_2TiO_5 reaction at higher temperature. This allows the possibility of a solid solution formation between Al_2TiO_5 and the isostructural Fe_2TiO_5 by a cation replacement mechanism.

From the ionic radii concept, the structural stabilization might be explained by the incorporation of Fe^{+3} (r=0.67Å) which decreases the structure distortion, caused by the Ti^{+4}: Al^{+3} radii difference (Shannon, R., 1969).

In the case of ilmenite addition, but from the concentrated mineral (Fig.19), the structural stabilization effect is evidenced only for the composition with 9% addition. This behavior might be attributed to the SiO_2 contamination which, on the other hand, benefits body densification by a liquid phase sintering mechanism.

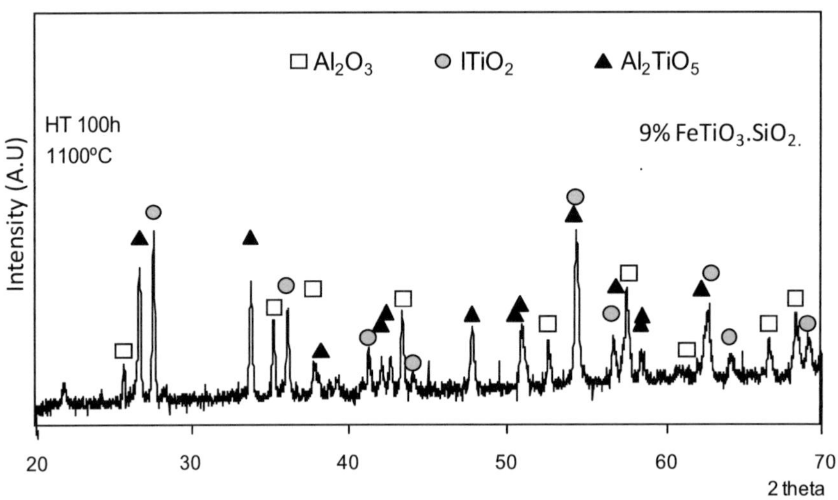

Fig. 19. X Ray Diffraction of Al_2TiO_5 with $FeTiO_3.SiO_2$ 9wt% addition, heat treated for 100hours at 1100°C.

The microstructure study by SEM-EDS corroborated the X-ray analysis evaluation (Fig. 20). All heat treated samples showed the characteristic elongated grain shape, typical of the Al_2TiO_5 decomposition into its original precursors raw materials Al_2O_3 and TiO_2. However, stabilized Al_2TiO_5 phase is observed in the compositions with 6% pure ilmenite, 9% concentrated mineral and in lower proportion in the samples with 6% ferrosilicon.

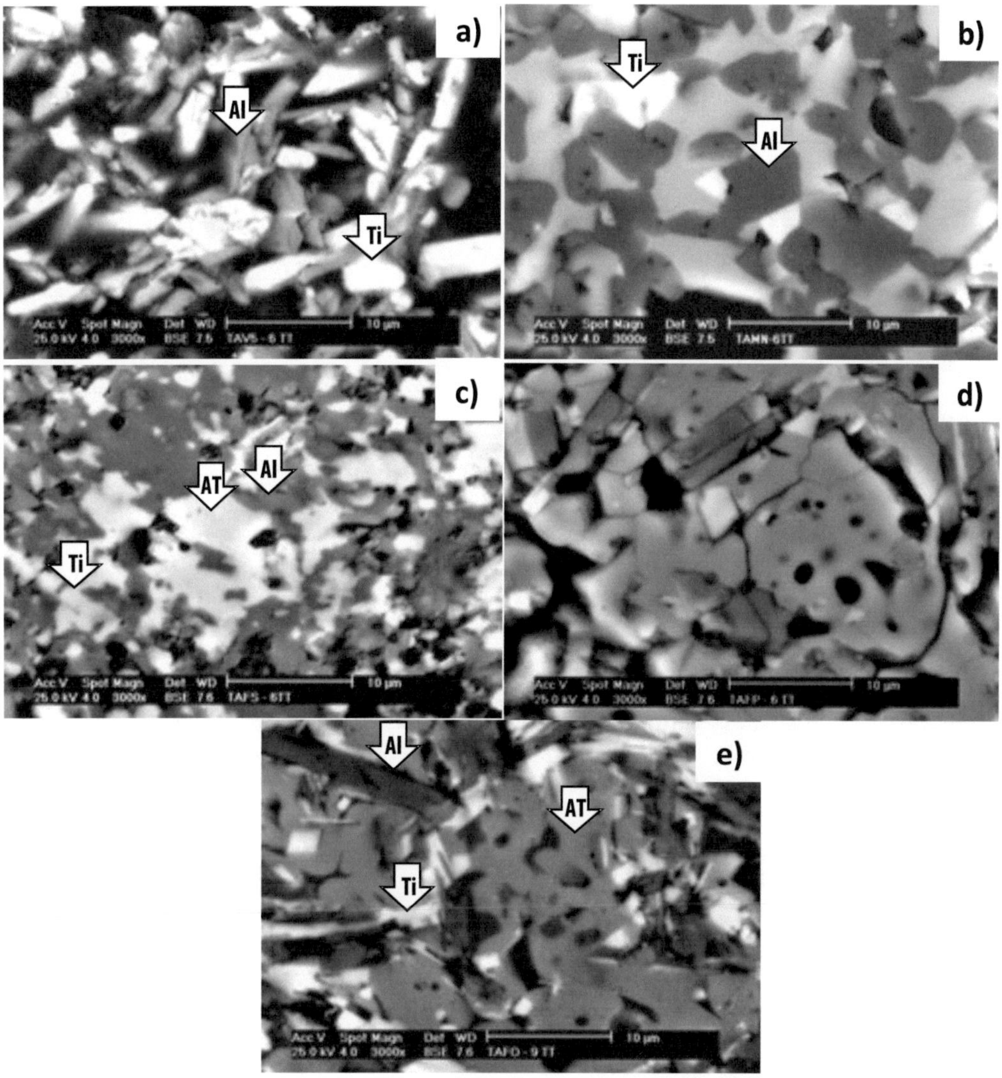

Fig. 20. a) BSE Microstructural detail x3000, Al_2TiO_5 with : a) V_2O_5 6wt% ; b) MnO 6wt%; c) $FeSi_2.SiO_2$ 6wt% d) $FeTiO_3$ 6wt% and e) $FeTiO_3.SiO_2$ 9wt% addition, heat treated at 1100°C for 100 hours. (AT: Aluminum titanate; Ti: Titania; Al: Alumina).

6.5 Al_2TiO_5 decomposition phase quantification

The quantification of Al_2TiO_5 decomposed after heat treatment was also determined using the internal standard method being the results showed in Table 3. The values obtained corroborate BSE image analysis, observations. The stabilization addition effect being higher in the compositions with pure and concentrated mineral, the MnO and the ferrosilicon stabilize slightly while vanadium oxide have not any effect.

AT+Additive (%)	%	%TiO_2 decomposed	% Al_2O_3 decomposed	%Al_2TiO_5 left
V_2O_5	3	43.86	55.82	0.32
	6	43.29	55.10	1.61
	9	43.71	55.63	0.66
MnO	3	35.69	45.41	18.90
	6	32.40	41.22	26.38
	9	32.90	41.84	25.26
$FeTiO_3 .SiO_2$	3	26.53	26.97	46.50
	6	23.06	29.30	47,64
	9	23.10	30.12	48.78
$FeTiO_3$	3	17.17	21.78	61.05
	6	10.28	13.01	76.71
	9	8.65	10.93	80.42
$FeSi_2$	3	38.01	48.36	13.63
	6	36.26	46.13	17.61
	9	33.20	42.24	24.56

Table 3. Al_2TiO_5 % Phase decomposition after heat treatment at 1100°C/ 100hours, by internal standard quantification method.

6.6 Grain size

Since the best stabilyzing behaviour after heat treatment was achieved by the ilmenites addition, these samples in the as sintered condition, were selected to determine the effect of additive contens on grains size. There were no significant variations in the grain size obtained with the two additives; however, there is a slight decrease in size as the percentage of the additive increases (in both cases), determined by image analysis of grain size (Table 4). The grain size varies between 9 and 12μm. In other words, the presence of second phase slightly inhibits the growth of grain.

$FeTiO_3$ (%)	Tg (μm)	$FeTiO_3.SiO_2$(%)	Tg (μm)
3	11.96	3	11.18
6	11.60	6	10.52
9	9.18	9	9,31

Table 4. Effect of $FeTiO_3$ (pure and concentrated mineral) on the grain size of sintered samples at 1450°C/ 3hours.

6.7 Mössbauer spectroscopy

It has confirmed the presence of the Fe^{+3} ions in all compositions with Fe added, i.e., the ilmenites and ferrosilicon.

The Mössbauer spectrum for material with 6% of ferrosilicon addition can be adjusted to two doublets (Fig.21a). The first doublet corresponds to the ferrous cation (Fe^{+2}) with a resonance that fits the hyperfine splitting with an isomer shift: IS = 1.01 ± 0.002 mm/s and a quadrupole splitting: QS= 0.664 ± 0.003 mm/s. The second doublet corresponds to the resonance of the ferric cation (Fe^{+3}) with a IS = 0. 323 ± 0.003 mm/s and a QS = 0. 520 ± 0.004 mm/s. For composition with 6% pure ilmenite addition (Fig. 21b.), it is revealed a consistent doublet with the ferric state (Fe^{+3}), with a IS = 0. 323 + 0.003 mm/s and QS = 0.520 ± 0.004 mm/s. The spectrum for the sample with 6% of mineral ilmenite (Fig. 19c), one could guess the doublet corresponds to both states ferrous (Fe^{+2}) and ferric (Fe^{+3}), however should be noted that results are not accurate, with considerable dispersion.

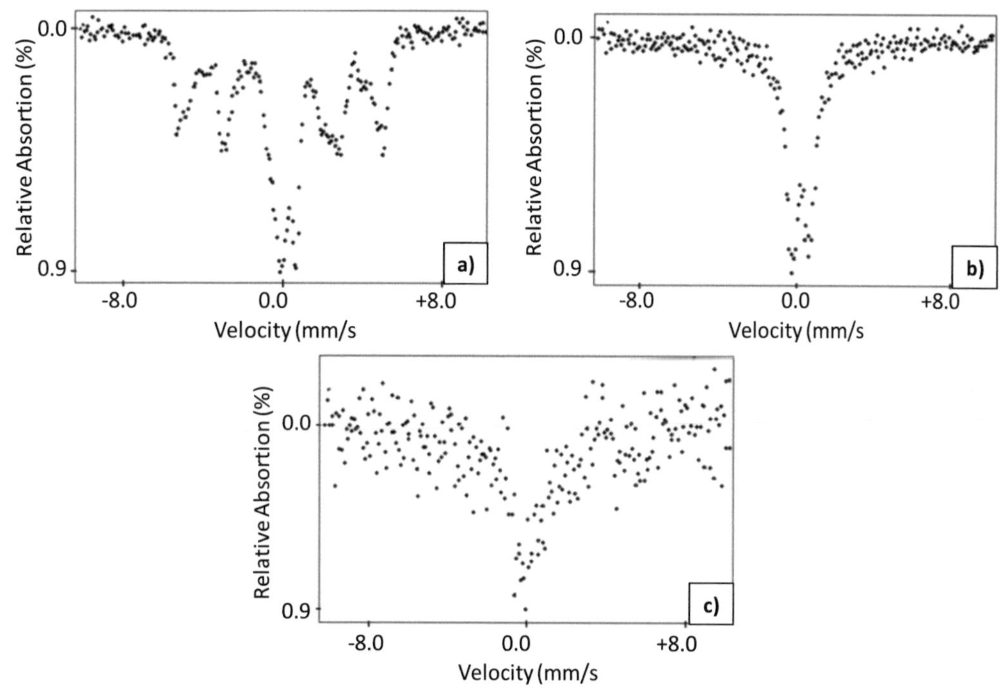

Fig. 21. Mössbauer Spectroscopy of Al_2TiO_5 with: a) 6wt% $FeSi_2.SiO_2$; b) 6wt% $FeTiO_3$ and c) 6wt% $FeTiO_3.SiO_2$ additions.

These results corroborate the possible replacement of the Al^{+3} ions by ion Fe^{+3}; there is higher stabilization in samples with pure ilmenite addition, where all Fe ions are in ferric state. However, as it was found in previous research (Barrios de Arenas & Cho, 2010) , the presence of Fe^{+2} ions, also represent the possibility of Al^{+3} ions substitution, with the creation of defects, which in turn promotes the diffusion in solid state.

6.8 Thermal expansion

Some authors have directly related the area of hysteresis in thermal expansion curves to sample microcracks density (Lingenberg W. 1985). In this study, after comparison of the ilmenite added and pure Al_2TiO_5 materials results (Fig.22), show an evident reduction in the area of hysteresis in the formers, being even more important in the samples with concentrated mineral $FeTiO_3$ addition (fig.22 b).

In both $FeTiO_3$-added samples, the property values are antagonist to addition contents.

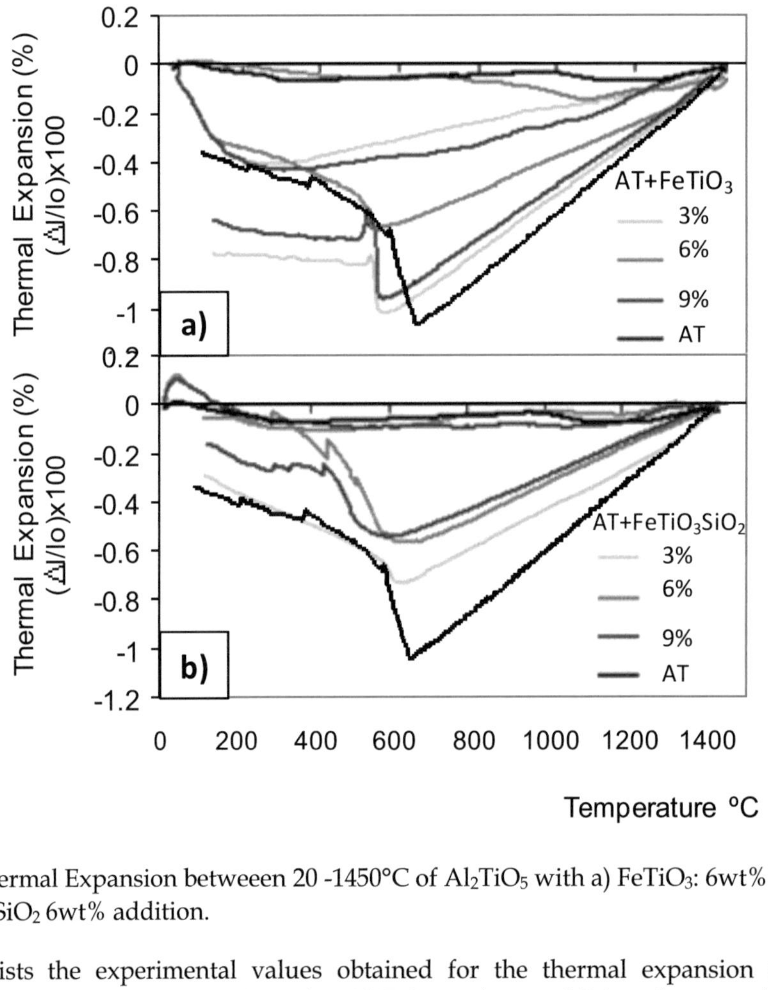

Fig. 22. Thermal Expansion betweeen 20 -1450°C of Al_2TiO_5 with a) $FeTiO_3$: 6wt% and b) $FeTiO_3.SiO_2$ 6wt% addition.

Table 5, lists the experimental values obtained for the thermal expansion coefficients between 25-1000°C and 25-1450°C. The Al_2TiO_5 without additives has a value slightly negative $\alpha_{25\text{-}1000°C}$= -0.55 x 10^{-6}°C^{-1}. Both additions turn this characteristic value into very low positive ones augmenting with the additive content in each case. This behavior could be explained by the reduction in grain size which causes a microcracking inhibition, as grains boundaries surface increases, i.e., it is necessary a higher energy for cracking (Yoleva et al. 2010).

Composition	$\alpha_{25\text{-}1000°C}$ x10^{-6} °C^{-1}	$\alpha_{25\text{-}1450°C}$ x10^{-6} °C^{-1}
Al_2TiO_5	-0.55	0.87
$FeTiO_3$ (%)		
3	0.62	0.96
6	0.76	1.05
9	0.86	1.16
$FeTiO_3.SiO_2$ (%)		
3	0.83	1.11
6	0.94	1.22
9	1.02	1.40

Table 5. Effect of $FeTiO_3$ (pure and concentrated mineral) addition, on Al_2TiO_5 Thermal Expansion.

7. Conclusion

The evident effect of $FeTiO_3$ additions on Al_2TiO_5 was established. An increase of the addition leads to a sensitive decrease on decomposition, this is due to the expected solid solution formed between Al_2TiO_5 and the isostructural Fe_2TiO_5, the latter being the product of ilmenite decomposition in the oxidizing conditions used.

The ilmenites additions produce a slightly increase in the thermal expansion coefficients being more important with the concentrated mineral addition. Although the values remain acceptable.

The MnO increases densification by the presence of a localized liquid phase, which allows the rearrangement of particles in the first stage of sintering and also slightly decomposition controls.The ferrosilicon ($FeSi_2.SiO_2$), SiO_2 reacts with Al_2O_3 and the TiO_2 forming a liquid phase, allowing rearrangement of particles; however, it lacks of stabilizing effect.

8. References

Asbrink, G. & Magneli, A. (1967) "X-Ray Studies on Some Mixed Oxides Systems of Pseudobrookite Structure", *Acta Chem. Scand* 21.

Bachmann, J. L. (1948) "Investigations of Properties of Aluminium Oxide and Some Aluminuos Materials"; *Ph.D Thesis, Pennsylvania State University.*

Barrios de Arenas I. & Cho S-A. (2011) "Efecto de la adición de ferrosilicio - $FeSi_2$ en la microestructura y estabilidad del titanato de aluminio- Al_2TiO_5." *Revista Latinoamericana de Metalurgia y Materiales* 2011; 31 (1): 11-19.

Barrios de Arenas I. (2010) "Estudio de materiales Compuestos de titanato de aluminio Al_2TiO_5 estabilizado con ilmenita Fe_2TiO_5, reforzado con Al_2O_3 y TiO_2", *Revista Latinoamericana de Metalurgia y Materiales;* 30 (2): 210-22.

Bayer, G. (1971) "Thermal Expansión Characteristics and Stability of Pseudobrookite Compounds, Me_3O_5". *Less Common Metals*, 24,[3],129-138.

Bayer, G. (1973) "Thermal Expansion Anisotropy of Oxide Compounds", *Proc. Brit. Ceram. Soc.* 22, de. D. J. Godfrey. Brit.Cer.Soc., Stoke on Trent, 39-53.

Brown, I. & Mc Gavin, D. (1994) "Effect of Iron Oxides Additives on Al_2TiO_5 Formation" Fourth Euroceramics. Vol 4. 487-492. Faenza editors. Italy

Buessem, W. R .& Lange, F. F.: "Residual Stresses in Anisotropic Ceramics", *Interceram.* 15 [3] (1966) 229-231.

Buscaglia, V.; Alvazzi, M.; Nanni, P.; Leoni, M. & Bottino, C. (1995) "Factors Affecting Microstucture Evolution During Reaction Sintering of Al_2TiO_5 Ceramics" Ceramics Charting The Future. Edit. P. Vicenzini, Techna Srl. 1867-1875.

Buscaglia, V.; Carracciolo, F.; Leoni, M.; Nanni, P.; Viviani, M. & Lematre, J. (1997) "Synthesis, Sintering and Expansion of $Al_{0.8}Mg_{0.6}Ti_{2.1}O_5$ Low Thermal Expansion Material Resistent to Thermal Decomposition" *J. Mater. Sci. 32* 6525-6531.

Buscaglia, V.; Musenich R.; Nanni, P. & Leoni, M. (1996) "Solid State Reactions in Ceramics Systems" pp. 123-26 in Proceedings of International Workshop on Advanced Ceramics´96, March 12-14, Inuyama, Japan.

Cleveland J. J. & Bradt R. C. (1978) "Grain Size Microcracking Relation for Pseudobrookite Oxides", *J. Am. Ceram. Soc.* 61 [11-12] 478-481.

Cleveland, J.J. (1977) Critical Grain Size for Microcraking in the Pseudobrookite Structure; M.S. Thesis, The Pennsylvania State University, March.

Epicier, T.; Thomas, G.; Wohlfromm, H.; Moya, J. (1991) "High Resolution Electron Microscopy of the Cationic Disorder in Al_2TiO_5". *J. Mater. Res.* 6 [1] 138-145.

Fonseca, A. P & Baptista, J. (2003) "Efecto de la estequiometria y la temperatura de cocción en el desarrollo de la fase Al_2TiO_5". Bol. Soc. Esp. Ceram. Vidrio. 42 [2] 65-68.

Freudenberg, B.: "Etude de la reaction à l'état solide: $Al_2O_3 + TiO_2 - Al_2TiO_5$", Tesis Doctoral, Eĉole Polytécnique, Lausanne 1987.

Golberg, D. (1968) "On the Systems Formed by Alumina with Several Trivalent and Tetravalent Metal Oxides, in particular Titanium Oxide" *Rev. Int. Hautes Temp. Refract.*, 5, 181-94.

Hasselman, D. P. & Singh, J. P. (1979) "Analysis of Thermal Stress Resistance of Microcracked Brittle Ceramics"; *Bull. Am. Cer. Soc.* 58, [9] 856-860.

Holcombe, C. E. & Coffey, A .L. (1973) "Calculated X-Ray Powder Diffraction Data for β-Al_2-TiO_5"; *J. Am. Ceram. Soc.* 56 [4] 220 -221.

Huber J. & Heinrich, J. (1988) Keramic im Motor; en ENVICERAM ´88, 1st. Int. Symp. And Exposition of Ceramics for Environmental Protection, 7-9 Dic. 1988, Köln, FRG.

Ishitsuka, M., Sao, T., Endo, T. & Shimada, M. (1987) "Synthesis and Thermal Stability of Aluminium Titanate Solid Solutions" *J. Am. Ceram. Soc.*, 70 69-71.

Jung, J.; Feltz, A.& Freudenberg, B. (1993) "Improved Thermal Stability of Al - Titanate Solid Solutions", *Cfi/ Ber. DKG* 70 No. 6 299-301.

Jürgen, H. & Aldinger, F. (1996) "Applied Phase Studies", *Z. Metallkd.* 87[11] 841-848.

Kato, E., Daimon, K. & Takahashi, J. (1980) "Decomposition Temperature of β- Al_2TiO_5", *J. Am. Ceram. Soc.* 63 (1980) 355.

Klug, H. & Alexander, L. (1954) "X Ray Diffraction Procedures" Edit. John Wiley, p. 410, New York.

Kolomietsev, V.; Suvorov, S.; Makarov, V. & Butalov, S.(1981) "Sintering and Some Problems of Composite Materials in the Al_2O_3 -Al_2TiO_5 System" *Refractories* 22, 627-631.

Kuszyk J. A. & Bradt, R. C.(1973) "Influence of Grain Size on Effects of Thermal Expansion Anisotropy in $MgTi_2O_5$", *J. Am. Ceram. Soc.*, 56 [8] 420-23.

Lang, S.; Fillmore, C. & Maxwell, L. (1952) "The System Berillia-Alumina-Titania: Phase Relations and General Physical Properties of Three Components Porcelains", *J. Res. Natl. Bur. Stand.*, 48 [4] 301-321.

Lingenberg, W. (1985) "Werkstoffeigenschaften von Al_2TiO_5 unter besonderer Berücksichtigung von Bildungs und Zerfallsreaktionen; Disertation, TU Clausthal.

Liu, Z.; Zhao, O. &Yuan J. J. (1996) "The Effects of Additives on Properties a Structure of Hot Pressed Aluminium Titanate Ceramics". *J. Mater. Sci. Lett.* 31 90-94.

Milosevski, M. (1997) "Thermal Diffusivity of Al_2TiO_5, $CaTiO_3$ and $BaTiO_3$" *Science of Sintering* 29 [2] 105-112.

Milosevski, M.; Ondracek, O.; Milisevska, R.; Spaseska, D. & Dimeska, A. (1995) "Thermal Expansion and Mechanical Properties of Al_2TiO_5 – SiO_2 System" Ceramics Charting The Future. Edit. P. Vicenzini, Techna Srl. 1875-1882.

Morishima, H.; Kato, Z.; Uematsu, K.; Saito, K.; Yano, T.; Ootsuka, N. (1987) "Synthesis of Aluminium Titanate - Mullite Composite having High Thermal Shock Resistance"; *J. Mater. Sci. Lett.* 6 389-390.

Morosin, B. & Lynch, R. W. (1972) "Structure Studies on Al_2TiO_5 at Room Temperature and 600°C"; *Acta Crystallogr.* Sect. B 28 1040.

Navrostky, A. (1975) "Thermodynamics of Formation of Some Compounds with the Pseudobrookite Structure and of the $FeTi_2O_5$-Ti_3O_5 So lid-Solutions Series"; *American Mineralogist* 60 249-256.

Ohya, Y. & Nakagawa, Z. (1996) "Measurement of crack volume due to thermal expasión anisotropy in aluminium titanate ceramics", *J. Mater. Sci.* 31 1555-1559.

ShannoN, R.D. & Pask, J.A. (1965) "Kinetics of the Anatase-Rutile Transformation"; *J. Am. Ceram. Soc.* 48 391-398.

Sheppard L. M. (1989) "A Global Perspective of Advanced Ceramics", *Am. Ceram. Soc. Bull.* 68 (9) 1624-1633.

Sleepetys, R. & Vaughan, P. (1969) "Solid Solution of Aluminium Oxide in Rutile Titanium Dioxide" *J. Phys. Chem.*, 73, , 2157 - 62.

Stingl, P.; Heinrich, J. & Huber, J. (1986) Proceedings of the 2nd International Symposium of ceramic Materials and Component Engines, Lübeck - *Travemunde (FRG)* April 1986. Edited by W. Bunk and H. Hausner. DKG Bad Honned 369.

Thomas, H. & Stevens, R. (1989) "Alumina Titanate - A Literature review. Part. 2 Engineering Properties and Thermal Stability"; *Br. Cer. Trans. J.* 88 184-190.

Tilloca, G. (1991) "Thermal Stabilization of Aluminium Titanate and Properties of Aluminium Solid Solutions" *J. Mater. Sci.* 26 2809-2814.

Woermann, E. (1985) Die Thermische Stabilität von Pseudobrookit-Mischkristallen; DFG-Abschlu-bericht Tialit (AZ. W. 81/23).

Wohlfromm, H.; Moya T. S. & Pena P. (1990) "Effect of $ZrSiO_4$ and MgO Additions on Reaction Sintering and Properties of Al_2TiO_5 Based Materials". *J. Mater. Sci* 25 3753-3764.

Yoleva, A.; Djambazov, S.; Arsenov, D. & Hristov, V. (2010) "Effect of SiO_2 addition on thermal hysteresis of Aluminum titanate" J. Univ. of Chem. Tech. and Met. 45 3 269-274.

23

Sintering to Transparency of Polycrystalline Ceramic Materials

Marta Suárez[1,2], Adolfo Fernández[1,2],
Ramón Torrecillas[2] and José L. Menéndez[2]
[1]*ITMA Materials Research*
[2]*Centro de Investigación en Nanomateriales y Nanotecnología (CINN).*
Consejo Superior de Investigaciones Científicas (CSIC) –
Universidad de Oviedo (UO) – Principado de Asturias
Spain

1. Introduction

There is currently a high demand for advanced materials for different types of applications (see Fig. 1.1) in which besides a high mechanical performance, a partial or total transparency in a given spectral range is required. Transparent ceramics become more and more important for applications in which materials are subject to extremely high mechanical and thermal stress in combination with optical properties. More recently, interest has focused on the development of transparent armor materials (ceramic) for both military and civil applications. Also, the development of new optoelectronic devices has extended the use of ordinary optical materials to new applications and environments such as temperature (IR) sensor, optical fiber communications, laser interferometers, etc. A considerable fraction of these new devices operates in aggressive environments, such as ovens, radiation chambers and aerospace sensors. In such cases, the sensitive electronic component must be preserved from the extreme external condition by a transparent window. Transparent and coloured ceramics are also often used as wear and scratch resistant parts such as bearings and watch glasses as well as for their aesthetic Properties in synthetic opals and rubies.

In science and technology the word transparent is used for those components that show clear images regardless of the distance between the object and the transparent window. Clear transparency is achieved when after transmission through the window the light does not undergo noticeable absorption or scattering. This applies, for example, to some glasses, single-crystalline and polycrystalline transparent ceramics.

Most of ordinary optical transparent materials, glasses, polymers or alkali hydrides are soft, weak and/or brittle. Figure 1.2 shows the transmittance spectrum of BK7, a commercial glass widely used for visible optics. However, this material presents a strong absorption in the IR range making it of no use for this spectral range. Furthermore, this material shows a very low melting point (559ºC) so it cannot be used at high temperature.

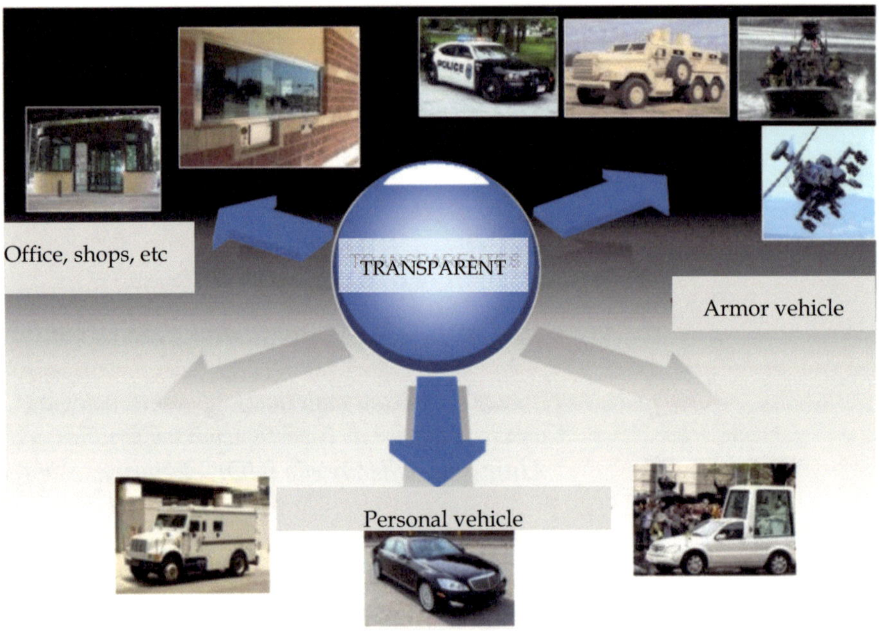

Fig. 1.1. Applications in which transparent materials are required.

However, some ceramic materials, such as corundum, spinel, yttria and YAG, do not show any absorption in a large range between IR and UV (see Fig. 1.2) and are suitable to work under extreme conditions due to their chemical stability and high mechanical performance. Many transparent ceramics are single crystal materials grown from the melt or by flame processes. However, the growth and machining of single crystals is an expensive task, which largely limits their scale up of production and, therefore, their range of applications.

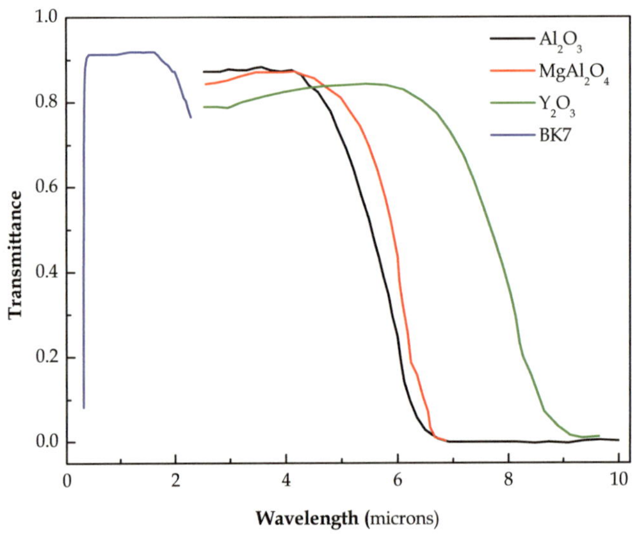

Fig. 1.2. Transmittance spectra of different materials.

Most of these problems can be solved with the use of polycrystalline materials as they show similar mechanical, chemical and thermal stability compared to single crystals and can be produced in complex shapes and are not size limited as single crystals. Among them, only a small number of materials such as MgF_2, ZnS and ZnSe are used for optical applications, but generally their use is limited to coatings and thin films and their mechanical behavior is not appropriate. For this reason, the current trend moves towards the production of polycrystalline transparent ceramics which, by providing more versatility to properties and to the production of complex shapes, will broaden the fields of application. Compared to single crystals, polycrystalline materials have a complex microstructure consisting on grains, grain boundaries, secondary phases and pores. These factors have a great influence on the optical properties, resulting in transparent, translucent or opaque materials, as can be seen in figure 1.3.

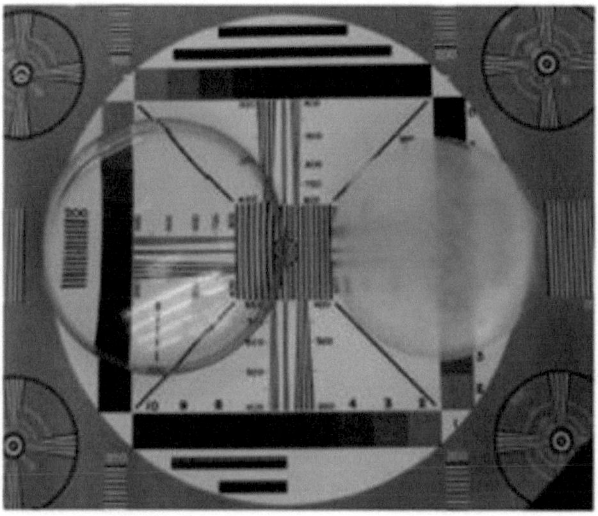

Fig. 1.3. Light scattering on polycrystalline ceramics.

In order to become transparent, the ceramic should be poreless and have optically perfect crystal boundaries and crystals. The surface of a pore is a boundary between phases with sharply different optical properties, and therefore, it intensely reflects and refracts light. A large number of pores makes ceramics opaque. Pores may be intercrystalline or intracrystalline. The elimination of intracrystalline pores, even if they are submicron in size, is a much longer process than elimination of closed intercrystalline pores. Intercrystalline pores occur at crystal boundaries which are sinks of vacancies, and this makes their removal easier. The presence of a second phase on the crystal boundaries which has different optical properties from the main crystalline phase leads to reflection and refraction of light and makes the ceramic less transparent. For this reason, transparent ceramics are obtained from raw material of high purity and the amount of additive is chosen so that they completely dissolve in the solid solution with the main phase.

The crystals in ceramics should be optically perfect, this is the absence of optical defects: pores, inclusions of the second solid phases, aggregate boundaries, and dislocations. In ceramics from optically anisotropic crystals an additional scattering of light arises on the boundaries because of their arbitrary crystallographic orientation.

The amount of light scattering therefore depends on the wavelength of the incident radiation, or light. For example, since visible light has a wavelength scale on the order of hundreds of nanometers, scattering centers will have dimensions on a similar spatial scale. Most ceramic materials, such as alumina and its compounds, are formed from fine powders, yielding a fine grained polycrystalline microstructure which is filled with scattering centers comparable to the wavelength of visible light. Thus, they are generally opaque as opposed to transparent materials. The degree of transparency also depends on the thickness of the component. These losses increase with growing thickness and result in translucent behavior. Transparency, which is independent of the thickness, is only possible for materials with an in-line transmission close to the theoretical maximum. These materials do not cause losses by absorption or scattering. These considerations lead us to the requirements to be observed by manufacturers of transparent ceramics.

2. Ceramic materials

Different materials have been proposed to prepare transparent ceramics such as alumina, spinel, yttria, AlON, YAG, etc (see Fig. 2.1).

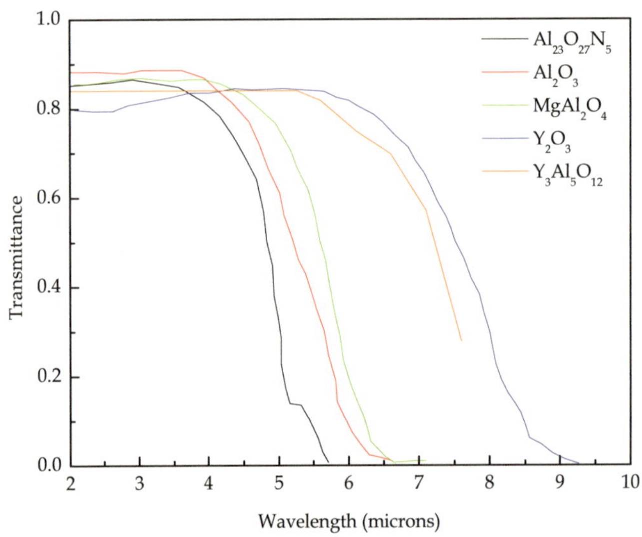

Fig. 2.1. Transmittance spectra of ceramic materials.

Aluminium oxynitride spinel ($Al_{23}O_{27}N_5$, AlON) is a cubic material and one of the leading candidates for transparent armor. This material is very useful for aircraft, missiles, IR and laser windows, etc (Mccauley et al., 2009) as it is stable up to 1200°C, relatively light weight and resistance to damage due to oxidation or radiation. AlON is optically transparent (≥80%) in the near ultraviolet, visible and near infrared regions of the electromagnetic spectrum (see Fig. 2.1). It is four times harder than fused silica glass, 85% as hard as sapphire and nearly 15% harder than magnesium aluminate spinel (table 1). The incorporation of nitrogen into aluminium oxide stabilizes a crystalline spinel phase, which due to its cubic crystal structure and unit cell, is an isotropic material which can be produced as a transparent ceramic nanomaterial. T. M. Hartnett et al (Hartnett et al., 1998) have obtained transparent AlON in the IR region by sintering at 1880ºC in a graphite oven. However, this material is not stable under low N_2 pressure and at temperatures higher than 1640ºC. Furthermore, this material is susceptible to oxide in air atmosphere at temperatures higher than 1200ºC making it not useful as projection lamps (Wei, 2009).

Material	Strength (MPa)	Knoop hardness (Kg/mm2)	Modulus (GPa)	Density (g/ml)	Melting point (K)
AlON	300	1950	323	3.68	2425
Sapphire	700	1500-2200	345-386	3.98	2300
Yttria	150	720	174	5.03	2430
Spinel	100-200	1400	273	3.58	2400
YAG	200-250	1215	300	4.55	1950
ZrO_2	210	1100	180-200	6.52	2128

Table 1. Physical properties of ceramic materials.

Zircona (ZrO_2) is a material with a cubic structure that can be used as a transparent material in different fields: domestic, industrial and military. It is not amongst the hard ceramics (table 1), and its high refractive index near 2.2 is similar to some perowskites. This may explain why few investigations have been published on sintered (polycrystalline) transparent ZrO_2. Single crystals of cubic zirconia are, however, important as artificial gemstones. Their high refractive index gives rise to a high degree of brilliance that comes close to diamonds. In the future the flexibility of powder technology in producing more complex shapes may stimulate the substitution of zirconia single crystals by sintered transparent decorative and optical products. Peuchert et al., (2009) have obtained a transparent material with 65% transmittance at 600 nm using vacuum atmosphere followed by hot isostatic pressing using titanium oxide as sintering aid.

Yttrium oxide (Y_2O_3) is another material with a cubic structure that can be used on pipes for discharge lamps or heat-resistant windows due to its corrosion resistance, thermal stability and transparency in a wide range of the electromagnetic spectrum (figure 2.1). Similarly, when introducing rare earths in its structure, it can be used as the active medium in solid-state lasers. Lukin et al., (1980) have obtained transparent yttria after sintering at 1900ºC under vacuum. Anderson (1998) has developed a ceramic $Nd:Y_2O_3$ by conventional sintering. Greskovish et al., (1973) have developed a laser $Y_2O_3:Nd^{3+}$ doped with ThO_2. La_2O_3 doped Y_2O_3 is of interest for IR applications because it is one of the longest wavelength transmitting oxides. It is a refractory with a melting point of 2430ºC and has a moderate thermal expansion coefficient (table 1). A major consideration is the low emissivity of yttria, which limits background radiation upon heating. In particular, lasers with ytterbium as dopant allow the efficient operation both in CW (Kong et al., 2005) and in pulsed regimes (Tokurakawa et al., 2007).

Magnesium aluminate spinel ($MgAl_2O_4$) is a transparent ceramic with a cubic crystal structure. This material shows an excellent optical transmission from 0.2 to 5.5 micrometers and high values of hardness (table 1) making it useful for a wide range of optical applications, including electronic and structural windows in the IR region (see Fig. 2.1). Optical quality transparent spinel has been produced by sinter/HIP, hot pressing, and hot press/HIP operations, and it has been shown that the use of a hot isostatic press can improve its optical and physical properties (Patel et al., 2000). Several authors have worked on obtaining materials of transparent spinel, as the case of Cook et al., (2005). These authors have synthesized spinel powders which subsequently formed by cold isostatic pressing and sintered in a hot pressing machine by applying a mechanical pressure of 4 MPa obtaining a transparent material. Lu et al., (2006) have obtained a transparent spinel material after sintering at low temperature and high pressure (2-5 GPa). Morita et al., (2008) have obtained a transparent spinel material after sintering at 1300°C in plasma sintering equipment (Spark Plasma Sintering, SPS), reaching 47% of transmittance at 550nm. Spinel offers some processing advantages over AlON such as the fact that spinel is capable of being processed at much lower temperatures than AlON, and has been shown to possess superior optical properties within the infrared (IR) region.

Yttrium aluminium garnet ($Y_3Al_5O_{12}$, YAG) is a polycrystalline material with has a cubic structure that can be used in structural (table 1) and functional applications. YAG has been considered a prime candidate for its use as a matrix in oxide-oxide composites in gas turbine engines (Mah et al., 2004). Also, it can be used as substrate for dielectric components, prisms and mirrors, as a part of discharge lamps, high intensity lamps as well as an active medium for the production of lasers (Ikesue et al., 1995), since it is able to accept trivalent cations in its structure, especially rare earths and transition metals. In this sense, YAG can be doped with Yb^{3+} for diode-laser (Takaichi et al., 2003), with Er^{3+} (Qin et al., 2004) widely used in medical applications or Eu^{3+} (Shikao et al., 2001) used as cathode ray tube. Finally, Nd^{3+} doped YAG is one of the most popular laser materials (Savastru et al., 2004). Due to its high thermal conductivity has been widely used in commercial, medical (ophthalmology, cosmetic, dental, etc), military and industry since its discovery in 1964. Three stable phases exist in the Y_2O_3–Al_2O_3 system: an orthorhombic perowskite$YAlO_3$ ("YAP") with an Y/Al ratio 1:1, an alumina-rich cubic garnet $Y_3Al_5O_{12}$ ("YAG"), and an yttria rich monoclinic phase with composition $Y_4Al_2O_9$ ("YAM"). Wen et al., (2004) have obtained a transparent polycrystalline material with a transmittance of 63% in the visible range and 70% in the IR region, using a solid state reaction of alumina and yttria and after sintering at 1750°C in vacuum atmosphere (see Fig. 2.1 for transparency in the IR). Reverse-strike coprecipitation is a common synthesis route for multicationic systems (Li et al., 2000a). Li et al., (2000b) have synthesized a material by the reverse coprecipitation route using aluminum and yttrium nitrates as precursors. The material was sintered at 1700°C in vacuum.

Traditionally, alpha alumina (α-Al_2O_3) has been considered one of the most widely used structural ceramics as basic matrix in many industrial applications, due to its good mechanical properties (table 1), refractory character and its chemical stability in harsh environments. Also, α-Al_2O_3 presents no absorption (see Fig. 2.1) from the near UV (>0.2 μm) until the IR range (> 5 μm). However, due to the birefringent character of alumina, an additional light scattering directly related to the grain size takes place in polycrystalline materials. This light scattering is originated at the boundaries between two alumina grains with different crystalline orientations. Two approaches can be followed to minimize/reduce this scattering: (1) increasing the texture of the polycrystalline material during the forming stage to minimize the misorientations between grains (2) decreasing the grain size as much as possible, which usually implies starting with nanometer sized alumina particles and often making use of non conventional sintering techniques, such as microwave (Jiping Cheng et al., 2002) or spark plasma sintering. However, since Coble´s work in the 60´s (Coble, 1962), the possibility of using other ceramic materials than the cubic ones for optical applications was demonstrated.

One way to avoid grain growth during sintering and obtain denser materials with reduced porosity is based on the modification of the diffusion mechanisms at high temperature by the introduction of second phases in grain boundaries or by doping with various elements that change the state of the charges at the grain boundary. In the case of alumina many additives have been proposed, mainly metal oxides (NiO, CoO, ZnO, SnO_2, Y_2O_3, MgO, ...) (Budworth, 1970). Magnesia (MgO) (Shuzhi et al., 1999) is one of the most used doping agents in the case of alumina as it inhibits the grain growth and promotes the development of a more uniform grain structure. The role of MgO has been widely studied since the work of Coble in the 60's, indicating that small additions of MgO ($\leq$ 0.25%) allowed obtaining alumina with a density close to theoretical. Bennison and Harmer (1990a) have also studied the role of MgO in the sintering of alumina, highlighting the effect it has on all the parameters that control the sintering of alumina. Handwerker et al. (1989) have also proposed that the addition of magnesia to alumina reduces the chemical heterogeneity due to the presence of impurities, reducing or eliminating the liquid phase generated by the abnormal growth of grains. Jiping Cheng et al., (2002) studied the influence of the presence of MgO on the transparency of alumina and observed that using MgO as sintering aid allows obtaining alumina samples with a more uniform grain size (equiaxed) and no porosity after microwave sintering.

3. Processing and sintering of transparent ceramic materials

Although many transparent ceramics are single crystal materials, transparent polycrystalline ceramics have different advantages such as low price, ease of manufacture, mass-production and more versatility in properties. Furthermore, the capability of producing complex shapes could broaden the fields of application.

The classical issues of ceramic processing are related with powder quality, purity, defect free processing and elimination of minor defects and pores. When the aim is to reach transparency, requirements will be similar from a qualitative point of view but noticeably more restrictive from a quantitative point of view. In addition, microstructural features such as maximum pore and grain size (the latter for birefringent materials) are critical in scattering process. The requirements for these parameters depend on the wavelength at which transparency is desired. In general, as it will be discussed in section 4, it is assumed that defect size must be $< \lambda/10$ in order to obtain transparent materials.

Then, in this chapter the processing and sintering techniques most widely used in ceramic manufacturing will be revised, describing the key parameters to be controlled and their effect on the optical properties of the final material.

Ceramic fabrication methods can be classified in different categories depending on the starting materials involved (gas, liquid or solid phase). Polycrystalline ceramics are usually manufactured by compacting powder to a body which is then sintered at high temperatures. The geometry, production volume and characteristic requirements for the component govern the choice of manufacturing process. Alternatively, ceramic materials can be simultaneously formed and sintered when pressure assisted sintering techniques are used (see Fig. 3.1).

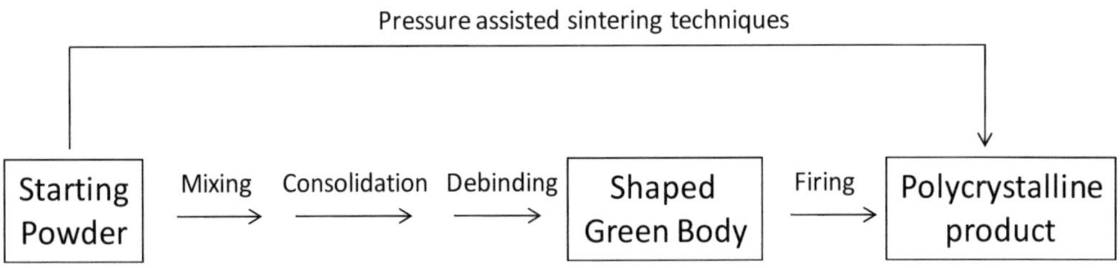

Fig. 3.1. Flow chart for the production of polycrystalline ceramics.

Polycrystalline ceramic processing

The processing steps involved in conventional fabrication of ceramics can be divided in two main parts; formation of green body and firing, and both of them must be carefully controlled in order to avoid residual porosity in the final material.

In the first part, the preparation of shaped green body from ceramic powders, three operations can be identified: mixing, consolidation and debinding. Mixing process includes the preparation of stable slurries from ceramic powders by addition of dispersants or pH controllers and the incorporation of binders and other additives. Consolidation can be done directly from wet slurries or after preparing conditioned powders. Finally, a debinding process for removing the additives used in previous actions before sintering is required.

Although the properties of the starting ceramic powders also play an important role in their behaviour during shaping and sintering, in this review the different ceramic powder synthesis methods will not be described. The description will be focused on the influence of the different processing steps starting from commercial raw materials.

Wet processing of ceramic powders is usually done in order to avoid the formation of undesired powder aggregates that could be responsible of microstructural defects in the final material. As it has been previously mentioned, when the aim is to obtain a transparent material, the critical flaw size above which an important loss of transmittance is observed is very small.

In the processing of ceramics, colloidal suspensions, consisting of a dispersion of solid particles in a liquid, are of particular interest. They are being used increasingly in the consolidation of ceramic powders to produce the green body. Compared to powder consolidation in the dry state, colloidal methods can lead to better packing uniformity in the green body which, in turn, leads to a better microstructural control during firing. Moreover, colloidal suspensions are usually prepared in order to obtain conditioned powders to be consolidated by dye pressing.

Stability of colloidal suspensions depends on particle size and their surface properties. The particles must not be too large otherwise gravity will produce rapid sedimentation. On the other hand, if the attractive force between the particles is large enough, the particles will collide and stick together, leading to rapid sedimentation of particle clusters.

Flocculation will therefore occur unless some process is promoted to produce repulsion between the particles which is sufficiently strong to overcome the attractive force. There are several ways for achieving this, but the most commonly used are:

1. Electrostatic stabilization in which the repulsion between the particles is based on electrostatic charges on the particles
2. Steric stabilization in which the repulsion is produced by uncharged polymer chains adsorbed onto to the particle surfaces
3. Electrosteric stabilization, consisting of a combination of electrostatic and steric repulsion, achieved by the adsorption of charged polymers (polyelectrolytes) onto the particle surfaces.

Rheological measurements are widely used to characterize the properties of colloidal suspensions. They can be used as a method of analysis as, for example, in determining the optimum concentration of dispersant required to stabilize a suspension by measuring the viscosity of the suspension as a function of the concentration of dispersant added.

Once the stable colloidal suspension is prepared, two alternative processes can be distinguished; direct consolidation in order to form a green body or drying under controlled conditions in order to produce a ready to press powder.

In the first case, the processes are known as colloidal forming techniques and their interest is due to the complete deagglomeration of starting powders, reducing the generation of defects. However, other limitations such as difficulties for obtaining larger or complex parts with simultaneously thin and thick cross sections due to density gradients in the green body or problems like the differential sedimentation due to the particle size distribution in the starting powders are found. Slip-casting will be described as an example of colloidal forming technique.

Slip-Casting

Slip casting, with or without pressure, constitutes an ideal combination of dewatering and shaping. As much of the slurry liquid at or near the mould surface is absorbed in the pores in the mould, a layer of solid is formed by the interlocking solid particles in the region near the mould surface. As the process continues, this solid layer increases so long as the mould pores continue to absorb the liquid of the slurry.

The biggest advantage of this process is its versatility in terms of shape, size and materials applicability. It can also accommodate a range of particle sizes, working with typical particle sizes from a fraction of a micron (green densities of ~ 40-50%) to several microns. It can also work fairly well with finer, nanoscale particles. A comparison between slip casting and uniaxial pressing of yttria ceramics show how the minimization of density gradients in green bodies prepared by slip casting allows obtaining more homogeneous materials in terms of translucency and microstructure (J. Mouzon et al., 2008).

However, slip casting has limitations, which, from an operational point of view, are primarily its slow casting rates, with thickness proportional to the square root of casting time, and hence increased cost of casting and of drying to avoid cracking, as well as costs for preparing and maintaining a large mould inventory and facilities for mould storage and drying. Using pressure to accelerate the dewatering process improves the productivity of the process. The main difference in comparison with slip casting is that the water is not removed by capillary suction (negative pressure in the plaster mould) but by pressurisation (positive slip pressure). Control of the filtration process is based on four parameters; the pressure differential on the body, the liquid-medium viscosity, the specific surface area of the slip's solids content and the body porosity (body formation is dependent on the permeability of the layer of body that has already formed from filtered material).

In order to solve those problems a great diversity of advanced forming techniques has been developed. Amongst them, aqueous injection moulding, centrifugal slip casting, direct coagulation casting, electrophoretic casting, gelcasting, hydrolysis assisted solidifications, etc can be mentioned. In all cases, the objective is to obtain a very homogenous green body in order to facilitate the preparation of defect free materials.

However, in other cases, stable suspensions of ceramic powders are prepared in order to produce dry powders especially conditioned for forming by pressing. In this case, the most critical issue is to avoid the formation of hard agglomerates that could lead to formation of defects in the final material which could not be removed during firing. Two drying techniques are especially suitable for this purpose; spray drying and freeze-drying.

Spray-Drying vs. freeze drying

The two main drying techniques for obtaining ready to press powders are spray drying and freeze drying. In spray drying process, the solvent is eliminated by evaporation when the slurry is passed through a chamber at a temperature over the solvent boiling point, whereas during freeze drying the suspension is previously frozen and then water is removed by sublimation. The two processes are similar, except for energy flow. In the case of spray drying, energy is applied to the droplet, forcing evaporation of the medium resulting in both energy and mass transfer through the droplet. In spray freeze drying, energy only is removed from the droplet, forcing the melted to solidify. Both techniques are schematized in figure 3.2.

Spray drying is the most widely used industrial process involving particle formation and drying. It is highly suited for the continuous production of dry solids in either powder, granulate or agglomerate form from liquid feedstocks as solutions, emulsions and pumpable suspensions.

There are three fundamental steps involved in spray drying.

1. Atomization of a liquid feed into fine droplets.
2. Mixing of these spray droplets with a heated gas stream, allowing the liquid to evaporate and leave dried solids.
3. Dried powder is separated from the gas stream and collected.

Spray drying involves the atomization of a liquid feedstock into a spray of droplets and contacting the droplets with hot air in a drying chamber.

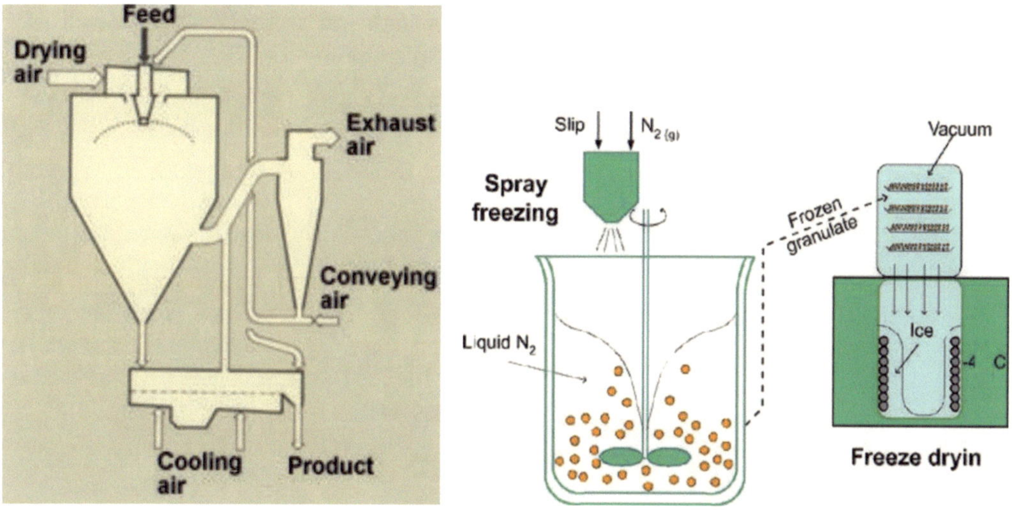

Fig. 3.2. Spray-drying and freeze-drying processes.

The sprays are produced by either rotary (wheel) or nozzle atomizers. Evaporation of moisture from the droplets and formation of dry particles proceed under controlled temperature and airflow conditions. Powder is discharged continuously from the drying chamber. Operating conditions and dryer design are selected according to the drying characteristics of the product and powder specification.

Spray formation is usually coupled to freeze drying process. This technique is named spray freeze drying. This process consists of

1. Atomization of liquid solutions or suspension using ultrasound, one-or two fluid nozzles or vibrating orifice droplet generators
2. Freezing of the droplets in a cryogenic liquid or cryogenic vapour
3. Ice sublimation at low temperature and pressure or alternatively atmospheric freeze-drying using a cold desiccant gas stream

The advantage of using conditioned powder for obtaining transparent ceramic materials is known (I. Amato et al., 1976). The enhanced behaviour during compaction of spray dried or freeze dried powders leads to a more homogeneous distribution of particles in the green body and finally a reduction in the residual porosity of the material. Nevertheless, the incorporation of additives during slurry preparation or binders for favouring soft granulation of powders makes an additional process before firing necessary. This process is named debinding.

Binders, which are used in the slip casting process or in pressing, give the green body a sufficient strength by gluing together particles at their boundary surfaces. Usually those binders are based on polyvinyl alcohols (PVA), polyacrylate or cellulose. High-polymeric compounds such as cellulose and polysaccharides work as plastification agents. They make the flow of ceramic masses during extruding possible.

The thermal treatment of the debinding process destroys the polymers by oxidation or combustion in oxygen containing atmosphere. Very often it is an uncontrolled reaction of high reaction rate inside the shaped part creating a high gas pressure, which can lead to ruptures of the compact. Debinding process is schematized in figure 3.3.

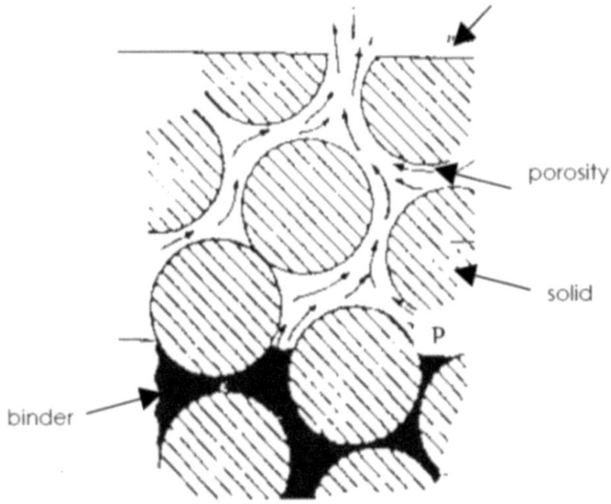

Fig. 3.3. Description of debinding process.

Debinding process is a critical step. In order to overcome the problems in thermal debinding, solvent debinding has been widely adopted by industry. In this process, a portion of the binder can be chemically removed by using solvents like acetone, trichloroethane or heptane. A large amount of open porosities, after solvent debinding, allows the degraded products to diffuse to the surface easily. A more environmentally friendly method is given by binder compositions containing water-soluble components, like polyethylene glycol.

Sintering

Finally, the green body is sintered by heating at high temperature in order to eliminate porosity and obtain the desired microstructure. The driving force for sintering is the reduction in surface free energy of the consolidated mass of particles. This reduction in energy can be accomplished by atom diffusion processes that lead to either densification of the body (by transport matter from inside the grains into the pores) or coarsening of the microstructure (by rearrangement of matter between different parts of the pore surfaces without actually leading to a decrease in the pore volume). The diffusion paths for densification and coarsening are shown in figure 3.4 for an idealized situation of two spherical particles in contact. The densification processes remove material from the grain boundary region leading to shrinkage whereas coarsening processes produce microstructural changes without causing shrinkage.

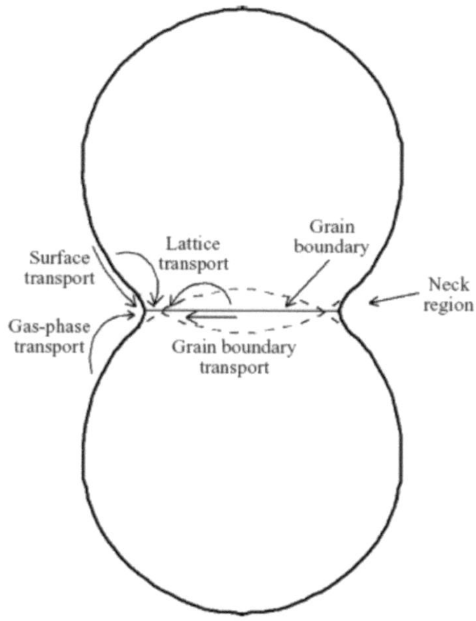

Fig. 3.4. Schematic representation of phenomena involved during firing of ceramics.

When the objective is to obtain transparent materials, it is necessary to control the microstructure evolution during ceramic sintering. The grain growth that occurs during sintering reduces the optical transmittance due to two reasons. Firstly, the pore size distribution in the sintered material is proportional to average grain size. Secondly, in the case of birefringent materials as alumina, the higher the grain size the more important the light scattering.

Grain growth inhibition can be attained by different methods. The most common strategies for reducing grain growth during sintering are the design of complex sintering cycles as two step sintering or the use of dopants for blocking coarsening mechanisms. Two-step sintering has been reported for preparing nanograin ceramics. In the process, the sample is first heated to a higher temperature to achieve an intermediate relative density (above 75%), and then rapidly cooled down and held at a lower temperature until it is fully dense (Z. Chen et al., 2008).

On the other hand, additives can impede the grain growth of ceramics during sintering. In particular, there are many studies treating the alumina grain growth inhibition using different dopants (S.J. Bennison et al., 1990a) (I. Alvarez et.al, 2010). An innovative method for controlling alumina grain growth is described by M. Suárez et al., (2011). In this case, the dopant used is an alumina precursor in order to obtain pure alumina after sintering. The effect of alumina precursor doping on process kinetics and microstructure evolution is discussed.

However, the residual porosity of materials after conventional sintering is usually high enough to reduce noticeably their transmittance or even lead to opaque materials. Thus, an additional process for removing that porosity is usually needed. An effective method for attaining complete densification is the post-HIP treatment. In hot isostatic pressing, the sample predensified, until only closed porosity is present, is placed in a pressure vessel. The pressurizing gas is introduced by means of a compressor to achieve a given initial gas pressure, and the sample is heated to the sintering temperature. During this time the gas pressure increases further to the required value and collapses around the sample, thereby acting to transmit the isostatic pressure to the sample. The use of this technique for improving transparency has been shown by different authors (K. Itatani et al., 2006 and M. Suárez et al., 2010).

Pressure assisted sintering techniques

The simultaneous application of pressure and heat is also used in the Hot-Pressing and Spark Plasma Sintering, methods that are known as pressure assisted sintering techniques. In this case, a mechanical uniaxial pressure is applied to the sample placed in a dye by a vertical piston while the system is heated. Graphite is the most common die material because it is relatively inexpensive, is easily machined, and has excellent creep resistance at high temperatures. The rate of densification can be deduced by following the piston displacement. Typically, in Hot- Pressing, the sintering temperature is chosen to achieve full densification in 30 min to 2 h. Some guidance for selecting the appropriate hot pressing temperature may be obtained from pressure sintering maps, but trial and error is usually done. Pressure is often maintained during the cooldown step.

A sintering method with a configuration similar to Hot-Pressing has been recently developed. It is named Spark Plasma Sintering (SPS) and its main characteristic is that a pulsed DC current is directly passed through the graphite die while uniaxial pressure is applied. The characteristics include (a) high heating rate, (b) the application of a pressure, and (c) the effect of the current. The main advantage in comparison with other sintering techniques is the high heating rates that can be applied during sample sintering. The description of SPS process and a comparison of cycle duration with Hot Pressing are shown in figures 3.5 and 3.6

These features have a great influence on the characteristics of the materials obtained. The extremely short processing times allow obtaining special microstructures in the final material that are unattainable by other sintering techniques. Thus, it is possible to fabricate a dense material with an average grain size similar to starting powders. There are many scientific works showing the potential of this technique for obtaining a wide diversity of transparent polycrystalline materials (B. N. Kim et al., 2007), (C. Wang et al., 2009), (G. Zhang et al., 2009), (R. Chaim et al., 2006). Nowadays, one of the challenges related with this technique is the scaling-up in order to obtain large samples or complex shape components.

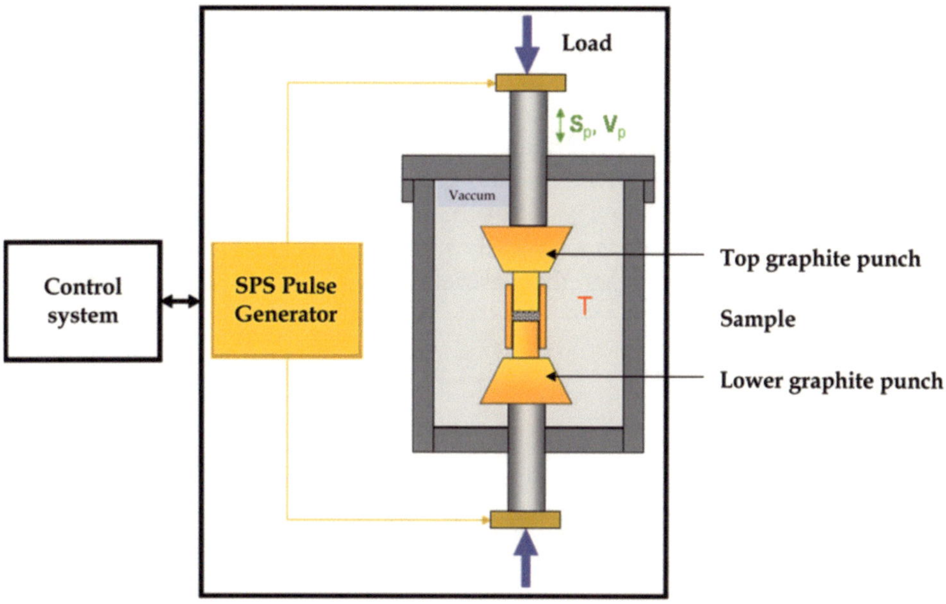

Fig. 3.5. Description of SPS process.

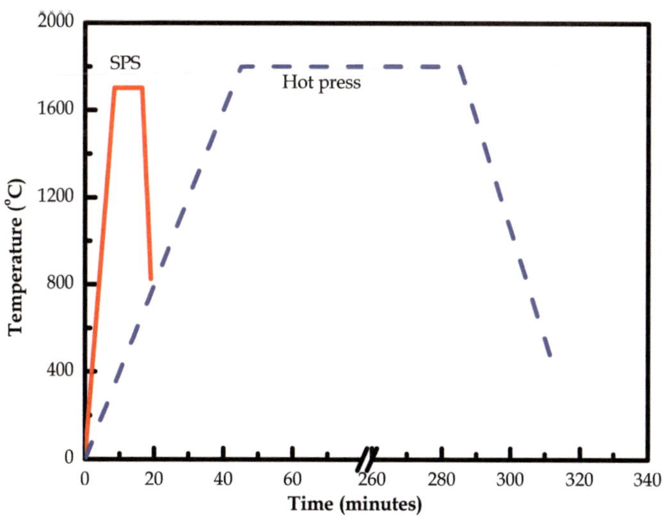

Fig. 3.6. Comparison of cycle duration between Hot-Pressing and Spark Plasma Sintering

4. Characterization of transparent materials

Transparency represents the ability of materials to allow the transmission of light through them. When traversing a transparent material, a light beam propagates along the same direction before and after traversing the material (see Fig. 4.1a). However, perfectly transparent materials are rare in nature and most of them present the so-called scattering centres. When the concentration of scattering centres is sufficiently large, the material still allows the transmission of light, but the transmitted light beam does not only propagate along the incident direction, but diffuse, off-scattered, beams can be detected; in this case the material becomes translucent (see Fig. 4.1b). In this work, we shall refer only to the transparency of a material, which is measured as the real in-line transmittance. When the amount of scattering centres is higher, the materials scatter so much light that they become opaque (see Fig. 4.1c).

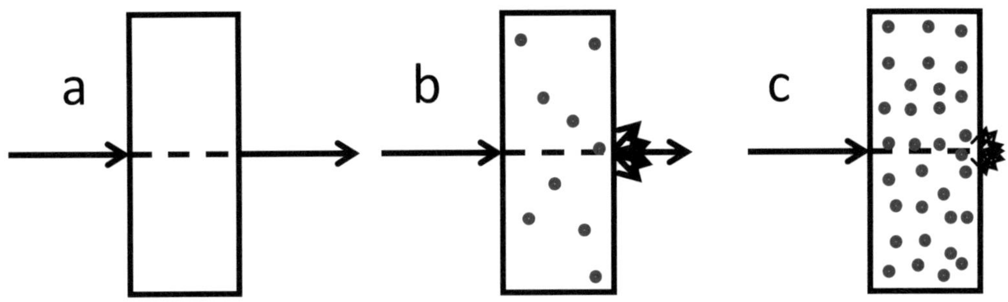

Fig. 4.1. Light propagation through a transparent (a), a translucent (b) and a highly dispersive, almost opaque (c) material.

Therefore, it becomes necessary to define a measurement technique that allows distinguishing between translucency and transparency. In this case, the relevant parameter is the real in line transmittance (Appetz & Bruggen, 2003) or RIT. In this case, a collimated light beam impinges on the sample and the detector is placed far away, usually 1 m, from the sample. This way, for ordinary detectors and light beams, the light scattered > 0.5° will not be detected. Any non-absorbing material must be spatially homogeneous with respect to its dielectric properties in order to become transparent. From an electromagnetic point of view, a defect, or scattering centre, is a spatial region in which a difference in the diagonal dielectric constant (refractive index) is present. The effect of dielectric heterogeneities is manifested through light scattering phenomena, which leads to losses both in the optical quality of the materials and in the total transmitted energy. In ceramics, the main scattering sources are given by the presence of pores, second phase inclusions, rough interfaces… (see Fig. 4.2). Whereas the roughness at interfaces can be minimized by an adequate polishing of the surfaces and the presence of second phases can be also made negligible by selecting pure materials, the presence of pores and the effects of grain boundaries can only be minimized by an adequate processing and sintering. In most cases, and particularly in cubic materials such as yttria, YAG, AlON, spinel, pores are by far the main source of scattering and most of the efforts in the literature have been devoted to minimization of this kind of scattering. In non-cubic materials, of which alumina is the most intensively studied material, there is an additional scattering given by the dependence of the value of the refractive index with the orientation of the crystalline grain, which we shall refer to hereafter as scattering due to grains.

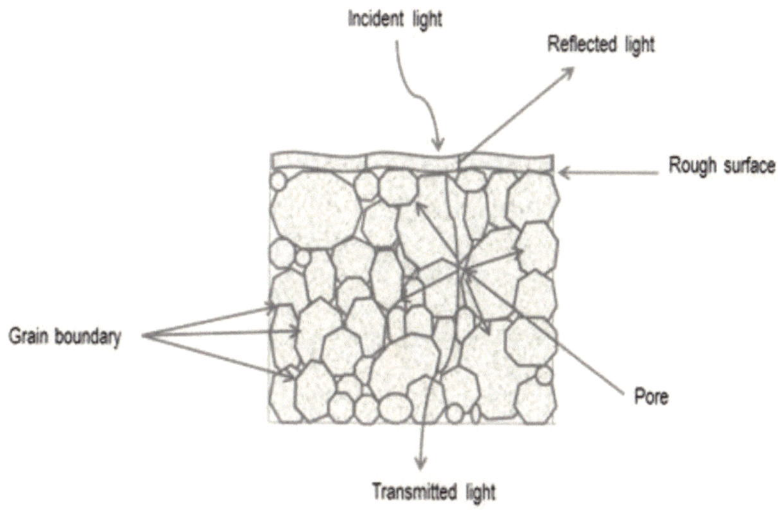

Fig. 4.2. Light scattering sources in polycrystalline ceramics.

Whenever a light beam reaches a surface separating two media with different refractive indices (see Figure 4.3), it deviates according to the Snell's law:

$$n_1 \cdot \sin\theta_1 = n_2 \cdot \sin\theta_2 \tag{1}$$

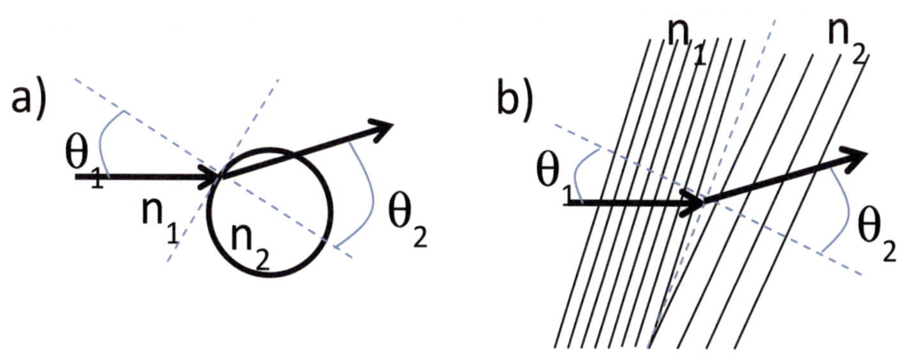

Fig. 4.3. Light refraction at (a) a matrix-pore boundary and (b) between two grains with different crystallographic orientations in a birefringent polycrystalline material.

A model based on geometrical optics was first produced by A. L. Dalisa and R. J. Seymour in which a spread function describing the effect on an incident collimated beam of randomly oriented spatial regions with a difference in the refractive index between zero and $\pm\, \Delta n_{max}$

was developed for ferroelectric ceramics (Dalisa & Seymour, 1973). This model was further applied by R. Appetz et al. to study the off-specular transmittance in polycrystalline alumina (Appetz & Bruggen, 2003), a birefringent material for which n_1 can be taken as 1.768 and n_2 as 1.76. Considering a grain boundary placed at 45°, the deviation of the light beam at this interface is calculated to be around 0.26°. If a beam light traverses a 1 mm thick polycrystalline alumina sample, in which the grain size is around 0.5 μm, the number of refractions that a light beam undergoes is not below several thousands. This gives an idea of the large scattering undergone by the beam. This discussion also illustrates why the larger the number of pores (for the same pore size), the smaller the real in line transmittance. However, following the same reasoning, one would expect the real in line transmittance to be higher for increasing grain size, whereas the experimental observations (Appetz & Bruggen, 2003) indicate that the amount of light scattered out of the normal incidence is larger in a sample with a grain size of 20 μm than in one with a grain size of 1 μm (Figure 4.4). This indicates that the geometrical model used above is not valid anymore to study the scattering of objects with dimensions of the order of the wavelength of the light.

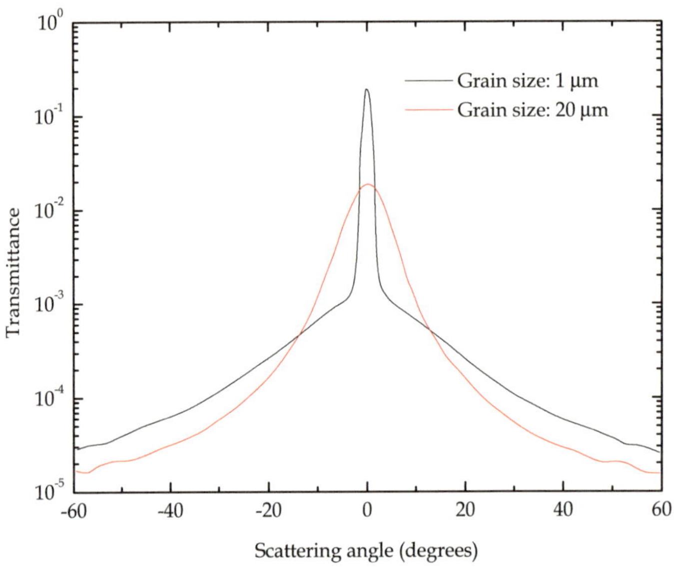

Fig. 4.4. Angular scattering profiles corresponding to 0.8 mm thick polycrystalline alumina samples with 1 μm (solid line) and 20 μm (dotted line) grain size (taken from Appetz & Bruggen, 2003).

In systems with a few defects and by using the Mie equation (Mie, 1908), it is possible to determine the intensity and directionality of the scattered radiation. This formulation is derived from the Maxwell equations and it is, therefore, valid for arbitrary defect sizes. However, as the Mie equation is complex to handle, it becomes helpful to use approximations. Two approximations are often used (Bohren & Huffmann, 2010; van de Hulst, 1957): Rayleigh-Gans-Debye scattering for large particles: $2\pi r >> \lambda$ and Rayleigh scattering for small particles: $2\pi r << \lambda$, with r being the radius of the scattering centre. When dealing with transparent materials or materials close to transparency, the number of pores and their dimensions verifies the conditions for the second approach (Rayleigh scattering) and, therefore, we'll focus our analysis under that simplification. According to this well-known approach, the intensity of electromagnetic radiation with wavelength λ scattered by an object of diameter *d* and refractive index *n*, at a distance *R* and angle of scattering *θ* is given by:

$$I = I_0 \frac{1+cos^2\theta}{2R^2}\left(\frac{2\pi}{\lambda}\right)^4\left(\frac{n^2-1}{n^2+2}\right)^2\left(\frac{d}{2}\right)^6 \quad (2)$$

The Rayleigh scattering cross section for a single particle can be expressed as:

$$\sigma_s = \frac{2\pi^5}{3}\frac{d^6}{\lambda^4}\left(\frac{n^2-1}{n^2+2}\right)^2 \quad (3)$$

In case a group of N scattering particles is considered, the scattering cross section is given by N times the single scattering cross section. This implies that the scattering due to pores of diameter d in a ceramic is much larger, by a factor of 2^6, than that due to pores of diameter d/2. If the size of the pores is reduced to half the initial size, the individual scattering is reduced by a factor of 2^6, whereas the number of scattering centres must be increased by a factor 2^3 if the total porosity is kept constant. This implies that, in total, by reducing the pore size by a factor of 2, the total dispersion is reduced by a factor 2^3, almost an order of magnitude. Therefore, in order to improve the real in line transmittance, two approaches can be followed: the porosity is strongly reduced or, as shown above, the size of the pores is kept much smaller than the wavelength of the radiation used.

Since the pioneering works by Peelen and Metselaar (Peelen and Metselaar, 1974), lots of efforts have been devoted to analyse and model the effects of the different scattering sources on the real in line transmittance. It is well known that the real in line transmittance decays exponentially with thickness, *d*, according to:

$$RIT = (1 - R_s)^2 \cdot \exp(-\gamma d) \quad (4)$$

where γ, the total scattering, is the sum of the scattering coefficients due to grains and pores: $\gamma = \gamma_{gb} + \gamma_p$. On the other hand, R_s accounts for the total reflected light at the air-material and material-air boundaries. It is well known that the amount of light reflected from a surface between two media (Born & Wolf, 2005) with refractive indices n_1 and n_2 at normal incidence is given by $R_s=(n_1-n_2)^2/(n_1+n_2)^2$. In this case, one of the media is air (n_1=1) and the other medium is the material under study (n_2=n). R. Appetz and M. P. B. van Bruggen calculated the light scattering coefficients due to pores and different grain orientations in non-cubic materials, considering just one pore size and one grain size (Appetz & Bruggen, 2003). The expressions given for those coefficients are:

$$\gamma_{gb} = \frac{3r_{gb}\pi^2}{\lambda_0^2}\Delta n_{gb}^2 \text{ and } \gamma_p = \frac{p}{\frac{4}{3}\pi r_p^3}\frac{3r_p\pi^2}{\lambda_0^2}\Delta n_p^2 \quad (5)$$

where r_{gb} (r_p) is the radius of the grain (pore); Δn^2_{gb} (Δn^2_p) is the difference in the refractive indices between grains in birefringent materials (pore and matrix); p is the total porosity of the material and λ_0 is the wavelength of light used in vacuum. More recently, this analysis was extended to a distribution of pore and particle sizes (Suarez et al., 2009). In Figure 4.5, the effect of considering a distribution of pores instead of a single pore is given. For this simulation, a Poisson pore size distribution with an average radius of 30 nm was taken into account. It can be observed that the main differences take place at short wavelengths, that is, in the visible part of the spectrum for the numbers considered in this example.

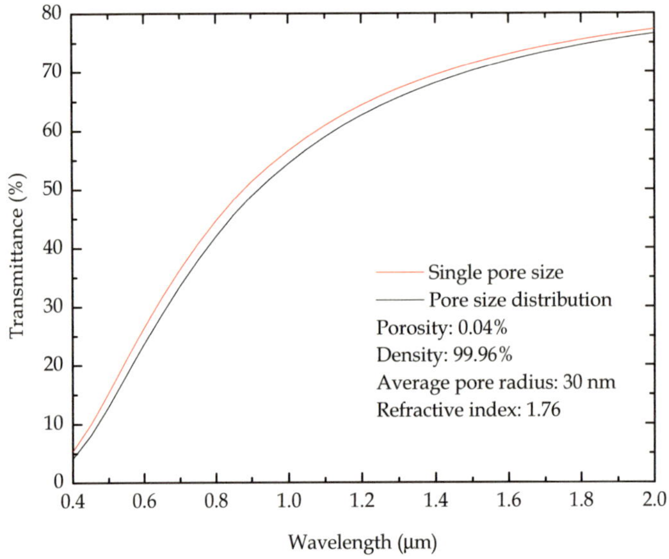

Fig. 4.5. Transmittance spectra in a material with a single pore size or a Poisson distribution of pore sizes.

In Figure 4.6, the effect of changing both the porosity for a fixed pore size (4.6.a) and the pore size at a fixed porosity (4.6.b) is simulated according to the formalism developed by Suarez et al. (Suarez et al., 2009). Some remarkable features should be highlighted: as shown in Figure 4.6.a, for a pore radius of 10 nm, in a 1 mm thick material and at a wavelength of 600 nm, i. e., in the visible range, a porosity of 0.5%, which corresponds to a density of the material of 99.5%, implies that the RIT decays to almost zero. Actually, values of the porosity over 0.3% lead to RIT values so small that their detection becomes very complicated for usual detectors. On the other hand, at the same wavelength and considering a porosity of only 0.05%, which corresponds to a density of 99.95%, it is shown that pore sizes over, approximately, 50 nm lead to very small RIT values. These two results combined imply that achieving a high density (>99.9%) is not usually enough to obtain a transparent material. It is also necessary to keep the values of pores below a given value that, according to these simulations, can be established below 30 nm.

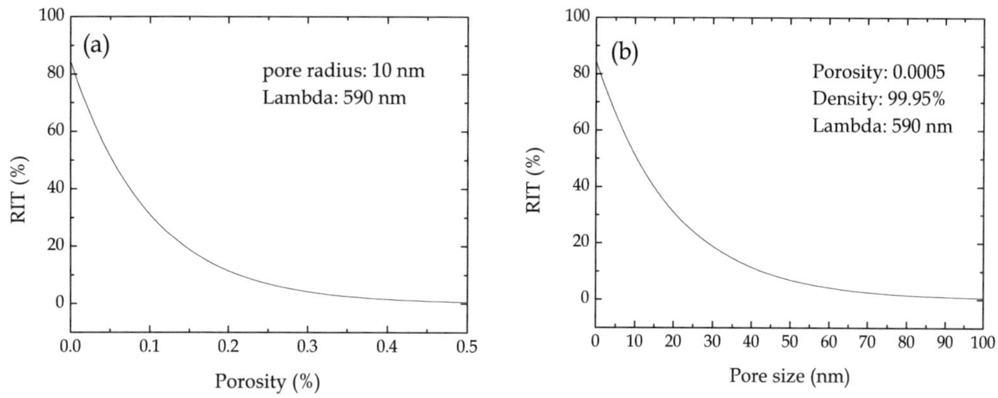

Fig. 4.6. Simulations considering different porosities and pore sizes.

The simulations shown above indicate the critical role of pores when transparent ceramics are being pursued. However, the situation can become more complex when the crystalline structure of the material considered is not cubic. In this case, the diagonal elements of the dielectric tensor will be, in general, different. In other words, the refractive index will depend on the crystalline orientation of each individual grain. For example, due to symmetry reasons, in alumina, two of the diagonal elements of the dielectric tensor are equal, but the third one is different ($\varepsilon_{xx}=\varepsilon_{yy}\neq\varepsilon_{zz}$). In alumina, this leads to a difference in the refractive index of only 0.008, but enough to induce a large scattering, particularly in the visible range. As shown in Figure 4.7a, the larger the grain size, the larger the scattering and, therefore, the smaller the RIT. Also, analogously to the porosity, the scattering becomes more critical at shorter wavelengths. It is shown in Figure 4.7a that whereas the RIT at 400 nm for a polycrystalline material does not reach 10%, it is close to 40% at 700 nm. Therefore, the grain size will be very important when materials are developed for the visible range and will not be so critical in devices operating in the infrared range. Following the same formalism, it can be shown that the RIT only falls to 60% at 1.5 microns and 70% at 2 microns when the average grain size is 2 microns in polycrystalline alumina. All RIT values shown so far have been calculated considering texture less materials. However, different routes have shown the possibility to induce some texture in the polycrystalline materials (Salem et al., 1989; Mao et al., 2008; Uchikoshi et al., 2004). Figure 4.7b shows how at two wavelengths, 400 and 700 nm, the RIT increases with increasing texture. In this simulation, texture = 0 corresponds to a perfectly polycrystalline material whereas texture = 1 corresponds to a fully textured, analogous to a single crystal, material. It is worth noting that no porosity has been considered in the simulations and that even so, at 400 nm, the RIT strongly decays from its maximum, around 80%, to only 20% due to a random orientation of the crystallites forming the material. The behaviour at 700 nm is similar, but the decay is not so abrupt, indicating again the strong wavelength dependence of the phenomenon. For these simulations, a grain size of only 500 nm has been considered, which is not easy to obtain in conventionally sintered alumina materials. Usual grain sizes are of the order of several μm, which leads to even more drastic reductions in the RIT, as can be deduced from Figure 4.7a.

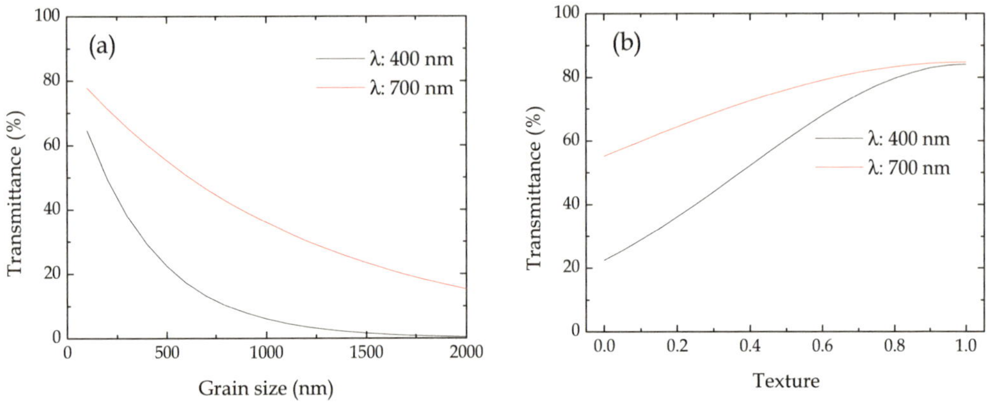

Fig. 4.7. Simulations for birefringent materials with different textures and grain sizes.

C. Pecharroman et al. developed a theoretical model (Pecharroman et al., 2009) for light scattering due to polycrystalline aggregates of uniaxial spheres within the Rayleigh-Gans-Debye approximation. This model shows that the scattering efficiency of each individual grain depends linearly on the grain size and on the texture of the samples. As indicated above, the most critical scattering is that due to the pores and this explains why common technical alumina ceramics which are considered dense at relative density over 99.5 % are opaque (white) when sintered in air, even if high purity raw materials are used. However, provided that the scattering due to pores decreases as $(d/\lambda)^4$, in order to make a material transparent it is not necessary to completely remove all the scattering centres, which is a cumbersome task, but it is enough to keep them all below a certain critical size. As an approximate rule, it is considered that the defects should be smaller than $\lambda/10$. In particular, for the visible range, in which the considered wavelengths range from 400 to 750 nm, the pore size should not be over a few tens of nm. Even so, the total porosity should never be over 0.05%. For this reason care must be taken not only during sintering, but also during processing and preparation of the green bodies. A uniform packing of the powders in green bodies has been shown in the literature to be critical to obtain low porosity ceramics. This is of particular importance when working with nanoceramics, as the attractive Van der Waals forces between particles are responsible for the high tendency of the nanoparticles to agglomerate. The presence of agglomerates leads to local differences in density which changes the sintering behaviour of the body leading to microcracks and microstructural effects such as residual pores and local coarsening and, therefore, to a large scattering. On the one hand, the raw powder should be very fine-grained in order to increase the sintering activity resulting in the elimination of residual porosity. Also, the structural homogeneity of sintering bodies is most important for a minimum of flaws that are detrimental to both optical and mechanical properties. This request for homogeneity makes the use of nanopowders < 100 nm still difficult. Casting methods are particularly suited to process raw powders with particle sizes between 100 and 150 nm.

Finally, it is obvious that preparing materials with the same thickness for a set of measurements is extremely difficult. For this reason, it is important to normalize measurements to a given thickness. The expression that relates the transmittance at a normalized thickness d_1 considering a measurement performed on a material with a thickness d_2 is given by

$$\frac{T(d_1)}{(1-R_s)^2} = e^{-\gamma d_1} = e^{-\gamma d_1 \cdot \frac{d_2}{d_2}} = \left[e^{-\gamma d_2}\right]^{d_1/d_2} = \left[\frac{T(d_2)}{(1-R_s)^2}\right]^{d_1/d_2} \quad (6)$$

5. Conclusions

In conclusion, the interest on transparent ceramic materials relies on the extraordinary combination of high mechanical performance, chemical resistance/inertness... and little or none absorption in different ranges of the electromagnetic spectrum from infrared to near ultra violet. It has been shown that obtaining a transparent ceramic material implies sintering to a density, very close to the theoretical density (≥99.9%), and keeping the few pores left with a pore size below 30 nm. In order to achieve this aim, attention must be paid not only to the raw materials: adequate granulometry, particle size... but it becomes also critical to control the packing of the powders in the green body. Usually, large defects present in the green bodies cannot be removed during the sintering process, and therefore, forming step is decisive for the preparation of transparent materials. Moreover, it has been shown how an adequate choice of the sintering technique combined with a tailoring of the starting powder is critical when an accurate control of the microstructural features is required. Finally, different analysis techniques of the transmittance spectra have been given from which average structural parameters such as pore size, porosity and grain size and texture in birefringent ceramics can be extracted.

6. References

Álvarez-Clemares I., Mata-Osoro G., Fernández A., López-Esteban S., Pecharromán C., Palomares J., Torrecillas R. & Moya J.S. (2010). Transparent alumina/ceria nanocomposites by spark plasma sintering. Advanced Engineering Materials, 12, 11, 1154-1160

Amato I., Baudrocco F. & Martorana D. (1976). Evaluation of freeze drying and spray drying processes for preparing transparent alumina. Material Science and Engineering, 26, 1, 73-78.

Anderson R.C. (1998). Transparent Yttria-based ceramics and method for producing same. U. S. Patent 3545987.

Appetz R. & Van Bruggen M.P.B. (2003). Transparent alumina: a light scattering model. Journal of the American Ceramic Society, 86, 3, 480-486.

Bennison S.J. & Harmer M.P. (1990a). A history of the role of MgO in the sintering of alfa-alumina. Ceramic Transactions, American Ceramic Society, 7, 13-49.

Bennison S.J. & Harmer M.P. (1990b). Effect of magnesia on surface diffusion in sapphire and the role of magnesia in the sintering of alumina. Journal of the American Ceramic Society, 73, 4, 833-837

Bohren C. F. & Huffmann D.R. (2010). Absorption and scattering of light by small particles, Wiley-Interscience, ISBN 3527406646, New York

Born M. & Wolf E. (2005). Principles of Optics-Electromagnetic theory of propagation, interference and diffraction of light, Cambridge University Press, ISBM 0521642221, Cambridge, UK.

Budworth D.W. (1970). The selection of grain-growth control additives for the sintering of ceramics. Mineralogical Magazine, 37, 291, 833-838.

Chaim R., Marder-Jaeckel R. & Shen J.Z. (2006). Transparent YAG ceramics by surface softening of nanoparticles in spark plasma sintering. Material Science and Engineering, 429, 1-2, 74-78

Cheng J., Agrawal D., Zhang Y. & Roy R. (2002). Microwave sintering of transparent alumina. Materials Letters, 56, 4, 587– 592.

Chen Z., Li J., Xu J. & Hu Z. (2008). Fabrication of YAG transparent ceramics by two-step sintering. Ceramics International, 34, 7, 1709-1712

Coble R. L. (1962). U.S. Pat. No.3 026 210.

Cook R., Kochis M., Reimanis I., Kleebe H. J. (2005). A new powder production route for transparent spinel windows properties: powder synthesis and windows properties. Proceedings of the Defense and Security Symposium.

Dalisa, A. L. & Seymour R. J. (1973). Convolution Scattering Model for Ferroelectric Ceramics and Other Display Media, Proceedings of the IEEE, 61, 7, 981–991.

Greskovich C. & Chernoch J.P. (1973). Polycrystalline ceramic lasers. Journal of Applied Physics, 44, 4599-4607.

Handwerker C.A., Morris P. A. & Coble R. L. (1989). Effects of chemical inhomogeneities on grain growth and microstructure in Al_2O_3. Journal of The American Ceramic Society, 72, 1, 130-136.

Hartnett T. M., Bernstein S. D., Maguire E. A., & Tustison R. W. (1998). Optical properties of AlON (aluminum oxynitride). Infrared Physics & Technology, 39, 4, 203–211.

Ikesue A., Furusato I. (1995). Fabrication of polycrystalline transparent YAG ceramics by a solid-state reaction method. Journal of the American Ceramic Society, 78, 1 225-228.

Itatami K., Tsujimoto T. & Kishimoto A. (2006). Thermal and optical properties of transparent magnesium oxide ceramics fabricated by post hot-isostatic pressing. Journal of the European Ceramic Society, 26, 4-5, 639-645

Kim B. N., Hiraga K., Morita K. & Yoshida H. (2007). Spark Plasma Sintering of transparent alumina. Scripta Materialia, 57, 7, 607-610

Kong, J., Tang, D.Y., Zhao B., Lu J., Ueda K., Yagi H. & Yanagitani T. (2005). 9.2-W diode-pumped $Yb:Y_2O_3$ ceramic laser. Applied Physics Letters, 86, 16, 16116-16119.

Mccauley J. W., Patel P., Chen M., Gilde G., Strassburger E., Paliwal B., Ramesh K.T. & Dandekar D.P (2099). AlON: A brief history of its emergence and evolution, Journal of the European Ceramic Society, 29, 2, 223-236.

Li J. G., Ikegami T., Lee J. H. & Mori T. (2000a). Low-temperature fabrication of transparent yttrium aluminum garnet (YAG) ceramics without additives. Journal of the American Ceramic Society, 83, 4, 961-963.

Li J. G., Ikegami T., Lee J. H., Mori T. & Yajima Y. (2000b). Co-precipitation synthesis and sintering of yttrium aluminum garnet (YAG) powders: the effect of precipitant. Journal of The European Ceramic Society, 20, 14-15, 2395-2405.

Lu T. C., Chang X. H., Qi J. Q., Lu X. J., Wei Q. M., Zhu S., Sun K., Lian J. & Wanga L. M. (2006). Low-temperature high-pressure preparation of transparent nanocrystalline $MgAl_2O_4$ ceramics. Applied Physics Letters, 88, 21, 213120-213123.

Lukin E. S., Vlasov A. S., Zubakhina M.A. & Datsenko A.M. (1980). Effect of γ radiation on the optical properties of transparent yttrium-oxide ceramics. Glass and Ceramics, 37, 5, 255-258.

Mah T. I., Parthasarathy T. A. & Lee H. D. (2004). Polycristalline YAG; structural of functional?. Journal of Ceramic Processing Research, 5, 4, 369-379.

Mao X., Wang S., Shimai S. & Guo J. (2008). Transparent Polycrystalline Alumina Ceramics with Orientated Optical Axes, Journal of the American Ceramic Society. 94, 10, 3431-3433.

Mie G. (1908). Beiträge zur Optik trüber Medien, speziell kolloidaler Metallösungen, Annalen der Physik. 330, 377–445.

Morita K., Kim B.N., Hiraga K., Yoshida H. (2008). Fabrication of transparent $MgAl_2O_4$ spinel polycrystal by spark plasma sintering processing. Scripta Materialia, 58, 12, 1114–1117.

Mouzon, J., Glowacki E. & Odén M. (2008). Comparison between slip-casting and uniaxial pressing for the fabrication of translucent yttria ceramics. Journal of Material Science, 43, 8, 2849-2856

Patel P.J., Gary A. G., Dehmer P. G. & McCauley J.W. Transparent Armor, The AMPTIAC Newsletter. Advanced Materials and Processes Technology, 4, 3, 1-24.

Pecharroman C., Mata-Osoro G., Diaz L. A., Torrecillas R. & Moya J. S. (2009). On the transparency of nanostructured alumina: Rayleigh-Gans model for anisotropic spheres, Optics Express 17, 8, 6899-6912.

Peelen J. G. J. & Metselaar R. (1974). Light Scattering by Pores in Polycrystalline Materials: Transmission Properties of Alumina, Journal of Applied Physics, 45, 1, 216-220.

Peuchert U., Okano, Y., Menke Y., Reichel S. & Ikesue, A. (2009). Transparent cubic-ZrO_2 ceramics for application as optical lenses. Journal of the European Ceramic Society, 29, 2, 283-291.

Qin G., Lu J., Bisson J. F., Feng Y., Ueda K. I., Yagi H. & Yanagitani T. (2004). Upconverion luminescence of Er^{3+} in highly transparent YAG ceramics. Solid State Communications, 132, 2, 103-106.

Salem J. A., Shannon Jr J. L. & Brad R. C. (1989). Crack Growth Resistance of Textured Alumina, Journal of the American Ceramic Society. 72, 1, 20-27.

Savastru D., Miclos S., Cotîrlan C., Ristici E., Mustata M., Mogîldea M., Mogîldea G., Dragu T. & Morarescu R. (2004). Nd:YAG Laser system for ophthalmology: Biolaser-1. Journal of Optoelectronics and Advanced Materials, 6, 2, 497–502.

Shikao S. & Jiye W. (2001). Combustion synthesis of Eu activated $Y_3Al_5O_{12}$ phosphor nanoparticles. Journal of Alloys and Compounds, 327, 1-2, 82–86.

Shuzhi L., Bangwei Z., Xiaolin S., Yifang O., Haowen X. & Zhongyu X. (1999). The structure and infrared spectra of nanostructured MgO-Al_2O_3 solid solution powders prepared by the chemical method. Journal of Materials Processing Technology, 89-90, 405-409.

Suarez M., Fernandez A., Menendez J.L. & Torrecillas R. (2009). Grain growth control and transparency in spark plasma sintered self-doped alumina materials. Scripta Materialia 61, 10, 931-934.

Suárez M., Fernández A., Menéndez J.L., Nygren M., Torrecillas R. & Zhao Z. (2010). Hot Isostatic pressing of optically active Nd:YAG powders doped by a colloidal processing route. Journal of the European Ceramic Society, 30, 6, 1489-1494

Suárez M., Fernández A., Menéndez J.L. & Torrecillas R. (2011). Blocking of grain reorientation in self-doped alumina materials. Scripta Materialia, 64, 6, 517-520

Takaichi K., Yahi H., Lu J., Shirakawa A., Ueda K., Yanagitani T. & Kaminskii A.A. (2003). Yb^{3+} doped $Y_3Al_5O_{12}$ ceramics- A new solid-state laser material. Physics state solid A, 200, 1, R5-R7.

Tokurakawa, M.T., Kazunori S., Akira U., Ken-ichi Y., Hideki Y., Takagimi K. & Alesander A. (2007). Diode-pumped 188 fs mode-locked Yb^{3+}:Y_2O_3 ceramic laser, Applied Physics Letters, 90, 7, 071101-071104.

Uchikoshi T., Suzuki T.S., Okuyama H. & Sakka Y. (2004). Control of crystalline texture in polycrystalline alumina ceramics by electrophoretic deposition in a strong magnetic field, Journal of Materials Research, 19, 5, 1487-1491.

Van de Hulst H. C. (1957). Light scattering by small particles, John Wiley and Sons, ISBN 0486642283, New York.

Wang C. & Zhao Z. (2009). Transparent $MgAl_2O_4$ ceramic produced by spark plasma sintering. Scripta Materialia, 61, 2, 193-196

Wei C. G. (2009). Transparent ceramis for lighting. Journal of the European Ceramic Society, 29, 2, 237-244.

Wen L., Sun X., Xiu Z., Chen S. & Tsai C .T. (2004). Synthesis of nanocrystalline yttria powder and fabrication of transparent YAG ceramics. Journal of the European Ceramic Society, 24, 2681-2688.

Zhang G., Wang Y., Fu Z., Wang H., Wang W., Zhang J., Lee S.W. & Niihara K. (2009). Transparent mullite ceramic from single-phase gel by spark plasma sintering. Journal of the European Ceramic Society, 29, 13, 2705-2711

24

Effects of the Microstructure Induced by Sintering on the Dielectric Properties of Alumina

Zarbout Kamel[1], Moya Gérard[2],
Si Ahmed Abderrahmane[2], Damamme Gilles[3] and Kallel Ali[1]
[1]*Sfax University, LaMaCoP, BP 1171, Sfax 3000,*
[2]*Aix-Marseille University, Im2np, UMR-CNRS 6242, Marseille,*
[3]*Commissariat à l'Energie Atomique, DAM Ile-de-France, Bruyère-le-Châtel*
[1]*Tunisie*
[2,3]*France*

1. Introduction

The dielectric properties of undoped corundum, α-alumina, of different kinds (single crystal or polycrystalline obtained by solid state sintering) have been the subject of numerous studies concerning, in particular, the breakdown strength and the charging behavior (Haddour et al., 2009; Liebault et al., 2001; Si Ahmed et al., 2005; Suharyanto et al., 2006; Thome et al., 2004; Touzin et al., 2010; Zarbout et al., 2008, 2010). The common feature pointed out by most of these investigations is the conspicuous role played by the microstructure and the impurities. It is also established that the microstructure induced by the sintering process (grain size and porosity) goes concomitantly along with impurities segregation at grain boundaries and/or the development of defects in the lattice (Chiang et al., 1996; Gavrilov et al., 1999; Lagerlöf & Grimes, 1998; Moya et al., 2003). To some extent, for a given composition, these evolutions can be governed by the sintering conditions, for instance the firing cycle in the case of solid state sintering (Chiang et al., 1996).

The breakdown strength is a key parameter for the reliability of dielectrics and in particular of microelectronic insulator components. This parameter is intimately linked to the charging properties as breakdown originates from the enhancement of the density of trapped charges, which stems from the competition between charge trapping and conduction (Blaise & Le Gressus, 1991; Le Gressus et al., 1991; Liebault et al., 2001; Haddour et al., 2009). The charges can be either generated by irradiation or injected through interfaces via an applied voltage. Charge trapping can occur around intrinsic point defects, defects induced by the dissolution of impurities, defects associated with grain boundaries interfaces and dislocations (Kolk & Heasell, 1980). We must also keep in mind that trapping in insulators gives rise to polarization and lattice deformation allowing energy accumulation within the material (Blaise & Le Gressus, 1991; Le Gressus et al., 1991; Stoneham, 1997). As a result, if some critical density of trapped charges (or some critical electric field) is reached, external stresses (thermal, electrical or mechanical) can trigger a collective relaxation process corresponding to a release of stored energy. If the amount of this energy is sufficient, breakdown could set in causing irreversible damages of the material (Blaise & Le Gressus, 1991; Moya & Blaise, 1998; Stoneham, 1997).

It appears that an improvement of the breakdown strength would require that conduction, which tends to decrease density of trapped charges, be favored to some extent without, however, substantially altering the insulating properties. Conduction will also be referred as the ability of the material to spread charges. Therefore, the control of the competition between charge accumulation (trapping) and spreading (conduction), via the fabrication processes, is a key technological concern. The foregoing arguments motivate the need to develop methods for the characterization of charge conduction (conversely charge trapping) and underscore furthermore the importance of controlling the microstructural development during sintering of ceramic insulators. The purpose of this chapter is to provide the physical background for a more comprehensive understanding of the effects of the microstructure (and the various defects) induced by the sintering conditions on charge conduction in alumina. This understanding, which could be generalized to other ceramics, appears as prerequisite for the fabrication of insulators of improved dielectric breakdown strength.

2. Defects in α-alumina

The α-alumina, exhibits the hexagonal corundum structure. In this structure, Al^{3+} cations occupy only two-thirds of the available sites and an interstitial unoccupied site arises between alternate pairs of Al^{3+}. Charge trapping in alumina may take place around defects that can be intrinsic in nature or stemming from the dissolution of impurities (i.e., the foreign cations and their charge compensating defects). In sintered materials, one has also to take into account the effect of grain boundaries, segregation of impurities and defects at interfaces. These defects are characterized by energy levels within the wide band gap (of about 9 eV in alumina).

2.1 Point defects

2.1.1 Intrinsic point defects (Schottky and Frenkel defects)

The Schottky defects consist of pairs of negatively charged cationic vacancies $V_{Al}^{'''}$ and positively charged anionic vacancies $V_O^{\bullet\bullet}$. The vacancies must be formed in the stoichiometric ratio (two aluminium for three oxygen) in order to preserve the electrical neutrality of the crystal. Using the Kröger-Vink notation, the formation of Schottky defects obey to the reaction:

$$2\,Al_{Al}^{x} + 3\,O_O^{x} \Leftrightarrow 2\,V_{Al}^{'''} + 3\,V_O^{\bullet\bullet} + Al_2O_3 \tag{1}$$

The Frenkel defects are formed when the Al^{3+} cation (Eq. 2) or the O^{2-} anion (Eq. 3) is displaced from its normal site onto an interstitial site giving a vacancy and an interstitial pair:

$$Al_{Al}^{x} \Leftrightarrow V_{Al}^{'''} + Al_i^{\bullet\bullet\bullet} \tag{2}$$

$$O_O^x \Leftrightarrow V_O^{\bullet\bullet} + O_i^{''} \quad (3)$$

Simulation results show that the formation energies of intrinsic point defects in α-alumina are relatively high (Atkinson et al., 2003). They are estimated respectively for Schottky defects, cation Frenkel and anion Frenkel at 5.15, 5.54 and 7.22 eV.

2.1.2 Extrinsic point defects

Extrinsic point defects are entailed by the dissolution of foreign elements. The solubility of an impurity depends mainly on its cation size (generally, small size elements exhibit high solubility). The charge compensating defects accompanying the dissolution of aliovalent impurities (i.e, defects that are required for ensuring the electrical neutrality) are determined not only by their valence (charge) but also by their position (interstitial or substitutional) in the host lattice.

In the case of a cation (M) greater in valence than the host cation (Al^{3+}), the dissolution mode in substitutional position is most likely the cationic vacancy compensation mechanism (Atkinson et al., 2003). Accordingly, for tetravalent cations, in MO_2 (such as SiO_2 or TiO_2), the dissolution reaction is:

$$3\,MO_2 + 4\,Al_{Al}^x \Leftrightarrow 3\,M_{Al}^{\bullet} + V_{Al}^{'''} + 2\,Al_2O_3 \quad (4)$$

This compensation by a cationic vacancy is somewhat corroborated by experiments involving solution of Ti^{4+} in α-Al_2O_3 (Mohapatra & Kröger, 1977; Rasmussen & Kingery, 1970).

For divalent cations, in MO (such as MgO or CaO), the anionic vacancy compensation of substitutional $M_{Al}^{'}$, is suggested (Atkinson et al., 2003):

$$2\,MO + 2\,Al_{Al}^x + O_O^x \Leftrightarrow 2\,M_{Al}^{'} + V_O^{\bullet\bullet} + Al_2O_3 \quad (5)$$

The interstitial dissolution of monovalent elements, in M_2O (such as Na_2O or Ag_2O), can be governed by a host cationic vacancy compensation mechanism. However, a self-compensating dissolution mode, involving both interstitial and substitutional positions of M, is also expected (Gontier-Moya et al., 2001).

In the case of isovalent elements, in M_2O_3 (such as Cr_2O_3 or Y_2O_3), the dissolution will not create charged defects in the lattice but can induce a stress field due to the misfit arising from the difference in cation radii.

As previously pointed out, the formation energies of intrinsic defects are very high in α-alumina. Consequently, a few ppm of impurities will make the concentrations of extrinsic defects higher than those of the intrinsic ones, even at temperatures near the melting point (Kröger, 1984; Lagerlöf & Grimes, 1998).

2.1.3 Point defects association

Isolated point defects can be associated, at appropriate temperature, to form neutral or charged defect clusters. This association leads to a substantial reduction in the solution energy due to strong coulombic interaction and lattice relaxation. A typical example of defect clustering is the association of defects induced by the dissolution of tetravalent cations (Eq. 4):

$$3\,M_{Al}^{\bullet} + V_{Al}^{'''} \Leftrightarrow (3\,M_{Al}^{\bullet} : V_{Al}^{'''})^{x} \tag{6}$$

Mass action calculations (Lagerlöf & Grimes, 1998) have shown that the relative concentrations of extrinsic defects (point defects and defect associations) depend on the equilibrium temperature under which they are created. In sintered alumina, they can be determined by the sintering temperature and time, i.e. isothermal part of the firing schedule.

2.1.4 Association of point defects with charges

Anionic vacancies $V_O^{\bullet\bullet}$ (or cationic vacancies $V_{Al}^{'''}$) can be associated with electrons (or holes) to form F centers (or V centers). In fact, upon trapping one electron (or two electrons), $V_O^{\bullet\bullet}$ becomes a F^+ center (or a F center). Anionic vacancies can also be associated with a substitutional divalent cation $M_{Al}^{'}$ to form F_{cation} center (such as F_{Mg} or F_{Ca}). The F, F^+ and F_{cation} act as a donor centers. The energy levels of F and F^+ are respectively estimated around 3 and 3.8 eV below the edge of the conduction band (Kröger, 1984).

2.2 Extended defects: Grain boundaries

Grain boundaries are the interfaces between like crystals, at which atomic planes are always disrupted to some extent. The atomic order of the lattice is preserved up to within approximately a unit cell of the dividing plane. Thus, the disordered region of a grain boundary is typically only 0.5–1 nm wide, although it does vary somewhat with the type of boundary and the crystal lattice periodicity. Grain boundaries provide segregation sites for impurities and defects.

3. Sintering of α-alumina

3.1 Sintering of α-alumina and induced microstructure

The polycrystalline alumina samples (of 0.2 cm thickness and 1.6 cm diameter) were processed, at Ecole Nationale Supérieure des Mines "ENSM" of Saint Etienne (France), by sintering from two types of powders of different purities (Table 1). The first referred below as "pure", elaborated by the Exal process and provided by CRICERAM Co., contains about 150 ppm of different impurities (with 90 ppm of silicon). The second referred as "impure", elaborated by the Bayer process and provided by REYNOLDS Co., has an impurity content near 4000 ppm (with 1497 ppm of silicon). In Table 1, the compositions of single crystals, which will be considered as reference material, are also given (cf. next section).

Before sintering, the powders were prepared according to conventional procedures involving successively aqueous dispersion, adding of organic binders, spray drying, uniaxial die forming and cold isostatic pressing. Sintering near theoretical density was performed in air with a firing schedule (Fig. 1), which comprises:

- a binder burnout stage where samples were heated slowly (1 K/min) up to 773 K in three steps during 5 hours,
- a heat-up stage, at 2 K/min, from 773 K to the required sintering temperature T_s,
- an isothermal heat treatment at T_s with a dwelling time t_s,
- a cooling stage at rate of about 10 K/min.

	SiO_2	CaO	MgO	Na_2O	Fe_2O_3	K_2O	Cr_2O_3	TiO_2
Alumina powders "pure" (Criceram)	90	5	< 5	40	12	---	---	---
Alumina powders "impure" (Reynolds)	1497	686	723	404	415	---	---	---
Single crystal (RSA)	290	16	< 10	19	48	---	---	---
Single crystal (Pi-Kem)	---	0.6	0.2	1.5	9	2.5	0.4	0.3

Table 1. Composition of alumina materials (impurities in ppm).

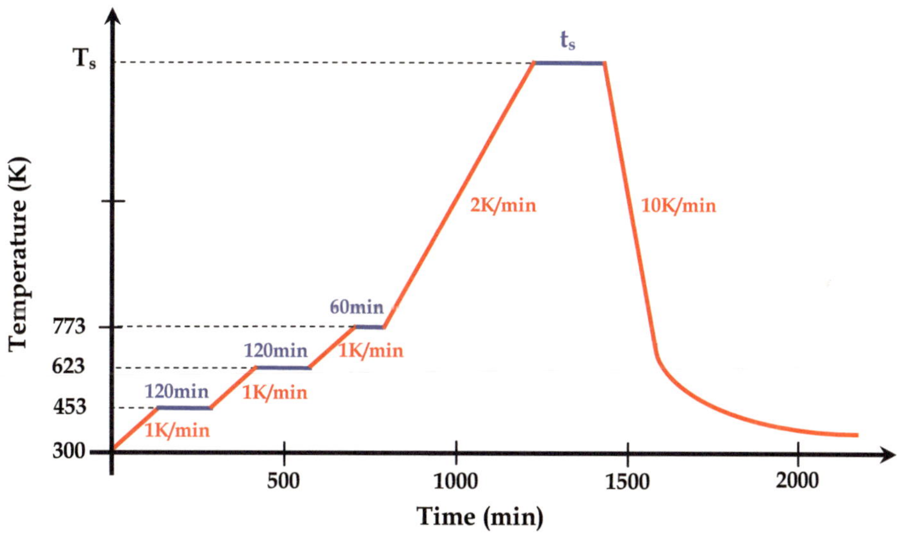

Fig. 1. Schematic description of the firing schedule of the sintering process.

The grain diameters, d, and densities of the sintered samples, which were achieved via the control of the sintering temperature, T_s, and the dwelling time, t_s, are given in Table 2.

The grain sizes were determined by the intercept method. The sample surfaces were first polished and then thermally etched to reveal the grain boundaries. Etching was performed by holding the sample at 50 K below the sintering temperature T_s (during 15 to 30 min) after a rapid heating. The average grain sizes were calculated from the values of 100 to 200 grain size measurements from Scanning Electron Microscope "SEM" images of the surface using different magnifications. Figure 2 shows, as example, the microstructure of the 1.2 μm grain diameter of the impure polycrystalline alumina sample.

	grain diameter, d (μm)	sintering temperature, T_s (K)	dwelling time, t_s (min)	density (in % of theoretical)
"pure" samples	1.7	1863	100	98.7
	2.7	1893	120	98.9
	4.5	1923	120	99.1
"impure" samples	1.2	1773	180	95.7
	2	1823	180	98
	4	1923	360	95.4

Table 2. Sintering temperature, dwelling time at the sintering temperature and corresponding grain diameters and densities (Liebault, 1999).

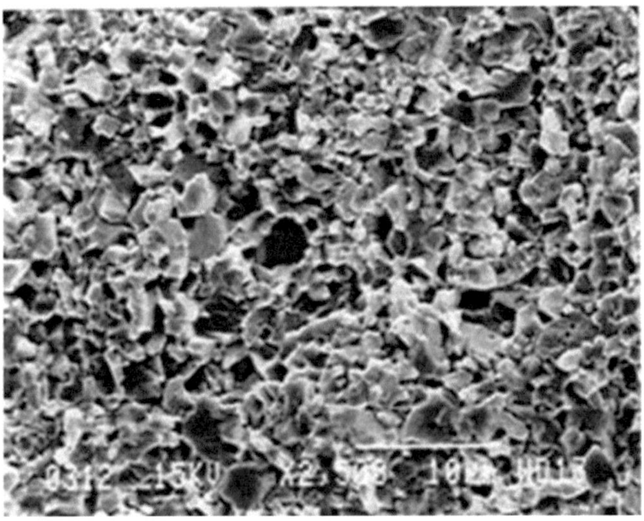

Fig. 2. Microstructure of the 1.2 μm grain diameter of the impure polycrystalline alumina sample (SEM image).

3.2 Single crystals (as reference materials)

The single crystals are selected for the purpose of providing reference materials, which will be compared to the sintered alumina. Two types of single crystals taken from a Verneuil-grown rod are considered (Table 1). The first of very low impurity content (about 15 ppm) is provided by Pi-Kem Co. (U.K.). The second, manufactured by RSA Co. (France), contains about 380 ppm of various impurities (with 290 ppm of silicon). The samples were polished to flat mirror surface finishes, using successively finer grades of diamond pastes down to 1 μm. In order to anneal the defects induced by machining and polishing, a thermal treatment in air at 1773 K during 4 hours was performed.

3.3 Characterization of defects in sintered alumina using positron annihilation lifetime spectroscopy

In this section, complementary investigations concerning the use of Positron Annihilation Lifetime Spectroscopy "PALS" for the characterization of defects in alumina will be summarized (Moya et al., 2003; Si Ahmed et al., 2005). After a short description of the procedure, we will spotlight the results relevant to the charging properties.

3.3.1 Experimental procedure

The positron lifetime spectra of the samples, identical to those described in Table 1, were recorded at room temperature using a conventional fast-fast coincidence system with γ detectors consisting of plastic scintillators characterized by a time resolution of 270 ps. The ^{22}Na positron source, 10 μCi sealed by 0.75 μm thick nickel foils, was sandwiched between two identical samples (of 20 mm diameter and 2 mm thickness). The spectra were measured in 2000 channels, with a calibration of 27 ps/channel, collecting 6.9×10^6 to 20×10^6 counts.

3.3.2 Spectra analysis

The experimental spectra were fitted via a LTV.9 program (Kansy, 1996), in which a three-state trapping model was introduced into the source code (Krause-Rehberg & Leipner, 1999).

In the case of single crystal Pi-Kem, the spectra analysis reduces to only one lifetime component (τ_b = 117 ± 1 ps) associated with reasonable confidence to annihilation in bulk free defects. The quite high intensity of this component, (98.4 ± 1.6) %, appears as an unequivocal justification of the absence of discernable defects that are able to trap positrons (e.g., dislocations, negatively charged vacancies, neutral complexes). This result is also interpreted as a confirmation of the very low impurity content, which therefore justifies its choice as reference material for assessing the effects of both the microstructure achieved by the sintering process and the impurity content. For the sintered samples, three lifetime components were deduced. The shortest lifetime (τ_b = 122 ± 4 ps) was attributed to annihilation in the bulk free defects as it is very close to the one of the reference material, the intermediate (τ_g = 137 ± 2 ps) to trapping in defects within the grains and the longest (τ_{gb} = 397 ± 22 ps) to trapping in clusters located at grain boundaries.

Since Silicon is the main impurity in sintered samples, the possible defects felt by positrons within the grains are isolated vacancies V_{Al}''' and $(Si_{Al}^{\bullet}:V_{Al}''')''$, $(2Si_{Al}^{\bullet}:V_{Al}''')'$, $(3Si_{Al}^{\bullet}:V_{Al}''')^x$ clusters. However, the neutral cluster $(3Si_{Al}^{\bullet}:V_{Al}''')^x$ is more likely than the others due to the sintering temperatures and dwelling times (Lagerlöf & Grimes, 1998). In any case, the positron lifetime in all these defects is expected to have about the same value (τ_g = 137 ± 2 ps) because it is primarily determined by V_{Al}''' (Moya et al., 2003). The nature of clusters at grain boundaries is more difficult to ascertain because of the competition between all impurities for the segregation sites. However, one can speculate that the lifetime of about 400 ps (τ_{gb} = 397 ± 22 ps) reflects positron trapping in neutral or negatively charged clusters of charge compensating native vacancies V_{Al}''' and $V_O^{\bullet\bullet}$, which are induced by the impurities that have the strongest tendency for segregation, for instance SiO_2, MgO and CaO. In particular, $V_O^{\bullet\bullet}$ could stem from the dissolution of CaO (Eq. 5), which, incidentally displays by far the highest enrichment ratio, about 1300 (Dörre & Hübner, 1984).

4. Method for the characterization of the charging state of dielectrics

In insulator material, the charges can be injected through interfaces, via an applied voltage, or created by irradiation via energetic particles. For instance, incident electrons, as they slowdown, can generate pairs of electrons and holes. These charge carriers can recombine, be trapped or move as a result of diffusion and/or field conduction. Concurrently, some of the electrons can be emitted from the sample surface and a distribution of trapped charges may develop within the irradiated volume. In our case, experiments are carried out using the electron beam of a SEM. The experimental set up, described below, provides means for measuring the evolution of the net amount of trapped charges Q_t, which characterize the charging state of the insulator. For this purpose, we use the Induced Current Measurements "ICM" method (called also the Displacement Current Measurements), which we have recently improved (Zarbout et al., 2008, 2010).

4.1 Experimental set up

The experiments are performed using a SEM (LEO 440), which is specially equipped (Fig. 3) to inject a controlled amount of charges Q_{inj} with appropriate conditions (energy of incident electrons, charge and current beam densities, temperature). The electron beam is monitored by a computer system allowing the control of the beam parameters:

- beam energy, E_p (varying between 300 and 40000 eV),
- current beam intensity, I_p (ranging from a few pA to several hundred nA, with the possibility of reaching a few μA),
- irradiated beam area of the sample (varying between a few nm when the beam is focused and few hundred μm when it is defocused).

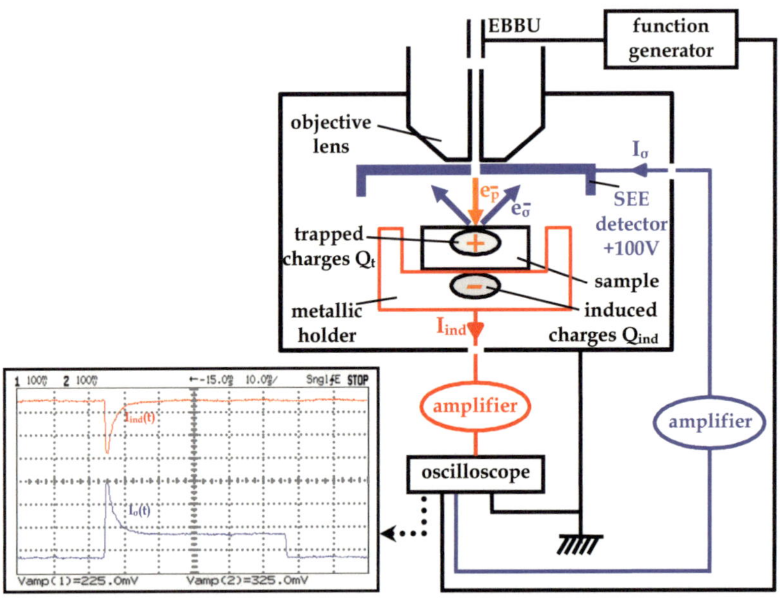

Fig. 3. Setting up for the measurement of the secondary electron current I_σ and induced current I_{ind}. The secondary electron detector (diameter 15 cm) is positioned at a distance of about 2 cm from the sample surface.

Furthermore, the injection time, t_{inj} (ranging from 10^{-3} to 1 s), is adjusted by the Electron Beam Blanking Unit "EBBU" using a function generator, allowing the turns off on the spot over the specimen surface with no electron track outside the investigated area when the beam is blanked.

The metallic sample holder is attached to a cooling-heating stage (temperature range 93 – 673 K). Hence, in situ thermal sample cleaning under vacuum (at T = 663 K during 180 minutes) and sample characterization at different temperatures are possible. Prior to electron irradiations, after the cooling that follows the cleaning step, the sample is held during 180 minutes at the testing temperature, so that the thermal equilibrium between the sample and the metallic holder is approached.

4.2 The improved Induced Current Measurement method

The ICM method is based on the measurement of the current I_{ind}, produced by the variation of the induced charges Q_{ind} (in the sample holder) due to the trapped charges in the sample Q_t (Liebault et al., 2001, 2003; Song et al., 1996). Since the influence coefficient in our experimental set up is close to one (Zarbout et al., 2008), the amount of the net trapped charges is given by:

$$Q_t(t) = -Q_{ind}(t) = -\int_0^t I_{ind}(t)\,dt \tag{7}$$

The improvement brought to the ICM method consists in the concurrent measurement of I_{ind} and the total secondary electron current I_σ due to the sole electrons emitted by the sample. For this purpose, as shown in Fig. 3, the SEM is specially equipped with a secondary electron low-noise collector located under the objective lens just above the sample. A biased voltage of 100 V is applied to it in order to collect all the electrons escaping from the sample surface.

During charge injection, the currents, I_σ and I_{ind}, are simultaneously amplified (Keithley 428) and observed on an oscilloscope (HP 54600B) where the material response is displayed after a short lag time. The primary current beam I_p (which is adjusted in a Faraday cage) and the current I_σ are always positive whereas I_{ind} can be positive or negative depending on the Secondary Electron Emission "SEE" yield σ, which is equal to I_σ/I_p. Then if I_σ is higher than I_p ($\sigma > 1$), the sample charges positively and I_{ind} is negative. In the other case ($\sigma < 1$), the sample charges negatively and I_{ind} is positive. The general variation of σ with E_p is shown in Fig. 4.

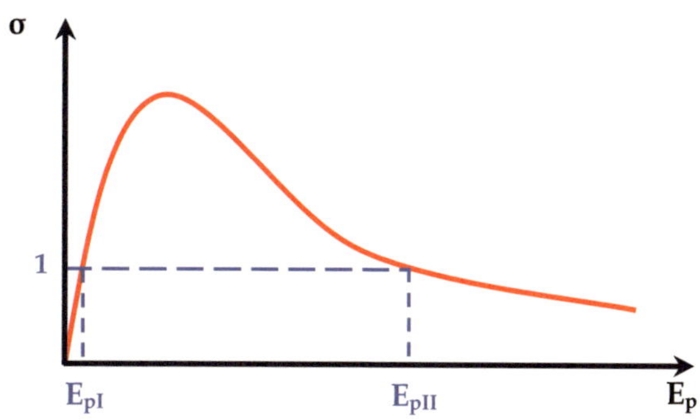

Fig. 4. Schematic evolution of the SEE yield σ with primary beam energy E_p for uncharged insulator materials.

The sign of the net sample charges (i.e., net trapped charges) depends mainly, for a given material, on the primary beam energy E_p and the primary current beam density J. In the case of α-alumina, a sign inversion of the net trapped charges is observed for J greater than 7×10^6 pA/cm² (Thome et al., 2004).

Experiments are performed with a 10 keV primary electron beam energy which is located, for the α-alumina materials, between the two crossover energies of primary electrons E_{pI} and E_{pII} (about 20 keV) for which σ is equal to 1. To probe a zone representative of the material we use a defocused beam over an area of 560 μm diameter Φ, which has been accurately measured using an electron resist (Zarbout et al., 2005). With the utilized value I_p = 100 pA, the primary current beam density is J = 4×10^4 pA/cm². These experimental conditions give rise to net positive trapped charges (σ > 1) and then a negative induced current, as shown in the curves recorded by the oscilloscope in Fig. 3.

The positive surface potential developed by the trapped charges Q_t does not exceed a few volts, which is very low to produce any disturbance of the incident electron beam and of the measurement process (Zarbout et al., 2008). Then, the good stability and reliability of the SEM ensure (for a biased voltage applied to the secondary electron collector greater than about 50 V) the current complementarity according to the Kirchhoff law:

$$I_p(t) = I_{ind}(t) + I_\sigma(t) \tag{8}$$

Hence, the presence of the collector with a sufficient biased voltage allows accurate measurements of the currents and therefore the precise determination of the quantity of trapped charges. Incidentally, this makes also possible a precise calculation of the SEE yield σ(t):

$$\sigma(t) = \frac{I_\sigma(t)}{I_p(t)} = \frac{I_p(t) - I_{ind}(t)}{I_{ind}(t) + I_\sigma(t)} = 1 - \frac{I_{ind}(t)}{I_{ind}(t) + I_\sigma(t)} \tag{9}$$

The SEE yield could be deduced directly from the secondary electron current $I_\sigma(t)$ and the primary one $I_p(t)$. However, the use of $I_\sigma(t)$ and $I_{ind}(t)$ in Eq. 9 is more appropriate as it provides the opportunity to circumvent any uncontrolled fluctuation of the primary electron current during the different phases of the experimental process.

It is worth noting that after the fabrication process of polycrystalline samples (with firing temperatures above 1863 K) or the thermal treatment of single crystals (at 1773 K for 4 hours), the final thermal cleaning stage under vacuum of all the samples prior to electron irradiation will completely remove initial charges and surface contamination that could interfere with the generated charges.

4.3 The charging kinetic

The measurement of the foregoing currents gives means to follow the net quantity of trapped charges during irradiation. The current curves $I_\sigma(t)$ and $I_{ind}(t)$ of Fig 5 are a typical example of the recorded currents. As pointed out in the experimental conditions, irradiation is performed with a 10 keV electron beam energy, which ensure a net amount of positive charges in the sample.

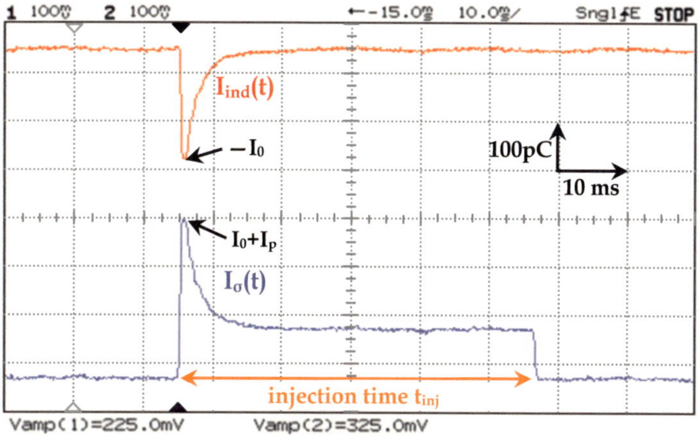

Fig. 5. Current curves recorded during the first injection in polycrystalline alumina sample (d = 4.5 μm) at 473 K. The currents, $-I_0 = -225$ pA and $I_0 + I_p = 325$ pA, are measured after a short lag time. The irradiation conditions are: $E_p = 10$ keV, $I_p = 100$ pA, $t_{inj} = 5$ ms, $Q_{inj} = 5$ pC and $J = 4\times10^4$ pA/cm^2.

At the beginning of injection, the current curves are affected by the rise time of the amplifiers. Therefore, a short lag time is required to display the actual current values $-I_0$ and I_0+I_p. As irradiation proceeds, the induced current, I_{ind}, increases from $-I_0$ to zero. Concurrently and since the current complementarity conditions are verified (as revealed by Eq. 8), the total secondary electron current I_σ decreases from $(I_0 + I_p)$ to I_p. This decrease has been associated with a progressive accumulation of positive charges that are distributed over a depth of about the escape length of secondary electrons λ. As suggested, this could be assigned to recombination of electrons with holes (in this zone), which should be otherwise emitted (Cazaux, 1986) or to field effect (Blaise et al., 2009). The other important feature is indeed that during injection, a negative charge distribution develops in the vicinity of the penetration depth of primary electrons R_p. As result, an internal electric field is established between the two charge distributions (holes near the surface and electrons around R_p) whose direction is oriented towards the bulk. Since the diameter Φ of the irradiated area is much larger than the penetration depth, a planar geometry can be used to evaluate this field (Aoufi & Damamme, 2008) from the Gauss theorem:

$$E(z,t) = \frac{-1}{\varepsilon_0\varepsilon_r(1+\varepsilon_r)}\left(1+\frac{\Phi^2}{4(1+\varepsilon_r)e^2}\right)^{-1}\int_0^e \rho(z,t)dz + \frac{1}{\varepsilon_0\varepsilon_r}\int_0^z \rho(z,t)dz \qquad (10)$$

In this expression, where the contribution of image charges has been taken into account, e is the sample thickness, ε_0 is the vacuum permittivity, ε_r the relative permittivity (taken equal to 10 for α-alumina) and ρ(z, t) is the density of charges at the depth z. The first term corresponds to the surface electric field E (0, t).

When I_{ind} reaches a zero value ($I_\sigma(t) = I_p(t)$), a steady state, which corresponds in fact to a self regulated flow regime, is achieved. There, on the average, for each incident electron one secondary electron is emitted. This state is characterized by some constant value of the electric field as well as by a maximum amount of net trapped charge in the sample, Q_{st}, equal to $-\int_0^{t_{inj}} I_{ind}(t)\,dt$. The steady state is interesting because it will be taken as a reference state and then the practical choice of the injection time t_{inj} is determined by the achievement of this state. The evolution of the net trapped charges Q_t during injection, which can be deduced from Eq. 7, is shown in Fig. 6.

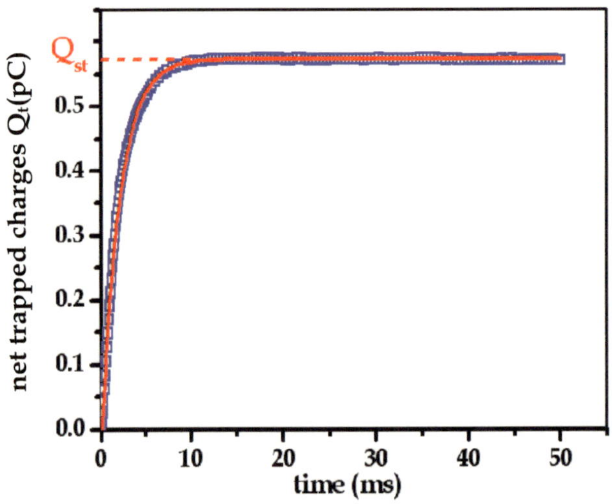

Fig. 6. Evolution with time, at 473 K, of the net charge, Q_t, during the first pulse injection in polycrystalline alumina sample (d = 4.5 μm). The quantity Q_t is derived from the currents of Fig. 5 via Eq. 7. The solid line represents the exponential fit of the data, as given by Eq. 11.

The best fit of data represented in Fig. 6, leads to an exponential time evolution:

$$Q_t(t) = Q_{st}\left[1 - \exp\left(-\frac{t}{\tau_c}\right)\right] \qquad (11)$$

where τ_c is the charging time constant (found equal to 2.07×10^{-3} s), which characterizes the charging kinetic of the material in the used experimental conditions.

5. Measurement of the ability to spread charges

The method described above gives the opportunity to evaluate the quantity of trapped charges during irradiation. If the probed zone is initially uncharged, $Q_t(t)$ characterizes the quantity of charges that accumulates after an irradiation time t. Depending on the insulator conduction properties, the accumulated charges can either remain localized or partially (or even totally) spread out of the irradiated volume (discharge phenomenon). Hence, the measurement of the ability of the material to spread charges is of technological interest. In this paragraph, we give details of the experimental procedure developed for the measurement of this ability and we define a recovery parameter allowing the quantitative evaluation of the extent of discharge.

5.1 Experimental protocol

To characterize the degree of spreading (of discharge), we set up a protocol allowing its evaluation. The procedure consists in analysing the states of charging deduced from two pulse electron injections over the same area, separated by some lapse of time as explained in the three following stages.

5.1.1 The first charging stage (first pulse injection)

The first stage is intended to achieve a charging reference state, which is attained once σ reaches the constant value of 1, i.e. the situation where for each electron entering one is emitted. This charging reference state is associated with the trapping of a maximum quantity of charges Q_{st} (cf. Fig. 6). The embedded charges can be either indefinitely trapped or just localised for some time lapse (seconds, minutes, hours or more). If all the charges were trapped in a stable way, the foregoing reference state would stay unchanged when irradiation is turned off. In the case of partially localised charges, some fraction of Q_{st} could manage to spread out from irradiated region, through conduction. To take into account this fact, a pause time long enough to appraise the degree of charge spreading is chosen for the pause stage.

5.1.2 The pause stage

During the pause time Δt, the stability of the amount of charges Q_{st} is determined by the efficiency of traps associated with impurities and lattice defects as well as the charge transport properties. Then, if one wants to evaluate the extent of discharging, one has to perform, with identical conditions, a second injection over the same area as the first one for the purpose of restoring the reference steady state.

5.1.3 The third stage (second pulse injection)

At the inception of the second injection, if the recorded currents are in the continuity of those of the first injection, all the charge Q_{st} have remained trapped in the irradiated volume during the pause time Δt. If it is not the case, a fraction of the charge Q_{st} has been removed from the irradiated zone as a result of discharging (i.e., detrapping and transport) during the pause time.

For the purpose of illustration, the curves of Fig. 7 display the recorded currents during the second injection, performed over the same area as the first injection and with identical experimental conditions, after a pause time Δt = 900 s. Here, the current curves are not in the continuity of those obtained at the end of the first injection (Fig. 5). As the second injection proceeds, the reference steady state is reached again. The ensuing quantity of net charge introduced during this stage is interpreted as the amount Q_d that has been removed, given by:

$$Q_d = \left[-\int_0^{t_{inj}} I_{ind}(t)\,dt \right]_{\text{second injection}} \qquad (12)$$

In other words, this interpretation means that any loss of charges, during the pause, can be compensated by those introduced during the second irradiation to restore Q_{st}.

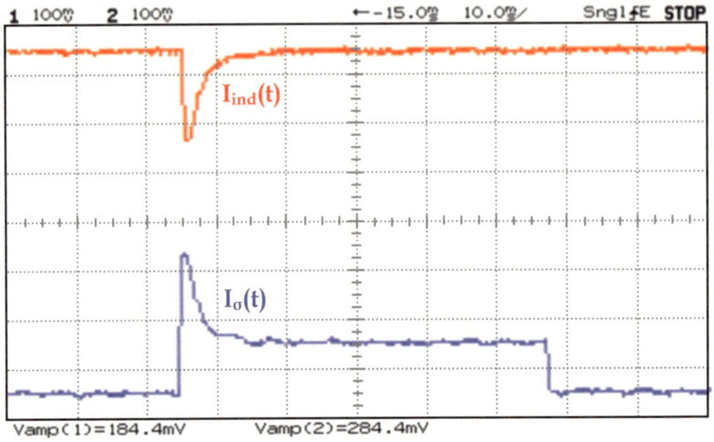

Fig. 7. Current curves obtained during the second pulse injection performed after a pause time Δt = 900 s. Irradiation is carried out over the same area as in the first injection under identical experimental conditions (given in Fig. 5).

The three stages are summarized in Fig. 8, which illustrates the evolution of the amounts of charges remaining in the irradiated volume in the polycrystalline sample of grain diameter d = 4.5 μm at T = 473 K. The data concern the example of Figs. 5 and 7, for which some discharging occurs during the pause Δt. During the pause time t_p ($0 \leq t_p \leq \Delta t$), the net charge that still remains in the irradiated volume $Q_l(t_p)$ evolves from its initial value, $Q_l(t_p = 0) = Q_{st}$, to the final one, $Q_l(t_p = \Delta t) = Q_f$. The form of the curve representing $Q_l(t_p)$ will be justified below.

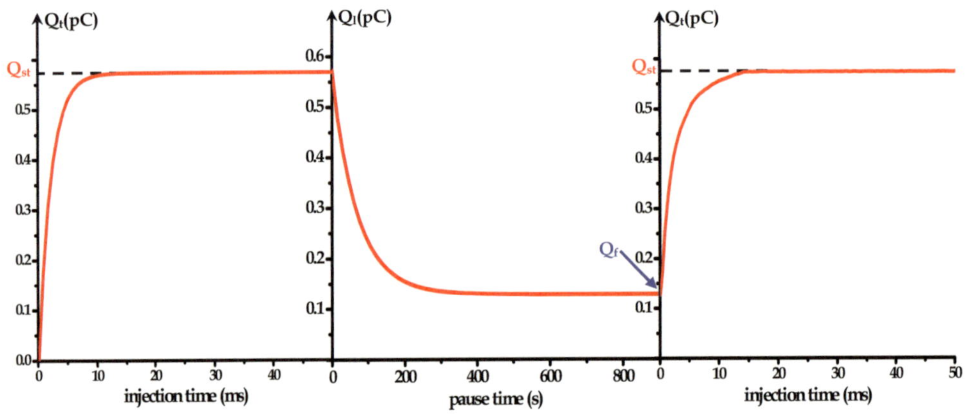

Fig. 8. Illustration of the time evolution of the amounts of charges remaining in the irradiated volume during the three stages. The first and the third curves correspond to Figs. 5 and 7 respectively. The net charge $Q_l(t_p)$ during the pause evolves from Q_{st} to Q_f (which remains in the irradiated volume after a pause of Δt =900 s) as justified below by Eq. (15).

5.2 Definition of a recovery parameter for the evaluation of discharge

We first try to find a kinetic description of the charges that remain in the irradiated volume after a pause time t_p. This amount is given by:

$$Q_l(t_p) = Q_{st} - Q_d(t_p) \tag{13}$$

where, $Q_d(t_p)$ is given by Eq. 12.

As an example, in Fig. 9, we report the best fit of the evolution of Q_d with t_p, for data obtained, at 473 K in polycrystalline alumina sample (d = 4.5 μm), by performing the measurements over different zones that are sufficiently distant from one another to avoid any overlaps of irradiated volumes.

The fit is well described by an exponential with a time constant τ_r:

$$Q_d(t_p, T) = Q_\infty(T)\left[1 - \exp\left(-\frac{t_p}{\tau_r(T)}\right)\right] \tag{14}$$

In this equation, Q_∞ is the asymptotic value reached by Q_d. In fact, it stands for the maximum amount of charges going out of the irradiated volume in the discharging process at temperature T. In the example of Fig. 9 where T = 473 K, Q_∞ is found equal 0.46 pC corresponding to about 80 % of Q_{st} (hence, 20 % of the net charge is still inside the irradiated volume). This is an indication of the existence of different trapping sites and that the temperature of 473 K is too low to provoke detrapping from the deeper ones.

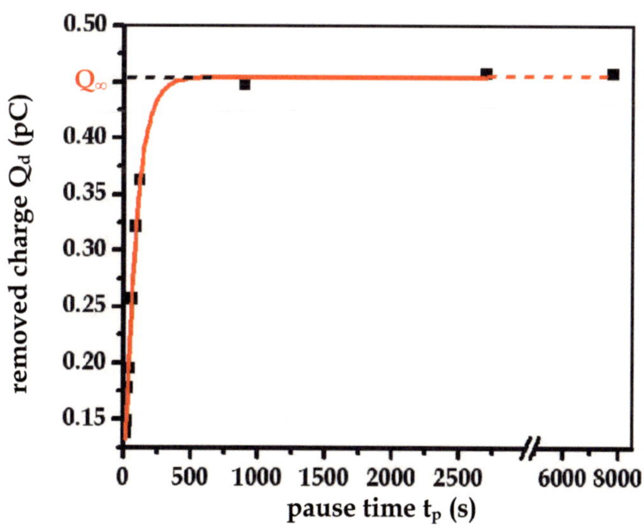

Fig. 9. Evolution with the pause time at T = 473 K of the amount of charges that is removed from the irradiated volume Q_d (in polycrystalline sample of 4.5 μm grain size). The solid line is the exponential fit of the data (Eq. 14). The asymptotic value $Q_\infty(T)$ is attained after a pause time of only 300 s.

From Eq. 14, we can associate to $Q_\infty(T)$ the amount of charges $Q_f(T)$ that still remain in the irradiated volume, $Q_\infty(T) = Q_{st} - Q_f(T)$. Therefore, the remaining quantity of charges $Q_l(t_p, T)$ can be obtained from Eqs. 13 and 14:

$$Q_l(t_p, T) = (Q_{st} - Q_f(T)) \exp\left(-\frac{t_p}{\tau_r(T)}\right) + Q_f(T) \quad (15)$$

The first term in this expression, which corresponds to the curve of the pause stage in Fig. 8, expresses charge decay under the internal electric field. Generally, the time constant τ_r can be set equal to ε/γ, where ε is the dielectric permittivity ($\varepsilon = \varepsilon_r\varepsilon_0$) and γ is the electric conductivity of the material (Adamiak, 2003; Cazaux, 2004). The value of τ_r deduced from Fig. 9 is 82 s giving a conductivity of about $10^{-14}\ \Omega^{-1}cm^{-1}$ which can be expected for this material, in agreement with the experimental value of the resistivity obtained for this sample in our laboratory ($1.2\ 10^{14}\ \Omega cm$).

The asymptotic value Q_∞ at T = 473 K is reached after only 300 s (Fig. 9). Therefore, one can anticipate that at temperatures within the range of interest (300 – 663 K), the condition for reaching the asymptotic value $Q_\infty(T)$ are met for the chosen pause time Δt of 900 s.

The measured value of $Q_\infty(T)$ is the result of detrapping of charges and their subsequent transport under the internal electric field. During this transport, a fraction of the detrapped charges can undergo a retrapping in deeper traps in the irradiated volume and eventually a recombination. The overall effect is a variation of charge distribution in the volume of interest, which affects the electric field. Consequently, since in the considered temperature range the experimental results do not reveal any significant dependence of Q_{st} on temperature, the ratio $R(T) = Q_\infty(T)/Q_{st}$ can be expressed in terms of the measured currents:

$$R(T) = \frac{Q_\infty(T)}{Q_{st}} = \frac{-\left[\int_0^{t_{inj}} I_{ind}(t)dt\right]_{second\ injection}}{-\left[\int_0^{t_{inj}} I_{ind}(t)dt\right]_{first\ injection}} \quad (16)$$

This experimental parameter, which measures the fraction of charge removed from the irradiated volume, also characterizes the extent of discharging.

The ratio R(T), which can vary between 0 and 1, corresponds to an evolution from either the dominance of stable charge trapping (low values of R) or of charge spreading (high values of R, with R = 1 for a complete recovery of the uncharged state). The rate at which charges are detrapped depends usually on an attempt escape frequency and an activation energy linked to the trap depth. As a result, one can expect that, the variation of R with temperature could shed some light on the discharge process.

6. Effect of microstructure induced by sintering on the ability of a dielectric to trap or spread charges

As it will be pointed out, trapping and spreading are intimately linked to the microstructure and defects. Sintering not only leads to the creation of new interfaces but also to important phenomena such as segregation at grain boundaries and defects association.

6.1 Charge spreading in reference materials (single crystals)

The results of the two types of single crystals are reported in Fig. 10. The values of R in both Pi-Kem and RSA are zero up to 473 K, indicating a perfect stable trapping behavior. Above 473 K, R increases but the enhancement is significant only for Pi-Kem.

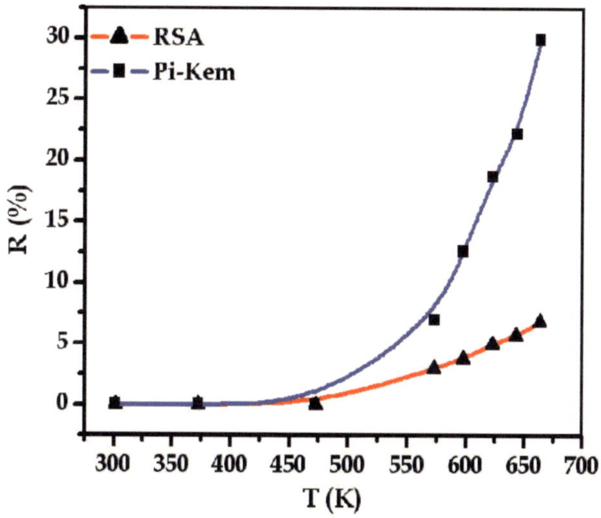

Fig. 10. Fraction R of charges removed from the irradiated volume as a function of temperature for the two types of single crystals (Pi-Kem and RSA samples).

In order to interpret these behaviors we can consider the following results:

i. Cathodoluminescence spectra obtained in similar Pi-Kem samples (Jardin et al., 1995) have identified mainly the F and F^+ centers. Stable trapping in these centers is expected due to their deep level in the band gap (3 eV, and 3.8 eV for the F and F^+, respectively). The increase of R above 473 K might be due to the intervention of excited F centers whose energy levels are believed very near the edge of the conduction band (Bonnelle & Jonnard, 2010; Jonnard et al. 2000; Kröger, 1984).
ii. The RSA material displays a more stable trapping ability than the Pi-Kem one above 473 K. This stable trapping raises queries about the role of the defects induced by the impurities (Table 1) and in particular those by the dominant silicon. With such amount of silicon (290 ppm), the concentration of the defects associated with Si exceeds the others. Consequently, one can deduce that stable trapping may occur on these defects.

As mentioned in paragraph 2, the dissolution of silicon into Al_2O_3, is expected to be compensated by a negatively charged cationic vacancy, $V_{Al}^{'''}$ (Eq. 4). In this context, the positively charged substitutional silicon $Si_{Al}^{\bullet}$ may act as electron trapping site while the cationic vacancy $V_{Al}^{'''}$ as hole trap. Upon trapping one electron during irradiation, $Si_{Al}^{\bullet}$ induces a donor level associated to $Si_{Al}^{\times}$, which is estimated at 1.59 eV below the edge of the conduction band (Kröger, 1984). With regard to $V_{Al}^{'''}$, hole trapping will give an acceptor level (associated to $V_{Al}^{''}$) located at 1.5 eV above the valence band (Kröger, 1984).

Accordingly, with such relatively deep trap levels, it is very unlikely that detrapping of charge carriers occurs at the temperatures of our experiments. However, the contribution of the other impurities, such as the ones of smaller valence than the host, cannot be ruled out because trapping depends not only on the defect concentration but also on their specific trapping properties such as the capture cross section of traps which is mainly determined by their charge state. Indeed, the cathodoluminescence spectra have also detected (Jardin et al., 1995) in similar RSA samples (which like ours were annealed at 1773 K during 4 hours) the deep centers F_{cation}, such as F_{Mg}^{X} with an energy level at 4 eV below the edge of the conduction band (Kröger, 1984).

We alternatively tried to shed some more light on the trapping behaviors of single crystals by using the Scanning Electron Microscope Mirror Effect "SEMME" method, which requires net negative charging (Liebault et al., 2003; Vallayer et al. 1999). Thus, we performed electron injection with 30 keV (energy greater than E_{pII}) focused beam and an injected dose of 300 pC followed by a scanning of the sample surface, with low electron beam energy (some hundred of eV). In RSA samples we find a mirror image at 300 K, which remains stable up to about 663 K. In contrast, in Pi-Kem materials no mirror image formation was achievable even at 300 K.

It must be reminded that in the SEMME method the observation of a mirror image, immediately after irradiation, is due to the presence of a sufficiently high concentration of stable traps for electrons. Thus, no mirror image is observed in Pi-Kem samples as they do not contain enough traps. The fact that in RSA the mirror is maintained at high temperatures confirms the presence of deep traps and somewhat supports the assumption, discussed above in point (ii), that stable trapping can be assigned to the defects induced by the dissolution of the dominant silicon impurity.

In contrast with SEMME method, the ICM method can give, during charging, an induced current whatever the trap concentration is. So, the parameter R gives indications on the stability of trapped charges, after a pause Δt, independently of the trap concentration. As a result, at temperatures lower than 573 K, for which trapped charges are stable in RSA and Pi-Kem samples, the R values are very low ($R = 0$ at room temperature for both kinds of single crystals). Above 573 K, RSA samples have R values lower than those of Pi-Kem. This is a supplementary indication of the existence of deeper traps in RSA, which is one of the requirements for the mirror image formation at high temperatures in the SEMME method.

6.2 Charge spreading in sintered alumina of various microstructures and impurities

6.2.1 Charge spreading in pure sintered alumina

The values of R(T) for the three polycrystalline alumina pure samples as function of temperature are shown in Fig. 11. The comparison with single crystals (R close to zero up to 573 K) indicates that the presence of grain boundaries makes trapping less stable in polycrystalline alumina (R is substantially above zero). This interpretation agrees with the fact that, with SEMME method, we do not detect any mirror image in the polycrystalline samples.

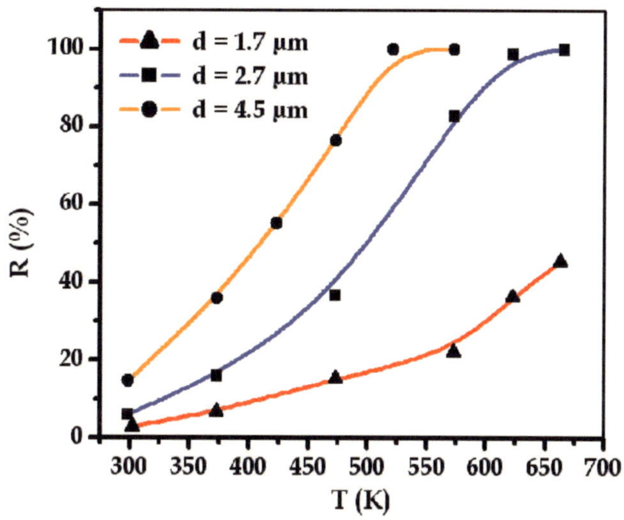

Fig. 11. Fraction R of charges removed from the irradiated volume as function of temperature for pure polycrystalline alumina samples of different grain diameters d.

In the RSA single crystal and pure polycrystalline alumina, silicon is the dominant impurity. Therefore, one can suggest that atomic disorder introduced by grain boundaries gives rise to states closer to the edge of the conduction band than those of $Si_{Al}^{\times}$ and $V_{Al}^{'''}$, which are most likely responsible for the stable trapping in the bulk.

For fine grains (d = 1.7 µm), R spans from 0 (at 300 K) to 45 % (at 663 K) whereas for large grains (d = 4.5 µm) R varies between 15 % (at 300 K) to near complete discharge (about 100 %) above 473 K. One would expect the behaviour of polycrystalline material to tend towards that of single crystals when the grain size increases. However, the reverse is clearly seen. The explanation of this apparent contradiction can be found by considering the distribution of impurities in the polycrystalline samples. We have to bear in mind that the microstructural evolution during the sintering process, with in particular the achievement of a given grain size, is concomitant with segregation to grain boundaries of impurities, which corresponds to a purification of the grain (interpreted as an internal gettering). In fact, the large grain size of 4.5 µm has required both a higher sintering temperature and a longer dwelling time than those used to attain 1.7 µm. Hence, the enhancement of gettering effect in large grain sample may have lowered the concentration of deep traps (such as $Si_{Al}^{\times}$ and $V_{Al}^{'''}$) in the bulk. In this assumption, the impurities (mainly Si) segregates at grain boundaries where they can be associated to other defects to form clusters, as suggested by positron measurements (cf. paragraph 3), which may be less efficient in trapping charges.

This agrees with the experimental facts:

i. The R values are always higher for the larger grain (for example at 573 K, R = 100 % for d = 4.5 µm and only 22 % for d = 1.7 µm). This means that the density of deep traps in large grains is substantially lower.

ii. The foregoing assumption is somewhat confirmed in Fig. 12, where below 573 K the semi-logarithmic plot of R versus reciprocal temperature exhibits an Arrhenius law leading to an activation energy about 0.12 eV whatever the grain size is. This same activation energy means that we are dealing with detrapping from similar trapping sites (i.e., grain boundary traps). The continuous variation in R over a large temperature range, as shown in Fig 11, indicates a detrapping from a density of continuous trapping states rather than from a single trapping level. This aspect characterizes disordered solids in which hopping conduction mechanism can occur with the same order of magnitude of activation energy (Blaise, 2001).

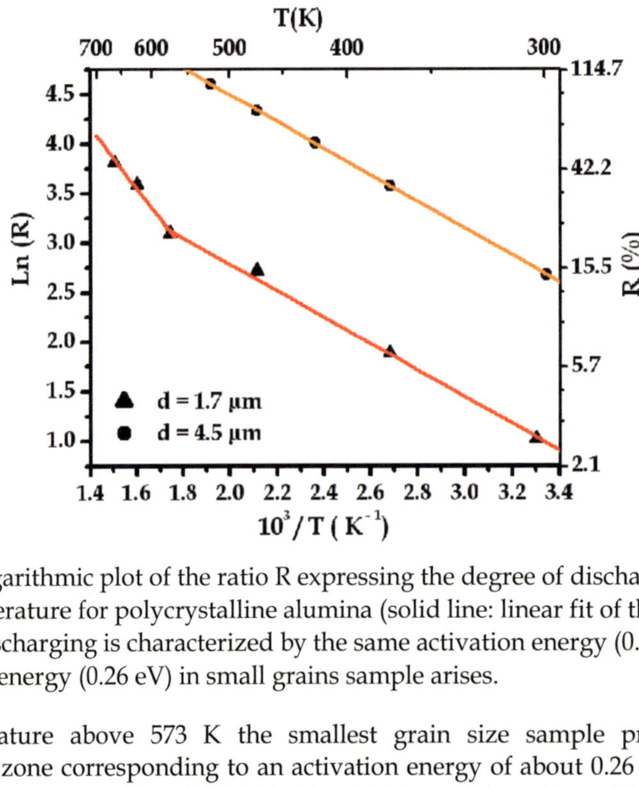

Fig. 12. Semi-logarithmic plot of the ratio R expressing the degree of discharge versus reciprocal temperature for polycrystalline alumina (solid line: linear fit of the data). For T below 573 K, discharging is characterized by the same activation energy (0.12 eV). Above 573 K, a second energy (0.26 eV) in small grains sample arises.

iii. for temperature above 573 K the smallest grain size sample presents a second detrapping zone corresponding to an activation energy of about 0.26 eV (i.e., twice the energy at lower temperature). This fact is a further confirmation of the presence in the smaller grain size of a higher density of deeper traps located within the grain (likely $Si_{Al}^{\times}$ and V_{Al}''') in accordance with a less efficient gettering effect.

6.2.2 Charge spreading in impure sintered alumina

The results of the three polycrystalline alumina impure samples are reported in Fig. 13. The comparison of the pure and impure polycrystalline samples reveals a more stable trapping behavior in impure samples in the whole temperature range. Furthermore, two opposite behaviors arise: stable trapping increases with the grain diameter in impure samples (Fig.13) and the contrary is obtained in the pure ones (Fig. 11). In the impure material, the contents of impurities are much higher and expected above the solubility limits. Hence, gettering might be less efficient due to a possible saturation of grain boundaries. In addition, interactions between the various defects generated by the foreign elements are expected (Gavrilov et al., 1999). It must be pointed out that the great variety of impurities and their substantial content make difficult the interpretation of the results due to a possible interference of co-segregation, which is difficult to predict when more than three elements are involved. The actual situation is even more complicated by the fact that segregation leads to the creation of a space charge at grain boundaries (Tiku & Kröger, 1980) with a sign that depends on the segregated impurities, which may interfere with the charging process.

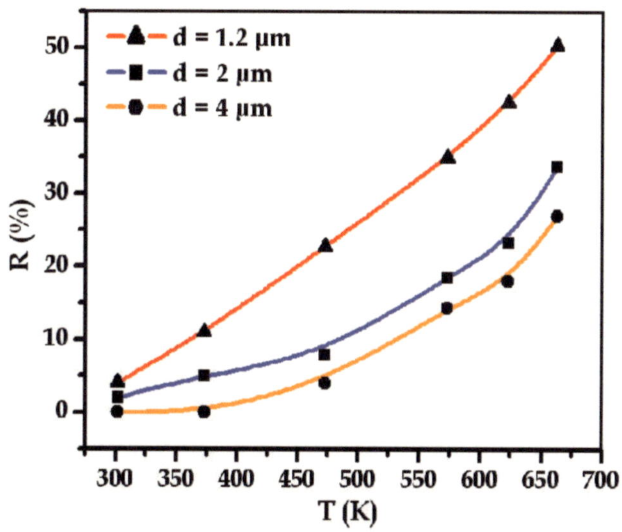

Fig. 13. Fraction R of charges removed from the irradiated volume as a function of temperature for impure polycrystalline alumina samples of different grain diameters d.

The semi-logarithmic plot of the degree of discharge R versus reciprocal temperature exhibits, as for the pure samples, an Arrhenius type law leading to an activation energy of about 0.12 eV for the smallest grain size sample (d = 1.2 µm) and about 0.28 eV for the largest one (d = 4 µm). These activation energies are an indication of detrapping from trapping sites located within the grain boundary.

It is worth noting that, at room temperature and in the same impure polycrystalline samples, breakdown strength increases with the grain sizes (Liebault, 1999; Si Ahmed et al., 2005), which is also the case for the fraction of removed charges R. Therefore this correlation confirms the importance of charge spreading to prevent breakdown.

6.3 Charge spreading in alumina of sub-micrometric grain size

The evolution of the properties with the grain size raises queries about the effect of changing the grain size from micron to nanometer scales. The study of charging properties of nanostructured alumina is beyond the scope of the present chapter. However, there is interest in trying to verify whether they can be obtained by simple extrapolation from the results of micrometric grain size materials.

Nanopowders, with grain diameter of about 27 nm, have been synthesized by the gaseous phase method and compacted via magnetic compaction process at Ural State Technical University, Russia (Kortov et al., 2008). Next, sintering of the compacts has been carried out at Institut National des Sciences Appliquées (INSA) of Lyon (France). The sintering temperature of 1473 K (dwelling time 60 min) was reached at a rate of 3 K/min. As expected, with such heating rate, a substantial grain growth occurred (i.e., from 27 to about 100 nm, cf. Fig. 14). Indeed, grain growth could have been reduced by using faster heating rates that are made possible by different sintering techniques such as spark plasma sintering.

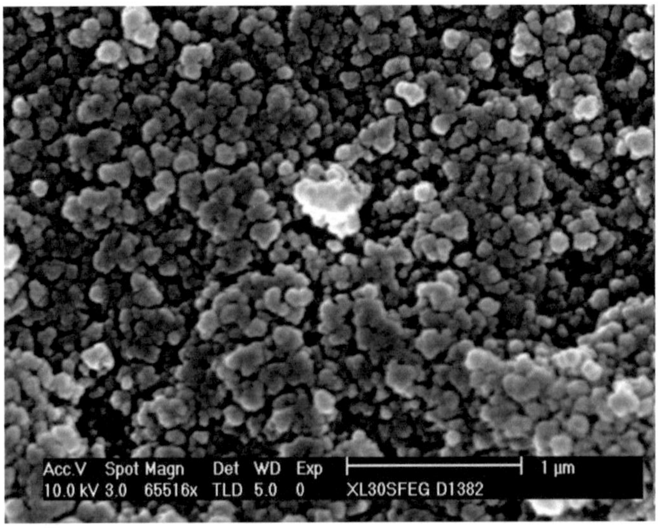

Fig. 14. Microstructure of the sub-micrometric grain size of the impure polycrystalline alumina sample after sintering. This picture is the SEM observation of fracture surface. The average grain diameter has grown after sintering to about of 100 nm (the initial particle diameter prior to sintering was about 27 nm).

The overall purity of this material is about the same as the one of impure polycrystalline alumina. In Fig. 15, the fraction R of charges removed from the irradiated volume as a function of temperature is given for the 0.1 and 1.2 µm samples. The manifest difference is the sharp enhancement of R between 600 and 663 K for the 0.1 µm sample (the value of R increases from 30 to 90 %), which contrasts with the continuous behaviour of the other polycrystalline samples. The activation energy that arises from the semi-logarithmic plot of the recovery parameter versus reciprocal temperature in the range 600-663 K is 0.53 eV (Moya et al., 2007). These results are an indication that detrapping occurs, at about 600 K, from a dominant efficient trap having a well-defined energy level in the gap as in the case of silver doped single crystal (Zarbout et al. 2010).

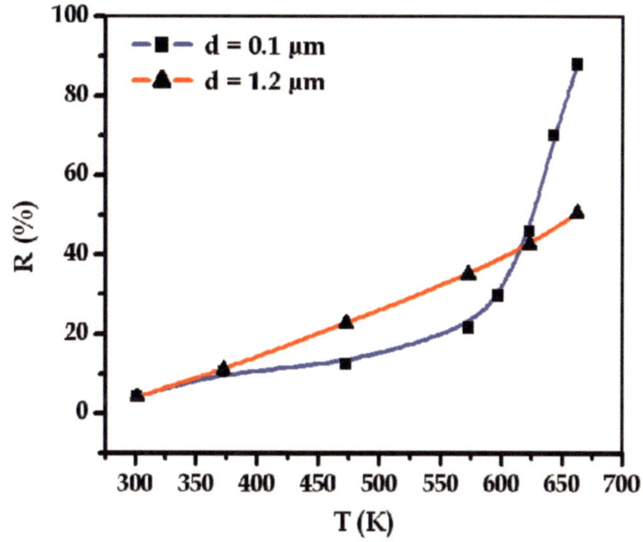

Fig. 15. Fraction R of charges removed from the irradiated volume as a function of temperature for the sub-micrometric and 1.2 µm grain sizes alumina samples.

7. Conclusion

This chapter provides a method for the characterization of charge trapping and spreading in dielectrics. A quantitative recovery parameters reflecting the relative degree of the two competing processes is accurately derived. The experimental set up makes possible the assessment of the effect of temperature (in the range 300-700 K). The ability of polycrystalline alumina to trap or, conversely, to spread charges depends strongly on the grain size and segregation of impurities at interfaces. The results suggest that the grain boundary interfaces can be associated to shallow traps whereas the defects within the grains to deeper ones. The strong tendency for segregation of the main impurities implies that an internal gettering effect can also intervene. It appears therefore that the control of the microstructural development, during the conventional sintering process, is of importance as it provides ways to influence the insulator properties in technological applications of oxide ceramics, for instance, the breakdown strength. Further investigations dealing with the properties of nanostructured materials, processed by sintering techniques that reduce grain growth, could bring more understanding of the role of interfaces.

8. Acknowledgments

The authors are grateful to Dr. Goeuriot D. and Dr. Liébault J. (E.N.S.M.), to Prof. Kortov V.S. (Ural State Technical University) for the supply of some samples. Fruitful discussions with Dr. Bernardini J. (Im2np) and Prof. Fakhfakh Z. (LaMaCoP) were highly appreciated. The first author gratefully acknowledges financial support from the Ministry of Higher Education and Scientific Research of Tunisia and the French Institute of Cooperation.

9. References

Adamiak, K. (2003). Analysis of charge transport in high resistivity conductors under different conduction models. *Journal of Electrostatics*, Vol. 57, No. 3-4, (March 2003), pp. 325-335, ISSN 0304-3886

Aoufi, A. & Damamme, G. (2008). Analysis and numerical simulation of secondary electron emission of an insulator submitted to an electron beam. *Proceedings of ISDEIV 2008 23th International Symposium on Discharges and Electrical Insulation in Vacuum*, pp. 21-24, ISBN 978-973-755-382-9, Bucharest, Romania, September 15–19, 2008

Atkinson, K. J. W.; Grimes, R. W.; Levy, M. R.; Coull, Z. L. & English, T. (2003). Accommodation of impurities in α-Al_2O_3, α-Cr_2O_3 and α-Fe_2O_3. *Journal of the European Ceramic Society*, Vol. 23, No. 16, (December 2003), pp. 3059-3070, ISSN 0955-2219

Blaise, G. & Le Gressus, C. (1991). Charging and flashover induced by surface polarization relaxation process. *Journal of Applied Physics*, Vol. 69, No. 9, (May 1991), pp. 6334-6339, ISSN 0021-8979

Blaise, G. (2001). Charge localization and transport in disordered dielectric materials. *Journal of Electrostatics*, Vol. 50, No. 2, (January 2001), pp. 69-89, ISSN 0304-3886

Blaise, G.; Pesty, F.; & Garoche, P. (2009). The secondary electron emission yield of muscovite mica: Charging kinetics and current density effects. *Journal of Applied Physics*, Vol. 105, No. 3, (February 2009), pp. 034101(1-12), ISSN 0021-8979

Bonnelle C. & Jonnard P. (2010). Dynamics of charge trapping by electron-irradiated alumina. *Physical Review B.*, Vol. 82, No. 7, (August 2010), pp. 075132(1-8), ISSN 1098-0121

Cazaux, J. (1986). Some considerations on the electric field induced in insulators by electron bombardment. *Journal of Applied Physics*, Vol. 59, No. 5, (March 1986), pp. 1413-1430, ISSN 0021-8979

Cazaux, J. (2004). Charging in Scanning Electron Microscopy "from Inside and Outside". *Scanning*, Vol. 26, No. 4, (July/August 2004), pp. 181-203, ISSN 1932-8745

Chiang, Y. -M.; Birnie III, D. & Kingery, W. D. (1996). *Physical Ceramics: Principles for Ceramic Science and Engineering*, (Wiley), John Wiley & Sons, ISBN 0-471-59873-9, New York

Dörre, E. & Hübner, H. (1984). *Alumina: processing, properties, and applications*, Springer-Verlag, ISBN 3-540-13576-6, Berlin

Gavrilov, K. L.; Bennison, S. J.; Mikeska, K. R. & Levy-Setti, R. (1999). Grain boundary chemistry of alumina by high-resolution imaging SIMS. *Acta Materialia*, Vol. 47, No. 15-16, (November 1999), pp. 4031-4039, ISSN 1359-6454

Gontier-Moya, E. G.; Bernardini, J. & Moya, F. (2001). Silver and Platinum diffusion in alumina single crystals. *Acta Materialia*, Vol. 49, No. 4, (autumn 2001), pp. 637-644, ISSN 1359-6454

Haddour, L.; Mesrati, N.; Goeuriot, D. & Tréheux, D. (2009). Relationships between microstructure, mechanical and dielectric properties of different alumina materials. *Journal of the European Ceramic Society*, Vol. 29, No. 13, (October 2009), pp. 2747-2756, ISSN 0955-2219

Jardin, C.; Durupt, P.; Bigarre, J. & Le Gressus, C. (1995). The surface potential and defects of insulating materials probed by electron and photon emissions. *Annual Report Conference on Electrical Insulation and Dielectric Phenomena*, pp. 548-551, ISBN 0-7803-2931-7, Virgina Beach, VA, USA, October 22-25, 1995

Jonnard, P.; Bonnelle, C.; Blaise, G.; Remond, G. & Roques-Carmes, C. (2000). F^+ and F centers in α-Al_2O_3 by electron-induced x-ray emission spectroscopy and cathodoluminescence. *Journal of Applied Physics*, Vol. 88, No. 11, (December 2000), pp. 6413-6417, ISSN 0021-8979

Kansy, J. (1996). Microcomputer program for analysis of positron annihilation lifetime Spectra. *Nuclear Instruments and Methods in Physics Research Section A: Accelerators, Spectrometers, Detectors and Associated Equipment*, Vol. 374, No. 2, (May 1996), pp. 235-244, ISSN 0168-9002

Kolk, J. & Heasell, E. L. (1980). A study of charge trapping in the Al-Al_2O_3-Si, MIS system. *Solid State Electron*, Vol. 23, No. 2, (February 1980), pp. 101-107, ISSN 0038-1101

Kortov, V. S.; Ermakov, A. E.; Zatsepin, A. F. & Nikiforov S. V. (2008). Luminescence properties of nanostructured alumina ceramic. *Radiation Measurements*, Vol. 43, No. 2-6, (February-June 2008), pp. 341 - 344, ISSN 1350-4487

Krause-Rehberg, R. & Leipner, H. S. (1999). *Positron Annihilation in Semi-conductors: defect studies*, Springer-Verlag, ISBN 3-540-64371-0, Berlin

Kröger, F. A. (1984). Electrical properties of α-Al_2O_3, In: *Advances in Ceramics, Vol. 10: Structure and Properties of MgO and Al_2O_3 Ceramics*, W. D. Kingery, pp. 1-15, American Ceramic Society, ISBN 0916094626, Columbus, OH

Lagerlöf, K. P. D. & Grimes, R. W. (1998). The defect chemistry of sapphire (α-Al_2O_3). *Acta Materialia*, Vol. 46, No. 16, (October 1998), pp. 5689-5700, ISSN 1359-6454.

Le Gressus, C.; Valin, F.; Henriot, M.; Gautier, M.; Duraud, J. P.; Sudarshan, T. S.; Bommakanti, R. G. & Blaise G. (1991). Flashover in wide-band-gap high-purity insulators: Methodology and mechanisms. *Journal of Applied Physics*, Vol. 69, No. 9, (May 1991), pp. 6325-6333, ISSN 0021-8979

Liebault, J. (1999). Ph. D. thesis. *Behavior of alumina materials during injection of charges. Relation between microstructure, dielectric breakdown and image current measurement (The SEMM method)*, Ecole Nationale Supérieure des Mines de Saint Etienne, France, INIST T 126061

Liebault, J.; Vallayer, J.; Goeuriot, D.; Tréheux, D. & Thévenot, F. (2001). How the trapping of charges can explain the dielectric breakdown performance of alumina ceramics. *Journal of the European Ceramic Society*, Vol. 21, No. 3, (March 2010), pp. 389-397, ISSN 0955-2219

Liebault, J.; Zarbout, K.; Moya, G. & Kallel A. (2003). Advanced measurement techniques of space-charge induced by an electron beam irradiation in thin dielectric layers. *Journal of Non-Crystalline Solids*, Vol. 322, No. 1-3, (July 2003), pp. 213-218, ISSN 0022-3093

Mohapatra, S. K. & Kröger, F. A., (1977). Defect structure of α-Al_2O_3 doped with titanium. *Journal of the American Ceramic Society*, Vol. 60, No. 9-10, (September 1977), pp. 381-387, ISSN 1551-2916

Moya, G. & Blaise, G. (1998). Charge trapping-detrapping induced thermodynamic relaxation processes, In: *Space Charge in Solids Dielectrics*, J.C. Fothergill & L.A. Dissado, pp. 19-28, Dielectric Society, ISBN 0 9533538 0 X, Leicester, UK

Moya, G.; Kansy, J.; Si Ahmed, A.; Liebault, J.; Moya F. & Gœuriot, D. (2003). Positron lifetime measurements in sintered alumina. *Physica Status Solidi (a)*, Vol. 198, No. 1, (July 2003), pp. 215-223, ISSN 1862-6319

Moya, G.; Zarbout, K.; Si Ahmed, A.; Bernardini, J.; Damamme, G. & Kortov, V. (2007). Grain size effect on electron transport properties of poly and nano-crystalline alumina, *First International Meeting on Nano-materials,* Belfast, January 13-15, 2007

Rasmussen, J. J. & Kingery, W. D., (1970). Effect of dopants on the defect structure of single-crystal aluminium oxide. *Journal of the American Ceramic Society*, Vol. 53, No. 8, (August 1970), pp. 436–440, ISSN 1551-2916

Si Ahmed, A.; Kansy, J.; Zarbout, K.; Moya, G.; Liebault, J. & Goeuriot, D. (2005). Microstructural origin of the dielectric breakdown strength in alumina: A study by positron lifetime spectroscopy. *Journal of the European Ceramic Society,* Vol. 25, No. 12, (2005), pp. 2813-2816, ISSN 0955-2219

Song, Z. G.; Ong, C. K. & Gong, H. (1996). A time-resolved current method for the investigation of charging ability of insulators under electron beam irradiation. *Journal of Applied Physics*, Vol. 79, No. 9, (May 1996), pp. 7123-7128, ISSN 0021-8979

Stoneham, A. M. (1997). Electronic and defect processes in oxides. The polaron in action. *IEEE Transactions on Dielectrics and Electrical Insulation*, Vol. 4, No. 5 (October 1997), pp. 604-613, ISSN 1070-9878

Suharyanto, Yamano, Y.; Kobayashi, S.; Michizono, S. & Saito, Y. (2006). Secondary electron emission and surface charging evaluation of alumina ceramics and sapphire. *IEEE Transactions on Dielectrics and Electrical Insulation*, Vol. 13, No. 1 (February 2006), pp. 72-78, ISSN 1070-9878

Thome, T.; Braga, D. & Blaise, G. (2004). Effect of current density on electron beam induced charging in sapphire and yttria-stabilized zirconia. *Journal of Applied Physics*, Vol. 95, No. 5, (March 2004), pp. 2619-2624, ISSN 0021-8979

Tiku, S. K. & Kröger, F. A. (1980). Effects of space charge, grain-boundary segregation, and mobility differences between grain boundary and bulk on the conductivity of polycrystalline Al_2O_3. *Journal of the American Ceramic Society*, Vol. 63, No. 3-4, (March 1980), pp. 183-189, ISSN 1551-2916

Touzin, M.; Goeuriot, D.; Guerret-Piécourt, C.; Juvé, D. & Fitting, H. -J. (2010). Alumina based ceramics for high-voltage insulation. *Journal of the European Ceramic Society,* Vol. 30, No. 4, (March 2010), pp. 805-817, ISSN 0955-2219

Vallayer, B.; Blaise, G. & Treheux, D. (1999). Space charge measurement in a dielectric material after irradiation with a 30 kV electron beam: Application to single-crystals oxide trapping properties. *Review of Scientific Instruments*, Vol. 70, No. 7, (July 1999), pp. 3102-3112, ISSN 0034-6748

Zarbout, K.; Moya, G. & Kallel, A. (2005). Determination of the electron beam irradiated area by using a new procedure deriving from the electron beam lithography technique. *Nuclear Instruments and Methods in Physics Research Section B: Beam Interactions with Materials and Atoms,* Vol. 234, No. 3, (June 2005), pp. 261-268, ISSN 0168-583X

Zarbout, K.; Si Ahmed, A.; Moya, G.; Bernardini, J.; Goeuriot, D. & Kallel, A. (2008). Stability of trapped charges in sapphires and alumina ceramics: Evaluation by secondary electron emission. *Journal of Applied Physics,* Vol. 103, No. 5, (March 2008), pp. 054107(1-7), ISSN 0021-8979

Zarbout, K.; Moya, G.; Si Ahmed, A.; Damamme, G. & Kallel, A. (2010). Study of discharge after electron irradiation in sapphires and polycrystalline alumina. *Journal of Applied Physics*, Vol. 108, No. 9, (November 2010), pp. 094109(1-8), ISSN 0021-8979

25

Evaluation of Dielectric Properties from the Cakes of Feldspathic Raw Material for Electrical Porcelain Production

V.P. Ilyina
Establishment the Karelian Centre of Science of the Russian Academy of Sciences Institute of Geology of the Russian Academy of Science
Russia

1. Introduction

Feldspars are anhydrous alumosilicates containing alkaline (Na^+, K^+) and alkaline-earth (Ca^{2+}) cations. The basic types of feldspars used in ceramic production are: potassic feldspar (microcline) $K_2O \bullet A1_2O_3 \bullet 6SiO_2$, sodic feldspar (albite) $Na_2O \bullet Al_2O_3 \bullet 6SiO_2$ and calcium feldspar (anorthite) $CaO \bullet A1_2O_3 \bullet 2SiO_2$. The dielectrical properties of electrical porcelain depend on the mineralogical composition, alkali ratio and total alkali content of feldspathic raw material [1].

To examine new sources of raw materials for electrical porcelain production, we have evaluated nonconventional types feldspathic rocks of Karelian. Feldspathic rocks such as alkaline and nepheline syenites from the Elisenvaara and Yeletozero deposits, aplite-like granite from the Louhi area and volcanics such as Kostomuksha halleflinta and Roza-Lambi quartz porphyry. On a chemical compound they represent Alumosilikate K, Na, Ca, less often Ba. Form isomorphic numbers, including Plagioclase [3]. In world practice, the above rocks are a common source of mineral products which has some advantages over pegmatite.

Nonconventional types of feldspathic raw materials were evaluated on cakes because the dielectrical and other properties of natural feldspars and their cakes are much the same, while the chemical bonds of natural feldspars persist after their melting and are inherited by the glass phase of the resulting ceramics [1,2].

Cakes (material in vitreous state, as in ceramics) were prepared from finely ground (particle size 0.063 mm) powders of deferrized feldspathic rocks by caking them in crucibles at 1350° C for 3 hr.

We have assessed the electrical properties (dielectrical permeability -ε, dielectrical loss angle tangent -tgδ and electrical resistance -lgρ), the thermal coefficient of linear expansion (TCLE) and pH of nonconventional types of feldspathic raw materials such as potassic halleflinta, quartz porphyry, syenite and granite-aplite and compared them with those of pegmatite, a common paw material for ceramics production. The compositions and physico-technical properties of feldspathic rocks are shown in Table 1.

2. Electrical characteristics of feldspathic raw material

To measure electrical properties, we prepared specimens of cakes. They were 20 - 25 mm in diameter and 2 - 3 mm in height. ε-, lgρ- and tg δ-values were measured using a bridge with a capacity E of 7-8 and a working frequency of 1000 Hz at 20°C.

Electrical properties were estimated by introducing an additional coefficient to account for additional capacity on the specimen zones not covered by electrodes using the formulas: dielectric permeability: $\varepsilon = \kappa_1 \kappa_2 c$, where $\kappa_1 = 1.14$ is the coefficient of the sensor, κ_2 is the coefficient of specimen thickness, c is specimen capacity; specific electrical resistance: $\lg\rho = \kappa/g \cdot l$, where $\kappa = 33.55$ is the sensor coefficient, g is conductivity, l is specimen thickness; dielectrical loss: $tg\,\delta = 0.175 \cdot g/s$, where c is specimen capacity and g is conductivity.

The results of the measurement of the dielectrical properties of feldspathic rocks are shown in Table 1, and the dependence of variations in ε, lgρ and tgδ on the quantities of microcline and plagioclase in the cake is shown in Figures 2.

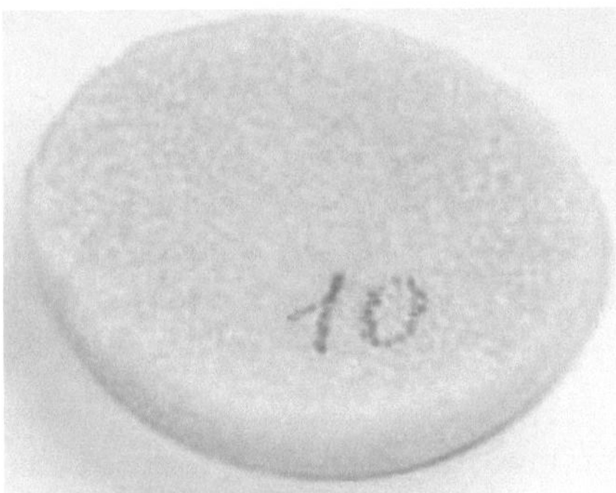

Fig. 1. Cake from feldspathic raw material.

Dielectrical permeability, ε, is the electrical parameter of a mineral showing its ability to polarize in an electrical field. According to experimental data, the dielectrical permeability of the cakes of the rocks evaluated (Nos. 1 – 12) varies from 3.26 to 8.1. The dielectrical permeabilities of feldspathic rocks from other deposits (Nos. 13 – 18), shown in Table 1, range from 5.6 to 7.7, and according to E.V. Rozhkova's results presented in the literature [4,5] /(Golod et al., 1975), they vary from 5.6 – 6.3. This means that the variation range of ε of the cakes of nonconventional feldspathic rocks is within known values.

Analysis of the dielectrical permeability values obtained has shown that ε depends largely on mineralogical composition. Microcline and quartz were shown to have the greatest effect on dielectrical permeability. A high percentage of quartz (44 mass.%), in contrast to that of other rocks, in volcanics from the Roza-Lambi deposit (cake no. 7) contributes to a decline in ε of the cake to 5.4 dtn/cm^2.

The ε dielectrical permeability of plagioclase rock cakes (Nos. 10 – 12, Table) is lower than that microcline cakes (3.26-4.02), which seems to be due to the presence of large quantities of quartz and Ca^{2+} and Mg^{2+} ions in cakes 11-12 and Ba^{2+} ions in cakeв no. 10, because the electrical conductivity of K^+ and Na^+ ions in dielectrics is higher than that of Ca^{2+}, Mg^{2+} and Ba^{2+} ions [1,4].

Type of breeds	Feldspar deposits		Mineral structure, *mass. %*				Dielectric properties			TKL R⊙10^{-6} 1/ grad	pH
	Area	Deposits	№пп	Q	Mi	Pl	E	ρ⊙10^{10} Ohm⊙ cm	tg δ	400^0C	
Pegmatite	Chaupino-Louhi	Hetolambino, Uracco, Civ-guba	1 2 3	2,5 28.6 24.2	70.0 64.1 61.3	27.5 7.3 14.5	7.70 7.53 7.35	- 0,94 1.00	- 0.148 0.062	7.70 7.93 7.72	9.67 9.08 9.85
	Priladozhye	Jccima Lypikko	4 5	27.2 27.8	49.8 46.6	23.0 25.6	6.20 6.73	0.95 1.00	0.900 0.127	8.10 8.51	9.83 9.82
	Uljlega	Cjryla	6	7.1	58.0	34.9	7.00	2.10	0.027	7.90	8.84
Volcanic rock	White Sea	Roza-Lambi	7	44.0	44.2	11.8	5.40	0.85	0.160	8.08	7.10
	Kalevala	Kostomuksha	8	24.2	68.0	7.8	7.50	1.23	0.050	9.15	10.0
Syenite	Louhi	Yeletozero	9	23.1	53,5	31.4	6.65	1.51	0.032	7.83	9.80
	Priladozhye	Elisenvaara	10	2.8	43,0	54.2	3.26	2.00	0.025	8.00	8.25
Granite-aplite	Louhi	Yeletozero	11	29.1	Or-30,8	40.4	3.87	0.85	0.207	7.89	7.98
		Slyudozero	12	22.0	Or-11,1	66.9	4.02	1.32	0.034	7.84	9.76
Pegmatite-granite	Mamsko-Chauiscoe	Mamskoe,B. Northern	13	9.0	71,0	20.0	7.5	-	-	7.78	-
	Enskoe	Rikolatva	14	10.1	69,5	26.4	7.0	-	-	7.52	8.4
	Central Asia	Ljngarskoe	15	15.5	47.7	36.8	5.6	-	-	7.82	-
	Kazakhstan	Karaotse-lgskoye	16	2.5	58,7	38.8	7.7	-	-	7.76	-
	Uralskoe	Malache-vskoye	17	43.6	29,3	27.1	5.6	-	-	7.80	
	Finland	Kemio (FFF)	18	8.0 50	37,0 50,0	55.0 450	-	-	-	-	8.3 83

Table 1. Mineral structure and properties of feldspar breeds.

Figure 2 shows that high microcline and plagioclase concentrations in feldspathic rocks are responsible for a low angle tangent of dielectrical loss in cakes: (no. 8) potassic halleflinta - 0,050, (no. 10) syenite - 0,025 – 0.032, (no. 6) and pegmatite (Cjryla) – 0.027- and increases their electrical resistance (Fig.3) -(1,23, 2,0, 2,10⊙10^{10} Ohm⊙cm), which is consistent with GOST parameters 2484-85 for electrical porcelain: dielectrical permeability not more than 7, the angle tangent of dielectrical loss – 0.030 and electrical resistance 1-2⊙10^{12}. Low dielectrical loss (0.025) and high electrical resistance (2.0⊙10^{10} Ohm⊙cm) are characteristic of Elisenvaara syenites which have a high total percentage of microcline and plagioclase (97 mass.%).

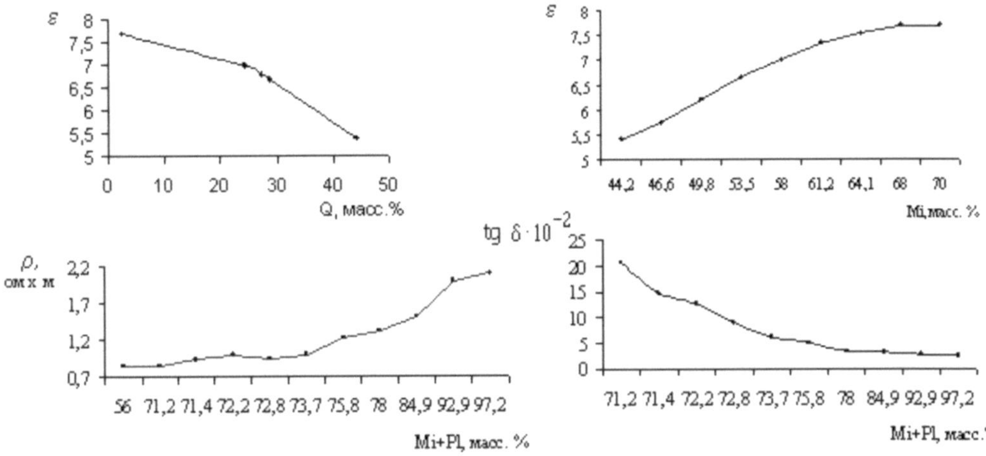

Fig. 2. Dependence of dielectric properties on feldspathic rock composition.

Elisenvaara syenites and Kostomuksha volcanics are most efficient feldspathic raw materials for electrical porcelain production in terms of their electrical properties.

3. Thermal Coefficient of Linear Expansion (TCLR)

Thermal coefficient of linear expansion (TCLR) is a parameter which shows thermal expansion upon burning of ceramic products.

TCLR was measured on a DKV- 4 quartz dilatometer on specimens, 55⊙ × 5 m in size, produced by pressing ground rocks to a grain size of 0.063 mm and subsequent burning in a silitic furnace at 1350°C to reach the maximum density of the specimens. TCLR was estimated using the formula: $\alpha = \frac{1}{l_0}\times\frac{l_t - l_0}{t_1 - t_0} + 55,5\times10^{-6}1/\text{deg.}$, where t_0 is initial measurement temperature, 20°C , t_l is final measurement temperature, 400, 700°C; l_0 is the original length of the specimen, mm, l_t is specimen lengthening, mm; 5.5⊙(10^{-6}- 1/deg. of the TCLR of quartz glass.

The highest TCLR values (9.15 and 11.75⊙10^{-6} 1/deg., respectively) are characteristic of halleflinta (cake no. 8) at 400° and 700°C, respectively. Halleflinta differs structurally from pegmatite in having finely dispersed quartz, which contributes to intense dissolution of quartz in a glass phase. Figure 3 shows that curves 1, 2 and 5 for the cakes of the rocks studied exhibit a non-uniform course of thermal expansion. A rapid increase in TCLR in the range 600 – 700° C is typical of compositions with a minimum quartz concentration and high microcline and plagioclase concentrations which form a less viscous liquid phase. Large quantities of quartz in rocks are responsible for the rectilinear course of the TCLR curves (3, 4, 6 and 7). A minimum TCLE value (7.10⊙10^{-6} 1/deg) in the range 600-700° C is observed in cake 7 with a minimum quartz concentration (44.0 mass.%), and a maximum TCLR value (8.9⊙10^{-6} 1/deg) in cake 1 with a quartz concentration of 2.5 mass. %.

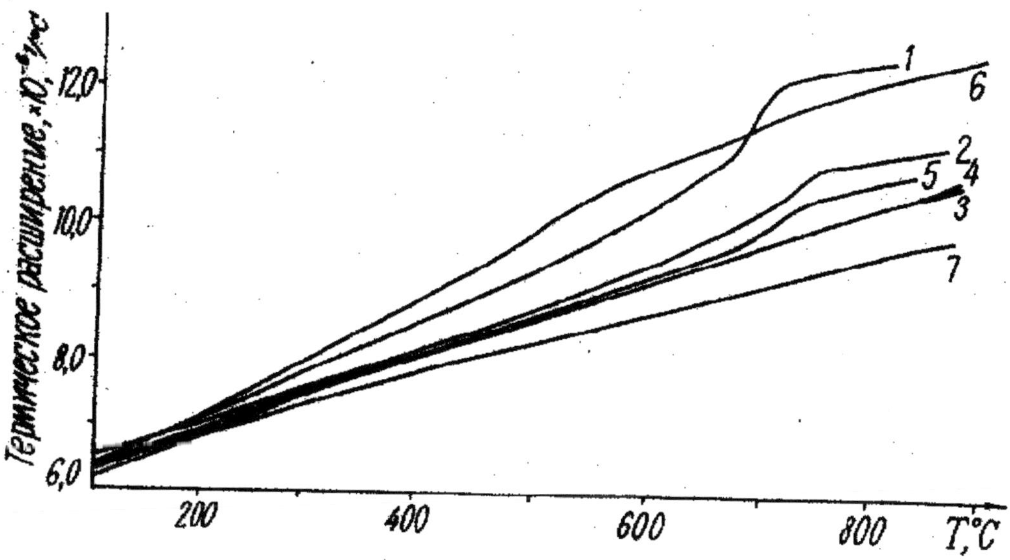

Fig. 3. Temperature dependence of the thermal coefficient of linear expansion: (numbers as in Table 1).

According to the literature [5] the dependence of TCLR on quartz concentration is due to a difference in the density of minerals. It is known that as mineral density increases, TCLR decreases. As the density of quartz (2.65 g/cm^2) is higher than that of albite (2.61 g/cm^2) and that of microcline (2.55 g/cm^2), low TCLR values in cakes with high quartz concentrations can well be expected. The dependence of TCLR on microcline-plagioclase ratio shows a more complex pattern. As quartz concentration rises with a decline in feldspar concentration [1], also observed in Roza-Lambi quartz porphyry, the porcelain burning expansion interval will expand to 100 °C without deformation of products upon burning.

Quartz has thus been shown to decrease the TCLR of feldspathic rocks, as indicated by the absence of rapid changes in the curves (Fig.3) in the range 600 – 700 °C. High concentrations of feldspathic minerals (microcline, plagioclase - 95 – 97.5 mass. %), which form a less viscous glass phase, increase the TCLR of feldspathic rocks.

4. pH of feldspar suspension

pH is a parameter showing the acidity of feldspathic rocks and the engineering properties of clay slurry (viscosity, the density of clay slurry, tixotropy for molding ceramic products. As pH rises, the tixotropy of the clay slurry increases. Interacting with water, feldspathic minerals change the acidity of ceramic shlicker, depending on their constituent cations and anions. The optimum viscosity of clay slurry for refractory products is achieved at pH 5.5 and that for porcelain at pH 7.5–9. According to GOST 21119.3-91 for feldspathic fillers in varnish and paint production, pH is 6-9.

The pH of suspensions was measured using the potentiometer method on an I-120.1-type ionometer. Suspensions were prepared from из finely dispersed samples (grain size 0.063 mm) at solid:fluid ratio of 1:2.5 in accordance with the procedure described in [6,7]. The results of measurements of the pH of feldspar rock suspensions are presented in Table 1, and Figure 4 shows the dependence of pH on total microcline and plagioclase concentration in the rock.

Our results have shown that a common pattern of variations in the pH of feldspathic rock suspensions is a rise in pH with increasing alkaline oxide concentration.

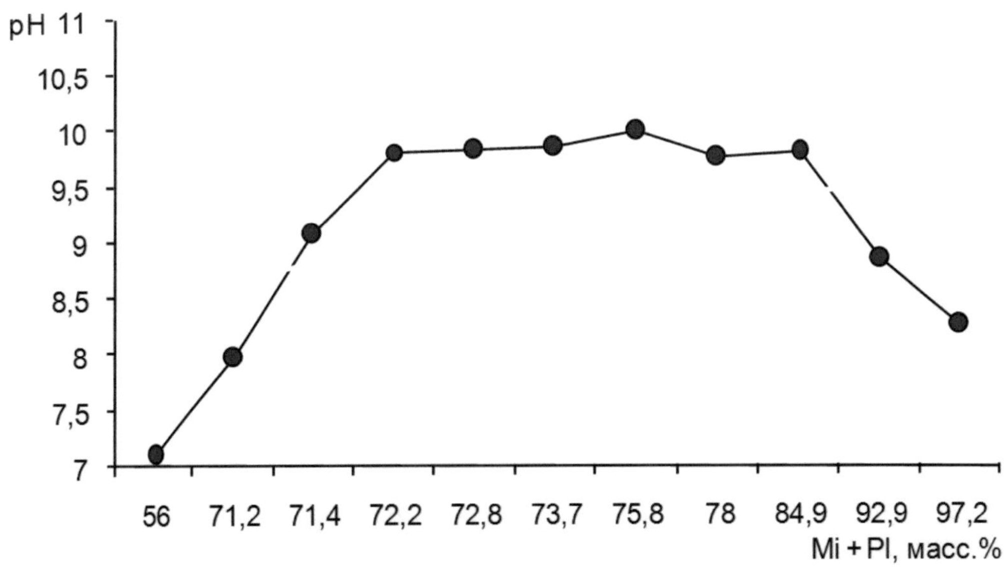

Fig. 4. Dependence of variations in pH on microcline and plagioclase.

Table 1 shows that suspensions of Khetolambino and Lupikko microcline pegmatites have the highest pH values: 9.67 and 9.82, respectively. Potassic halleflinta, a nonconventional type of feldspathic raw material, has a high pH (10.0). Comparison of the pH of quartz-feldspathic rock suspensions has shown that large quantities of quartz (pH of quartz is 6.4-6.9) decrease their pH considerably, as in Roza-Lambi volcanics (pH 7.1).

Figure 4 shows that rocks that contain a total microcline and plagioclase concentration of 75.8–78.0 mass. % and quartz concentration of 22.0–24.2 mass. % have the highest pH.

Thus, the pH of feldspar suspensions depends on total microcline and plagioclase concentrations and quartz concentration in the rock. The pH of Elisenvaara syenite is 8.25 is consistent with feldspathic raw material according to GOST 21119.3-91 (pH 6-9) and Kemio feldspar from Finland (pH 8.3).

5. Conclusions

The electrical properties ε , lg ρ, tg δ of the cakes of nonconventional types of feldspathic rocks and conventional types, such as pegmatite, are basically affected by alkaline oxides and quartz. Dielectrical permeability depends more on microcline and quartz concentrations in rocks. The angle tangent of dielectrical loss and electrical resistance depend on total microcline and plagioclase concentration. Elisenvaara syenites, which contain high microcline and plagioclase concentrations (97 mass.%), typically show lower tg δ values (0.025) and higher electrical resistance ($2.0\odot10^{-12}$ Ohm$\odot$cm) than other rocks. Their TCLR is consistent with that of microcline pegmatite. The pH of the suspensions of nonconventional feldspathic raw material depends on total microcline and plagioclase concentration and quartz concentration. Elisenvaara alkaline syenites are a promising feldspathic raw material for electrical porcelain production.

6. References

[1] Avgustinnik, A.I. Ceramics. L.: Stroyizdat, 1975. 590 p.

[2] Avgustinnik, A.I., Kholodok, N.I, Golod, M.I. et al. On the thermal-physical properties of feldspathic minerals from pegmatite veins. // Geophysical study of the eastern Baltic Shield. Petrozavodsk, 1980. P.108-117.

[3] Ilyina, V.P. Technological evaluation of feldspathic rocks from the Louhi district // Geological and technological evaluation of minerals and rocks from the Republic of Karelia and some regions of the European continent. Petrozavodsk, 1997. P.57-59.

[4] Golod, M.I., Grodnitsky, L.L., Klabukov, B.N. On the dielectrical permeability of plagioclase from pegmatite veins of the Kola Peninsula// Indicator minerals of the characteristics of their host environment. L., "Наука",1975. P. 47-49.

[5] Kholodok, N.I., Golod, M.I., Popova, I.A., Klabukov, B.N. On the dielectrical permeability of potassic feldspars from pegmatite veins // Pegmatites from Karelia and the Kola Peninsula. Petrozavodsk, 1977. P.160-164.

[6] Karyukina, V.N. Determination of minerals from the pH of their suspension // Modern methods for mineralogical study of rocks, ores and minerals. M.: Gosgeotekhizdat, 1957. P.208-230.

[7] Kopchenova, E.V., Karyukina, V.N. Estimation of the pH of mineral suspension and the implication of this index in mineralogical studies // Modern methods for mineralogical study. M.: Nedra, 1969. P.148-155.

26

Sintering of Transparent Conductive Oxides: From Oxide Ceramic Powders to Advanced Optoelectronic Materials

Guido Falk
Saarland University, Chair Powder Technology of Glass and Ceramics, Saarbruecken, Germany

1. Introduction

Since Cadmium thin films have been sputtered and thermally oxidized for the first time in 1907 (Chopra, et al., 1983) the technological innovation in transparent conductive oxides (TCO) has developed rapidly and substantially during the last years. Indium-tin-oxide (In:90:Sn:10, ITO) is today the most important TCO material due to the potential combinations of high electrical conductivity in the range of 10^4 S/cm and high transparency in the visible range of 90 % with a layer thickness of 100 nm.

Today TCO thin films are processed by chemical or physical vapour deposition, especially vacuum based sputtering and evaporation processes. Due to these elaborate and complex manufacturing, especially for large-size TCO applications direct structuring of TCO layers by printing, sol-gel coating and other powder- and paste-based manufacturing processes have been the subject of many investigations (Hyatt, 1989, Straue, et al., 2009).

These as-processed TCO materials serve as transparent electrodes in liquid crystal displays, thin-film electroluminescence displays, electrochromic displays, transparent conductive coatings of highly sensitive radiation detectors, ferroelectric photoconductors and memory devices, transparent conductive oxidic films as gate-electrodes for injection and charge coupled devices and are used in products as flat panel displays, touchscreens, organic light emitting diodes, electroluminescence lamps as well as numerous components of solar technology.

Resulting from the increased indium consumption and the expected supply gap of indium raw materials within the next decades (Carlin, 2007) an increasing priority is attributed to the development of suitable substitute materials, whereas currently aluminium doped zinc oxide (ZnO:Al, AZO) have consistently attached the highest importance (Ellmer, et al., 2008). Beside the TCO applications mentioned above AZO is used in highly promising developments such as blue and ultraviolet lasers, components of improved light amplification of GaN-based LEDs, transparent thin film transistors, photo detectors, varistors, catalysts and optical current transformers.

Sputtered Al:ZnO thin films have already been used in commercially available flat panel displays und thin layer solar cells with an electrical resistance of 1-3 $\cdot 10^{-4}$ Ω cm at aluminium doping concentrations between 1.6 to 3.2 at.-% (Anders, et al., 2010).

Among others current research is focussing on sputtering process stability based on ITO target materials, on AZO target materials with improved electrical conductivity, thermal and mechanical stability as well as highest transparency of TCO thin layers by enhanced optimization of chemical composition, microstructure tailoring and advanced sintering and densification methodologies of TCO target materials.

2. Microstructure-property relationship of TCO materials

2.1 Nodule formation

Nodules, also called „black growths" or „black crystals", are conical defects formed during the sputtering process on the surface of the target material. The influence of nucleation on nucleus growth of nodules was investigated by camera monitoring (Schlott, et al., 1996). It was found, that especially impurities originating from the sputtering process, i.e. SiO_2 or Al_2O_3 particles , were collected on the target material and act as nucleus in the nodule formation process.

There are also other important impurities in the form of inclusions that originates from the target manufacturing process. Target materials that happen to feature the increased densification, are more suitable for sputtering processes since they exhibit fewer nodule failures (Schlott, et al., 1995, Gehman, et al., 1992). The shape of nodules has been investigated by SEM analysis and partly a distinct peak is observed, partly a flattened conical shape is formed (Schlott, et al., 1996). From time to time spherical globule like particles are observed at the nodules peak being identified as SiO_2 and Al_2O_3. In this case the particles could be assigned to impurities of the sputtering process being directly related to the nodule formation process. It was found by further detailed investigations that the external layer of the nodules contains pure ITO phase, the thickness of the layers being up to several tens of microns. Additionally these layers containing fewer oxygen concentrations compared to the bulk target material are supposed to be formed by self-sputtering. Self-sputtering means that the target itself was coated by the sputtering process as it is shown schematically in Figure 1. The oxygen deficiency can be explained by the fact that the atmosphere nearby the target surface contains less oxygen.

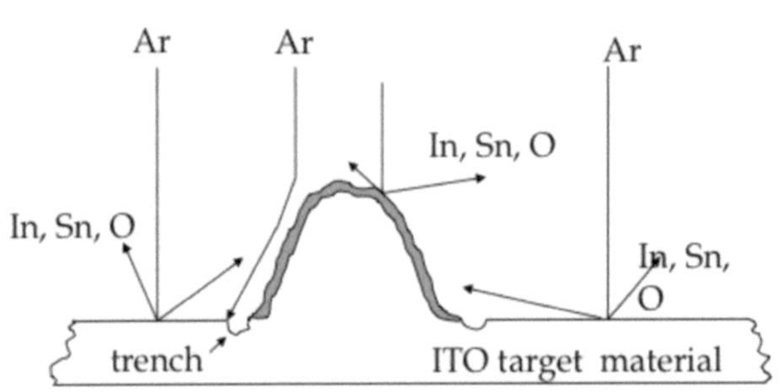

Fig. 1. Cross section scheme of a nodule formed on the surface of an ITO sputtering target according to (Schlott, et al., 1996).

Nodules have to be avoided, since they modify the sputtering process and it is therefore necessary to interrupt the process and to clean or even to exchange the target material. Nodule formation is a severe problem since nodules show a reduced sputtering voltage compared to the surrounding material. Thereby the nodules prevent the sputtering of the material being covered by the nodule layers. Furthermore when the nodule formation is avoided, arc discharge is not needed and the sputtering process is excecuted at increased sputtering voltages enabling to operate the facility at higher efficiencies (Nadaud, et al., 1995). The nodule formation was observed for metallic indium-tin targets during the reactive sputtering process (Schlott, et al., 1996) and also for sputtering of oxide ceramic ITO targets (Schlott, et al., 1995). The elimination of inclusions and metallic phases is predominant for the effective avaoidance of nodule formation (see Figure 2).

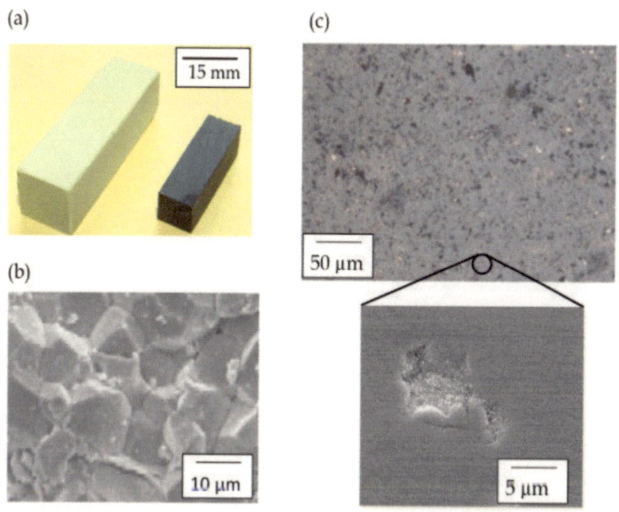

Fig. 2. Cold isostatically pressed ITO body (a, left) and sintered ITO body with a shrinkage of 15 % (a, right), SEM topography of fractured ITO surface after sintering with a mean grain size of 25 µm (b) and outbreak (dark) and segregations (bright) analysed by SEM at polished ITO surfaces (c).

A trouble-free microstructure is important to prevent the chipping (Schlott, et al., 1996). Impurties and metallic InSn eutectica have to be avoided in order to guarantee sufficient thin film qualities (Schlott, et al., 1996).

Once being formed the nodules grow contineously during the sputtering process. They do not dissolve unless they explode due to thermal stresses or due to the power of micro arc discharge effects (Schlott, et al., 1996). If that is done the desintegrated nodule particles are scattered and form nucleus of new nodules or they form small holes, so called pinholes, in the growing layer contributing to a significant quality degradation of the sputtered thin film (Kukla, et al., 1998).

Nodule formation is observed particularly frequently when the target material consists of several tiles being composed to enlarge the target surface. The split between the tiles are considered as collecting sites for impurites and dust. Furthermore the split could be coated by self-sputtering. If an Ar atom is smashed onto these coated splits, the scattered particles composed of In, Sn and O are able to be deposited onto the sputtered thin films (Schlott, et al., 1996). A key component for the suppression of undesired nodule formation is the basic understanding of most important ITO characteristics as well as specific sintering techniques and its effectiveness to tailor distinct microstructural properties of TCO target materials. In the following sections an overview of the state of the art and future trends of the related topics are given.

2.2 The system ITO and related defect structures

Enoki (Enoki, et al., 1991) proposed the following In_2O_3-SnO_2 phase diagram shown in Figure 3.

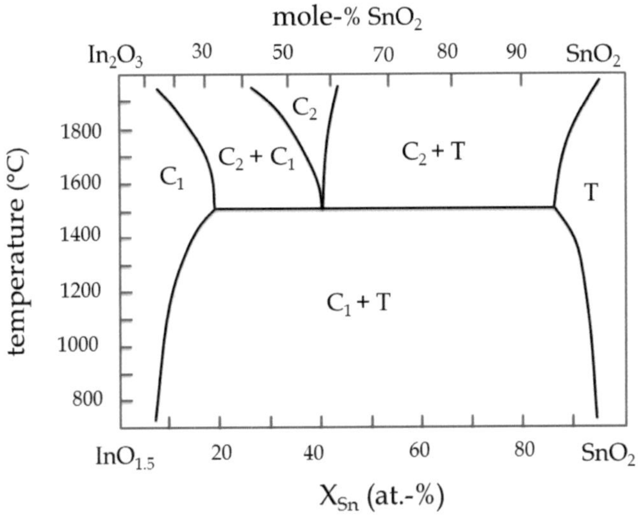

Fig. 3. Phase diagram of the pseudo-binary system In_2O_3-SnO_2 (χ_{Sn} = atom concen-tration of tin (%)).

Hereby the abbriviations C_1 and C_2 represent the cubic ITO phase and the orthorhombic intermediate respectively, and T is the rutile type SnO_2. It was found that the C_2 phase is formed in the concentration range between 47.9 and 59.3 mole-% Sn at temperatures exceeding 1573 K.

Morover it was oberserved that the intermediate (C_2) falls apart into In_2O_3 (C_1) and the T-phase according to the following eutectic reaction:

$$C_2 \leftrightarrow C_1 + T \quad (1)$$

Finally it was observed that the solubility limit of SnO_2 in In_2O_3 phase is between about 12.4 to 15.0 mole-% which is independent of the temperature (Enoki, et al., 1991). For some basic understanding it is helpful to consider the crystal structure of solid In_2O_3. The C type rare earth sesquioxide-or bixbyite crystal structure of Indium oxide is a variation of the cubically body-centred crystal structure (Marezio, 1966).

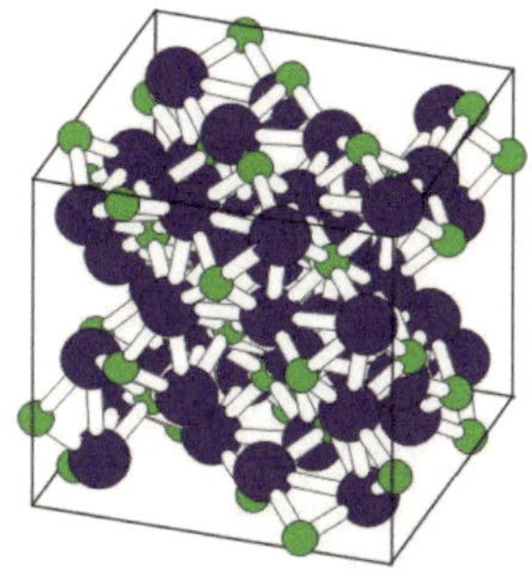

Fig. 4. Cubic In_2O_3 unit cell.

The unit cell has 80 atoms or 16 molecular weights (Warschkow, et al., 2003) and has two unequal cation positions (Granqvist & Hultaker, 2002). The lattice constant is 10.117 ± 0.001 Å (Marezio, 1966). The space group is referred to as Ia^3 or Th^7. Pure In_2O_3 has a density of 7.17 g/cm³ (Bates, et al., 1986) and the theoretical density of the cubic structure is 7.12 g/cm³ (Marezio, 1966). The electron subshell configuration of both atoms indium and oxygen is as follows according to (Chopra, et al., 1983):

$$\text{O:}\quad 1s^2 \mid 2s^2\, 2p^4 \,\text{- -} \mid \tag{2}$$

$$\text{In:}\quad 1s^2 \mid 2s^2\, 2p^6 \mid 3s^2\, 3p^6\, 3d^{10} \mid 4s^2\, 4p^6\, 4d^{10} \mid 5s^2\, 5p \tag{3}$$

The oxygen atoms need two more p-electrons to reach a stable 8-electron configuration. The indium atoms have three electrons in addition to a stable electron configuration. As a result the stoichiometry of the oxide is In_2O_3 resulting in a transition of electrons from In to O and a crystal structure with In^{3+} and O^{2-} ions in the lattice (Mayr, 1998). The unit cell has two unequal In-positions, the first space-diagonally (d position) and the other surface-diagonally (b position). With the d positions the oxygen atoms are located in the corners of an slightly distorted cube with two space-diagonal empty sites. In the second cube the oxygen atoms are located in a slightly differently distorted cube with two surface-diagonal empty sites. The characteristics of the defect structure are determined by this special arrangement in atoms (Hwang, et al., 2000). In_2O_3 exhibit a cubic bixbyite structure (Marezio, 1966), SnO_2 in contrast has a tetragonal structure (see Fig. 4) similarly to the rutile structure (Enoki, et al., 1991). The density of SnO_2 is 6.95 g/cm³ (Bates, et al., 1986).

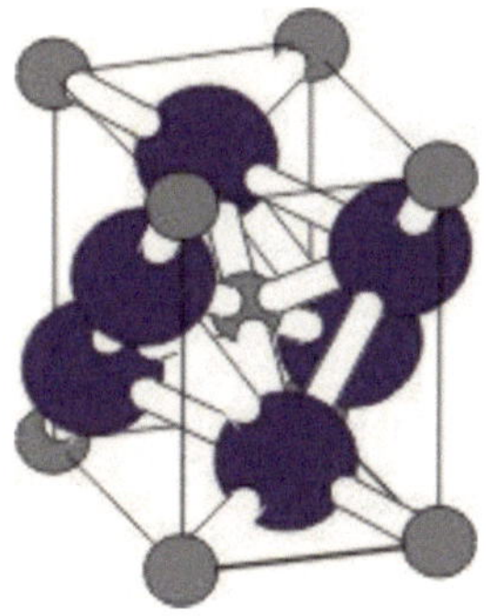

Fig. 5. Tetragonal SnO_2 unit cell.

Indium atoms are incorporated in the lattice as In^{3+} and the tin atoms as Sn^{2+}. The tin atoms have the following electron configuration (Mayr, 1998):

$$Sn : 1s^2 \mid 2s^2\, 2p^6 \mid 3s^2\, 3p^6\, 3d^{10} \mid 4s^2\, 4p^6\, 4d^{10} \mid 5s^2\, 5p^2 \tag{4}$$

It was shown by means of X-ray diffraction analysis (Frank & Köstlin, 1982) that the cubic In_2O_3 structure is preserved by doping with SnO_2 up to the solubility limit of Sn in In_2O_3. The exact solubility limit of Sn in In_2O_3 is not exactly known and varies between 6 ± 2 at.-% of Sn. Up to this concentration every tin atom is substituted by an indium atom. The solubility in thin layers is even higher (Hwang, et al., 2000). The maximum solubility of Indium in the SnO_2 lattice is as low as 1 at.-%. Thereby Sn^{4+} ions are substituted by In^{3+} ions significantly decreasing the electrical conductivity. The ion radius of the Sn^{4+} is 0.71 Å and should lead to a linear reduction of the latticed constant with increasing doping of Sn^{4+} according to the Vegard law (Nadaud, et al., 1998) since the ion radius of In^{3+} is 0.81 Å. However, this is not observed. Udawatte (Udawatte, et al., 2000) reports on a maximum lattice constant reached at 5 mole-% of Sn content dropping below the maximum solubility of 6 mole-% of Sn in the In_2O_3 lattice reported by Nadaud (Nadaud, et al., 1998). These authors have calculated a lattice constant of 10.1247 Å at the maxium solubility limit of Sn compared to a lattice constant of 10.1195 Å that is observed for pure In_2O_3. The presence of Sn strongly changes the behaviour of the oxygen ions. Due to Sn doping the general distance between oxygen and cation increases, but the distance between oxygen and Sn decreases (Nadaud, et al., 1998). Typical phase modifications of indium-tin oxide are listed in the following table.

phase	unit cell	Structure	lattice constant (Å)	density (g/cm³)
In_2O_3	cubic	Bixbyite	10.117	7.17
SnO_2	tetragonal	Rutile	-	6.95
$In_4Sn_3O_{12}$	rhomboedric	Fluorite	-	7.30

Table 1. ITO phases, unit cells, structures, lattice constants and theoretical desities.

The formation of Indium oxide and the subsequent reaction contributing, inter alia , to the defect structure can be described by Kröger-Vink notation (Rahaman, 1995) according to the following structure elements of the chemical reactions (see Table 2).

abbreviaton		subscript term		superscript term	
V	vacancy	In	In lattic site	x	neutral
In	In atom	O	O lattice site	•	positively charged
O	O atom	i	Interstitial	′	negatively charged
e	electron	g	gas phase		
h	hole				

Table 2. Structure elements of chemical reactions according to Kröger-Vink notation.

The formation of Indium oxide is described according to the following reaction equation:

$$In_2O_3 = 2In^x{}_{In} + 3O^x{}_O \leftrightarrow 2In_g + 3/2O_g \quad (5)$$

In parallel to this reaction the formation of gaseous In_2O can occur:

$$2In^g + (1/2O_2)_g \leftrightarrow (In_2O)_g \quad (6)$$

Furthermore In_2O_3 is produced from gaseous In_20^g in oxygen atmospheres according to the following reaction:.

$$In_2O_3 \leftrightarrow 2In^x{}_{In} + 3O^x{}_O \leftrightarrow (In_2O)_g + (O_2)_g \quad (7)$$

Taking account of mass action law it is obvious that the oxygen partial pressure controles the number of oxygen vacancies $V^x{}_O$. The number of charge carriers in In_2O_3 is thus depending very much on oxygen partial pressure. Theoretically the number of charge carriers would increase at decreasing oxygen partial pressures associated with an increase of electrical conductivity (Mayr, 1998). However, the mechanismen is more complicated in reality since scattering mechanisms can occur consequently decreasing the charge carrier mobility and ion conductivity. More detailed information about scattering mechanisms are given in (Chopra, et al., 1983).

Oxygen vacancy concentration is a function of the defect structure of indium tin oxide (Freeman, et al., 2000). At the beginning first oxygen vacancies start to form as soon as the oxygen atoms leave their interstitial lattice sites $((2Sn^{\bullet}{}_{In}O_i'')^x)$ and are transformed to gaseous oxygen phase $(1/2O_{2g})$ (Freeman, et al., 2000). The following equation describes the formation of oxygen vacancies and the release of two electrons:

$$(2Sn^{\bullet}{}_{In}O_i'')^x \leftrightarrow 2Sn^{\bullet}{}_{In} + (1/2O_2)_g + 2c' \quad (8)$$

Only under reducing conditions $(\sim pO_2{}^{-1/8})$ oxygen vacancies are formed by diffusing into the bulk from their former lattice sites $(O^x{}_o)$ (Hwang, et al., 2000). Under extreme reducing conditions ($pO_2 \sim 10^{-14}$ atm) non-reducing defects are formed such as $(2Sn^{\bullet}{}_{In}3O_OO_i'')^x$ (González, et al., 2001). The equation (9) decribes the transformation of gaseous species whereas two electrons for any oxygen vacancy. Equation (10) shows that the number of oxygen vacancies $(V^{\bullet\bullet}{}_o)$ formed and the number of doped lattice sites ($n = [D^{\bullet}{}_{In}]$) is a function of oxygen partial pressure $(pO_2{}^{1/2})$. This relation is described by the equilibrium constant K:

$$O^x{}_o \leftrightarrow (1/2O_2)_g + V^{\bullet\bullet}{}_o + 2e' \quad (9)$$

$$K = pO_2{}^{1/2}[V^{\bullet\bullet}{}_o]n^2 \quad (10)$$

If one oxygen atom is missing in the unit cell the valance electrons of the surroun-ding atoms have a reduced ionisation energy, which then is provided by thermal lattice vibrations. These electrons are in a quasi-free state and act as conduction electrons. That means that oxygen vacancies provide electrons for the conduction bands (Mayr, 1998). Even in the undoped state small oxygen deficencies can be detected. In this case oxygen vacancies appear in reduced concentrations compared to other defect structures (Hwang, et al., 2000).

system In_2O_3:Sn		
vacancy formation & self-compensation	$2In_{In}^{x} + 2SnO_2 \leftrightarrow (2Sn^{\bullet}_{In}O_i'')^{x} + In_2O_3$	(11)
donator formation	$(2Sn^{\bullet}_{In}O_i'')^{x} \leftrightarrow 2Sn^{\bullet}_{In} + (1/2O_2)_g + 2e'$	(12)
	$O^{x}_{o} \leftrightarrow (1/2O_2)_g + V^{\bullet\bullet}_{o} + 2e'$	(13)
	$2In_{In}^{x} + 2SnO_2 \leftrightarrow 2Sn^{\bullet}_{In} + 2e' + In_2O_3 + (1/2O_2)_g$	(14)

Table 3. Overview of reactions on formation and self-compensation of vacancies as well as formation of donators of the system In_2O_3:Sn according to Kröger-Vink notation.

Nadaud and co-workers investiated oxygen concentration of bulk-ITO by neutron diffraction and Rietveld analysis (Nadaud, et al., 1998). After sintering at 1400 °C in reducing nitrogen atmospheres stoiciometric oxygen concentrations were detected for both undoped and 6 at.-% Sn doped In_2O_3. On the other hand sintering in oxygen atmospheres resulted in bulk oxygen excess of about 3 %. This could be explained by the stoiciometry of the neutral $(2Sn^{\bullet}_{In}O''_{i})^{x}$ complexes. Another explanation is the formation of large Sn-based oxygen complexes being difficult to get reduced in the intermediate temperature regime. The same authors investigated these large complexes by Mössbauer, EXAFS and neutron diffraction.

ITO Mössbauer- and EXAFS data von ITO reveal a relaxation of the Sn-O shell similar to the observed relaxation in Sn rhich $In_4Sn_3O_{12}$ (Nadaud, et al., 1998). The analytical data allow to extend the explanation of inefficacy doping above 6 at.-% Sn to the effect that Sn atoms are incorporated at cation sites where they are inactive.

The scientific community is controversial discussing the precipitation of the rhomboedric phase $In_4Sn_3O_{12}$ at Sn concentrations exceeding 6 at.-% Sn. The $In_4Sn_3O_{12}$ structure has been analysed by neutron diffraction experiments and it was ascertaiend that the structure is similar to the Fluorite structure and that a sixfold occupation of M1-sites by Sn cations and sevenfold occupation of M2-sites by In- and Sn cations occurs (Nadaud, et al., 1998). The following figure shows the unit cell of the rhomboedric $In_4Sn_3O_{12}$ phase. The lattice constant is 6.2071 Å and the mismatch angle ø is 99.29°.

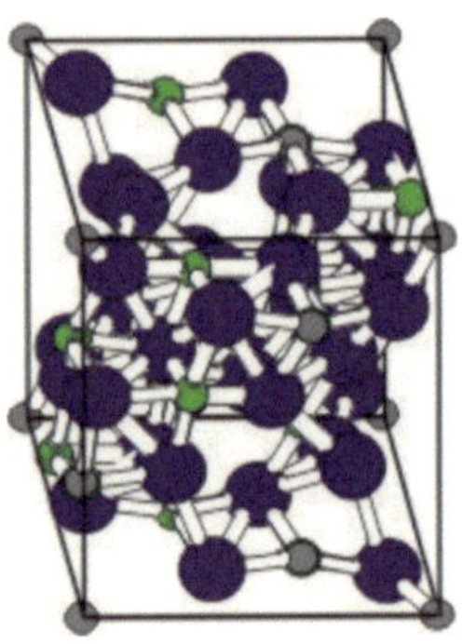

Fig. 6. Unit cell of the $In_4Sn_3O_{12}$ phase.

The atoms of the rhomboedric cells are more densely packed compared to the cubic structure. The rhomoedric phase was fist discovered by Bates et. al. and the density was calculated to 7.303 g/cm³ (Bates, et al., 1986). It was found by X-ray diffraction experiments that the conformation is a densely packed $M'_mM''_nO_{3m}$ defect structure typically observed for the compositions Yb_7O_2 and Pr_7O_{12}. The In_2O_3 as well as SnO_2 solubility in the $In_4Sn_3O_{12}$-phase is limited (Bates, et al., 1986).

3. Synthesis of nano- and microcrystalline ITO ($Sn:In_2O_3$) and AZO ($Al:ZnO_2$) powders

3.1 ITO powder synthesis

In most cases indium-tin-oxide powders are synthesized by hydrothermal processes. Gel formation is based on the co-precipitation of $InCl_3{\cdot}H_2O$ and $SnCl_2$ educts (Udawatte & Yanagisawa, 2000). The microcrystalline powder were homogeneous and reveal the composition of tin doped indium oxide [$In(OH)_3$:Sn] and tin doped indium hydroxide [InOOH:Sn]. Calcination of $In(OH)_3$:Sn at 300 °C resulted in cubic tin doped indium oxide [In_2O_3:Sn]. At calcination temperatures above 500 °C the InOOH:Sn phase is transformed to a solid solution of the formula ($In_2Sn_{1-x}O_{5-y}$). Both powders have been calcinated in air atmosphere. It was found by Mössbauer analysis that the Sn^{4+}-ion coordination number is 8 because each tin atom has 8 neighbouring oxygen atoms which bear opposite charge. A similar synthesis scheme is presented in (Yanagisawa, et al., 2000). In this case indium-tin-oxide has been synthesized from a In-Sn-hydrogel. The hydrothermal treatment of the gel at 300 °C resulted in the formation of InOOH:Sn with mean particle sizes in the range of 80 nm. The subsequent calcination of the product at different calcination temperatures led to different microstructural vacancy configurations of indium tin oxide solid solutions. Calcination at 700 °C lead to a powder with primary particles sizes in the range of about 160 nm. Another approach is the processing of aqueous $In(NO_3)_3{\cdot}H_2O$ solutions, subsequent heating and calcination at 500 °C in order to achieve nanosized ITO powders (Sorescu, et al., 2004). The solution of metallic tin and indium in HCl is precipitated by [NH_4OH] (Nam, et al., 2001) and the resulting indium-tin-hydroxide gel is dried, grinded and calcinated at 600 °C.

The processing of tin doped indium oxide crystalites from a direct indium-tin smelting enriched with oxygen is described in (Frank, et al., 1976). The hetero-geneous nucleation process was initiated by small In_2O_3 crystalites. Different compositions and concentration ranges of the smelting process resulted in different tin doping concentrations. Indium tin oxide synthesis by a chemical transport process is reported in (Werner, et al., 1996). Starting from metallic indium and tin and dissolution of the educts in nitric acid the solution is dried and calcinated at 900 °C to achieve In_2O_3 and SnO_2. The mixture of both oxides were doped with transport media iodine, or sulfur, or chlorine. Indium oxide crystals doped with 8.2 mole-% tin and tin oxide crystals doped with 2.4 mole-% indium were attained. These powders have been characterized and the synthesis reactions based on chlorine transport media have been thermodynamically modeled (Patzke, et al., 2000).

In case the seperate oxides are provided as educt materials for the synthesis reaction, indium doped tin oxide, as reported in (Nadaud, et al., 1994) could be processed. In this case In_2O_3 und SnO_2 powder (purity of 99,99 wt.-%) have been mixed in ethanol and calcinated at

1380 °C for theree hours. Colouration from yellow to green and XRD peaks of indium-tin-oxide without exception proved the complete transforma-tion to indium tin oxide.

An alternative ITO synthesis process is given in (Stenger, et al., 1999). Here molten metallic indium tin alloy has been reacted with oxygen in a plasma arc furnace. The product is subsequently quenched by a gas stream at cooling rates between 10^5 K/s to 10^8 K/s to a final temperature between +50 °C and +400 °C. Thereby mixed indium tin oxide powders were produced with specific surface area of 3 m²/g at maximum. As the powder tends to formation of large agglomerates the size of different ramifications has been analysed by electron microscopy and documented as primary particle size being in the range of 0.03 µm up to 1.0 µm. Hot melt spraying of indium tin melts in oxydizing atmospheres and subsequent quenching is also suitable to process indium tin oxide powders. The agglomerated particle are porous and reveal particle sizes up to 170 µm (see Figure 4).

Compared to this the commercial powder ZY-ITO (Dow Chemical Co. Ltd.) is composed of a mixture of 90 wt.-% In_2O_3 and 10 wt.-% SnO_2, with a specific surface area of 32.0 m²/g and a mean primary particle size in the range of 3.4 µm.

3.2 AZO powder synthesis

Although various studies of synthesis, characterization and application of Al:ZnO thin layer systems are given in the literature (Stanciu, et al., 2001, Selmi, et al., 2008), there are only a few investigations referring to synthesis and processing of nanosized Al:ZnO particles. Flame spray pyrolysis of liquid Al- ad Zn precursors led to uncontrolled Al:ZnO stoichiometries and segregations of the spinel and wurzite phase combined with evaporation of the metallic Zn phase (Kim, 2008). Vapour condensation (Strachowski, et al., 2007) and spray pyrolysis (Hsiao, et al., 2007) as well as low-pressure spray pyrolysis (Hidayat, et al., 2008) resulted in highly agglomerated particles exhibiting poor redispersing characteristics.

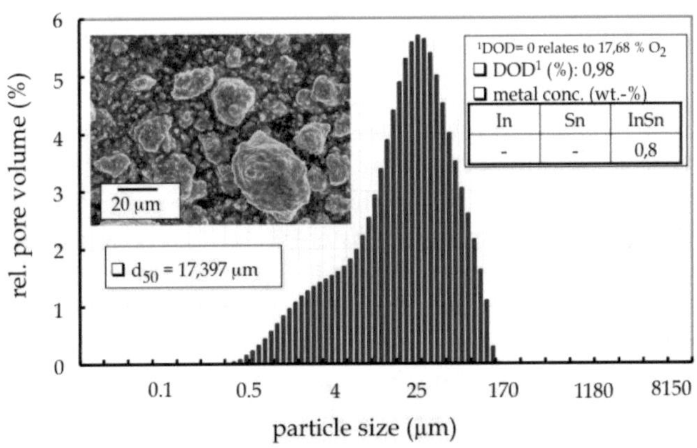

Fig. 7. Indium tin oxide powder produced by hot melt spraying and subsequent quenching: Pore volume of the powder as a function of particle size.

The processing of suitable AZO particles for optoelectronic applications is therefore still challenging. Several alternative processing routes have been elaborated with limited success. AZO particles synthesized by co-precipitation (Nayak, et al., 2007, Aimable, et al., 2010, Shui, et al., 2009) and solvothermal decomposition (Thu & Maenosono, 2010) are characterized by wide particle size distributions and heterogeneous Al doping concentrations of the microstructure. Hydrothermally (Piticescu, et al., 2006) and sol-gel based processing (Chen, et al., 2008) of AZO particles resulted in particles of high agglomeration degree and mean particles diameters considerably higher than 30 nm and undefined redispersing behaviour. The sintering and densification kinetics of Al:ZnO green bodies is slowed down with increasing Al doping concentrations and the grain growth effects are significantly reduced (Han, et al., 2001, Hynes, et al., 2002). Information about processability of synthesized AZO particles within technical and functional ceramic processing chains is still missing.

4. Sintering of TCO materials

4.1 Sintering and Degree of Oxygen Deficiency (DOD)

The degree of oxygen deficiency (DOD) is referred to oxygen deficient in relation to oxygen concentration of the fully oxidized powder (Schlott, et al., 1995). In (Weigert, et al., 1992) the degree of oxygen deficiency is defined as follows:

$$DOD = (a-b) / (a-c) \tag{15}$$

Here a is the oxygen concentration of stoiciometric oxidic compound, b is the oxygen concentration of the partially reduced oxidic compound and c is the oxygen concentration of the completely reduced metallic compound. The degree of oxygen deficiency is closely related to the oxygen vacancies contributing in parts to the electrical conductivity of ITO. Oxygen release is caused by thermolysis of the ITO powder and oxygen uptake is observed during cooling. The arrangement of oxygen vacancies is likely to be closely connected to the powder synthesis.

Thermal treatment in air leads to light-green colouring of the powder and thermal treatment in reducing atmospheres results in black colouring of the powder. Sputtering is causing a yellow colour of the target (Otsuka-Matsua-Yao, et al., 1997). After sintering the powder can also change colour to dark green or black(Udawatte, et al., 2000). The thermal history of the powder is of significant importance since the powder characteristics are closely correlated with the target quality having a profound impact on the characteristics of the sputtered thin layers (Weigert, et al., 1992).

Evidence proves that oxygen release and uptake of ITO follows a chemical hysteresis (Otsuka-Matsua-Yao, et al., 1997). In repeated heating and cooling cycles different quantities of oxygen are incorporated or released. According to the findings of experimental sintering investigations (Otsuka-Matsua-Yao, et al., 1997) the quantities of released oxygen below 773 K was marginal. A preliminar thermolysis at 1073 K was needed to release carbon monoxide absorbed on the surface. The authors came to the result, that the thermodynamic oxygen characteristics of ITO samples were changed during heating and cooling cycles. Hereby the total oxygen concentration is reduced during each cycle and the kinetics of oxygen release is slowed down cycle by cycle. The initiation temperaure of oxygen release was observed to increase on every iteration of heating and cooling cycle. It was assumed by the authors that there are several phases and several phase transformations inbetween these cycles causing the discontinuous oxygen release. The related phases exhibit similar chemical composition and resemble the CaF_2 structure. The phases are devided in coherent domaines and the surface energy could have a major impact on the *Gibb's free energy*. This means that chemical hysteresis appears and the oxygen concentration can not be determined explicitly as a function of the temperature and the oxygen partial pressure (Otsuka-Matsua-Yao, et al., 1997).

But there is evidence by experimental investigations that several microphases could act as barriers thereby preventing the oxygen release. These microstructures could originate from extended defects, these are sheared structures or domaines, which diverge from each other in therms of chemical compositions on a very small scale. However the exact analysis and qualitiative and quantitative identification of these structures by X-ray diffraction is not possible since they are charcterized by the structure of the rare earth elements, as for example In_4SnO_8, $In_4Sn_2O_{10}$, In_2O_3 etc. The oxygen release could therefore depend on the interfacial energy between these microphases. In sputtering processes partially reduced indium-tin-oxide target materials are preferred in order to achieve thin film characteristics of minimum specific electric resistance and maximum sputtering efficiencies. Partially reduced targets reveal improved electrical and thermal characteristics as well as optimized densification.

The reduction of the powder or sintered body can be achieved by several methods, i.e. sintering in vacuum or reducing atmospheres (H_2, CO, H_2-Ar oder H_2-N_2). Alternatively the specimen can be hot pressed in graphite moulds or carbon releasing materials or the carbon releasing materials can be doped with the specimen (Weigert, et al., 1992). However under these reducing conditions the caculated oxygenn stoichiometry can not be accurately controlled. Target materials with reproducible and uniform degree of oxygen deficiency are preferred, since they guarantee superior sputtering efficiencies and reduced operation and coating times (Weigert, et al., 1992).

Furthermore the adding of reducing agents and ingredients causes normally the formation of metal particles of different sizes considerably larger than 50 μm (Schlott, et al., 1995). The microstructure appears spotty and targets processed by the powder qualities reveal lower fracture toughness characteristics. From this reason the target specimen shall be reduced below 1000 °C and preferably even below 800 °C. The subsequent densification of the powder and/or the sintered body is realized usually by hot isostatic pressing at temperatures above the melting point of the metal indium-tin-phases (Schlott, et al., 1995).

The hydrophilic characteristics of In_2O_3 should be noted. The water uptake from the surrounding atmosphere leads to the transformation of In_2O_3 to $In(OH)_3$. The water absorption capacity is increasing with increased specific surface area (Lippens, 1996).

Ceramic ITO target materials show low thermal conductivities compared to the metallic target materials and are therefore very susceptible to thermal stresses resulting form non-uniform heating during the sputtering process. It is therefore preferred to increase the thermal conductivity within certain limits in order to guarantee improved thermal shock resistance characteristics (Schlott, et al., 2001).

Based on those optimized target materials the sputtering process is able to be performed at higher specific powder densities without resulting in target desintegration due to formation of thermal gradients.

Improved thermal shock resistance characteristics also allow increased sputtering rates and shortened processing and operation times and consequently reduced production cost. At comparable sputtering rate increased thermal conductivity causes decreased target surface temperature and the decreased temperature gradient ensures a reduction of thermal stresses as well as increased reliability and target life time. This is especially true for target materials working as cathodes under pulsed sputtering operation conditions (Schlott, et al., 2001). In the state-of-the-art the thermal conductivity of ITO target materials is in the range of 7 - 10 W/mK.

In principle a high densification of the target materials is desirable, that means the density should exceed 95 % of the theoretical density in order to guarantee exceptable thermal conductivity characteristics. Usually the thermal conductivity could also be increased by partially reduction (Schlott, et al., 2001).

The theoretical melting point of indium-tin-oxide is 2223 °C. However above the temperature of 1600 °C ITO is evaporated due to a critical vapour pressure regime of the indium and tin metal phases (Vojnovich & Bratton, 1975). At a temperature of 1000 °C the vapour pressure of indium is $6.6 \cdot 10^{-5}$ bar and that of tin is $6.6 \cdot 10^{-7}$ bar (Nichols, 1982).

4.2 Specific TCO sintering techniques

4.2.1 Thermal vacuum degassing

Water absorbed in the green body has to be thermally desorbed in order to avoid crack formation and specimen desintegration after sintering and subsequent cooling. This is especially true for capsuled and hot isostatically pressed samples. The gas formation during sintering has to be prevented by preheating of green bodies within controlled degassing atmospheres as for example N_2 or Ar (Lippens, 1996). The thermal degassing process purifies the grain boundaries and is resulting in a reduction of specific surface area. Hereby the sintering activity is reduced and the densification has to be achieved by pressure assisted sintering techniques (Schlott, et al., 2001).

As referred to the previous section, thermal vacuum degassing exerts a strong influ-ence on the dgree of oxygen deficiency. Other studies (Falk, 2008) came to the result that vauum degassing and combined hot isostatic pressure sintering of capsuled ITO resulted in optimized sintering densities if the vacuum degassing time and temperature is correlated with a defect free micro structure, adopted concentration of free metal species as well as suitable degrees of oxygen deficiency (see Figure 8).

4.2.2 Pre-sintering

Based on the findings of Son (Son & Kim, 1998) it was shown that pressure assisted pre-sintering of In_2O_3 at 5 MPa could increase the densification rate significantly. Dilatometer experiments have proven that maximum densification rates were achieved at 1130 °C and the pressure was increased for another 5 MPa at this temperature. Sintering in air (1 atm) at 1500 °C, however, only leads to a desification of 76 % of the theoretical density.

In (Son & Kim, 1998) it was found, that the agglomerated green structure was transformed in a homogeneous polycrystalline microstructure at temperatures above 1070 °C and the agglomerates showes increased densification rates compared to the surrounding matrix leading to pore formation, so called interagglomerate pores, inbetween the matrix and the densified microstructure regimes. These pore formations could be effectively prevented by pressure assisted sintering and rearrangement of densified areas and by moving the pores to the surface area.

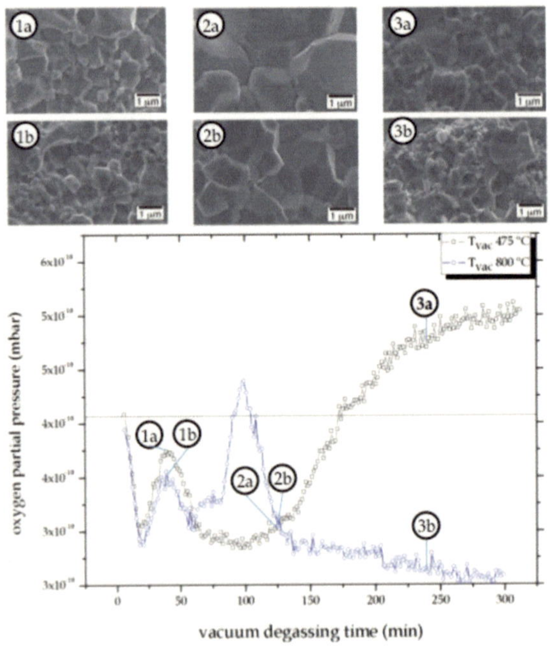

a)

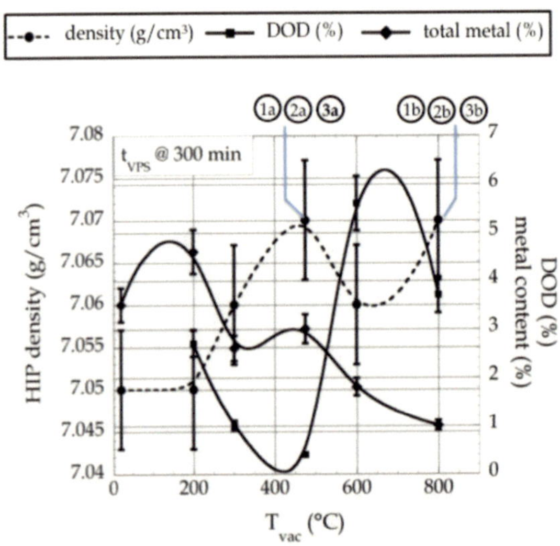

b)

Fig. 8. Oxygen release and oxygen capturing and attained microstructures during vacuum degassing at 475 °C and 800 °C as a function of vacuum degassing time (a), HIP densities, free total metal content and degree of oxygen deficiency (DOD) as a function of vaccum degassing temperature of ITO sintered bodies (b) according to (Andersson, et al. 2005).

4.2.3 Sintering without additives

In (Stenger, et al., 1999) the avoidance of binders and/or dry pressing agents is proposd to prevent any contamination of the received ITO powder. Futhermore it is suggested to avoid the evaporation of gaseous species during pyrolysis of additives since these proceses are likely to reduce efficient pressure build up during hot isostatic pressing.

In (Udawatte, et al., 2000) the group reported about additive-free sintering of ITO powder compacts in air atmospheres. Starting from hydrothermally prepared and at 500 °C calcinated precursors ITO powders of the composition $In_2Sn_{1-x}O_{5-y}$ were attained. The authors have shown previously that pre-sintering of ITO and significant densification is achieved when the $In_2Sn_{1-x}O_5$ phase is transformed to cubic $In_{2-y}Sn_yO_3$ in the temperature range between 1000 °C and 1200 °C. Sintering necks were observed in this temperature range and by excceding the sintering temperature above 1250 °C sigificant grain growth was initiated. At 1300 °C a uniform grain size of 2 µm up to 3 µm and a sintering density of 65% related to the theoretical density was observed.

The maximum densification was achieved at 1450 °C correlated with a mean grain size of about 7 µm. In this case triple grain boundary pores arised more and more frequently compared to intragranular pores.

The conclusion of these experimental investigations were that the sintering is activated mainly in the temperature range between 1300 °C and 1400 °C. The maximum density of about 92 % of the theoretical density was achieved after sintering at 1450 °C for three hours. Furthermore it was concluded that tin doping results in higher densification rates. In (Udawatte, et al., 2000) it is mentioned that tetragonal SnO_2 phase formation counteracts ITO densification.

In (Udawatte & Yanagisawa, 2001) small dry pressed powder compacts (diameter 10 mm, thickness 1.5 mm) have been sintered at 1400 °C for three hours. Taking the theoretical density of 7.106 g/cm^3 as a basis a maximim density of 99.3 % of the theoretical density was achieved. The powder used had a mean particle size of about 80 nm.

Compared to conventional sintering elevated densities have been achieved by „spark plasma sintering" (SPS) (Takeuchi, et al., 2002). At a reduced dwell time of 5 minutes and high heating rates up to 50 K/min the SPS experiments resulted in considerably low sintering densities, probably due to inhomogeneous temperature distribution.

The sintering of cubic and rhomboedric nanosized ITO powders with mean particle sizes in the range of 50 to 100 nm were sintered up to 900 °C (Kim, et al., 2002) where the cubic phase was transformed. This transformation should theoretically results in a volume expansion of 2.1 % which was not observed since grain growth and pore formation were initiated. It was very complicated to eliminate these pores by subsequent sintering at elevated temperatures. The phase transformation promoted the active diffusion of atoms resulting in inhomogeneous grain growth with intragranular pore formation. It is therefore recommended to prevent phase transformation during sintering in order to achieve higher densification rates and more homogeneous microstructures (Kim, et al., 2002).

4.2.4 Sintering with additives

Vojnovich and co-workers (Vojnovich & Bratton, 1975) investigated the influence of imputities in terms of densification rates of In_2O_3 powders. Combinations of the impurities Si, Ca, Mg, Pb and Fe as well as the doping with Kaolin, Al_2O_3 and SiO_2 have been experimentally studied. It was found that the impurities forming liquid phases due to eutectic reactions even contribute to increased densification rates as for example by doping with Ca, Si and Mg as well as Kaolin. Impurities leading to low melting eutectic phase formations favour liquid phase sintering conditions and resulting in higher density values (Vojnovich & Bratton, 1975). It was found that TiO_2 doping causes an increase of sintering density and limits grain growth. From a concentration of 0.5 wt.-% TiO_2 as sintering additive the phase In_2TiO_5 is precipitated at the grain boundary resulting in increased grain boundary diffusion at reduced diffusion activation energies (Nadaud, et al., 1997).

The influence of Si and Zr impurities on ITO sintering was ivestigated in (Nadaud, et al., 1994). These studies have been motivated by the successful ITO-doping with tetravalent Ti (Nadaud, et al., 1997) and the attempt to achieve even better results with alternative tetravalent doping additives. It came out that a sigificant influence of SiO_2 on ITO densification was not observed. Zirconia, however, has a negative influence on densification and increase specific electrical resistance of ITO ceramics at room temperature. The approach to incorporate sintering additives in order to increase densification is being questioned in (Schlott, et al., 1996), since it was found that imputities, i.e. TiO_2, presented as inclusions or segregations in the microsructure, could cause the formation of nodules.

4.3 Sintering behavior and TCO microstructure

Yanai and Nakamura (Yanai & Nakamura, 2003) found that SnO_2 aggregates in the powder could negatively influence the sintering of ITO and could consequently result in higher porosity of the sintered body. In this case the walls of the pores are covered by SnO_2 segregations. In case of finely distributed SnO_2 segregations decreasing porosity is achieved. And finally in (Hsueh & Jonghe, 1984) it was agreed upon that inhomogeneous sintering takes place in case of heterogeneous density and stress distribution. The experimental evidence of these facts was given by measurements of heterogeneous stress distribution along the grain boundaries causing gradients of the chemical potential and acting as a driving force of the material transport.

Investigations of the microstructure of target materials processed at different processing conditions have shown that finely distributed metallic phases led to improved densification and significant recrystallization of the matrix during hot isostatic pressing.

This means that the liquid metal phase could further enhance the densification although under normal operation condition the wetting of indium oxide and tin oxide by the metallic phases of indium and tin is not observed (Schlott, et al., 1995).

The recrystallization leads to higher densities and and an essantial increase of fracture toughness (Falk, 2007, Falk, 2009) . This is unexpected too, since for ceramic materials a reduction of grain size leads to improved mechanical properties. Parallel to these findings it was also oberserved that the morphology of partially reduced powders have a major impact on densification and mechanical stability (Schlott, et al., 1995).

Studies of the fracture surface have shown that the intergranular bonding was much higher in recrystallized microstructures. Hereby the metallic phases take over a crack arresting function.

Good quality targets are attained at DOD values in the area of 0.02 to 0.2. If the degree of oxygen deficiency is too low metallic phases are rare and the positive influence of these phases on densification and mechanical strength is neglectable.

If the degree of oxygen deficiency is too high large area metal segregations act as microstructural failures and cause decreasing mechanical strength and worse densification (Schlott, et al., 1995). These authors processed targets of two different powders incoporting metal segregations of mean diameter between 1 µm to 10 µm and < 200 µm. Grain growth effects as a function of the seggregation size was detected and it was conclued that grain growth was especially pronounced in the case of large seggregated microstructures and consequently the maximum sputtering efficiency could not be achieved due to temporarily devating arc discharge at the surface of the target material.

In case of slightly reduced target materials especially tin segregation have been observed and in the event of substantially reduced targets the metallic phases are indentified as InSn alloys. The In-Sn eutectic phase with 48.3 at.-% Sn is melting at 120 °C according to (Shunk, 1969).

4.4 Sintering behavior and electrical conductivity

The electrical conductivity of both pure indium oxide and pure tin oxide is a result of stoichiometry disturbance due to formation of oxygen vacancies. In case of In_2O_3 this structure can be described by the complex $In_2O_{3-x}(V_o)_x e'_{2x}$ (Mayr, 1998). By tin doping, having a higher valence number compared to indium, negative charge carriers are incorporated into the lattice contributing to an additional increase of electrical conductivity.

Freeman and co-workers (Freeman, et al., 2000) have calculated the theoretical distribution of energy bands in tin doped and undoped In_2O_3. In the case of tin doped indium oxide the s-band in the lower section of the conduction band is broadened. Consequently a high mobility of electrons is achieved explaining the high electrical conductivity. Further investigation of electronic ITO structure was elaborated in (Odaka, et al., 2001).

In general the electrical conductivity of Sn doped ITO is lower than theoretically predicted. It was observed that high Sn doping concentrations even reduce the electrical conductivity. The $In_4Sn_3O_{12}$ phase mentioned before is not or slightly electrically conductive (Nadaud, et al., 1998). This fact can be explained by the inactivity of Sn-cations located in M1-positions due to the ternary coordiation of the surrounding oxygen ions. This is similar to the structure of SnO_2 which is neutral. Enventhough In and Sn are located in close vicinity of the periodic system of elements the additional electron of the Sn atom causes a higher affinity to oxygen compared to indium.

The fact that Sn is deteriorating the electrical conductivity becomes more reasonable when high doping levels exceeding 9 at.-% Sn doping concentrations at increased oyxgene partial pressures ($p0_2$ = 1 atm) are applied as it was experimentally proven by Nadaud (Nadaud, et al., 1998).

With the exception of the $In_4Sn_3O_{12}$-phase formation there are mainly two reasons explaining the decrease of electrical conductivity of ITO as soon as doping concentrations exceeding limiting values of 6 at.-% of tin. First the electrical conductivity is reduced due to the formation of the neutral irreducable clusters (Frank & Köstlin, 1982) as for example $(2Sn^{\bullet}_{In}O_i)^x$ complex or the strongly attached and neutral $(Sn_2O_4)^x$ complex (Frank & Köstlin, 1982). Second the lattice is progressively distored as soon as the doping concentration are increased. The atoms are displaced from their original positions and In_2O_3 is similar to SnO_2 crystalline structure. Consequently Sn_2O_i''-cluster are formed acting as neutral lattice defects (Nadaud, et al., 1998). Furthermore these formed clusters are able to restrict doped charge carrier concentrations und decrease charge carrier mobility by causing *„un-ionized impurity scattering"* (Hwang, et al., 2000).

At increasing oxygen vacancy concentrations a compensation process is initiated resulting in a further decrease of the electrical conductivity according to the following equation (Mayr, 1998):

$$O_i'' + V_o \Leftrightarrow O^x_o \quad (16)$$

The electrical characteristics of ITO -phases have been experimentally investigated by Bates and co-workers (Bates, et al., 1986). After sintering of ITO the cubic body centred In_2O_3 phase, being able to incorporate tin concentrations up to 20 mole-% by solid solution process, the rhomboedric $In_4Sn_3O_{12}$ as well as the tetragonal SnO_2 have been detected. It was found that the electrical conductivity is increasing with increased In_2O_3 phase content up to a phase concentration of about 30 mole-%, passing a constant conductivity level up to about 50 mole-% In_2O_3 and reaching a maximum conductivity level at In_2O_3 phase concentration of about 80 mole-%.

This maximum electrical conductivity is a factor of 20 to 25 times higher than the electrical conductivity of pure In_2O_3 (1.6 up to $2.7 \cdot 10^3$ /Ωcm) and a factor of 6 to 20 times higher than the electrical conductivity of $In_4Sn_3O_{12}$ (100 up to 300 /Ωcm). These values are clearly lower than those of thin layers which are in the range of up to 10^4 /Ωcm (Nadaud, et al., 1994).

Studies of the cyclical heating and subsequent cooling of 10 up to 70 mole-% In_2O_3 in air resulted in a reproducible hysteresis of the electrical conductivity and the thermoelectric power. These characteristics are connected to the formation of the high temperature $In_4Sn_3O_{12}$ phase according to the authors (Bates, et al., 1986).

It is also claimed that there are other phase transformations at elevated temperatures, contributing to additional explanations of the fluctuations in electrical conductivity. In continuous thermogravimetrical studies of ITO targets a hysteresis of oxygen uptake and release was observed from multiple cyclical heating and cooling in atmospheres with controlled oxygen partial pressure (Otsuka-Matsua-Yao, et al., 1997). The conclusion is that the microstructural transformation of ITO in the temperature range between 1273 K and

1773 K is correlated with the oxygen content. These results also indicate that the typical ITO characteristics not only depend on a specific oxygen concentration range since a hysteresis is present. Investigations of electrical In_2O_3 characteristics at elevated temperatures have been elaborated by De Wit (Wit, 1975). Hwang (Hwang, et al., 2000) proposed three different regimes of electrical conductivity as a function of oxygen partial pressure and tin concentration. These three regimes have been ascertained experimentally by electrical conductivity and thermoelectric power measurements. The thermoelectric power is a measurement for the thermal diffusion current which is achieved by a temperature gradient. The first regime is characterized by low oxygen partial pressure ($\sim pO_2^{-1/6}$) and low tin doping concentrations. The second regime is distinguished by mean oxygen partial pressure ($\sim pO_2^{0}$) and mean tin concentrations and the third regime by high oxygen partial pressure ($\sim pO_2^{-1/8}$) and high tin doping concentrations. The different doping concentrations result in different defect structures.

The electrical conductivity of bulk nano-ITO is significantly lower compared to the electrical conductivity of bulk μ-ITO. The explanation seems to be that the charge carrier density and the mobility of charge carriers is much lower in nano-ITO (Hwang, et al., 2000). Modeling of optical and electrical characteristics of ITO-thin films made of nano-ITO have been elaborated by Granqvist and co-workers (Ederth, et al., 2003).

5. Conclusions and future trends

Ceramic transparent conductive oxides are widely used for the processing of thin transparent conductive oxides films by vacuum sputtering techniques. These thin film layers are used in liquid crystal display technologies and various application fields such as energy conservation, information storage, electrophotography, electromagnetic radiation shielding and optoelectronic industry. In order to achieve maximum electrical and thermal conductivities and high sputtering efficiencies usually TCO target materials with distinct degrees of oxygen deficiencies are being used. The transparent semiconductor indium-tin-oxide with its high transmission for visible light, its high electrical conductivity and its strong plasma reflection in the near infrared is one of the most common transparent conductive materials. A simplified description of basic understanding of most important ITO characteristics are given and correlated to the desired microstructural properties of sintered TCO target materials.

Specific sintering techniques, i.e. hot isostatic pressing of vacuum pre-sintered, compacted and capsuled ITO bodies results in a distinct consolidation of microstructure and a homogenisation of ITO phase and thus to an increase in HIP sintered density close to the theoretical density. In order to achieve homogeneous microstructures with small mean grain sizes of the pure ITO phase elaborate demands on specific sintering methodologies and adopted sintering processing chains have to be applied. It is important to note that the characteristics of industrially available raw powders in terms of phase composition, particle size distribution, powder density, degree of oxygen deficiency and concentrations of free metal species such as In, Sn and InSn intermetallics could have considerable impact on the quality of the later ITO product. The assessment of superior powder quality in the framework of stream lined and rationalized powder synthesis and processing can therefore be seen as a way forward to realize further optimization of ITO target materials for sputtering applications. Facing the shortage of indium raw material resources it will be of increasing importance within the coming decades to develop suitable substitute TCO materials. At present aluminium doped zinc oxide (AZO) is one of the most promising candidates. Although the attempts of development of flexible transparent electrode coatings on the base of TCO nano powders for large scale applications have yet to yield any spectacular breakthroughs, new cost-effective niche applications, i.e. flexible displays, are likely to rise up and compete with the state-of-the-art sputtering thin film methodologies in the near future. It is therefore expected that new modern sintering techniques of nano particulate TCO materials, i.e. laser treatment, will be developed in order to significantly improve electrical conductivity of TCO particulate materials on polymer substrates.

6. References

Aimable, A., et al. (2010). Comparison of two innovative precipitation systems for ZnO and Al-doped ZnO nanoparticle synthesis. *Processing and Application of Ceramics,* Vol. 4, No. 3, pp. 107-114, ISSN 1820-6131

Anders, A., et al. (2010). High quality ZnO:Al transparent conducting oxide films synthesized by pulsed filtered cathodic arc deposition. *Thin Solid Films,* Vol. 518, No. pp. 3313-3319, ISSN 0040-6090

Andersson, L., et al. (2005). On a Correlation Between Chemical Hysteresis and Densification Behavior of SnO_2-In_2O_3 Powder Compacts *cfi/Ber. Dtsch. Keram. Ges.,* Vol. 13, No. pp. 208-211, ISSN 0173-9913

Bates, J. L., et al. (1986). Electrical conductivity, Seebeck coefficient, and structure of In_2O_3-SnO_2. *Am. Ceram. Soc. Bull.,* Vol. 65, No. 4, pp. 673-678, ISSN 0002-7812

Carlin, J. F. (2007). Indium, In: *Minerals Yearbook: Metals and Minerals,* J. H. DeYoung, J., pp. 78-79, U.S. Department of the Interiour, U.S. Geological Survey, ISBN 978-1-4113-3015-3, Washington

Chen, K. J., et al. (2008). The crystallization and physical properties of Al-doped ZnO nanoparticles. *Appl. Surf. Sci.,* Vol. 254, No. pp. 5791–5795, ISSN 0169-4332

Chopra, K. L., et al. (1983). Transparent conductors - a status review. *Thin solid films,* Vol. 102, No. pp. 1-46, ISSN 0040-6090

Ederth, J., et al. (2003). Indium tin oxide films made from nanoparticles; models for the optical and electrical properties. *Thin Solid Films,* Vol. 445, No. pp. 199-206, ISSN 0040-6090

Ellmer, K., et al. (2008). *Transparent Conductive Zinc Oxide,* Springer Verlag, ISBN 978-3-540-73611-9, Berlin, Heidelberg, New York

Enoki, H., et al. (1991). The intermediate compound in the In_2O_3-SnO_2 system. *J. Mater. Sci.,* Vol. 26, No. 15, pp. 4110-4115, ISSN 0022-2461

Falk, G. (2008). Densification of $(In_{0,9}Sn_{0,1})_2O_3$ by vacuum pre-sintering and hot isostatic pressing. *cfi/Ber. Dtsch. Keram. Ges.,* Vol. 85, No. 13, pp. 35-38, ISSN 0173-9913

Falk, G. (2009). Heißisostatisches Pressen nanokristalliner Oxidkeramiken, In: *Technische Keramische Werkstoffe,* Kriegesmann, J., pp. 1-37, HvB-Verlag, ISBN 978-3-938595-00-8, Ellerau

Falk, G. (2007). Optimierung von ITO-Targets durch Hippen. *Fortschrittsberichte der DKG: Verfahrenstechnik,* Vol. 21, No. 1, pp. 102-111, ISSN 0173-9913

Frank, G., et al. (1976). The solubilities of Sn in In_2O_3 and of In in SnO_2 crystals grown from Sn-In melts. *J. Cryst. Growth,* Vol. 36, No. 1, pp. 179-180, ISSN 0022-0248

Frank, G. & Köstlin, H. (1982). Electrical properties and defect model of tin-doped indium oxide layers. *Appl. Phys. A,* Vol. 27, No. pp. 197-206, ISSN 1432-0630

Freeman, A. J., et al. (2000). Chemical and thin-film strategies for new transparent conducting oxides. *MRS Bulletin,* Vol. No. August, pp. 45-51, ISSN 0883-7694

Gehman, B. L., et al. (1992). Influence of manufacturing process of indium tin oxide sputtering targets on sputtering behavior. *Thin Solid Films,* Vol. No. 220, pp. 333-336, ISSN 0040-6090

González, G. B., et al. (2001). Neutron diffraction study on the defect structure of indium-tin-oxide. *J. Appl. Phys.,* Vol. 89, No. 5, pp. 2550-2555, ISSN 0021-8979

Granqvist, C. G. & Hultaker, A. (2002). Transparent and conducting ITO films: New developments and applications. *Thin solid films,* Vol. 411, No. 2002, pp. 1-5, ISSN 0040-6090

Han, J., et al. (2001). Densification and grain growth of Al-doped ZnO. *J. Mater. Res.,* Vol. 16, No. 2, pp. 459-468, ISSN 2044-5326

Hidayat, D., et al. (2008). Single crystal ZnO:Al nanoparticles directly synthesized using low-pressure spray pyrolysis. Mat. Sci. Eng. B, Vol. 151, No. pp. 231–237, ISSN 0921-5107

Hsiao, K.-C., et al. (2007). Synthesis, characterization and photocatalytic property of nanostructured Al-doped ZnO powders prepared by spray pyrolysis. *Mat. Sci. Eng. A,* Vol. 447, No. pp. 71-76, ISSN 0921-5093

Hsueh, H. H. & Jonghe, L. C. D. (1984). Particle rotation in early sintering. *J. Am. Ceram. Soc.,* Vol. 67, No. 10, pp. C215-C217, ISSN 0002-7820

Hwang, J. H., et al. (2000). Point defects and electrical properties of Sn-doped In-based transparent conducting oxides. *Solid State Ionics,* Vol. 129, No. pp. 135-144, ISSN 0167-2738

Hyatt, E. P. (1989). Continuous Tape Casting for Small Volumes. *Ceramic Bulletin,* Vol. 68, No. 4, pp. 869-870, ISSN 0002-7812

Hynes, A. P., et al. (2002). Sintering and characterization of nanophase zinc oxide. *J. Am. Ceram. Soc.* , Vol. 85, No. 8, pp. 1979-1987, ISSN 0002-7820

Kim, B. C., et al. (2002). Effect of phase transformation on the densification of coprecipitated nanocrystalline indium tin oxide powders. *J. Am. Ceram. Soc.,* Vol. 85, No. 8, pp. 2083-2088, ISSN 0002-7820

Kim, M. (2008). Mixed-metal oxide nanopowders by liquid-feed flame spray pyrolysis (LF-FSP): Synthesis and processing of core-shell nanoparticles, Ph.D., The University of Michigan, Michigan

Kukla, R., et al. (1998). Sputtered ITO-layers: new approaches for high-quality, low-cost-production, *Proceedings of twelth international conference on vacuum web coating conference,* pp. 104-111, Reno, Nevada

Lippens, P. (1996). Integration of target manufacturing in the sputtering plant, *Proceedings of 39th Annual Technical Conference Proceedings,* pp. 424-430, Albuquerque, ISSN 0737-5921

Marezio, M. (1966). Refinement of the crystal structure of In_2O_3 at two wavelengths. *Acta Cryst.*, Vol. 20, No. pp. 723-728, ISSN 1600-5724

Mayr, M. (1998). High vacuum sputter roll coating: a new large-scale manufacturing technology for transparent conductive ITO layers, Leybold-Heraeus, Report

Nadaud, N., et al. (1995), Matériau pour cible de pulverisation cathodique, EU 0679731, Patent, Saint Gobain Vitrage International

Nadaud, N., et al. (1997). Titania as a sintering additive in indium oxide ceramics. *J. Am. Ceram. Soc.*, Vol. 80, No. 5, pp. 1208-1212, ISSN ISSN 0002-7820

Nadaud, N., et al. (1998). Structural studies of tin-doped indium oxide (ITO) and $In_4Sn_3O_{12}$. *J. Solid State Chem.*, Vol. 135, No. 1, pp. 140-148, ISSN 1095-726X

Nadaud, N., et al. (1994). Sintering and electrical properties of titania- and zirconia-containing In_2O_3-SnO_2 (ITO) ceramics. *J. Am. Ceram. Soc.*, Vol. 77, No. 3, pp. 843-46, ISSN 0002-7820

Nam, J. G., et al. (2001). Synthesis and sintering properties of nanosized In_2O_3-10 wt.-% SnO_2 powders. *Scripta Mater.*, Vol. 44, No. 8-9, pp. 2047-2050, ISSN 1359-6462

Nayak, J., et al. (2007). Yellowish-white photoluminescence from ZnO nanoparticles doped with Al and Li. *Superlattices and Microstructures*, Vol. 42, No. pp. 438–443, ISSN 0749-6036

Nichols, D. R. (1982). ITO films: adaptable to many applications. *Photonics Spectra*, Vol. No. pp. 57-60, ISSN 0731-1230

Odaka, H., et al. (2001). Electronic structure analyses of Sn-doped In_2O_3. *Jpn. J. Appl. Phys.*, Vol. 40, No. 5A, pp. 3231-3235, ISSN 0021-4922

Otsuka-Matsua-Yao, S., et al. (1997). Chemical hysteresis on the release and uptake of oxygen by SnO_2-doped In_2O_3 powders. *J. Electrochem. Soc.*, Vol. 144, No. 4, pp. 1488-1494, ISSN 0013-4651

Patzke, G. R., et al. (2000). Chemischer Transport fester Loesungen. 8. Chemischer Transport und Sauerstoffionenleitfaehigkeit von Mischkristallen im System In_2O_3/SnO_2. *Z. Anorg. Allg. Chem.*, Vol. 626, No. 11, pp. 2340-2346, ISSN 1521-3749

Piticescu, R. R., et al. (2006). Synthesis of Al-doped ZnO nanomaterials with controlled luminescence. *J. Eur. Ceram. Soc.*, Vol. 26, No. pp. 2979–2983, ISSN 0955-2219

Rahaman, M. N. (1995). *Ceramic processing and sintering*, Marcel Dekker, Inc., ISBN 0-8247-9573-3, New York

Schlott, M., et al. (2001), Method of preparing indium oxide/tin oxide target for cathodic sputtering, US 6187253, Leybold Materials GmbH, Patent

Schlott, M., et al. (1996). Nodule formation on indium-oxide tin-oxide sputtering targets, *Proceedings of 1996 SID International Symposium*, Santa Anna, CA, USA

Schlott, M., et al. (1995), Target für die Kathodenzerstäubung zur Herstellung transparenter, leitfähiger Schichten und Verfahren zu seiner Herstellung, DE 4407774, Patent, Leybold Materials GmbH

Selmi, M., et al. (2008). Studies on the properties of sputter-deposited Al-doped ZnO films. *Superlattices and Microstructures*, Vol. 44, No. pp. 268–275, ISSN 0749-6036

Shui, A., et al. (2009). Preparation and properties for aluminum-doped zinc oxide powders with the coprecipitation method. *J. Ceram. Soc. Japan*, Vol. 117, No. 5, pp. 703-705, ISSN 1882-1022

Shunk, F. A. (1969). *Constitution of binary alloys* (2 supplement), McGraw-Hill, Inc., ISBN 07-057315-8, Chicago

Son, J. W. & Kim, D. Y. (1998). Enhanced densification of In_2O_3 ceramics by presintering with low pressure (5 MPa). *J. Am. Ceram. Soc.,* Vol. 81, No. 9, pp. 2489-2492, ISSN 0002-7820

Sorescu, M., et al. (2004). Nanocrystalline rhombohedral In_2O_3 synthesized by hydrothermal and postannealing pathways. *J. Mater. Sci.,* Vol. 39, No. pp. 675-677, ISSN 1573-4838

Stanciu, L. A., et al. (2001). Effects of Heating Rate on Densification and Grain Growth during Field-Assisted Sintering of α-Al_2O_3 and $MoSi_2$ Powders. *Metall. Mater. Trans. A,* Vol. 32, No. pp. 2633-2638, ISSN 1073-5623

Stenger, B., et al. (1999), Verfahren zum Herstellen eines indium-Zinn-Oxid-Formkörpers, DE 19822570, Patent, W.C. Heraeus GmbH

Strachowski, T., et al. (2007). Morphology and luminescence properties of zinc oxide nanopowders doped with aluminum ions obtained by hydrothermal and vapor condensation methods. *J. Appl. Phys.,* Vol. 102, No. 7, pp. 073513 - 073513-9 ISSN 0021-8979

Straue, N., et al. (2009). Preparation and soft lithographic printing of nano-sized ITO-dispersions for the manufacture of electrodes for TFTs. *J. Mater. Sci.,* Vol. 44, No. 22, pp. 6011-6019, ISSN 1573-4838

Takeuchi, T., et al. (2002). Rapid preparation of indium tin oxide sputtering targets by spark plasma sintering. *J. Mater. Sci. Lett.,* Vol. 21, No. pp. 855-857, ISSN 0261-8028

Thu, T. V. & Maenosono, S. (2010). Synthesis of high-quality Al-doped ZnO nanoink *J. Appl. Phys.* , Vol. 107, No. 1, pp. 014308-014308-6, ISSN 0021-8979

Udawatte, C. P. & Yanagisawa, K. (2001). Fabrication of low-porosity indium tin oxide ceramics in air from hydrothermally prepared powders. *J. Am. Ceram. Soc.,* Vol. 84, No. 1, pp. 251-53, ISSN 0002-7820

Udawatte, C. P. & Yanagisawa, K. (2000). Hydrothermal preparation of highly sinterable tin doped indium oxie powders: the effect of the processing parameters, *Proceedings of Ceramic Processing Science VI,* Inuyama City, Japan,

Udawatte, C. P., et al. (2000). Sintering of additive free hydrothermally derived indium tin oxide powders in air. *J. Solid State Chem.,* Vol. 154, No. 2, pp. 444-450, ISSN 0022-4596

Vojnovich, T. & Bratton, R. J. (1975). Impurity effects on sintering and electrical resistivity of indium oxide. *Am. Ceram. Soc. Bull.,* Vol. 54, No. 2, pp. 216-217, ISSN 0002-7812

Warschkow, O., et al. (2003). Defect structures of tin-doped indium oxide. *J. Am. Ceram. Soc.,* Vol. 86, No. 10, pp. 1700-1706, ISSN 0002-7820

Weigert, M., et al. (1992), Target für die Kathodenzerstäubung und Verfahren zu dessen Herstellung, DE 4124471, Patent, Degussa AG

Werner, J., et al. (1996). Chemical transport of restricted solid solutions of In_2O_3 and SnO_2: experiments and thermodynamic process analysis. *J. Cryst. Growth,* Vol. 165, No. 3, pp. 258-267, ISSN 0022-0248

Wit, J. H. W. d. (1975). The high temperature behaviour of In_2O_3. *J. Solid State Chem.,* Vol. 13, No. pp. 192-200, ISSN 0022-4596

Yanagisawa, K., et al. (2000). Preparation and characterization of fine indium tin oxide powders by a hydrothermal treatment and postannealing method. *J. Mater. Res.*, Vol. 15, No. 6, pp. 1404-1408, ISSN 0884-2914

Yanai, Y. & Nakamura, A. (2003), Manufacturing method of ITO powder with tin dissolved in indium oxide, and manufacturing method of ITO target, US 2003/0039607, Patent,